Contemporary Clinical Neuroscience

Series Editor
Mario Manto, Department of Neurology, CHU-Charleroi, Charleroi, Belgium and
Department of Neurosciences, University of Mons, Mons, Belgium

Contemporary Clinical Neurosciences bridges the gap between bench research in the neurosciences and clinical neurology work by offering translational research on all aspects of the human brain and behavior with a special emphasis on the understanding, treatment, and eradication of diseases of the human nervous system. These novel, state-of-the-art research volumes present a wide array of preclinical and clinical research programs to a wide spectrum of readers representing the diversity of neuroscience as a discipline. Volumes in the series have focused on Attention Deficit Hyperactivity Disorder, Neurodegenerative diseases, G Protein Receptors, Sleep disorders, and Addiction issues.

More information about this series at http://www.springer.com/series/7678

Hidehiro Mizusawa • Shinji Kakei
Editors

Cerebellum as a CNS Hub

Springer

Editors
Hidehiro Mizusawa
Neurology, NCNP Hospital
National Center of Neurology and Psychiatry (NCNP)
Kodaira, Tokyo, Japan

Shinji Kakei
Department of Anatomy and Physiology
Faculty of Life Sciences
Jissen Women's University
Tokyo, Japan

ISSN 2627-535X ISSN 2627-5341 (electronic)
Contemporary Clinical Neuroscience
ISBN 978-3-030-75816-5 ISBN 978-3-030-75817-2 (eBook)
https://doi.org/10.1007/978-3-030-75817-2

Cover illustration by Katsunori Goto

This Springer imprint is published by the registered company Springer Nature Switzerland AG
The registered company address is: Gewerbestrasse 11, 6330 Cham, Switzerland

Cerebellum as a CNS Hub

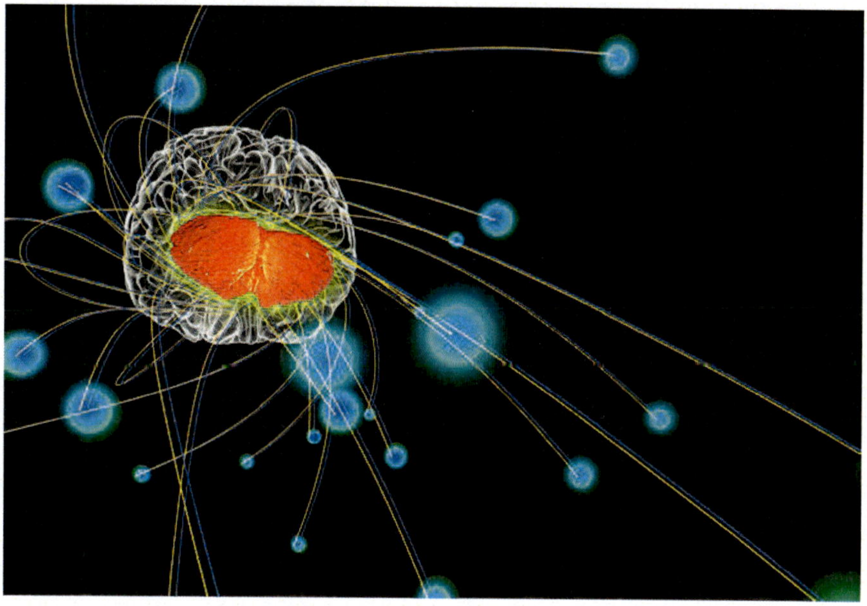

Illustration by Katsunori Goto

Preface

This book originated in 2017 when late Prof. Masao Ito, one of the greatest leaders of cerebellar research, encouraged us to organize an international symposium for research on the cerebellum. He stimulated us by emphasizing the importance of a multidisciplinary meeting for both cutting-edge neuroscientists and clinicians to boost cerebellar research in the future. Fortunately, thanks to the grant from the Fujihara Foundation of Science, our dream came true as the 75th Fujihara Seminar "The Cerebellum as a CNS hub – From evolution to therapeutic strategy" (Dec 1–4, 2018 at Tokyo Medical and Dental University, Tokyo, Japan). In the 4-day seminar, we tried to address a new point of view about the cerebellum, which will guide both basic and clinical researches about the cerebellum for years to come. Behind this ambitious title lay our primary question: What is the cerebellum for?

A prototype of the cerebellum emerged in the alar plate (i.e., sensory domain) of the rhombencephalon of jawless fish such as lampreys to collect multimodal inputs, which included both exteroceptive (lateral-line, vestibular, acoustic, visual) and somatosensory inputs (Olof Larsell, 1967, *The Comparative Anatomy and Histology of the Cerebellum.* Volumes 1 and 2, The University of Minnesota Press). It was literally located at the pivotal point in the lamprey brain. The cerebellum further gained access, in mammals, from the cerebral cortex, which include association and limbic areas as well as motor and sensory areas. In humans, the cerebellum is estimated to share no less than 80% of the CNS neurons in no more than 10% of the brain volume. Overall, the cerebellum has been, throughout its long history, a unique "hub" in the CNS to collect inputs from the entire brain, and probably has contributed to optimize adaptive behaviors of its owners.

Despite diversity of functional repertoires and input-output organizations in different cerebellar regions, the local neuron circuitry is characterized by its superb homogeneity, sometimes expressed as "crystal-like." Therefore, it is widely believed that a variety of inputs to the cerebellum are transformed with common dynamics. In other words, it is possible to discuss the diverse cerebellar functions on a common ground. And in the 75th Fujihara Seminar, we tried to approach the ground from the following viewpoints:

1. Evolution and development of the cerebellum
2. Neurocircuitry of the cerebellum
3. Information processing in the cerebellar neurocircuitry and its model
4. Complex spikes and plasticity of the cerebellar neurocircuitry
5. Cerebro-cerebellar loop and its contribution of the cerebellum to higher brain functions
6. Cerebellar disorders and their evaluation
7. Mechanisms and models of spinocerebellar ataxia and new treatments

Therefore, in this book, we arranged the chapters to conform to the organization.

We thank all the contributors to this book and all the participants of the 75th Fujihara Seminar. We also thank the Fujihara Foundation of Science for supporting the 75th Fujihara Seminar. We also thank Prof. Nobutaka Hirokawa, the chairperson of the foundation, for his invaluable advices and encouragements for the seminar.

We would like to acknowledge the staff at Springer Nature who provided excellent service throughout this project. We particularly wish to acknowledge William Lamsback, Jefferey Newton, Mallaigh Nolan, and Joseph Quatela (in alphabetical order) for their constant support and high degree of professionalism. They have been invaluable in helping to bring this work to completion during this historically hard period of the COVID-19 pandemic.

Finally, we wish to express our greatest thanks to late Prof. Ito. He was looking forward to the 75th Fujihara Seminar, while we were looking forward to celebrating his 90th birthday during the seminar. Unfortunately, he was not able to attend the seminar due to his deteriorating health condition. To our great sorrow, he passed away on December 18, 2018, 2 weeks after the seminar. Therefore, we wish to dedicate this book to late Prof. Masao Ito, whose ideas will ignite our imagination for decades to come.

National Center of Neurology and Psychiatry (NCNP) Hidehiro Mizusawa
Tokyo, Japan

Department of Anatomy and Physiology Shinji Kakei
Faculty of Life Sciences,
Jissen Women's University
Tokyo, Japan

Contents

Part I
Evolution and Development of the Cerebellum

Chapter 1
Evolutionary and Developmental Perspectives on the Origin and Diversification of the Vertebrate Cerebellum

Yasunori Murakami and Fumiaki Sugahara

1.1 The Origin of Vertebrates

Vertebrates are characterized by several morphological traits, such as paired sense organs, W- or V-shaped myotomes, segmented pharyngeal arches, notochords that only transiently appear in the developmental period in some vertebrates, and a hollow neural tube. Based on these morphological characters, recent paleontological studies identified the earliest vertebrates, namely, *Myllokunmingia fengjiaoa*, *Haikouichthys ercaicunensis*, and *Metaspriggina walcotti*, from sediment of the Cambrian period (Shu et al., 2003; Butler & Hodos, 2005; Morris & Caron, 2014). Thus, the origin of vertebrates may date back 540 million years. Because these animals lack jaws, they are grouped into agnathans (jawless animals). However, agnathans are not a monophyletic group; rather, they are a mixture of several jawless animals (see below). Although we could not identify the morphology of the brains of these early vertebrates, we speculated on the presence of some brain centers: their paired eyes may indicate the presence of a forebrain that gives rise to an optic stark, while the single nostril indicates the presence of an olfactory epithelium.

Many extant vertebrates are gnathostomes (jawed vertebrates), which are characterized by upper and lower jaws derived from the anterior pharyngeal arch (Fig. 1.1). However, in the early period of vertebrate evolution, many gnathostomes did not possess movable jaws. Thus, these vertebrates are referred to as agnathan grade and are often called "jawless gnathostomes." Given the fact that many extant vertebrates possess jaws, the establishment of jaws may be a key innovation in the evolution of vertebrates. However, two groups of jawless animals, lampreys and hagfishes, exist

Y. Murakami (✉)
Graduate School of Science and Engineering, Ehime University, Matsuyama, Ehime, Japan
e-mail: muakami.yasunori.mu@ehime-u.ac.jp

F. Sugahara
Division of Biology, Hyogo College of Medicine, Nishinomiya, Hyogo, Japan

H. Mizusawa, S. Kakei (eds.), *Cerebellum as a CNS Hub*,
Contemporary Clinical Neuroscience, https://doi.org/10.1007/978-3-030-75817-2_1

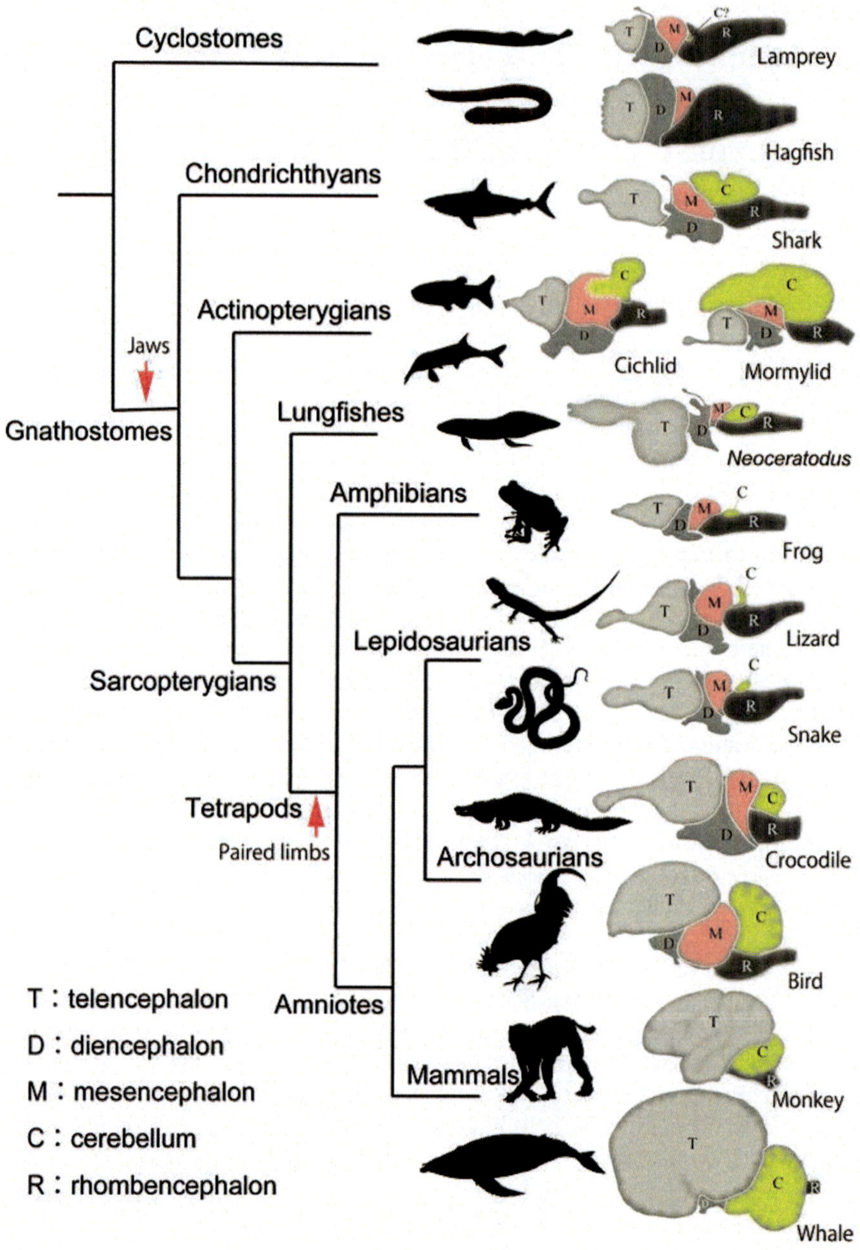

Fig. 1.1 The lineage of extant vertebrates with schematic drawings of their brains. Brain anatomy is based on Nieuwenhuys (1998)

currently in fresh- and saltwater environments. Although they possess a complex and excellent oral apparatus, as observed in gnathostomes, the morphology and the developmental mechanism of their facial region are quite different from those of gnathostomes (Ota & Kuratani, 2006; Ota et al., 2007, 2011; Oisi et al., 2013). Recent morphological and molecular phylogenetic studies indicate that lampreys and hagfishes are cyclostomes (Fig. 1.1; Mallatt & Sullivan, 1998; Kuraku & Kuratani, 2006; Heimberg et al., 2010; Oisi et al., 2013). Because cyclostomes are thought to have diversified from the other vertebrate lineage approximately 500 million years ago (Kuraku & Kuratani, 2006), a comparative study between cyclostomes and gnathostomes may identify ancestral traits that were established before the diversification between cyclostomes and gnathostomes. It is important to note that extant cyclostomes are not "primitive" animals because many of their morphological and physiological characters have been changing progressively toward their evolution.

1.2 Cyclostomes

Extant lampreys include 40 species, 29 of which are confined to freshwater environments (Nelson, 2016). Adult animals breed in shallow rivers, and their hatched larvae are called ammocoetes. The adult lamprey possesses an oral funnel with rings of horny teeth, a single nostril, a single pair of eyes, and seven pairs of gill openings (Fig. 1.2). In the vestibular organ, lampreys have anterior and posterior semicircular canals, unlike gnathostomes, which possess three canals (anterior, posterior, and horizontal canals). They lack paired appendages, although they have unpaired dorsal and tail fins.

Extant hagfishes include 78 species and live in a marine environment (Nelson, 2016). They possess a single nostril and lack paired appendages, as observed in lampreys. Unlike lampreys, however, extant hagfishes have only rudimentary eyes and lack larval stages in their life cycle. They possess three pairs of short tentacles around the mouth that function as mechanosensory and chemosensory apparatuses (Fig. 1.2). Their skin contains a large number of Schreiner's organs, which consist of multiple sensory cells (Braun, 1998; Finger, 2009). They have 1–6 pairs of pharyngeal openings and a mucous gland on the body surface, to prevent attacks by enemies. Their semicircular canal appears to be single, but comparative developmental data indicate that vertebrates shared the early stages of semicircular canal development, whereas hagfishes exhibit simplified later stages of canal development (Jorqensen et al., 1998; Higuchi et al., 2019).

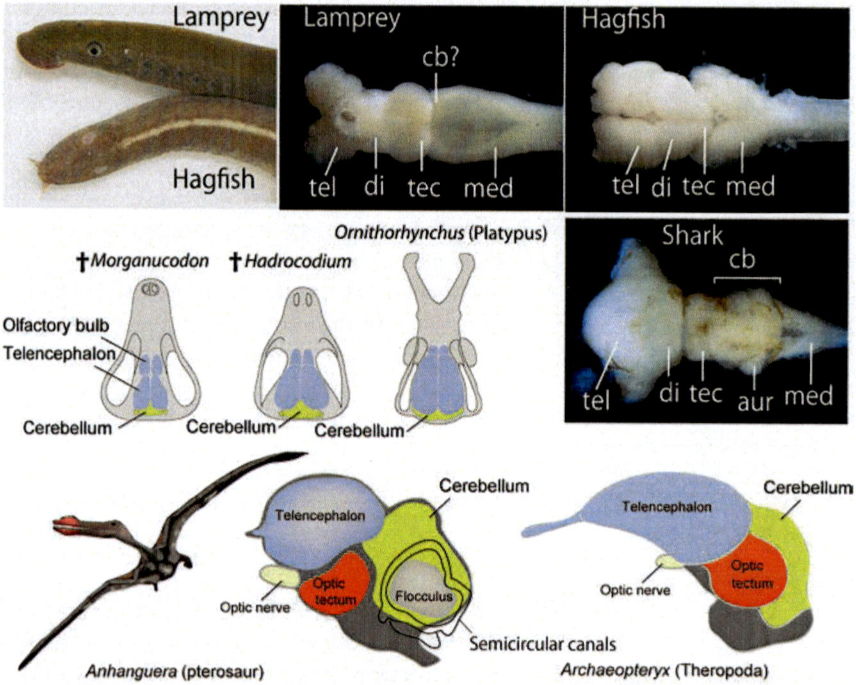

Fig. 1.2 Morphology and brains of cyclostomes, sharks, Mammaliaformes (*Morganucodon and Hadrocodium*), *Anhanguera*, and *Archaeopteryx*. Abbreviations: aur auricle, cb cerebellum, di diencephalon, med medulla, tel telencephalon, tec tectum. Animal drawing is based on Witton (2013). Brain illustrations of *Anhanguera* and *Archaeopteryx* were based on Alonso et al. (2004) and Witmer et al. (2003), respectively. (The illustration of mammals was based on Luo et al. (2001) and Rowe et al. (2011))

1.3 Gnathostomes

In the course of evolution, extant gnathostomes have been divided into chondrichthyans (Elasmobranchii and Holocephali), actinopterygians (ray-finned fishes), and sarcopterygians (lungfishes, coelacanthiformes, amphibians, and amniotes) (Fig. 1.1; Zhu et al., 2013). Among these groups, actinopterygians gave rise to teleosts that diversified into a variety of forms in the course of the evolution; extant teleosts comprise about 30,000 species. Tetrapods, including amphibians and amniotes, possess fore and hind limbs, which have allowed animals to walk on land. Importantly, paired limbs were lost subsequently in some lineages, such as snakes. Extant amphibians are included in Lissamphibia, which are divided into anurans, urodeles, and gymnophiones. Traditionally, amniotes have been divided into reptiles, birds, and mammals. However, in terms of modern phylogeny, extant amniotes are divided into synapsids and sauropsids; the former include mammals and the latter are further subdivided into lepidosaurians (tuatara, lizards, and snakes), turtles, and archosaurians (crocodiles and birds). In this context, birds should be included in archosaurians because of morphological and molecular similarities (Donoghue & Benton, 2007; Crawford et al., 2012; Green et al., 2014; Tzika et al., 2015). Thus, traditional "reptilian" species are paraphyletic group that are sometimes referred to as "nonavian sauropsids."

1.4 The Vertebrate Brain

The vertebrate brain can be divided into several regions, including the telencephalon, diencephalon, mesencephalon, and rhombencephalon (pons and medulla oblongata) (Fig. 1.1). The anteroposterior arrangement of the brain is highly conserved among vertebrates, and the neuronal subtypes of those regions and the stereotyped framework of longitudinal or commissural tracts are also observed in many vertebrate lineages (Nieuwenhuys, 1998). Conversely, vertebrate brains exhibit a remarkable morphological diversity. For example, mammals possess a well-developed telencephalon and cerebellum, whereas teleosts and birds have a large optic tectum, which is part of the midbrain (Fig. 1.1). The morphological and physiological adaptation of animals to various environments may lead to the diversification of brain regions that act as centers of sensory perception and motor coordination. As in other brain regions, the vertebrate cerebellum, which receives input from many parts of the body and brain, exhibits a remarkable diversity and plays an essential role in the coordination of body movement and balance (Fig. 1.1; Nieuwenhuys, 1967).

Although the word cerebellum means "small brain," this brain region possesses a large number of neurons. In addition, the cerebellum might be the most variable brain region in vertebrates, namely, in *Mormyridae* (a family of teleosts sometimes called elephantfish), the cerebellum is the largest region of the brain (Fig. 1.1). Conversely, among cyclostomes, lampreys possess undifferentiated corpus cerebelli, and hagfishes appear to lack a cerebellum, although lampreys possess "cerebellar-like" structures that include the dorsal and medial octavolateralis nuclei, which are involved in the processing of electroreceptive and lateral-line sensory information, respectively (Figs. 1.1 and 1.2; Nieuwenhuys, 1967; Yopak et al., 2017). Therefore, it seems that a true cerebellum may have been acquired in the gnathostome lineage after the split between gnathostomes and cyclostomes (Montgomery et al., 2012). However, as mentioned below, our recent study using several brain marker genes suggests the ancient origin of the developmental plan of the cerebellum.

1.5 Diversity of the Vertebrate Cerebellum

In many gnathostomes, the cerebellum is located in the dorsal part of the hindbrain, in which a single bulge called the corps cerebelli can be observed. In many gnathostomes, except for mammals, this region represents the main area for the function of the cerebellum. In mammals, the vermis receives proprioceptive input from the spinocerebellar tract, as observed for the corpus cerebelli in many vertebrates, and the lateral hemisphere receives input from the telencephalon via the pontine nucleus. Thus, the vermis is thought to be a homolog of the corpus cerebelli. However, the corpus cerebelli in teleosts receives input from the telencephalon, as observed for the mammalian lateral cerebellar hemisphere (Imura et al., 2003, 2017), suggesting that the corpus cerebelli in fishes may not be fully homologous to the mammalian vermis. Actinopterygians possess a well-developed valvula cerebelli in the anterior part of the cerebellum, which often intrudes into the ventricle of the mesencephalon. In actinopterygians, except for *Polypterus*, one of the cerebellum-like regions, called the torus longitudinalis, lies adjacent to the dorsal midline of the optic tectum (Northmore, 2017). The cerebellum in gnathostomes also possesses the auricle, which receives input from the vestibular system (Fig. 1.2). The auricle is homologous to the flocculus in amniotes. The flocculus, which is involved in the integration of visual and vestibular information, is well developed in flying amniotes, such as birds and pterosaurs (Fig. 1.2; Witmer et al., 2003), and may represent a parallel evolution for flying ability (see below).

In general, a large cerebellum is present in fishes with high swimming performance and birds, which have adapted to the oceanic and aerial environments (Fig. 1.1). Conversely, animals that move mainly in two-dimensional space, such as benthic fishes or amphibians, possess cerebellum with a relatively small size (Fig. 1.1; Nieuwenhuys, 1967; Yopak et al., 2017). This is consistent with the fact that the cerebellum functions as a center for the vestibular system and motor control. The most interesting example of cerebellar evolution can be observed in mormyrids, a group of freshwater fishes that comprise the family *Mormyridae*; they possess well-developed valvula cerebelli, which occupy about 80% of the total brain (Nieuwenhuys & Nicholson, 1967; Bell & Szabo, 1986; Meek et al., 2008; Hibi et al., 2017).

In sarcopterygians, which include coelacanth, lungfish, and tetrapods, the brains of coelacanth and *Neoceratodus* (a kind of lungfish) possess an extensive cerebellum and distinct laterally paired auricles. An endocast analysis of the fossil dipnoan *Rhinodipteus kimberleyensis* from the late Devonian showed that the shape of the mesencephalic–metencephalic cavity suggests the presence of large auricles, as in coelacanth and *Neoceratodus* (Clement & Ahlberg, 2014). In terrestrial amphibians, auricles and the cerebellum-like nuclei that are associated with the lateral line or vestibular function have become less prominent. Thus, in the course of the adaptive radiation to the terrestrial field, the morphology of the cerebellum appears to have changed.

Among nonavian sauropsids, which include lepidosaurians (tuatara, lizards, and snakes), turtles, and crocodiles, the crocodilian cerebellum is relatively large compared with other lineages, which is consistent with the high motor abilities of crocodiles (Fig. 1.1). For example, freshwater crocodiles (*Crocodylus johnsoni*) perform asymmetrical gaits (gallops) at high speeds. However, the cerebellum of crocodiles is still smaller than that of birds and mammals.

1.6 The Cerebellum in Mammals

The mammalian cerebellum is characterized by a large volume and remarkable foliation. It is composed of the vermis, paravermis, and the cerebellar hemisphere. The mammalian cerebellum also contains a large amount of afferent and efferent axons. The vermis receives input from the spinal cord, whereas the cerebellar hemispheres receive strong projections from the neocortex via the pontine nuclei. The mammalian cerebellum regulates motor systems by monitoring the motor programs that are generated in the motor centers in the telencephalon. In addition, it integrates several types of sensory information and is involved in various cognitive functions, such as emotion, language, and memory.

A recent study showed that the fibroblast growth factor 8 (FGF8) is required for the development of the mammalian vermis and paravermis, while the avian cerebellum develops independently of FGF8 (Butts et al., 2014). These results suggest that the role of FGF8 in the development of the cerebellar vermis and paravermis changed during amniote evolution.

Regarding the origin of the mammalian cerebellum, an endocast analysis revealed that the brains of cynodonts, a clade originating in the late Permian, represent a narrow and featureless forebrain with a small olfactory bulb; however, the cerebellum was wider than the forebrain (Rowe et al., 2011). The brain of *Morganucodon*, the basal-most member of Mammaliaformes (a clade originating in the late Triassic), shows an expansion of the telencephalon and the olfactory bulb (Fig. 1.2). The cerebellum of *Morganucodon* is also enlarged, implying the expansion of the basal ganglia, thalamus, and pontine nuclei. The cerebellum of *Hadrocodium*, the closest known extinct relative of mammalia, bulges backward, bending the occipital plate. Because this animal possesses a thick spinal cord, the cerebellum exhibited an enhanced ability of sensory-motor integration. Thus, the Mammaliaformes may have evolved under a selective pressure for increasing the cerebellum size.

1.7 The Cerebellum in Flying Vertebrates

In the course of vertebrate evolution, only three groups (i.e., pterosaurs, birds, and Chiropteran mammals) successfully acquired an ability to fly using their muscles and the skeletal unit, which is called powered flight. It is important to note that these groups tend to have a well-developed cerebellum. Among archosaurs, the cerebellum in birds and pterosaurs is remarkably larger than that of crocodiles (Figs. 1.1 and 1.2). Because such a large cerebellum could not be observed in the ancestors of birds and pterosaurs, the size of the cerebellum may have increased independently in both lineages. Thus, the brains of these flying reptiles have evolved under the selective pressure that accelerated the expansion of the cerebellum. Endocast analyses revealed that the cerebellum in pterosaurs shows highly expanded features with a quite enlarged flocculus, which might receive input from well-developed semicircular canals (Fig. 1.2; Witmer et al., 2003). Because of this specialized cerebellum, pterosaurs, which have an extremely large body size, were able to fly. However, it may be premature to conclude that the enlargement of the flocculus is closely linked to flying ability because a recent analysis using microcomputed tomography in 60 extant bird species showed that relative "flocculus" size (as indexed by the relative volume of the floccular fossa) and flying ability were not significantly correlated (Walsh et al., 2013).

All vertebrate groups that perform powered flight appear to have an ability called endothermy, i.e., the capacity to generate and maintain a high body temperature by themselves without depending on an external heat source. Birds and mammals are endothermic, and pterosaurs probably also had endothermic ability because their bodies were covered in dense hair-like "pycnofibers," which may have functioned to maintain a high body temperature (Witton, 2013). These high physiological functions are necessary for the maintenance and development of the brain, which requires a high level of energy. Therefore, flying ability, endothermy, and brain development are closely related to each other.

Regarding the origin of the avian cerebellum, an endocast analysis of *Archaeopteryx* revealed that this animal had a relatively small cerebellum with no traces of a foliar structure (Fig. 1.2), suggesting that the cerebellum of *Archaeopteryx* was more primitive than that of any modern bird (Alonso et al., 2004).

1.8 The Origin of the Cerebellum: A Paleontological Study

Jawless animals in the early evolutionary stage include lineages such as conodonts, the presumptive ancestor of modern cyclostomes, and several jawless gnathostomes, including osteostracans (Donoghue et al., 2000; Goudemand et al., 2011). Among the basal gnathostomes, some lineages acquired jaws that possess a skeletal joint made by upper and lower jaws that are derived from pharyngeal arch components. Placoderms are one of the earliest jawed fishes and are characterized by the presence of bony plates in their head and thorax (Fig. 1.3). Among the jawed fishes, *Climatius* (Acanthodii) and *Andreolepis* (Actinopterygii) have been identified from sediment from around 420 million years ago (Janvier, 2002; Chen et al., 2016). Furthermore, *Megamastax*, a sarcopterygian, lived in the late Silurian period (about 423 million years ago) (Choo et al., 2014). A recent study showed that placoderms may have diversified before the diversification of acanthodians, chondrichthyes,

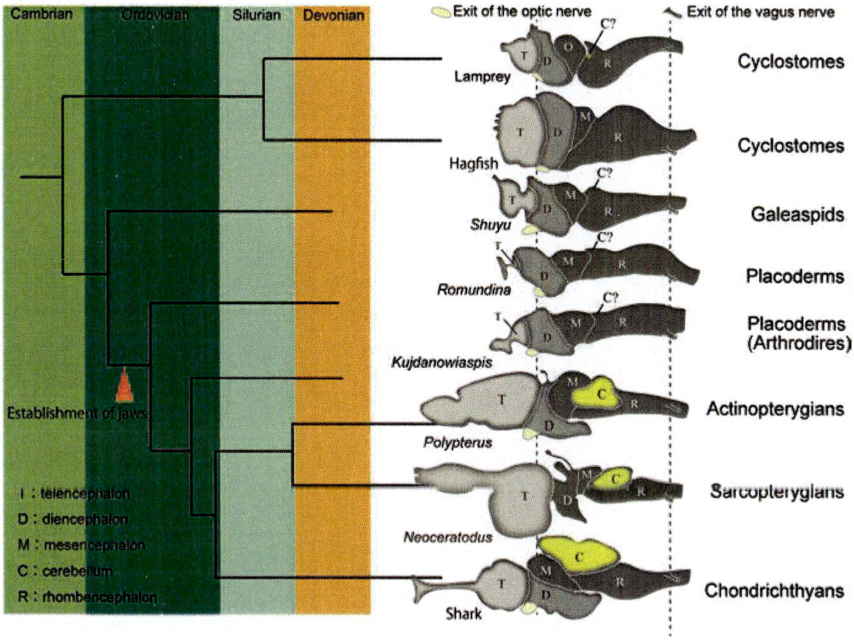

Fig. 1.3 The brain of early vertebrates. Schematic drawings of the brains of cyclostomes and gnathostomes (osteostracans, placoderms, actinopterygians, sarcopterygians, and chondrichthyans). Brains are aligned based on the position of the optic and vagus nerve exits. The rostral position is to the left (after Donoghue & Keating, 2014; Dupret et al., 2014)

sarcopterygians, and actinopterygians (Fig. 1.3; Zhu et al., 2013). Subsequently, the descendants of jawed fishes spread their habitat to the freshwater and saltwater environments, while many of the jawless gnathostomes disappeared in the vertebrate history (Fig. 1.3).

Some fossil osteostracans, which possess a skeletal head shield, have preserved casts of many internal organs, including the brain. However, generally, it is difficult to identify the brain morphology from the fossils of anamniotes because the brain in most anamniotes does not completely fill the endocranial cavity, in which fat tissues often occupy the space between the brain and the internal surface of the skull. However, some good endocasts from basal cartilaginous and bony fishes indicate the morphologies of brain regions. Furthermore, recent advanced techniques using synchrotron radiation X-ray tomography images revealed the brain cavity in the skull with high resolution and allowed the reconstruction of the external morphology of the ancient brain. Using this new technique, the brain morphologies of *Romundina*, a type of placoderm, have been reported (Fig. 1.3). Although *Romundina* is a jawed vertebrate, its brain retains a jawless fish-type morphology in terms of the relatively small size of the telencephalon (Dupret et al., 2014). Unfortunately, unlike that observed for the telencephalon, the morphology of the cerebellum cannot be clearly identified because the boundary between the midbrain and the cerebellum is sometimes invisible, and, as mentioned above, the brain incompletely fills the brain cavity in the skull. However, some fossil specimens from the Devonian period exhibit signs of central and peripheral nervous systems. The brains of the Galeaspid *Shuyu* and the Osteostracan *Cephalaspis* or *Norselaspis* show a bulge on the dorsal hindbrain, which might represent a sign of a cerebellum (Fig. 1.3; Stencio, 1927; Janvier, 2002; Gai et al., 2011). These results suggest that early gnathostomes had already acquired a cerebellum, albeit with a small size. It is important to note that the telencephalon and the cerebellum appear to be enlarged in the crown gnathostomes, which include the actinopterygian or chondrichthyan lineages (Fig. 1.3), suggesting that the developmental program underlying the formation of these brain regions may have been modified in the common ancestor of these lineages.

1.9 The Origin of the Cerebellum: A Comparative Study

As mentioned above, endocast analyses provide valuable insights into the size and external morphology of the cerebellum in fossil specimens. However, we could not identify the cytoarchitecture or axonal connections of ancient cerebellums. Comparative analysis of extant animals is another approach that can be used for the study of brain evolution. If particular characters are shared by two groups, they may be derived from their common ancestor as plesiomorphic characters. Using these morphological characters, we could reconstruct the internal structure of the ancient brain. Recent studies using mammals, birds, and teleosts have shown that various transcription factors and morphogens are expressed in embryonic mouse brains in a region-specific manner. The cerebellum emerges from the dorsal side of the hindbrain, which corresponds to the first rhombomere segment and is marked by several transcription factors and signaling molecules. Therefore, the comparison of those molecular cues among various species may provide insights into the origins and diversification processes of the vertebrate cerebellum.

1.10 Cerebellar Cytoarchitecture

The gnathostome cerebellum shows a tremendous diversity with respect to its relative size and structure. The cerebellar cortex is divided into three layers in all extant gnathostomes, namely, the granule cell layer, the Purkinje cell layer, and the molecular layer, which includes Purkinje cell dendrites and granule cell axons (Fig. 1.4; Butler & Hodos, 2005; Yopak et al., 2017). However, there are some exceptions. In Chondrichthyes, the cell bodies of granule cells are clustered in two longitudinal ridges called the prominentiae granulares, which are located on either side of the median plane (Fig. 1.4; Nieuwenhuys, 1967, 1998). The Purkinje cell layer is

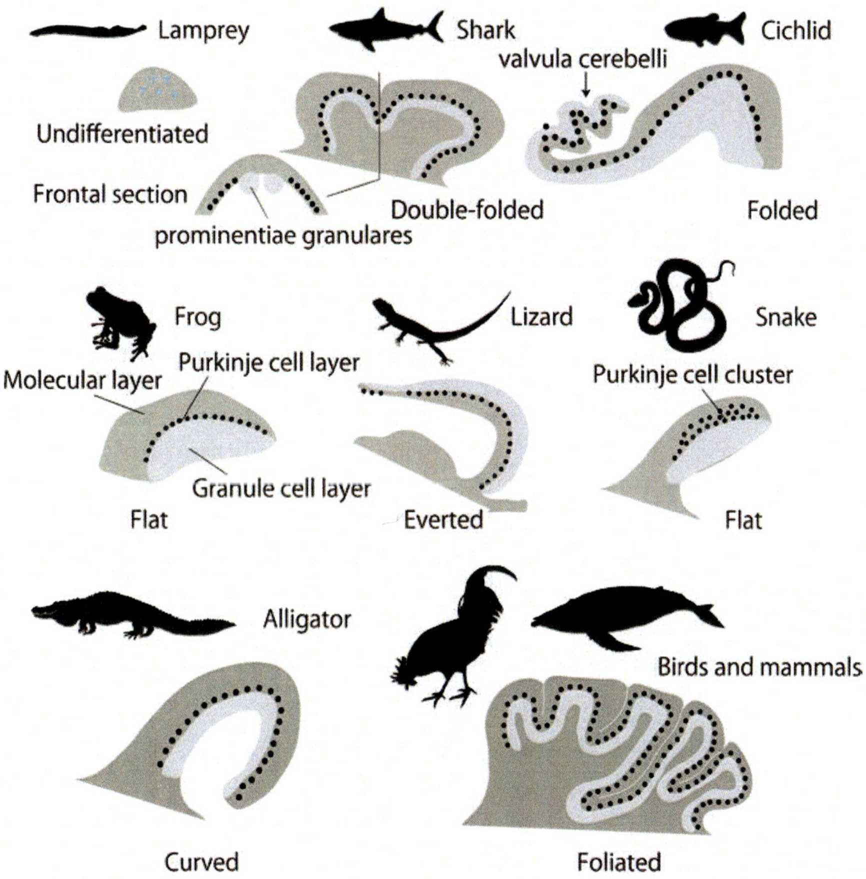

Fig. 1.4 Morphology of the corpus cerebelli in vertebrates. In each example, the rostral position is to the left (after Butler & Hodos, 2005)

usually a monolayer; however, in snakes, Purkinje cells do not form the monolayer typically observed in other amniotes; rather, Purkinje cell somata form several clusters in the molecular layer (Fig. 1.4; Aspden et al., 2015). In lampreys, granule cell-like neurons were observed, while Purkinje cells could not be identified in the corpus cerebelli (see below). In many gnathostomes, the cerebellum possesses one or more deep cerebellar nuclei. In apes, the dentate nucleus, which is linked to the well-developed lateral hemisphere, is enlarged and complexly folded. However, teleosts lack cerebellar nuclei; instead, they have eurydendroid cells (Murakami & Morita, 1987). Although their axonal projection is similar to those of the deep cerebellar nuclei, the expression property of eurydendroid cells is different from that of deep cerebellar nuclei, suggesting that eurydendroid cells are not homologous to the deep cerebellar nuclei (Kani et al., 2010).

1.11 The Origin of Cerebellar Neurons

To elucidate the evolutionary processes of the vertebrate cerebellum, cyclostomes, which diversified from the gnathostome lineage in the early evolutionary period, stand at the key position in terms of phylogeny. As cyclostomes bifurcated from gnathostomes in the early evolutionary stage, we could hopefully identify the ancestral (plesiomorphic) or derivative (apomorphic) traits by comparing the cerebella of cyclostomes and gnathostomes. Extant lampreys possess an undifferentiated corpus cerebelli, in which Purkinje-like cells can be observed (Johnston, 1902; Larsell, 1967). However, these cells are intermingled with granule-like cells; moreover, morphologically, lamprey Purkinje-like cells show poor ramification, unlike the gnathostome Purkinje cells, and they are not stained by antibodies against zebrin, which recognize gnathostome Purkinje cells (Lannoo & Hawkes, 1997). Furthermore, hagfishes have only an uncertain sign of a cerebellum in the hindbrain (Fig. 1.2; Nieuwenhuys, 1967; Butler & Hodos, 2005). Cyclostomes appear to lack deep cerebellar nuclei and also some cerebellum-related nuclei, such as the red nucleus and precerebellar nuclei, both of which are present in sharks (Wicht, 1996; Pose-Mendez et al., 2014). Thus, the organization of a cerebellum that possesses molecular, Purkinje-cell, and granular layers with a precerebellar system may have been established after the split between cyclostomes and gnathostomes.

1.12 Combinatory Expression of Regulatory Genes in the Anterior Rhombencephalon

In all vertebrates studied to date, *Otx2*, which encodes a type of transcription factor, is expressed in the anterior neural tube, with a sharp boundary at the midbrain/hindbrain junction, and is absent in the rhombencephalon, which gives rise to the cerebellum (Fig. 1.5). Ectopic expression of Otx2 in the anterior rhombencephalon results in the induction of the midbrain, instead of the cerebellum (Broccoli et al., 1999; Katahira et al., 2000). Conversely, the most anterior expression domain of Hox-family genes appears to be rhombomere 2 (r2), where the transcripts of *Hoxa2* were detected (Fig. 1.5). Because the vertebrate cerebellum appeared in the most anterior part of rhombomeres, the developmental position of the cerebellum should correspond to rhombomere 1 (r1), where the expression of the *Otx* and *Hox* genes is absent (Fig. 1.5). Thus, an Otx/Hox-free rhombomere may be a prerequisite for the patterning of the cerebellum. This hypothesis is supported by the fact that the *Hoxa2* knockout mice, which lost the r1/r2 boundary, exhibit posterior expansion of the cerebellum (Butts et al., 2014).

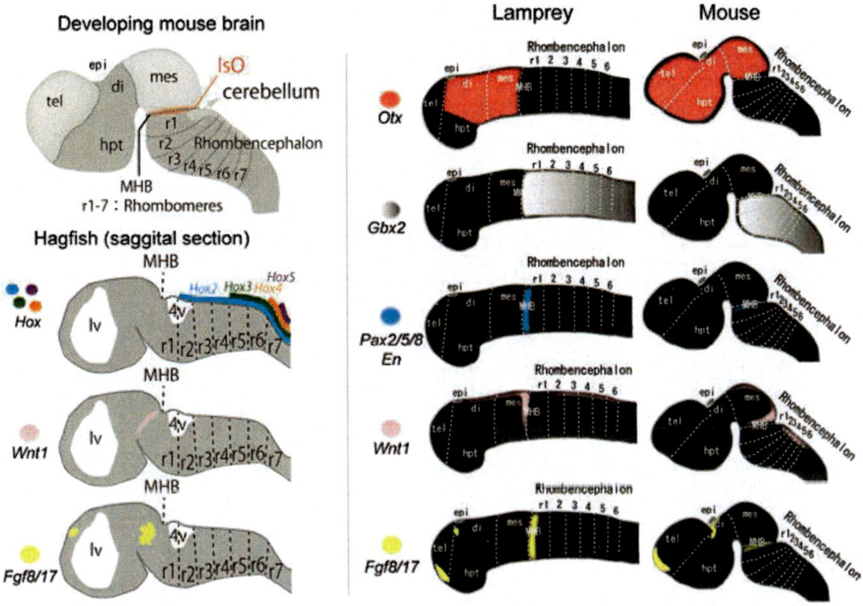

Fig. 1.5 Developmental plan of the vertebrate cerebellum. Expression patterns of regulatory genes along the anteroposterior neuraxis in the developing mouse, lamprey, and hagfish are shown. The rostral position is to the left. Abbreviations: di diencephalon, epi epiphysis, hpt hypothalamus, IsO isthmic organizer, lv lateral ventricle, med medulla, mes mesencephalon, MHB mid-hindbrain boundary, tel telencephalon, 4v fourth ventricle (after Takio et al., 2007; Sugahara et al., 2011, 2016)

1.13 Organizing Centers

During the development of the vertebrate brain, some specific regions act as important signaling centers called organizers, one of which is the isthmic organizer (IsO), located at the boundary between the mesencephalon and the metencephalon (Fig. 1.5). The IsO plays a crucial role in the formation of the cerebellum, and it is characterized by the combinatorial expression of transcription factors and signaling molecules (Fig. 1.5). Namely, the IsO is patterned in the expression boundary between *Otx2* and *Gbx2* (Millet et al., 1996; Broccoli et al., 1999; Katahira et al., 2000). Subsequently, the expression domains of the signaling factors *Wnt1* and *Fgf8* appeared within the *Otx2*- and *Gbx2*-expressing neuroepithelia, respectively (Hidalgo-Sanchez et al., 1999). Zebrafish embryos with inactivation of Fgf8 function exhibit an *acerebellar* (cerebellum absent) phenotype (Reifers et al., 1998). In chicks, ectopic expression of Fgf8b, a splicing variant of Fgf8, changes the fate of the mesencephalic alar plate to the cerebellum (Sato et al., 2001). Based on these results, Fgf8 is now accepted as one of the most important organizing signals (Harada et al., 2016). Importantly, the gene expression pattern in the IsO has been partially identified in cephalochordates and urochordates. Curiously, the combinatory gene expression that characterizes the IsO is also found in the surface ectoderm of hemichordates (Pani et al., 2012). This finding suggests that the developmental basis that gives rise to the IsO might have emerged in an ancestor of hemichordates as an ectodermal-patterning mechanism. Subsequently, in chordates, these organizers might have been diverted to a brain-patterning program. However, chordates, including amphioxus and ascidians (a member of Tunicata), appear to lack some of the molecular cues that are necessary for the patterning of the IsO. For example, tunicates exhibit loss of *Gbx* from their genome (Wada et al., 2003). In amphioxus, cognates of *En* and *Fgf8* are not expressed at the expression boundary between *Gbx* and *Otx* (Castro et al., 2006; Meulemans & Bronner-Fraser, 2007; Bertrand et al., 2011). Amphioxus also lack *Dmbx* and *pax2/5/8* expression rostral to the anterior expression boundary of *Gbx* (Kozmik et al., 1999; Takahashi & Holland, 2004). These results suggest that the deep deuterostome ancestry of genetic programs for brain development appears to have degenerated secondarily in the amphioxus and ascidians (Yao et al., 2016).

1.14 IsO in Cyclostomes

Adult lampreys exhibit only an uncertain sign of the cerebellum (Fig. 1.2). Recent studies using lamprey embryos found that an *Otx2* ortholog was expressed anterior to the midbrain/hindbrain boundary and that a *Gbx2* ortholog was expressed posterior to this boundary (Fig. 1.5; Murakami et al., 2001; Takio et al., 2007). These expression domains are similar between lampreys and chicks. A recent study also showed that the lamprey homolog of *Hoxa2* was expressed posterior to r2, as well as in other vertebrates (Takio et al., 2007; Parker et al., 2014), suggesting that an Otx/Hox-free rhombomere, which is a prerequisite for the patterning of the cerebellum, is present in the lamprey. In addition, the lamprey orthologs of *Pax2/5/8*, *Wnt1*, and *Fgf8/17* are restricted to the midbrain/hindbrain boundary, as well as the mouse brain (Fig. 1.5; Murakami et al., 2001; Sugahara et al., 2011, 2016). These results suggest that an IsO is present in the lamprey. Moreover, recent studies using hagfish embryos (see below) showed that one of the *Otx2* homologous genes is expressed up to the posterior end of the midbrain and that *Wnt1*, *Fgf8/17* cognates in the hagfish are also expressed in the mid-hindbrain boundary, which positionally corresponds to the IsO (Fig. 1.5; Sugahara et al., 2016; Sugahara and Murakami in prep.). These results strongly suggest that the IsO is present in cyclostomes, i.e., it was already established in the last common ancestor of vertebrates.

1.15 The Origin of the Cerebellar Cytoarchitecture

Extant hagfishes and lampreys lack a cerebellar organization that includes molecular, Purkinje-cell, and granular layers (Fig. 1.4). Thus, it has been thought that these neurons and their connections were established in the gnathostome lineage after the split from cyclostomes. In gnathostomes, the cerebellum contains two major neuronal types, namely, excitatory granule neurons and inhibitory Purkinje cells. These cells originate from the dorsal side of the rhombencephalon (Watson et al., 2011). The precursors of cerebellar neurons are marked by several transcription factors. In mammals, *Ptf1a*-expressing cells are generated from the ventricular zone in r1 (Fig. 1.6a). Subsequently, these cells migrate dorsally and differentiate into GABAergic inhibitory cerebellar neurons, including Purkinje cells (Butts et al., 2014). Conversely, *Atoh1*-positive cells originate from the two bilateral dorsal-most edges of the hindbrain, the "rhombic lip," and migrate tangentially in the rhombic lip to form an external granular layer (Wingate & Hatten, 1999; Butts et al., 2014). These cells then downregulate *Atoh1* and upregulate the transcription factor

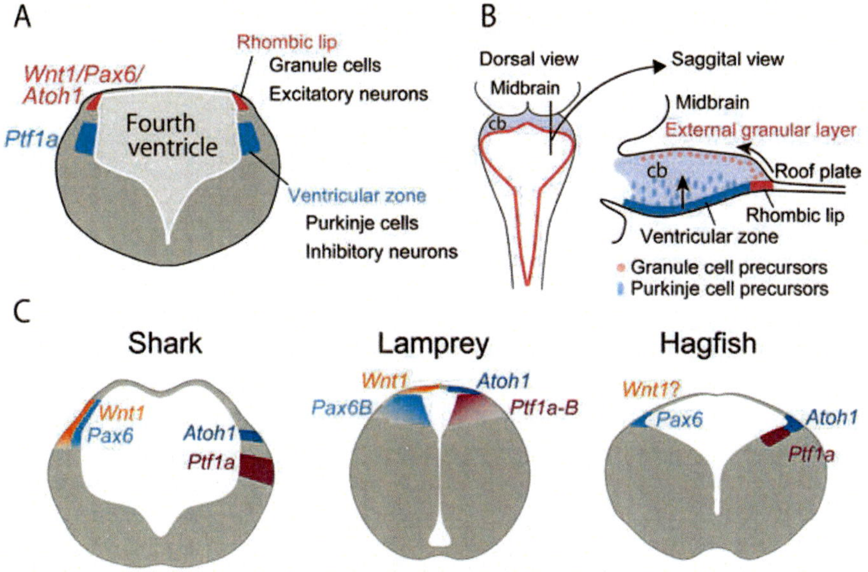

Fig. 1.6 Schematic drawing of the development of the vertebrate cerebellum. (**a**) The expression domains of regulatory genes and primordia of cerebellar neurons are shown in the coronal section of the anterior rhombencephalon of mammals. (**b**) Dorsal and sagittal views of the cerebellar primordium. Granule cells and Purkinje cells migrate from the rhombic rip and the ventricular zone, respectively. (**c**) Expression domains of regulatory genes in the shark, lamprey, and hagfish in a coronal section of the anterior rhombencephalon. Abbreviations: cb cerebellum (after Butts et al., 2014; Sugahara et al., 2016)

NeuroD. Finally, these cells migrate ventrally through the Purkinje-cell layer and form an internal granular layer (Fig. 1.6b).

The cellular organization and expression of *Atoh1* and *Ptf1a* orthologs in the cerebellar primordium have been observed in many gnathostome species, including amniotes, teleosts, and sharks (Chaplin et al., 2010), suggesting that the developmental plan that produced a cerebellar cytoarchitecture could date back at least to the common ancestor of gnathostomes. However, Sugahara et al. (2016) found that presumptive lamprey cognates of *Atoh1* are expressed in the rhombic lip throughout the rhombencephalon (Fig. 1.6c). Interestingly, lamprey *Atoh1* expression in the posterior rhombic lip is downregulated in the pharyngeal stage, while the expression in the most anterior rhombic lip is maintained until the respiration stage (Sugahara et al., 2016). This implies that the nature of *Atoh1*-positive cells in r1 is different from those of caudal rhombomeres. Moreover, the lamprey homolog of *NeuroD* is expressed adjacent to the *Atoh1*-positive cells in the anterior rhombic lip (Sugahara and Murakami in prep.). Because *NeuroD* is expressed in the postmitotic stage of cerebellar granule cells in gnathostomes, it is likely that lampreys possess the molecular consequence that covers a whole stage of granule cell differentiation.

Ptf1a in the lamprey could be identified in the genome, and its expression domains are located just ventral to the rhombic lip, which might correspond to the cerebellar ventricular zone in jawed vertebrates (Sugahara et al., 2016). Moreover, lamprey neurons that express glutamic acid decarboxylase, an enzyme required for GABA synthesis, were found slightly lateral to the *Ptf1a*-positive ventricular zone (Sugahara and Murakami in prep.). This indicates that *Ptf1a*-positive neural progenitor cells migrate laterally to differentiate into GABA-producing inhibitory neurons. In gnathostomes, cerebellar inhibitory neurons include Golgi, basket, and stellate cells in addition to Purkinje cells, and all of these cells are generated from the *Ptf1a*-positive precursor cells located in the ventricular zone. Thus, further analysis is necessary to determine the fate of *Ptf1a*-positive inhibitory neurons in the lamprey ventricular zone.

Unlike that observed in lampreys, hagfishes have only a few fibers of the vestibular lateral-line commissure at the posterior end of the midbrain (Larsell, 1947). Thus, many textbooks state that the cerebellum is absent or doubtful in the hagfish (Yopak et al., 2017). As mentioned above, if hagfishes had bifurcated from the other vertebrates before the diversification of lampreys, it would be consistent to assume that the cerebellum was acquired after the divergence of the hagfishes. However, such a simple scenario is now unlikely because, as noted above, cyclostomes are assumed to be a monophyletic group. Alternatively, the ancestral cerebellum seems to have been established before the divergence of cyclostomes, to then degenerate secondarily in the hagfish lineage. The commissural fibers of the spinocerebellar tract, which could be observed in gnathostomes and lampreys, are absent in hagfish (Larsell, 1947), suggesting that the tracts might have degenerated in hagfishes. In contrast, the vestibular lateral line commissures were retained in both cyclostome lineages. Therefore, to identify the origin of the cerebellum, the development of the hagfish hindbrain is quite important. However, hagfish embryology has been hampered because of the difficulty in obtaining fertilized eggs. Nevertheless, recent studies have successfully obtained eggs and living embryos from the Japanese hagfish *Eptatretus burgeri* (Ota & Kuratani, 2006; Ota et al., 2007) and have revealed that they share a common body plan with lampreys (Ota et al., 2007, 2011; Oisi et al., 2013).

Using hagfish embryos, Sugahara et al. (2016) identified the anteroposterior arrangement of the CNS from the telencephalon to rhombencephalon and also found differential expression patterns of some transcription factor-encoding genes in the hagfish hindbrain. Recent studies further identified that an *Otx*/*Hox*-free rhombomere is present in the hagfish rhombencephalon as well as in lamprey and that the expression domains of hagfish *Pax6* and *Atoh1* are restricted to the dorsal tip of the rhombencephalon, which appeared to be a rhombic lip (Fig. 1.6c; Sugahara et al., 2016; Pascual-Anaya et al., 2018). In turn, the expression of *Pax6* in the rhombic lip is a prerequisite for the migration of granular cells into the cerebellar cortex, to form the internal granule cell layer (Engelkamp et al., 1999). Moreover, in the mammalian cerebellum, *Pax6* is necessary for the development of parallel fibers, which are formed by branches of granule cell axons (Yamasaki et al., 2001). Thus, the presence of hagfish *Pax6* in the rhombic lip indicates that hagfishes may retain the molecular signals that are necessary for the axon guidance of granule cells. Curiously, *Pax6* was not detected in the lamprey rhombic lip (Murakami et al., 2001; Sugahara et al., 2016), although *Pax6B*, a paralog of lamprey *pax6*, was indeed expressed in a deeper layer of the dorsal rhombencephalon and overlapped with *Ptf1a* expression, unlike that observed in the hagfish and gnathostomes (Fig. 1.6c; Sugahara et al., 2016). Thus, lampreys might have lost the function of *Pax6* secondarily in the rhombic lip. We also found that the hagfish homolog of *Ptf1a* is expressed in the ventral region of the *Atoh1*/*Pax6* expression domain in r1 (Sugahara and Murakami in prep.). Taken together, these findings suggest that the gene expression patterns in the presumptive cerebellar primordium are conserved between gnathostomes and hagfish and further indicate that hagfishes may possess developmental programs that are equivalent to the cerebellum-induced rhombic lip and the ventricular zone, both of which are sources of the cerebellum granule cells and Purkinje cells in gnathostomes.

1.16 The Origin and Diversification of the Cerebellum

In summary, many brain compartments may have been established in the common ancestor of vertebrates, and the developmental architecture of the cerebellum may have been acquired as an evolutionary novelty in the vertebrate lineage before the divergence of cyclostomes. In other words, the specific genetic background that underlies the acquisition of the cerebellum might have already been established in the latest common ancestor of vertebrates, although the cerebellum proper is less prominent in extant cyclostomes.

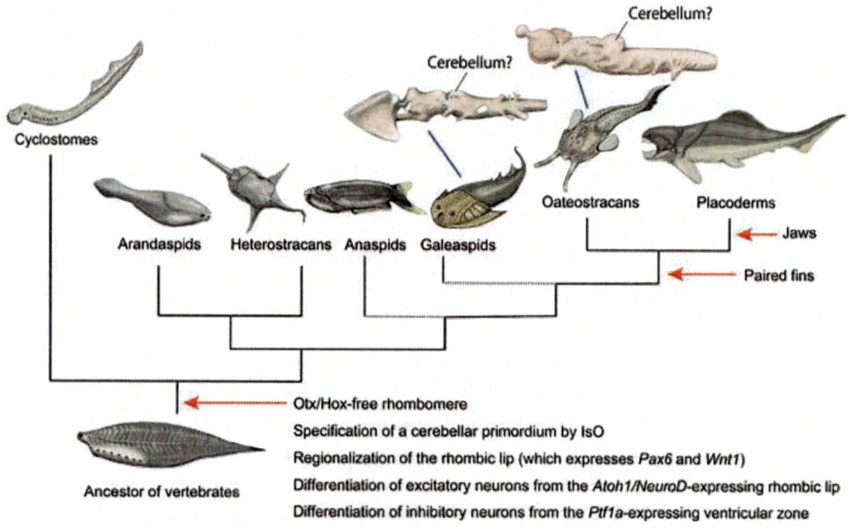

Fig. 1.7 Evolutionary scenario of the developmental patterning in the vertebrate cerebellum. Animal and brain drawings based on Janvier (2002) and Gai et al. (2011). Phylogeny adapted from Donoghue and Keating (2014) and Morris and Caron (2014)

If so, how early did vertebrates acquire a cerebellum? As mentioned above, fossil Galeaspid *Shuyu* have a small bulge on the dorsal metencephalon, and the Osteostraci *Norselaspis*, which is closely related to jawed vertebrates, possess a relatively remarkable bulge in the metencephalon compared with that of *Shuyu* (Fig. 1.7). It is important to note that *Shuyu* have a well-developed trunk region and that *Norselaspis* possess paired fins (Fig. 1.7). In gnathostomes, these structures are innervated by spinal nerves and the cerebellum receives strong input from the vestibular and spinocerebellar tracts. The evolution of these structures, which enabled animals to move accurately in three-dimensional space, may have accelerated the evolution of the cerebellum as an evolutionary novelty in gnathostomes. Subsequently, the flexibility of the brain developmental program may have resulted in the morphological and physiological diversity of the cerebellum and enabled vertebrates to adapt to complex environments. However, knowledge of the evolutionary process of the cerebellum remains fragmental. Further advances in evolutionary developmental biology will provide insights into the origin and evolution of the vertebrate cerebellum.

Acknowledgments We thank Dr. Makiko Fukui for technical support and valuable discussions. We thank all past and present members and collaborators of the YM and FS laboratories for support and constructive discussions. This project has been supported by RIKEN, Kobe, Japan, Ehime University (Research unit) and the Japan Society for the Promotion of Science (JSPS; grant number 20K06892 to YM, and grant number 19K06794 to FS).

References

Alonso, P. D., Milner, A. C., Ketcham, R. A., Cookson, M. J., & Rowe, T. B. (2004). The avian nature of the brain and inner ear of archaeopteryx. *Nature, 430*, 666–669.

Aspden, J. W., Armstrong, C. L., Gutierrez-Ibanez, C. I., Hawkes, R., Iwaniuk, A. N., Kohl, T., Graham, D. J., & Wylie, D. R. (2015). Zebrin II / aldolase C expression in the cerebellum of the western diamondback rattlesnake (Crotalus atrox). *PLoS One, 10*, e0117539.

Bell, C. C., & Szabo, T. (1986). Electroreception in mormyrid fish: Central anatomy. In T. H. Bullock & W. Heiligenberg (Eds.), *Electroreception*. Wiley.

Bertrand, S., Camasses, A., Somorjai, I., Belgacem, M. R., Chabrol, O., Escande, M. L., Pontarotti, P., & Escriva, H. (2011). Amphioxus FGF signaling predicts the acquisition of vertebrate morphological traits. *Proceedings of the National Academy of Sciences of the United States of America, 108*, 9160–9165.

Braun, C. B. (1998). Schreiner organs: A new craniate chemosensory modality in hagfishes. *The Journal of Comparative Neurology, 392*, 135–163.

Broccoli, V., Boncinelli, E., & Wurst, W. (1999). The caudal limit of Otx2 expression positions the isthmic organizer. *Nature, 401*, 164–168.

Butler, A. B., & Hodos, W. (2005). *Comparative vertebrate neuroanatomy: Evolution and adapta tion* (2nd ed.). Wiley-Interscience.

Butts, T., Green, M. J., & Wingate, R. J. (2014). Development of the cerebellum: Simple steps to make a 'little brain'. *Development, 141*, 4031–4041.

Castro, L. F., Rasmussen, S. L., Holland, P. W., Holland, N. D., & Holland, L. Z. (2006). A Gbx homeobox gene in amphioxus: Insights into ancestry of the ANTP class and evolution of the midbrain/hindbrain boundary. *Developmental Biology, 295*, 40–51.

Chaplin, N., Tendeng, C., & Wingate, R. J. (2010). Absence of an external germinal layer in zebrafish and shark reveals a distinct, anamniote ground plan of cerebellum development. *The Journal of Neuroscience, 30*, 3048–3057.

Chen, D., Blom, H., Sanchez, S., Tafforeau, P., & Ahlberg, P. E. (2016). The stem osteichthyan Andreolepis and the origin of tooth replacement. *Nature, 539*, 237–241.

Choo, B., Zhu, M., Zhao, W., Jia, L., & Zhu, Y. (2014). The largest Silurian vertebrate and its palaeoecological implications. *Scientific Reports, 4*, 5242.

Clement, A. M., & Ahlberg, P. E. (2014). The first virtual cranial endocast of a lungfish (sarcopterygii: dipnoi). *PLoS One, 9*, e113898.

Crawford, N. G., Faircloth, B. C., McCormack, J. E., Brumfield, R. T., Winker, K., & Glenn, T. C. (2012). More than 1000 ultraconserved elements provide evidence that turtles are the sister group of archosaurs. *Biology Letters, 8*, 783–786.

Donoghue, P. C., & Benton, M. J. (2007). Rocks and clocks: Calibrating the Tree of Life using fossils and molecules. *Trends in Ecology & Evolution, 22*, 424–431.

Donoghue, P. C., & Keating, J. N. (2014). Early vertebrate evolution. *Paléo, 57*, 879–893.

Donoghue, P. C., Forey, P. L., & Aldridge, R. J. (2000). Conodont affinity and chordate phylogeny. *Biological Reviews of the Cambridge Philosophical Society, 75*, 191–251.

Dupret, V., Sanchez, S., Goujet, D., Tafforeau, P., & Ahlberg, P. E. (2014). A primitive placoderm sheds light on the origin of the jawed vertebrate face. *Nature, 507*, 500–503.

Engelkamp, D., Rashbass, P., Seawright, A., & van Heyningen, V. (1999). Role of Pax6 in development of the cerebellar system. *Development, 126*, 3585–3596.

Finger, T. E. (2009). The evolution of taste systems. In J. H. Kaas (Ed.), *Evolutinary neuroscience* (pp. 459–478). Academic Press.

Gai, Z., Donoghue, P. C., Zhu, M., Janvier, P., & Stampanoni, M. (2011). Fossil jawless fish from China foreshadows early jawed vertebrate anatomy. *Nature, 476*, 324–327.

Goudemand, N., Orchard, M. J., Urdy, S., Bucher, H., & Tafforeau, P. (2011). Synchrotron-aided reconstruction of the conodont feeding apparatus and implications for the mouth of the first vertebrates. *Proceedings of the National Academy of Sciences of the United States of America, 108*, 8720–8724.

Green, R. E., Braun, E. L., Armstrong, J., Earl, D., Nguyen, N., Hickey, G., Vandewege, M. W., St John, J. A., et al. (2014). Three crocodilian genomes reveal ancestral patterns of evolution among archosaurs. *Science, 346*, 1254449.

Harada, H., Sato, T., & Nakamura, H. (2016). Fgf8 signaling for development of the midbrain and hindbrain. *Development, Growth & Differentiation, 58*, 437–445.

Heimberg, A. M., Cowper-Sal-lari, R., Semon, M., Donoghue, P. C., & Peterson, K. J. (2010). microRNAs reveal the interrelationships of hagfish, lampreys, and gnathostomes and the nature of the ancestral vertebrate. *Proceedings of the National Academy of Sciences of the United States of America, 107*, 19379–19383.

Hibi, M., Matsuda, K., Takeuchi, M., Shimizu, T., & Murakami, Y. (2017). Evolutionary mechanisms that generate morphology and neural-circuit diversity of the cerebellum. *Development, Growth & Differentiation, 59*, 228–243.

Hidalgo-Sanchez, M., Millet, S., Simeone, A., & Alvarado-Mallart, R. M. (1999). Comparative analysis of Otx2, Gbx2, Pax2, Fgf8 and Wnt1 gene expressions during the formation of the chick midbrain/hindbrain domain. *Mechanisms of Development, 81*, 175–178.

Higuchi, S., Sugahara, F., Pascual-Anaya, J., Takagi, W., Oisi, Y., & Kuratani, S. (2019). Inner ear development in cyclostomes and evolution of the vertebrate semicircular canals. *Nature, 565*, 347–350.

Imura, K., Yamamoto, N., Sawai, N., Yoshimoto, M., Yang, C. Y., Xue, H. G., & Ito, H. (2003). Topographical organization of an indirect telencephalo-cerebellar pathway through the nucleus paracommissuralis in a teleost, Oreochromis niloticus. *Brain, Behavior and Evolution, 61*, 70–90.

Imura, K., Yamamoto, N., Yoshimoto, M., Endo, M., Funakoshi, K., & Ito, H. (2017). Fiber connections of the caudal corpus cerebelli, with special reference to the intrinsic circuitry, in a teleost (Oreochromis niloticus). *Brain, Behavior and Evolution, 89*, 15–32.

Janvier, P. (2002). *Early vertebrates*. Oxford University Press.

Johnston, J. B. (1902). The brain of petromyzon. *The Journal of Comparative Neurology, 12*, 1–82.

Jorqensen, J. M., Shichiri, M., & Genese, F. A. (1998). Morphology of the hagfish inner ear. *Acta Zoologica, 79*, 251–256.

Kani, S., Bae, Y. K., Shimizu, T., Tanabe, K., Satou, C., Parsons, M. J., Scott, E., Higashijima, S., et al. (2010). Proneural gene-linked neurogenesis in zebrafish cerebellum. *Developmental Biology, 343*, 1–17.

Katahira, T., Sato, T., Sugiyama, S., Okafuji, T., Araki, I., Funahashi, J., & Nakamura, H. (2000). Interaction between Otx2 and Gbx2 defines the organizing center for the optic tectum. *Mechanisms of Development, 91*, 43–52.

Kozmik, Z., Holland, N. D., Kalousova, A., Paces, J., Schubert, M., & Holland, L. Z. (1999). Characterization of an amphioxus paired box gene, AmphiPax2/5/8: Developmental expression patterns in optic support cells, nephridium, thyroid-like structures and pharyngeal gill slits, but not in the midbrain-hindbrain boundary region. *Development, 126*, 1295–1304.

Kuraku, S., & Kuratani, S. (2006). Time scale for cyclostome evolution inferred with a phylogenetic diagnosis of hagfish and lamprey cDNA sequences. *Zoological Science, 23*, 1053–1064.

Lannoo, M. J., & Hawkes, R. (1997). A search for primitive Purkinje cells: Zebrin II expression in sea lampreys (Petromyzon marinus). *Neuroscience Letters, 237*, 53–55.

Larsell, O. (1947). The cerebellum of myxinoids and petromyzonts including developmental stages in the lampreys. *The Journal of Comparative Neurology, 86*, 395–445.

Larsell, O. (1967). *The comparative anatomy and histology of the cerebellum*. University of Minnesota Press.

Luo, Z. X., Crompton, A. W., & Sun, A. L. (2001). A new mammaliaform from the early Jurassic and evolution of mammalian characteristics. *Science, 292*, 1535–1540.

Mallatt, J., & Sullivan, J. (1998). 28S and 18S rDNA sequences support the monophyly of lampreys and hagfishes. *Molecular Biology Evolution, 15*, 1706–1718.

Meek, J., Yang, J. Y., Han, V. Z., & Bell, C. C. (2008). Morphological analysis of the mormyrid cerebellum using immunohistochemistry, with emphasis on the unusual neuronal organization of the valvula. *The Journal of Comparative Neurology, 510*, 396–421.

Meulemans, D., & Bronner-Fraser, M. (2007). Insights from amphioxus into the evolution of vertebrate cartilage. *PLoS One, 2*, e787.

Millet, S., Bloch-Gallego, E., Simeone, A., & Alvarado-Mallart, R. M. (1996). The caudal limit of Otx2 gene expression as a marker of the midbrain/hindbrain boundary: A study using in situ hybridisation and chick/quail homotopic grafts. *Development, 122*, 3785–3797.

Montgomery, J. C., Bodznick, D., & Yopak, K. E. (2012). The cerebellum and cerebellum-like structures of cartilaginous fishes. *Brain, Behavior and Evolution, 80*, 152–165.

Morris, S. C., & Caron, J. B. (2014). A primitive fish from the Cambrian of North America. *Nature, 512*, 419–422.

Murakami, T., & Morita, Y. (1987). Morphology and distribution of the projection neurons in the cerebellum in a teleost, Sebastiscus marmoratus. *The Journal of Comparative Neurology, 256*, 607–623.

Murakami, Y., Ogasawara, M., Sugahara, F., Hirano, S., Satoh, N., & Kuratani, S. (2001). Identification and expression of the lamprey Pax6 gene: Evolutionary origin of the segmented brain of vertebrates. *Development, 128*, 3521–3531.

Nelson, J. S. (2016). *Fishes of the world Fifth edition*. Wiley.

Nieuwenhuys, R. (1967). Comparative anatomy of the cerebellum. *Progress in Brain Research, 25*, 1–93.

Nieuwenhuys, R. (1998). *The central nervous system of vertebrates* (Nieuwenhuys, R., TenDonkelaar, H. J., Nicholson, C., ed.), Springer-Verlag Berlin Heidelberg.

Nieuwenhuys, R., & Nicholson, C. (1967). Cerebellum of mormyrids. *Nature, 215*, 764–765.

Northmore, D. P. (2017). Holding visual attention for 400millionyears: A model of tectum and torus longitudinalis in teleost fishes. *Vision Research, 131*, 44–56.

Oisi, Y., Ota, K. G., Kuraku, S., Fujimoto, S., & Kuratani, S. (2013). Craniofacial development of hagfishes and the evolution of vertebrates. *Nature, 493*, 175–180.

Ota, K. G., & Kuratani, S. (2006). The history of scientific endeavors towards understanding hagfish embryology. *Zoological Science, 23*, 403–418.

Ota, K. G., Kuraku, S., & Kuratani, S. (2007). Hagfish embryology with reference to the evolution of the neural crest. *Nature, 446*, 672–675.

Ota, K. G., Fujimoto, S., Oisi, Y., & Kuratani, S. (2011). Identification of vertebra-like elements and their possible differentiation from sclerotomes in the hagfish. *Nature Communications, 2*, 373.

Pani, A. M., Mullarkey, E. E., Aronowicz, J., Assimacopoulos, S., Grove, E. A., & Lowe, C. J. (2012). Ancient deuterostome origins of vertebrate brain signalling centres. *Nature, 483*, 289–294.

Parker, H. J., Bronner, M. E., & Krumlauf, R. (2014). A Hox regulatory network of hindbrain segmentation is conserved to the base of vertebrates. *Nature, 514*(7523), 490–493.

Pascual-Anaya, J., Sato, I., Sugahara, F., Higuchi, S., Paps, J., Ren, Y., Takagi, W., Ruiz-Villalba, A., et al. (2018). Hagfish and lamprey Hox genes reveal conservation of temporal colinearity in vertebrates. *Nature Ecology & Evolution, 2*, 859–866.

Pose-Mendez, S., Candal, E., Adrio, F., & Rodriguez-Moldes, I. (2014). Development of the cerebellar afferent system in the shark Scyliorhinus canicula: Insights into the basal organization of precerebellar nuclei in gnathostomes. *The Journal of Comparative Neurology, 522*, 131–168.

Reifers, F., Bohli, H., Walsh, E. C., Crossley, P. H., Stainier, D. Y., & Brand, M. (1998). Fgf8 is mutated in zebrafish acerebellar (ace) mutants and is required for maintenance of midbrain-hindbrain boundary development and somitogenesis. *Development, 125*, 2381–2395.

Rowe, T. B., Macrini, T. E., & Luo, Z. X. (2011). Fossil evidence on origin of the mammalian brain. *Science, 332*, 955–957.

Sato, T., Araki, I., & Nakamura, H. (2001). Inductive signal and tissue responsiveness defining the tectum and the cerebellum. *Development, 128*, 2461–2469.

Shu, D. G., Morris, S. C., Han, J., Zhang, Z. F., Yasui, K., Janvier, P., Chen, L., Zhang, X. L., et al. (2003). Head and backbone of the Early Cambrian vertebrate Haikouichthys. *Nature, 421*, 526–529.

Stencio, E. A. (1927). "The Dountonian and Devonian vertebrates of Spitsbergen. I, Family Cephalaspidae", (I kommisjon hos J. Dybwad, 1927).

Sugahara, F., Aota, S., Kuraku, S., Murakami, Y., Takio-Ogawa, Y., Hirano, S., & Kuratani, S. (2011). Involvement of Hedgehog and FGF signalling in the lamprey telencephalon: Evolution of regionalization and dorsoventral patterning of the vertebrate forebrain. *Development, 138*, 1217–1226.

Sugahara, F., Pascual-Anaya, J., Oisi, Y., Kuraku, S., Aota, S., Adachi, N., Takagi, W., Hirai, T., et al. (2016). Evidence from cyclostomes for complex regionalization of the ancestral vertebrate brain. *Nature, 531*, 97–100.

Takahashi, T., & Holland, P. W. (2004). Amphioxus and ascidian Dmbx homeobox genes give clues to the vertebrate origins of midbrain development. *Development, 131*, 3285–3294.

Takio, Y., Kuraku, S., Murakami, Y., Pasqualetti, M., Rijli, F. M., Narita, Y., Kuratani, S., & Kusakabe, R. (2007). Hox gene expression patterns in Lethenteron japonicum embryos--insights into the evolution of the vertebrate Hox code. *Developmental Biology, 308*, 606–620.

Tzika, A. C., Ullate-Agote, A., Grbic, D., & Milinkovitch, M. C. (2015). Reptilian transcriptomes v2.0: An extensive resource for sauropsida genomics and transcriptomics. *Genome Biology and Evolution, 7*, 1827–1841.

Wada, S., Tokuoka, M., Shoguchi, E., Kobayashi, K., Di Gregorio, A., Spagnuolo, A., Branno, M., Kohara, Y., et al. (2003). A genomewide survey of developmentally relevant genes in Ciona intestinalis. II. Genes for homeobox transcription factors. *Development Genes and Evolution, 213*, 222–234.

Walsh, S. A., Iwaniuk, A. N., Knoll, M. A., Bourdon, E., Barrett, P. M., Milner, A. C., Nudds, R. L., Abel, R. L., et al. (2013). Avian cerebellar floccular fossa size is not a proxy for flying ability in birds. *PLoS One, 8*, e67176.

Watson, C., Paxinos, G., & Puelles, L. (2011). *The mouse nervous system*. Academic Press.

Wicht, H. (1996). The brains of lampreys and hagfishes: Characteristics, characters, and comparisons. *Brain, Behavior and Evolution, 48*, 248–261.

Wingate, R. J., & Hatten, M. E. (1999). The role of the rhombic lip in avian cerebellum development. *Development, 126*, 4395–4404.

Witmer, L. M., Chatterjee, S., Franzosa, J., & Rowe, T. (2003). Neuroanatomy of flying reptiles and implications for flight, posture and behaviour. *Nature, 425*, 950–953.

Witton, M. P. (2013). *Pterosaurs: Natural history, evolution, anatomy*. Princeton University Press.

Yamasaki, T., Kawaji, K., Ono, K., Bito, H., Hirano, T., Osumi, N., & Kengaku, M. (2001). Pax6 regulates granule cell polarization during parallel fiber formation in the developing cerebellum. *Development, 128*, 3133–3144.

Yao, Y., Minor, P. J., Zhao, Y. T., Jeong, Y., Pani, A. M., King, A. N., Symmons, O., Gan, L., et al. (2016). Cis-regulatory architecture of a brain signaling center predates the origin of chordates. *Nature Genetics, 48*, 575–580.

Yopak, K. E., Pakan, J. M. P., & Wylie, D. (2017). The cerebellum of nonmammalian vertebrates. In J. H. Kaas (Ed.), *Evolution of nervous systems* (2nd ed.). Academic Press.

Zhu, M., Yu, X., Ahlberg, P. E., Choo, B., Lu, J., Qiao, T., Qu, Q., Zhao, W., et al. (2013). A Silurian placoderm with osteichthyan-like marginal jaw bones. *Nature, 502*, 188–193.

Chapter 2
Cerebellum-Like Systems in Actinopterygian Fishes with a Special Focus on the Diversity of Cerebellum-Like System in the Mesencephalon

Naoyuki Yamamoto and Hanako Hagio

Abbreviation

AUR	Auricula cerebelli
CC	Corpus cerebelli
CC_4	Lobule 4 of corpus cerebelli
CCg	Granular layer of corpus cerebelli
CLS	Cerebellum-like system
CrC	Cerebellar crest
DCN	Dorsal cochlear nucleus
dlz	Dorsolateral zone of mormyrid ELL
DO	Descending octaval nucleus
DOdm	Dorsomedial zone of DO
DON	Dorsal octavolateral nucleus
EG	Eminentia granularis
EGp	Posterior part of EG
ELL	Electrosensory lateral line lobe
flm	Medial longitudinal fascicle
EOD	Electric organ discharge
GR	Corpus glomerulosum pars rotunda
GP	Granule population of McCormick and Hernandez (1996)
HB	Habenula
IP	Interpeduncular nucleus

N. Yamamoto (✉)
Laboratory of Fish Biology, Department of Animal Sciences, Graduate School of Bioagricultural Sciences, Nagoya University, Nagoya, Japan
e-mail: nyama@agr.nagoya-u.ac.jp

H. Hagio
Laboratory of Fish Biology, Department of Animal Sciences, Graduate School of Bioagricultural Sciences, Nagoya University, Nagoya, Japan

Institute for Advanced Research, Nagoya University, Nagoya, Japan

H. Mizusawa, S. Kakei (eds.), *Cerebellum as a CNS Hub*,
Contemporary Clinical Neuroscience, https://doi.org/10.1007/978-3-030-75817-2_2

l	Lateral fiber part of TL
L	Lateral granular cell part of TL
LC	Lobus caudalis
LCg	Granular layer of LC
LFN	Lateral funicular nucleus
LI	Inferior lobe
ll	Lateral lemniscus
m	Medial fiber part of TL
M	Medial granular cell part of TL
MLL	Mechanosensory lateral line lobe
mz	Medial zone of mormyrid ELL
NAT	Anterior tuberal nucleus
NLV	Lateral valvular nucleus
NLVpl	Posterolateral part of NLV of Ito and Yoshimoto (1990)
NLVpm	Posteromedial part of NLV of Ito and Yoshimoto (1990)
NM	Medial octavolateral nucleus
NPC	Nucleus paracommissuralis
NPE	Nucleus praeeminentialis
NPEl	Lateral part of NPE
NPEm	Medial part of NPE
NSE	Nucleus subeminentialis
NSV	Nucleus subvalvularis
nII	Optic nerve
NIII	Oculomotor nucleus
NVm	Trigeminal motor nucleus
nVIII	Octaval nerve
P	Pituitary
pc	Posterior commissure
RCLS	Rhombencephalic CLS
pLL	Posterior lateral line nerve
Rm	Middle reticular formation
rV	Central root of trigeminal nerve
rVII	Central root of facial nerve
SAC	Stratum album centrale
SFGS	Stratum fibrosum et griseum superficiale
SGC	Stratum griseum centrale
SGN	Secondary gustatory nucleus
SM	Stratum marginale
SO	Stratum opticum
SOP	Secondary octaval population
SPV	Stratum periventriculare
SV	Vascular sac
TE	Telencephalon

TL	Torus longitudinalis
TLa	Torus lateralis
TO	Optic tectum
TS	Torus semicircularis
TSc	Central nucleus of TS
TSl	Lateral nucleus of TS
TSvl	Ventrolateral nucleus of TS
VC	Valvula cerebelli
VCl	Lateral lobe of VC
VCm	Medial lobe of VC
vlz	Ventrolateral zone of mormyrid ELL

2.1 Introduction

We begin this chapter with a brief description of the cerebellum in teleosts, the largest group in actinopterygians, which is necessary to understand the cerebellum-like systems (CLSs). The cerebellum of teleosts is composed of the molecular, Purkinje cell, and granular layers, as in mammals. In teleosts, however, eurydendroid cells, which receive synaptic inputs from Purkinje cells and send fibers out of the cerebellum, also situate in the Purkinje cell layer (Fig. 2.1). The cerebellum of teleosts is composed of the corpus cerebelli (CC), valvula cerebelli (VC), lobus caudalis (LC), and eminentia granularis (EG) (Fig. 2.2).

CC is an unpaired cerebellar region (Fig. 2.2), and afferents to this cerebellar region include spinal projections (Finger, 2000) and telencephalic inputs relayed by the lateral valvular nucleus (NLV) (Yang et al., 2004). The corpus hence may include components comparable to the cerebellar hemisphere and vermis of mammals. VC is not situated dorsal to the pons but protrudes rostrally from CC into the mesencephalic ventricle (Fig. 2.2). Among vertebrates, VC is found only in actinopterygians and receives descending telencephalic inputs relayed by NLV (Yang et al., 2004) but does not appear to receive spinal inputs. LC is the caudalmost region of the cerebellum, which situates ventrocaudally adjacent to CC (Fig. 2.2). EG is a protrusion situated caudolaterally adjacent to CC; receives primary afferents carrying auditory, vestibular, and lateral line modalities (Yamamoto & Ito, 2005; Noro et al., 2007); and projects to the molecular layer of LC. LC and EG together are usually regarded as the vestibulolateralis cerebellum and might be comparable to the flocculo-nodular lobe of mammals. An important point is that the granule cells of LC and EG are cerebellar components that are, at the same time, part of CLSs, as will be detailed below.

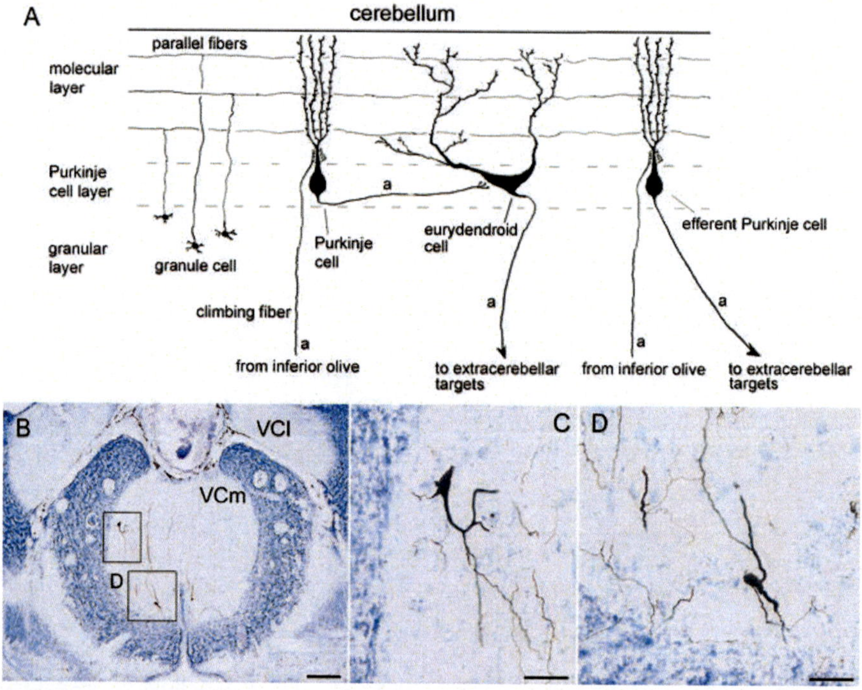

Fig. 2.1 Laminar organization, component cells, and their connections of the cerebellum in teleosts. (**a**) Schematic line drawings of cerebellar layers and cerebellar neurons. Note that eurydendroid cells, or cerebellar efferent neurons, are present in the Purkinje cell layer. The majority of Purkinje cells send the axon to eurydendroid cells, while a minor type of Purkinje cells project to extracerebellar targets. a: axon. (Modified from Yamamoto (2018)). (**b**) Transverse section through the valvula cerebelli of goldfish with a tracer injection into the tegementum (Nakase and Yamamoto, unpublished data). The section is counter-stained with Nissl staining. (**c**) and (**d**) Enlarged photographs of boxed areas in panel **b**, showing eurydendroid cells labeled from the tegmentum. For other abbreviations, see list. Scale bar = 200 μm in (**b**); 50 μm in (**c**, **d**)

2.2 Widespread Distribution of CLSs in Vertebrates (Figs. 2.1 and 2.2)

CLSs are found in most vertebrate taxa (Bell, 2002; Fig. 2.3). The name "cerebellum-like" comes from similarities to the cerebellum in terms of component cells, laminar organization, and neural circuitry. Cells that are similar to granule and Purkinje cells as well as molecular layer-like neuropil composed of apical dendrites of Purkinje-like cells and parallel fiber-like axons from granular cells are found in CLSs.

In mammals except monotremes, a cerebellum-like nucleus can be found in the rhombencephalon, which is the dorsal cochlear nucleus. There are two cerebellum-like nuclei in the rhombencephalon in lampreys, cartilaginous fishes, sarcopterygian

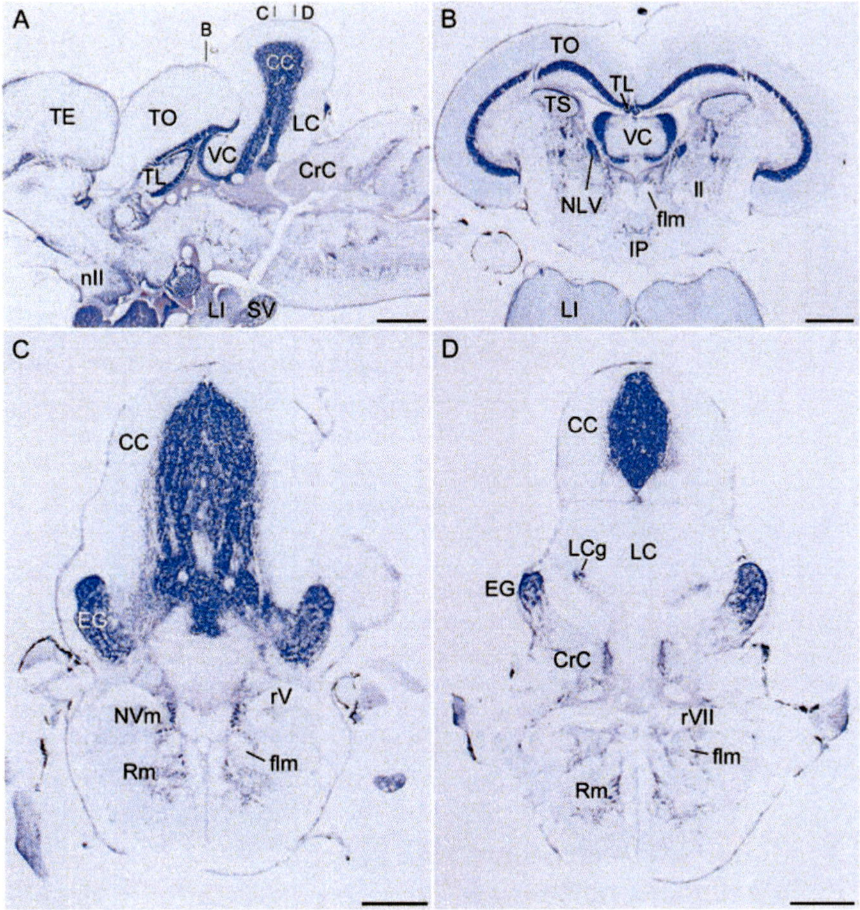

Fig. 2.2 Nissl-stained sections of yellowfin goby. (**a**) Sagittal section through the torus longitudinalis (TL), valvula cerebelli (VC), corpus cerebelli (CC), lobus caudalis (LC), and cerebellar crest (CrC). (**b**–**d**) Transverse sections through VC (**b**), CC and eminentia granularis (EG) (**c**), and LC (**d**). For other abbreviations, see list. Scale bar = 500 μm in (**a**–**d**)

fishes, and urodeles: the dorsal octavolateral nucleus (DON), or a primary electrosensory lateral line nucleus and the medial octavolateral nucleus (NM), or a primary mechanosensory lateral line nucleus. In anurans and some hagfish (myxinoids) only NM is present. The most diverse CLSs are found in actinopterygians, as detailed below. A CLS is missing in some hagfish (eptaretids) and reptiles including birds.

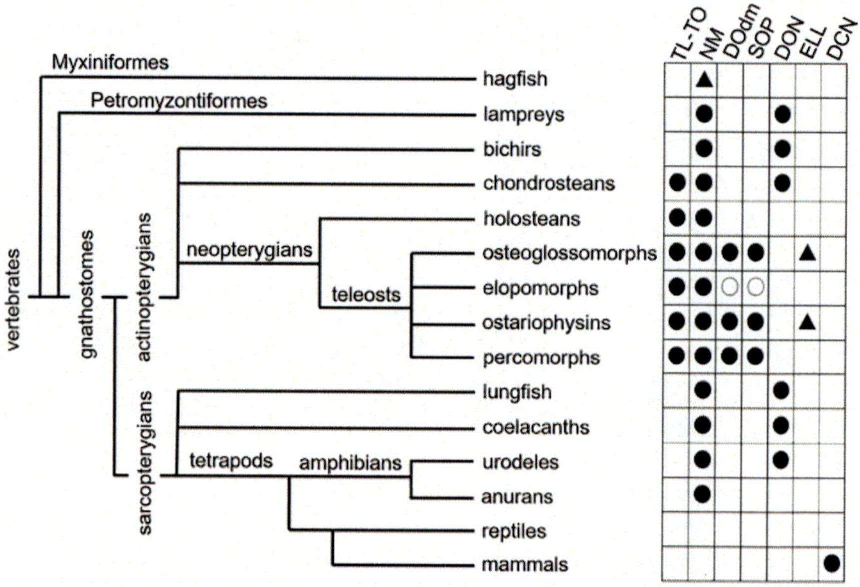

Fig. 2.3 Phylogenetic distribution of cerebellum-like systems in vertebrates. Closed circle, present; open circle, probably present (experimental investigation yet to be done); closed triangle, present in some species. DCN dorsal cochlear nucleus, DOdm dorsomedial zone of descending octaval nucleus, DON dorsal octavolateral nucleus, ELL electrosensory lateral line lobe, NM medial octavolateral nucleus, SOP secondary octaval population, TL-TO torus longitudinalis-optic tectum system

2.3 Multitude of Cerebellum-Like Systems in Actinopterygians

Actinopterygians or ray-finned fishes include cladistians (bichirs), chondrosteans (sturgeons and paddlefish), and neopterygians (holosteans [gar and amia] and teleosts). In all actinopterygians, NM is present in the rhombencephalon (Fig. 2.3). In addition, a CLS involved in electrosensensory processing is present in the rhombencephalon in electrosensory species. Cladistians and chondrosteans possess DON. An electrosensory lateral line system is lacking in gar, amia, and basal teleost taxa like eels, and it is widely accepted that the electrosensory system was once lost at the base of neopterygian radiation. However, the electrosensory lateral line system re-evolved in some teleosts. In these fishes, the primary electrosensory structure is cerebellum-like and called the electrosensory lateral line lobe (ELL). Furthermore, there are two additional cerebellum-like nuclei (Finger, 1984; McCormick & Hernandez, 1996; Yamamoto & Ito, 2005): the dorsomedial zone of descending octaval nucleus (DOdm) and the secondary octaval population (SOP).

In sharp contrast to other vertebrates, in which CLSs are present only in the rhombencephalon, a still different CLS, the torus longitudinalis (TL)-optic tectum

(TO) system, is present in the mesencephalon of actinopterygians (Fig. 2.3). A diencephalic structure, the nucleus rostrolateralis of Butler and Saidel (1991), is considered as a CLS by some authors (e.g., Bell, 2002). Much remains to be studied for this system, and it will not be dealt with further. Below, we will describe the organization, cellular morphology, and circuitry of CLSs in actinopterygians with a special focus on the TL-TO system.

2.4 Rhombencephalic CLSs (RCLSs) in Actinopterygians

2.4.1 Structural Organization and Neural Circuitry

Medial Octavolateral Nucleus (NM)

NM is covered dorsally by a neuropil called cerebellar crest (CrC) (Fig. 2.4). The nucleus is composed of large fusiform neurons and smaller round neurons. The large fusiform neurons line just beneath CrC. These cells are called crest cells, since they extend an apical dendrite into CrC (Fig. 2.5). Crest cells express glutamate receptor ∂2 as do Purkinje cells, indicating their similarity at the molecular level (Mikami et al., 2004). The apical dendrite bifurcates into a number of spiny branches in a fashion somewhat similar to that of Purkinje cell dendrites. However, unlike Purkinje cell dendrites that distribute in a two-dimensional plane, the dendritic arbor of crest cells forms a three-dimensional cone. The apical dendrites receive synaptic input from granule cells of EG and LC as well as from granular cells of nucleus praeeminentialis (NPE) (Yamamoto et al., 2010) (Figs. 2.6 and 2.7). Thin fibers from these afferent sources enter CrC and run side by side parallel to the surface of the crest (Fig. 2.7), similarly to parallel fibers of the cerebellum. This is not odd, since EG and granule cells of LC also give rise to parallel fibers in the vestibulolateralis cerebellum, an authentic cerebellar division. Some of the crest cells possess a basal dendrite (Fig. 2.5), and others do not (New et al., 1996). The crest cells presumably receive mechanosensory lateral line inputs directly on basal dendrites from primary afferents that terminate in NM or indirectly via interneurons in NM. Many, if not all, crest cells project to extracerebellar targets such as the torus semicircularis (TS; homologue of inferior colliculus), more specifically the ventrolateral nucleus of TS (TSvl; McCormick & Hernandez, 1996; Yamamoto et al., 2010) (Fig. 2.5).

Dorsomedial Zone of Descending Octaval Nucleus (DOdm)

The descending octaval nucleus (DO) receives vestibular, auditory, and minor lateral line inputs, with the primary auditory afferents terminating in the dorsomedial zone (DOdm) (McCormick & Braford Jr., 1994; O'Marra & McCormick, 1999; Yamamoto & Ito, 2005). DOdm is a paramedian nucleus situated medially adjacent

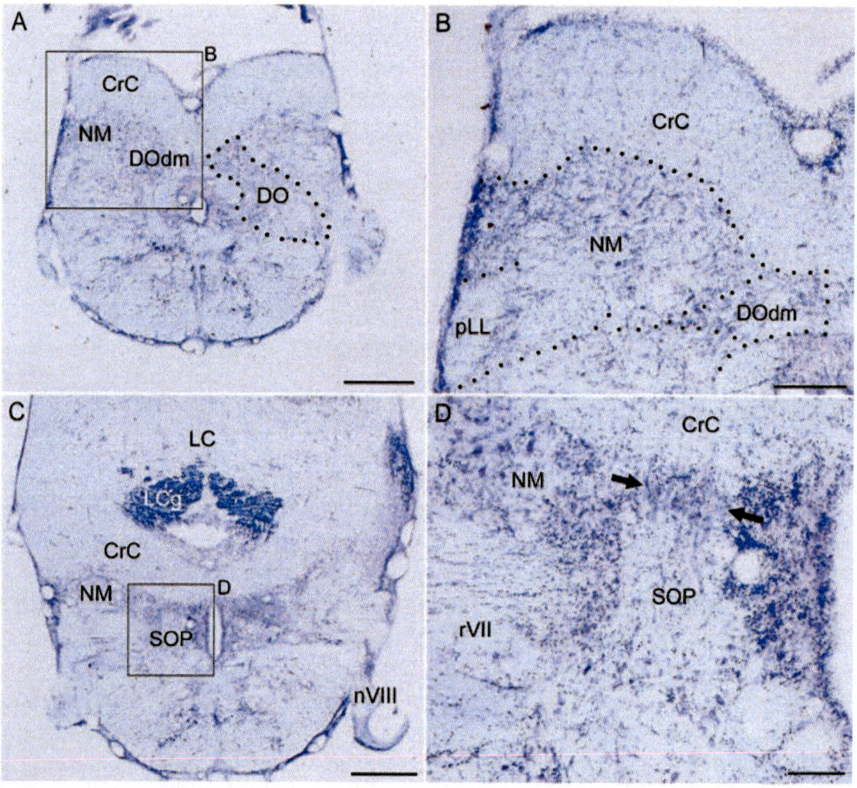

Fig. 2.4 Nissl-stained transverse sections of goldfish. (**a**) Section through medial octavolateral nucleus (NM) and dorsomedial zone of descending octaval nucleus (DOdm). (**b**) Enlarged photograph of boxed area in panel (**a**). (**c**) Section through medial octavolateral nucleus (NM) and secondary octaval population (SOP). (**d**) Enlarged photograph of boxed area in panel (**c**). Arrows indicate the layer of crest cells. For other abbreviations, see list. Scale bar = 500 μm in (**a, c**); 200 μm in (**b**); 100 μm in (**d**)

to NM (Fig. 2.4). Similarly to NM, DOdm is covered dorsally by the CrC, and crest cells are present along the dorsal margin of the nucleus. Crest cells in DOdm quite likely receive synaptic input on the apical dendrite from granular cells in EG, LC, and NPE, as is the case for NM. Crest cells presumably receive inputs directly or indirectly via interneurons from primary auditory afferents terminating in DOdm and project to the central nucleus of TS (TSc) (McCormick & Hernandez, 1996; Yamamoto & Ito, 2005).

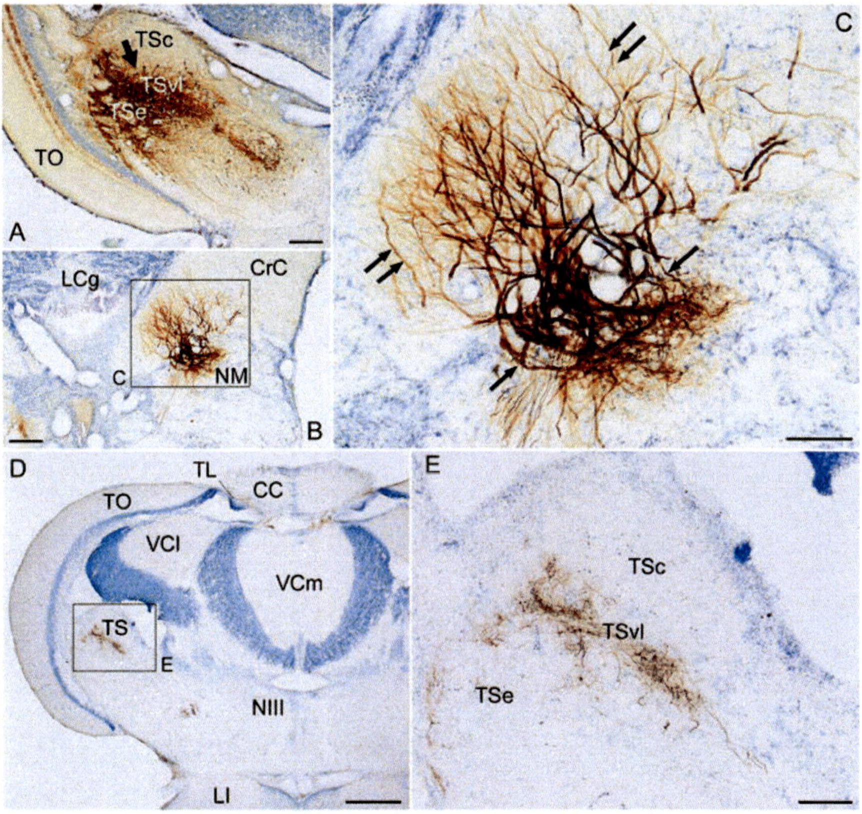

Fig. 2.5 Transverse sections of goldfish with injection of biotinylated dextran amine into torus semicircularis (TS) (**a–c**) and biocytin into medial octavolateral nucleus (NM) and cerebellar crest (CrC) (**d**, **e**). Sections are counter-stained with Nissl staining. (**a**) Injection site (arrow) into ventrolateral (TSvl) and external (TSe) nuclei of TS. (**b**) Retrogradely labeled crest cells of NM. (**c**) Enlarged photograph of the boxed area in panel (**b**). double arrows: apical dendrites of crest cells; arrow: basal dendrites of crest cells. (**d**) Labeled terminals in TS (injection site shown in Fig. 2.6a). (**e**) Enlarged photograph of the boxed area in panel (**d**), showing labeled terminals restricted in TSvl. For other abbreviations, see list. Scale bar = 200 μm in (**a**, **b**); 100 μm in (**c**, **e**); 500 μm in (**d**)

Secondary Octaval Population (SOP)

SOP, which is also called the superior olive, situates just rostral to DOdm. Similarly to the latter nucleus and NM, SOP is covered dorsally by CrC, and there are crest cells on the dorsal margin of the nucleus (Fig. 2.4). Ventral to the crest cell region is the spherical small cell zone, and fusiform neurons are present in the ventralmost zone of the nucleus. The morphology of crest cells is much the same as those in DOdm and NM (McCormick & Hernandez, 1996; Yamamoto & Ito, 2005). Afferents to the medial crest region covering SOP come from granule cells in EG and NPE as well as from the granule population (GP) of McCormick and Hernandez (1996) (Fig. 2.6). Many crest cells (if not all) project to TSc. The basal dendrites of crest

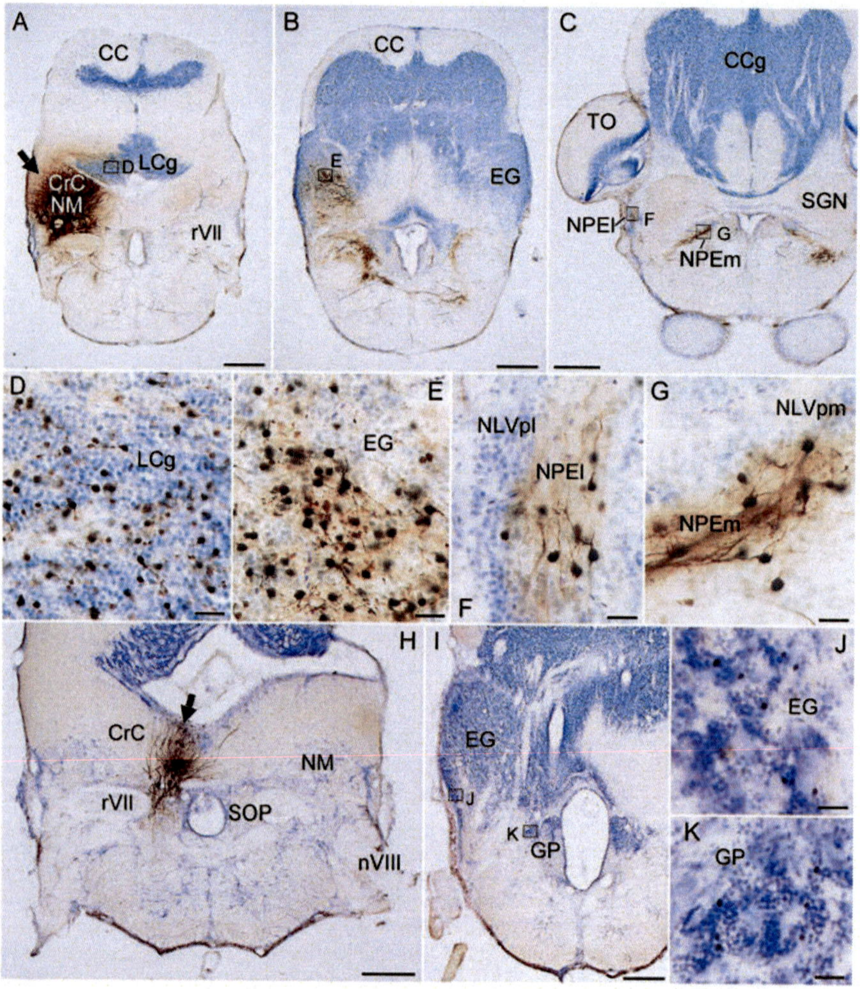

Fig. 2.6 Transverse sections showing labeled structures in the goldfish brain following a biocytin injection into the medial octavolateral nucleus (NM) (**a–g**) and injection of biotinylated dextran amine into the medial zone of cerebellar crest (CrC) covering the seconadry octaval population (SOP) (**h–k**). Sections are counter-stained with Nissl staining. (**a**) Injection site into NM and CrC (arrow). (**b**) Section through the eminentia granularis (EG). (**c**) Section through the isthmic region. (**d–g**) Retrogradely labeled cells in the granular layer of lobus caudalis (LCg), medial EG, and lateral part (NPEl) and medial part (NPEm) of nucleus preeminentialis, respectively. (**h**) Injection site restricted to the medial CrC (arrow). (**i**) Section through EG. (**j, k**) Enlarged photographs of boxed areas in panel I, showing retrogradely labeled cells in EG and granule population (GP), respectively. For other abbreviations, see list. Scale bar = 500 μm in (**a–c, h–i**); 20 μm in (**d–g, j–k**)

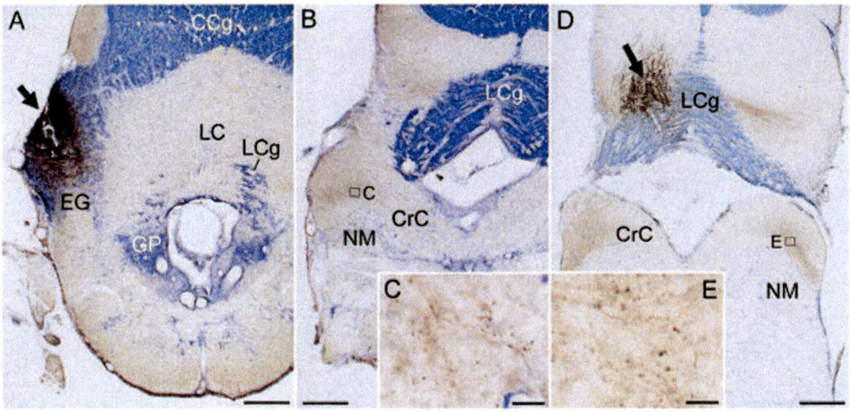

Fig. 2.7 Transverse sections of goldfish brain showing labeled structures following biocytin injections into the eminentia granularis (EG) (**a–c**) and lobus caudalis (LC) (**d**, **e**). Sections are counterstained with Nissl staining. (**a**) Biocytin injection site into EG (arrow). (**b**) Section through the cerebellar crest (CrC). (**c**) Higher magnification photograph of the boxed area in (**b**), showing labeled "parallel fibers" in CrC. (**d**) Biocytin injection site into LC (arrow), including the granular layer. (**e**) Higher magnification photograph of the boxed area in (**d**), showing labeled "parallel fibers" in CrC. For other abbreviations, see list. Scale bar = 500 μm in (**a–b**, **d**); 10 μm in (**c**, **e**)

cells quite likely receive secondary auditory inputs from DOdm neurons whose axons terminate in SOP (Yamamoto & Ito, 2005).

Dorsal Octavolateral Nucleus (DON)

Among actinopterygians, DON is present only in polypteryforms and chondrosteans. DON situates dorsomedial to CrC in polypteryiforms (Fig. 2.8) and dorsal to the crest in chondrosteans (McCormick, 1982), configurations different from that of NM, DOdm, and SOP in teleosts. DON receives primary electrosensory afferents through the lateral line nerves. In paddlefish large DON neurons projecting to the optic tectum possess multiple dendrites extending into CrC and a dendrite coursing to the opposite direction, which probably receives electrosensory input similarly to the basal dendrite of crest cells in other RCLSs (Hofmann et al., 2002). Granule cell populations present rostral or close to DON likely supply "parallel fibers" in CrC.

Electrosensory Lateral Line Lobe (ELL)

Among teleosts, only mormyrids, *Xenomystus* (African brown knifefish), gymnotids, and silurids possess the electrosensory system. Mormyrids and *Xenomystus* are closely related taxa of osteoglossiforms. Likewise, gymnotids and silurids are closely related taxa of ostariophysins. It is considered that an electrosensory system

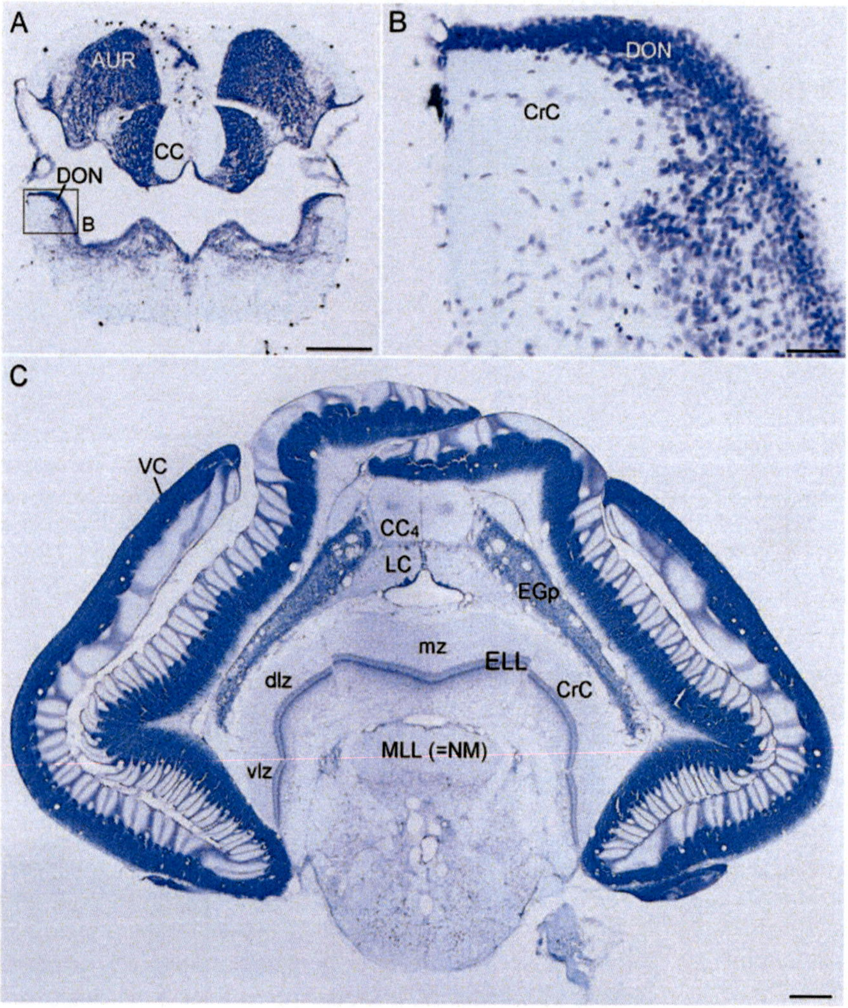

Fig. 2.8 Nissl-stained transverse sections of gray bichir (**a**, **b**) and elephantnose fish (**c**). (**a**) Section through the dorsal octavolateral nucleus (DON). (**b**) Higher magnification photograph of the boxed area in panel (**a**), showing DON and cerebellar crest (CrC). (**c**) Section through the electrosensory lateral line lobe (ELL). For other abbreviations, see list. Scale bar = 500 μm in (**a**, **c**); 50 μm in (**b**)

re-evolved twice independently in osteoglossiforms and ostariophysins (Bullock et al., 1983). Among the four teleost taxa, mormyrids and gymnotids are not only electrosensory but also weakly electric fish that produce weak electric currents by the electric organ. Electric organ discharges (EODs) are used to perceive objects by sensing the distortion of the electric field and to communicate with other individuals.

Like other cerebellum-like nuclei enumerated above, ELL is covered by the "molecular layer" or CrC that is composed of "parallel fibers" coming from populations of granular cells (Fig. 2.8). Also, there are crest cells in the lobe, together with other cell types. Although ELLs in gymonotids and mormyrids show several differences, there are a number of similarities as well, suggesting that the two taxa share similar molecular mechanisms responsible for making up of CLS.

Silurids Silurids respond only passively to electric signals from outside with receptors called the ampullary organ. ELL is comprised of four layers with the molecular layer being the most superficial. Just below the molecular layer, two types of crest cells are lined up. Both types are common in having an apical dendrite that bifurcates into spiny branches in the molecular layer. One of the crest cell type possesses a long basal dendrite and the other a shorter one. It is considered that the former type receives direct electrosensory inputs from primary afferents, and the latter indirectly relayed by GABAergic interneurons (Finger, 1986). Crest cells project to NPE and lateral nucleus of TS (TSl). The molecular layer receives parallel fibers from EG, granule cells of LC, and NPE, and it is known that fibers from the former structures terminate superficially and those from the latter deeper (Tong & Finger, 1983).

Mormyrids ELL of mormyrids extends rather dorsally and overlies NM that is also called mechanosensory lateral line lobe (Fig. 2.8). ELL of elephantnose fish *Gnathnemus petersi* is composed of three divisions: medial, dorsolateral, and ventrolateral (Fig. 2.8). The medial and dorsolateral zones receive primary afferents from different subtypes of tuberous receptor organs called mormyromasts, which are respectively used for active electroreception and encoding amplitudes of EODs. The ventrolateral zone receives primary afferents from ampullary receptors for passive electroreception. Primary afferent projections to each subdivision are topographically organized; there are three maps within the lobe reflecting receptor distributions on the skin. Mormyrids possess the third type of receptor, the knollenorgans, which are very sensitive to EODs. Information coded by the knollenorgan reaches the nucleus of lateral line lobe, not ELL. Among a variety of cell types in ELL, four cell types extend spiny apical dendrites into the molecular layer and may be regarded as crest cells. Two of them possess a medium-sized soma in the ganglionic layer, which is just below the molecular layer, and are called medium-sized ganglionic cells. They are GABAergic interneurons with basal dendrites (Meek et al., 1996, 1999). The other two cell types are large, glutamatergic, and project to TSl and NPE. One of the cell types is called large ganglionic cells whose soma situates in the ganglionic layer, while the other type is called large fusiform cells whose soma situates deeper. The four types appear to receive electrosensory inputs only indirectly via interneurons (Meek et al., 1999). One type of medium-sized ganglionic cells (type II) sends an axon to the soma and basal dendrites of large ganglionic cell, and the other type (type I) to large fusiform cell. The superficial zone of the molecular layer receives parallel fibers from the posterior part of EG (EGp) and deep zone from NPE.

Gymnotids Gymnotids possess ampullary organs for passive electric sense and tuberous organs to detect EODs for active electroreception. ELL is comprised of four zones: the medial, anteromedial, centrolateral, and lateral zones. The medial zone receives inputs from ampullary organs in a topographic manner. The latter three zones receive inputs from tuberous organs through primary afferents that give rise to collaterals to form a topographic map in each zone (Bell, 2002). There are three major types of crest cells: basilar pyramidal cells, non-basilar pyramidal cells, and polymorphic cells. Both pyramidal cell types extend spiny, apical dendrites into the molecular layer and project to TS and dorsal NPE, but the types differ in that basilar pyramidal cells receive electrosensory inputs on the basal dendrites, while basal dendrites are lacking for non-basilar pyramidal cells (Bastian, 1981). Polymorphic cells extend smooth apical dendrites into the cerebellar crest, are equipped with additional dendrites reaching deeper layers to receive electrosensory inputs, and give rise to an axon terminating on pyramidal cells. The molecular layer receives parallel fibers that originate from EGp and NPE.

2.4.2 *Functional Significances of RCLSs*

There are a number of common features in the circuitries of RCLSs. First of all, most of crest cell types or similar cells called under different names such as pyramidal cells (hereafter all will be called crest cells for simplicity) receive sensory inputs directly from the primary afferents upon the basal dendrite or indirectly via interneurons. Second, crest cells receive synaptic inputs from "parallel fibers" upon their apical dendrites (mostly spiny). The parallel fibers come from cerebellar or isthmic populations of granular cells such as EG, granule cells of LC, and NPE. Third, RCLSs send ascending fibers that terminate in TS and NPE. Crest cells are a major source of ascending projections, although there are crest cell types that are interneurons (e.g., polymorphic cells of gymnotids). In this regard, efferent crest cells are similar to eurydendroid cells in the cerebellum of teleosts, and crest cells with local projections are similar to Purkinje cells. Alternatively, efferent crest cells correspond to a minor type of Purkinje cells that project directly to extracerebellar targets (Brochu et al., 1990) (Fig. 2.1). Fourth, none of the RCLSs receive climbing fibers from the inferior olive or comparable nuclei, unlike the cerebellum. These similarities appear to reflect common functional features of RCLSs that are considered to serve for elimination of self-generated sensory inputs (Bell, 2001; Fig. 2.9).

The best known among RCLSs is the functions of ELL. Parallel fibers to ELL come from EGp and NPE. Afferent connections to EGp have been studied extensively in mormyrids and gymnotids. In mormyrids EGp receives information such as signals associated with the motor command for EODs (corollary discharge signals) and proprioceptive signals related to posture (Bell et al., 1992). The ampullary zone of ELL serves to detect extraneous electric inputs, and the sensory inputs that occur in response to EODs generated by the fish itself are distracting noises. EGp-mediated inputs reaching the ampullary zone serve to eliminate

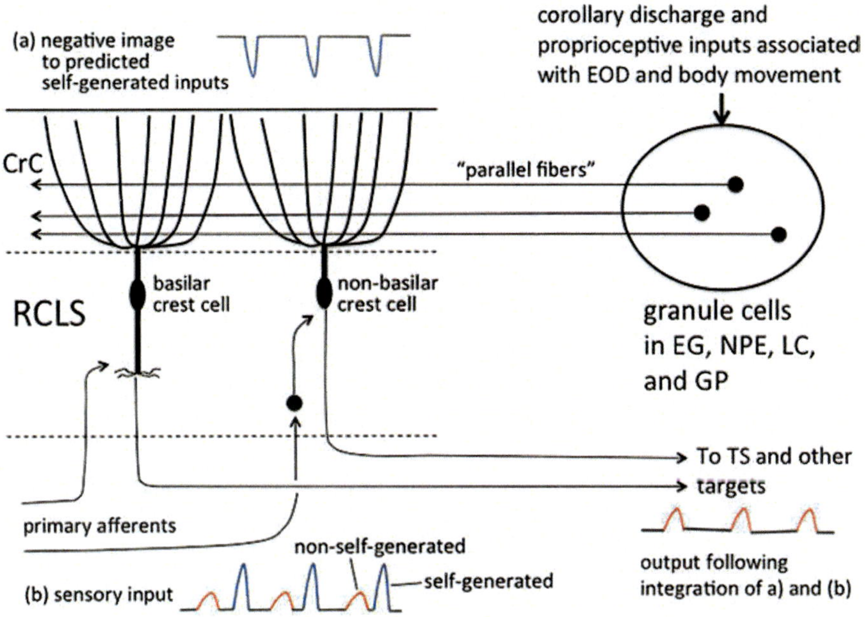

Fig. 2.9 Schematic diagram illustrating elimination or minimization of self-generated sensory inputs by the rhombencephalic cerebellum-like system (ECLS), making use of corollary discharge and proprioceptive inputs. For other abbreviations, see list

self-generated sensory inputs (Bell, 1981). Corollary discharge inputs arriving at the molecular layer are integrated with the sensory inputs from ampullary organs by crest cells, to learn to generate the negative image to predicted sensory responses to EODs from the fish itself. Thus, distracting inputs are eliminated or minimized (Bell, 2001). Similar processes work to eliminate unwanted noises caused by the changes in posture. Detailed studies in gymnotids have been done for ELL region serving for active electrolocation. Bending of the tail causes changes in the electric signal received by the electrosensory receptors, which has nothing to do with the extraneous stimuli. Proprioceptive information reaching the molecular layer mediated by EGp is utilized to eliminate or minimize such interfering noises, through the mechanism generating negative-image to predicted, self-generated sensory inputs (Bastian, 1986; Bell, 2001).

Subtraction of self-generated sensory inputs is not limited for ELL. Studies indicate that a similar mechanism also works for DON in passively electrosensory skates, in which respiratory movements induce strong stimulation of ampullary receptors (Bell, 2001). The same quite likely applies to the other cerebellum-like nuclei: NM, DOdm, and SOP. CrC in cyprinids receives "parallel fibers" from EG, granule cells of LC, NPE, and GP (McCormick & Hernandez, 1996; Yamamoto & Ito, 2005). EG receives primary afferents from the mechanosensory lateral line and octaval nerves (Noro et al., 2007), ascending fibers from NM (mechanosensory lateral line), lateral funicular nucleus (LFN; possibly proprioceptive), and spinal cord (possible proprioceptive and/or corollary discharge signal associated with swimming) (Yamamoto et al., 2010). NM, SOP, LFN, spinal cord, trigeminal nerve, and TSvl project to NPE (Xue et al., 2006; Yamamoto et al., 2010). Thus, EG and NPE likely provide the cerebellum-like nuclei with proprioceptive inputs and corollary discharge signals to eliminate self-generated mechanosensory lateral line and auditory inputs.

2.5 Mesencephalic Cerebellum-Like System in Acanthopterygians: TL-TO System

2.5.1 Structural Organization and Neural Circuitry

The mesencephalic CLS is composed of TL, and stratum marginale (SM) and pyramidal cells of the optic tectum (Figs. 2.10 and 2.11). The TL is a paramedian cell mass hanging down from the optic tectum into the mesencephalic ventricle. Near its rostral pole, the torus situates dorsally adjacent to the posterior commissure, where the cross-sectional area is the largest in most species. The torus is composed mainly of densely packed small neurons called granule cells (Fig. 2.11). Larger cells have

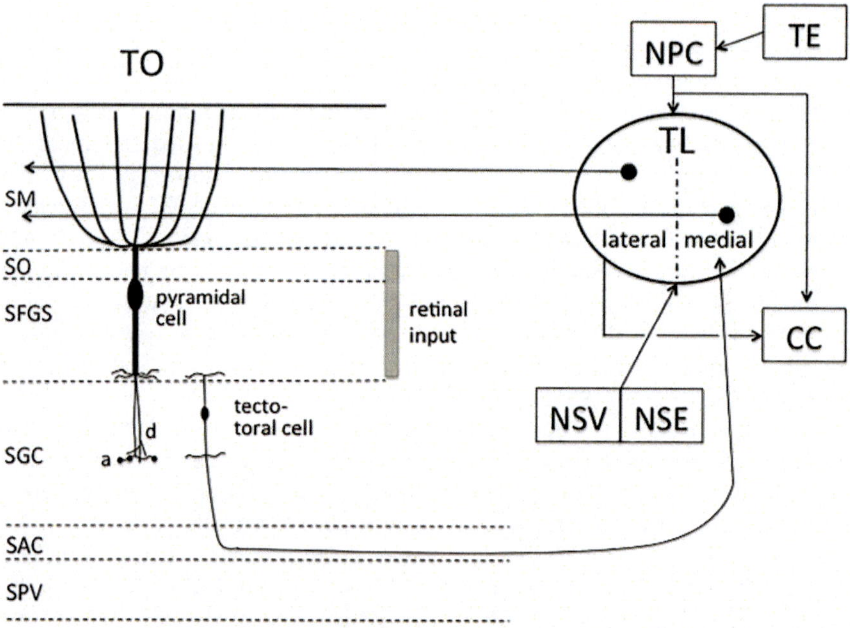

Fig. 2.10 Schematic diagram illustrating component cells and their connections of the mesencephalic cerebellum-like system in squirrelfish. A axon, d dendrite. For other abbreviations, see list

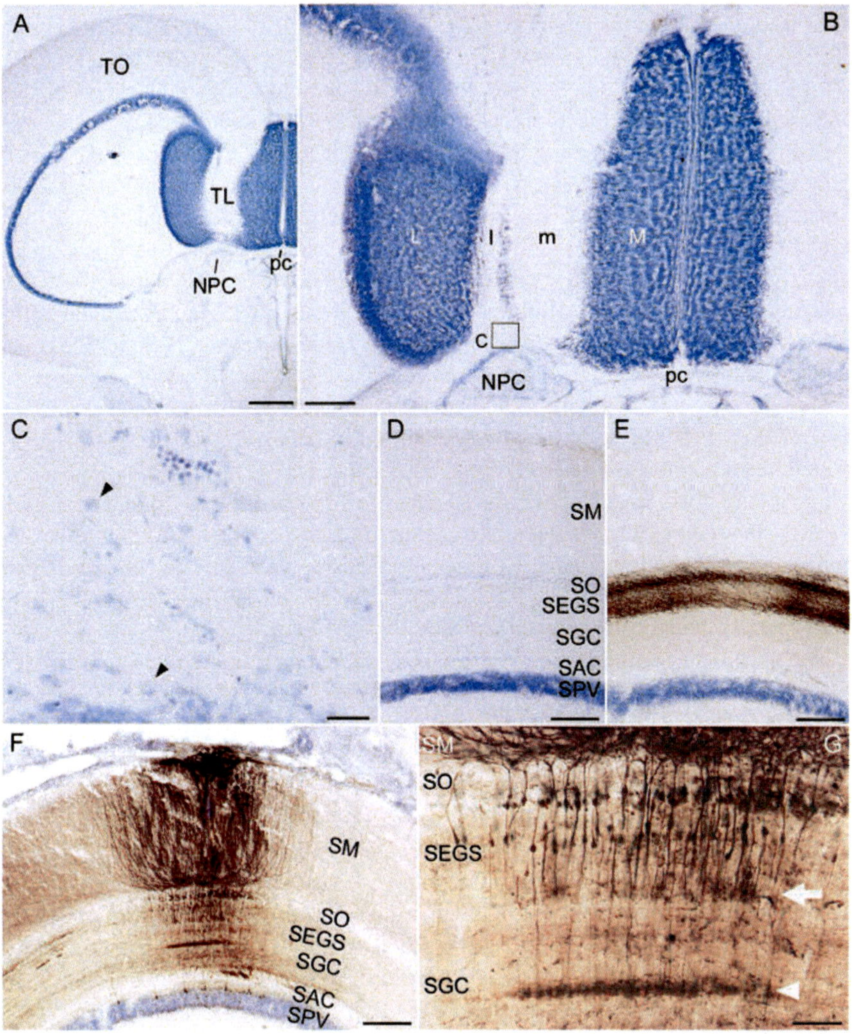

Fig. 2.11 Cytoarchitecture of the torus longitudinalis (TL) (**a–c**), laminar organization of the optic tectum (TO) (**d**, **e**), and morphology of pyramidal cells (**f**, **g**) in crown squirrelfish. (**a**) Transverse section through TL at the level of posterior commissure (pc). (**b**) Higher magnification of TL. TL of squirrelfish is of type IV, which can be divided into medial (granule cell zone, M; fiber zone, m) and lateral (granule cell zone, L; fiber zone, l) parts. (**c**) Higher magnification photograph of the boxed area in panel (**b**), showing larger cells (arrowhead) in the fiber zone. (**d**) Nissl-stained transverse section of TO. (**e**) Transverse section through TO of a squirrelfish with biocytin injections into the optic nerve. (**f**) Biocytin injection site into the stratum marginale (SM). Pyramidal cells are labeled, which took up the tracer from their dendrites in SM. (**g**) Higher magnification of pyramidal cells. Cell bodies situate in the boundary zone between the stratum opticum (SO) and stratum fibrosum et griseum superficiale (SFGS), and in the middle layer of SFGS. White arrow: level of arborization of horizontal branches from the basal dendrite; white arrowhead: level of arborization of a thin deep branch from the basal dendrite and termination of the pyramidal cell axon. For other abbreviations, see list. Scale bar = 500 μm in (**a**); 200 μm in (**b**, **d–f**); 20 μm in (**c**); 50 μm in (**g**)

been also identified in some species (e.g., carp, Ito, 1971; squirrelfish, Xue et al., 2003; rainbow trout, Folgueira et al., 2007; zebrafish, Folgueira et al., 2020) (Fig. 2.11). The granule cells send axons along the most superficial layer of the optic tectum or SM (Ito & Kishida, 1978) (Fig. 2.10). These "parallel fibers" are called marginal fibers. Deep to SM, bipolar cells called pyramidal cells (or type I cells of Meek & Schellart, 1978) are present in the stratum fibrosum et griseum superficiale (SFGS) (Figs. 2.10 and 2.11). Similarly to crest cells of RCLSs, the apical dendrite of pyramidal cells bifurcates into spiny branches forming a cone-shaped dendritic arbor in SM (Figs. 2.10 and 2.11), where marginal fibers make synaptic contacts with the dendrites (Ito, 1970; Ito et al., 1980). Pyramidal cells express glutamate receptor ∂2, as is the case for the crest cells in NM and Purkinje cells (Mikami et al., 2004). All pyramidal cells possess a basal dendrite, which penetrates through SFGS, where retinal axons terminate massively (Vanegas & Ito, 1983; Meek, 1981; Figs. 2.10 and 2.11). The basal dendrite gives rise to thin dendritic branches extending laterally for short distances at the deepest zone of SFGS, where retinal fibers terminate. The basal dendrite also gives rise to a thin branch and an axon that course further deep into the stratum griseum centrale (SGC), at the middle of which both the dendritic branch and axon give rise to fine branches that extend horizontally (Xue et al., 2003; Folgueira et al., 2007) (Figs. 2.10 and 2.11). Thus, the pyramidal cells are interneurons.

A tract-tracing study in carp suggested that the major afferent source to TL is VC (Ito & Kishida, 1978). Subsequent studies, however, failed to confirm this connection (*Pantodon buccholzi*, Wullimann & Roth, 1994; carp, Ito et al., 2003; squirrelfish, Xue et al., 2003; rainbow trout, Folgueira et al., 2007; zebrafish, Folgueira et al., 2020). In *Pantodon* TL receives fibers from the dorsal preglomerular nucleus, NLV, and oculomotor nucleus (Wullimann & Roth, 1994). In carp and squirrelfish, TL receives fibers from the optic tectum, nucleus subeminentialis (NSE), nucleus subvalvularis (NSV), and nucleus paracommissuralis (NPC) (carp, Ito et al., 2003; squirrelfish, Xue et al., 2003; Fig. 2.10). In rainbow trout and zebrafish, TL receives afferents from similar sources, although subeminential inputs to TL may be missing. The NSV is ventral to the NLV and laterally adjacent to the oculomotor nucleus and may correspond to the nucleus Q that sends fibers to TL in *Gnathonemus petersii* (Meek et al., 1986). Oculomotor neurons labeled from DiI application to the torus in *Pantodon* may correspond to neurons of NSV. NPC receives descending projections from the dorsal telencephalon (mainly its central part) and projects also to the cerebellum likely via collateral fibers (Ito et al., 1982; Imura et al., 2003; Xue et al., 2003) (Fig. 2.10). These connections are shared with the dorsal preglomerular nucleus of *Pantodon* (Wullimann & Roth, 1994), and the latter nucleus may well correspond to NPC. Among the afferent sources to TL enumerated above, the best understood is the tectotoral cells. Tectotoral cells in squirrelfish are bipolar with the soma in SGC. The apical and basal dendrites form horizontal dendritic arbors in the deepest zone of SFGS and middle zone of SGC, respectively (Xue et al., 2003; Fig. 2.10). The former dendrites receive retinal input (Northmore, 2017; Robles et al., 2021). The depth of the latter dendritic arbor coincides well with the terminal layer of the pyramidal cell axons. The tectotoral cells might integrate visual and pyramidal cell inputs and feedback to TL, although it may not be ruled out that tectotoral cells receive descending telencephalic inputs that reach the middle zone of the SGC (Murakami et al., 1983; Xue et al., 2003). In carp the somata of tectoral cells are situated in the middle of SGC with dendrites in deep SFGS and middle of SGC, although the somata are multipolar. Tectotoral cells are unipolar and possess only an apical dendrite reaching the deep zone of SFGS in rainbow trout, while those of zebrafish possess dendrites that distribute in the deep zone of SGC in addition to an apical dendrite, suggesting species differences in the morphology tectotoral cells. Tectotoral cells project only to the medial TL in squirrelfish (Xue et al., 2003), while NPC appears to project to the entire TL (Folgueira et al., 2007).

2.5.2 Functional Significances of TL-TO System

Kishida (1979) reported that teleost species with thick SM (e.g., squirrelfish), which would reflect importance of toral inputs to the tectum in comparison with other afferents, actively move from shallow to deep water or live in turbulent water, suggesting involvement of TL in equilibrium. Electrophysiological studies also provided clues to the functions of the mesencephalic CLS. Extracellular recordings revealed spiking discharges influenced by light from the dorsal region of TL, while saccade-related activities were recorded from the ventral region in mojarra, squirrelfish, and goldfish (Northmore et al., 1983; Northmore, 1984). The dorsal region presumably corresponds to medial TL, and the ventral region to lateral TL of Xue et al. (2003) in squirrelfish, and hereafter the former region will be called the dorsomedial TL and the latter ventrolateral TL after Northmore (2017). The former light-responsive (photic) activities are, in particular, sensitive to dimming with receptive fields along the equator or subequatorial zone of the visual field. The latter saccade-related activities persisted even after paralysis or in the dark and occurred in response to active, but not passive, eye movements, indicating that the activities are driven by corollary discharge inputs. The multiunit activity encodes the amplitude of saccade. The photic activities in the dorsomedial torus are quite likely induced by afferents from tectotoral neurons that respond to dimming stimuli (Robles et al., 2021); TL itself does not receive retinal projections. The source of saccade-related inputs to the ventrolateral TL might be NSV and/or NSE. NSV is close to the oculomotor nucleus and is in a position suitable for receiving oculomotor-related information. NSE is ventrally adjacent to EG, a region close to the site of electrical stimulations that induced eye movement (Demski, 1983).

Ablation of TL reduced dorsal light reflex, suggesting a role of TL in equilibrium control (Gibbs & Northmore, 1996). This finding appears to be in accord with photic responses recorded from the dorsomedial TL. However, the significance of saccade-related activities seems difficult to interpret from the point of view of controlling dorsal light reflex. There are other hypotheses on the functions of TL. Meek (1992) proposed that the torotectal system serves as a coincidence detector integrating lateral line and visual inputs to generate goal directed movements.

Lateral line afferents to TL, presumed to come from TS, are uncertain. Although labeled cells occurred in TS after tracer application to TL in carp and *Pantodon* (Wullimann & Roth, 1994; Ito et al., 2003), this is likely due to labeling of tectal commissure rather than TL (Folgueira et al., 2007). Recently, a Ca^{2+} imaging study confirmed responses of TL neurons to dimming stimuli (Robles et al., 2021). The authors of the latter study hypothesized that TL stimulated by an object primes the other side of the tectum to enhance detection of the object entering the contralateral visual field. The functions proposed in the studies enumerated above focus on either the photic or saccade-related activities. Northmore (2011) proposed a different hypothesis that takes account of both activities, as below. Pyramidal cells transmit photic information to tectotoral cells that in turn project back to the dorsomedial TL. It takes a significant delay time before the information comes back again from the dorsomedial TL onto the apical dendrites of pyramidal cells via marginal fibers. Thus, the apical dendrites receive visual information regarding the scene a while before (delayed visual input reflecting pre-saccadic eye position). The delayed inputs are integrated with the inputs from ventrolateral TL coding the magnitude of saccade. By learning through plasticity of parallel fiber-apical dendrite synapses, the pyramidal cells become able to make prediction of what the visual scene will be like after saccade. By analogy to the subtraction of self-generated sensory inputs as seen for RCLSs, pyramidal cells subtract saccade-driven changes in visual inputs so that the tectum can attend to non-self-generated changes. More recently, taking into account the presence of parallel fibers with higher conduction velocities (Vanegas et al., 1979), Northmore (2017) constructed neural network models and proposed a new hypothesis on the function of TL: visual attention shift and its learning. In the model, the pyramidal cells receiving delayed photic inputs from the dorsomedial TL serve formation of an attention locus for the one selected from multiple moving objects through winner-take-all circuitry in the tectum. The visual attention locus in the tectum can follow slowly moving objects through the abovementioned tectal circuitry but fail to follow rapid movements such as those by saccade. The saccade-related inputs from the ventrolateral TL are utilized to shift the locus of attention in the tectum to attend to the same object in spite of the eye position change after saccade. This is assumed to be achieved by combining excitatory drive influencing the pyramidal cells in tectal regions surrounding the current tectal locus of attention with inputs signaling the predicted magnitude of saccade (Fig. 2.12). The author suggested a third category of inputs onto the apical dendrites of pyramidal cells in addition to delayed photic and saccade-associated inputs. The third category of inputs with faster conduction velocities, which are supposed to occur from the dorsomedial TL (presumably larger granule cells), mediate photic information more rapidly than the delayed inputs; i.e. the rapid signals can mediate post-saccade photic information when the delayed inputs are still signaling pre-saccadic information to pyramidal cells. Making use of the three categories of information, precise shift of attention locus can be learned, which the author considers important for fish whose visual systems grow continuously. The hypothesis of Northmore (2017) reflects cell types, circuitry, and physiology of TL, and the TL may well serve such

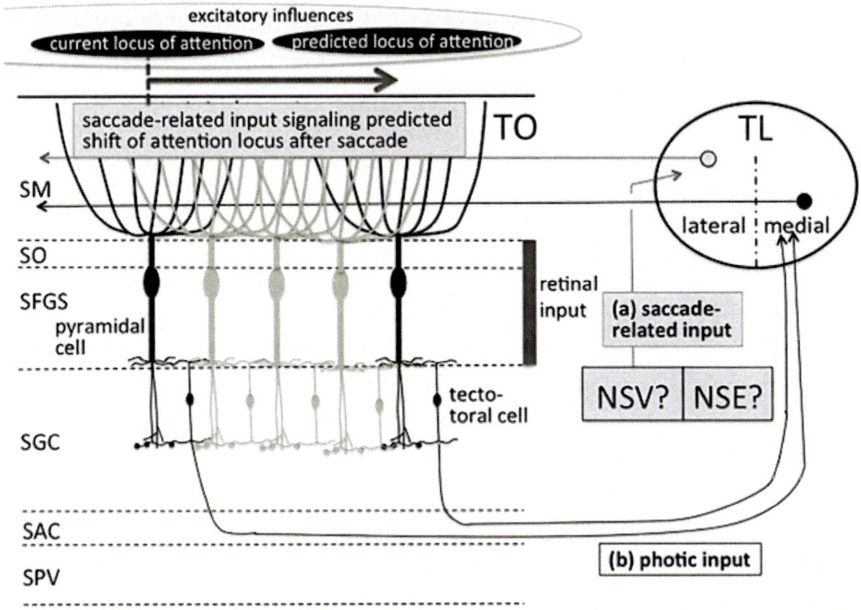

Fig. 2.12 Schematic illustration of the functional model of torus longitudinalis-optic tectum (TL-TO) system proposed by Northmore (2017). Pyramidal cells can "spotlight" a specific locus of the tectum that attends to a visual target of interest (the most salient visual item). TL-TO circuitry enables the tectum to track a single visual target continuously, in spite of the saccade-induced eye-position changes. This is made possible by shifting the attention locus to a new tectal position, upon which the visual target will be projected after the saccade. Corollary discharge inputs of oculomotor command signal, which are mediated by lateral (ventrolateral) TL to apical dendrites of pyramidal cells, can be used to predict the eye-position change by the occurring saccade

functions. However, there are a few points that remain to be studied further. The receptive field of dorsomedial TL is the equatorial visual field, the simplest interpretation for which is that tectotoral cells in the mediolaterally intermediate tectum with equatorial visual receptive field provide photic information to the dorsomedial TL. However, this does not appear to be the case. Not only intermediate but also medial or lateral tectal injection of tracer result in labeled terminals in TL in yellowfin goby (Reanalysis of the specimen of Hagio et al., 2021). Similarly, toral tracer injections label cells in a wide tectal zone in zebrafish (Folgueira et al., 2020). Some toral network mechanisms yet to be elucidated may serve to generate equatorial visual receptive field in the dorsomedial TL. Another remaining issue is the descending telencecphalic inputs to the TL mediated by NPC. Electrical stimulations of the central part of dorsal telencephalon brought about reproduction related behavior in a cichlid fish *Lepomis* (Demski, 1983). Therefore, this pathway possibly mediates TL with a copy of descending signal for intended behavior.

2.5.3 *Diversity of Torus TL-TO System in Actinopterygians.*

Functions of TL are yet to be definitively determined, although a number of hypotheses have been put forward as to its significances, as enumerated above. We therefore investigated the diversity of TL-TO system in search of the traits that may be connected to the functions of TL.

The cytoarchitecture of TL shows considerable diversity among different actinopterygians. First of all, a TL is not present in polypteryforms (*Polypterus senegalus*; Fig. 2.13). In chondrosteans, a tiny TL is present in sturgeons, *Acipenser transmontanus* and Bester (Fig. 2.13). Only small, granular cells are found in the sturgeon TL, and no clear cytoarchitectonic differentiation is apparent in different

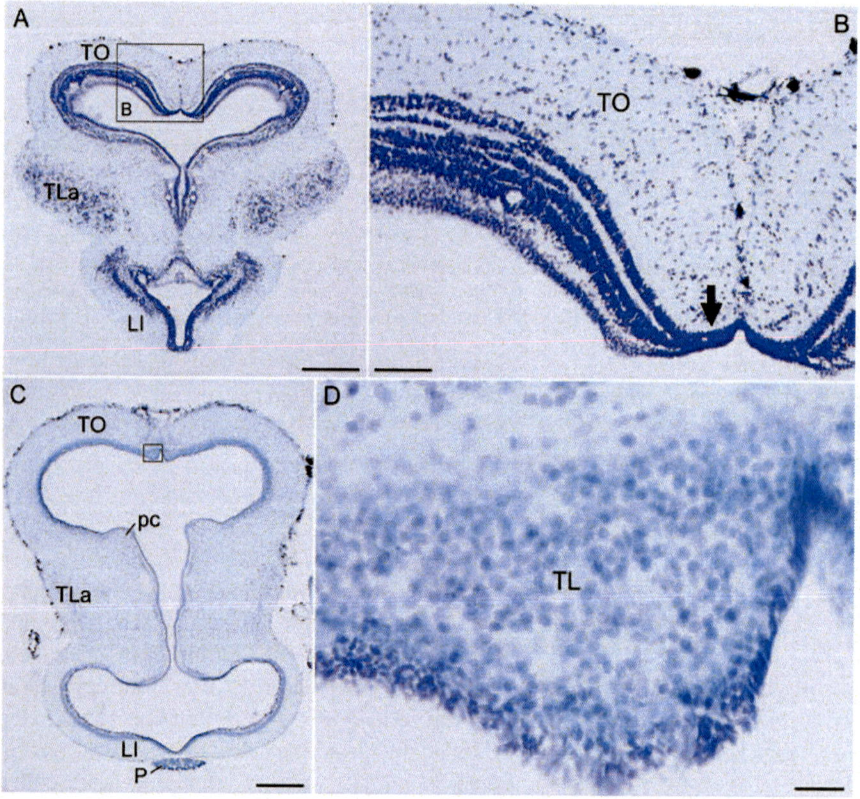

Fig. 2.13 Nissl-stained transverse sections of gray bichir (**a**, **b**) and Bester (hybrid of male *Huso huso* and female *Acipenser ruthenus*) (**c**, **d**). (**a**) Section through the optic tectum (TO). (**b**) Higher magnification photograph of the boxed area in panel (**a**). The arrow points to the site where the torus longitudinalis (TL) is expected to hang, if present. Only the deep tectal layer of small cells continues to this site, and no bulge is found that is indicative of TL. (**c**) Section through TO at a level slightly caudal to the posterior commissure (pc). (**d**) Higher magnification photograph of the boxed area in panel (**c**). Small cells populate in TL with only tiny cell-free zones that represent neuropils. For other abbreviations, see list. Scale bar = 500 μm in (**a**, **c**); 100 μm in (**b**); 20 μm in (**d**)

zones of TL. TL is present in all neopterygians, suggesting that TL evolved subsequent to the diversification of polypteryforms from other actinopterygian lineages. In longnose gar, small cells distribute very densely in the ventral zone of TL, with the density gradually being lower more dorsally. Larger cells are scattered in the dorsalmost zone of the torus. This cytoarchitecture is similar to type IA TL of teleosts, which will be explained in more detail below. The morphology of TL shows amazing species differences in teleosts. We observed Nissl- and Bodian-stained section series, and tract-tracing materials with Nissl-counterstaining of teleost brains in 51 species, and found that the cytoarchitecture of TL can be categorized at least into four types.

Type I is the cytoarchitectonic pattern seen in 34 species from a wide variety of teleost taxa (Table 2.1), such as silver arowana, goldfish (Fig. 2.14), carp, zebrafish, rainbow trout, pineconefish, flathead grey mullet (Fig. 2.14), Nile tilapia, and so on. Small granule cells occupy much of the extent of the torus. The granule cells distribute with almost even density in some species such as flathead grey mullet, while granule cells show uneven densities in other species such as goldfish (Fig. 2.14). Lower density zones in the latter species probably represent small neuropils. A larger neuropil or fibrous zone with sparser cells is present in the dorsal region of TL, where larger cells stained paler than granule cells are present. At least some of the larger cells are GABAergic (rainbow trout, Folgueira et al., 2007; zebrafish, Folgueira et al., 2020). In carp and zebrafish, cells in the dorsal TL were regarded as large cells, and medium-sized and small cells were identified among the granule cells (Ito, 1971; Folgueira et al., 2020). We failed to distinguish between small- and medium-sized granule cells, with casual observation with light microscopy in the present survey. Unlike the species explained above, the granule cells of TL in elephantnose fish can be clearly divided into two zones, dorsal and ventral (Fig. 2.15), although granule cells are very dense and relatively evenly distributed as in TL of other type I species. Therefore, TL of elephantnose fish is categorized as type IB, being distinguished from more commonly found type I TL that is categorized as type IA. The TL of elephantnose fish is also distinct from TL of other species in that a clear neuropil is present in the ventral TL above the posterior commissure (Fig. 2.15). There are large cells close to the ventral neuropil. It remains unknown whether or not they are GABAergic.

Type II is a pattern found only in seven species: Japanese eel, spotted garden-eel, Japanese rice fish, yellow striped flounder, striated frogfish, a sea toad *Chaunax abei*, and spotted green pufferfish. Densely packed granule cells of a few to several cell thickness form the medial, lateral, and ventral walls of the torus in these species. Granule cells continue inward in some portions of the torus, but they do not spread to occupy a significant area of the inner torus; i.e., cell-free/sparse neuropil zones are clearly appreciated in the TL. Larger cells are present in the dorsal zone of the TL as in the case of type IA torus. In the species except spotted garden ell and *Chaunax abei*, cells in the inner TL cannot be differentiated clearly from those forming the superficial walls of TL with casual microscopic observations. Tectal injections of neural tracer resulted in labeling almost all small cells in TL in Japanese rice fish (Fig. 2.16), suggesting that the vast majority of small cells are

Table 2.1 Phylogenetic distribution[a] of cyroarchitectonic types of torus longitudinalis in teleosts

Order	Family	English name	Japanese name	Scientific name	TL type
Anguilliformes	Congridae	Spotted garden-eel[b]	chin-anago	*Heteroconger hassi*	IIB
	Anguillidae	Japanese eel[b]	nihon-unagi	*Anguilla japonica*	IIA
Osteoglossiformes	Osteoglossidae	Arawana	n.a.	*Osteoglossum bicirrhosum*	IA
	Mormyridae	Elephantnose fish[b]	n.a.	*Gnathonemus petersii*	IB
Clupeiformes	Clupeidae	Japanese sardinella	sappa	*Sardinella zunasi*	IA
		Pacific herring	nishin	*Clupea pallasii pallasii*	IA
		South American pilchard	ma-iwashi	*Sardinops sagax*	IA
		Red-eye round herring	urume-iwashi	*Etrumeus teres*	IA
Cypriformes	Cyprinidae	Goldfish[b]	kingyo	*Carassius auratus*	IA
		Common carp[b]	koi	*Cyprinus carpio*	IA
		Zebrafish[b]	n.a.	*Danio rerio*	IA
Siluriformes	Siluridae	Amur catfish	namazu	*Silurus asotus*	IA
	Plotosidae	Striped eel catfish	gonzui	*Plotosus japonicus*	IA
Salmoniformes	Salmonidae	Rainbow trout[b]	niji-masu	*Oncorhynchus mykiss*	IA
Beryciformes	Holocentridae	Crown squirrelfish[b]	niji-ebisu	*Sargocentron diadema*	IV
		North Pacific squirrelfish	ittoudai	*Sargocentron spinosissimum*	IV
		Pinecone soldierfish	yogore-matsukasa	*Myripristis murdjan*	IV
	Monocentridae	Pineconefish	matsukawa-uo	*Monocentris japonica*	IA
Ophidiiformes	Ophidiidae	Armoured cusk	yoroi-itachi-uo	*Hoplobrotula armata*	IA
Gobiiformes	Oxudercidae	Yellowfin goby[b]	ma-haze	*Acanthogobius flavimanus*	III
		Black goby	tobi-haze	*Periophthalmus modestus*	III
		Shimofuri-goby	shimofuri-shima-haze	*Tridentiger bifasciatus*	III
		Chameleon goby	akaobi-shima-haze	*Tridentiger trigonocephalus*	III
	Gobiidae	n.a.	kawa-yoshinobori	*Rhinogobius flumineus*	III
		Whitebarred goby	sarasa-haze	*Amblygobius phalaena*	III

Mugiliformes	Mugilidae	Flathead grey mullet[b]	bora	*Mugil cephalus*	IA
Cichliformes	Cichlidae	Nile tilapia	n.a.	*Oreochromis niloticus*	IA
Beloniformes	Adrianichthyidae	Japanese rice fish[b]	medaka	*Oryzias latipes*	IIA
Anabantiformes	Osphronemidae	Dwarf gourami	n.a.	*Trichogaster lalius*	IA
Pleuronectiformes	Pleuronectidae	Yellow striped flounder[b]	ma-garei	*Pseudopleuronectes herzensteini*	IIA
	Paralichthyidae	Bastard halibut[b]	hirame	*Paralichthys olivaceus*	IA
Trachiniformes	Champsodontidae	n.a.	wani-gisu	*Champsodon snyderi*	IA
Labriformes	Labridae	n.a.	kyuusen	*Parajulis poecilepterus*	IA
Perciformes	Mullidae	Japanese goatfish[b]	himeji	*Upeneus japonicus*	IA
		Manybar goatfish[b]	ojisan	*Parupeneus multifasciatus*	IA
	Toxotidae	Banded archerfish	teppou-uo	*Toxotes jaculatrix*	IA
	Leiognathidae	Spotnape ponyfish[b]	hiiragi	*Nuchequula nuchalis*	IA
Scorpaeniformes	Scorpaenidae	Darkbanded rockfish[b]	mebaru	*Sebastes inermis*	IA
		False kelpfish[b]	kasago	*Sebastiscus marmoratus*	IA
	Platycephalidae	Crocodile flathead	ine-gochi	*Cociella crocodila*	IA
	Hexagrammidae	Spotty-bellied greenling[b]	kujime	*Hexagrammos agrammus*	IA
	Liparidae	Tanaka's snailfish	kusa-uo	*Liparis tanakae*	IA
Lophiiformes	Antennariidae	Striated frogfish	kaeru-ankou	*Antennarius striatus*	IIA
	Chaunacidae	n.a.	midori-fusa-ankou	*Chaunax abei*	IIB
Tetraodontiformes	Ostraciidae	Bluespotted boxfish[b]	hako-fugu	*Ostracion immaculatus*	IA
	Monacanthidae	Whitespotted pygmy filefish[b]	amine-hagi	*Rudarius ercodes*	IA
	Tetraodontidae	Panther puffer[b]	higan-fugu	*Takifugu pardalis*	IA
		n.a.[b]	kusa-fugu	*Takifugu niphobles*	IA
		Purple puffer[b]	ma-fugu	*Takifugu porphyreus*	IA
		Blunthead puffer	yorito-fugu	*Sphoeroides pachygaster*	IA
		Spotted green pufferfish	midori-fugu	*Dichotomyctere nigroviridis*	IIA

n.a. not applicable
[a]Phylogeny is based on Nelson et al. (2016)
[b]Species used for relative size measurements

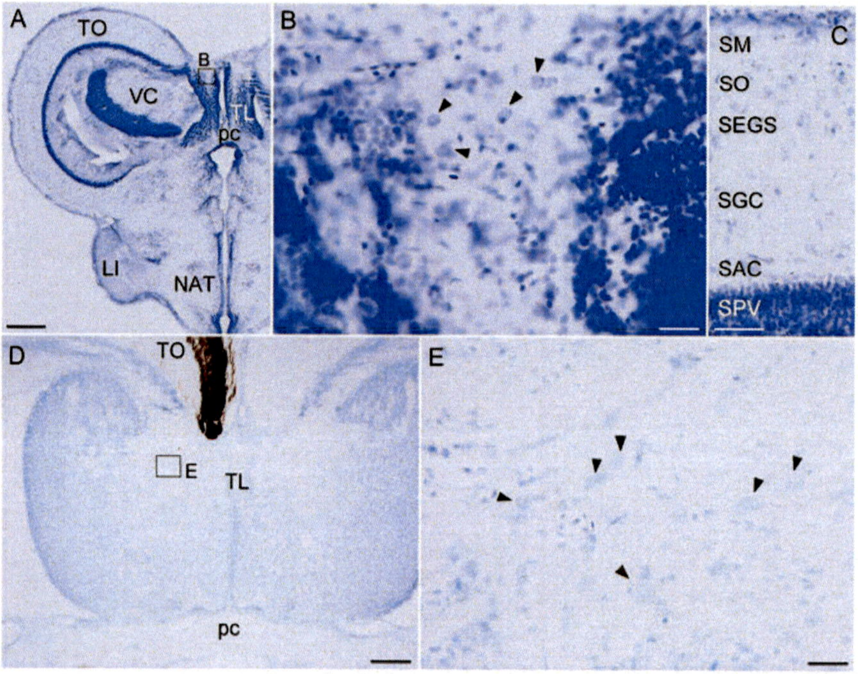

Fig. 2.14 Transverse sections through the mesenecphalon of goldfish (**a**–**c**) and flathead grey mullet (**d**, **e**), showing type IA torus in these species. (**a**) Nissl-stained section through the torus longitudinalis (TL) at the level of posterior commissure (pc). At this level TL is elongate dorsoventrally. TL is composed mostly of small granule cells except for the dorsocentral zone. (**b**) Higher magnification photograph of the boxed area in panel (**a**), showing cells in the dorsocentral zone of TL. Larger cells (arrowhead) are present in this specific region of the torus. Aggregated, small granule cells are present lateral and medial to the larger-celled zone. (**c**) Laminar organization of the optic tectum (TO). (**d**) Section through the TL of the mullet with biocytin injections into the optic nerve. Small granule cells occupy much of TL. (**e**) Higher magnification photograph of the boxed area in panel (**c**), showing larger cells (arrowhead) in the dorsal zone of TL. For other abbreviations, see list. Scale bar = 500 μm in (**a**); 20 μm in (**b**, **e**); 50 μm in (**c**); 200 μm in (**d**)

granule cells sending fibers to SM. In spotted garden-eel and *Chaunax abei*, inner cells are clearly larger and stained paler than those in the outer cell walls (Fig. 2.16). On the basis of this cytoarchitectonic difference, the pattern in the latter two species is categorized as type IIB and that in the other species as type IIA.

Quite differentiated cytoarchitecture characterizes type III TL. All six species of gobies investigated exhibit this type, and we failed to find this pattern in other teleosts, suggesting that this type represents a derived feature shared in gobies. Similarly to type II torus, lateral and ventral walls are formed by densely packed lamina of granule cells (Fig. 2.17). Unlike type II TL, however, the tori on both sides are fused at the midline to form an unpaired structure. Also different from type II, the inner torus is composed mostly of neuropil and fibrous zones. Deep to the superficial granule cell lamina, small cells are diffusedly distributed in the neuropil. Islands of

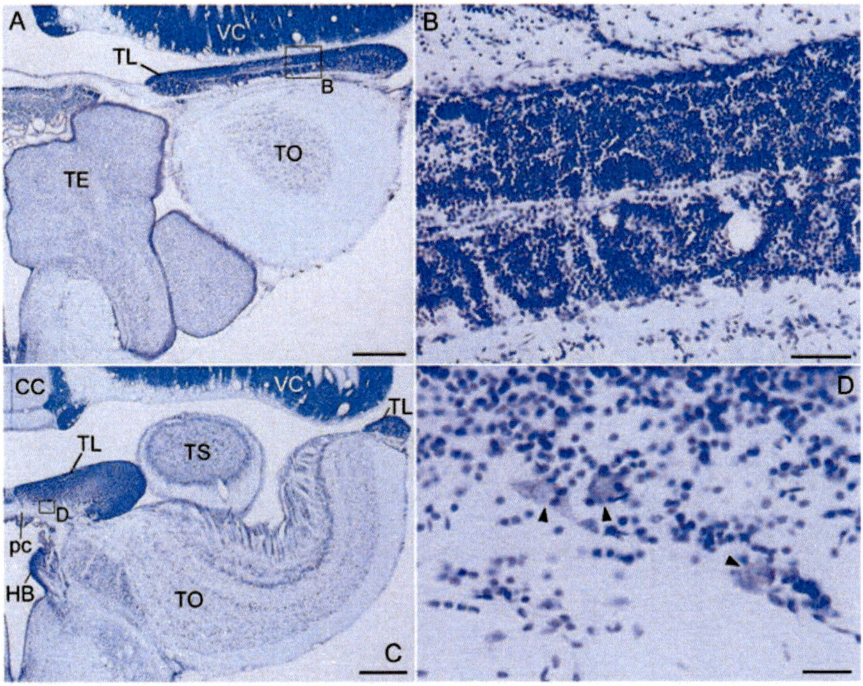

Fig. 2.15 Nissl-stained transverse sections through the type IB torus longitudinalis (TL) of elephantnose fish. (**a**) Section through TL lying above the optic tectum (TO). TL is composed mostly of small granule cells. TL at this level is flattened to become elongate mediolaterally, a configuration of TL not seen in other teleosts. (**b**) Higher magnification photograph of the boxed area in panel (**a**). TL can be divided into dorsal and ventral part at this level. (**c**) Section at the level of posterior commissure (pc), caudal to the one shown in panels (**a**) and (**b**). At this level, TL becomes separated into medial and lateral portions with the medial one attached onto pc and the lateral one onto the dorsolateral edge of TO. (**d**) Higher magnification photograph of the boxed area in panel (**c**). Large cells (arrowhead) are present near the boundary between the granule cell zone and neuropil zone. For other abbreviations, see list. Scale bar = 500 μm in (**a**, **c**); 50 μm in (**b**); 20 μm in (**d**)

small cells occupy near the midline of torus, presumably reflecting remnant of bilateral medial walls of torus after fusion during development. In the dorsolateral torus, many small cells form a dorsoventrally elongated island, and large cells are present dorsomedial to the island, which appear to be GABAergic (Fig. 2.17). Medium-sized to large cells were also encountered more ventrolaterally within the neuropil. These cells are stained paler than the large cells in the dorsomedial TL (Fig. 2.17).

Holocentrid fish exhibit a different type of well-differentiated TL. This type, regarded as type IV, can be found only in three species of holocentrids analyzed in the present survey (crown squirrelfish, North Pacific squirrelfish, and pinecone soldierfish). The TL of holocentrids is composed of medial and lateral parts, each with granule cell and fiber zones (Fig. 2.11). The medial and lateral parts are of similar sizes in squirrelfishes, while the lateral part is quite small in soldierfish. Larger cells

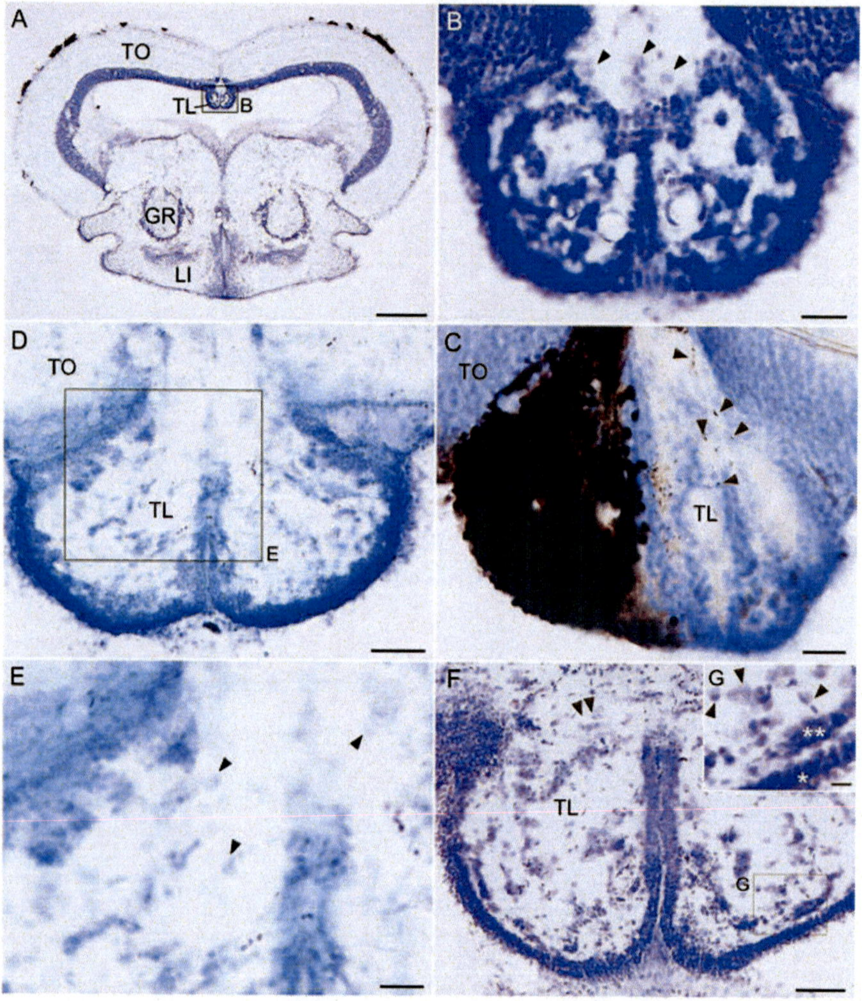

Fig. 2.16 Transverse sections through the type II torus longitudinalis (TL) in Japanese rice fish (**a–c**), Japanese eel (**d**, **e**), and spotted garden-eel (**f**, **g**). (**a**) Nissl-stained section through TL, which is of type IIA. (**b**) Higher magnification photograph of the boxed area in panel (**a**). Aggregated granule cells form lateral, medial, and ventral walls of TL. Cell free/sparse zones (neuropils) are clearly appreciated. Aggregates of small granule cells are also present deeper in the torus. Larger cells (arrowhead) are present only in the dorsal zone of TL. (**c**) Section through TL of rice fish with tracer injections into the optic tectum (TO). The vast majority of granule cells are retrogradely labeled. Note that the neuropils also become labeled, due to the presence of labeled dendrites of granule cells. Labeled terminals (arrowhead) appear in the neuropil of TL contralateral to the injected tectum. (**d**) Section through the eel torus of type IIA . Lateral, medial, and ventral walls of TL are formed by aggregated granule cells. Groups of small granule cells are present deeper, surrounded by neuropils. (**e**) Higher magnification photograph of the boxed area in panel (**d**). Larger cells are present in the dorsal TL. (**f**) Section through garden-eel TL of type IIB type. Densely packed small granule cells form lateral, medial, and ventral walls of TL, as in eel and rice fish. Islands of small granule cells and medium-sized cells are present in the central neuropil zone. Larger cells (double arrowheads) are present in the dorsal TL as in tori of types IA and IIA. (**g**) Higher magnification photograph of the boxed area in panel (**f**). Densely packed, small granule cells forming the wall of TL (white asterisk), a cell lamina of small granule cells (double white asterisks), and medium-sized cells (arrowhead) are visible. For other abbreviations, see list. Scale bar = 500 μm in (**a**); 20 μm in (**b**, **c**, **e**); 50 μm in (**d**, **f**); 10 μm in (**g**)

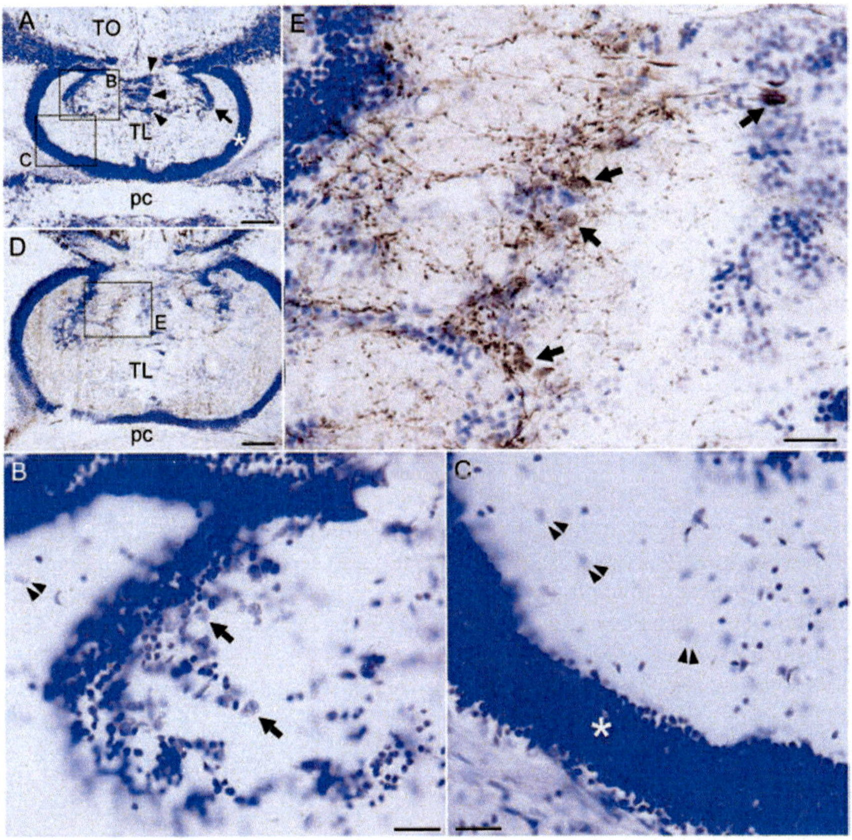

Fig. 2.17 Transverse sections through the torus longitudinalis (TL) of yellowfin goby. (**a**) Nissl-stained section. Lateral and ventral walls of TL are formed by a clear lamina of small granule cells (white asterisk). Deep to the walls is central neuropil zone. Within the central zone, there are islands of small cell aggregate at the midline (arrowheads). In the dorsolateral zone of TL is a curved dorsoventral lamina of small cells (arrow). (**b**) Higher magnification photograph of the boxed area in panel (**a**). Large cells are present medial to the curved dorsoventral lamina of small cells (arrow). There are medium-sized pale cells in the neuropil zone closer to small cell laminae (double arrowheads). (**c**) Higher magnification photograph of the boxed area in panel (**a**). Medium-sized pale cells (double arrowheads) are scattered in the neuropil zone close to the small cell lamina (white asterisk). Small cells are scattered in the neuropil. (**d**) Section through TL processed for immunohistochemistry against GAD65/67. Labeled puncta and fibers appear throughout the neuropil zone. The level of section corresponds to that of panel **a**. (**e**) Higher magnification photograph of the boxed area in panel (**d**). Lightly labeled somata (arrow) appear only in the dorsal TL. For other abbreviations, see list. Scale bar = 100 μm in (**a**, **d**); 20 μm in (**b**, **c**, **e**)

are present near or within the fiber layers and also within the granule cell region. Although these larger cells have been categorized into medium-sized and large cells (Xue et al., 2003), it is not really clear if such a clear-cut classification is possible, as assessed by casual observation in the present survey.

Interpretation of the functions and evolution TL in teleosts from the point of view of cytoarchitecture may be a difficult question with the present analysis of limited number of species. However, the plesiomorphic cytoarchitecture may be type I, considering (1) uniform distribution of granular cells in sturgeons and gar, representing non-teleost actinopterygian lineages that diverged prior to the emergence of teleosts, and (2) widespread distribution of type I in various teleost taxa; types II, III, and IV emerge only sporadically. Among these subtypes types III and IV show quite well-differentiated cytoarchitecture. We found type III TL only in gobies and type IV only in holocentrids. The majority of gobies dwell relatively shallow water and stay on or close to the bottom, and many species sensitively respond to moving objects. The yellowfin goby possesses as many as three high cell density regions in the retina (Miyazaki et al., 2019). Kishida (1979) pointed that squirrelfish live in turbulent water and moves from shallow to deep and interpreted well-developed TL-SM systems as connected to equilibrium. It should be pointed, however, that the holocentrid fish possess vary large eyes. The unique cytoarchitecture of TL in gobies and holocentrids may well reflect well-developed visual system, which may be in favor of the hypothesis proposed by Northmore (2017). It should be also noted, however, that TL of pinecone fish is of type I. Large eyes are equipped with pinecone fish, which is a beryciform fish like squirrelfishes and soldierfish but does not belong to holocentrids. An alternative interpretation, therefore, may be that the cytoarchitectonic characteristics of TL are under strong influences of phylogeny rather than ecology or differential development of visual system.

TL is not only diverse in cytoarchitecture but also in size. For example, TL is amazingly large in squirrelfish (Fig. 2.11), while it is tiny in Japanese rice fish (Fig. 2.16). A previous study investigated the relative thickness of SM to the total thickness of the tectum (Kishida, 1979). We have done a similar analysis in 24 species of teleosts including species that were not analyzed in that previous study. The relative thickness of SM, however, only relates to the magnitude of toral output, neglecting the input channels to the torus. Therefore, as a casual analysis of the toral size, we measured the frontal sectional area of TL at or close to the level of posterior commissure: i.e., the level where TL is largest as seen in transverse sections. To compensate for the overall brain size differences between species, we divided the toral area by the sectional area of the telencephalon at the level of anterior commissure. The telencephalon was selected for this standardization, since the telencephalon receives all major sensory input (Yamamoto et al., 2007) and possibly provides a relatively standardized indication for brain size.

The relative thickness of SM to the tectum and relative sectional size of TL to the telencephalon showed similar trends (Fig. 2.18). The largest relative thickness of SM was 45.7% in crown squirrelfish, which is larger than 33% for North Pacific squirrelfish reported by Kishida (1979). Crown squirrelfish also showed the largest relative size of TL (24.6%). Second and third largest values of relative SM thickness were 28.7% and 24.4% for elephantnose fish and flathead grey mullet, respectively. These species were not studied by Kishida (1979). Flathead grey mullet had the second largest relative TL size (22.9%). Common carp showed the third largest relative TL size (18.8%). Very thick relative thickness of SM in squirrelfishes may

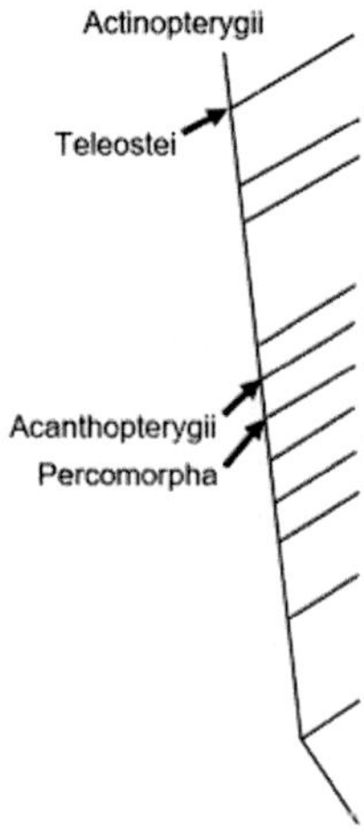

Order	Species	TL/TE (%)	SM/TO (%)
Anguilliformes	*Heteroconger hassi* (spotted garden-eel)	5.8	8.2
	Anguilla japonica (Japanese eel)	2.1	4.3
Osteoglossiformes	*Gnathonemus petersii* (elephantnose fish)	11.8	28.7
Cypriniformes	*Carassius auratus* (goldfish)	16.2	12.5
	Cyprinus carpio (common carp)	18.8	14.6
	Danio rerio (zebra danio)	7.4	7.4
Salmoniformes	*Oncorhynchus mykiss* (rainbow trout)	10.3	12.9
Beryciformes	*Sargocentron diadema* (crown squirrelfish)	24.6	45.7
Gobiiformes	*Acanthogobius flavimanus* (yellowfin goby)	8.4	15.2
Mugiliformes	*Mugil cephalus* (flathead grey mullet)	22.9	24.4
Beloniformes	*Oryzias latipes* (Japanese rice fish)	2.6	5.9
Pleuronectiformes	*Pseudopleuronectes herzensteini* (yellow striped flounder)	1.9	4.0
	Paralichthys olivaceus (bastard halibut)	4.5	9.7
Perciformes	*Upeneus japonicus* (Japanese goatfish)	3.9	7.3
	Parupeneus multifasciatus (manybar goatfish)	3.8	4.8
	Nuchequula nuchalis (spotnape ponyfish)	15.2	9.8
Scorpaeniformes	*Sebastes inermis* (darkbanded rockfish)	5.5	9.2
	Sebastiscus marmoratus (false kelpfish)	3.3	11.8
	Hexagrammos agrammus (spotty-bellied greenling)	3.6	7.6
Tetraodontiformes	*Ostracion immaculatus* (bluespotted boxfish)	6.1	12.4
	Rudarius ercodes (whitespotted pygmy filefish)	5.8	8.3
	Takifugu pardalis (panther puffer)	2.1	5.9
	Takifugu niphobles (kusafugu)	2.8	6.6
	Takifugu porphyreus (purple puffer)	2.0	7.0

Fig. 2.18 Phylogenetic tree of 24 teleost species, in which development of torus-tectal system was assessed by measurements. Relative thickness of the stratum marginale to the total thickness of tectum (SM/TO) and relative sectional area of the torus longitudinalis to the sectional area of the telencephalon (TL/TE) are indicated in the right of the figure

reflect importance of equilibrium for these fishes as suggested by Kishida or excellent visual ability with large eyes. The thick relative SM of elephantnose fish was totally unexpected. Elephantnose fish live in muddy rivers or lakes with slow water current and mainly feed at night. Visually related functions of TL unlikely explain the hypertrophy of TL-TO system in elephantnose fish. So far as the knowledge of the authors, flathead grey mullet dwells relatively shallow water and does not migrate much from shallow to deep or vice versa, and the same is true to common carp. The eyes of flathead grey mullet are quite large, while the eyes of common carp are, at most, of ordinary size. Thus, vision may be important for flathead grey mullet, while neither equilibrium nor vision appear to be very important for common carp.

Both the relative SM thickness and TL size were small for yellow striped flounder, Japanese rice fish, Japanese eel, and three species of puffer fish. In spotted garden-eel, a diurnal species unlike nocturnal Japanese eel, both measurements show ordinary values. This apparently seems a species difference in line with the involvement of TL-TO system in vision-related functions. However, yellow striped flounder vigorously responds to moving items, Japanese rice fish swim on the surface searching for food with large eyes, and puffer fish have large eyes and respond well to shiny or fluorescent items.

As enumerated above, teleost species with quite different ecology can show similarly developed or underdeveloped TL-TO systems. Therefore, the present comparative analyses on relative thickness of SM and sectional size of TL failed to provide a straightforward answer to the question of the roles played by TL-TO system. However, the present survey remained rough analyses using brain sections and crude evaluations on the importance of equilibrium and vision for the species in question. More precise analyses using volumetric data on relative size of TL to the whole brain and assessments on more specific functions such as the abilities for maintaining visual attention target (after Northmore's hypothesis) may provide more fruitful data that could unravel better the functional significances of TL. Last but not the least, the unexpected finding of well-developed TL-TO system in elephantnose fish strongly suggests a non-visual function of the system in this species or perhaps flexible, evolutionary function changes (electric sense-related function?) of the system as an adaptation to muddy water.

References

Bastian, J. (1981). Electrolocation II. The effects of moving objects and other electric stimuli on the activity of two categories of posterior lateral line lobe cells in *Apteronotus albifrons*. *Journal of Comparative Physiology, 144*, 481–494.

Bastian, J. (1986). Electrolocation: Behavior, anatomy, and physiology. In T. H. Bullock & W. Heiligenberg (Eds.), *Electroreception* (pp. 577–612). Wiley.

Bell, C. C. (1981). An efference copy which is modified by reafferent input. *Science, 214*, 450–453.

Bell, C. C. (2001). Memory-based expectation in electrosensory systems. *Current Opinion in Neurobiology, 11*, 481–487.

Bell, C. C. (2002). Evolution of cerebellum-like structures. *Brain, Behavior and Evolution, 59*, 312–326.

Bell, C. C., Grant, K., & Serrier, J. (1992). Corollary discharge effects and sensory processing in the mormyrid electrosensory lobe: I. Field potentials and cellular activity in associated structures. *Journal of Neurophysiology, 68*, 843–858.

Brochu, G., Maler, L., & Hawks, R. (1990). Zebrin II: A polypeptide antigen expressed exclusively in Purkinje cells reveals compartments in rat and fish cerebellum. *The Journal of Comparative Neurology, 291*, 538–552.

Bullock, T. H., Bodznick, D. A., & Northcutt, R. G. (1983). The phylogenetic distribution of electroreception: Evidence of convergent evolution of a primitive vertebrate sense modality. *Brain Research Reviews, 6*, 25–46.

Butler, A. B., & Saidel, W. N. (1991). Retinal projections in the freshwater butterfly fish, *Pantodon buchholzi* (Osteoglossoidei). *Brain, Behavior and Evolution, 38*, 127–153.

Demski, L. S. (1983). Behavioral effects of electrical stimulation of the brain. In R. E. Davis & R. G. Northcutt (Eds.), *Fish neurobiology* (Vol. 2, pp. 317–359). The University of Michigan Press.

Finger, T. E. (1984). Central organization of eighth nerve and mechanosensory lateral line systems in the brainstem of ictalurid catfish. *The Journal of Comparative Neurology, 229*, 129–151.

Finger, T. E. (1986). Electroreception in catfish: Anatomy and electrophysiology. In T. H. Bullock & W. Heiligenberg (Eds.), *Electroreception* (pp. 287–317). Wiley.

Finger, T. E. (2000). Ascending spinal systems in the fish, *Prionotus carolinus*. *The Journal of Comparative Neurology, 422*, 106–122.

Folgueira, M., Riva-Mendoza, S., Ferreño-Galmán, N., Catro, A., Bianco, I. H., Anadón, R., & Yáñez, J. (2020). Anatomy and connectivity of the torus longitudinalis of the adult zebrafish. *Frontier in Neural Circuits, 14*, 8.

Folgueira, M., Sueiro, C., Rodríguez-Moldes, I., Yáñez, J., & Anadon, R. (2007). Organization of the torus longitudinalis in the rainbow trout (*Oncorhynchus mykiss*): An immunohistochemical study of the GABAergic system and a DiI tract-tracing study. *The Journal of Comparative Neurology, 503*, 348–370.

Gibbs, M. A., & Northmore, D. P. M. (1996). The role of torus longitudinalis in equilibrium orientation measured with the dorsal light reflex. *Brain, Behavior and Evolution, 48*, 115–120.

Hagio, H., Kawaguchi, M., Abe, H., & Yamamoto, N. (2021). Afferent and efferent connections of the nucleus prethalamicus in the yellowfin goby *Acanthogobius falvimanus*. *The Journal of Comparative Neurology, 529*(1), 87–110.

Hofmann, M. H., Wojtenek, W., & Wilkens, L. A. (2002). Central organization of the electrosensory system in the paddlefish (*Polyodon spathura*). *The Journal of Comparative Neurology, 446*, 25–36.

Imura, K., Yamamoto, N., Sawai, N., Yoshimoto, M., Yang, C.-Y., Xue, H.-G., & Ito, H. (2003). Topographic organization of an indirect telencephalo-cerebellar pathway through the nucleus paracommissuralis in a teleost, *Oreochromis niloticus*. *Brain, Behavior and Evolution, 61*, 70–90.

Ito, H. (1970). Fine structures of the carp tectum opticum. *Journal für Hirnforschung, 12*, 325–354.

Ito, H. (1971). Fine structure of the carp torus longitudinalis. *Journal of Morphology, 135*, 153–164.

Ito, H., Butler, A. B., & Ebbesson, S. O. E. (1980). An ultrastructural study of the normal synaptic organization of the optic tectum and the degenerating tectal afferent from retina, telencephalon, and contralateral tectum in a teleost, *Holocentrus rufus*. *The Journal of Comparative Neurology, 191*, 639–659.

Ito, H., & Kishida, R. (1978). Afferent and efferent fiber connections of the carp torus longitudinalis. *The Journal of Comparative Neurology, 181*, 465–476.

Ito, H., Murakami, T., & Morita, Y. (1982). An indirect telencephalocerebellar pathway and its relay nucleus in teleosts. *Brain Research, 249*, 1–13.

Ito, H., Yamamoto, N., Yoshimoto, M., Sawai, N., Yang, C.-Y., Xue, H.-G., & Imura, K. (2003). Fiber connections of the torus longitudinalis in a teleost: *Cyprinus carpio* re-examined. *The Journal of Comparative Neurology, 457*, 202–211.

Ito, H., & Yoshimoto, M. (1990). Cytoarchitecture and fiber connections of the nucleus lateralis valvulae in the carp (*Cyprinus carpio*). *The Journal of Comparative Neurology, 298*, 385–399.

Kishida, R. (1979). Comparative study on the teleostean optic tectum. Lamination and cytoarchitecture. *Journal für Hirnforschung, 20*, 57–67.

McCormick, C. A. (1982). The organization of the octavolateralis area in actinopterygian fishes: A new interpretation. *Journal of Morphology, 171*, 159–181.

McCormick, C. A., & Braford, M. R., Jr. (1994). Organization of inner ear endorgan projections in the goldfish, *Carassius auratus*. *Brain, Behavior and Evolution, 43*, 189–205.

McCormick, C. A., & Hernandez, D. V. (1996). Connections of octaval and lateral line nuclei of the medulla in the goldfish, including the cytoarchitecture of the secondary octaval population in goldfish and catfish. *Brain, Behavior and Evolution, 47*, 113–137.

Meek, J. (1981). A Golgi-electron microscopic study of goldfish optic tectum. II. A quantitative aspects of synaptic organization. *The Journal of Comparative Neurology, 199*, 175–190.

Meek, J. (1992). Why run parallel fibers parallel? Teleostean Purkinje cells as possible coincidence detectors, in a timing device subserving spatial coding of temporal differences. *Neuroscience, 48*, 249–283.

Meek, J., Grant, K., & Bell, C. (1999). Structural organization of the mormyrid electrosensory lateral line lobe. *Journal of Experimental Biology, 202*, 1291–1300.

Meek, J., Grant, K., Sugawara, Y., Hafmans, T. G. M., Veron, M., & Denizot, J. P. (1996). Interneurons of the ganglionic layer in the mormyrid electrosensory lateral line lobe: Morphology, immunohistochemistry, and synaptology. *The Journal of Comparative Neurology, 375*, 43–65.

Meek, J., Nieuwenhuys, R., & Elsevier, D. (1986). Afferent and efferent connections of cerebellar lobe C_3 of the mormyrid fish *Gnathonemus petersi*: An HRP study. *The Journal of Comparative Neurology, 245*, 342–358.

Meek, J., & Schellart, N. A. M. (1978). A Golgi study of goldfish optic tectum. *The Journal of Comparative Neurology, 182*, 89–122.

Mikami, Y., Yoshida, T., Matsuda, N., & Mishina, M. (2004). Expression of glutamate receptor ∂2 in neurons with cerebellum-like wiring. *Biochemical and Biophysical Research Communications, 322*, 168–176.

Miyazaki, T., Kato, A., Ikenaga, T., Hagio, H., & Yamamoto, N. (2019). A lambda-shaped retractor lentis muscle in the yellowfin goby *Acanthogobius flavimanus*. *Journal of Morphology, 280*, 526–533.

Murakami, T., Morita, Y., & Ito, H. (1983). Extrinsic and intrinsic fiber connections of the telencephalon in a teleost, *Sebastiscus marmoratus*. *The Journal of Comparative Neurology, 216*, 115–131.

Nelson, J. S., Grande, T. C., & Wilson, M. V. H. (2016). *Fishes of the world* (5th ed.). Wiley.

New, J. G., Coombs, S., McCormick, C. A., & Oshel, P. E. (1996). Cytoarchitecture of the medial octavolateralis nucleus in the goldfish, *Carassius auratus*. *The Journal of Comparative Neurology, 366*, 534–546.

Noro, S., Yamamoto, N., Ishikawa, Y., Ito, H., & Ijiri, K. (2007). Studies on the morphology of the inner ear and semicircular canal endorgan projections of *ha*, a medaka behavior mutant. *The Fish Biology Journal MEDAKA, 11*, 31–41.

Northmore, D. P. M. (1984). Visual and saccadic activity in the goldfish torus longitudinalis. *Journal of Comparative Physiology-A, 155*, 333–340.

Northmore, D. P. M. (2011). The optic tectum. In A. P. Farrell (Ed.), *Encyclopedia of fish physiology: From genome to environment* (Vol. 1, pp. 131–142). Elsevier.

Northmore, D. P. M. (2017). Holding visual attention for 400 million years: A model of tectum and torus longitudinalis in teleost fishes. *Vision Research, 131*, 44–56.

Northmore, D. P. M., Williams, B., & Vanegas, H. (1983). The teleostean torus longitudinalis: Responses related to eye movements, visuotopic mapping, and functional relations with the optic tectum. *Journal of Comparative Physiology-A, 150*, 39–50.

O'Marra, S. K., & McCormick, C. A. (1999). Organization and connections of the dorsal descending nucleus and other presumed acoustic areas in the brainstem of the teleost fish, *Astronotus ocellatus*. *Hearing Research, 129*, 7–19.

Robles, E., Fields, N. P., & Baier, H. (2021). The zebrafish visual system transmits dimming information via multiple segregated pathways. *The Journal of Comparative Neurology, 529*, 539–552.

Tong, S.-L., & Finger, T. E. (1983). Central organization of the electrosensory lateral line system in bullhead catfish *Ictalurus nebulosus*. *The Journal of Comparative Neurology, 217*, 1–16.

Vanegas, H., & Ito, H. (1983). Morphological aspects of the teleostean visual system: A review. *Brain Research Reviews, 6*, 117–137.

Vanegas, H., Williams, B., & Fresman, J. A. (1979). Response to stimulation of marginal fibers in the teleostean optic tectum. *Experimental Brain Research, 34*, 335–349.

Wullimann, M. F., & Roth, G. (1994). Descending telencephalic information reaches longitudinal torus via the dorsal preglomerular nucleus in the teleost fish *Pantodon buchholzi*: A case of neural preaptation. *Brain, Behavior and Evolution, 44*, 338–352.

Xue, H.-G., Yamamoto, N., Yang, C.-Y., Kerem, G., Yoshimoto, M., Imura, K., & Ito, H. (2003). Fiber connections of the torus longitudinalis and optic tectum in holocentrid teleosts. *The Journal of Comparative Neurology, 462*, 194–212.

Xue, H.-G., Yang, C.-Y., Ito, H., Yamamoto, N., & Ozawa, H. (2006). Primary and secondary trigeminnal projections in a cyprinid teleost, carp *Cyprinus carpio*. *The Journal of Comparative Neurology, 499*, 626–644.

Yamamoto, N. (2018). The brain of fish. In S. Shigeno, T. Nomura, & Y. Murakami (Eds.), *The strange world of brain – As revealed from genes. (In Japanese)* (pp. 289–331). Isshiki Shuppan.

Yamamoto, N., Ishikawa, Y., Yoshimoto, M., Xue, H.-G., Bahaxar, N., Sawai, N., Yang, C.-Y., Ozawa, H., & Ito, H. (2007). A new interpretation on the homology of the teleostean telencephalon based on hodology and a new eversion model. *Brain, Behavior and Evolution, 69*, 96–104.

Yamamoto, N., & Ito, H. (2005). Fiber connections of the central nucleus of semicircular torus in cyprinids. *The Journal of Comparative Neurology, 491*, 186–211.

Yamamoto, N., Kato, T., Okada, Y., & Somiya, H. (2010). Somatosensory nucleus in the torus semicircularis of cyprinid teleosts. *The Journal of Comparative Neurology, 518*, 2475–2502.

Yang, C.-Y., Yoshimoto, M., Xue, H.-G., Yamamoto, N., Imura, K., Sawai, N., Ishikawa, Y., & Ito, H. (2004). Fiber connections of the lateral valvular nucleus in a percomorph teleost, *Tilapia (Oreochromis) niloticus*. *The Journal of Comparative Neurology, 474*, 209–226.

Chapter 3
Modeling of Human Cerebellar Development and Diseases with Pluripotent Stem Cell-Derived Brain Organoids

Atsushi Tamada and Keiko Muguruma

3.1 Introduction

Our basic knowledge of the structure, functions, and development of the cerebellum has been established by extensive physiological and anatomical studies. Rapid progress in molecular biology during the last few decades has further accelerated the investigation of the molecular mechanisms of the cerebellum. In particular, phenotypic analyses of spontaneous or genetically engineered animals have revealed the developmental mechanisms of the cerebellum at the molecular and cellular levels. In addition, recent advances in comprehensive genetic expression analysis and noninvasive image analysis have enabled the identification of cerebellar diseases and developmental malformation in humans. Although cerebellar development appears to be fundamentally conserved among vertebrates including humans (Butts et al., 2014), human cerebellar diseases cannot be fully explained by currently existing animal disease models (Haldipur & Millen, 2013). In such situations, the emergence of induced pluripotent stem cells (iPSCs) and brain organoid techniques that can be used to differentiate PSCs into the cerebellar cells and tissues has provided an alternative approach for investigating human cerebellar development and diseases. In this review, we first provide an overview of the fundamental processes of cerebellar development. Then, we introduce the organoid techniques that can be used to construct cerebellum from pluripotent stem cells. Finally, we discuss the current status and future perspectives on the applications of these techniques to the investigation of cerebellar development and diseases.

A. Tamada · K. Muguruma (✉)
Department of iPS Cell Applied Medicine, Faculty of Medicine, Kansai Medical University, Hirakata, Osaka, Japan
e-mail: muguruke@hirakata.kmu.ac.jp

H. Mizusawa, S. Kakei (eds.), *Cerebellum as a CNS Hub*,
Contemporary Clinical Neuroscience, https://doi.org/10.1007/978-3-030-75817-2_3

3.2 Cerebellar Development

The cerebellum lies at the dorsal part of the rostral hindbrain in adults. Developmentally, it originates in a specific region defined by the anterior-posterior and the dorsal-ventral axes of the embryonic brain (Fig. 3.1). The primordium of the cerebellum is anteroposteriorly located in the first rhombomere, which is a transient embryonic swelling of the rhombencephalon, and dorsoventrally in the alar plate, the dorsal half of the neural tube (Martinez & Alvarado-Mallart, 1989a, b; Hallonet & Le Douarin, 1993; Wingate & Hatten, 1999; Joyner et al., 2000; Martinez et al., 2013; Wurst & Bally-Cuif, 2001). The molecular mechanisms of cerebellar development have been extensively studied and largely elucidated in vertebrate animals including rodents (Fig. 3.2). At the neural plate stage, the presumptive areas for the midbrain and the hindbrain start to differentially express two homeobox transcription factors, *Otx2* (Orthodenticle Homeobox 2) and *Gbx2* (Gastrulation Brain Homeobox 2) (Fig. 3.2a). *Otx2* is expressed anteriorly in the forebrain and the midbrain with a caudal limit at the presumptive boundary between the midbrain and the hindbrain. *Gbx2* is expressed posteriorly in the hindbrain with a rostral limit at the mid-hindbrain boundary. *Otx2* and *Gbx2* are key regulatory molecules at the determination of anterior-posterior regions. Deletion of the *Otx2* gene anteriorly shifts the mid-hindbrain boundary and expands the cerebellar primordium to the midbrain region. On the contrary, a *Gbx2* mutation posteriorly shifts the mid-hindbrain regions with an expansion of the midbrain region at the expense of the cerebellum

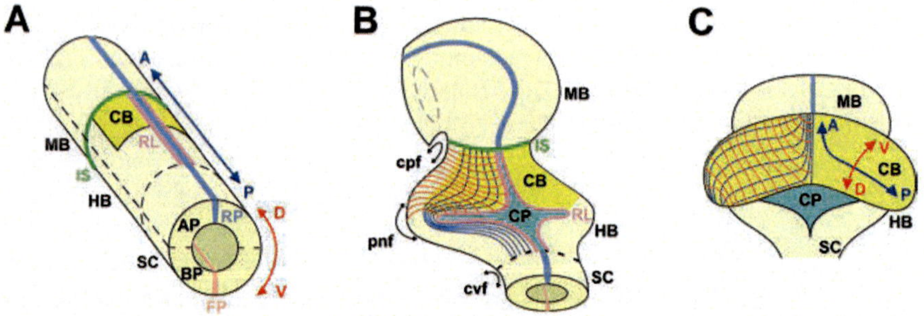

Fig. 3.1 Morphogenesis of the cerebellum. (**a**) At the neural tube stage, the CNS differentiates specific regions according to the positions along the anterior-posterior (A-P) and the dorsal-ventral (D-V) axes. The hindbrain (HB) contains the rhombic lip (RL) structure adjacent to the dorsal midline roof plate (RP). The presumptive cerebellum (CB) appears in the alar plate (AP) of the rostral HB and anteriorly demarcated by the isthmus (IS), the mid-hindbrain boundary. (**b**) As development proceeds, the neural tube is distorted by inhomogeneous cell growth and formation of flexures, including the cephalic flexure (cpf), the pontine flexure (pnf), and the cervical flexure (cvf). The RP is expanded to form the choroid plexus (CP). Note that the A-P axis (blue lines) and the D-V axis (red lines) of the neural tube are also distorted in the primordium of the CB. (**c**) During maturation, the anterior part of the CB is fused to form a continuous cerebellar tissue. As a result of the transformation, the A-P and the medial-lateral axes of the anatomical terminology correspond to the ventral-dorsal and the A-P axes of the neural tube, respectively. MB midbrain, SC spinal cord, FP floor plate, BP basal plate

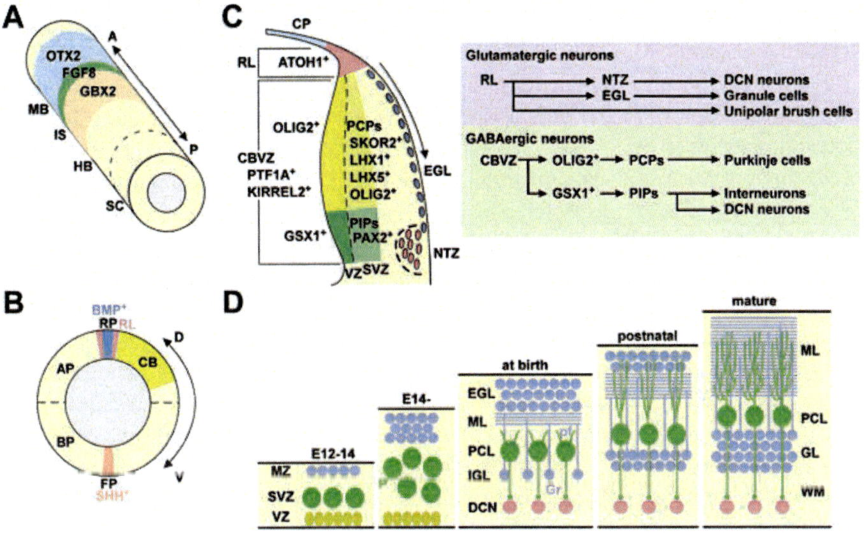

Fig. 3.2 Molecular and cellular mechanisms of the cerebellar development. (**a**) Regional specification along the A-P axis. OTX2 in the MB and GBX in the HB trigger and maintain the expression of FGF8 at their boundary and forms the organizing center IS. The anterior HB differentiates in GBX2 domain under control of FGF8. (**b**) Regional specification along the D-V axis at the level of the anterior HB. Counteracting influences of the dorsal BMP and the ventral SHH differentiate specific regions along the D-V axis. The primordial CB differentiates in the dorsal alar plate (AP). (**c**) Neurogenesis in the cerebellum. ATOH1^{+} RL generates glutamatergic neuronal progenitors that migrate tangentially to form the external granular layer (EGL) and the nuclear transition zone (NTZ). OLIG2^{+} VZ generates Purkinje cell progenitors (PCPs), whereas GSX1^{+} VZ generates PAX2^{+} interneuron progenitors (PIPs) that give rise to interneurons of the cerebellar cortex and deep cerebellar nuclei (DCN). (**d**) Development of cytoarchitecture in the CB of the mouse. SVZ subventricular zone, MZ marginal zone, ML molecular layer, PCL Purkinje cell layer, IGL internal granular layer, P Purkinje cell, Gr granule cell, pf parallel fiber, GL granule cell layer, WM white matter

(Hidalgo-Sanchez et al., 2005). Mutually exclusive expression of *Otx2* and *Gbx2* generates a signaling center called the isthmic organizer at their expression boundary. The isthmic organizer gives rise to an anatomical structure called the isthmus, which forms a constricted boundary between the midbrain and the hindbrain. At embryonic day (E)8-11.5 in mice, the isthmic organizer regulates cell survival and pattern formation in the midbrain and the hindbrain by secreting two signaling molecules: FGF8 (Fibroblast Growth Factor 8) and WNT1 (Sato & Joyner, 2009; Harada et al., 2016). Deletion of *Fgf8* or *Wnt1* leads to a lack in mid-hindbrain regions. Manipulation of these molecular activities affects the pattern formation of the cerebellum. While these isthmic organizing molecules specify the anterior limit of the cerebellum, *Hoxa2* demarcates the cerebellum by its expression posterior to rhombomere 1. Expression of *Hoxa2* is suppressed by the isthmic *Fgf8*, resulting in its absence in rhombomere 1 (Irving & Mason, 2000; Mason et al., 2000). Deletion of *Hoxa2* results in the expansion of the cerebellar territory (Gavalas et al., 1997), whereas ectopic expression of *Hoxa2* in rhombomere 1 perturbs the development of cerebellar neurons (Mason et al., 2000; Eddison et al., 2004).

Following the area specification along the anteroposterior axis, the cerebellar primordium appears in the dorsal alar plate region (Fig. 3.2b). The acquisition of dorsal properties is largely achieved by the roof plate. The roof plate is a monolayered structure formed at the dorsal midline of the neural tube. It primarily functions as a signaling center by secretion of BMP (Bone Morphogenetic Protein) and WNT. It later differentiates into the choroid plexus of the fourth ventricle. Secreted from the roof plate at the level of rhombomere 1, WNT accelerates the proliferation of the cerebellar primordial cells in the ventricular zone. Secreted BMP induces the cerebellar rhombic lip structure, which is the germinal source of cerebellar excitatory neurons, and also induces the expression of the transcription factor PTF1A (pancreas specific transcription factor 1a) in the cerebellar ventricular zone (Mishima et al., 2009; Millen et al., 2014; Yamada et al., 2014). PTF1A specifies the ventral boundary of the cerebellar territory in rhombomere 1, and its deletion abnormally transforms the cerebellar ventricular cells into cells with more ventral brainstem properties (Millen et al., 2014).

The neural tube literally takes a simple tubular structure when it is formed, but it is subsequently distorted into a complex structure. The cerebellar plate, which is the thickened cerebellar primordium at dorsal rhombomere 1, also undergoes a complex transformation at around E9-12.5 in mice (Fig. 3.1b). The neural tube starts to fold at the cephalic flexure around the isthmus and at the pontine or rhombic flexure around the boundary between the metencephalon (rostral hindbrain) and the myelencephalon (caudal hindbrain). The proliferating pons, cerebellum, and medulla occupy the large mass, whereas the less-proliferating monolayered choroid plexus is laterally expanded at the level of pontine flexure, forming a rhomboid-like structure. The lateral expansion and the folding of the neural tube in the hindbrain rotate the anteroposterior axis of the cerebellar plate by 90°, aligning it along the anatomical mediolateral axis (Fig. 3.1c). Then the left and the right cerebellar plates are fused at the anterior region to form the cerebellar vermis at the midline (Sgaier et al., 2005).

Cells in the cerebellum are derived from two germinal sources, the cerebellar plate and the upper rhombic lip (URL) (Fig. 3.2c). At around E10.5-14.5, GABAergic inhibitory neurons of the cerebellum are differentiated from PTF1A-positive neural stem cells located at the cerebellar ventricular zone (Hoshino et al., 2005). GABAergic neurons of the cerebellar nuclei are born first. Subsequently, GABAergic Purkinje cells, the sole output neurons of the cerebellar cortex, are born and start to migrate outward along the radial glial fibers running from the ventricular zone to the pial surface. Lastly, other GABAergic interneurons including basket cells, stellate cells, Golgi cells, chandelier cells, and Lugaro cells are born (Sudarov et al., 2011). After neural stem cells are born in the ventricular zone, the remaining radial glial cells are thought to lose their apical contacts, extend pial endfeet, and differentiate into Bergmann glial cells. Among the PTF1A-positive neural stem cells in the cerebellar ventricular zone, the dorsal populations express a beta-helix-loop-helix (bHLH)-type transcription factor OLIG2 (oligodendrocyte transcription factor 2) and generate Purkinje cell progenitors (PCPs). On the other hand, the ventral populations express a homeobox gene GSX1 and generate PAX2-positive interneuron progenitors (PIPs). OLIG2-positive PCP-generating cells predominate in the PTF1A-positive ventricular zone at E12.5. However, they are replaced with GSX1-positive PIP-generating cells at E14.5, due to suppression of OLIG2 expression by GSX1 (Leto et al., 2016; Yamada et al., 2014). These may underlie the sequential generation of Purkinje cells and other interneurons. In addition to OLIG2, PCPs express cell adhesion molecules KIRREL2 (Kirre like nephrin family adhesion molecule 2) and E-Cadherin (Mizuhara et al., 2010). PCPs later express a transcriptional corepressor SKOR2 (SKI family transcriptional corepressor 2) (Minaki et al., 2008).

The URL, the other germinal source of cerebellar cells, is formed until E10.5 at the dorsal margin of the cerebellar plate ventricular zone beneath the roof plate. The URL is induced by BMP secreted from the roof plate and the choroid plexus and comes to express the bHLH-type transcription factor ATOH1 (Atonal bHLH transcription factor 1) (Alder et al., 1999). Neural stem cells in the URL generate all of the excitatory glutamatergic neurons in the cerebellum (Machold & Fishell, 2005; Wang et al., 2005; Carletti & Rossi, 2008; Leto et al., 2016; Marzban et al., 2014). Initially, glutamatergic cerebellar nuclei neurons are born at the URL, tangentially migrate along the pial surface of the cerebellar primordium, and assemble in a cellular mass called the nuclear transitory zone (NTZ). The cells migrate out of the NTZ and move deeply into the cerebellum to form the cerebellar nucleus with the inhibitory neurons derived from the ventricular zone (Fink et al., 2006). Subsequently at E11.5, a larger number of granule cell progenitors are born in the URL, similarly migrate along the pial surface, but spread over the whole surface of the cerebellar plate just beneath the meninges, forming a layer called the external granular layer (EGL). The granule cell progenitors rapidly proliferate in the EGL by the mitogenic activity of SHH (Sonic hedgehog) secreted from Purkinje cells (Dahmane & Ruiz i Altaba, 1999) (Fig. 3.2d). The proliferation in the EGL peaks at around postnatal day (P) 7 and continues until around P15. The explosive proliferation of the granule cell progenitors increases the volume of the cerebellum and induces the formation of the folia (Sudarov & Joyner, 2007). The foliation may be caused by an anisotropic expansion of the EGL along the anteroposterior axis (Legue et al., 2015). The granule cell progenitors extend long bipolar processes that later become the parallel fibers along the mediolateral axis. The granule cell progenitors exit the proliferation cycle at the inner EGL. The postmitotic granule cells radially translocate their cell bodies inward along the fibers of Bergmann glial cells, but leave the trailing processes along the radial axis and the parallel fibers in the EGL, representing a T-shaped cellular morphology. The migrating granule cell bodies pass by Purkinje cells and form the internal granular layer (IGL) under the Purkinje cell layer. After completion of the granule cell migration, the EGL disappears and the IGL forms the mature granule cell layer. The parallel fibers from the granule cells come to occupy the surface of the cerebellum and form the molecular layer. To this layer, Purkinje cells extend fan-shaped dendrites that are aligned along the anteroposterior axis and make synapses with the parallel fiber axon terminals. Thus, a series of precisely regulated events during development underlie the formation of the highly-ordered layer structures found in the mature cerebellum (Millen & Gleeson, 2008; Butts et al., 2014; Marzban et al., 2014).

The molecular mechanisms of the perinatal development of the cerebellum have been extensively investigated. It is known that spontaneous mutant mice called Reeler show ataxia and cerebellar malformation due to abnormal cell migration. The causative gene codes a huge secreted glycoprotein Reelin (RELN), which is expressed by the URL-derived cerebellar nucleus and granule cell progenitors migrating beneath the meninges. Importantly, a lissencephaly caused by RELN mutation shows cerebellar hypoplasia (Hong et al., 2000). In addition, the radial migration of the granule cells could be explained by various mechanisms including the intracellular calcium dynamics of granule cells, the scaffolding of Bergmann glial fibers, and SDF1 (Stromal-cell-derived factor 1) secreted from the meninges. However, the mechanisms of the cerebellar development still remain largely unknown. This may be because cerebellar development consists of multiple complex steps from the regional specification of the cerebellum at the neural tube stage to the acquisition of cellular identity and because these steps often proceed concurrently and interact with each other. These complex developmental mechanisms may be revealed by combining the analytical methods of classical molecular, cellular, and developmental biology with a newly emerging strategy that recapitulates ontogenetic events in vitro.

3.3 Recapitulation of Cerebellar Development with PSCs

It was revealed that PSCs can be differentiated into various cell types by recapitulating ontogenetic processes (Muguruma & Sasai, 2012; Suzuki & Vanderhaeghen, 2015). With this notion, mouse ESCs (mESCs) were differentiated to cerebellar neurons by a combinatory treatment of positional signals (Salero & Hatten, 2007; Su et al., 2006). mESCs was first treated with secreted molecules (FGF8, WNT1 and retinoic acid; RA) that are known to induce cerebellar primordium, then with BMPs and GDF7, which induce granule cell progenitors and mitogens (SHH and JAGGED 1). $ATOH1^+$ granule cell progenitors were differentiated at relatively high efficiency (>15%) with this culture protocol, but Purkinje cells were rarely generated (less than 1%). A similar but modified protocol was developed for hESC culture (Erceg et al., 2010). This protocol, requiring a large amount of growth factors, somewhat improved the efficiency of Purkinje cell differentiation, but it does not appear to induce the characteristic morphology, such as dendritic extension and large soma. Another group reported an improved protocol for mESCs that uses a co-culture system with feeder cells to promote the generation, survival, and maturation of Purkinje cells (Tao et al., 2010). This protocol differentiates Purkinje cells with morphological and electrophysiological properties at higher efficiency (twofold to ninefold), but it remains unknown whether this protocol can be applied to hPSCs.

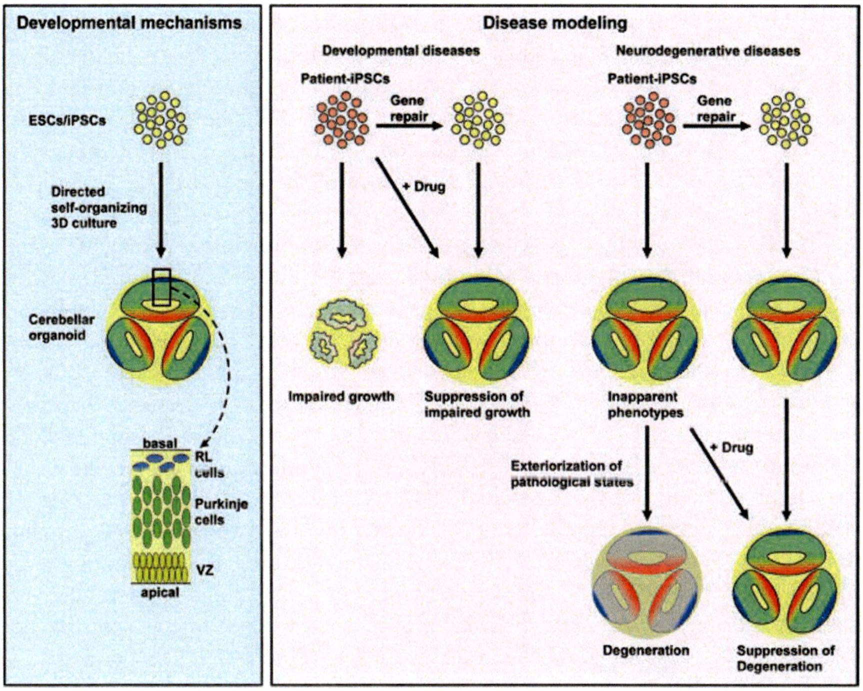

Fig. 3.3 Applications of cerebellar organoids to investigation of human cerebellum. Human ESC/iPSC-derived cerebellar organoids provide a way to access human cerebellar tissues and a platform to experimentally investigate the developmental mechanisms. Combination of the organoid culture with patient-derived iPSCs accelerates the pathogenic investigation of developmental and neurodegenerative cerebellar diseases. Organoid culture systems can be used to identify disease phenotypes, pathogenic mechanisms, effective drug treatment or gene repair to alleviate the diseases

An innovation in neural differentiation of PSCs was brought by introduction of 3D culture systems that recapitulate ontogenetic tissue formation utilizing self-organizing principles (Sasai et al., 2012; Sasai, 2013) (Fig. 3.3). A 3D floating culture method called serum-free culture of the embryoid body-like aggregates with quick reaggregation (SFEBq) was developed for 3D tissue formation (Eiraku et al., 2008; Wataya et al., 2008). The 3D culture system, also called brain organoid culture (Lancaster & Knoblich, 2014), allows researchers to form 3D tissues that strikingly resemble mouse or human brain. In the case of cerebellar development, differentiation starts from the formation of the isthmic organizer as described in the previous section. The 3D SFEBq culture of mESCs was modified to recapitulate the innate program of initial cerebellar development associated with the isthmic organizer (Muguruma et al., 2010). Treatment of the culture with moderate caudalizing factors (FGF2 and insulin) achieved robust differentiation (>80% efficacy) of EN2^{+} progenitors, representing the caudal midbrain to rostral hindbrain regions. A majority of the induced progenitors come to express rostral hindbrain markers. Further treatment with an inhibitor of SHH, a ventralizing signaling molecule, directed the tissue to acquire dorsal properties with expression of early cerebellar markers PTF1A and KIRREL2, suggesting efficient generation of cerebellar neuroepithelium. The induced KIRREL2⁺ progenitors subsequently differentiate into Purkinje cells at high efficiency when co-cultured with mouse URL-derived granule cells. Thus, self-organizing 3D organoid culture combined with treatment with appropriate signaling molecules is an efficient method for differentiation of PSCs into cerebellar cells and tissues.

This 3D organoid culture based on a self-formation principle is applicable to the differentiation of human cerebellar neurons from hPSCs, with some modifications (Muguruma et al., 2015). The modifications include treatment with TGFß inhibitor, which inhibits mesenchymal differentiation and promotes neuroectodermal differentiation, together with the Rho-associated coiled-coil forming kinase (ROCK) selective inhibitor Y-27632 (Watanabe et al., 2007). These hPSC-derived neural progenitors come to express EN2 and GBX2 after 21 days in culture. Around 35 days in culture, substantial populations of hPSC-derived cells express the Purkinje cell progenitor markers LHX5, KIRREL2, PTF1A, and SKOR2. As in the case of mESCs, hPSC-derived cerebellar progenitors come to express L7, Calbindin, GABA, and Aldolase C (specific markers for mature Purkinje cells) after long-term co-culture with mouse URL-derived granule cells. $L7^+$/$Calbindin^+$ hPSC-derived Purkinje cells extend elaborate dendritic branches with spines that are positive for the Purkinje cell-specific glutamate receptor GRID2 (GluRδ2). GRID2 was associated with CBLN1, which is expressed at the presynaptic termini of the parallel fibers. The electrophysiological properties of hPSC-derived Purkinje cells showed typical characteristics as seen in rodents (Muguruma et al., 2015).

Differentiation studies with mESCs and hPSCs revealed substantial differences between mouse and human cerebellar organoids. It is known that BMP signals from the roof-plate promote RL formation (Alder et al., 1999). BMP signals are necessary for the induction of RL-derived $ATOH1^+$ progenitors in mESC culture (Muguruma et al., 2010). In contrast, $ATOH1^+$ cells, $ATOH1^+$/$BARHL1^+$ granule cells, and $TBR1^+$/$SMI\text{-}32^+$ glutamatergic DCN projection neurons were generated from hPSCs even in the absence of BMP4 (Muguruma et al., 2015). Furthermore, unlike mESC-derived organoids, the differentiated human cerebellar organoids come to contain hollow neural tube-like neuroepithelial structures with dorsoventral polarity. The outer or superficial portion of the tube acquired a dorsal cerebellar fate and contained both Purkinje and granule cell precursors, whereas the inner or lumen side acquired the ventral properties of the midline floor plate and the basal plate of the brainstem at the level of the cerebellum.

The formation of polarized cerebellar neuroepithelium seems to be a somewhat stochastic event. Sequential treatment with FGF19 and SDF1 (CXCL2) efficiently promoted the formation of larger neuroepithelial tissues with RL-like structures. In this culture condition, hPSC-derived organoids display a neuroepithelial structure with an apico-basally layered arrangement of multiple cell types. It consists of the innermost cerebellar ventricular zone, an intermediate layer containing precursors of Purkinje neurons, and an outermost layer occupied by derivatives of the RL, which have the characteristics of granule cell precursors. Thus, hPSC-derived organoids appear to have more potential to differentiate into cerebellar tissues than those from mouse PSCs.

The 3D cerebellar organoid culture enables us to differentiate the tissues up until the first or second trimester. However, differentiation beyond the third trimester has not yet been accomplished. Other protocols reported by several groups seem to reach similar stages (Erceg et al., 2010; Salero & Hatten, 2007; Higuera et al., 2017; Wang et al., 2015; Watson et al., 2018; Silva et al., 2020). Further maturation of the cerebellar organoids may require some environmental changes like vascularization or inclusion of non-neuronal cells such as astrocyte or microglia.

3.4 Investigation of Cerebellar Diseases with PSCs

In recent years, many labs have developed brain organoids from hPSCs and constructed in vitro models for investigating human brain diseases, development, and evolution (Hong & Do, 2019; Chhibber et al., 2020). Conventional approaches using human cell line models or animal models encounter substantial problems in investigating human diseases, such as insufficient phenotypic recapitulation in cultured cells that are irrelevant to the disease targets and differential drug responses between human and animal species. There are ethical regulations and technical difficulties in using human tissues for studying human development and evolution, and there are also ethical considerations when using species that are genetically close to humans like primates, for experimental investigations. Therefore, the methods for generating iPSCs from humans including patients, and the in vitro differentiation of iPSCs into various target cells and organoid tissues provide an effective platform to investigate human diseases and development. In the following subsections, we introduce recent progress and future perspectives on the investigations of cerebellar diseases with a focus on the application of the hiPSCs and cerebellar organoids derived from them (Fig. 3.3).

3.5 Spinocerebellar Ataxias

Spinocerebellar ataxias (SCAs), which belong to a large family of cerebellar ataxias, are characterized by progressive neurodegeneration of the cerebellum and its related areas. In the cerebellum, massive neurodegeneration occurs in Purkinje cells. The clinical phenotype includes progressive ataxia, dysarthria, and oculomotor deficits. SCAs are classified into sporadic diseases and inherited diseases with genetic mutations. The inherited SCAs are chronologically numbered after identification of the causative gene, and 48 SCAs have been identified to date. The majority of the inherited SCAs are caused by autosomal-dominant mutations, such as nucleotide repeat expansions in coding or non-coding sequences, or by point mutations (Ashizawa et al., 2018). At least seven subtypes (SCA 1, 2, 3, 6, 7, 8, and 17) are caused by expansion of CAG repeats encoding polyQ tracts. iPSCs have been generated from SCA2, SCA3, SCA6, and SCA7 patients, and some of them have been subjected to phenotypic analysis (Naphade et al., 2019; Tamada et al., 2020). Phenotypic analysis can be done with patient-derived iPSCs themselves or partially differentiated neurons without cell-type specification. However, for in vitro recapitulation of a disease phenotype, it is desirable to perform the analysis with fully differentiated target cells or tissues that are affected in the disease. In this direction, Purkinje cells differentiated from SCA6 patient-derived iPSCs were used for identification of disease-specific phenotypes (Ishida et al., 2016; Naphade et al., 2019; Tamada et al., 2020).

3.6 Cerebellar Malformations

In humans, the development of the cerebellum starts as early as the fourth gestational week. The development continues over a protracted period after birth, which creates vulnerability in the cerebellum. Cerebellar malformations often accompany defects in other regions including the brainstem. Here, we focus on the frequently occurring malformations that could be the targets of PSC-based investigations.

Dandy-Walker malformation (DWM) is a heterogeneous disorder defined by a hypoplastic, upwardly rotated vermis, an enlarged fourth ventricle, and an enlarged posterior fossa (Aldinger & Doherty, 2016). The cerebellar vermis is more severely affected than the hemispheres. There is variance among patients in the clinical features and developmental outcomes. The patients exhibit various symptoms from intellectual disability to autism. Some patients do not notice any symptoms until diagnosed. The empiric recurrence risk is low at an estimated 1–5% (Murray et al., 1985), suggesting de-novo, somatic mosaic, or complex genetic causes (Aldinger & Doherty, 2016; Haldipur & Millen, 2013). The genetic cause of DWM remains largely unknown, but few genes are involved in rare cases. Heterozygous deletions of chromosome 3q24 including the zinc finger in cerebellum (ZIC)1 and ZIC4 genes are involved in some cases (Grinberg & Millen, 2005; Blank et al., 2011). Additionally, heterozygous deletions of chromosome 6p25 encompassing the Forkhead Box 1 (FOXC1) genes are also associated with DWM (Aldinger et al., 2009; Delahaye et al., 2012).

Cerebellar hypoplasia is characterized as an underdevelopment of the cerebellum. It is frequently associated with other brain malformations including lissencephaly. Lissencephaly is caused by defects in either the RELN pathway or microtubule formation. RELN is transiently expressed by neurons within superficial layers in both the cerebral cortex and cerebellum and regulates radial neuronal migration. In a similar way to Reeler mutant mice, human RELN mutations cause cerebellar hypoplasia accompanied with abnormality in Purkinje cell and granule cell migration. Patients with RELN mutation have pachygyria and an extremely hypoplastic cerebellum (Hong et al., 2000). Patients with a mutation in the reelin receptor VLDLR exhibit mild pachygyria and a slightly small cerebellum (Boycott et al., 2009). Loss of α- or β-tubulins is also a major cause of brain malformations, such as lissencephaly, pachygyria, and microgyria. This may be due to the importance of tubulins in neuronal migration (Kato, 2015). It is also known that underglycosylation of α-dystroglycan with O-linked carbohydrates causes severe cerebellar hypoplasia and/or dysmorphology (Martin, 2005).

Currently there is no cure to prevent cerebellar malformations. However, if accurate human disease models are established with 3D organoid technology using patient-derived iPSCs, that would promote a mechanistic understanding that could lead to development of effective treatments or drugs against cerebellar malformations.

3.7 Future Perspective

About 10 years have passed since the brain organoid techniques were first developed, and they are still being developed and improved. So far, the protocols have only recapitulated human brain development up until the second trimester in the context of gene expression and tissue construction. Because current brain organoids lack circulation systems and are poor in extracellular matrices, they require trial-and-error improvement of the culture environment such as testing of various media components and efficient gas exchange. Extensive research has been aimed at developing methods for formation of vascular networks by assembling vascular endothelial cells or pericytes, on-tip culture of organoids with bioengineering of microfluidic devices. In addition, inclusion of non-neuronal cells such as glial cells could help maintain the brain environment and functional expression in the organoids. The brain organoid techniques, which are currently in rapid progress, have potential for various applications other than pathological investigation and regenerative medicine (Koo et al., 2019; Kim et al., 2020). Recently, comparative transcriptome analysis of Purkinje cells from mice and those from human iPS cells have confirmed conserved genetic expression patterns among species and have further revealed novel cell-specific markers (Buchholz et al., 2020). Brain organoids can be applied to cancer research, such as development of in vivo medulloblastoma models with cerebellar organoids (Ballabio et al., 2020). Thus, we expect that brain organoids will become a fundamental technique in embryology and stem cell biology research and be used in other fields as well.

Acknowledgments This work is supported by MEXT/JSPS KAKENHI Grant Numbers JP18H02535 and JP19K22983 to K.M. and JP18H04762 to A.T.

References

Alder, J., Lee, K. J., Jessell, T. M., & Hatten, M. E. (1999). Generation of cerebellar granule neurons in vivo by transplantation of BMP-treated neural progenitor cells. *Nature Neuroscience, 2*(6), 535–540. https://doi.org/10.1038/9189

Aldinger, K. A., & Doherty, D. (2016). The genetics of cerebellar malformations. *Seminars in Fetal & Neonatal Medicine, 21*(5), 321–332. https://doi.org/10.1016/j.siny.2016.04.008

Aldinger, K. A., Lehmann, O. J., Hudgins, L., Chizhikov, V. V., Bassuk, A. G., Ades, L. C., et al. (2009). FOXC1 is required for normal cerebellar development and is a major contributor to chromosome 6p25.3 Dandy-Walker malformation. *Nature Genetics, 41*(9), 1037–1042. https://doi.org/10.1038/ng.422

Ashizawa, T., Oz, G., & Paulson, H. L. (2018). Spinocerebellar ataxias: Prospects and challenges for therapy development. *Nature Reviews. Neurology, 14*(10), 590–605. https://doi.org/10.1038/s41582-018-0051-6. 10.1038/s41582-018-0051-6 [pii].

Ballabio, C., Anderle, M., Gianesello, M., Lago, C., Miele, E., Cardano, M., et al. (2020). Modeling medulloblastoma in vivo and with human cerebellar organoids. *Nature communications, 11*(1), 583. https://doi.org/10.1038/s41467-019-13989-3

Blank, M. C., Grinberg, I., Aryee, E., Laliberte, C., Chizhikov, V. V., Henkelman, R. M., et al. (2011). Multiple developmental programs are altered by loss of Zic1 and Zic4 to cause Dandy-Walker malformation cerebellar pathogenesis. *Development, 138*(6), 1207–1216. https://doi.org/10.1242/dev.054114

Boycott, K. M., Bonnemann, C., Herz, J., Neuert, S., Beaulieu, C., Scott, J. N., et al. (2009). Mutations in VLDLR as a cause for autosomal recessive cerebellar ataxia with mental retardation (dysequilibrium syndrome). *Journal of Child Neurology, 24*(10), 1310–1315. https://doi.org/10.1177/0883073809332696

Buchholz, D. E., Carroll, T. S., Kocabas, A., Zhu, X. D., Behesti, H., Faust, P. L., et al. (2020). Novel genetic features of human and mouse Purkinje cell differentiation defined by comparative transcriptomics. *Proceedings of the National Academy of Sciences of the United States of America, 117*(26), 15085–15095. https://doi.org/10.1073/pnas.2000102117

Butts, T., Green, M. J., & Wingate, R. J. (2014). Development of the cerebellum: Simple steps to make a 'little brain'. *Development, 141*(21), 4031–4041. https://doi.org/10.1242/dev.106559

Carletti, B., & Rossi, F. (2008). Neurogenesis in the cerebellum. *The Neuroscientist, 14*(1), 91–100. 1073858407304629 [pii]. https://doi.org/10.1177/1073858407304629

Chhibber, T., Bagchi, S., Lahooti, B., Verma, A., Al-Ahmad, A., Paul, M. K., et al. (2020). CNS organoids: An innovative tool for neurological disease modeling and drug neurotoxicity screening. *Drug Discovery Today, 25*(2), 456–465. https://doi.org/10.1016/j.drudis.2019.11.010

Dahmane, N., & Ruiz i Altaba, A. (1999). Sonic hedgehog regulates the growth and patterning of the cerebellum. *Development, 126*(14), 3089–3100.

Delahaye, A., Khung-Savatovsky, S., Aboura, A., Guimiot, F., Drunat, S., Alessandri, J. L., et al. (2012). Pre- and postnatal phenotype of 6p25 deletions involving the FOXC1 gene. *American Journal of Medical Genetics. Part A, 158A*(10), 2430–2438. https://doi.org/10.1002/ajmg.a.35548

Eddison, M., Toole, L., Bell, E., & Wingate, R. J. (2004). Segmental identity and cerebellar granule cell induction in rhombomere 1. *BMC Biology, 2*, 14. https://doi.org/10.1186/1741-7007-2-14

Eiraku, M., Watanabe, K., Matsuo-Takasaki, M., Kawada, M., Yonemura, S., Matsumura, M., et al. (2008). Self-organized formation of polarized cortical tissues from ESCs and its active manipulation by extrinsic signals. *Cell Stem Cell, 3*(5), 519–532. https://doi.org/10.1016/j.stem.2008.09.002. S1934-5909(08)00455-4 [pii].

Erceg, S., Ronaghi, M., Zipancic, I., Lainez, S., Rosello, M. G., Xiong, C., et al. (2010). Efficient differentiation of human embryonic stem cells into functional cerebellar-like cells. *Stem Cells and Development, 19*(11), 1745–1756. https://doi.org/10.1089/scd.2009.0498

Fink, A. J., Englund, C., Daza, R. A., Pham, D., Lau, C., Nivison, M., et al. (2006). Development of the deep cerebellar nuclei: Transcription factors and cell migration from the rhombic lip. *The Journal of Neuroscience, 26*(11), 3066–3076. https://doi.org/10.1523/JNEUROSCI.5203-05.2006

Gavalas, A., Davenne, M., Lumsden, A., Chambon, P., & Rijli, F. M. (1997). Role of Hoxa-2 in axon pathfinding and rostral hindbrain patterning. *Development, 124*(19), 3693–3702.

Grinberg, I., & Millen, K. J. (2005). The ZIC gene family in development and disease. *Clinical Genetics, 67*(4), 290–296. https://doi.org/10.1111/j.1399-0004.2005.00418.x

Haldipur, P., & Millen, K. J. (2013). Cerebellar malformation: Deficits in early neural tube identity found in CHARGE syndrome. *eLife, 2*, e01873. https://doi.org/10.7554/eLife.01873

Hallonet, M. E., & Le Douarin, N. M. (1993). Tracing neuroepithelial cells of the mesencephalic and metencephalic alar plates during cerebellar ontogeny in quail-chick chimaeras. *The European Journal of Neuroscience, 5*(9), 1145–1155. https://doi.org/10.1111/j.1460-9568.1993.tb00969.x

Harada, H., Sato, T., & Nakamura, H. (2016). Fgf8 signaling for development of the midbrain and hindbrain. *Development, Growth & Differentiation, 58*(5), 437–445. https://doi.org/10.1111/dgd.12293

Hidalgo-Sanchez, M., Millet, S., Bloch-Gallego, E., & Alvarado-Mallart, R. M. (2005). Specification of the meso-isthmo-cerebellar region: The Otx2/Gbx2 boundary. *Brain Research. Brain Research Reviews, 49*(2), 134–149. https://doi.org/10.1016/j.brainresrev.2005.01.010

Higuera, G. A., Iaffaldano, G., Bedar, M., Shpak, G., Broersen, R., Munshi, S. T., et al. (2017). An expandable embryonic stem cell-derived Purkinje neuron progenitor population that exhibits in vivo maturation in the adult mouse cerebellum. *Scientific Reports, 7*(1), 8863. https://doi.org/10.1038/s41598-017-09348-1. 10.1038/s41598-017-09348-1 [pii].

Hong, S. E., Shugart, Y. Y., Huang, D. T., Shahwan, S. A., Grant, P. E., Hourihane, J. O., et al. (2000). Autosomal recessive lissencephaly with cerebellar hypoplasia is associated with human RELN mutations. *Nature Genetics, 26*(1), 93–96. https://doi.org/10.1038/79246

Hong, Y. J., & Do, J. T. (2019). Neural lineage differentiation from pluripotent stem cells to mimic human brain tissues. *Frontiers in Bioengineering and Biotechnology, 7*, 400. https://doi.org/10.3389/fbioe.2019.00400

Hoshino, M., Nakamura, S., Mori, K., Kawauchi, T., Terao, M., Nishimura, Y. V., et al. (2005). Ptf1a, a bHLH transcriptional gene, defines GABAergic neuronal fates in cerebellum. *Neuron, 47*(2), 201–213. doi:S0896-6273(05)00485-X [pii]. https://doi.org/10.1016/j.neuron.2005.06.007

Irving, C., & Mason, I. (2000). Signalling by FGF8 from the isthmus patterns anterior hindbrain and establishes the anterior limit of Hox gene expression. *Development, 127*(1), 177–186.

Ishida, Y., Kawakami, H., Kitajima, H., Nishiyama, A., Sasai, Y., Inoue, H., et al. (2016). Vulnerability of Purkinje cells generated from spinocerebellar ataxia type 6 patient-derived iPSCs. *Cell Reports, 17*(6), 1482–1490. doi:S2211-1247(16)31429-2 [pii]. https://doi.org/10.1016/j.celrep.2016.36026

Joyner, A. L., Liu, A., & Millet, S. (2000). Otx2, Gbx2 and Fgf8 interact to position and maintain a mid-hindbrain organizer. *Current Opinion in Cell Biology, 12*(6), 736–741. doi:S0955-0674(00)00161-7 [pii]. https://doi.org/10.1016/s0955-0674(00)00161-7

Kato, M. (2015). Genotype-phenotype correlation in neuronal migration disorders and cortical dysplasias. *Front Neurosci-Switz, 9*, 181. https://doi.org/10.3389/Fnins.2015.00181

Kim, J., Koo, B. K., & Knoblich, J. A.. Human organoids: Model systems for human biology and medicine. Nature reviews. *Molecular Cell Biology*. 2020. https://doi.org/10.1038/s41580-020-0259-3

Koo, B., Choi, B., Park, H., & Yoon, K. J. (2019). Past, present, and future of brain organoid technology. *Molecules and Cells, 42*(9), 617–627. https://doi.org/10.14348/molcells.2019.0162

Lancaster, M. A., & Knoblich, J. A. (2014). Organogenesis in a dish: Modeling development and disease using organoid technologies. *Science, 345*(6194), 1247125. https://doi.org/10.1126/science.1247125

Legue, E., Riedel, E., & Joyner, A. L. (2015). Clonal analysis reveals granule cell behaviors and compartmentalization that determine the folded morphology of the cerebellum. *Development, 142*(9), 1661–1671. https://doi.org/10.1242/dev.120287

Leto, K., Arancillo, M., Becker, E. B., Buffo, A., Chiang, C., Ding, B., et al. (2016). Consensus paper: Cerebellar development. *Cerebellum, 15*(6), 789–828. https://doi.org/10.1007/s12311-015-0724-2. 10.1007/s12311-015-0724-2 [pii].

Machold, R., & Fishell, G. (2005). Math1 is expressed in temporally discrete pools of cerebellar rhombic-lip neural progenitors. *Neuron, 48*(1), 17–24. https://doi.org/10.1016/j.neuron.2005.08.028

Martin, P. T. (2005). The dystroglycanopathies: The new disorders of O-linked glycosylation. *Seminars in Pediatric Neurology, 12*(3), 152–158. https://doi.org/10.1016/j.spen.2005.10.003

Martinez, S., & Alvarado-Mallart, R. M. (1989a). Rostral cerebellum originates from the caudal portion of the so-called 'Mesencephalic' vesicle: A study using chick/quail chimeras. *The European Journal of Neuroscience, 1*(6), 549–560. https://doi.org/10.1111/j.1460-9568.1989.tb00362.x

Martinez, S., & Alvarado-Mallart, R. M. (1989b). Transplanted mesencephalic quail cells colonize selectively all primary visual nuclei of chick diencephalon: A study using heterotopic transplants. *Brain Research. Developmental Brain Research, 47*(2), 263–274. https://doi.org/10.1016/0165-3806(89)90181-8

Martinez, S., Andreu, A., Mecklenburg, N., & Echevarria, D. (2013). Cellular and molecular basis of cerebellar development. *Frontiers in Neuroanatomy, 7*, 18. https://doi.org/10.3389/fnana.2013.00018

Marzban, H., Del Bigio, M. R., Alizadeh, J., Ghavami, S., Zachariah, R. M., & Rastegar, M. (2014). Cellular commitment in the developing cerebellum. *Frontiers in Cellular Neuroscience, 8*, 450. https://doi.org/10.3389/fncel.2014.00450

Mason, I., Chambers, D., Shamim, H., Walshe, J., & Irving, C. (2000). Regulation and function of FGF8 in patterning of midbrain and anterior hindbrain. *Biochemistry and Cell Biology, 78*(5), 577–584.

Millen, K. J., & Gleeson, J. G. (2008). Cerebellar development and disease. *Current Opinion in Neurobiology, 18*(1), 12–19. https://doi.org/10.1016/j.conb.2008.05.010

Millen, K. J., Steshina, E. Y., Iskusnykh, I. Y., & Chizhikov, V. V. (2014). Transformation of the cerebellum into more ventral brainstem fates causes cerebellar agenesis in the absence of Ptf1a function. *Proceedings of the National Academy of Sciences of the United States of America, 111*(17), E1777–E1786. https://doi.org/10.1073/pnas.1315024111

Minaki, Y., Nakatani, T., Mizuhara, E., Inoue, T., & Ono, Y. (2008). Identification of a novel transcriptional corepressor, Corl2, as a cerebellar Purkinje cell-selective marker. *Gene Expression Patterns, 8*(6), 418–423. https://doi.org/10.1016/j.gep.2008.04.004

Mishima, Y., Lindgren, A. G., Chizhikov, V. V., Johnson, R. L., & Millen, K. J. (2009). Overlapping function of Lmx1a and Lmx1b in anterior hindbrain roof plate formation and cerebellar growth. *The Journal of Neuroscience, 29*(36), 11377–11384. https://doi.org/10.1523/JNEUROSCI.0969-09.2009

Mizuhara, E., Minaki, Y., Nakatani, T., Kumai, M., Inoue, T., Muguruma, K., et al. (2010). Purkinje cells originate from cerebellar ventricular zone progenitors positive for Neph3 and E-cadherin. *Developmental Biology, 338*(2), 202–214. https://doi.org/10.1016/j.ydbio.2009.11.032

Muguruma, K., Nishiyama, A., Kawakami, H., Hashimoto, K., & Sasai, Y. (2015). Self-organization of polarized cerebellar tissue in 3D culture of human pluripotent stem cells. *Cell Reports, 10*(4), 537–550. https://doi.org/10.1016/j.celrep.2014.12.051. S2211-1247(14)01104-8 [pii].

Muguruma, K., Nishiyama, A., Ono, Y., Miyawaki, H., Mizuhara, E., Hori, S., et al. (2010). Ontogeny-recapitulating generation and tissue integration of ES cell-derived Purkinje cells. *Nature Neuroscience, 13*(10), 1171–1180. https://doi.org/10.1038/nn.2638. nn.2638 [pii].

Muguruma, K., & Sasai, Y. (2012). In vitro recapitulation of neural development using embryonic stem cells: From neurogenesis to histogenesis. *Development, Growth & Differentiation, 54*(3), 349–357. https://doi.org/10.1111/j.1440-169X.2012.01329.x

Murray, J. C., Johnson, J. A., & Bird, T. D. (1985). Dandy-Walker malformation: Etiologic heterogeneity and empiric recurrence risks. *Clinical Genetics, 28*(4), 272–283. https://doi.org/10.1111/j.1399-0004.1985.tb00401.x

Naphade, S., Tshilenge, K. T., & Ellerby, L. M. (2019). Modeling polyglutamine expansion diseases with induced pluripotent stem cells. *Neurotherapeutics, 16*(4), 979–998. https://doi.org/10.1007/s13311-019-00810-8

Salero, E., & Hatten, M. E. (2007). Differentiation of ES cells into cerebellar neurons. *Proceedings of the National Academy of Sciences of the United States of America, 104*(8), 2997–3002. doi:0610879104 [pii]. https://doi.org/10.1073/pnas.0610879104

Sasai, Y. (2013). Cytosystems dynamics in self-organization of tissue architecture. *Nature, 493*(7432), 318–326. https://doi.org/10.1038/nature11859

Sasai, Y., Eiraku, M., & Suga, H. (2012). In vitro organogenesis in three dimensions: Self-organising stem cells. *Development, 139*(22), 4111–4121. https://doi.org/10.1242/dev.079590

Sato, T., & Joyner, A. L. (2009). The duration of Fgf8 isthmic organizer expression is key to patterning different tectal-isthmo-cerebellum structures. *Development, 136*(21), 3617–3626. https://doi.org/10.1242/dev.041210

Sgaier, S. K., Millet, S., Villanueva, M. P., Berenshteyn, F., Song, C., & Joyner, A. L. (2005). Morphogenetic and cellular movements that shape the mouse cerebellum; insights from genetic fate mapping. *Neuron, 45*(1), 27–40. https://doi.org/10.1016/j.neuron.2004.12.021

Silva, T. P., Bekman, E. P., Fernandes, T. G., Vaz, S. H., Rodrigues, C. A. V., Diogo, M. M., et al. (2020). Maturation of human pluripotent stem cell-derived cerebellar neurons in the absence of co-culture. *Frontiers in Bioengineering and Biotechnology, 8*, 70. https://doi.org/10.3389/fbioe.2020.00070

Su, H. L., Muguruma, K., Matsuo-Takasaki, M., Kengaku, M., Watanabe, K., & Sasai, Y. (2006). Generation of cerebellar neuron precursors from embryonic stem cells. *Developmental Biology, 290*(2), 287–296. doi:S0012-1606(05)00787-6 [pii]. https://doi.org/10.1016/j.ydbio.2005.11.010

Sudarov, A., & Joyner, A. L. (2007). Cerebellum morphogenesis: The foliation pattern is orchestrated by multi-cellular anchoring centers. *Neural Development, 2*, 26. https://doi.org/10.1186/1749-8104-2-26

Sudarov, A., Turnbull, R. K., Kim, E. J., Lebel-Potter, M., Guillemot, F., & Joyner, A. L. (2011). Ascl1 genetics reveals insights into cerebellum local circuit assembly. *The Journal of Neuroscience, 31*(30), 11055–11069. https://doi.org/10.1523/JNEUROSCI.0479-11.2011

Suzuki, I. K., & Vanderhaeghen, P. (2015). Is this a brain which I see before me? Modeling human neural development with pluripotent stem cells. *Development, 142*(18), 3138–3150. https://doi.org/10.1242/dev.120568

Tamada, A., Watanabe, S., & Muguruma, K. (2020). Investigating developmental and disease mechanisms of the cerebellum with pluripotent stem cells. *Molecular and Cellular Neurosciences, 107*, 103530. https://doi.org/10.1016/j.mcn.2020.103530

Tao, O., Shimazaki, T., Okada, Y., Naka, H., Kohda, K., Yuzaki, M., et al. (2010). Efficient generation of mature cerebellar Purkinje cells from mouse embryonic stem cells. *Journal of Neuroscience Research, 88*(2), 234–247. https://doi.org/10.1002/jnr.22208

Wang, S., Wang, B., Pan, N., Fu, L., Wang, C., Song, G., et al. (2015). Differentiation of human induced pluripotent stem cells to mature functional Purkinje neurons. *Scientific Reports, 5*, 9232. https://doi.org/10.1038/srep09232. srep09232 [pii].

Wang, V. Y., Rose, M. F., & Zoghbi, H. Y. (2005). Math1 expression redefines the rhombic lip derivatives and reveals novel lineages within the brainstem and cerebellum. *Neuron, 48*(1), 31–43. https://doi.org/10.1016/j.neuron.2005.08.024

Watanabe, K., Ueno, M., Kamiya, D., Nishiyama, A., Matsumura, M., Wataya, T., et al. (2007). A ROCK inhibitor permits survival of dissociated human embryonic stem cells. *Nature Biotechnology, 25*(6), 681–686. doi:nbt1310 [pii]. 10.1038/nbt1310.

Wataya, T., Ando, S., Muguruma, K., Ikeda, H., Watanabe, K., Eiraku, M., et al. (2008). Minimization of exogenous signals in ES cell culture induces rostral hypothalamic differentiation. *Proceedings of the National Academy of Sciences of the United States of America, 105*(33), 11796–11801. https://doi.org/10.1073/pnas.0803078105. 0803078105 [pii].

Watson, L. M., Wong, M. M. K., Vowles, J., Cowley, S. A., & Becker, E. B. E. (2018). A simplified method for generating Purkinje cells from human-induced pluripotent stem cells. *Cerebellum, 17*(4), 419–427. https://doi.org/10.1007/s12311-017-0913-2. 10.1007/s12311-017-0913-2 [pii].
Wingate, R. J., & Hatten, M. E. (1999). The role of the rhombic lip in avian cerebellum development. *Development, 126*(20), 4395–4404.
Wurst, W., & Bally-Cuif, L. (2001). Neural plate patterning: Upstream and downstream of the isthmic organizer. *Nature Reviews. Neuroscience, 2*(2), 99–108. https://doi.org/10.1038/35053516
Yamada, M., Seto, Y., Taya, S., Owa, T., Inoue, Y. U., Inoue, T., et al. (2014). Specification of spatial identities of cerebellar neuron progenitors by ptf1a and atoh1 for proper production of GABAergic and glutamatergic neurons. *The Journal of Neuroscience, 34*(14), 4786–4800. https://doi.org/10.1523/JNEUROSCI.2722-13.2014

Chapter 4
mGluR1 Is a Molecular "Hub" for Synapse Elimination in the Developing Cerebellum

Masanobu Kano, Takaki Watanabe, and Naofumi Uesaka

4.1 Prologue

In 1982, an author of this chapter, Masanobu Kano, entered the PhD course at Graduate School of Medicine, The University of Tokyo, and became a graduate student of late Masao Ito. In this year, Ito published the first paper demonstrating long-term depression (LTD) of responsiveness of Purkinje cells in the cerebellar flocculus to vestibular nerve stimulation after conjunctive stimulation with climbing fibers in decerebrate rabbit (Ito et al., 1982). However, vestibular nerve stimulation activates floccular Purkinje cells di-synaptically via vestibular mossy fiber to granule cell synapses and parallel fiber (i.e., granule cell axon) to Purkinje cell synapses; it was not conclusive whether LTD occurs at the latter synapses. Therefore, Ito started with Kano to examine whether LTD is induced at parallel fiber to Purkinje cell synapses by examining field potentials elicited by direct parallel fiber stimulation, which reflect excitatory postsynaptic potentials (EPSPs) in Purkinje cells, in the dorsal paraflocculus of decerebrate rabbit. They showed that LTD was induced at parallel fiber to Purkinje cell synapses after conjunctive stimulation with climbing fibers (Ito & Kano, 1982). Then, Kano worked with Carl-Frederik Ekerot who came from University Lund and stayed in Ito's lab as a JSPS fellow for 10 months from 1982 to 1983. They recorded single unit activity from Purkinje cells in response to parallel fiber stimulation and induced LTD by conjunctive parallel fiber and

M. Kano (✉) · T. Watanabe
Department of Neurophysiology, Graduate School of Medicine, The University of Tokyo, Tokyo, Japan

International Research Center for Neurointelligence (WPI-IRCN), The University of Tokyo Institutes for Advanced Study (UTIAS), The University of Tokyo, Tokyo, Japan
e-mail: mkano-tky@m.u-tokyo.ac.jp

N. Uesaka
Graduate School of Medical and Dental Sciences, Tokyo Medical and Dental University, Tokyo, Japan

H. Mizusawa, S. Kakei (eds.), *Cerebellum as a CNS Hub*,
Contemporary Clinical Neuroscience, https://doi.org/10.1007/978-3-030-75817-2_4

climbing fiber stimulation in the dorsal paraflocculus of decerebrate rabbit. They found "off-beam" parallel fiber stimulation, which presumably activate molecular layer interneurons and induce feedforward inhibition to Purkinje cells under recording, effectively blocked LTD (Ekerot & Kano, 1985). This study suggests that sufficient depolarization of Purkinje cells during conjunctive parallel fiber and climbing fiber stimulation is required for the induction of LTD.

After Ekerot returned to Lund, Kano started to examine with Makoto Kato which subtype(s) of glutamate receptor at parallel fiber to Purkinje cell synapses is involved in LTD. In the middle of 1980s, molecular identity of glutamate receptor was not known, and they were pharmacologically classified into three subtypes, namely, "quisqualate-preferring," "kainite-preferring," and "NMDA-preferring" receptors (Foster & Fagg, 1988). It should also be noted that LTD was yet to be successfully induced in cerebellar slice preparation, which awaited the report by Sakurai in 1887 (Sakurai, 1987). Therefore, Kano had to use in vivo decerebrate rabbit preparation as he used in the previous two studies. Kano and Kato recorded extracellular spike activity from a Purkinje cell in the paraflocculus, stimulated parallel fibers, and placed an electrode for ionotropic application of a glutamate agonist onto the dendrite of the Purkinje cell under recording. They examined whether ionophoretic application of a glutamate agonist in conjunction with climbing fiber stimulation induced LTD of Purkinje cell responsiveness to parallel fiber stimulation. They found that quiaqualate and glutamate effectively induced LTD, but kainite and aspartate failed to do so (Kano & Kato, 1987). Quisqualate appeared to be more potent than glutamate to induce LTD. Ionophoresis of NMDA inhibited spontaneous activity of Purkinje cells, suggesting that Purkinje cells have very low NMDA sensitivity and NMDA preferentially activated molecular layer interneurons (Kano & Kato, 1987). Thus, they concluded "quisqualate receptors (i.e., glutamate receptors with agonist preference of quisqualate > glutamate >>> kainite, aspartate)" are specifically involved in LTD (Kano & Kato, 1987). However, this agonist preference was clearly different from that for any of the three glutamate receptors reported at that time (Foster & Fagg, 1988).

Soon after the publication of Kano and Kato's paper, Sugiyama et al. reported a new type of glutamate receptor, namely, "metabotropic glutamate receptor," that couples heterotrimeric GTP-binding protein and induces calcium mobilization from internal calcium store mediated by inositol trisphosphate (IP3) (Sugiyama et al., 1987). Subsequently, Masu et al. (1991) cloned and characterized the first metabotropic glutamate receptor that induces IP3-mediated calcium release (Masu et al., 1991). This receptor, termed metabotropic glutamate receptor 1 (mGluR1), is strongly expressed in cerebellar Purkinje cells and has a unique agonist preference of quisqualate > glutamate >>> kainite (Masu et al., 1991; Aramori & Nakanishi, 1992). Thus, "qusqualate receptors" responsible for LTD induction reported by Kano and Kato (1987) appeared to correspond to mGluR1.

4.2 mGluR1 Is Essential for LTD at Parallel Fiber to Purkinje Cell Synapse

In 1991, David Linden reported that quiqualate and glutamate, but not aspartate, induced LTD of glutamate responsiveness of cultured Purkinje cells (Linden et al., 1991), confirming the results of Kano and Kato. Crepel et al. (1991) reported that application of an mGluR agonist to cerebellar slices induced long-lasting suppression of parallel fiber to Purkinje cell synaptic transmission analogous to LTD (Crepel et al., 1991). In 1994, Shigemoto et al. demonstrated that antibodies against mGluR1, which blocked mGluR1-mediated calcium mobilization, blocked LTD of glutamate responsiveness in cultured Purkinje cells (Shigemoto et al., 1994).

In January of 1994, Kano started collaboration with Atsu Aiba who generated mGluR1 knockout mice at Susumu Tonegawa's lab at MIT. Although mGluR1 knockout mice were severely ataxic indicative of defect in the cerebellum, gross anatomy of the cerebellum, morphology of Purkinje cell, basic properties of parallel fiber to Purkinje cell transmission, and those of climbing fiber to Purkinje cell transmission were apparently normal in mGluR1 knockout mice. However, LTD at parallel fiber to Purkinje cell synapse was deficient and classical eyeblink conditioning, a form of discrete motor learning known to be dependent on the cerebellum (McCormick & Thompson, 1984; Lee et al., 2015; Nakao et al., 2019), was severely impaired in mGluR1 knockout mice (Aiba et al., 1994). In the same year, Conquet et al. (1994) reported deficient LTD at parallel fiber to Purkinje cells synapses and motor deficit in another line of mGluR1 knockout mice they generated (Conquet et al., 1994). These results clarified that mGluR1 is essential for LTD at parallel fiber-Purkinje cell synapses and provided a supportive evidence that LTD is a cellular basis of cerebellum-dependent discrete motor learning.

4.3 mGluR1 Is Essential for Climbing Fiber Synapse Elimination in the Developing Cerebellum

It is well known that Purkinje cells in the mature animal are innervated by single climbing fibers (mono-innervation). However, Purkinje cells are innervated by multiple climbing fibers at birth (Crepel et al., 1976; Crepel, 1982; Lohof et al., 1996). Then massive elimination of redundant climbing fibers occurs during postnatal development, and the mono-innervation pattern is established by the end of the third postnatal week (Crepel et al., 1976; Crepel, 1982; Lohof et al., 1996). After publication of the first paper on the cerebellar abnormalities of mGluR1 knockout mice (Aiba et al., 1994), Kano tackled the issue together with Kouichi Hashimoto whether mGluR1 is required for climbing fiber synapse elimination during postnatal development of the cerebellum. Kano and his colleagues examined climbing fiber to Purkinje cell synapse rigorously with electrophysiological technique. Meanwhile, they were examining the cerebellum of protein kinase Cγ (PKCγ) knockout mice that were generated also in Susumu Tonegawa's lab and show ataxia and motor discoordination. PKCγ is a classical PKC that is strongly expressed in Purkinje cells and considered to be activated at the downstream of mGluR1. They made whole-cell recordings from Purkinje cells and stimulated climbing fibers in cerebellar slices prepared from mGluR1 or PKCγ knockout mice at postnatal day 20 (P20) or older and from age-matched wild-type mice. When the stimulus strength was gradually increased, multiple steps of excitatory postsynaptic currents (EPSCs) were elicited at distinct thresholds in more than 30% of Purkinje cells from mGluR1 knockout mice (Fig. 4.1) or PKCγ knockout mice (Kano et al., 1995, 1997). In contrast, single steps of climbing fiber-mediated EPSCs (CF-EPSCs) were elicited in the majority of Purkinje cells from wild-type mice (Fig. 4.1) (Kano et al., 1995, 1997). Levenes et al. also reported multiple climbing fiber innervation in adult Purkinje cells from the other line of mGluR1 knockout mice (Levenes et al., 1997). These electrophysiological data clearly show that substantial percentage of Purkinje cells after P20

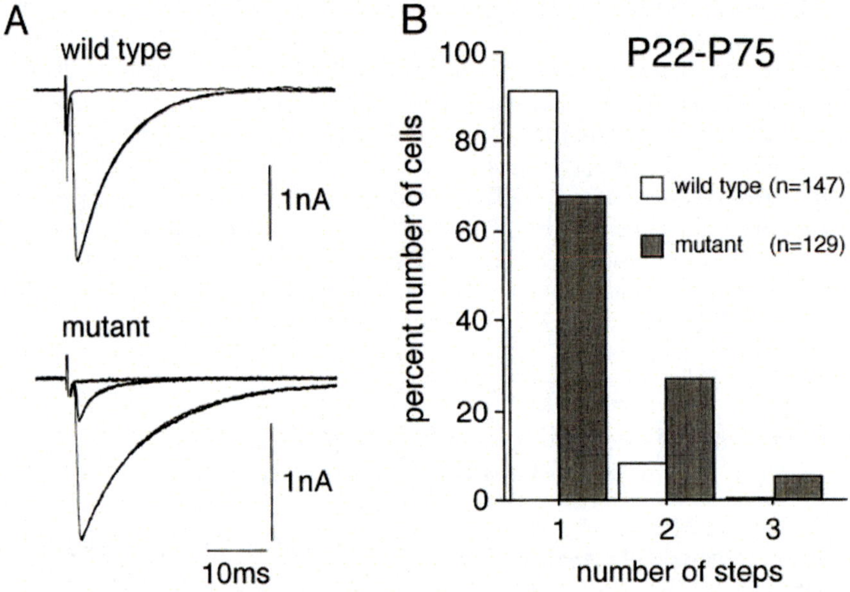

Fig. 4.1 Persistent multiple climbing fiber innervation of Purkinje cells of mGluR1 knockout mice at P22-P75. (**a**) EPSCs elicited by stimulation of climbing fibers in the granule cell layer in a wild-type (P24) and a mGluR1 mutant (P27) Purkinje cell. With gradually increasing stimulus intensities, EPSCs of the wild type were obtained in an all-or-none fashion, while those of the mutant occurred at two discrete stimulus intensity steps, indicating that at least two climbing fibers innervate this mutant Purkinje cell. Stimuli were applied at 0.1 Hz. Holding potentials were −20 mV for both the wild-type and mutant Purkinje cells to inactivate voltage-dependent channels. (**b**) Summary histograms showing number of discrete steps of CF-EPSCs of the wild-type (open columns) and mGluR1 mutant (hatched columns) Purkinje cells. Data obtained from mice at P22–P75. Numbers of tested Purkinje cells are n = 147 (from 15 mice, 107 cells studied blind to the mouse genotype) for the wild type and n = 129 (from 13 mice, 97 cells studied blind) for the mGluR1 mutant. (From Kano et al., 1997)

were innervated by multiple climbing fibers in both mGluR1 knockout and PKCγ knockout mice, whereas most Purkinje cells were innervated by single climbing fibers in wild-type Purkinje cells. Then they examined postnatal development of climbing fiber innervation in these two lines of knockout mice. They found that climbing fiber innervation during the first and the second postnatal weeks were normal, but significantly higher percentage of Purkinje cells were innervated by multiple climbing fibers during the third postnatal week in both mGluR1 knockout and PKCγ knockout mice (Kano et al., 1995, 1997).

Since mGluR1 is coupled to Gq protein and is considered to activate phospholipase Cβ (PLCβ), Kano and his colleagues examined climbing fiber synapse elimination in mutant mice deficient in Gq and those lacking PLCβ4, a major PLC isoform expressed in Purkinje cells (Offermanns et al., 1997; Kano et al., 1998). They found that Gq knockout and PLCβ4 knockout mice exhibit essentially the same phenotypes in terms of developmental climbing fiber synapse elimination as those seen in mGluR1 knockout and PKCγ knockout mice. Climbing fiber innervation is normal during the first two postnatal weeks, but significantly higher percentage of Purkinje cells are innervated by multiple climbing fibers in the third postnatal week, which persists into adulthood (Offermanns et al., 1997; Kano et al., 1998). Thus, the results of the four strains of knockout mice indicate that the mGluR1-Gq-PLCβ4-PKCγ signaling cascade is not involved in climbing fiber synapse development during the first two postnatal weeks but it is crucially required for climbing fiber synapse elimination in the third postnatal week (Kano et al., 1995, 1997, 1998; Offermanns et al., 1997; Hashimoto & Kano, 2013; Kano & Watanabe, 2017).

Purkinje cells are known to express two splice variants of mGluR1, namely, mGluR1a and mGluR1b. mGluR1a has a long C-terminal domain that interacts with scaffolding proteins in the postsynaptic density, but mGluR1b lacks such a C-terminal domain (Ferraguti et al., 2008). To clarify whether mGluR1 in Purkinje cells is responsible for various cerebellar phenotypes of mGluR1 knockout mice, Aiba generated transgenic mice in which mGluR1a or mGlu1b was introduced into Purkinje cells of mGluR1 knockout mice (mGluR1a-rescue or mGluR1b-rescue mice) (Ichise et al., 2000; Ohtani et al., 2014). In collaboration with the Kano lab, Aiba and his colleagues demonstrated that the impairment of climbing fiber synapse elimination was almost completely restored in mGluR1a-rescue mice (Ichise et al., 2000). In marked contrast, the impaired climbing fiber synapse elimination was not rescued by Purkinje cell-specific expression of mGluR1b into mGluR1 knockout mice (Ohtani et al., 2014). These results clearly show that signaling from mGluR1a to PKCγ in Purkinje cells but in no other cell types is crucial for climbing fiber synapse elimination.

Another question was how and at which synapse mGluR1 is activated for climbing fiber synapse elimination. mGluR1 is densely expressed at perisynaptic regions of Purkinje cell dendritic spines facing parallel fiber synaptic terminals (Baude et al., 1993; Nusser et al., 1994; Petralia et al., 1998; Lopez-Bendito et al., 2001). However, mGluR1 is also expressed at perisynaptic regions of climbing fiber to Purkinje cell synapses (Baude et al., 1993; Nusser et al., 1994; Petralia et al., 1998;

Lopez-Bendito et al., 2001). Brief burst stimulation of parallel fibers readily activates mGluR1 and induces slow EPSCs or IP3-mediated localized calcium release in Purkinje cell dendrites (Batchelor et al., 1994; Batchelor & Garthwaite, 1997; Finch & Augustine, 1998; Takechi et al., 1998). In contrast, climbing fiber stimulation does not readily activate mGluR in Purkinje cells unless glutamate concentration around synapse is elevated by blocking glutamate transporters (Dzubay & Otis, 2002). These results suggest that mGluR1 is likely to be activated by parallel fiber inputs rather than climbing fiber inputs in the cerebellum in vivo. Kakizawa et al. showed that chronic blockade of NMDA receptors in the cerebellum at P14 to P16 caused persistent multiple climbing fiber innervation of Purkinje cells in adulthood (Kakizawa et al., 2000). This effect was not observed when NMDA receptors were blocked earlier than P14 or later than P17, suggesting that NMDA receptors are involved in climbing fiber synapse elimination during the same stage of climbing fiber synapse elimination in which mGluR1 is involved (Kakizawa et al., 2000). However, Purkinje cells are devoid of functional NMDA receptors at least during the second and third postnatal weeks (Kakizawa et al., 2000). In contrast, granule cells richly express NMDA receptors at mossy fiber to granule cell synapses during postnatal development (Ebralidze et al., 1996; Kadotani et al., 1996; Takahashi et al., 1996). Therefore, one possibility is that neural activity along mossy fiber – granule cell – parallel fiber pathway activates mGluR1 and its downstream signaling in Purkinje cell to cause elimination of redundant climbing fiber synapses (Kakizawa et al., 2000).

Meanwhile, Kouichi Hashimoto intensively studied postnatal development of climbing fiber innervation of Purkinje cells and clarified that it consists of four distinct phases (Hashimoto & Kano, 2003; Hashimoto et al., 2009a, b; Kano et al., 2018). At birth, each Purkinje cell is innervated by more than five climbing fibers that produces EPSCs with similar amplitudes, indicating that neonatal Purkinje cells receive more than five climbing fiber inputs with similar synaptic strengths. From P3 to P7, synaptic input from a single climbing fiber selectively becomes stronger among multiple climbing fibers in each Purkinje cell (Functional differentiation). From around P9, only the strongest climbing fiber extends its synaptic territory along dendrites of Purkinje cell (Climbing fiber translocation). In parallel, redundant climbing fiber synapses on the soma are eliminated from around P7 to P11, which proceeds independently from parallel fiber to Purkinje cell synapse formation (Early phase of climbing fiber elimination) and from around P12 to P17, which requires normal parallel fiber to Purkinje cell synapse formation (Late phase of climbing fiber elimination) (Hashimoto & Kano, 2003; Hashimoto et al., 2009a, b; Kano et al., 2018). Thus, the aforementioned phenotypes of mGluR1, Gq, PLCβ4, and PKCγ knockout mice indicate that mGluR1 to PKCγ signaling is required rather specifically for the late phase of climbing fiber elimination (see Fig. 4.2).

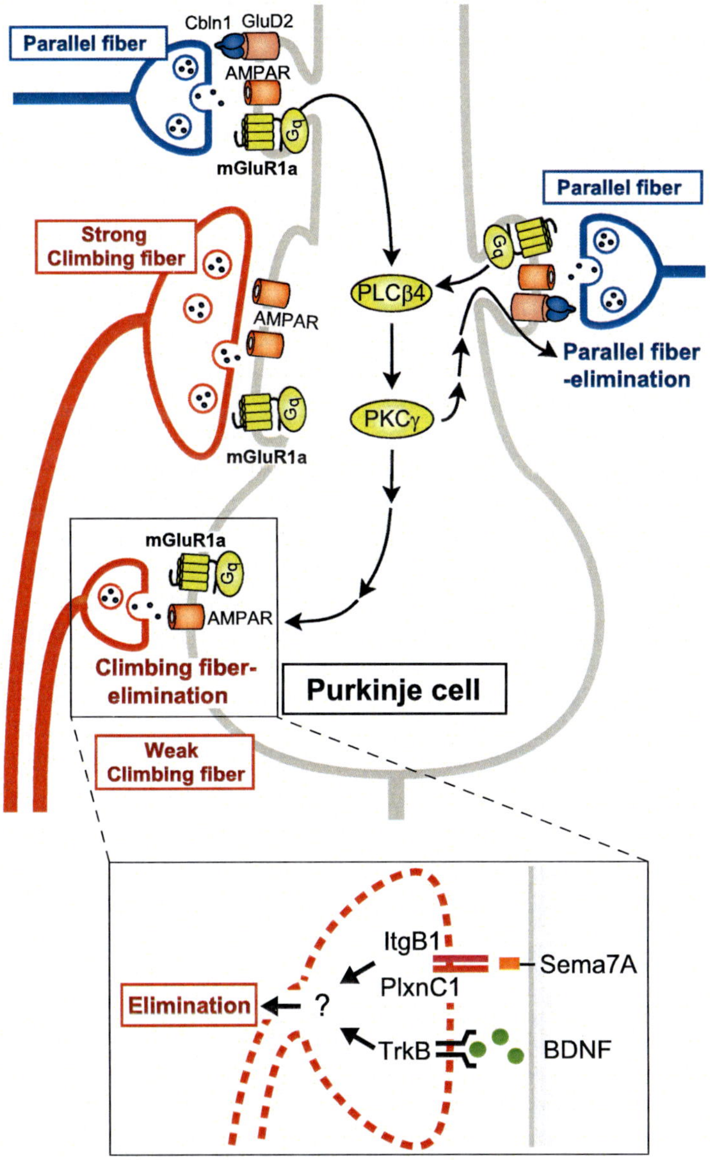

Fig. 4.2 Schematic diagram of mGluR1 signaling in Purkinje cells required for developmental climbing fiber synapse elimination. Parallel fiber synaptic inputs activate mGluR1 and its downstream signaling (Gq –PLCβ4 –PKCγ) in Purkinje cells. Sema7A retrogradely acts on its Plexin C1 (PlxnC1)/Integrin B1 (ItgB1) receptor on "weak" climbing fibers and eliminates them from the soma during P15 to P17. BDNF also acts retrogradely on TrkB on "weak" climbing fibers and eliminates them from the soma during P15 to P17. The Sema7A signal and the BDNF signal appear to converge within climbing fiber terminals. The mGluR1 to PKCγ signaling eliminates parallel fiber synapses from proximal dendrites during P15 to P30. The long C-terminal domain of mGluR1a is required for climbing fiber synapse elimination. (Figure 3 of Kano & Watanabe, F1000Res 6: 416, 2017, is modified)

4.4 Downstream Molecules of mGluR1 Involved in Climbing Fiber Synapse Elimination

Since mGluR1 mediate signals within Purkinje cells and causes elimination of redundant climbing fiber synaptic terminals from the soma, there must be mechanisms that transmit signals retrogradely from postsynaptic Purkinje cells to presynaptic climbing fiber terminals. Uesaka et al. performed screening of candidate retrograde signaling molecules by using micro-RNA-mediated knockdown of target molecules in Purkinje cells in the developing cerebellum (Uesaka et al., 2012; Uesaka et al., 2014). They found that microRNA-mediated knockdown of semaphorin7A (Sema7A) specifically in Purkinje cells of neonatal mice caused an increase of the percentage of Purkinje cells with multiple climbing fiber innervation, which manifested at P15, indicating that Sema7A is required for the late phase of climbing fiber elimination (Uesaka et al., 2014). To test whether Sema7A functions at the downstream of mGluR1, they compared the effect of double knockdown of mGluR1 and Sema7A and that of mGluR1 knockdown alone. They found that the effect of double knockdown was the same as that of single mGluR1 knockdown, indicating that mGluR1 and Sema7A function along the same signaling pathway for climbing fiber synapse elimination (Uesaka et al., 2014). Moreover, overexpression of Sema7A into Purkinje cells with mGluR1 knockdown restored normal level of climbing fiber synapse elimination, indicating Sema7A functions at the downstream of mGluR1 to eliminate surplus climbing fiber synapses (Uesaka et al., 2014). Uesaka et al. searched receptors at climbing fibers that bind Sema7A derived from Purkinje cells and exert elimination of those climbing fibers. They found that knockdown of putative Sema7A receptors, Plexin C1 and Integrin B1, in inferior olivary neurons of neonatal mice impaired climbing fiber synapse elimination to the extent similar to Sema7A knockdown in Purkinje cells (Uesaka et al., 2014). They also demonstrate that overexpression of constitutively active cofilin in climbing fibers occluded the effect of Plexin C1 knockdown and that knockdown of focal adhesion kinase in climbing fibers occluded the effect of Integrin B1 knockdown (Uesaka et al., 2014; Uesaka & Kano, 2018). These results collectively indicate that, at the downstream of mGluR1, retrograde Sema7A to Plexin C1/Integrin B1 signaling from Purkinje cells to climbing fibers facilitates climbing fiber synapse elimination after P15 by regulating cofilin and focal adhesion kinase in climbing fibers (Uesaka et al., 2014; Uesaka & Kano, 2018) (see Fig. 4.2).

Besides Sema7A, Choo et al. (2017) demonstrated that brain-derived neurotrophic factor (BDNF) mediate retrograde signal from Purkinje cells to climbing fibers to facilitate their elimination (Choo et al., 2017). Previous studies showed that climbing fiber synapse elimination was impaired in mice with global or cerebellum-specific knockout of TrkB, a high affinity receptor for BDNF (Bosman et al., 2006; Johnson et al., 2007). Consistent with these previous reports, Choo et al. found that climbing fiber synapse elimination after P15 was impaired in Purkinje cell-specific BDNF knockout mice and in Purkinje cells with microRNA-mediated BDNF knockdown. They found similar impairment of climbing fiber synapse elimination when TrkB was knocked down in climbing fibers by injecting lentivirus carrying microRNA against TrkB into the neonatal inferior olive. The effect of mGluR1 knockdown in Purkinje cells of wild-type mice on the impairment of climbing fiber synapse elimination was the same as that of Purkinje cell-specific BDNF knockout mice, suggesting that mGluR1 and BDNF function along the same signaling pathway (see Fig. 4.2). Furthermore, Choo et al. showed that the effect of Sema7A knockdown in Purkinje cells of wild-type mice was similar to that of Purkinje cell-specific BDNF knockout mice, suggesting that BDNF and Sema7A share a common signaling pathway for the late phase climbing synapse elimination, which presumably functions within climbing fiber terminals (Choo et al., 2017) (see Fig. 4.2).

4.5 mGluR1 Mediates Parallel Fiber Synapse Elimination Following the Late Phase of Climbing Fiber Elimination

Ichikawa et al. discovered that massive elimination of parallel fiber to Purkinje cell synapses from proximal dendrites occurs from around P15 to P30 (Ichikawa et al., 2016). Climbing fibers and parallel fibers are known to innervate proximal and distal portions of Purkinje cell dendrites, respectively. Between these two dendritic portions, there are intermediate dendritic portions with mixed innervation by both climbing fibers and parallel fibers. Ichikawa et al. performed serial electron microscopic analysis of Purkinje cell dendrites during postnatal development. They found that the dendritic regions with mixed innervation were steadily expanded from around P7 to P15, and then parallel fiber synapses are massively eliminated from proximal dendrites occurs from around P15 to around P30. Consequently, the dendritic regions with mixed climbing fiber and parallel fiber innervation become markedly smaller and the climbing fiber and parallel fiber territories become well segregated (Ichikawa et al., 2016). Importantly, they found that the massive parallel fiber synapse elimination from P15 to around P30 was deficient in mGluR1 and PKCγ knockout mice (Ichikawa et al., 2016). These results indicate that mGluR1 to PKCγ cascade is essential not only for the late phase of CF synapse elimination but also for the subsequent PF synapse elimination (see Fig. 4.2).

4.6 Conclusion: mGluR1 Is a Molecular "Hub" for Purkinje Cell Synaptic Development and Function

In this chapter, we have made an overview of how mGluR1 plays crucial roles in climbing fiber synapse elimination in the developing cerebellum. We also referred to the role of mGluR1 in LTD at parallel fiber to Purkinje cell synapses in the cerebellum. However, mGluR1 is known to be involved in many other functions in

Purkinje cells. For example, brief parallel fiber bursts readily activate mGluR1 and induce IP3 receptor-mediated local calcium release in dendrites (Finch & Augustine, 1998; Takechi et al., 1998) and slow EPSCs involving TRPC3 (Hartmann et al., 2008). Dysregulation of such mGluR1 signaling in Purkinje cells has been found in several clinically relevant mouse models of human cerebellar ataxias, and mutations of mGluR1 and related molecules have been reported in certain types of human ataxias (Meera et al., 2016; Power et al., 2016; Kano & Watanabe, 2017). Furthermore, mGluR1 is crucial for release of the endogenous cannabinoid 2-arachidonoylglycerol that transmits information retrogradely from Purkinje cells to presynaptic terminals and suppresses neurotransmitter release (Maejima et al., 2001, 2005; Kano et al., 2008; Tanimura et al., 2010). Thus, mGluR1 is a molecular "hub" in Purkinje cell essential for refinement of synaptic wiring during development, modulation of synaptic transmission and neuronal excitability, and synaptic plasticity that underlies motor learning (Kano et al., 2008; Kano & Watanabe, 2017).

Acknowledgments This work was supported in part by Grants-in-Aid for Scientific Research 18H04012 and 20H05915 to M.K. from JSPS, Japan.

References

Aiba, A., Kano, M., Chen, C., Stanton, M. E., Fox, G. D., Herrup, K., Zwingman, T. A., & Tonegawa, S. (1994). Deficient cerebellar long-term depression and impaired motor learning in mGluR1 mutant mice. *Cell, 79*, 377–388.

Aramori, I., & Nakanishi, S. (1992). Signal transduction and pharmacological characteristics of a metabotropic glutamate receptor, mGluR1, in transfected CHO cells. *Neuron, 8*, 757–765.

Batchelor, A. M., & Garthwaite, J. (1997). Frequency detection and temporally dispersed synaptic signal association through a metabotropic glutamate receptor pathway. *Nature, 385*, 74–77.

Batchelor, A. M., Madge, D. J., & Garthwaite, J. (1994). Synaptic activation of metabotropic glutamate receptors in the parallel fibre-Purkinje cell pathway in rat cerebellar slices. *Neuroscience, 63*, 911–915.

Baude, A., Nusser, Z., Roberts, J. D., Mulvihill, E., McIlhinney, R. A., & Somogyi, P. (1993). The metabotropic glutamate receptor (mGluR1α) is concentrated at perisynaptic membrane of neuronal subpopulations as detected by immunogold reaction. *Neuron, 11*, 771–787.

Bosman, L. W., Hartmann, J., Barski, J. J., Lepier, A., Noll-Hussong, M., Reichardt, L. F., & Konnerth, A. (2006). Requirement of TrkB for synapse elimination in developing cerebellar Purkinje cells. *Brain Cell Biology, 35*, 87–101.

Choo, M., Miyazaki, T., Yamazaki, M., Kawamura, M., Nakazawa, T., Zhang, J., Tanimura, A., Uesaka, N., Watanabe, M., Sakimura, K., & Kano, M. (2017). Retrograde BDNF to TrkB signaling promotes synapse elimination in the developing cerebellum. *Nature Communications, 8*, 195.

Conquet, F., Bashir, Z. I., Davies, C. H., Daniel, H., Ferraguti, F., Bordi, F., Franz-Bacon, K., Reggiani, A., Matarese, V., Conde, F., et al. (1994). Motor deficit and impairment of synaptic plasticity in mice lacking mGluR1. *Nature, 372*, 237–243.

Crepel, F. (1982). Regression of functional synapses in the immature mammalian cerebellum. *Trends in Neurosciences, 5*, 266–269.

Crepel, F., Daniel, H., Hemart, N., & Jaillard, D. (1991). Effects of ACPD and AP3 on parallel-fibre-mediated EPSPs of Purkinje cells in cerebellar slices in vitro. *Experimental Brain Research, 86*, 402–406.

Crepel, F., Mariani, J., & Delhaye-Bouchaud, N. (1976). Evidence for a multiple innervation of Purkinje cells by climbing fibers in the immature rat cerebellum. *Journal of Neurobiology, 7*, 567–578.

Dzubay, J. A., & Otis, T. S. (2002). Climbing fiber activation of metabotropic glutamate receptors on cerebellar Purkinje neurons. *Neuron, 36*, 1159–1167.

Ebralidze, A. K., Rossi, D. J., Tonegawa, S., & Slater, N. T. (1996). Modification of NMDA receptor channels and synaptic transmission by targeted disruption of the NR2C gene. *The Journal of Neuroscience, 16*, 5014–5025.

Ekerot, C. F., & Kano, M. (1985). Long-term depression of parallel fibre synapses following stimulation of climbing fibres. *Brain Research, 342*, 357–360.

Ferraguti, F., Crepaldi, L., & Nicoletti, F. (2008). Metabotropic glutamate 1 receptor: Current concepts and perspectives. *Pharmacological Reviews, 60*, 536–581.

Finch, E. A., & Augustine, G. J. (1998). Local calcium signalling by inositol-1,4,5-trisphosphate in Purkinje cell dendrites. *Nature, 396*, 753–756.

Foster, A. C., & Fagg, G. E. (1988). Acidic amino acid receptor nomenclature: Time for change. *Trends in Neurosciences, 11*, 17–18.

Hartmann, J., Dragicevic, E., Adelsberger, H., Henning, H. A., Sumser, M., Abramowitz, J., Blum, R., Dietrich, A., Freichel, M., Flockerzi, V., Birnbaumer, L., & Konnerth, A. (2008). TRPC3 channels are required for synaptic transmission and motor coordination. *Neuron, 59*, 392–398.

Hashimoto, K., Ichikawa, R., Kitamura, K., Watanabe, M., & Kano, M. (2009a). Translocation of a "winner" climbing fiber to the Purkinje cell dendrite and subsequent elimination of "losers" from the soma in developing cerebellum. *Neuron, 63*, 106–118.

Hashimoto, K., & Kano, M. (2003). Functional differentiation of multiple climbing fiber inputs during synapse elimination in the developing cerebellum. *Neuron, 38*, 785–796.

Hashimoto, K., & Kano, M. (2013). Synapse elimination in the developing cerebellum. *Cellular and Molecular Life Sciences, 70*, 4667–4680.

Hashimoto, K., Yoshida, T., Sakimura, K., Mishina, M., Watanabe, M., & Kano, M. (2009b). Influence of parallel fiber-Purkinje cell synapse formation on postnatal development of climbing fiber-Purkinje cell synapses in the cerebellum. *Neuroscience, 162*, 601–611.

Ichikawa, R., Hashimoto, K., Miyazaki, T., Uchigashima, M., Yamasaki, M., Aiba, A., Kano, M., & Watanabe, M. (2016). Territories of heterologous inputs onto Purkinje cell dendrites are segregated by mGluR1-dependent parallel fiber synapse elimination. *Proceedings of the National Academy of Sciences of the United States of America, 113*, 2282–2287.

Ichise, T., Kano, M., Hashimoto, K., Yanagihara, D., Nakao, K., Shigemoto, R., Katsuki, M., & Aiba, A. (2000). mGluR1 in cerebellar Purkinje cells essential for long-term depression, synapse elimination, and motor coordination. *Science, 288*, 1832–1835.

Ito, M., & Kano, M. (1982). Long-lasting depression of parallel fiber-Purkinje cell transmission induced by conjunctive stimulation of parallel fibers and climbing fibers in the cerebellar cortex. *Neuroscience Letters, 33*, 253–258.

Ito, M., Sakurai, M., & Tongroach, P. (1982). Climbing fibre induced depression of both mossy fibre responsiveness and glutamate sensitivity of cerebellar Purkinje cells. *The Journal of Physiology, 324*, 113–134.

Johnson, E. M., Craig, E. T., & Yeh, H. H. (2007). TrkB is necessary for pruning at the climbing fibre-Purkinje cell synapse in the developing murine cerebellum. *The Journal of Physiology, 582*, 629–646.

Kadotani, H., Hirano, T., Masugi, M., Nakamura, K., Nakao, K., Katsuki, M., & Nakanishi, S. (1996). Motor discoordination results from combined gene disruption of the NMDA receptor NR2A and NR2C subunits, but not from single disruption of the NR2A or NR2C subunit. *The Journal of Neuroscience, 16*, 7859–7867.

Kakizawa, S., Yamasaki, M., Watanabe, M., & Kano, M. (2000). Critical period for activity-dependent synapse elimination in developing cerebellum. *The Journal of Neuroscience, 20*, 4954–4961.

Kano, M., Hashimoto, K., Chen, C., Abeliovich, A., Aiba, A., Kurihara, H., Watanabe, M., Inoue, Y., & Tonegawa, S. (1995). Impaired synapse elimination during cerebellar development in PKCγ mutant mice. *Cell, 83*, 1223–1231.

Kano, M., Hashimoto, K., Kurihara, H., Watanabe, M., Inoue, Y., Aiba, A., & Tonegawa, S. (1997). Persistent multiple climbing fiber innervation of cerebellar Purkinje cells in mice lacking mGluR1. *Neuron, 18*, 71–79.

Kano, M., Hashimoto, K., & Tabata, T. (2008). Type-1 metabotropic glutamate receptor in cerebellar Purkinje cells: A key molecule responsible for long-term depression, endocannabinoid signalling and synapse elimination. *Philosophical Transactions of the Royal Society of London. Series B, Biological Sciences, 363*, 2173–2186.

Kano, M., Hashimoto, K., Watanabe, M., Kurihara, H., Offermanns, S., Jiang, H., Wu, Y., Jun, K., Shin, H. S., Inoue, Y., Simon, M. I., & Wu, D. (1998). Phospholipase Cβ4 is specifically involved in climbing fiber synapse elimination in the developing cerebellum. *Proceedings of the National Academy of Sciences of the United States of America, 95*, 15724–15729.

Kano, M., & Kato, M. (1987). Quisqualate receptors are specifically involved in cerebellar synaptic plasticity. *Nature, 325*, 276–279.

Kano, M., & Watanabe, T. (2017). Type-1 metabotropic glutamate receptor signaling in cerebellar Purkinje cells in health and disease. *F1000Res, 6*, 416.

Kano, M., Watanabe, T., Uesaka, N., & Watanabe, M. (2018). Multiple phases of climbing fiber synapse elimination in the developing cerebellum. *Cerebellum, 17*, 722–734.

Lee, K. H., Mathews, P. J., Reeves, A. M., Choe, K. Y., Jami, S. A., Serrano, R. E., & Otis, T. S. (2015). Circuit mechanisms underlying motor memory formation in the cerebellum. *Neuron, 86*, 529–540.

Levenes, C., Daniel, H., Jaillard, D., Conquet, F., & Crepel, F. (1997). Incomplete regression of multiple climbing fibre innervation of cerebellar Purkinje cells in mGluR1 mutant mice. *Neuroreport, 8*, 571–574.

Linden, D. J., Dickinson, M. H., Smeyne, M., & Connor, J. A. (1991). A long-term depression of AMPA currents in cultured cerebellar Purkinje neurons. *Neuron, 7*, 81–89.

Lohof, A. M., Delhaye-Bouchaud, N., & Mariani, J. (1996). Synapse elimination in the central nervous system: Functional significance and cellular mechanisms. *Reviews in the Neurosciences, 7*, 85–101.

Lopez-Bendito, G., Shigemoto, R., Lujan, R., & Juiz, J. M. (2001). Developmental changes in the localisation of the mGluR1alpha subtype of metabotropic glutamate receptors in Purkinje cells. *Neuroscience, 105*, 413–429.

Maejima, T., Hashimoto, K., Yoshida, T., Aiba, A., & Kano, M. (2001). Presynaptic inhibition caused by retrograde signal from metabotropic glutamate to cannabinoid receptors. *Neuron, 31*, 463–475.

Maejima, T., Oka, S., Hashimotodani, Y., Ohno-Shosaku, T., Aiba, A., Wu, D., Waku, K., Sugiura, T., & Kano, M. (2005). Synaptically driven endocannabinoid release requires Ca^{2+}-assisted metabotropic glutamate receptor subtype 1 to phospholipase Cβ4 signaling cascade in the cerebellum. *The Journal of Neuroscience, 25*, 6826–6835.

Masu, M., Tanabe, Y., Tsuchida, K., Shigemoto, R., & Nakanishi, S. (1991). Sequence and expression of a metabotropic glutamate receptor. *Nature, 349*, 760–765.

McCormick, D. A., & Thompson, R. F. (1984). Cerebellum: Essential involvement in the classically conditioned eyelid response. *Science, 223*, 296–299.

Meera, P., Pulst, S. M., & Otis, T. S. (2016). Cellular and circuit mechanisms underlying spinocerebellar ataxias. *The Journal of Physiology, 594*, 4653–4660.

Nakao, H., Kishimoto, Y., Hashimoto, K., Kitamura, K., Yamasaki, M., Nakao, K., Watanabe, M., Kano, M., Kirino, Y., & Aiba, A. (2019). mGluR1 in cerebellar Purkinje cells is essential for the formation but not expression of associative eyeblink memory. *Scientific Reports, 9*, 7353.

Nusser, Z., Mulvihill, E., Streit, P., & Somogyi, P. (1994). Subsynaptic segregation of metabotropic and ionotropic glutamate receptors as revealed by immunogold localization. *Neuroscience, 61*, 421–427.

Offermanns, S., Hashimoto, K., Watanabe, M., Sun, W., Kurihara, H., Thompson, R. F., Inoue, Y., Kano, M., & Simon, M. I. (1997). Impaired motor coordination and persistent multiple climbing fiber innervation of cerebellar Purkinje cells in mice lacking Gαq. *Proceedings of the National Academy of Sciences of the United States of America, 94*, 14089–14094.

Ohtani, Y., Miyata, M., Hashimoto, K., Tabata, T., Kishimoto, Y., Fukaya, M., Kase, D., Kassai, H., Nakao, K., Hirata, T., Watanabe, M., Kano, M., & Aiba, A. (2014). The synaptic targeting of mGluR1 by its carboxyl-terminal domain is crucial for cerebellar function. *The Journal of Neuroscience, 34*, 2702–2712.

Petralia, R. S., Zhao, H. M., Wang, Y. X., & Wenthold, R. J. (1998). Variations in the tangential distribution of postsynaptic glutamate receptors in Purkinje cell parallel and climbing fiber synapses during development. *Neuropharmacology, 37*, 1321–1334.

Power, E. M., English, N. A., & Empson, R. M. (2016). Are Type 1 metabotropic glutamate receptors a viable therapeutic target for the treatment of cerebellar ataxia? *The Journal of Physiology, 594*, 4643–4652.

Sakurai, M. (1987). Synaptic modification of parallel fibre-Purkinje cell transmission in in vitro guinea-pig cerebellar slices. *The Journal of Physiology, 394*, 463–480.

Shigemoto, R., Abe, T., Nomura, S., Nakanishi, S., & Hirano, T. (1994). Antibodies inactivating mGluR1 metabotropic glutamate receptor block long-term depression in cultured Purkinje cells. *Neuron, 12*, 1245–1255.

Sugiyama, H., Ito, I., & Hirono, C. (1987). A new type of glutamate receptor linked to inositol phospholipid metabolism. *Nature, 325*, 531–533.

Takahashi, T., Feldmeyer, D., Suzuki, N., Onodera, K., Cull-Candy, S. G., Sakimura, K., & Mishina, M. (1996). Functional correlation of NMDA receptor epsilon subunits expression with the properties of single-channel and synaptic currents in the developing cerebellum. *The Journal of Neuroscience, 16*, 4376–4382.

Takechi, H., Eilers, J., & Konnerth, A. (1998). A new class of synaptic response involving calcium release in dendritic spines. *Nature, 396*, 757–760.

Tanimura, A., Yamazaki, M., Hashimotodani, Y., Uchigashima, M., Kawata, S., Abe, M., Kita, Y., Hashimoto, K., Shimizu, T., Watanabe, M., Sakimura, K., & Kano, M. (2010). The endocannabinoid 2-arachidonoylglycerol produced by diacylglycerol lipase α mediates retrograde suppression of synaptic transmission. *Neuron, 65*, 320–327.

Uesaka, N., & Kano, M. (2018). Presynaptic mechanisms mediating retrograde semaphorin signals for climbing fiber synapse elimination during postnatal cerebellar development. *Cerebellum, 17*, 17–22.

Uesaka, N., Mikuni, T., Hashimoto, K., Hirai, H., Sakimura, K., & Kano, M. (2012). Organotypic coculture preparation for the study of developmental synapse elimination in mammalian brain. *The Journal of Neuroscience, 32*, 11657–11670.

Uesaka, N., Uchigashima, M., Mikuni, T., Nakazawa, T., Nakao, H., Hirai, H., Aiba, A., Watanabe, M., & Kano, M. (2014). Retrograde semaphorin signaling regulates synapse elimination in the developing mouse brain. *Science, 344*, 1020–1023.

Part II
Neurocircuitry of the Cerebellum

Chapter 5
Cerebellar Lobules and Stripes, Viewed from Development, Topographic Axonal Projections, Functional Localization, and Interspecies Homology

Izumi Sugihara

5.1 Introduction

Cerebellar lobules and cerebellar longitudinal stripes are the landmark structures useful for locating various anatomical elements and physiological activities in the cerebellum. While the lobules are formed by the structural foliation of the cerebellar cortex, stripes are formed by the longitudinal distribution pattern of Purkinje cell subsets of specific molecular expression profiles. These structures are consistent among individuals and generally preserved among different mammalian species. Developmental studies have revealed that the stripe organization of the adult cerebellum arises from the reorganization of the embryonic Purkinje cell clusters (Fujita et al., 2012; Tran-Anh et al., 2020; Vibulyaseck et al., 2017). The afferent (climbing fiber axons and mossy fiber axons) and efferent (Purkinje cells) axonal projection patterns are tightly correlated to both the lobular and striped organizations in the cerebellar cortex. Through this relationship, the output neurons show activity specific to the lobule and stripe, which would lead to the formation of the functional localization in the cerebellar cortex. As a whole, cerebellar lobules and stripes are important fundamental anatomical structures of the cerebellar organization which has significance as a developmental scaffold to frame topographic axonal projections. Therefore, the first aim of this article is to summarize the morphological and functional significance of the lobular and striped organization of the cerebellum and how they frame axonal projection patterns and operation of Purkinje cell populations.

Among different mammalian species that are used in brain experiments (e.g., humans, non-human primates, and rodents), the shape of the cerebellum is

I. Sugihara (✉)
Department of Systems Neurophysiology, Graduate School of Medical and Dental Sciences and Center for Brain Integration Research, Tokyo Medical and Dental University, Tokyo, Japan
e-mail: isugihara.phy1@tmd.ac.jp

H. Mizusawa, S. Kakei (eds.), *Cerebellum as a CNS Hub*,
Contemporary Clinical Neuroscience, https://doi.org/10.1007/978-3-030-75817-2_5

significantly variable. However, to correlate the cerebellar functional localization among them, a correct understanding of the consistency of the lobular and striped organization across animal species is important. Therefore, the second aim of this article is to give special attention to the comparative aspects of the lobular and striped organization of the cerebellum. The striped pattern and axonal projection patterns, which have been clarified not only in rodents but also in primates, can support the identification of lobules. I summarize the interspecies lobule identification in two areas, the vermal central area (lobules VI-VII) and the hemispheric ansiform area (crus I in rodents and crus I + II in primates), in this article.

Notably, the ansiform area shows a significant volume increase in dexterous mammals. Particular parts of the cerebellar cortex including the ansiform area (crus I + II, vermal lobules VII, and lobule HIX) are deeply involved in non-motor (cognitive) function and neuropsychological/neurodevelopmental disorders in humans (Guell et al., 2018). To establish animal models about these disorders, the correct identification of homologous lobules and stripes among animal species is vital. If homologous lobules are identified, the rodent cerebellum can be a more useful model to investigate mechanisms for the cerebellar non-motor function. Correlates in striped and axonal projection patterns of these lobules for the non-motor function reported in rodents and primates are also summarized in this article. The contents of this article would hopefully provide some contribution in answering questions 6 and 7 among seven major remaining questions in research on the cerebellum of late Dr. Masao Ito (2012).

5.2 Cerebellar Lobules

The cerebellum has a large surface area (as large as half of the cerebral cortex), which is deeply foliated by transverse fissures. When unfolded, the entire cerebellar cortex is formed by a continuous single sheet of the cortex. Consequently, the entire cerebellar cortex can be shown schematically in an unfolded scheme in a two-dimensional space, in which lobules and stripes can be mapped.

Some deep fissures separate the cerebellum into several lobules. The lobular structure is well developed in the avian and mammalian cerebellums. While most of the fissures are continuous throughout the mediolateral extent of the cerebellum in aves (Vibulyaseck et al., 2015), medial fissures and lateral fissures are not simply continuous with each other in the posterior part of the mammalian cerebellum (Fujita et al., 2010). This is the morphological basis for the distinction between the lobular organization of the vermis and hemisphere in the mammalian cerebellum.

The lobular organization of the cerebellum is highly conserved among individuals of the same animal species and well conserved between phylogenetically close animal species such as rats and mice (rat: Larsell, 1952, 1970; Paxinos & Watson, 2007; mouse: Paxinos & Franklin, 2001). The major lobulation pattern in adult mammals and the developmental appearance of major fissures that separate individual lobules are similar across various mammalian species (Larsell, 1937, 1970; Larsell & Jansen, 1972).

The current definition of mammalian cerebellar lobules is owed primarily to comparative morphological studies by Larsell (1970), as well as Larsell and Jansen (1972), which examined the foliation of the cerebellar surface and in midsagittal sections. Subsequent publications adopted Larsell's definition (human, Schmahmann et al., 1999; macaque, Madigan & Carpenter, 1971; Paxinos et al., 2009; marmoset, Fujita et al., 2010; Paxinos et al., 2011; rat, Voogd, 2004; Paxinos & Watson, 2007; Swanson, 1998; mouse, Marani & Voogd, 1979; Paxinos & Franklin, 2001; Fujita et al., 2014).

5.2.1 *Lobules in the Vermis*

In the midsagittal section of the vermis, the lobular organization is relatively well conserved among mammals, supporting the idea that the basic lobular organization in the hemisphere is regarded as commonly shared by all mammals (Bolk, 1906; Larsell, 1970). Vermal lobules (Lobules I–II, III, IV–V, VI–VII, VIII, IX and X, Larsell, 1970, Larsell & Jansen, 1972) are classified into three groups by the two deepest fissures: lobules I–V, primary fissure, lobules VI–VIII, secondary fissure, and lobules IX–X from the anterior to the posterior regions. Then, four next deepest fissures divide each of the three groups of lobules into three, two and two (groups of) lobules (Fig. 5.1); lobules I–V is divided into lobules I–II, III, and lobules IV–V, lobules VI–VIII is divided into lobules VI–VII and VIII, lobules IX–X is divided into lobules IX and X. This lobule classification is applicable in various mammals including rodents, non-human primates, and humans (Luo et al., 2017).

Subdivisions of lobules VI–VII are complicated since fissures between them are not necessarily deep and lobular shape is not very consistent among mammals. It seems impossible to determine homologous sublobules only from the fissure structures (foliation pattern). By taking the zebrin expression pattern as well as other morphological characteristics into account between the rat and marmoset cerebellums, the subdivision correspondence has been proposed: rodent (sub)lobules VIa, VIb, VIc, and VII equivalent with primate (sub)lobules VI, VIIAa, VIIAb-d, VIIB (Fujita et al., 2010; upper half of Table 5.1), i.e., the boundary between lobules VI and VII is not consistent between rodents and primates in the Larsell nomenclature.

5.2.2 *Lobules in the Hemisphere*

As in the vermal lobules, the basic organization in the hemispheric lobules is regarded as commonly shared by all mammals (Bolk, 1906; Larsell, 1970). However, species-dependent differences in lobular structure are greater in the hemisphere than in the vermis.

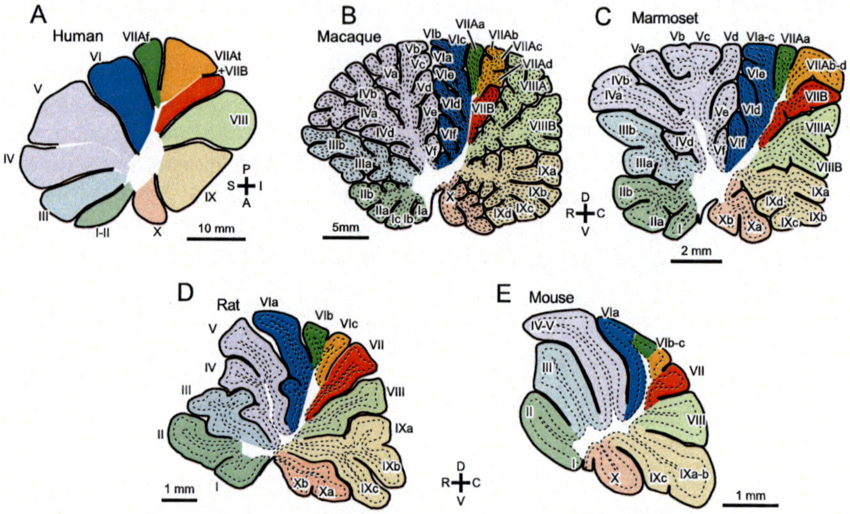

Fig. 5.1 Midsagittal section of the human (**a**), macaque (*Macaca mulatta*) (**b**), marmoset (*Callithrix jacchus*) (**c**), rat (Long-Evans) (**d**), and mouse (C57BL/6N) (**e**) cerebellums. (Putatively) homologous lobules are labeled in the same color in the central vermal area (lobules VI–VII). (Note different names in the area of lobules VI-VII between primates and rodents. Human and mouse section drawings are based on sections. Marmoset and rat drawings are based on the data used in figures of Fujita et al. (2010). The macaque section was redrawn from Larsell (1953). Abbreviations, *A* anterior, *C* caudal, *D* dorsal, *I* inferior, *P* posterior, *R* rostral, *S* superior; V ventral)

Vermal lobules I-V simply extend laterally to form the hemispheric parts of these lobules. Although lobule I hardly forms its hemispheric parts and lobule II extends to a limited extent, these lobules merge at their most lateral parts. Lobule VI (or Lobule VIa in rat and mouse) also simply extends laterally to its lateral part, hence designated as simple lobule (Bolk, 1906). The same hemispheric lobule is also designated as "hemispheric lobule VI (HVI)" (Larsell, 1970; Schmahmann et al., 1999) in primates and humans.

The relationship between the vermis and hemisphere is not simple in lobules caudal to lobules I–V. Bolk (Bolk, 1906) has drawn the sequential axis of cerebellar foliation in the hemisphere of the cerebellums of the lemur to classify hemispheric lobules. In the area caudal to the simple lobule (or lobule VI), the hemispheric sequential axis extends laterally (crus I) and returns medially (crus II) by making a loop (ansiform lobule) to the paramedian area (paramedian lobule) and curved laterally again toward the ventrolateral direction (paraflocculus and flocculus). Consequently, the continuity between vermal and hemispheric lobules has not been much considered in these names. However, posterior hemispheric lobules are continuous, beyond the paravermal fissure, to vermal lobules, as is apparent in rat and mouse cerebellums and as revealed by careful observation in primates (Fujita et al., 2010; Larsell, 1970). This continuity helps in the correct identification of hemispheric lobules.

Table 5.1 Putative homologous cerebellar lobules among humans, macaques, and rodents. Nomenclatures are not consistent among literature, since identification of homologous lobules is not simple. Shaded cells indicate the ansiform area

		Paxinos type nomenclature		Larsell/Brodal type nomenclature		
	Human	Macaque	Marmoset	Macaque	Marmoset	Rat/mouse
Vermis	Schmahmann et al., 1999	Paxinos et al., 2009	Paxinos et al., 2011	Larsell (1970), Larsell (1953)	Fujita et al., 2010	Larsell, 1970 (Albino rat chapter) and others[1]
	I-V	I-V	I-V	I-V	I-V	I-V
	VI	VI	VI	VI	VI	VIa
	VIIAf	VIIAa	VIIAa	VIIA	VIIAa	VIb
	VIIAt	VIIAb-d	VIIAb-d	VIIAb-d	VIIAb-d	VIc
	VIIB	VIIB	VIIB	VIIB	VIIB	VII
	VIII-X	VIII-X	VIII-X	VIII-X	VIII-X	VIII-X
Hemisphere	HVI	Sim	Sim	HVI (Sim)	Sim	Sim/HVI
	Crus I + Crus II + (?)	Crus I + Crus II	Crus I + Crus II	Crus I	Crus I	Crus I
	HVIIB (?)	Par	Par	Crus II + HVIIB (par a)	Crus II + Par a	Crus II + Par
	HVIII (?)	Cop	Cop	HVIIIA (par p)	Par p (cop)	Cop

Abbreviations: a, A, B, p, sublobule a, A, B, p; Cop, copula pyramidis; Par, paramedian lobule; Sim, simple lobule;
[1]Paxinos & Watson, 2007; Voogd, 2004; Paxinos & Franklin, 2001, Fujita et al., 2014

In the rodent cerebellum, the lateral extension of lobule VIb-c and VII, the combination of which is equivalent to VII in the primate cerebellum (Fujita et al., 2010), forms crus I and crus II/paramedian lobule, respectively. In the primate cerebellum, the continuity analysis was performed in the marmoset cerebellum (Fujita et al., 2010), which has a much smaller number of foliation than the macaque cerebellum. Later, we also performed continuity analysis in the macaque and human cerebellum by using MRI imaging and sections (Luo et al., 2017) (Fig. 5.2). These studies revealed nomenclature discrepancy in hemispheric lobules between rodents and primates and also among primate literature (cf. Luo et al., 2017 for detailed description). Here, I recapitulate our essential conclusion from our previous article (Luo et al., 2018). In the rat and mouse, the lateral extension of lobule VIb-c forms the lobule that extends most laterally, which is designated as crus I (Larsell, 1970). Although designated as crus I, not including crus II, this lobule by itself is equivalent to the whole ansiform area (Luo et al., 2018). "Ansiform area" here means the area that is defined by a particular striped pattern and axonal projection patterns and

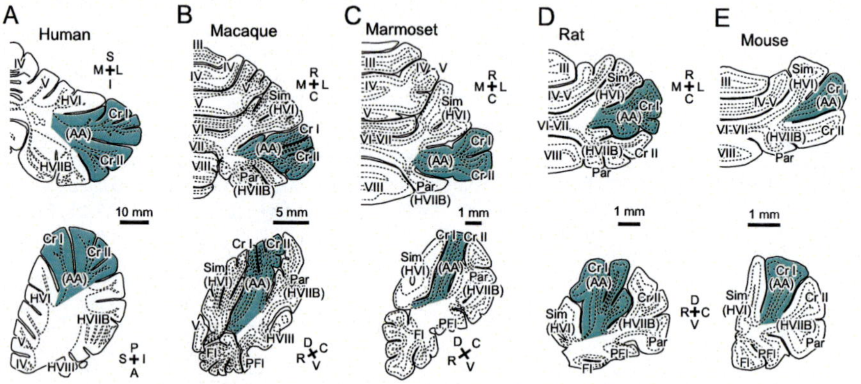

Fig. 5.2 Horizontal and sagittal sections of the human (**a**), macaque (*Macaca fuscata*) (**b**), marmoset (*Callithrix jacchus*) (**c**), rat (Long-Evans) (**d**), and mouse (C57BL/6N) (**e**) cerebellums. (Putatively) homologous lobules of the ansiform area (AA) are labeled in light blue. (Drawings are modified from Luo et al. (2017). Abbreviations, *A* anterior, *AA* ansiform area, *C* caudal, *Cop* copula pyramidis, *Cr I* crus I, *Cr II* crus II, *Fl* flocculus, *I* inferior, *L* lateral, *M* medial, *P* posterior, *Par* paramedian lobule, *PFl* Paraflocculus, *R* rostral, *Rt* right, *S* superior, *Sim* simple lobule)

is supposed to fit the ansiform lobule in the original Bolk's (1906) interpretation. In contrast, the lateral extension of lobule VII forms crus II and the paramedian lobule in the rodent cerebellum. In the non-human primates under Paxinos-type nomenclature (Macaque: Paxinos et al., 2009; Marmoset: Paxinos et al., 2011), the lateral extension of lobule VIIA (equivalent to rat/mouse lobule VIb-c) forms crus I and crus II of the ansiform lobule, fitting with the Bolk's interpretation. The lateral extension of lobule VIIB (equivalent to rat/mouse lobule VII) forms the rostral part of the paramedian lobule. In humans, foliation is complicated. Crus I (superior semilunar lobule) is the extension of lobule VIIAf, and crus II (inferior semilunar lobule) is the extension of lobule VIIAt. Caudal to these lobules, two other lobules, lobules HVIIB and HVIII, are defined and supposed to be the lateral extensions of lobule VIIB and VIII (Larsell & Jansen, 1972; Schmahmann et al., 1999). However, further analysis would be required to identify the exact extent of the ansiform area in the human cerebellum. The main question is whether lobule HVIIB is wholly or partly included in the ansiform area besides crus I and crus II. Axonal connection analysis by tractography from MRI image (Steele et al., 2016) would provide some information in the future. The description here is summarized in the lower half of Table 5.1.

The hemispheric extension of lobule VIII is recognizable in rat and mouse and designated as copula pyramidis (copula = connection, pyramidis = of lobule VIII, Larsell, 1970). The equivalent lobule, carefully identified in the marmoset cerebellum (Fujita et al., 2010), is the posterior part of the paramedian lobule, according to the primate cerebellum nomenclature (Paxinos et al., 2009).

The paraflocculus (lobule HIX) and flocculus (lobule HX) are generally regarded as the hemispheric part of lobules IX and X, respectively. However, the cortical structure is not continuous between the paraflocculus and lobule IX or between flocculus and lobule X in the mammalian cerebellum. The paraflocculus was located adjacent to crus II-paramedian lobule-copula pyramidis in the embryonic brain but shows remarkable lateral protrusion to fit into the arcuate fossa inside the anterior internal semicircular canal in the first postnatal week in the mouse (Panezai et al., 2020). The flocculus is located ventromedial to the paraflocculus.

5.3 Longitudinal Stripes and the Relationship between Lobules and Stripes

Purkinje cells, the sole output neurons of the cerebellar cortex, are composed of heterogeneous populations of different molecular expression profiles. A population of Purkinje cells that has a particular profile of molecular expression is arranged in longitudinally stripe-shaped areas. If focused on the expression of a particular molecule, a characteristic distribution pattern of Purkinje cell populations that have different molecular expression intensities emerges in the cerebellar cortex. Aldolase C or zebrin II is the gold standard of such molecules in the rat and mouse (Brochu et al., 1990; Fujita et al., 2014; Sugihara et al., 2004; Voogd & Glickstein, 1998). The striped distribution pattern of aldolase C (zebrin II)-positive and aldolase C (zebrin II)-negative Purkinje cells (zebrin pattern) have been clarified in the entire cerebellar cortex in the rat and mouse (Hawkes & Leclerc, 1987; Fig. 5.3b) as mapped in the unfolded scheme of the cerebellar cortex (rat: Sugihara & Shinoda, 2004; Ruigrok et al., 2015; Sugihara & Quy, 2007; mouse: Fujita et al., 2014; Sarpong et al., 2018). The striped distribution pattern is well conserved among individuals (Fujita et al., 2014; Hawkes & Leclerc, 1987) and between the wild type and mutants to express fluorescent proteins under the enhancer of aldolase C gene (Fujita et al., 2014). The zebrin pattern is highly correlated with the lobular organization and also with the topographic connections of Purkinje cell axons and climbing fiber axons (see Sect. 5.5.1).

Besides rat and mouse, all examined mammals seem to have more or less similar zebrin pattern at least in the anterior lobules (lobules I-V), where the three clear narrow positive stripes facilitate recognition (Sillitoe et al., 2005). Detailed examination of the zebrin expression pattern in the whole cerebellar cortex has been performed in marmoset (Fujita et al., 2010; Fig. 5.3a) and chick (Marzban et al., 2010; Vibulyaseck et al., 2015), besides in the rat and mouse so far. Whereas the patterns in the rat and mouse have high similarity (Fujita et al., 2014; Sugihara & Quy, 2007), the zebrin pattern in the marmoset shows basic characteristics shared with that in the rat and mouse as well as some unique characteristics (Fujita et al., 2010). Different type of complex zebrin pattern is observed in the avian cerebellum (Vibulyaseck et al., 2015). In vertebrates other than mammals and birds, the cerebellum does not seem to have any similar striped expression of zebrin.

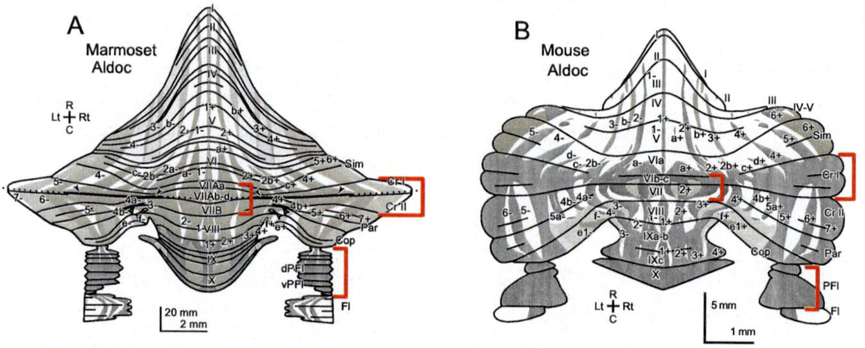

Fig. 5.3 Zebrin stripes, i.e., the longitudinal stripe-shaped distribution pattern of zebrin (aldolase C)-positive and zebrin (aldolase C)-negative Purkinje cells mapped in the unfolded scheme of the cerebellar cortex in the marmoset (**a**) and mouse. These schemes were based on the data used in figures of Fujita et al. (2010) (**a**) and Sarpong et al. (2018) (**b**). Red boxes indicate the homologous lobules in the central vermis, ansiform area, and paraflocculus. Paxinos-type nomenclature (Paxinos et al., 2011) was used in the marmoset cerebellum in (**a**), not the original nomenclature of Fujita et al. (2010). (Abbreviations, *C* caudal, *Cop* copula pyramidis, *Cr I* crus I, *Cr II* crus II, *dPFl* dorsal paraflocculus, *Fl* flocculus, *Lt* left, *Par* paramedian lobule, *PFl* Paraflocculus, *R* rostral, *Rt* right, *Sim* simple lobule, *vPFl* ventral paraflocculus)

The zebrin pattern is uniquely correlated with the lobular organization; the zebrin-striped patterns in different lobules are quite distinct from one another (Fujita et al., 2014). The change in the zebrin pattern has been conventionally correlated with the four groups of lobules in the rat and mouse vermal cerebellum (Hawkes & Leclerc, 1987; Ozol et al., 1999). In lobule I–V and the rostral part of lobule VIa (anterior zone of Ozol et al., 1999), zebrin-negative stripes are much wider than zebrin-positive stripes. In vermal lobule VI (posterior part) and lobule VII, zebrin-positive stripes are much wider and occupy most of the areas (central zone). In lobule VIII, anterior part of IX, crus II, paramedian lobule, and copula pyramidis (posterior zone), both zebrin-positive and zebrin-negative stripes occupy substantial areas. In the posterior part of lobule IX, X, paraflocculus, and flocculus, most of the areas are occupied by zebrin-positive Purkinje cells (nodular zone). However, if details of the zebrin pattern are compared in the entire cerebellar cortex (Fig. 5.3), the area-dependent change in the pattern is more complicated than the above four-group classification. In the hemisphere, the zebrin striped pattern changes significantly in different lobules. A remarkable feature of the striped pattern in the hemisphere is that all stripes shift laterally and positive stripes are wider and merge in crus I (in the rat and mouse) (Sugihara & Quy, 2007; Sugihara & Shinoda, 2004).

The nomenclature of zebrin stripes is complicated. Major positive stripes were named numerically from the medial to lateral areas in individual lobules as (P1+, P2+, P3+, P4+, P5+, P6+, P7+) in the original paper (Hawkes & Leclerc, 1987). Later, other newly recognized positive stripes were named alphabetically or alphanumerically (a+, b+, c+, d+, 4b+, 5a+, e+, Voogd et al., 2003; 2b+, 2b+s, 3d+, e1+, e2+, f+, Sugihara & Shinoda, 2004). Negative stripes are usually indicated by appending "-," in place of "+," in the name of the medially neighboring positive stripe. The analyses of olivary axons and Purkinje cell axons (cf. Sect. 5.5.1) showed that a zebrin stripe in the rostral lobules (lobules I–VIa in the vermis, lobules III–V and simple lobule in the hemisphere) and caudal lobules (lobules VII–X in the vermis, crus II, paramedian lobule and copula pyramidis in the hemisphere) make a pair that share the same axonal projection patterns: a pair of stripes are innervated divergently by branches of common olivary climbing fiber axons. Purkinje cell axons originating from a pair of stripes project convergently to the same area in the cerebellar nuclei (Sugihara et al., 2009; Sugihara & Quy, 2007; Sugihara & Shinoda, 2004, 2007). Such pair of zebrin stripes are regarded as linked stripes. Interestingly, linked zebrin stripes have been given different names between the rostral and caudal lobules in the above nomenclature. We have used "//" to indicate the linked stripes in the rostral and caudal cerebellum as in "4+//5+" (Sugihara & Shinoda, 2004). The paraflocculus is entirely zebrin-positive, whereas the flocculus is divided into the rostral part that has weaker expression of zebrin and the caudal part which is zebrin-positive (Fujita et al., 2014). No consistent names of zebrin stripes have been established in the paraflocculus and flocculus in multiple articles.

Expression of other molecules in Purkinje cell populations show various patterns, which has been examined mainly in rats and mouse. A group of molecules (e.g., EAAT4, PLCb3) show nearly the same distribution as zebrin (aldolase C), although critical observation has not been performed. Another group of molecules such as PLCb4 shows the pattern exactly complementary to that of zebrin (Sarna et al., 2006). Other molecular patterns are different from the zebrin pattern to various degrees. Detailed expression patterns of some of them have been reported in the mouse. HSP25 is expressed in multiple longitudinal stripes in lobule VII, IX, paraflocculus, and flocculus in the mouse (Fujita et al., 2014). Tyrosine hydroxylase is expressed mainly in parts of zebrin-positive stripes in lobules VII-X and copula pyramidis, paraflocculus, and flocculus (Locke et al., 2020). Pcdh10 is expressed mainly in some parts of zebrin-positive stripes (Sarpong et al., 2018).

The appearance of the striped patterns of zebrin (above) and other molecules such as Pcdh10 is special in crus I in the hemisphere of the mouse cerebellum. Pcdh10 is expressed in most parts of the crus I in the pars intermedia, whereas it is expressed only in several striped areas in other neighboring lobules (Sarpong et al., 2018; Fig. 5.4a–d), suggesting that crus I have an organization distinct from that of other neighboring lobules. Since the zebrin striped pattern is examined in animals other than the rat or mouse, the similarity of the zebrin-striped pattern can be utilized to identify homologous lobules among mammals. In the ansifrom area, the lateral shift of the pattern of stripes as well as widening and merging of zebrin-positive stripes are similarly observed in crus I in the mouse and the combination of crus I and II in the marmoset (Paxinos-type nomenclature, Fig. 5.3a, b, right red box; Luo et al., 2017). In the central vermis, the widening of zebrin-positive stripes is similarly observed in lobule VII in the marmoset and lobules VIb-c and VII in the mouse (Fig. 5.3a, b, left red box, Fujita et al., 2010). Besides, the paraflocculus contained zebrin-positive stripes nearly exclusively in both marmoset and mouse

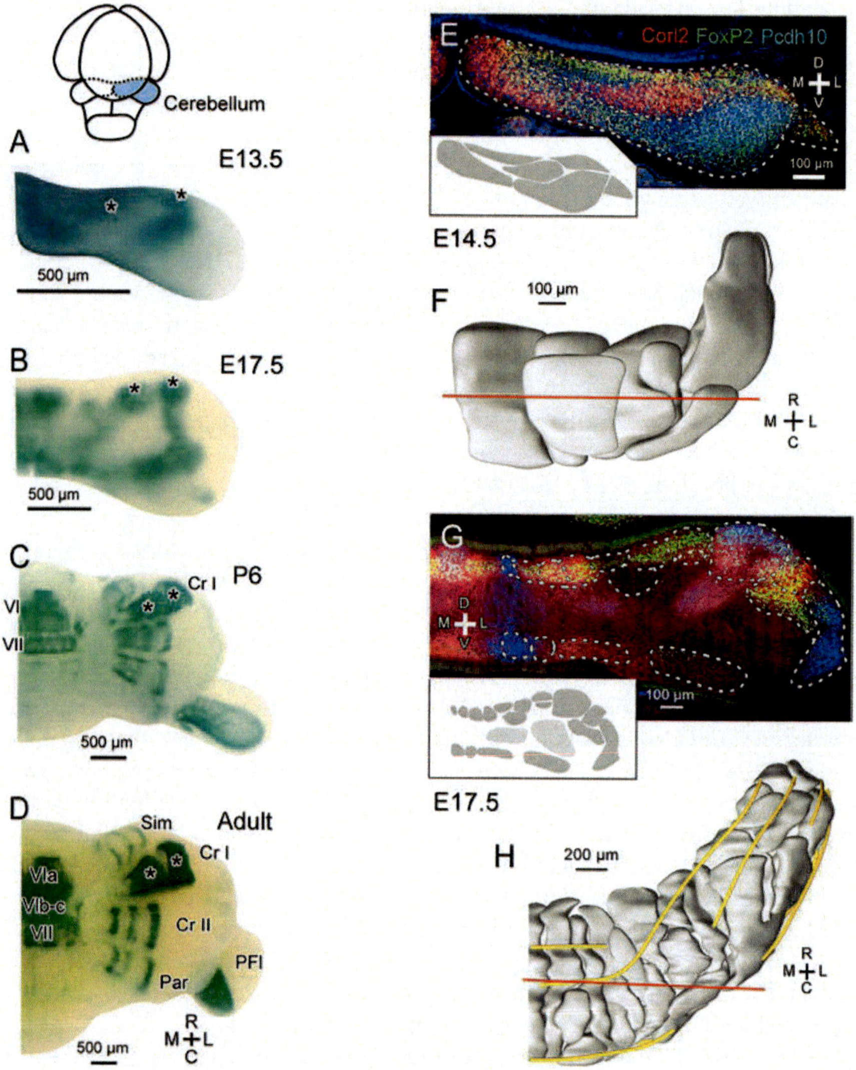

Fig. 5.4 Development of stripes and lobules. **(a–d)**, Development of protocadherin 10 (Pcdh10)-positive stripes, which have peculiar wide distributions in crus I, in the dorsocaudal view of whole-mount preparations of Pcdh10-reporter mouse (Modified from Vibulyaseck et al., 2017 and Sarpong et al., 2018). Asterisks track two major groups of Pcdh10-positive Purkinje cells that are eventually localized in crus I. **(e-h)**, Recognition of clusters of heterogeneous Purkinje cells in the E14.5 and E17.5 mouse embryos. Immunostaining of marker molecules (Corl2, FoxP2, and Pcdh10) in a coronal section **(e, g)** and three-dimensional reconstruction **(f, h)** are shown. Dashed lines circumscribe clusters in **(e)** and **(g)**. Red lines in **(f)** and **(h)** indicate the level of the section in E and H, respectively. Yellow lines in **(h)** indicate immature major fissures. (Redrawn from Tran-Anh et al. (2020). Abbreviations, *C* caudal, *Cr I* crus I, *Cr II* crus II, *D* dorsal, *L* lateral, *M* medial, *Par* paramedian lobule, *PFl* Paraflocculus, *R* rostral, *Rt* right, *Sim* simple lobule, *V* ventral)

(Fig. 5.3a, b, bottom red box). These three areas may correspond to the three major areas involved in the cognitive function in the human cerebellum (see Sect. 5.6).

5.4 Development of Cerebellar Lobules and Stripes

The developmental origins of the cerebellar hemisphere and vermis arise from the alar plate of rhombomere (r) 1 and r0 or isthmus (Martinez et al., 2013), respectively. The difference in origins of individual lobules in the vermis or hemisphere remains to be explored. Excitatory cerebellar nucleus neurons and Purkinje cells are the earliest neurons generated in the cerebellum. Purkinje cells are generated in the ventricular zone during the period from E10.5 to E12.5 or E13.5 (Hashimoto & Mikoshiba, 2003; Zhang et al., 2020). They migrate radially toward the cerebellar surface and also tangentially to the longitudinal direction. At E14.5, the entire cerebellar anlage forms a U-shaped curved bar- or dumbbell-like structure (Tran-Anh et al., 2020; Vibulyaseck et al., 2017). Purkinje cells are mostly organized into nine clusters that are positioned in distinct areas in the cerebellar anlage (Tran-Anh et al., 2020; Wizeman et al., 2019). Each of the clusters contains a subset of Purkinje cells that are born in a particular period between E10.5 and E13.5 and have a distinct profile of molecular expression are recognized in the E14.5 mouse cerebellum (Tran-Anh et al., 2020). Expansion (spread) of the external granule cell precursors originating from the "germinal triangle" or the rostral rhombic lip toward the entire surface of the cerebellar anlage is accompanied by the positioning of Purkinje cell clusters underneath the external granular layer. Genetic mechanisms in the developing cerebellum enable correct migration of granule cell precursors that give rise to lobulation (Sudarov & Joyner, 2007).

The earliest trace of lobule formation (or fissure formation) becomes apparent around E16.5 in the mouse cerebellum. At E17.5, a trace of major fissures appears. Thus, E17.5 seems a key timing in which the structures that lead to the adult lobules and stripes emerge (Fujita et al., 2012). The number of recognizable Purkinje cells increases from 9 at E14.5 to 18 at E15.5, and 28 and 37 at E17.5 (Fig. 5.4e–h). Since some clusters at E17.5 can be further subdivided into multiple clusters depending on the criteria of cluster recognition, our previous study counted 54 clusters at E17.5 (Fujita et al., 2012). This increase is brought about by divisions of clusters, i.e., partial change in the molecular expression profile in a part of a cluster and/or emergence of Purkinje cell-free gaps between these parts. Along with the increase of Purkinje cell clusters, spatial rearrangement of clusters occurs. Daughter clusters that belong to the lineage of the common original cluster often share similar axonal projection patterns and partly similar molecular expression profiles. Therefore, the spatial rearrangement process of embryonic Purkinje cell clusters before E17.5 is essential in understanding the striped and lobule-dependent compartmental organization of the cerebellum.

A typical rearrangement of the striped organization is seen in the case of the C1/C3 module (refer to the next section), which is demarcated as the multiple zebrin-negative and weakly positive stripes in dual paravermal areas in anterior and posterior lobules. The dual representation of the body (somatotopy) is conspicuous in the C1/C3 module. Among E14.5 Purkinje cell clusters, the c-l (central-lateral) cluster which lacks E10.5-born Purkinje cells and show high expression of EphA4 and Plcb4 are divided into five c-l lineage clusters. They separately migrate underneath other clusters and positioned far apart mediolaterally as well as rostrocaudally by E17.5. They are eventually transformed mainly into multiple separate zebrin-negative and weakly positive stripes, which together configure the adult C1/C3 module, in the anterior and posterior paravermal lobules (Tran-Anh et al., 2020). These results suggest that the spatial rearrangement of embryonic Purkinje cell clusters is involved in forming the functional localization including the dual somatotopic representation in the adult cerebellar cortex.

The lateral shift of striped arrangement in the central lobules (crus I and lobules IV–VII, in the mouse) also has its origin in the cluster development in the embryonic cerebellum. In the vermal lobule VI–VII, a lateral protrusion or expansion of Pcdh10-positive area occurs between E15.5 and E17.5 (Vibulyaseck et al., 2017). It is not clear how this area expands (migration of Pcdh10-positive neurons to this area or change of molecular expression pattern of neurons). In the lateral part of the central lobules which later forms crus I, Pcdh10-positive Purkinje cell clusters show remarkable lateral migration. This lateral migration of Pcdh10-positive cluster is concomitant with the rostrocaudal division of c-l lineage clusters (preceding paragraph). This is partly the cause of the lateral shift of zebrin (aldolase C) stripes in crus I and separation of zebrin-negative stripes into rostral and caudal lobules (Tran-Anh et al., 2020; Vibulyaseck et al., 2017).

Expansion of the cerebellar volume occurs in the postnatal period. Clusters that are positioned at variable depths in the embryonic cerebellum are all arranged on the surface, immediately under the external granular layer. Deep clusters intercalate superficial clusters, and all clusters are extended rostrolaterally concomitant with the longitudinal expansion of the postnatal cerebellum. As a result, the clustered organization of Purkinje cell subsets changes into the striped organization. At P6 in the mouse, the lobular foliation pattern and striped molecular compartmentalization pattern of the adult cerebellum roughly emerges (Fujita et al., 2012).

The formation of the paraflocculus may be noted as the remarkable case of the dynamic expansion of the postnatal cerebellar cortex. The paraflocculus is protruded laterally and longitudinal orientation within the paraflocculus is rotated within the short early postnatal period between P3 and P7 (Panezai et al., 2020).

Concerning the precise striped compartmental organization of the adult cerebellar cortex, details of the formation process have not been fully clarified yet. For example, how the alternating positioning of zebrin-positive and negative stripes is formed and how the individual zebrin-positive and negative stripes differentiate from embryonic Purkinje cell clusters have not been fully tracked yet.

5.5 The Relationship to Axonal Projections of Lobules and Stripes

Projection patterns of various neurons have a certain topographic relationship to the lobular and striped organization of the cerebellum. Some molecules expressed in a subset of Purkinje cells that form a cluster in the embryonic cerebellum may be involved in the pathfinding of axons of projection neurons. Candidates of such molecules include the cadherin family molecules (Sarpong et al., 2018) and ephrin/Eph receptor family molecules (Hashimoto et al., 2012; Lackey & Sillitoe, 2020). However, molecular mechanisms for particular topographic projections within the cerebellum have not been clarified yet.

The afferent and efferent projection patterns of axons are highly correlated with lobular organization of the cerebellum as axons consistently branch to and converge from specific preferred lobules (Fujita & Sugihara, 2013; Quy et al., 2011). The pattern is also highly correlated with the longitudinal striped patterns (Pijpers et al., 2006; Sugihara & Shinoda, 2004; Voogd et al., 2003).

5.5.1 Climbing Fiber and Purkinje Cell Projections

The cerebellar cortex sends its output to the cerebellar nuclei (CN) through the Purkinje cell inhibitory projection (Ito & Yoshida, 1966) and receives its climbing fiber input through the projection from the inferior olive (IO) (Llinás, 2014; Ruigrok et al., 2015). The morphology of these projections is tightly correlated to each other. A single inferior olive axon branches into seven climbing fibers on average in the rat and also gives rise to collaterals to terminate in the cerebellar nucleus (Sugihara et al., 1999). Climbing fibers originating from a single olivary axon usually terminate in a longitudinally band-shaped area in a single lobule or multiple neighboring or separate lobules (Sugihara et al., 2001). It is often observed that some branches project to rostral lobules and other branches project to caudal lobules from a single olivary axon (lobules I–VIa versus lobule VIII, or simple lobule versus crus II and paramedian lobule; Sugihara et al., 2001). When they terminate in multiple lobules, the longitudinal band-shaped termination areas in different lobules are located in a similar mediolateral position (Sugihara et al., 2001). Concerning this, a group of Purkinje cells located in similar mediolateral positions in different lobules convergently project to a common area within the cerebellar nucleus (Sugihara et al., 2009). Furthermore, the collateral of olivary axons that project to this group of Purkinje cells projects to the same area in the cerebellar nucleus (Sugihara & Shinoda, 2007). As a whole, olivary axons and Purkinje cell projections are organized into common longitudinal band-shaped areas in the cerebellar cortex. Note that there are exceptions to the above basic organization scheme. For example, some Purkinje cell axons project to structures outside of the cerebellar nucleus, such as the parabrachial nucleus (Hashimoto et al., 2018), and some olivary axons branch transversely to project to Purkinje cells located in mediolaterally separate stripes (Fujita & Sugihara, 2013).

Mass labeling studies have shown the longitudinal bands described above are classified into several discreet groups (or "modules" as designated later, Ruigrok et al., 2015) that has a clear topographical relationship between subareas of the inferior olive and cerebellar nuclei (Buisseret-Delmas & Angaut, 1993; Voogd & Bigaré, 1980). At the broadest level, five gross modules (A, B, C1/C3, C2 and D) have been defined in the vermis, lateral vermis, paravermis, paravermis, and hemisphere, respectively, each with a topographically connected subarea in the IO and CN plus the lateral vestibular nucleus (Ruigrok et al., 2015). Finer classification of the modules has been identified (A, AX, X, X-CX, B, A2, C1/C3, CX, C2, D0, D1, D2; Buisseret-Delmas & Angaut, 1993).

Later, the striped zebrin (=aldolase C) expression pattern was used as the positional reference to map axonal projection patterns of Purkinje cell axons and olivary axons (Sugihara & Quy, 2007; Sugihara & Shinoda, 2004; Voogd et al., 2003). These studies have clarified topographical projection patterns in virtually all zebrin stripes. Consequently, correspondence between the modules and zebrin stripes have been identified in the rat and mouse cerebellar cortex (Fujita et al., 2010; Sugihara et al., 2009; Sugihara & Shinoda, 2004; Voogd et al., 2003). Again, a similar analysis has been performed on the marmoset zebrin stripes (Fujita et al., 2010).

Compared to the longitudinal striped pattern, the lobular organization has not been much correlated to olivary and Purkinje cell projections. However, several pieces of evidence indicate that the Purkinje cell projections and olivary projections are both correlated with the lobular organization as well. First, some zebrin stripes do not extend across all lobules but are localized only in several neighboring lobules. For example, zebrin-negative and zebrin lightly positive stripes in the paravermal areas, which are equivalent to the C1/C3 module, are present in anterior and posterior lobules but absent in the apex of the central lobule or crus I (Fig. 5.5b). Consequently, the olivary and Purkinje cell projections for the C1/C3 module are absent in the apex of crus I. Second, in zebrin stripes that extend across all lobules, the topographic axonal projection pattern has subdivisions related to the lobular organization within a lobule. For example, the C2 module, which is present in all lobules from III, IV–V, simple lobule Crus I, crus II, paramedian lobule, and copula pyramidis, is innervated by olivary axons arising from the rostral part of the medial accessory olive subnucleus (rMAO) of the inferior olive and sends its Purkinje cell axons to the posterior interpositus nucleus (PIN) in the rat (Ruigrok et al., 2015; Sugihara et al., 2009). There are at least two lobule-related subdivisions within this topographic organization; the C2 module in crus I and the C2 module in the simple lobule, crus II, and paramedian lobule are each innervated by the rostral and caudal parts of the rMAO, respectively, and send their axons to the dorsal and ventral part of the PIN, respectively (Luo et al., 2017). Within the former subdivision, some axons branch into crus I and paraflocculus (Fujita & Sugihara, 2013), indicating that the C2 module in the paraflocculus also belongs to this subdivision and some functional link between crus I and the paraflocculus. In paravermal and hemispheric modules other than C1/C3 or C2 modules, some data in the published mapping

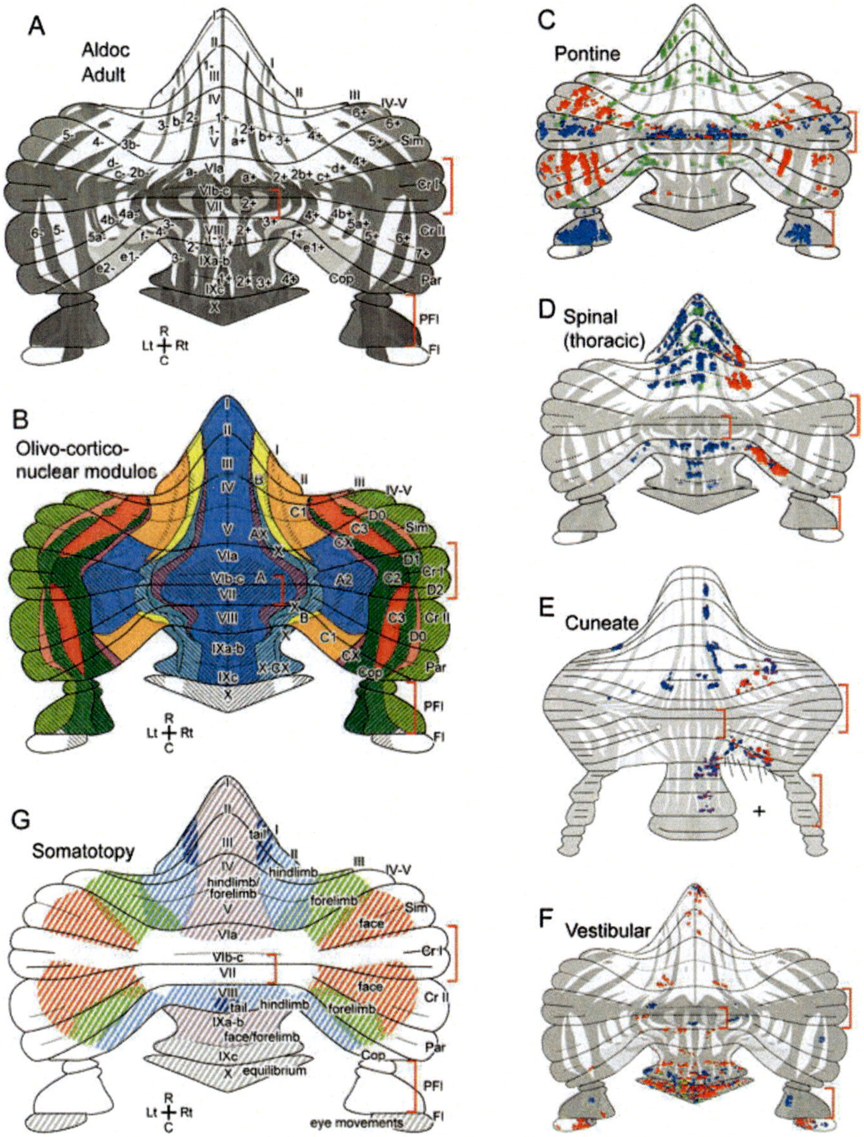

Fig. 5.5 Axonal projection patterns in cerebellar lobules and stripes. (**a**), Zebrin stripes mapped in the unfolded scheme of the mouse cerebellar cortex. Based on the data used in a figure in Sarpong et al. (2018). (**b**), Olivo-cortico-nuclear modules mapped on the zebrin stripes. Modified from Tran-Anh et al. (2020). (**c–f**), Mapping of reconstructed single mossy fiber axons originating from the pontine nucleus (**c**), thoracic spinal cord (**d**), cuneate and external cuneate nuclei (**e**, rat), and medial vestibular nucleus (**f**). Mossy fiber terminals of several axons belonging to the same group are mapped in an identical color. (Based on data used in figures of Biswas et al. (2020) (**c**), Luo et al. (2018), Quy et al. (2011), and Ando et al. (2020). (**g**), Schematic drawing of somatotopic representation in the cerebellar cortex. Modified from Tran-Anh et al. (2020). Red boxes indicate crus I (ansiform area), lobules VIb-c and VII (vermal central area), and paraflocculus, which are innvated by the blue group of pontine axons (**c**) and putatively homologous to the non-motor function lobules in the human cerebellum (Guell et al., 2018). Abbreviations, *C* caudal, *Cop* copula pyramidis, *Cr I* crus I, *Cr II* crus II, *Fl* flocculus, *Lt* left, *Par* paramedian lobule, *PFl* Paraflocculus, *R* rostral, *Rt* right, *Sim* simple lobule)

(Sugihara et al., 2001; Sugihara & Quy, 2007) also suggest subdivisions within a module. However, further studies would be needed to clarify the lobule-related subdivision organization in cerebellar modules. Lobular subdivisions are also observed in the vermal module; generally distinct groups of climbing fiber axons that branch into (1) lobule I–V (or VIa) and VIII, (2) lobule IX, (3) lobules (VIa), VIb-c, and VII have been observed (rat, Sugihara et al., 2001; Sugihara & Shinoda, 2004; mouse, Sugihara & Quy, 2007).

5.5.2 *Mossy Fiber Axons*

The topography of mossy fiber projection has been studied by mass labeling as well as single axon tracing. The latter technique revealed branching projection of single mossy fiber axons into multiple lobules and multiple stripes clearly, clarifying projection patterns of these axons significantly (Ando et al., 2020; Biswas et al., 2019; Luo et al., 2018; Luo et al., 2020; Na et al., 2019; Quy et al., 2011; Wu et al., 1999). A single mossy fiber axon shows abundant branching and terminates in multiple lobules and multiple stripes. Although mossy fiber axons terminate in the granule cell layer, their terminations have some correlation to stripes, which can be recognized by extending the zebrin-striped boundaries from the molecular and Purkinje cell layers to the granular layer. Mossy fiber projections are correlated with both lobules and stripes. For example, pontine mossy fibers terminate more in zebrin-positive stripes (or the granular layer under the zebrin-positive Purkinje cell layer) (Biswas et al., 2019; Na et al., 2019). Spinocerebellar and cuneocerebellar mossy fibers tend to terminate in complementarily striped subareas in the vermal lobules IV–V and VIII (Ji & Hawkes, 1994).

Mossy fibers axons originating from the precerebellar nuclei, such as the pontine nucleus and lateral reticular nucleus, and the spinal cord usually possess mossy fiber termination as the main axonal arbor (precerebellar type mossy fiber axons). On the contrary, some types of axons such as primary vestibular afferent axons (Ando et al., 2020), some axons originating from the ventral horn of the spinal cord (Luo et al., 2018), and some output axons from the cerebellar nucleus have projections to somewhere outside of the cerebellum as their main axonal arbor and possess a small number of mossy fiber terminals in their collaterals. In this article, we deal with only the precerebellar type mossy fiber axons. These axons seldom project to targets outside the cerebellum. In the cerebellar cortex, they usually branch first in the deep white matter to divide into collaterals terminating in multiple lobules and multiple mediolaterally separate areas. Collaterals then terminate frequently in the folial white matter and granular layer. Eventually, they have a variable number of terminals in multiple lobules and multiple stripes. The terminals originating from a single axon are often widely and sparsely spread but sometimes aggregated in a small cortical area (Luo et al., 2018). The number of mossy fiber terminals per axon generally ranges between 50 and 150 (54.1 on the average, for mouse vestibular nuclear axons; 66.5 for mouse pontine nucleus axons, Biswas et al., 2019; 99.4 and 91.0 for mouse non-crossing Clark's column axons and non-crossing marginal Clarke's column axons (Luo et al., 2018), 123.3 for rat dorsal column nucleus axons, Quy et al., 2011; 154.0 for rat lateral reticular nucleus axons, Wu et al., 1999). Some of the mossy fiber axons project to the cerebellar nucleus with some degree of topographic relationship, while some others do not project to the cerebellar nucleus.

In contrast to the inferior olive projection that covers the entire cerebellar cortex with the topographical relationship, no mossy fiber projection from a single source cover the entire cerebellar cortex. The spinocerebellar or cuneocerebellar projections do not cover vermal lobule VI–VII, X, or hemispheric areas of any lobules (Luo et al., 2018; Luo et al., 2020; Quy et al., 2011), the vestibulocerebellar projections preferentially project to lobule X, IXc, and flocculus (Ando et al., 2020). Among the precerebellar nuclei, the pontine nucleus sends its axons to the widest areas of the cerebellar cortex; the pontine nucleus projection covers all cerebellar lobules but for lobule X (Biswas et al., 2019). Therefore, the lobule-related or stripe-related topographic organization of the mossy fiber system, if there is any, can be well studied in the pontocerebellar mossy fiber projection. The topography of the pontocerebellar projection is simply summarized into three groups. Pontine nucleus axons originating from the rostra, medial, and lateral parts of the pontine nucleus mainly terminated in the paraflocculus, crus I, and lobules VIb-c and VII. Those originating from the central part of the pontine nucleus mainly terminated in the simplex lobule, crus II, and paramedian lobule. Those originating from the caudal part of the pontine nucleus mainly terminated in lobules II–VIa, VIII, and copula pyramidis. The interlobular branching pattern of pontocerebellar axons seems to determine the group of cerebellar lobules that are involved in a related functional localization of the cerebellum. It is noticeable that crus I (in the rat and mouse) and neighboring lobules (simple lobule, crus II, paramedian lobule) belong to different groups. This indicates that crus I in the hemisphere may be functionally distinct from neighboring lobules (simple lobule and crus II) in the mouse and rat cerebellum based on the pontocerebellar axonal projection pattern, supporting the idea that only crus I constitute the ansiform area in the rat and mouse. In mass retrograde labeling studies in macaque, crus I and crus II in Paxinos definition (=ansiform area) are innervated from the rostromedial pons, neighboring lobules (HVI, HIIB and HVIII) are innervated from the more central areas (Brodal, 1979; Brodal & Bjaalie, 1992). This agreed to the results in the rat and mouse if the proposed homology between the mouse and rat crus I and primate crus I and crus II is considered (Luo et al., 2017).

Concerning zebrin stripes, about 66% of terminals are located in zebrin-positive stripes. Mossy fiber projection is relayed by granule cells and granule cell axons (parallel fibers), which has a length of 2.5 mm or more to cross stripes. Therefore, the significance of stripe-specific projection of mossy fibers may be blurred by the parallel fiber running transversely across stripes, which produce a transversal spread of the excitation (Cramer et al., 2013). This question has not been fully solved.

5.5.3 Cerebellar Nucleus Projections

Subdivision organization of the cerebellar nucleus can be defined by the topographic projection patterns of the collaterals of the olivary axons and Purkinje cell axons (preceding section). The cerebellar nucleus has subdivisions which approximately correspond to the longitudinally striped organization of the cerebellar cortex (Sugihara & Shinoda, 2004, 2007), although the positional pattern of the correspondence is not simple. In the medial nucleus (MN), Purkinje cell axons from zebrin-positive stripes terminate in the ventral (in the most medial part of the MN) or caudoventral (in the rest of the medial nucleus) part of the medial nucleus, whereas Purkinje cell axons from zebrin-negative stripes terminate in the dorsal or rostrodorsal parts (Sugihara & Shinoda, 2007). Consequently, the medial nucleus has caudoventral zebrin-positive and rostrodorsal zebrin-negative subdivisions, because no other neurons than Purkinje cell axons express zebrin in the cerebellar nuclei. The boundary between the zebrin-positive and zebrin-negative areas is not as clear as in the cerebellar cortex since the projection of Purkinje cell axons show some spread beyond the boundary more or less. In the interpositus nucleus, zebrin-negative (and faintly-positive) Purkinje cell axons project to the rostrodorsal part, i.e., the anterior interpositus nucleus (AIN), and to the medial part of the PIN, while zebrin-positive Purkinje cell axons project to the ventral and lateral parts of the PIN. Thus, the striped arrangement of zebrin-positive and zebrin-negative stripes in the cerebellar cortex is transformed into the rostrodorsal vs. caudoventral segregation of zebrin-positive and zebrin-negative areas in the medial and interpositus cerebellar nuclei.

Furthermore, there seems to be another type of subdivision in the cerebellar nuclei which connects to different lobules. For example, in the lateral part of the PIN, Purkinje cells in crus II and simple lobules project dorsally while those in crus I project ventrally (Luo et al., 2017). However, the lobule-linked subdivisions of the cerebellar nuclei have not been fully clarified.

The cerebellar nuclei contain various types of excitatory and inhibitory neurons regarding the morphology, projection, neurotransmitter, and molecular expression (Fujita et al., 2020; Uusisaari et al., 2007). The distribution of types of neurons in the cerebellar nuclei is partly related to the compartmentalization determined by the topography of the Purkinje cell axonal projections. Concerning the output from the cerebellar nuclei, topographical connection to different target areas is mainly formed according to the molecular expression differences among output neurons (Fujita et al., 2020). However, output neurons of different molecular expression are localized in particular areas in the cerebellar nuclei (Fujita et al., 2020). As a whole, cerebellar output originating from different cortical compartments defined stripes and lobules are preserved more or less in the nuclear output connections.

5.6 Functional Localization in Lobules and Stripes

It would be quite reasonable to think that the lobular and striped organization of the cerebellum is well related to the function of the cerebellum. The question is how they are related to the function of the cerebellum. First of all, the topographic afferent and efferent axonal connections, which are correlated with lobules and stripes (refer to the preceding section), can give specific functions in particular lobules and stripes. Second, the different molecular expression profiles in Purkinje cell subsets in stripes and lobules can give different physiological properties in Purkinje cells and their synapses (Nguyen-Minh et al., 2019). Such different properties facilitate and modulate the function of particular stripes, affecting the function of the cerebellum. Concerning the topographic axonal connections, mossy fibers, climbing fibers, and the axonal connection is supposed to be the most essential factor to determine the functional localization. Among these points, the topographic axonal connection pattern is supposed to determine the gross functional localization.

To understand the functional aspects of the cerebellar organization, data from recording and imaging studies in awake animals and humans are needed. But these techniques have some limitations. Imaging studies can be applied in varieties of protocols in humans with the resolution of lobular foliation, but no information about stripes can be extracted from the data. In animal experiments with electrical and optical recordings, the recording area can be readily related to the lobule. However, such data can be related to the stripes only when immunostaining is performed later or when animals with stripe reporter expression are used (Cramer et al., 2013; Shimuta et al., 2020; Tsutsumi et al., 2019). Therefore, information about the involvement of stripes in the cerebellar function is limited. Consequently, the functional localization of the cerebellum is mostly described in terms of lobules.

Concerning lobules, functional localization has been long studied. Briefly, the vermal area in lobule I–VIa and VIII is involved in the adaptation of locomotion (Jahn et al., 2008; Luo et al., 2020; Ozden et al., 2012). Vermal area in lobule VIb-c and VII is involved in the non-motor function and oculomotor control (Catz & Thier, 2007; Suzuki et al., 2012; Watson et al., 2014). Lobule IXa-b is presumably involved in the orientation of the face and head (Sugihara, 2005; Waespe et al., 1985; Welker, 1987). The nodules, lobule IXc (or ventral uvula), and flocculus are involved in the control of adaptation of reflexive eye movements (macaque, Lisberger & Fuchs, 1978; cat, Sato & Kawasaki, 1984; rabbit, Ito et al., 1977; Barmack et al., 1992; mouse, Koekkoek et al., 1997). In the paravermal and medial hemispheric areas in lobule III-VIa (HVI) and in crus II, paramedian lobule and copula pyramidis (HVIIB-HVIII) are both involved somatosensorimotor control of fine motor activity of body parts in a somatotopic manner (Manni & Petrosini, 2004; Thickbroom et al., 2003; Tran-Anh et al., 2020). The body area is dually represented in the above two areas in the cerebellum. The rest of the cerebellar cortex (crus I and paraflocculus) will be discussed in a separate paragraph.

Then, how the stripes are related to the functional localization? The interpositus nucleus is divided into the AIN and PIN. The AIN and the cortical areas that project to the AIN (C1/C3 module) is involved in the control of fine body movements such as grasping, limb cutaneous reflexes, and eyeblink reflex (Horn et al., 2010; Pijpers et al., 2008). The C1/C3 module occupies the majority of the zebrin-negative and zebrin-positive stripes in the paravermal area in anterior and posterior cerebellar lobules (lobules III-V and simple lobule; crus II, paramedian lobule and copula pyramidis; Fig. 5.5b). Both the anterior and posterior parts of this area are topographically innervated by the climbing fiber axons originating from the dorsal accessory olive (Cerminara et al., 2013; Ekerot et al., 1997; Low et al., 2018; Ruigrok et al., 2015). Climbing fiber activity in the zebrin stripes of the C1/C3 module and other zebrin stripes occurs in a different particular context in crus II of behaving mice (Tsutsumi et al., 2019).

In the MN, involvement in eight different functions (posturomotor, oromotor, orienting, positional-autonomic, and vigilance) has been linked to excitatory output neurons that are localized in different subareas and express different marker molecules (Fujita et al., 2020). The functional localization in the cerebellar vermis would be considered based on the topographic projection patterns of Purkinje cells (Sugihara et al., 2009), which is related to the stripes primarily and to lobules secondary.

Functional localization in the rest of the cerebellar cortex is less clearly understood. In terms of stripes, the function of modules AX, X-CX, C2, D1 and D2, zebrin-positive stripes in the lateral vermis, lateral pars intermedia, and hemisphere are included in less clear areas. In terms of lobules, entire crus I (or the ansiform area, crus I + II in primates) and entire paraflocculus and the most lateral areas of all hemispheric lobules are included. However, recent studies including human imaging studies have produced significant progress in the understanding of the function of these lobules (Batson et al., 2015; D'Mello & Stoodley, 2015; Stoodley & Schmahmann, 2009).

In terms of the ansiform area (crus I in rodents, crus I and II in primates, Luo et al., 2018), a significant increase in its volume relative to the whole cerebellar volume is observed in more dextrous primates (Balsters et al., 2010; Luo et al., 2018; Fig. 5.6). The volume increase of the ansiform area culminates in the human cerebellum. This indicates that the ansiform area is essential in cognitive and executive functions. Imaging studies in humans have shown that crus I and crus II are mainly involved in cognitive, executive, language processing, and saccadic functions (Batson et al., 2015; D'Mello & Stoodley, 2015; Stoodley & Schmahmann, 2009). Non-human primate studies characterize crus I/II by their connectivity to the prefrontal cortex underlying non-motor functions (Strick et al., 2009). Similarly, crus I is also involved in cognitive functions such as ASD-like behavior (Kelly et al., 2020), besides somatosensory or somatomotor-related function.

Besides the ansiform area (crus I and II in primates and humans), human imaging studies showed two other major areas of the cognitive function in the cerebellum, vermal lobules VI–VII, and the paraflocculus (Guell et al., 2018). These three areas match with the divergent projection pattern of the pontocerebellar axons originating from the rostral, medial, and lateral parts of the pontine nucleus (Biswas et al., 2019; Na et al., 2019). The olivary projection also partly matches with these areas

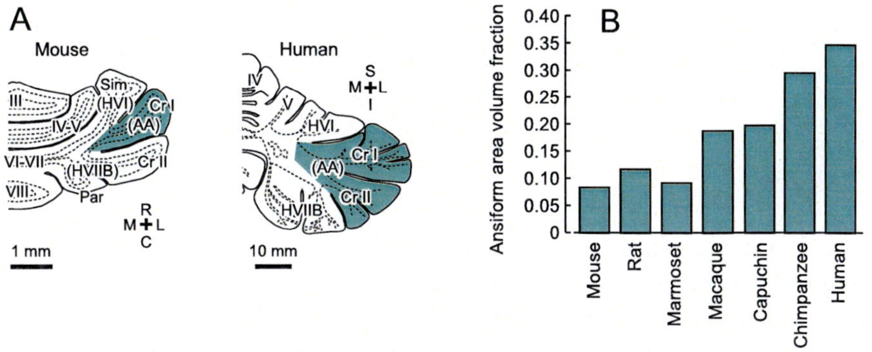

Fig. 5.6 Enlargement of the ansiform area in dexterous animals. (**a**). Drawings of cerebellar sections of the mouse (left, horizontal section) and human (right, coronal section = equivalent to the mouse coronal section) to show the ansiform area (light blue). Crus I in the mouse and crus I and crus II in the human are supposed to be homologous (Luo et al., 2017). (**b**), Volume fraction of the ansiform area measured in various animals (Luo et al., 2017). (Modified from Luo et al. (2017). Data for capuchin and chimpanzee are from Balsters et al., 2010. Abbreviations, *AA* ansiform area, *C* caudal, *Cr I* crus I, *Cr II* crus II, *I* inferior, *L* lateral, *M* medial, *Par* paramedian lobule, *R* rostral, *S* superior, *Sim* simple lobule)

since it shows branching between ansiform area (crus I in rodents) and paraflocculus (Fujita & Sugihara, 2013). In animals, the paraflocculus is a peculiar bulb-like protrusion in the ventrolateral part of the cerebellum (Panezai et al., 2020). Functional localization of the paraflocculus is not much clarified besides the control of smooth pursuit eye movements in primates (Shidara & Kawano, 1993), tinnitus in rats (Bauer et al., 2013). The possible involvement in non-motor function may be further studied in the animal paraflocculus.

5.7 Concluding Remarks

Cerebellar lobules and stripes have been frequently considered as independent structures for the cerebellar compartmentalization like X and Y axes of the two-dimensional coordinates. However, recent findings show that they are mutually tightly related to each other. This article summarized such relations with regard to cerebellar developments and axonal projections. In the end, several findings on how lobules and stripes are correlated with functional localization of the cerebellum have been described. Recent findings seem to identify longitudinal stripes as the structure for the functional localization, which is supported by the anatomical properties of stripes. In the comparative aspects, identifying equivalent lobules and stripes beyond species is important. The lobules in the ansiform area are differently named between rodents and primates. If this fact is taken into account, rodent models can be better utilized to understand the involvement of the cerebellum in various motor and non-motor functions.

Acknowledgments This study was supported by Grant-in-Aid for Scientific Research (KAKENHI) from the Japan Society for the Promotion of Science (19K06919). The author thanks Dr. Luo Yuanjun and Mr. Richard Nana Abankwah Owusu Mensah for comments and proofreading.

References

Ando, T., Ueda, M., Luo, Y., & Sugihara, I. (2020). Heterogeneous vestibulocerebellar mossy fiber projections revealed by single axon reconstruction in the mouse. *The Journal of Comparative Neurology, 528*, 1775–1802. https://doi.org/10.1002/cne.24853

Balsters, J. H., Cussans, E., Diedrichsen, J., Phillips, K. A., Preuss, T. M., Rilling, J. K., & Ramnani, N. (2010). Evolution of the cerebellar cortex: The selective expansion of prefrontal-projecting cerebellar lobules. *NeuroImage, 49*, 2045–2052. https://doi.org/10.1016/j.neuroimage.2009.10.045

Barmack, N. H., Baughman, R. W., Eckenstein, F. P., & Shojaku, H. (1992). Secondary vestibular cholinergic projection to the cerebellum of rabbit and rat as revealed by choline acetyltransferase immunohistochemistry, retrograde and orthograde tracers. *The Journal of Comparative Neurology, 317*, 250–270.

Batson, M. A., Petridou, N., DWJ, K., Frens, M. A., & Neggers, S. F. W. (2015). Single session imaging of cerebellum at 7 tesla: Obtaining structure and function of multiple motor subsystems in individual subjects. *PLoS One, 10*, e0134933. https://doi.org/10.1371/journal.pone.0134933

Bauer, C. A., Kurt, W., Sybert, L. T., & Brozoski, T. J. (2013). The cerebellum as a novel tinnitus generator. *Hearing Research, 295*, 130–139.

Biswas, M. S., Luo, Y., Sarpong, G. A., & Sugihara, I. (2019). Divergent projections of single pontocerebellar axons to multiple cerebellar lobules in the mouse. *The Journal of Comparative Neurology, 527*(12), 1966–1985. https://doi.org/10.1002/cne.24662

Bolk, L. (1906). *Das Cerebellum der Saugetiere, Eine Vergleichend Anatomische Untersuchung*. Verlag von Gustav Fischer.

Brodal, P. (1979). The pontocerebellar projection in the rhesus monkey: An experimental study with retrograde axonal transport of horseradish peroxidase. *Neuroscience, 4*, 193–208.

Brochu, G., Maler, L., & Hawkes, R. (1990). Zebrin II: A polypeptide antigen expressed selectively by purkinje cells reveals compartments in rat and fish cerebellum. *The Journal of Comparative Neurology, 2915*, 538–552. https://doi.org/10.1002/cne.902910405

Brodal, P., & Bjaalie, J. G. (1992). Organization of the pontine nuclei. *Neuroscience Research, 13*, 83–118.

Buisseret-Delmas, C., & Angaut, P. (1993). The cerebellar olivo-corticonuclear connections in the rat. Buisseret-Delmas C, Angaut P. *Progress in Neurobiology, 40*, 63–87. https://doi.org/10.1016/0301-0082(93)90048-w

Cerminara, N. L., Aoki, H., Loft, M., Sugihara, I., & Apps, R. (2013). Structural basis of cerebellar microcircuits in the rat. *Journal of Neuroscience, 33*, 16427–16442.

Cramer, S. W., Gao, W., Chen, G., & Ebner, T. J. (2013). Reevaluation of the beam and radial hypotheses of parallel fiber action in the cerebellar cortex. *The Journal of Neuroscience, 33*(28), 11412–11424. https://doi.org/10.1523/JNEUROSCI.0711-13.2013

Catz, N., & Thier, P. (2007). Neural control of saccadic eye movements. *Developments in Ophthalmology, 40*, 52–75.

D'Mello, A. M., & Stoodley, C. J. (2015). Cerebro-cerebellar circuits in autism spectrum disorder. *Frontiers in Neuroscience, 9*, 408. https://doi.org/10.3389/fnins.2015.00408

Ekerot, C. F., Garwicz, M., & Jörntell, H. (1997). The control of forelimb movements by intermediate cerebellum. *Progress in Brain Research, 114*, 423–429. https://doi.org/10.1016/s0079-6123(08)63378-6

Fujita, H., Oh-Nishi, A., Obayashi, S., & Sugihara, I. (2010). Organization of the marmoset cerebellum in three-dimensional space: Lobulation, aldolase C compartmentalization and axonal projection. *The Journal of Comparative Neurology, 518*, 1764–1791. https://doi.org/10.1002/cne.22301

Fujita, H., Morita, N., Furuichi, T., & Sugihara, I. (2012). Clustered fine compartmentalization of the mouse embryonic cerebellar cortex and its rearrangement into the postnatal striped configuration. *The Journal of Neuroscience, 32*, 15688–15703. https://doi.org/10.1523/JNEUROSCI.1710-12.2012

Fujita, H., & Sugihara, I. (2013). Branching patterns of olivocerebellar axons in relation to the compartmental organization of the cerebellum. *Front Neural Circuits, 7*, 3. https://doi.org/10.3389/fncir.2013.00003

Fujita, H., Aoki, H., Ajioka, I., Yamazaki, M., Abe, M., Oh-Nishi, A., Sakimura, K., & Sugihara, I. (2014). Detailed expression pattern of aldolase C (Aldoc) in the cerebellum, retina and other areas of the CNS studied in Aldoc-Venus knock-in mice. *PLoS One, 9*, e86679. https://doi.org/10.1371/journal.pone.0086679

Fujita, H., Kodama, T., & du Lac, S. (2020). Modular output circuits of the fastigial nucleus for diverse motor and nonmotor functions of the cerebellar vermis. *eLife, 9*, e58613. https://doi.org/10.7554/eLife.58613

Guell, X., Schmahmann, J. D., Gabriell, J. D. E., & Ghosh, S. S. (2018). Functional gradients of the cerebellum. *eLife, 7*, e36652. https://doi.org/10.7554/eLife.36652

Hashimoto, M., & Mikoshiba, K. (2003). Mediolateral compartmentalization of the cerebellum is determined on the "birth date" of Purkinje cells. *Journal of Neuroscience, 23*, 11342–11351.

Hashimoto, M., Ito, R., Kitamura, N., Namba, K., & Hisano, Y. (2012). Epha4 controls the midline crossing and contralateral axonal projections of inferior olive neurons. *The Journal of Comparative Neurology, 520*(8), 1702–1720. https://doi.org/10.1002/cne.23008

Hashimoto, M., Yamanaka, A., Kato, S., Tanifuji, M., Kobayashi, K., & Yaginuma, H. (2018). Anatomical evidence for a direct projection from Purkinje cells in the mouse cerebellar vermis to medial parabrachial nucleus. *Front Neural Circuits, 12*, 6. https://doi.org/10.3389/fncir.2018.00006

Hawkes, R., & Leclerc, N. (1987). Antigenic map of the rat cerebellar cortex: The distribution of parasagittal bands as revealed by monoclonal anti-Purkinje cell antibody mobQ113. *The Journal of Comparative Neurology, 256*, 29–41.

Horn, K. M., Pong, M., & Gibson, A. R. (2010). Functional relations of cerebellar modules of the cat. *Journal of Neuroscience, 30*, 9411–9423.

Ito, M. (2012). *The cerebellum: Brain for an implicit self.* Person.

Ito, M., & Yoshida, M. (1966). The origin of cerebellar-induced inhibition on Deiters neurons. I. Monosynaptic initiation of the inhibitory postsynaptic potentials. *Experimental Brain Research, 2*, 330–349.

Ito, M., Nisimaru, N., & Yamamoto, M. (1977). Specific patterns of neuronal connexions involved in the control of the rabbit's vestibulo-ocular reflexes by the cerebellar flocculus. *The Journal of Physiology, 265*, 833–854.

Jahn, K., Deutschländer, A., Stephan, T., Kalla, R., Wiesmann, M., Strupp, M., & Brandt, T. (2008). Imaging human supraspinal locomotor centers in brainstem and cerebellum. *NeuroImage, 39*, 786–792.

Ji, Z., & Hawkes, R. (1994). Topography of purkinje cell compartments and mossy fiber terminal fields in lobules II and III of the rat cerebellar cortex: Spinocerebellar and cuneocerebellar projections. *Neuroscience, 61*(4), 935–954.

Kelly, E., Meng, F., Fujita, H., Morgado, F., Kazemi, Y., Rice, L. C., Ren, C., Escamilla, C. O., Gibson, J. M., Sajadi, S., Pendry, R. J., Tan, T., Ellegood, J., Basson, M. A., Blakely, R. D., Dindot, S. V., Golzio, C., Hahn, M. K., Katsanis, N., Robins, D. M., Silverman, J. L., Singh, K. K., Wevrick, R., Taylor, M. J., Hammill, C., Anagnostou, E., Pfeiffer, B. E., Stoodley, C. J., Lerch, J. P., du Lac, S., & Tsai, P. T. (2020). Regulation of autism-relevant behaviors by cerebellar-prefrontal cortical circuits. *Nature Neuroscience, 23*(9), 1102–1110. https://doi.org/10.1038/s41593-020-0665-z

Koekkoek, S. K., Alphen, A. M., Burg, J., Grosveld, F., Galjart, N., & De Zeeuw, C. I. (1997). Gain adaptation and phase dynamics of compensatory eye movements in mice. *Genes and Function, 1*, 175–190.

Lackey, E. P., & Sillitoe, R. V. (2020). Eph/ephrin function contributes to the patterning of spinocerebellar mossy Fibers into parasagittal zones. *Frontiers in Systems Neuroscience, 14*, 7. https://doi.org/10.3389/fnsys.2020.00007

Larsell, O. (1937). The cerebellum. A review and interpretation. *Archives of Neurology & Psychiatry, 38*, 580–607.

Larsell, O. (1952). The morphogenesis and adult pattern of the lobules and fissures of the cerebellum of the white rat. *The Journal of Comparative Neurology, 97*, 281–356.

Larsell, O. (1953). The cerebellum of the cat and the monkey. *The Journal of Comparative Neurology, 99*, 135–199.

Larsell, O. (1970). *The comparative anatomy and histology of the cerebellum from Monotremes through apes*. The University of Minnesota Press.

Larsell, O., & Jansen, J. (1972). *The comparative anatomy and histology of the cerebellum, the human cerebellum, cerebellar connections and cerebellar cortex*. The University of Minnesota Press.

Lisberger, S. G., & Fuchs, A. F. (1978). Role of primate flocculus during rapid behavioral modification of vestibuloocular reflex. I. Purkinje cell activity during visually guided horizontal smooth-pursuit eye movements and passive head rotation. *Journal of Neurophysiology, 41*, 733–763.

Llinás, R. R. (2014). The olivo-cerebellar system: A key to understanding the functional significance of intrinsic oscillatory brain properties. *Front Neural Circuits, 7*, 96. https://doi.org/10.3389/fncir.2013.00096

Locke, T. M., Fujita, H., Hunker, A., Johanson, S. S., Darvas, M., du Lac, S., Zweifel, L. S., & Carlson, E. S. (2020). Purkinje cell-specific knockout of tyrosine hydroxylase impairs cognitive Behaviors. *Frontiers in Cellular Neuroscience, 14*, 228. https://doi.org/10.3389/fncel.2020.00228

Low, A. Y., Thanawalla, A. R., Yip, A. K., Kim, J., Wong, K. L., Tantra, M., Augustine, G. J., & Chen, A. I. (2018). Precision of discrete and rhythmic forelimb movements requires a distinct neuronal subpopulation in the interposed anterior nucleus. *Cell Reports, 22*, 2322–2333.

Luo, Y., Fujita, H., Nedelescu, H., Biswas, M. S., Sato, C., Ying, S., Takahashi, M., Akita, K., Higashi, T., Aoki, I., & Sugihara, I. (2017). Lobular homology in cerebellar hemispheres of humans, non-human primates and rodents: A structural, axonal tracing and molecular expression analysis. *Brain Structure & Function, 2017*(222), 2449–2472. https://doi.org/10.1007/s00429-017-1436-9

Luo, Y., Patel, R. P., Sarpong, G. A., Sasamura, K., & Sugihara, I. (2018). Single axonal morphology and termination to cerebellar aldolase C stripes characterize distinct spinocerebellar projection systems originating from the thoracic spinal cord in the mouse. *The Journal of Comparative Neurology, 526*, 681–706. https://doi.org/10.1002/cne.24360

Luo, Y., Onozato, T., Wu, X., Sasamura, K., Sakimura, K., & Sugihara, I. (2020). Dense projection of Stilling's nucleus spinocerebellar axons that convey tail proprioception to the midline area in lobule VIII of the mouse cerebellum. *Brain Structure and Function Brain Structure Function, 225*, 621–638. https://doi.org/10.1002/cne.24360

Madigan, J. C., & Carpenter, M. B. (1971). *Cerebellum of the rhesus monkey, atlas of lobules, laminae, and folia, in sections*. University Park Press.

Manni, E., & Petrosini, L. (2004). A century of cerebellar somatotopy: A debated representation. *Nature Reviews. Neuroscience, 5*, 241–249.

Marani, E., & Voogd, J. (1979). The morphology of the mouse cerebellum. *Acta Morphologica Neerlando-Scandinavica, 17*, 33–52.

Martinez, S., Andreu, A., Mecklenburg, N., & Echevarria, D. (2013). Cellular and molecular basis of cerebellar development. *Frontiers in Neuroanatomy, 7*, 18. https://doi.org/10.3389/fnana.2013.00018

Marzban, H., Chung, S. H., Pezhouh, M. K., Feirabend, H., Watanabe, M., Voogd, J., & Hawkes, R. (2010). Antigenic compartmentation of the cerebellar cortex in the chicken (Gallus domesticus). *The Journal of Comparative Neurology, 518*, 2221–2239.

Na, J., Sugihara, I., & Shinoda, Y. (2019). The entire trajectories of single pontocerebellar axons and their lobular and longitudinal terminal distribution patterns in multiple aldolase C-positive compartments of the rat cerebellar cortex. *The Journal of Comparative Neurology, 527*, 2488–2511. https://doi.org/10.1002/cne.24685

Nguyen-Minh, V. T., Tran-Anh, K., Luo, Y., & Sugihara, I. (2019). Electrophysiological excitability and parallel fiber synaptic properties of zebrin-positive and -negative Purkinje cells in lobule VIII of the mouse cerebellar slice. *Frontiers in Cellular Neuroscience, 12*(513), 1–11. https://doi.org/10.3389/fncel.2018.00513

Ozden, I., Dombeck, D. A., Hoogland, T. M., Tank, D. W., & Wang, S. S. (2012). Widespread state-dependent shifts in cerebellar activity in locomoting mice. *PLoS One, 7*, e42650. https://doi.org/10.1371/journal.pone.0042650

Ozol, K., Hayden, J. M., Oberdick, J., & Hawkes, R. (1999). Transverse zones in the vermis of the mouse cerebellum. *The Journal of Comparative Neurology, 412*, 95–111.

Panezai, S. K., Luo, Y., Vibulyaseck, S., Sarpong, G. A., Nguyen-Minh, V. T., Nedelescu, H., Hirano, S., & Sugihara, I. (2020). Reorganization of longitudinal compartments in the laterally protruding paraflocculus of the postnatal mouse cerebellum. *The Journal of Comparative Neurology, 528*(10), 1725–1741. https://doi.org/10.1002/cne.24849

Paxinos, G., Huang, X. F., Petrides, M., & Toga, A. W. (2009). *The rhesus monkey brain, in Steriotaxic coordinates* (2nd ed.). Academic-Elsevier.

Paxinos, G., Watson, C., Petrides, M., Rosa, M., & Tokuno, H. (2011). *The marmoset brain in stereotaxic coordinates*. Academic Press.

Paxinos, G., & Franklin, K. B. J. (2001). *The mouse brain in stereotaxic coordinates* (2nd ed.). Academic Press.

Paxinos, G., & Watson, C. (2007). *The rat brain in stereotaxic coordinates* (6th ed.). Academic Press.

Pijpers, A., Apps, R., Pardoe, J., Voogd, J., & Ruigrok, T. J. H. (2006). Precise spatial relationships between mossy fibers and climbing fibers in rat cerebellar cortical zones. *The Journal of Neuroscience, 26*, 12067–12080.

Pijpers, A., Winkelman, B. H., Bronsing, R., & Ruigrok, T. J. (2008). Selective impairment of the cerebellar C1 module involved in rat hind limb control reduces step-dependent modulation of cutaneous reflexes. *Journal of Neuroscience, 28*, 2179–2189.

Quy, P. N., Fujita, H., Sakamoto, Y., Na, J., & Sugihara, I. (2011). Projection patterns of single mossy fiber axons originating from the dorsal column nuclei mapped on the compartments in the rat cerebellar cortex. *The Journal of Comparative Neurology, 519*, 874–899. https://doi.org/10.1002/cne.22555

Ruigrok, T. J., Sillitoe, R. V., & Voogd, J. (2015). Cerebellum and cerebellar connections. In G. Paxinos (Ed.), *The rat nervous system* (4th ed., pp. 133–205). Amsterdam.

Sarna, J. R., Marzban, H., Watanabe, M., & Hawkes, R. (2006). Complementary stripes of phospholipase C beta 3 and C beta 4 expression by Purkinje cell subsets in the mouse cerebellum. *The Journal of Comparative Neurology, 496*, 303–313.

Sarpong, G. A., Vibulyaseck, S., Luo, Y., Biswas, M. S., Fujita, H., Hirano, S., & Sugihara, I. (2018). Cerebellar modules in the olivo-cortico-nuclear loop demarcated by pcdh10 expression in the adult mouse. *Journal of Comparative Neurology, 526*(15), 2406–2427. https://doi.org/10.1002/cne.24499

Sato, Y., & Kawasaki, T. (1984). Functional localization in the three floccular zones related to eye movement control in the cat. *Brain Research, 290*, 25–31.

Schmahmann, J. D., Doyon, J., McDonald, D., Holmes, C., Lavoie, K., Hurwitz, A. S., Kabani, N., Toga, A., Evans, A., & Petrides, M. (1999). Three-dimensional MRI atlas of the human cerebellum in proportional stereotaxic space. *NeuroImage, 10*, 233–260.

Shidara, M., & Kawano, K. (1993). Role of Purkinje cells in the ventral paraflocculus in short-latency ocular following responses. *Experimental Brain Research, 93*, 185–195.

Shimuta, M., Sugihara, I., & Ishikawa, T. (2020). Multiple signals evoked by unisensory stimulation converge onto cerebellar granule and Purkinje cells in mice. *Communications Biology, 3*, 381. https://doi.org/10.1038/s42003-020-1110-2

Sillitoe, R. V., Marzban, H., Larouche, M., Zahedi, S., Affanni, J., & Hawkes, R. (2005). Conservation of the architecture of the anterior lobe vermis of the cerebellum across mammalian species. *Progress in Brain Research, 148*, 283–297.

Steele, C. J., Anwander, A., Bazini, P. L., Trampel, R., Schaefer, A., Turner, R., Ramnani, N., & Villringer, A. (2016). Human cerebellar sub-millimeter diffusion imaging reveals the motor and non-motor topography of the dentate nucleus. *Cerebral Cortex, 27*, 4537–4548. https://doi.org/10.1093/cercor/bhw258

Stoodley, C. J., & Schmahmann, J. D. (2009). Functional topography in the human cerebellum: A meta-analysis of neuroimaging studies. *NeuroImage, 44*, 489–501.

Strick, P. L., Dum, R. P., & Fiez, J. A. (2009). Cerebellum and nonmotor function. *Annual Review of Neuroscience, 32*, 413–434. https://doi.org/10.1146/annurev.neuro.31.060407.125606

Sudarov, A., & Joyner, A. L. (2007). Cerebellum morphogenesis: The foliation pattern is orchestrated by multi-cellular anchoring centers. *Neural Development, 2*, 26.

Sugihara, I., Wu, H., & Shinoda, Y. (1999). Morphology of single olivocerebellar axons labeled with biotinylated dextran amine in the rat. *The Journal of Comparative Neurology, 414*, 131–148.

Sugihara, I. (2005). Microzonal projection and climbing fiber remodeling in single olivocerebellar axons of newborn rats at postnatal days 4-7. *The Journal of Comparative Neurology, 487*, 93–106. https://doi.org/10.1002/cne.20531

Sugihara, I., & Shinoda, Y. (2004). Molecular, topographic and functional organization of the cerebellar cortex: A study with combined aldolase C and olivocerebellar labeling. *The Journal of Neuroscience, 24*, 8771–8785.

Sugihara, I., & Quy, P. N. (2007). Identification of aldolase C compartments in the mouse cerebellar cortex by olivocerebellar labeling. *The Journal of Comparative Neurology, 500*, 1076–1092.

Sugihara, I., & Shinoda, Y. (2007). Molecular, topographic and functional organization of the cerebellar nuclei: Analysis by three-dimensional mapping of the olivonuclear projection and aldolase C labeling. *The Journal of Neuroscience, 27*, 9696–9710.

Sugihara, I., Wu, H. S., & Shinoda, Y. (2001). The entire trajectories of single olivocerebellar axons in the cerebellar cortex and their contribution to cerebellar compartmentalization. *The Journal of Neuroscience, 21*, 7715–7723.

Sugihara, I., Ebata, S., & Shinoda, Y. (2004). Functional compartmentalization in the flocculus and the ventral dentate and dorsal group y nuclei: An analysis of single olivocerebellar axonal morphology. *The Journal of Comparative Neurology, 470*, 113–133.

Sugihara, I., Fujita, H., Na, J., Quy, P. N., Li, B. Y., & Ikeda, D. (2009). Projection of reconstructed single Purkinje cell axons in relation to the cortical and nuclear aldolase C compartments of the rat cerebellum. *The Journal of Comparative Neurology, 512*, 282–304. https://doi.org/10.1002/cne.21889

Suzuki, L., Coulon, P., Sabel-Goedknegt, E. H., & Ruigrok, T. J. (2012). Organization of cerebral projections to identified cerebellar zones in the posterior cerebellum of the rat. *The Journal of Neuroscience, 32*, 10854–10869. https://doi.org/10.1523/JNEUROSCI.0857-12.2012

Swanson, L. W. (1998). *Brain maps: Structure of the rat brain* (2nd ed.). Elsevier.

Thickbroom, G. W., Byrnes, M. L., & Mastaglia, F. L. (2003). Dual representation of the hand in the cerebellum: Activation with voluntary and passive finger movement. *NeuroImage, 18*, 670–674.

Tran-Anh, K., Zhang, J., Nguyen-Minh, V. T., Fujita, H., Hirata, T., & Sugihara, I. (2020). A common origin of the cerebellar dual somatotopic areas revealed by tracking embryonic Purkinje cell clusters with birthdate tagging. *eNeuro*. 0251-20.2020. https://doi.org/10.1523/ENEURO.0251-20.2020

Tsutsumi, S., Hidaka, N., Isomura, Y., Matsuzaki, M., Sakimura, K., Kano, M., & Kitamura, K. (2019). Modular organization of cerebellar climbing fiber inputs during goal-directed behavior. *eLife, 8*, e47021. https://doi.org/10.7554/eLife.47021

Uusisaari, M., Obata, K., & Knöpfel, T. (2007). Morphological and electrophysiological properties of GABAergic and non-GABAergic cells in the deep cerebellar nuclei. *Journal of Neurophysiology, 97*, 901–911.

Vibulyaseck, S., Luo, Y., Fujita, H., Oh-Nishi, A., Ohki-Hamazaki, H., & Sugihara, I. (2015). Compartmentalization of the chick cerebellar cortex based on the link between the striped expression pattern of aldolase C and the topographic olivocerebellar projection. *Journal of Comparative Neurology, 523*, 1886–1912. https://doi.org/10.1002/cne.23769

Vibulyaseck, S., Fujita, H., Luo, Y., Tran, A. K., Oh-Nishi, A., Ono, Y., Hirano, S., & Sugihara, I. (2017). Spatial rearrangement of Purkinje cell subsets forms the transverse and longitudinal compartmentalization in the mouse embryonic cerebellum. *The Journal of Comparative Neurology, 525*, 2971 2990.

Voogd, J., & Bigaré, F. (1980). Topographical distribution of olivary and cortico nuclear fibers in the cerebellum: A review. In J. Courville, C. de Montigny, & Y. Lamarre (Eds.), *The inferior olivary nucleus. anatomy and physiology* (pp. 207–234). Raven Press.

Voogd, J., & Glickstein, M. (1998). The anatomy of the cerebellum. *Trends in Neurosciences, 21*(9), 370–375. https://doi.org/10.1016/s0166-2236(98)01318-6

Voogd, J., Pardoe, J., Ruigrok, T. J. H., & Apps, R. (2003). The distribution of climbing and mossy fiber collateral branches from the copula pyramidis and the paramedian lobule: Congruence of climbing fiber cortical zones and the pattern of zebrin banding within the rat cerebellum. *The Journal of Neuroscience, 23*, 4645–4656.

Voogd, J. (2004). Cerebellum. In G. Paxinos (Ed.), *The rat nervous system* (3rd ed., pp. 205–242). Elsevier Academic Press.

Waespe, W., Cohen, B., & Raphan, T. (1985). Dynamic modification of thevestibulo-ocular reflex by the nodulus and uvula. *Science, 228*, 199–202.

Watson, T. C., Becker, N., Apps, R., & Jones, M. W. (2014). Back to front: Cerebellar connections and interactions with the prefrontal cortex. *Frontiers in Systems Neuroscience, 8*, 4. https://doi.org/10.3389/fnsys.2014.00004

Welker, W. (1987). Spatial organization of somatosensory projections to granule cell cerebellar cortex: Functional and connectional implications of fractured somatotopy (summary of Wisconsin studies). In J. S. King (Ed.), *New concepts in cerebellar neurobiology* (pp. 239–280). Liss.

Wizeman, J. W., Guo, Q., Wilion, E. M., & Li, J. Y. (2019). Specification of diverse cell types during early neurogenesis of the mouse cerebellum. *eLife, 8*, e42388.

Wu, H. S., Sugihara, I., & Shinoda, Y. (1999). Projection patterns of single mossy fibers originating from the lateral reticular nucleus in the rat cerebellar cortex and nuclei. *The Journal of Comparative Neurology, 411*, 97–118.

Zhang, J., Tran-Anh, K., Hirata, T., & Sugihara, I. (2020). Striped distribution pattern of Purkinje cells of different birthdates in the mouse cerebellar cortex studied with the Neurog2-CreER transgenic line. *Neuroscience*. https://doi.org/10.1016/j.neuroscience.2020.07.028

Chapter 6
Olivocerebellar Somatotopy Revisited

Takayuki Michikawa and Atsushi Miyawaki

6.1 Introduction

Two major theories have been proposed for the information processing in the brain (Kanwisher, 2010). One is the theory of modularity, which argues that the brain is organized into discrete modules, each of which processes a particular type of information. The other is the theory of distributed processing, which argues that any information is processed by multiple regions of the brain and that any brain region represents multiple types of information. The discussion on these theories has been one of the prominent topics in the history of the field of neuroscience (Glickstein, 2014).

Point-to-point correspondence between an area of the body and an area of the central nervous system is called somatotopy (Saladin, 2021). A widely accepted assumption is that at least basic sensory and motor functions reside in specialized regions of the brain, such as somatosensory areas and motor areas of the cerebral cortex, respectively. The concept of functional somatotopic organization is consistent with the theory of modularity and has been well-documented in a number of textbooks. However, recent studies provide evidence for alternate views regarding the relations between the body parts and the central nervous system, in particular, both the cerebrum and the cerebellum.

T. Michikawa (✉) · A. Miyawaki
Biotechnological Optics Research Team, RIKEN Center for Advanced Photonics, Hirosawa, Saitama, Japan

Laboratory for Cell Function Dynamics, RIKEN Center for Brain Science, Hirosawa, Saitama, Japan
e-mail: takayuki.michikawa@riken.jp

H. Mizusawa, S. Kakei (eds.), *Cerebellum as a CNS Hub*,
Contemporary Clinical Neuroscience, https://doi.org/10.1007/978-3-030-75817-2_6

6.2 Penfield's Homunculus and Woolsey's Simiusculus in the Cerebral Cortex

In 1937, Wilder Penfield and Edwin Boldrey published an everlastingly influential paper describing a neocortical homunculus (Fig. 6.1a), a distorted representation of the human body so called grotesque creature, based on a neurological "map" of the areas and proportions of the human cerebral cortex dedicated to processing sensory functions, or motor functions, for different parts of the body (Penfield & Boldrey, 1937). They went through experiments with electrical stimulation of different neocortical areas of patients undergoing open brain surgery to control epilepsy and

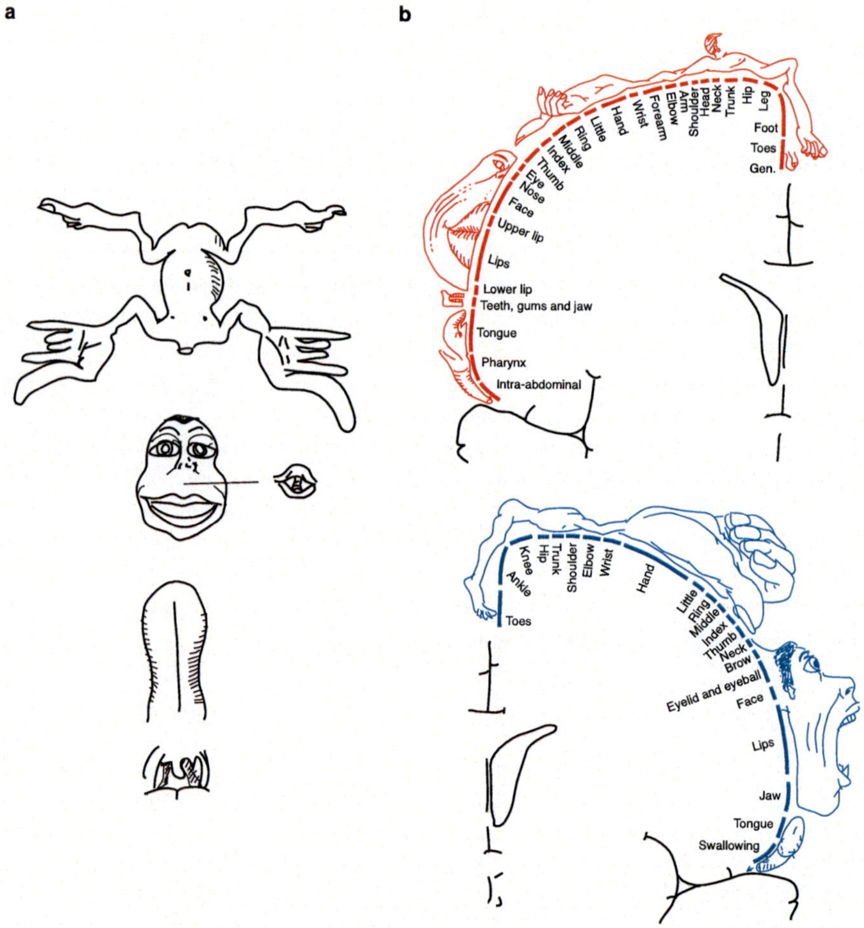

Fig. 6.1 Penfield's homunculus. (**a**) Sensory and motor homunculus. Modified from (Penfield & Boldrey, 1937). (**b**) Sensory (red) and motor (blue) homunculi. (Modified from Penfield & Rasmussen, 1950)

analyzed the sensation or movement produced by stimulation from approximately 400 operations performed from 1928 to 1946 on the Royal Victoria Hospital and later in Montreal Neurological Institute (Penfield & Rasmussen, 1950). They discovered that both sensory and motor responses were elicited primarily from the cerebral cortex adjacent to the central fissure. Approximately 75% of the points giving sensory responses were in the postcentral gyrus and remaining 25% were in the precentral gyrus. The authors concluded that the major cortical representation of somatic sensation (proprioceptive and discriminatory) is in the postcentral gyrus and that there is also a minor representation in the precentral gyrus. On the other hand, approximately 80% of these points giving motor responses were precentral in the location and 20% postcentral. Postcentral gyrus is known as the primary somatosensory cortex (S1) and precentral gyrus as the primary motor cortex (M1). To illustrate the order and comparative extent of the representation of the elements in the sensory and motor sequences, they figured out the combined sensory and motor homunculus (Fig. 6.1a), drawn by Mrs. Hortense P. Cantlie (Penfield & Boldrey, 1937). The figure was revised as a sensory homunculus (Fig. 6.1b) and a motor homunculus (Fig. 6.1c) in the later publication (Penfield & Rasmussen, 1950). These drawings have become probably the most well-known conceptual maps in modern neuroscience.

Around the same time, Clinton N. Woolsey at the Johns Hopkins University analyzed and reported detailed studies of the localization of cortical functions in many mammalian species. Woolsey and his mentor A. Philip Bard first relied on ablation to study cortical function, which involved creating cortical lesions in animals and observing which body parts lost motor function. Once Wade Marshall arrived at the Johns Hopkins Physiology department with a cathode ray oscilloscope, they used electrodes to record cerebral cortical potentials that were evoked in response to stimulus of specific body regions (Marshall et al., 1937). Woolsey's reverse strategy of stimulating the body rather than stimulating or damaging cortex yielded more precise, accurate functional mapping information of the associated sensory areas. He mapped and analyzed with the cerebral cortices of many mammals, including humans (Woolsey & Erickson, 1950). He produced the ape analog of Penfield's homunculus, referred to as a "simiusculus" (Fig. 6.2).

A cartoon summarizing cortical somatotopy, such as Penfield's homunculus or Woolsey's simiusculus, provided an appealing notion of how voluntary motor control or sensory inputs could be represented in the cerebral cortex and were reproduced in any relevant textbooks. These cartoons imply strict somatotopic representations in the primary sensory and motor cortex, but the authors greatly simplified the complexity of their experimental findings. For example, Penfield occasionally observed movements of multiple body parts, such as hand, arm, and shoulder, evoked by the stimulation of a single cortical region, but such information was ignored in the homunculus for the sake of simplicity. Penfield stated that "The motor homunculus may be used as an aid to memory in regard to movement sequence and the relative extent of cortex in which such movement finds representation. It is a cartoon of representation in which scientific accuracy is impossible" (Penfield & Rasmussen, 1950). Woolsey investigated receptive fields for individual

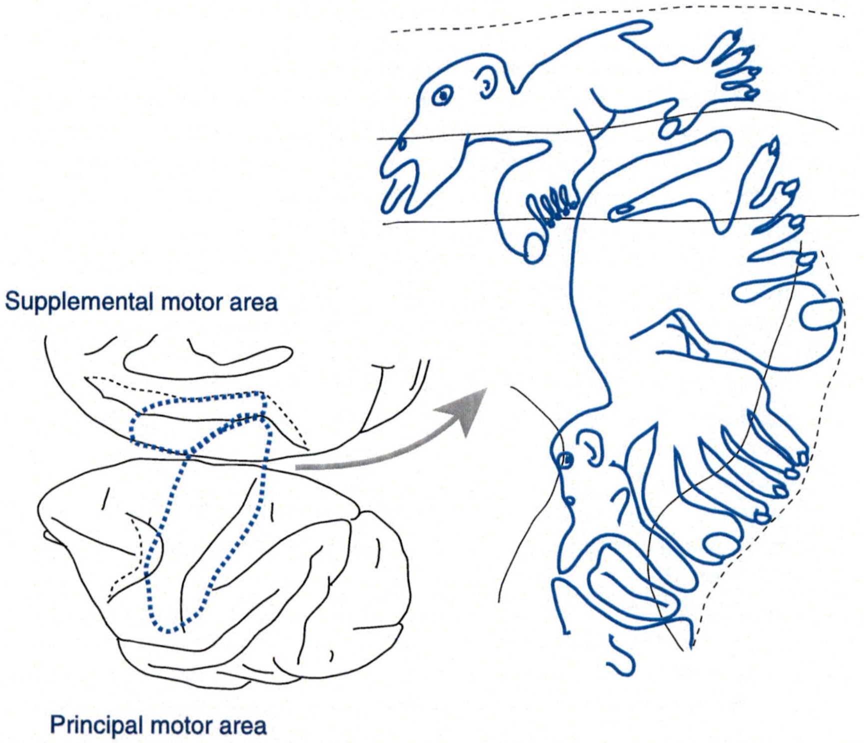

Fig. 6.2 Woolsey's simiusculus. Motor simiusculi for precentral and supplementary motor areas. (Modified from Woolsey et al., 1952)

cortical regions, but such one-to-many relationships were also excluded from the drawing of the simiusculus. Woolsey stated that "It must be emphasized, however, that this diagram is an inadequate representation of the localization pattern, since in a line drawing one can not indicate the successive overlap which is so characteristic a feature of cortical representation, not only in the motor but also in the sensory areas" (Woolsey et al., 1952). The following statement by Penfield well expresses the role of the cartoons, "[The homunculus] was one of a number of illustrations which we used to try to illustrate the truth. Of course, there is nothing like the homunculus as far as cortical representation is concerned, but it seems to be the only sort of thing that people in general understand" (Gandhoke et al., 2019).

Prior to and after the publication of the Penfield's homunculus, more realistic descriptions of the cerebral functional localization have been reported in particular in the motor cortex. In 1870, the British neurologist John Hughlings Jackson first described what are today called Jacksonian seizures. He was intrigued by the fact that partial seizures seemed to start in the hands and then move systematically up the body towards the face. Based on careful observations of such movements, Jackson was proposing a weighted distribution of function, in which each "unit" controls multiple body parts but controls some more than others in executing movements, rather than a simple one-to-one relationship between the brain region and the body part as proposed as the homuncular somatotopy (Berkowitz, 2018). In line with the Jackson's proposal, distributed and overlapped localizations of neuronal activity over the cerebral cortex accompanied with the movement of different body parts, such as digits and limbs, have been observed in the monkey motor cortex (Schieber & Hibbard, 1993) and the human premotor cortex (Willett et al., 2020). These single- and multi-unit recordings demonstrated that individual cortical neurons are responsible for the movement of multiple body parts.

In 2002, Michael Graziano and colleagues discovered that the primate motor cortex encompasses a map of ethologically relevant actions, which can be achieved via a variety of movements of multiple body parts, such as hand-to-mouth actions, defensive actions, reaching actions, climbing and leaping postures, hand-in-lower-space actions, manipulation in central space, and chewing and licking movements (Graziano et al., 2002). They applied a train of electrical pulses with a behaviorally relevant time scale, i.e., 500 ms, which was traditionally used for analyzing eye movements, perceptual decision-making, and controlling motivated states (Graziano, 2016). However, such elongated stimulations have not typically been used for mapping functional localization in the motor cortex, since researchers have explicitly (or implicitly) assumed that the motor cortex operates through a descending pathway from cortex to muscles in accordance with the homuncular somatotopy hypothesis. Hence, brief pulses of electrical stimulation (no more than 50 ms) have typically been used in order to evoke muscle twitches. After the discovery in the monkey, ethological action maps were observed in the motor cortex of various species, including rats, mice, cats, and humans (Graziano, 2016). The action map may seem incompatible with the classical homuncular somatotopy. However, as described above, motor cortical areas have been shown to contain overlapped functions, such as being responsible for the movements of multiple body parts. Therefore, the possibility of the coexistence of two maps in the motor cortex has been discussed (Graziano, 2016). Further studies will be necessary for solving the exact relationship between the body map (Penfield & Rasmussen, 1950; Woolsey & Erickson, 1950) and the action map (Graziano, 2016) in the motor cortex.

6.3 The Somatotopy in the Cerebellum

So far, as we describe above, somatotopy in the cerebral cortex is still under debate. The researches conducted for exploring the somatotopic organization in the cerebellum will be briefly summarized in this section.

Cerebellar Purkinje cells are the sole output neurons in the cerebellar cortex that receives two types of excitatory inputs: climbing fibers and mossy fibers relayed by granule cells (Eccles et al., 1967). Climbing fibers arise exclusively from the inferior olive, which is anatomically divided into several subdivisions (Brodal & Kawamura, 1980; Kooy, 1917). The topographical distribution of climbing (olivocerebellar) fibers in the contralateral cortex from each of the inferior olive subdivisions was characterized as a parasagittally elongated "zone" (Brodal & Kawamura, 1980; Voogd et al., 2013) (see below). The single input of a climbing fiber induces a complex spike (Thach, 1967), a short burst of impulses, in a postsynaptic Purkinje cell. On the other hand, a Purkinje cell receives inputs from granule cells and exhibits changes in the firing rate of simple spikes (Eccles et al., 1967).

Sensations arising from the skin, such as the sense of touch, pressure, cold, heat, and pain, and from the muscles, tendons, and joints, such as the sense of the position and movement of limbs and trunk, effort, force, and heaviness, are known as somatic sensations (Smith, 2008). The former is mediated by the tactile system and the latter the proprioceptive system. These multiple types of information are transmitted to the cerebrum and cerebellum separately via different ascending paths in the spinal cord. Both climbing and mossy fibers transmit sensory signals from peripheral tissues through spino-olivocerebellar and spinocerebellar pathways, respectively (Oscarsson, 1965, 1980), but afferent somatotopy, the point-to-point correspondence of an area of the body to a specific point on the cerebellar cortex, has not yet been well established (Manni & Petrosini, 2004).

6.3.1 Snider's Cerebellar Homunculi

In 1944, 7 years after the publication of the Penfield's homunculus (Penfield & Boldrey, 1937), Ray S. Snider and Averill Stowell at the Johns Hopkins University published a paper describing the cerebellar afferent somatotopy representing separate bodily parts in the cat and monkey cerebellum (Snider & Stowell, 1944). They recorded potentials from the surface of the cerebellar cortex evoked by tactile stimulations to arm, leg, or vibrissae under the anesthesia with barbiturate, chloralosan, or a mixture of various anesthetics. They started this research project in 1941 (Snider, 1950). At that time, researchers believed that the cerebellum receives only proprioceptive inputs, as Sir Charles S. Sherrington stated that the cerebellum as "the head ganglion of proprioceptive system" (Sherrington, 1920). The authors realized that clinicians had long hold that certain tactile functions passed through the gracile and cuneate nuclei, which are connected with the cerebellum by arcuate fibers. Hence, they investigated the existence of tactile projections to the cerebellum and actually discovered it (Snider & Stowell, 1944). In addition, they found that when the intensity of the stimulation was carefully controlled, localized responses were observed in the cerebellum (Fig. 6.3a). They found three responsive areas for stimulations of hairs around pads of left forelimb: an anterior area centered at the posterolateral corner of the ipsilateral anterior lobe and the adjacent parts of crus I and two posterior areas, one in each paramedian lobule (Fig. 6.3b). Two responding areas were identified for stimulations of hairs around pads of left hindlimb: one situated ipsilaterally in the anterior lobe covers the lateral borders of the fifth, seventh, and eighth folia rostral to the primary fissure; the other located in the ipsilateral paramedian lobule and the two most medial folia of crus II (Fig. 6.3b). Single area

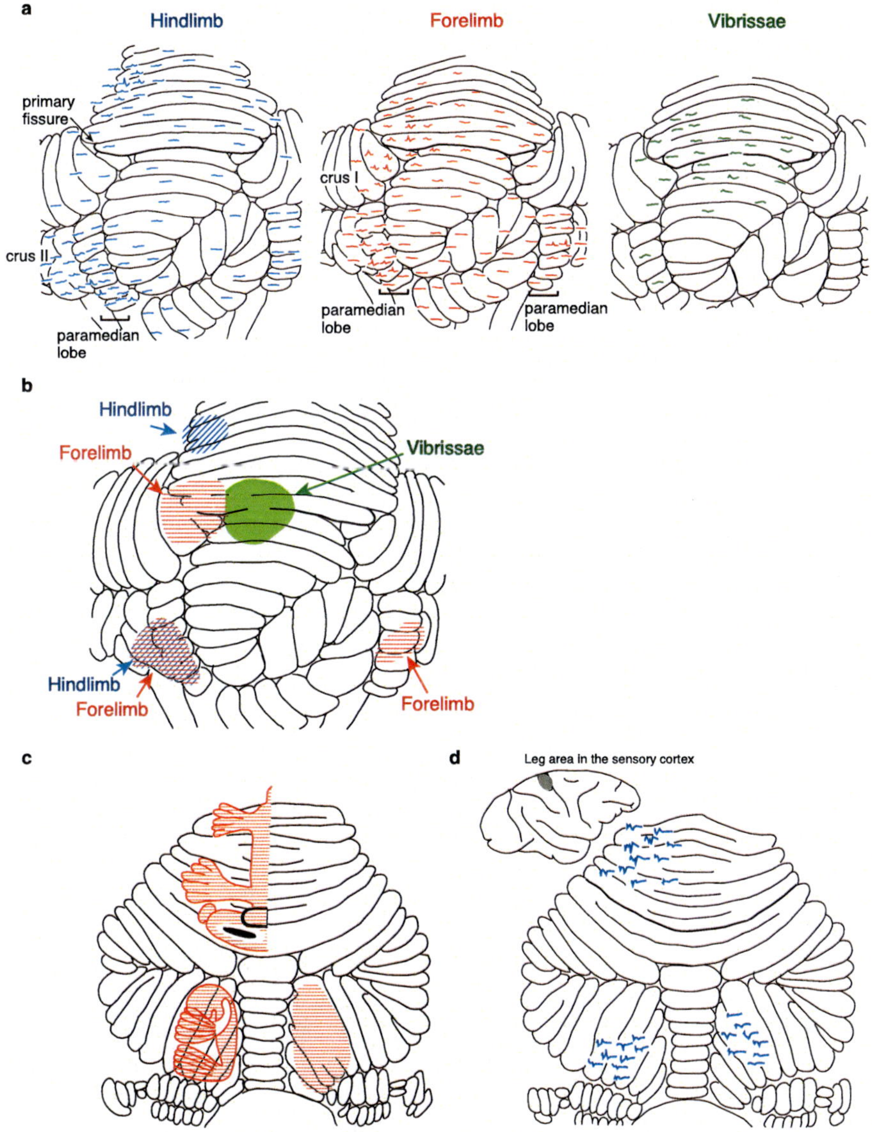

Fig. 6.3 Snider's cerebellar homunculus. (**a**) Cerebellar surface potentials evoked by tactile stimulation to hindlimb, forelimb and vibrissae. Recordings were made from the cat cerebellum under pentobarbital- or chloralosan-anesthesia. Modified from (Snider & Stowell, 1944). (**b**) Summary of receiving areas identified by tactile stimulations. Modified from (Snider & Stowell, 1944). (**c**) Cerebellar homuncular somatotopic maps. Modified from (Snider & Eldred, 1952). (**d**) Cerebellar surface potentials evoked by electrical stimulations to the cerebral sensory area. (Modified from Snider & Eldred, 1952)

was identified for the stimulation of vibrissae in left side: the area extends on both sides of the primary fissure (Fig. 6.3b). Six years after his publication, Snider looked back that "During the era of 1930 to 1940 there was a growing tendency to question of the unitarian concept of the cerebellar function" (Snider, 1950). The unitarian concept, "the cerebellum functions as a whole," was proposed by a French physiologist Jean Pierre Flourens in the early nineteenth century (Flourens, 1960). Snider stated that "... does it answer definitely and finally the long-standing question of whether or not there is localization of function within the cerebellum, ..." (Snider, 1950).

During an experiment in which the cerebellar tactile area of the face was being mapped, they noticed that "click" sound generated by the stimulating equipment evoked surface responses in the vermis of posterior lobe (Snider, 1950). Lesions in the inferior colliculus abolished the responses, suggesting that auditory signals elicited cerebellar responses. Then, they searched cerebellar cortical areas for other sensory modalities, such as gustatory, olfactory, and visual areas, and found the existence of the visual projection to the area which is coextensive with the auditory area. The visual responses occurred in the absence of extraocular muscles and cerebral cortex but were abolished by the lesions of superior colliculus. Based on these results, the authors concluded that projections of certain sensory systems, such as tactile, visual, and auditory projections, were localized within the cerebellar cortex (Snider & Stowell, 1944).

In his publication in 1944 (Snider & Stowell, 1944), Snider cited the Woolsey's work showing the representation of cutaneous tactile sensibility in the monkey cerebral cortex (Woolsey et al., 1942). He next examined the interrelationship between the newly discovered tactile, visual, and auditory areas in the cerebellum and those in the cerebral cortex in the monkey and the cat. Snider and Earl Eldred recorded cerebellar responses evoked by the electrical stimulation in various cerebral cortical areas, such as sensory, motor, visual, and auditory areas. They described that responsive areas in the cerebellar cortex identified by the stimulation of motor and sensory areas in the cerebral cortex were consistent with the cerebellar tactile areas identified in their previous work (Snider & Eldred, 1952). Similar interrelationship was reported by Edgar D. Adrian at the Cambridge University in 1943 (Adrian, 1943). On the basis of these results, they summarized their results by depicting homunculi on the cerebellar cortex (Fig. 6.3c) in this research paper (Snider & Eldred, 1952) and the preceding review article (Snider, 1950). Such an impressive caricature has been illustrated in textbooks of neuroscience (Purves et al., 2018; Purves et al., 2007) and articles reviewing cerebellar somatotopy (Guell et al., 2018; Iorio-Morin & Mathieu, 2020; Manni & Petrosini, 2004), providing non-negligible influence in the field. Nevertheless, Snider and Eldred's paper perplexes the reader as they failed to show most of the data in which the neocortical sensory and motor areas were stimulated. They showed cathode-ray traces obtained by stimulating only the somatic leg area (Fig. 6.3d), even though the cerebellar homunculus possesses the leg, arm, and face (Fig. 6.3c). They described that "If the cerebral motor areas are electrically excited, a type of projection pattern can be outlined upon the various cerebellar lobules which is almost identical to the one just described for the somatic systems," but readers cannot confirm the underlying data at all. In addition, there was the inconsistency between the cerebellar cortical homunculi (Fig. 6.3c) and the tactile areas in the cerebellar cortex identified in their previous work (Fig. 6.3b). They portrayed an inverted homunculus in the anterior lobe (Fig. 6.3c), but the arm and head (vibrissae) areas were not aligned sagittally when the responses evoked by the tactile stimulus were recorded (Fig. 6.3b). An upright homunculus was depicted at the ipsilateral paramedian lobule (Fig. 6.3c), but the arm and leg areas were almost entirely coextensive and the head area was not detected (Fig. 6.3b). Lastly, at the time the distinction between simple spikes and complex spikes was not made, and so it was unclear whether homuncular somatotopy was suggested for mossy fiber or climbing fiber inputs.

In 1954, C. Murphy Combs at the Northwestern University published a paper describing cerebellar localization of afferent inputs (Combs, 1954). It was demonstrated that somatotopical localization was not detected in the decerebrated unanesthetized cats by electrical stimulation of peripheral nerves or tactile stimulations (Fig. 6.4a). These results were well-consistent with those obtained in previous studies of the mapping both spinocerebellar (Anderson, 1943; Beck, 1927; Horrax, 1915) and spino-olivocerebellar (Brodal et al., 1950; Dow, 1942; Dow & Anderson, 1942) connections. It was found, interestingly, that when the nembutal was injected into decerebrated cats, spatially localized surface responses were observed in the same cats (Fig. 6.4b), as demonstrated by Snider and Stowell (Snider & Stowell,

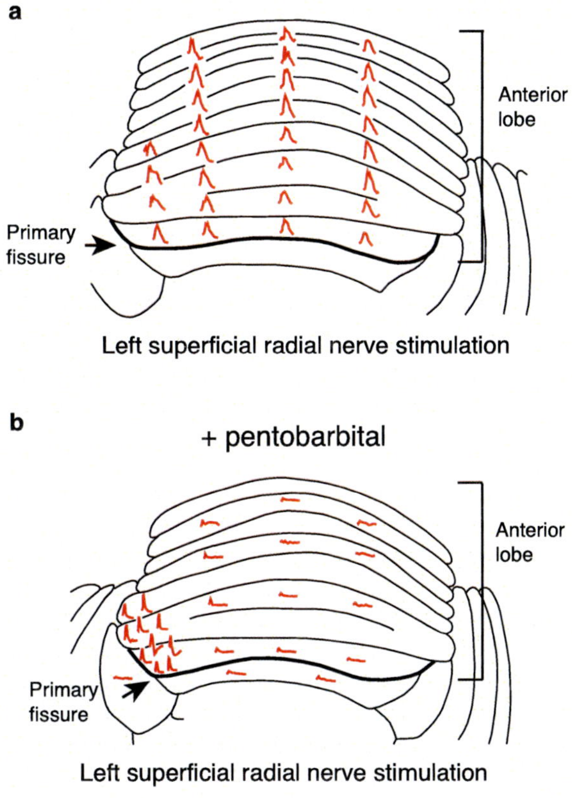

Fig. 6.4 The absence of somatotopy in the cerebellum in decerebrated unanesthetized cats. (**a**) Surface responses recorded in decerebrated unanesthetized cats. (**b**) Localized responses were observed only under pentobarbital anesthesia. (Modified from Combs, 1954)

1944). Combs speculated that there are two types of afferent inputs to the cerebellar cortex, one is distributed over the cortex and the other is localized, and that the nembutal specifically inhibits the former type of inputs. Combs described in his 1954 paper that "The authors would like to express his appreciation to Dr. Ray S. Snider for the advice and counsel which he so freely gave throughout the course of this work", suggesting that Snider was well aware of Combs' work at the time. Nevertheless, Snider published an article for general readers in 1958 with a revised version of cerebellar homunculi without any description of the effect of anesthesia (Snider, 1958). After reviewing the arguments both for and against the cerebellar somatotopy in their book published in 1958, Robert S. Dow and Giuseppe Moruzzi concluded that "the somatotopic arrangement is by far less pronounced on the cerebellar than on the cerebral cortex" (Dow & Moruzzi, 1958).

6.3.2 Oscarsson's Longitudinal Somatotopy

Olov Oscarsson at the University of Lund conducted a series of experiments to identify and characterize the pathways from spinal cord to cerebellum in cat since the mid 1950s initially with Anders Lundberg (University of Lund) and John C. Eccles (Australian National University). Oscarsson and his colleagues analyzed both spinocerebellar (Oscarsson, 1965) and spino-olivocerebellar pathways (Oscarsson, 1980). One of the distinctive techniques they used was the transection of the spinal cord except for a certain funiculus to analyze the spino-olivocerebellar pathways individually. Oscarsson reported in 1968 a "longitudinal" somatotopic organization of climbing fiber projections into the cerebellar anterior lobe via ventral funiculus (Fig. 6.5) (Oscarsson, 1968), which was marked in contrast with the previous "transverse" somatotopic organization of afferent projections shown by Adrian (Fig. 6.5) (Adrian, 1943), Snider (Fig. 6.3c) (Snider & Stowell, 1944), and others (Carrea & Grundfest, 1954; Combs, 1954; Grant, 1962; Provini et al., 1967). Indeed, in this study, Oscarsson analyzed climbing fiber responses by recording from single Purkinje cells and recording mass activity at the cerebellar surface or at the molecular layer. It was shown at that time that climbing fibers conveyed surface responses recorded from the ipsilateral side of the anterior lobe of the cat cerebellum following electrical stimulation of hindlimb afferents (Armstrong & Harvey, 1968). Oscarsson considered that Adrian recorded mossy fiber responses via the dorsal spinocerebellar tract (DSCT) and the cuneocerebellar tract (CCT) , rather than the climbing fiber responses, since recorded potentials showed short latency (Oscarsson, 1968). In contrast, Oscarsson considered that the transverse organization described by Snider and Stowell (Snider & Stowell, 1944) and by Combs (Combs, 1954) represented climbing fiber projections via dorsal- and dorsolateral-spino-olivocerebellar pathways based on the long latency surface potentials recorded by these authors (Oscarsson, 1968), consistent with the interpretation made by Eccles (Eccles et al., 1967) (see below). Oscarsson unfortunately did not make further discussion of the discrepancy between longitudinal and transverse somatotopic organizations proposed for climbing fiber projection (Oscarsson, 1968).

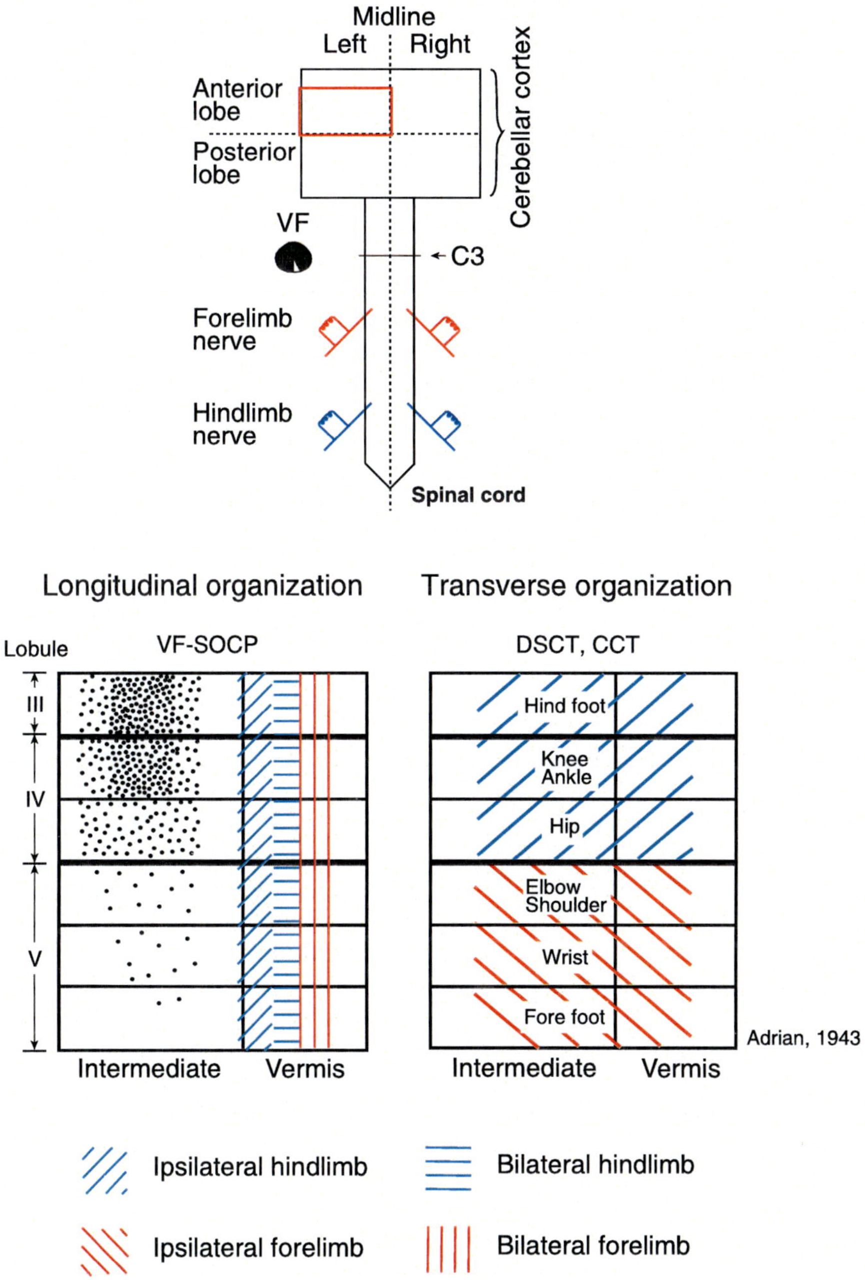

Fig. 6.5 Longitudinal versus transverse somatotopic organization in the anterior lobe of the cerebellum. (**a**) Projection areas of the spino-olivocerebellar path ascending through the ventral funiculus (VF-SOCP). The projection to the vermis in sagittal bands is indicated by hatching and the long-latency, presumably indirect, projection to the intermediate part is indicated by stippling. Density of dots shows the usual amplitude distribution of the positive potentials, which are evoked exclusively from ipsilateral nerves. (**b**) Transverse organization as given by the termination areas of the dorsal spinocerebellar tract (DSCT) and the cuneocerebellar tract (CCT). The detailed transverse organization described by Adrian is indicated. (**a, b**) The diagrams represent right halves of lobules V and IV and part of lobule III. Interrupted vertical line indicates mid line. (Modified from Oscarsson, 1968)

Concurrently, Jan Voogd at the State University of Leiden showed longitudinal zonal projections of climbing fibers based anatomically (Fig. 6.6) (Voogd, 2011). In 1969, Oscarsson and Voogd first met in a meeting organized in Chicago, and they immediately recognized that they studied an identical system (Voogd, 2011). After that, Oscarsson and his colleagues studied the olivocerebellar projections using electrophysiological methods based on the anatomically defined cerebellar cortical zones. A typical analysis which they carried out, for example, is shown in Fig. 6.7. The electrical stimulation was applied for the ipsilateral or the contralateral forelimb or hindlimb nerves (Fig. 6.7a). Surface potentials along the transverse axis of each folia of the anterior lobe (lobules III, IV and V) have been recorded (Fig. 6.7b). For each experiment, the maximal amplitudes of the evoked potentials were plotted against the mediolateral distance from the midline by ignoring negative potentials (Fig. 6.7c). Oscarsson stated that "Negative potentials were recorded from extensive areas and presumably represent electrical spread of fields generated in the projection areas." (Oscarsson, 1968). They counted the number of animals showing positive potentials for each zone and produced a map showing the somatotopical organization of these longitudinal zones (Fig. 6.7d). The cerebellar cortical zones were determined in accordance with the definition Voogd's scheme. In cases where positive potentials have been observed in a particular zone in all animals, the corresponding zone has been indicated by the hatching. However, when the positive responses were observed only in some animals, the sparse hatching was used. The hatching density in their maps therefore represented the animal-to-animal variability in the projection of climbing fibers.

Oscarsson and his colleagues analyzed the ventral (Oscarsson, 1968; Oscarsson & Sjolund, 1977), dorsal (Ekerot & Larson, 1979a; Oscarsson, 1969), dorsolateral (Larson et al., 1969b), and lateral (Larson et al., 1969a) spino-olivocerebellar pathways. In the latter versions of the somatotopic maps published in the 1970s, they distinguished direct pathways (hatching) and indirect pathways (stippling) depending on the latency of the evoked surface potentials (Fig. 6.8), instead of anima-to-animal variability. They considered that inputs from hindlimb were represented in the medial parasagittal zones of the anterior lobe, while inputs from forelimb were represented in the lateral parasagittal zones in both the ventral and dorsal spino-olivocerebellar pathways. They stated that "the hindlimb and forelimb areas in the anterior lobe are separated by an oblique line rather than by a transverse one as assumed before" (Ekerot & Larson, 1979a).

6.3.3 Eccles's Patchy Distribution of Climbing Fiber Projections

The longitudinal olivocerebellar somatotopy has been criticized by several researchers (Bloedel, 1973; Eccles, 1970; Llinás & Simpson, 1981), although the parasagittal arrangement of climbing fiber inputs has been observed by others as well (Armstrong et al., 1974; VanGilder & O'Leary, 1970). Eccles and his colleagues

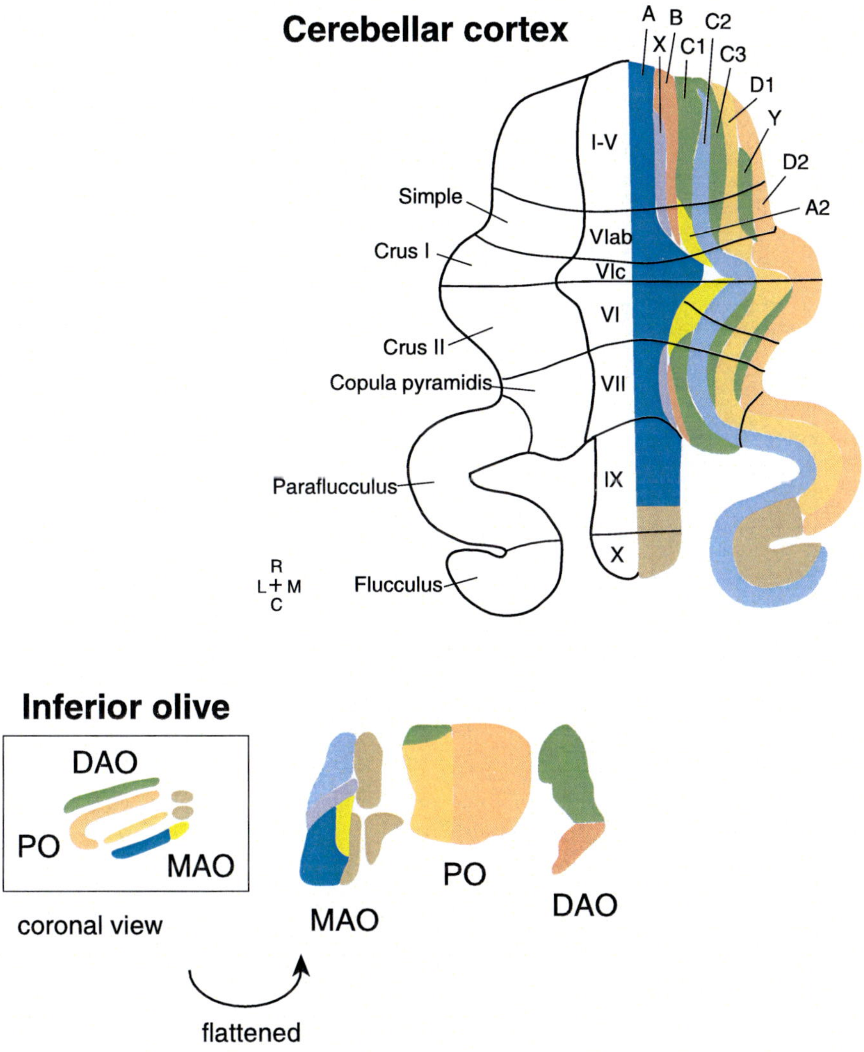

Fig. 6.6 Parasagittal zones in the cerebellum. Parasagittal zones (A, A2, B, C1, C2, C3, D1, D2, X, and Y) were defined according to the anatomical arrangements of climbing fibers and Purkinje cell axons (Voogd et al., 2013). DAO dorsal accessory olive, MAO medial accessory olive, PO principal olive. (Modified from Voogd et al., 2013)

evaluated the climbing fiber projections by applying arrays of microelectrode tracks ordered in an oblique transverse plane, penetrating to the right through the entire depth of the anterior lobe of the cat cerebellum (Eccles et al., 1968b). Beforehand, they analyzed electrical responses evoked by mossy fibers and climbing fibers at varying depths in the molecular layer of the cat cerebellar cortex (Eccles et al., 1968a). Based on these analyses, Eccles interpreted that the surface response

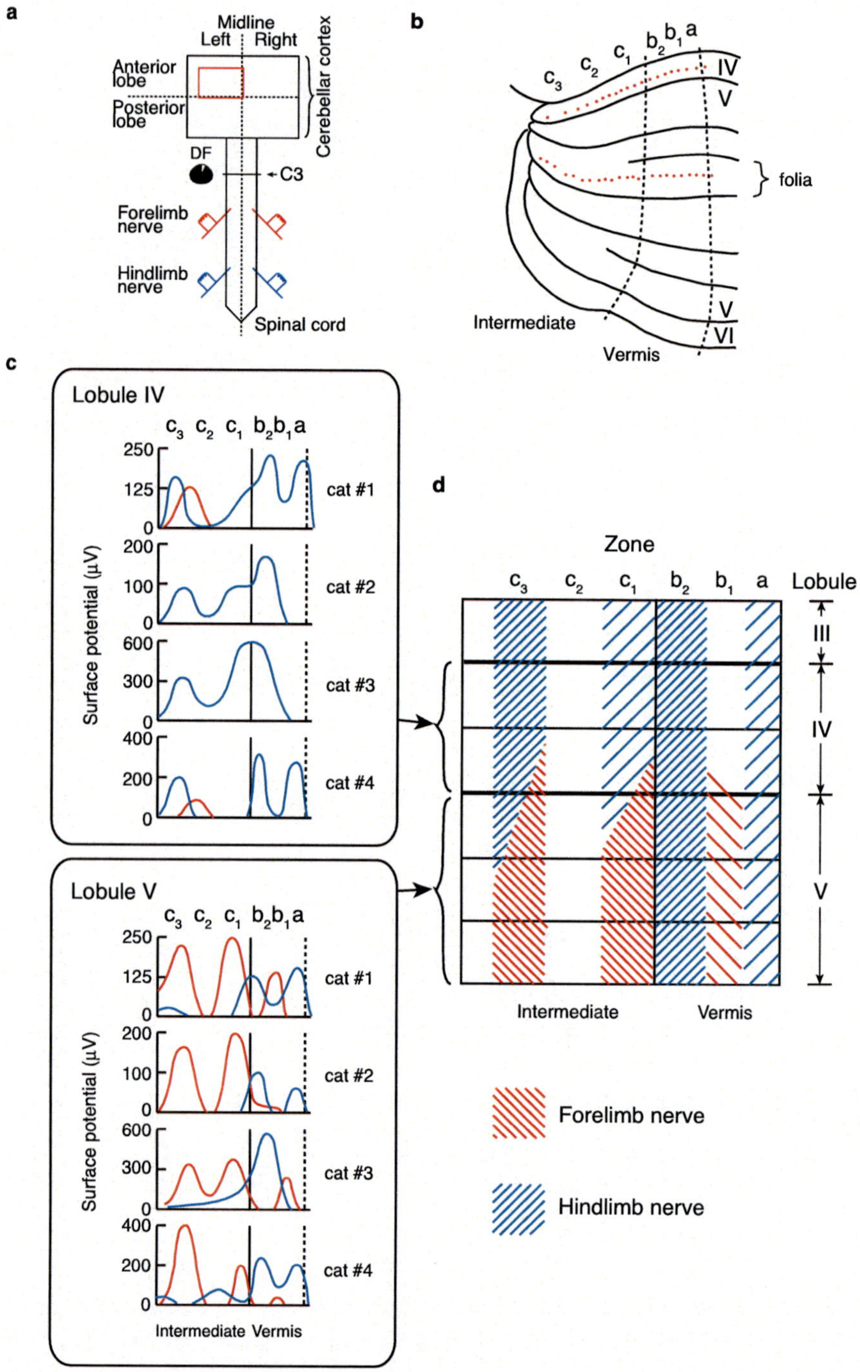

Fig. 6.7 (**a–d**) Cerebellar cortical mapping. Maximal amplitudes of surface potentials evoked by the stimulation of deep radial or hamstring in folia of lobule V and IV are plotted against the mediolateral distance from the midline of the recording position. The results of four different experiments are shown. Positive potentials were found in the sagittal zones indicated by hatching. Sparse hatching indicates that the zones were found in some experiments only. (Modified from Oscarsson, 1969)

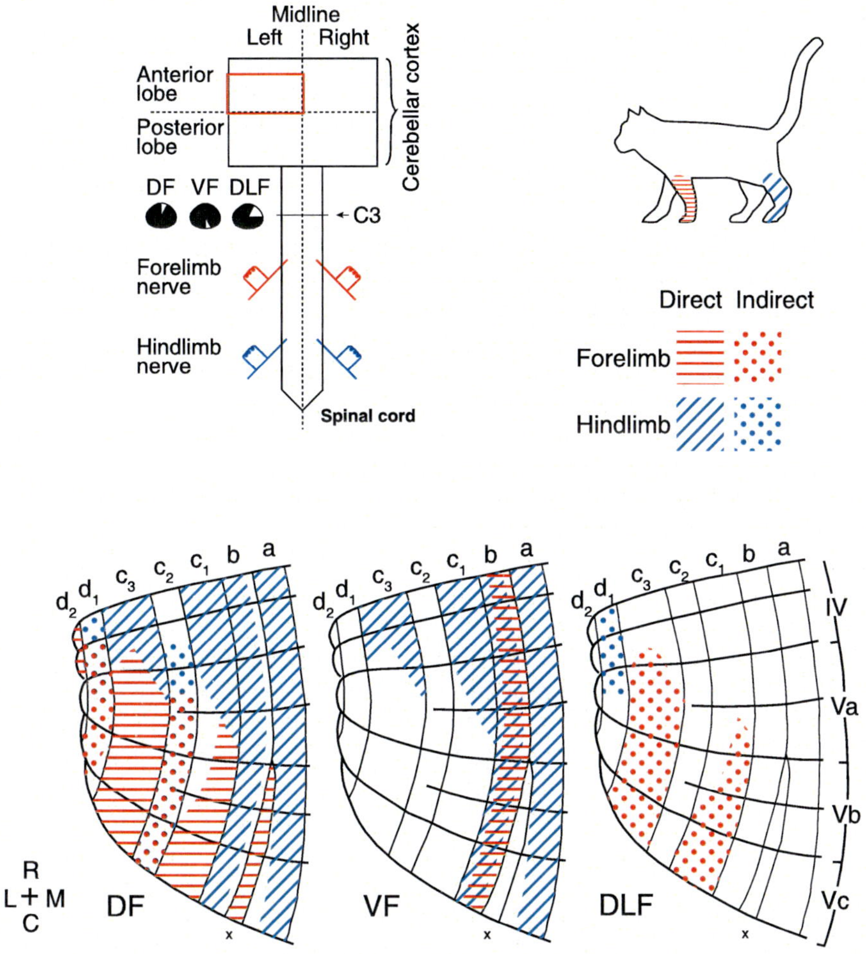

Fig. 6.8 Somatotopic organization of the termination zones of spino-olivocerebellar pathways. The parts of lobules IV and V in the left half of the cerebellar cortex are shown. The projection from the dorsal funiculus to c2 zone and the projection from the ventral funiculus to b zone are bilateral, whereas all other projections are ipsilateral. Surface potentials evoked by electrical stimuli of peripheral nerves were recorded in pentobarbital- or alpha chloralose-anesthetized cats with the spinal cord transected at the C3 level with examined funiculus maintained: dorsal (DF), ventral (VF) or dorsolateral (DLF) funiculus. (Modified from Ekerot & Larson, 1979b)

recorded by Snider and Stowell (Snider & Stowell, 1944) corresponded to the climbing fiber responses, rather than the mossy fiber responses (Eccles et al., 1967; Eccles et al., 1968b). Eccles used extracellular negative potential fields in the molecular layers for evaluating the distribution of the climbing fiber in his study, since negative potentials were generated by climbing fiber synaptic actions in immediate proximity, while positive potentials were generated at a distance. Eccles found that there is a sharp forelimb-hindlimb separation for the climbing fiber-evoked responses in the pars intermedia of the anterior lobe with a transverse delimitation usually by the fissure between the fifth and fourth lobules. In the vermis there is much more admixture, with the hindlimb-evoked responses tending to dominate in the lateral vermis of the fifth lobule, and the forelimb more medially. More detailed examination of climbing fiber responses evoked by a large variety of hindlimb and forelimb nerves demonstrated that there is a fine grain of climbing fiber inputs in the anterior lobe. Every grain (recording site) received different combinations of effective climbing fiber inputs from a variety of limb nerves. Eccles described the property of climbing fiber distribution as an ill-posed "patchy" character (Eccles et al., 1968b) that was markedly different from parasagittal strips reported by Oscarsson. Since patches were unordered somatotopically on the cerebellar cortex, they were also called mosaic. The patchy or mosaic distribution of climbing fiber projections was observed by other researchers (Eccles et al., 1972; Hiss et al., 1977; Leicht et al., 1973; Miles & Wiesendanger, 1975; Robertson & Laxer, 1981; Robertson et al., 1982; Rubia & Tandler, 1981; Rushmer et al., 1980).

Eccles discussed the factors responsible for the discrepancy between the Oscarsson's longitudinal somatotopy and the patchy distribution observed by himself (Eccles et al., 1968b). One of the factors he pointed out was the effect of the spinal cord transection applied by Oscarsson since Eccles used animals with intact spinal cord. On this issue, his colleague, Robert F. Schmidt, further investigated complex spike responses evoked by hindlimb and forelimb stimulations using single-unit recordings from unanesthetized cats with intact neuraxis (Leicht & Schmidt, 1977). Schmidt showed that responsive Purkinje cells were distributed throughout the anterior lobe irrespective of the stimulated limb and that no somatotopic organization of the climbing fiber projections was observed in unanesthetized cats (Fig. 6.9) (Leicht & Schmidt, 1977), consistent with previous studies using animals with intact neuraxis (Allen et al., 1974; Eccles et al., 1968b; Eccles et al., 1972; Ishikawa et al., 1972; Leicht et al., 1973; Provini et al., 1968). Carl-Frederik Ekerot, who used to work with Oscarsson, on the other hand, described their interpretation for the possible cause for the discrepancy that "The failure of detect the sagittal organization of the projection areas in these studies is presumably largely explained by the mapping technique used. Recording were obtained from caudorostrally oriented tracks which would cut obliquely across the sagittal zones. It should be emphasized that the orientation of the zones is not strictly in the sagittal plane but rather in the plane orthogonal to the folia and that medio-lateral shifts may occur between folia separated by deep fissure" (Ekerot & Larson, 1979b).

In the last chapter of his book entitled *The Cerebellum as a Neuronal Machine* published in 1967, Eccles described the possibility of the function of the cerebellum as a computer on the basis of the known input and output of the cerebellum citing the works done by Oscarsson (Eccles et al., 1967). He analyzed the afferent and efferent connections of the cerebellum by himself to construct models for the mode of operation of the neuronal machinery of the cerebellar cortex and discovered "patchy" distribution of climbing fiber projections to the cerebellar cortex. However,

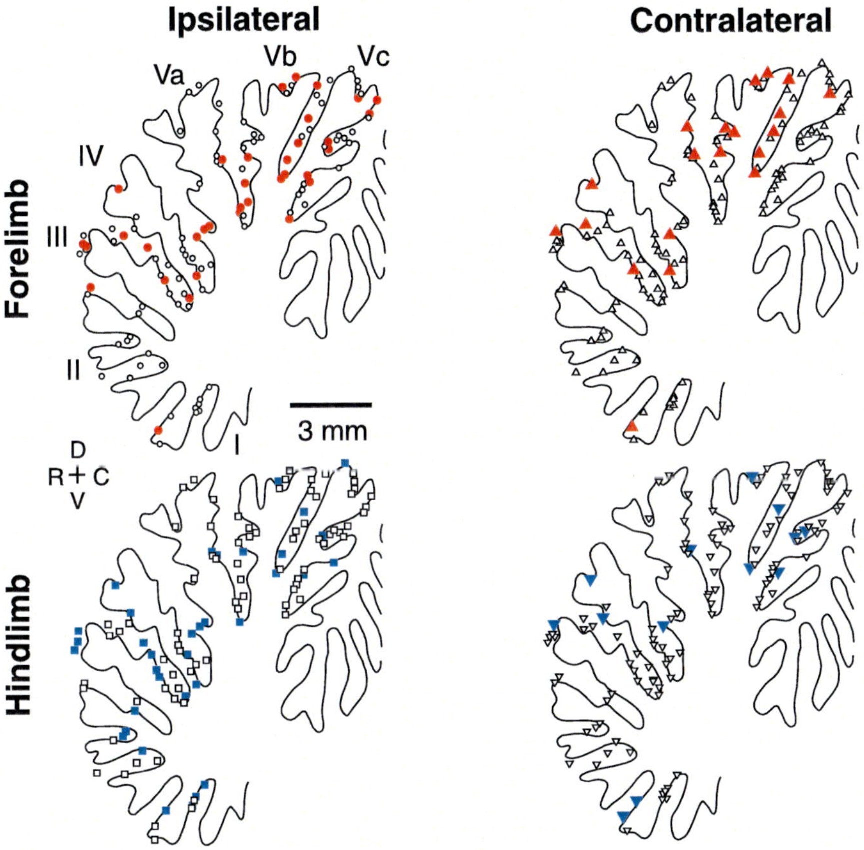

Fig. 6.9 The absence of somatotopy in unanesthetized cats with intact neuraxis. Distribution of Purkinje cells in the cerebellar anterior lobe with (filled symbols) or without (open symbols) the climbing fiber inputs from cutaneous mechanoreceptors of the ipsilateral forelimb (circle), ipsilateral hindlimb (square), contralateral forelimb (triangle), and contralateral hindlimb (inverted triangle) (left). (Modified from Leicht & Schmidt, 1977)

the argument concerning the olivocerebellar somatotopy stopped gradually after the retirement of Eccles in 1975 (Eccles, 1977). In the case where there is no somatotopic projection on the cerebellar cortex as reported by Eccles and others, it is still unclear how the sensory information is encoded in the olivocerebellar system.

Existence of the somatotopic organization of climbing fiber inputs is still controversial, even though the discussion is limited to the anterior lobe of the cat cerebellum, since the proposed models, i.e., the transverse somatotopy along lobules (homunculi), the longitudinal somatotopy along parasagittal zones, and the nonsomatotopic arrangement of ill-posed patches, are contradictory.

6.4 Large-Scale Optical Measurements of Cerebellar Complex Spikes

Considering that surface potentials reflect the mass activity of cortical neurons, this can be used for exploring large areas, but spatial resolution is not sufficient for the fine-grained mapping of climbing fiber projections. Single-unit recording of complex spikes of individual Purkinje cells holds an ideal spatial resolution for mapping, but the analysis of wide areas with dense sampling is definitely difficult. The comprehensive understanding of olivocerebellar somatotopy is waiting for a new method for wide mapping of climbing fiber projections with high spatiotemporal resolution. Recently, we introduced a method for real-time in vivo optical measurements of complex spike activity from more than 10,000 Purkinje cells at the same time (Michikawa et al., 2020). For this purpose, we used a genetically encoded Ca^{2+} sensor based on fluorescent resonance energy transfer (FRET), yellow cameleon 2.60 (YC2.60) (Miyawaki et al., 1997; Nagai et al., 2004). YC2.60 can be selectively expressed in cerebellar Purkinje cells using adenovirus expression system (Horikawa et al., 2010; Yamada et al., 2011) or crossing YC2.60 transgenic mice with L7-Cre mice (Kuroki et al., 2018; Michikawa et al., 2020). Simultaneous measurements of dendritic signals of YC2.60 and extracellular single-unit records to confirm that every increase in the FRET signals exclusively reflected occurrence of a complex spike (Michikawa et al., 2020).

Indeed, our large-scale measurements uncovered a contiguous transition of ever-changing patterns of both local and global complex spike synchrony over the cerebellar cortex. In order to analyze the complex patterns of spontaneous activity, we employed principal component analysis and spatiotemporal independent component analysis to isolate clusters of Purkinje cell that exhibit synchronous complex spikes. Applying this method, we found that the dorsal surface of the mouse cerebellar cortex is divided into approximately 200 segments (Michikawa et al., 2020).

We investigated the representation pattern of climbing fiber inputs to the cerebellar cortex. In order to do this, we visualized complex spike responses elicited by the application of a weak and brief electrical pulse on the flexor muscle of ipsilateral and contralateral forelimb and hindlimb. Results show that a single muscle stimulus elicited complex spike responses which were distributed widely over the cortex almost in a symmetrical way. It was found that electrical stimulations of four different limb muscles generated almost identical patterns of evoked complex spike activity over the cerebellar cortex and that each olivocerebellar segment received inputs largely from all four limbs. The Bayesian inference has shown that the timing and location of stimulations can be encoded from the ensemble activity of whole olivocerebellar segments. Those findings suggest that climbing fiber inputs originated from a limb muscle are distributed over the cerebellar cortex and that different limb signals are represented in almost all of the olivocerebellar segments investigated (Michikawa et al., 2020).

6.5 Distributed Sensory Representation in Unit of Olivocerebellar Segments

Results of the large-scale optical measurements suggest that ensemble patterns of complex spike activity of olivocerebellar segments, rather than the cortical position or identity of responsive Purkinje cells, constitute moment-by-moment sensory signals. Hence, we propose distributed and highly overlapped organization of climbing fiber projections in units of olivocerebellar segments, which is consistent with the previous proposals of non-somatotopic organization (Brodal et al., 1950; Combs, 1954; Dow, 1942; Dow & Anderson, 1942; Eccles et al., 1968b; Leicht & Schmidt, 1977), stemmed from the pioneering ideas of Flourens (Flourens, 1960) and Jackson (Berkowitz, 2018).

Every movement needs coordination of multiple body parts. The measurements of our system revealed that each olivocerebellar segment receives inputs from multiple limb muscles, suggesting that convergence of sensory information takes place within a single olivocerebellar segment. By this critical finding, we will provide a clue for analyzing the role of the cerebellum especially in motor coordination. Elucidation of the mechanism underlying decoding of sensory information represented by a population of complex spikes in unit of the olivocerebellar segment is a next challenge for better understanding of the operational principle of the neuronal machinery of the cerebellar cortex.

Acknowledgments This work was supported by grants from the Japan Ministry of Education, Culture, Sports, Science, and Technology Grant-in-Aid for Scientific Research on Innovative Areas: Resonance Bio (15H05948).

References

Adrian, E. D. (1943). Afferent areas in the cerebellum connected with the limbs. *Brain, 66*, 289–315.

Allen, G. I., Azzena, G. B., & Ohno, T. (1974). Somatotopically organized inputs from fore- and hindlimb areas of sensorimotor cortex to cerebellar Purkyne cells. *Experimental Brain Research, 20*, 255–272.

Anderson, R. F. (1943). Cerebellar distribution of the dorsal and ventral spino-cerebellar tracts. *The Journal of Comparative Neurology, 79*, 415–423.

Armstrong, D. M., & Harvey, R. J. (1968). Responses to a spino-olivo-cerebellar pathway in the cat. *The Journal of Physiology, 194*, 147–168.

Armstrong, D. M., Harvey, R. J., & Schild, R. F. (1974). Topographical localization in the olivo-cerebellar projection: An electrophysiological study in the cat. *The Journal of Comparative Neurology, 154*, 287–302.

Beck, G. M. (1927). The cerebellar terminations of the spinocerebellar fibers of the lower limbar and sacral segments of the cat. *Brain, 50*, 60–98.

Berkowitz, A. (2018). You can observe a lot by watching: Hughlings Jackson's underappreciated and prescient ideas about brain control of movement. *The Neuroscientist, 24*, 448–455.

Bloedel, J. R. (1973). Cerebellar afferent systems: A review. *Progress in Neurobiology, 2*, 3–68.

Brodal, A., & Kawamura, K. (1980). Olivocerebellar projection: A review. *Advances in Anatomy, Embryology and Cell Biology, 64*(IVIII), 1–140.
Brodal, A., Walberg, F., & Blackstad, T. (1950). Termination of spinal afferents to inferior olive in cat. *Journal of Neurophysiology, 13*, 431–454.
Carrea, R. M., & Grundfest, H. (1954). Electrophysiological studies of cerebellar inflow. I. Origin, conduction and termination of ventral spino-cerebellar tract in monkey and cat. *Journal of Neurophysiology, 17*, 208–238.
Combs, C. M. (1954). Electro-anatomical study of cerebellar localization; stimulation of various afferents. *Journal of Neurophysiology, 17*, 123–143.
Dow, R. S. (1942). Cerebellar action potentials in response to stimulation of the cerebral cortex in monkeys and cats. *Journal of Neurophysiology, 5*, 122–136.
Dow, R. S., & Anderson, R. (1942). Cerebellar action potentials in response to stimulation of proprioceptors and exteroceptors in the rat. *Journal of Neurophysiology, 5*, 363–371.
Dow, R. S., & Moruzzi, G. (1958). *The physiology and pathology of the cerebellum*. The University of Minnesota Press.
Eccles, J. C. (1970). The topography of the mossy and climbing fiber inputs to the anterior lobe of the cerebellum. In W. S. Fields & J. W. D. Willis (Eds.), *The cerrebellum in health and disease* (pp. 231–266). St. Louis, Missouri.
Eccles, J. C. (1977). My scientific odyssey. *Annual Review of Physiology, 39*, 1–18.
Eccles, J. C., Ito, M., & Szentágothai, J. (1967). *The cerebellum as a neuronal machine*. Springer.
Eccles, J. C., Provini, L., Strata, P., & Taborikova, H. (1968a). Analysis of electrical potentials evoked in the cerebellar anterior lobe by stimulation of hindlimb and forelimb nerves. *Experimental Brain Research, 6*, 171–194.
Eccles, J. C., Provini, L., Strata, P., & Taborikova, H. (1968b). Topographical investigations on the climbing fiber inputs from forelimb and hindlimb afferents to the cerebellar anterior lobe. *Experimental Brain Research, 6*, 195–215.
Eccles, J. C., Sabah, N. H., Schmidt, R. F., & Taborikova, H. (1972). Integration by Purkyne cells of mossy and climbing fiber inputs from cutaneous mechanoreceptors. *Experimental Brain Research, 15*, 498–520.
Ekerot, C. F., & Larson, B. (1979a). The dorsal spino-olivocerebellar system in the cat. I. Functional organization and termination in the anterior lobe. *Experimental Brain Research, 36*, 201–217.
Ekerot, C. F., & Larson, B. (1979b). The dorsal spino-olivocerebellar system in the cat. II. Somatotopical organization. *Experimental Brain Research, 36*, 219–232.
Flourens, P. (1960). Investigations of the properties and the functions of the various parts which compose the cerebral mass. In G. von Bonin (Ed.), *Some papers on the cerebral cortex*. Charles C Thomas.
Gandhoke, G. S., Belykh, E., Zhao, X., Leblanc, R., & Preul, M. C. (2019). Edwin Boldrey and Wilder Penfield's homunculus: A life given by Mrs. Cantlie (in and out of realism). *World Neurosurgery, 132*, 377–388.
Glickstein, M. (2014). *Neuroscience: A historical introduction*. The MIT Press.
Grant, G. (1962). Spinal course and somatotopically localized termination of the spinocerebellar tracts. An experimental study in the cat. *Acta Physiologica Scandinavica. Supplementum, 56*, 1–61.
Graziano, M. S., Taylor, C. S., & Moore, T. (2002). Complex movements evoked by microstimulation of precentral cortex. *Neuron, 34*, 841–851.
Graziano, M. S. A. (2016). Ethological action maps: A paradigm shift for the motor cortex. *Trends in Cognitive Sciences, 20*, 121–132.
Guell, X., Gabrieli, J. D. E., & Schmahmann, J. D. (2018). Triple representation of language, working memory, social and emotion processing in the cerebellum: Convergent evidence from task and seed-based resting-state fMRI analyses in a single large cohort. *NeuroImage, 172*, 437–449.
Hiss, E., Leicht, R., & Schmidt, R. F. (1977). Cutaneous receptive fields of cerebellar Purkyne cells of unanesthetized cats. *Experimental Brain Research, 27*, 319–333.

Horikawa, K., Yamada, Y., Matsuda, T., Kobayashi, K., Hashimoto, M., Matsu-ura, T., Miyawaki, A., Michikawa, T., Mikoshiba, K., & Nagai, T. (2010). Spontaneous network activity visualized by ultrasensitive Ca^{2+} indicators, yellow Cameleon-Nano. *Nature Methods, 7*, 729–732.

Horrax, G. (1915). A study of the afferent fibers of the body wall and of the hind legs to the cerebellum of the dog by the method of degeneration. *The Anatomical Record, 9*, 307–321.

Iorio-Morin, C., & Mathieu, D. (2020). Perspective on the homunculus, the history of cerebral localization, and evolving modes of data representation. *World Neurosurgery, 135*, 42–47.

Ishikawa, K., Kawaguchi, S., & Rowe, M. J. (1972). Actions of afferent impulses from muscle receptors on cerebellar Purkyne cells. I. Responses to muscle vibration. *Experimental Brain Research, 15*, 177–193.

Kanwisher, N. (2010). Functional specificity in the human brain: A window into the functional architecture of the mind. *Proceedings of the National Academy of Sciences of the United States of America, 107*, 11163–11170.

Kooy, F. H. (1917). The inferior olive in vertebrates. *Folia Neurobiol, 10*, 205–369.

Kuroki, S., Yoshida, T., Tsutsui, H., Iwama, M., Ando, R., Michikawa, T., Miyawaki, A., Ohshima, T., & Itohara, S. (2018). Excitatory neuronal hubs configure multisensory integration of slow waves in association cortex. *Cell Reports, 22*, 2873–2885.

Larson, B., Miller, S., & Oscarsson, O. (1969a). A spinocerebellar climbing fibre path activated by the flexor reflex afferents from all four limbs. *The Journal of Physiology, 203*, 641–649.

Larson, B., Miller, S., & Oscarsson, O. (1969b). Termination and functional organization of the dorsolateral spino-olivocerebellar path. *The Journal of Physiology, 203*, 611–640.

Leicht, R., Rowe, M. J., & Schmidt, R. F. (1973). Cutaneous convergence on to the climbing fibre input to cerebellar Purkyne cells. *The Journal of Physiology, 228*, 601–618.

Leicht, R., & Schmidt, R. F. (1977). Somatotopic studies on the vermal cortex of the cerebellar anterior lobe of unanaesthetized cats. *Experimental Brain Research, 27*, 479–490.

Llinás, R., & Simpson, J. I. (1981). Cerebellar control of movement. In A. L. Towe & E. S. Luschei (Eds.), *Handbook of behavioral neurobiology, 5 motor coordination* (pp. 231–302). New York.

Manni, E., & Petrosini, L. (2004). A century of cerebellar somatotopy: A debated representation. *Nature Reviews. Neuroscience, 5*, 241–249.

Marshall, W. H., Woolsey, C. N., & Bard, P. (1937). Cortical representation of tactile sensibility as indicated by cortical potentials. *Science, 85*, 388–390.

Michikawa, T., Yoshida, T., Kuroki, S., Ishikawa, T., Kakei, S., Itohara, S., & Miyawaki, A. (2020). Distributed sensory coding by cerebellar complex spikes in units of cortical segments. bioRxiv. Doi. https://doi.org/10.1101/2020.09.18.301564

Miles, T. S., & Wiesendanger, M. (1975). Organization of climbing fibre projections to the cerebellar cortex from trigeminal cutaneous afferents and from the SI face area of the cerebral cortex in the cat. *The Journal of Physiology, 245*, 409–424.

Miyawaki, A., Llopis, J., Heim, R., McCaffery, J. M., Adams, J. A., Ikura, M., & Tsien, R. Y. (1997). Fluorescent indicators for Ca^{2+} based on green fluorescent proteins and calmodulin. *Nature, 388*, 882–887.

Nagai, T., Yamada, S., Tominaga, T., Ichikawa, M., & Miyawaki, A. (2004). Expanded dynamic range of fluorescent indicators for Ca^{2+} by circularly permuted yellow fluorescent proteins. *Proceedings of the National Academy of Sciences of the United States of America, 101*, 10554–10559.

Oscarsson, O. (1965). Functional Organization of the Spino- and Cuneocerebellar Tracts. *Physiological Reviews, 45*, 495–522.

Oscarsson, O. (1968). Termination and functional organization of the ventral spino-olivocerebellar path. *The Journal of Physiology, 196*, 453–478.

Oscarsson, O. (1969). Termination and functional organization of the dorsal spino-olivocerebellar path. *The Journal of Physiology, 200*, 129–149.

Oscarsson, O. (1980). Functional organization of olivary projection to the cerebellar anterior lobe. In J. Courville & Y. Lamarre (Eds.), *The inferior olivary nucleus, anatomy and physiology, C.d.M* (pp. 279–289). Raven Press.

Oscarsson, O., & Sjolund, B. (1977). The ventral spino-olivocerebellar system in the cat. I. Identification of five paths and their termination in the cerebellar anterior lobe. *Experimental Brain Research, 28*, 469–486.
Penfield, W., & Boldrey, E. (1937). Somatic motor and sensory representation in the cerebral cortex of man as studied by electrical stimulation. *Brain, 60*, 389–443.
Penfield, W., & Rasmussen, T. (1950). *The cerebral cortex of man*. Macmillan.
Provini, L., Redman, S., & Strata, P. (1967). Somatotopic organization of mossy and climbing fibres to the anterior lobe of cerebellum activated by the sensorimotor cortex. *Brain Research, 6*, 378–381.
Provini, L., Redman, S., & Strata, P. (1968). Mossy and climbing fibre organization on the anterior lobe of the cerebellum activated by forelimb and hindlimb areas of the sensorimotor cortex. *Experimental Brain Research, 6*, 216–233.
Purves, D., Augustine, G. J., Fitzpatrick, D., Hall, W. C., LaMantia, A.-S., Mooney, R. D., Platt, M. L., & White, L. E. (2018). *Neuroscience* (6th ed.). Sinauer Associates.
Purves, D., Brannon, E. M., Cabeza, R., Huettel, S. A., & LaBar, K. S. (2007). *Principles of cognitive neuroscience*. Sinauer Associates.
Robertson, L. T., & Laxer, K. D. (1981). Localization of cutaneously elicited climbing fiber responses in lobule V of the monkey cerebellum. *Brain, Behavior and Evolution, 18*, 157–168.
Robertson, L. T., Laxer, K. D., & Rushmer, D. S. (1982). Organization of climbing fiber input from mechanoreceptors to lobule V vermal cortex of the cat. *Experimental Brain Research, 46*, 281–291.
Rubia, F. J., & Tandler, R. (1981). Spatial distribution of afferent information to the anterior lobe of the cat's cerebellum. *Experimental Brain Research, 42*, 249–259.
Rushmer, D. S., Woollacott, M. H., Robertson, L. T., & Laxer, K. D. (1980). Somatotopic organization of climbing fiber projections from low threshold cutaneous afferents to pars intermedia of cerebellar cortex in the cat. *Brain Research, 181*, 17–30.
Saladin, K. D. (2021). *Anatomy & physiology: The unity of form and function* (9th ed.). McGraw-Hill Education.
Schieber, M. H., & Hibbard, L. S. (1993). How somatotopic is the motor cortex hand area? *Science, 261*, 489–492.
Sherrington, C. S. (1920). *The integrative action of the nervous system*. Yale University Press.
Smith, C. U. M. (2008). *Biology of sensory systems* (2nd ed.). Wiley.
Snider, R. S. (1950). Recent contributions to the anatomy and physiology of the cerebellum. *Archives of Neurology and Psychiatry, 64*, 196–219.
Snider, R. S. (1958). The cerebellum. *Scientific American, 199*, 84–90.
Snider, R. S., & Eldred, E. (1952). Cerebrocerebellar relationships in the monkey. *Journal of Neurophysiology, 15*, 27–40.
Snider, R. S., & Stowell, A. (1944). Receiving areas of the tactile, auditory and visual systems in the cerebellum. *Journal of Neurophysiology, 7*, 331–358.
Thach, W. T., Jr. (1967). Somatosensory receptive fields of single units in cat cerebellar cortex. *Journal of Neurophysiology, 30*, 675–696.
VanGilder, J. C., & O'Leary, J. L. (1970). Topical projection of the olivocerebellar system in the cat: An electrophysiological study. *The Journal of Comparative Neurology, 140*, 69–80.
Voogd, J. (2011). Cerebellar zones: A personal history. *Cerebellum, 10*, 334–350.
Voogd, J., Shinoda, Y., Ruigrok, T. J., & Sugihara, I. (2013). Cerebellar nuclei and the inferior olivary nuclei: Organization and connections. In M. Manto, D. L. Gruol, J. D. Schmahmann, N. Koibuchi, & F. Rossi (Eds.), *Handbook of the cerebellum and cerebellar disorders* (pp. 377–436). Springer.
Willett, F. R., Deo, D. R., Avansino, D. T., Rezaii, P., Hochberg, L. R., Henderson, J. M., & Shenoy, K. V. (2020). Hand knob area of premotor cortex represents the whole body in a compositional way. *Cell, 181*, 396–409.
Woolsey, C. N., & Erickson, T. C. (1950). Study of the postcentral gyrus of man by the evoked potential technique. *Transactions of the American Neurological Association, 51*, 50–52.

Woolsey, C. N., Marshall, W. H., & Bard, P. (1942). Representation of cutaneous tactile sensibility in cerebral cortex of monkey as indicated by evoked potentials. *Bulletin of the Johns Hopkins Hospital, 70*, 399–441.

Woolsey, C. N., Settlage, P. H., Meyer, D. R., Sencer, W., Pinto Hamuy, T., & Travis, A. M. (1952). Patterns of localization in precentral and "supplementary" motor areas and their relation to the concept of a premotor area. *Research Publications – Association for Research in Nervous and Mental Disease, 30*, 238–264.

Yamada, Y., Michikawa, T., Hashimoto, M., Horikawa, K., Nagai, T., Miyawaki, A., Häusser, M., & Mikoshiba, K. (2011). Quantitative comparison of genetically encoded Ca indicators in cortical pyramidal cells and cerebellar Purkinje cells. *Frontiers in Cellular Neuroscience, 5*, 18.

Chapter 7
Purkinje Cell Dendrites: The Time-Tested Icon in Histology

Yukari H. Takeo and Michisuke Yuzaki

Proper regulation of the cell-type-specific shape of dendrites is vital for neuronal function (Lefebvre et al., 2015). The cerebellar Purkinje cells (PCs) exhibit a distinctive elaborate dendritic morphology (Fig. 7.1a$_1$), which mostly develops postnatally and serves as an excellent model to study the mechanisms underlying the development of dendrites. Recent technical advances in cell-type-specific, mosaic, or temporally regulated genetic manipulation have allowed quantitative and mechanistic analyses of the dendritic morphology of single PCs. Here, we review the latest findings on the molecular mechanisms underlying PC dendrite formation and maintenance. First, we summarize the geometrical and anatomical features of the dendritic morphology of PC during development through adulthood. Next, we describe recent technical strategies for the mechanistic study of dendritic morphology and discuss several key findings.

7.1 Anatomical and Geometrical Features of Developing and Mature PC Dendrites

7.1.1 Unique Morphology of PC Dendrites and their Synaptic Inputs

PCs are easily detectable in the cerebellar cortex by their large nuclei with a pear-shaped cell body, aligned to form a single cell layer, the PC layer (PCL), along the cerebellar folia. Their dendritic structures are characterized by thick dendritic trunks with highly branched dendrites that outspread in the translobular plane in the

Y. H. Takeo (✉) · M. Yuzaki (✉)
Department of Physiology, Keio University School of Medicine, Tokyo, Japan
e-mail: yhayashi@a3.keio.jp; myuzaki@keio.jp

H. Mizusawa, S. Kakei (eds.), *Cerebellum as a CNS Hub*,
Contemporary Clinical Neuroscience, https://doi.org/10.1007/978-3-030-75817-2_7

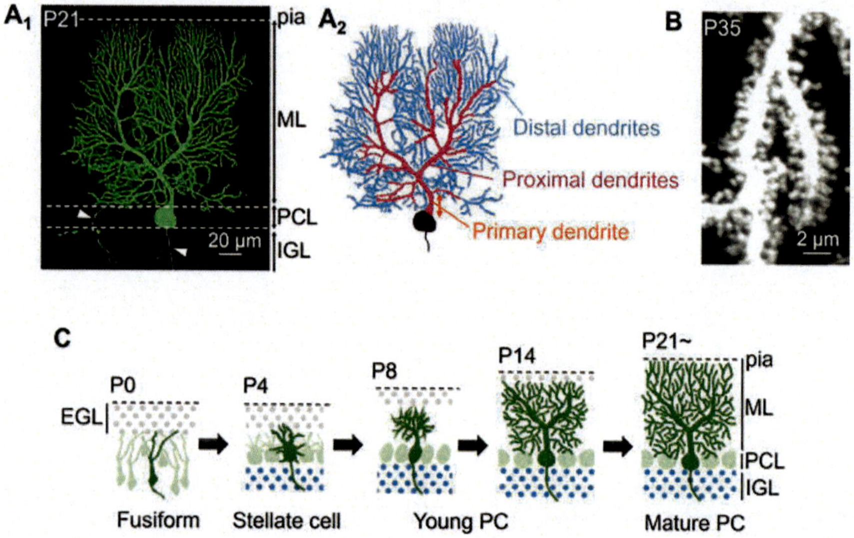

Fig. 7.1 Dendritic morphology of developing and mature cerebellar Purkinje cells. (**a**) A representative confocal image of a mouse Purkinje cells (PC) labeled by GFP at P21 (**a**$_1$). Schematic representation of a single primary dendrite, proximal (magenta) and distal (blue) dendrites is shown in **a**$_2$. White arrowheads indicate the axon. (**b**) A representative image of GFP-labeled distal dendrites of a mature PC at P35. For **a** and **b**, GFP was expressed via IUE at E11.5. (**c**) Schematic summary of postnatal development of mouse PCs. (*ML* molecular layer, *PCL* Purkinje cell layer, *IGL* internal granular layer, *EGL* external granular layer, *pia* pial surface)

molecular layer (ML). In mice, most PCs have single primary dendrites arising from the soma (Fig. 7.1a$_2$), except for those in the lobule IX. The primary dendrite, a thick, relatively long dendritic trunk arising from the soma, bifurcates once or twice in the ML, giving rise to hundreds of thinner, shorter, equally sized dendritic branches. These uniformly branched dendrites seldom overlap with each other and are arranged in a two-dimensional rectangular plane, reaching the pia mater.

The dendritic branches at higher branching orders have numerous mushroom-shaped dendritic spines (Fig. 7.1b), innervated by parallel fibers (PFs), axons of the cerebellar granule cells. The spiny dendritic branches are called distal dendrites, whereas the thick dendritic trunks with some branches at lower branching orders, which are associated with a smaller number of mushroom-shaped spines, are called proximal dendrites (Fig. 7.1a$_2$). The proximal dendrites mainly receive inputs from single climbing fibers (CF), the excitatory afferents from the inferior olive. The intermediate regions between the proximal and distal dendrites are innervated by both PFs and CFs (Ichikawa et al., 2016). The dendritic territories innervated by PFs or CFs are plastic and modified by synaptic activity; decreased PF or increased CF activities extend CF territories to more distal regions of the dendrites (Bosman & Konnerth, 2009). Although developing PCs are temporarily innervated by excitatory mossy fibers until around the third postnatal week (Kalinovsky et al., 2011),

PFs and CFs become the only excitatory afferents to the mature PCs. PCs also receive inhibitory inputs from the molecular layer interneurons (MLIs) around the soma and the axon initial segment.

7.1.2 Multistep Morphological Development of Dendrites

Developmental morphological changes in dendrites are commonly observed in many types of neurons across species. The most dramatic morphological transformation of dendrites has been observed in invertebrates during metamorphosis. During circuit remodeling at the pupal stage, *Drosophila* dendritic arborization (da) neurons lose all elaborate dendrites and completely reconstruct mature dendrites (Williams & Truman, 2005). In mammals, although they do not undergo metamorphosis, many neurons mature through a series of scrap-and-build processes involving pre-existing and new dendrites. For example, cortical pyramidal neurons retract their primitive dendrites during postmitotic migration and cortical layer formation (Hatanaka & Hirata, 2020). Spiny stellate neurons in layer IV of the barrel cortex retract and reassemble dendrites during circuit formation (Callaway & Borrell, 2011; Nakazawa et al., 2018). Dendrites of PCs also undergo multistep developmental stages accompanied by the most drastic morphological features during the first three postnatal weeks (Fig. 7.1c). At birth, post-mitotic PCs, which have migrated to the cerebellar cortex, exhibit a bipolar "fusiform" shape, characterized by a few long and poorly branched primitive dendrites arising from the spindle-shaped soma. At this fusiform stage, the PC somata are scattered in a relatively broader space below the external granular layer. By postnatal day (P) 4, these fusiform PCs lose their primitive dendrites and line up in a single cell layer to form the PCL. PCs begin to develop numerous short dendrites in all directions around the soma, exhibiting the "stellate-cell" shape. During this stellate-cell stage, the multiple apical (oriented toward the pia mater in the ML) primary dendrites form branches with filopodia-like protrusions, whereas the basolateral short primary dendrites around the soma gradually regress. Then, by around P8, all the excess dendrites disappear except for a single primary dendritic trunk, and the PCs exhibit small, but roughly two-dimensional, uniformly branched dendritic trees, which are characteristic of young PCs. From this stage on, PCs further grow and position hundreds of dendritic branches to form flat, rectangular dendritic territories as they mature both morphologically and functionally. Although the formation of the cerebellar circuit is classically thought to be completed in three postnatal weeks (Sotelo & Dusart, 2009; Weiss & Pysh, 1978), recent precise quantification revealed that PCs continue to grow dendrites until the fifth week (Kaneko et al., 2011).

7.1.3 PCs as a Model System for the Study of Dendritic Morphogenesis

Although developmental morphological changes of dendrites have been reported in many types of neurons, the underlying mechanisms remain largely unclear in the mammalian central nervous system in vivo. This is mainly due to the difficulty in observing and manipulating the rapid and non-synchronous changes in dendritic shapes during development. In contrast, the dendritic development of PCs follows the stereotypical stages, as described in the previous section, slowly over several postnatal weeks. Thus, PCs have provided an excellent model to investigate the mechanisms by which dendrites grow and mature during development. Taking advantage of recent genetic tools, studies in the past decades have identified a wide array of molecular events that specify the dendritic morphology of PCs.

7.2 Strategies for a Mechanistic Understanding of PC Dendritic Morphogenesis

7.2.1 In Vitro *Culture Systems*

Time-lapse imaging of dendrites during the development of cultured neurons has provided insights into the underlying molecular mechanisms. One caveat of using cultured neurons is that the developmental process may not be completely recapitulated in vitro. For example, PCs isolated from embryonic or neonatal cerebella grow more than two primary dendrites without undergoing the fusiform-like shape in dissociated cultures. Despite such differences, individual dendritic trees arising from the primary dendrites show relatively long trunks and short spiny branchlets with a fixed length, a stereotyped branching pattern observed in PCs in vivo (Fujishima et al., 2012). The discrepancy between the in vivo and in vitro morphology of dendrites suggests that certain developmental steps require extrinsic factors that are not available in vitro. Indeed, the PCs in organotypic slice cultures prepared from P0 pups show the fusiform and stellate-cell shapes similar to PCs in vivo, indicating that organotypic cultures contain environmental cues necessary for proper dendritic development (Boukhtouche et al., 2006). Nevertheless, not all PCs exhibit single primary dendrites in organotypic cultures, even after 10 days in vitro. Similarly, the dendritic arborization of PCs in organotypic cultures is not as elaborate or flat as PCs in vivo. Although research using dissociated or organotypic cultures have provided key insights into the molecular mechanisms underlying dendritic branching and spine formation (Boukhtouche et al., 2006; Fujishima et al., 2012; Fukumitsu et al., 2015; Gao et al., 2011; Hatsukano et al., 2017; Heuer & Mason, 2003; Shima et al., 2004), it remains unclear why and how PCs acquire single primary dendrites in a single plane in vivo.

7.2.2 *Genetic Strategies to Label and Manipulate PCs* in Vivo

To understand how dendrites develop in the mammalian brain in vivo, tools to manipulate gene expression in specific types of neurons are essential. Recent genetic tools have also allowed the visualization of single PC dendrite morphology in vivo. Sparse labeling of PCs by expressing fluorescent proteins using recombinant virus vectors, such as adeno-associated virus and lentivirus, has been used to clarify the development of PC dendrites in vivo (Chen et al., 2013; Gibson et al., 2014; Ing-Esteves et al., 2018; Kaneko et al., 2011; Lefebvre et al., 2012). The L7/Pcp-2 promoter (Oberdick et al., 1990; Smeyne et al., 1991; Tomomura et al., 2001) is often used to drive PC-specific transgene expression. A drawback of viral vectors is that they take at least several days to express transgenes. Thus, although the development of PCs in the second postnatal week can be studied by injecting AAV vectors into the neonatal cerebellum (Gibson et al., 2014; Kawabata Galbraith et al., 2018), it is not easy to follow the early developmental events during the first postnatal week. Furthermore, the L7/Pcp-2 promoter is known to drive transgene expression after the second postnatal week. Finally, the maximum length of allowable transgenes in AAV and lentivirus is limited to approximately 5 kb and 8 kb, respectively.

To overcome these limitations, in utero electroporation (IUE) of transgenes has served as a powerful complementary method. By injecting cDNAs into the fourth ventricle of embryos and applying electrical pulses over the uterus, multiple genes can be introduced into PCs that have just completed terminal differentiation near the fourth ventricle. An advantage of this technique is that multiple plasmids can be expressed in a cell type-specific manner, whose expression lasts for at least 3 months (Nishiyama et al., 2012, personal communication). In addition, there are theoretically no limits to the size of the transgenes used for gene transfer by electroporation. Indeed, plasmids as large as 12 kb (Nishiyama et al., 2012), and even bacterial artificial chromosomes (Barnabé-Heider et al., 2008), were reported to be introduced into neurons by IUE and in vivo electroporation into adult mice, respectively. Furthermore, using drug-inducible systems, such as Tet-on/off and CreER^{T2}/loxP, the expression of transgenes can be temporally controlled to target specific developmental stages (Nishiyama et al., 2012; Takeo et al., 2015). The drawback of IUE is that the learning curve is relatively steep because cDNAs should be injected into the fourth ventricle of early embryos (embryonic days 10.5–12.5) to be expressed in PCs (Nishiyama et al., 2012; Takeo, 2016).

Mosaic analysis, which does not label cells in a specific manner, also serves as a powerful method to investigate single-cell morphogenesis in vivo. Traditionally, Golgi staining has been used to sparsely label wild-type and mutant PCs (Berry & Bradley, 1976a; Bradley & Berry, 1978; Kim et al., 2011). Mosaic analysis with double marker (MADM) mutant mice has allowed the sparse labeling of both mutant and control PCs in the same animals (Joo et al., 2014; Takeo et al., 2021). Chimeric mice from induced pluripotent stem cells have revealed a cell-autonomous mechanism of dendritic patterning of PCs (Toyoda et al., 2014).

7.3 Molecular Mechanisms Underlying Stepwise Dendritogenesis

7.3.1 *The Transition from the Fusiform to Stellate-Cell Stages: Roles of RORα*

More than half a century ago, the intrinsic mechanism that regulates the transformation of PCs from the fusiform stage to the stellate-cell stage was attributed to a single transcriptional factor, retinoic acid-related orphan receptor α (RORα). *RORα*, which is highly expressed in PCs, is responsible for spontaneous *staggerer* mutant mice (Hamilton et al., 1996; Sidman et al., 1962). The *staggerer* mice are characterized by severe ataxia, which becomes obvious during the third postnatal week. The *staggerer* mice have a strikingly small cerebellum lacking most granule cells due to loss of sonic hedgehog (Shh), a trophic factor essential for the proliferation and differentiation of granule cells. Shh is transactivated by RORα and secreted from PCs (Gold et al., 2003), indicating RORα expressed in PCs non-cell-autonomously regulates the development of granule cells. On the other hand, PCs in the *staggerer* cerebella are atrophic and typically characterized by long, thin, spineless dendrites and spindle-shaped soma, which is reminiscent of immature PCs at the fusiform stage (Fig. 7.1c). Mosaic analyses using *staggerer* ↔ wild-type chimeric mice demonstrated that PCs require cell-autonomous *RORα* for normal dendritic morphogenesis (Soha & Herrup, 1995). Overexpression of RORα in PCs at P0 by lentivirus accelerated the dendritic transformation from a fusiform-like shape to a stellate-cell-like shape in organotypic slice cultures (Boukhtouche et al., 2006). More recently, we showed that knockdown of RORα in sparse PCs at E11.5, via IUE, resulted in fusiform-like long branchless dendrites with spindle-shaped somata at P14 (Takeo et al., 2015) (Fig. 7.2$a_{1,2}$). In addition, PCs in which RORα was knocked down often showed mislocalization of somata within the internal granular layer (IGL) (Takeo et al., 2015). These results indicate two cell-autonomous roles of RORα during the transition from the fusiform to the stellate-cell stage: retraction of primitive dendrites and disposition of the soma in the PCL.

How does RORα regulate these processes? Post-mitotic PCs migrate along the radial fibers and then move tangentially to form a few-cell layer called the Purkinje plates by E14.5, in a manner dependent on reelin-dab1 signaling (Miyata et al., 2010). Interestingly, the knockdown effect of RORα at E11.5 on the postnatal positioning of PCs was not detected at P0 but only became apparent at P4 (Takeo et al., 2015), indicating that RORα is unlikely to be involved in PC migration and the formation of Purkinje plates. Rather, RORα is necessary for PCs to form a single layer in the PCL from P0 to P4. Because PCs mislocalized within the IGL tend to show longer neurites, the RORα-dependent regression of primitive dendrites may be a prerequisite for forming a single layer of PCs in the PCL. Basolateral short dendrites oriented in all directions around the soma, a characteristic feature of the stellate-cell stage PCs, may also be necessary to detect extracellular cues to form a single layer. It is necessary to identify downstream genes regulated by RORα (Gold

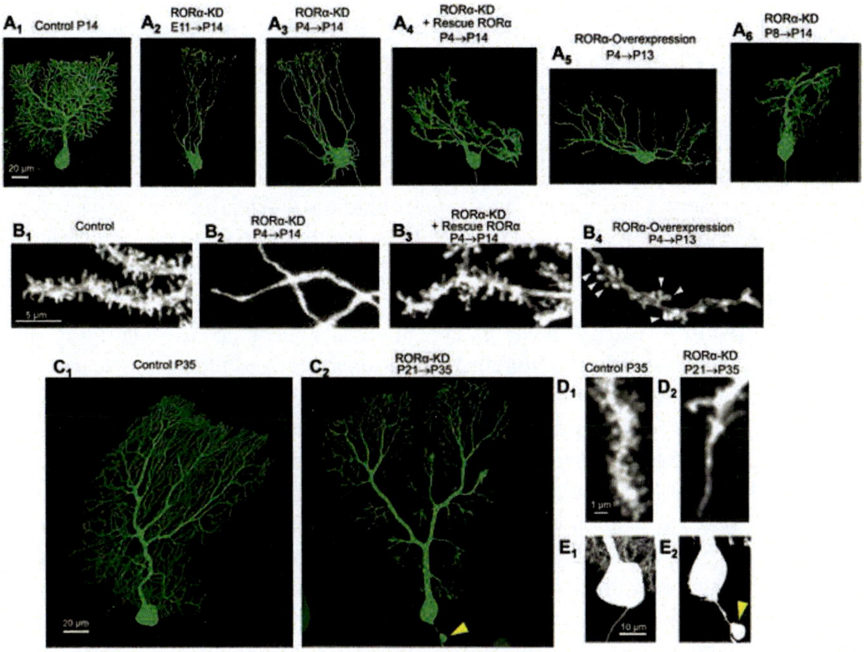

Fig. 7.2 Cell-autonomous roles of RORα for the development and maintenance of PC morphology (Takeo et al., 2015). (**a**, **b**) Representative images of GFP-labeled control PCs ($\mathbf{a}_1$, $\mathbf{b}_1$) and PCs expressing microRNA against RORα (KD) and/or RNAi-resistant RORα at P13-14. For $\mathbf{a}_{3-6}$ and $\mathbf{b}_{2-4}$, gene expression was temporarily induced by tamoxifen at P4 ($\mathbf{a}_{3-5}$ and $\mathbf{b}_{2-4}$) or P8 ($\mathbf{a}_6$). (**c**–**e**) Tamoxifen-induced conditional knockdown of RORα in mature PCs at P35. Representative images of dendrite morphology (**c**), distal dendritic spines (**d**) and somata (**e**) of wild-type control ($\mathbf{c}_1$, $\mathbf{d}_1$, $\mathbf{e}_1$) and RORα knockdown ($\mathbf{c}_2$, $\mathbf{d}_2$, $\mathbf{e}_2$) PCs are shown. Yellow arrowheads indicate the atrophic axonal torpedo

et al., 2003) that mediate the retraction of primitive dendrites and disposition of soma in the PCL at this stage.

Unexpectedly, when RORα was overexpressed at E11.5, PCs showed long neurites and were mislocalized in the white matter (Takeo et al., 2015), indicating that appropriate expression levels of RORα in PCs are important for the migration of PCs. Long and disoriented neurites may physically interfere with the migration of PCs. Because IUE at E11.5 could introduce transgenes into a small population of the deep cerebellar nucleus and other cell types (Nishiyama et al., 2012) in addition to PC precursors, another possibility is that overexpression of RORα may have changed the cell fate of these non-PCs. Indeed, RORα has been reported to induce the expression of several PC-specific genes, such as *Car8* and *Calbindin* (Gold et al., 2003). In any case, the effect of transgene overexpression in early postmitotic neurons should be carefully interpreted.

7.3.2 *The Transition from Stellate-Cell to Young PC Stage: Selective Growth of Single Primary Dendrites*

The timing when PCs at the stellate-cell stage lose superfluous dendrites and form single thick primary dendrites to become young PCs (Fig. 7.1c) depends on the cerebellar lobules. In addition, the timing could vary from cell to cell because stellate-cell stage PCs and young PCs are occasionally observed in the same lobules (Sotelo & Dusart, 2009). These findings indicate that PCs establish single primary dendrites in a non-synchronized manner in a relatively short period. This situation makes it difficult to clarify the steps and molecular mechanisms underlying this phenomenon by classical morphological analysis using fixed tissues.

Single primary dendrites can be established by several possible mechanisms as follows: 1) excess dendrites are removed except for the future single primary dendrites, 2) all dendrites retract, and the single primary dendrites newly emerge, and 3) single primary stem dendrites are formed by fusing multiple apical dendrites, perhaps by the translocation of the PC soma (Sotelo & Dusart, 2009). PCs in cultured cerebellar slices (Tanaka et al., 2003) have been shown to lose excess primary dendrites during maturation, supporting the first hypothesis. However, how the single "winner" dendrites are selected and escape from the elimination step in vivo remains unclear.

RORα, Thyroid Hormones, and aPKC

In addition to the role of RORα in the transition from the fusiform to the stellate-cell stage, RORα plays a distinct role at the stellate-cell stage to establish single primary dendrites (Fig. 7.1c). Tamoxifen-induced conditional knockdown of RORα at the stellate-cell stage between P4 and P6 caused PCs to retain multiple thin apical primary dendrites with a small number of branches and numerous short basolateral dendrites around the soma, reminiscent of those of the stellate-cell PCs (Fig. 7.2a_3) (Takeo et al., 2015). The somatic basolateral dendrites observed in the RORα-knockdown PCs were eliminated by co-expression of the knockdown-resistant RORα (Fig. 7.2a_4), suggesting a role for RORα in the elimination of basolateral dendrites at this stage. Meanwhile, the increased number of apical primary dendrites was not completely restored by the expression of the knockdown-resistant RORα. Indeed, tamoxifen-induced overexpression of RORα at the stellate-cell stage of P4 also increased the number of primary apical dendrites with a small number of branches. These results indicate that the precise expression level of RORα is crucial not only for eliminating basal dendrites but also for the establishment of single apical primary dendrites at this stage.

The thyroid hormone 3,3′,5-triiodo-L-thyronine (T3) has long been thought to be important for PC dendritic growth during the first two postnatal weeks (Anderson, 2008; Faustino & Ortiga-Carvalho, 2014; Kapfhammer, 2004). T3 acts primarily through nuclear thyroid hormone receptors (TRs), TR-α1, TR-β1, and TR-β2, which mediate transcriptional control of the target genes. T3 upregulates the expression and transcriptional activity of RORα in PCs of organotypic cultures (Boukhtouche et al., 2010). T3 also induces the expression of peroxisome proliferator-activated receptor-gamma co-activator 1α (PGC-1α), a master regulator of mitochondrial biogenesis, around P7–P8, when PCs start to exit the stellate-cell stage. Knocking down *PGC-1α* by IUE caused PCs to retain multiple primary dendrites with a small number of dendritic branches (Hatsukano et al., 2017). The dendritic morphology was similar to that of PCs wherein RORα was knocked down at the stellate-cell stage, suggesting that T3 and RORα share downstream molecular mechanisms during the transition from the stellate-cell stage to the young PC stage.

Live imaging of zebrafish PCs has shown that the Golgi apparatus is localized near the root of one of the multiple apical dendrites during development and at the base of a single primary dendrite at later stages (Tanabe et al., 2010). Knockout of atypical protein kinase C (aPKC), *Prkci*, or mosaic expression of a kinase-inactive form of Prkci, caused mislocalization of the Golgi apparatus and retention of multiple primary dendrites. Although the regulation of aPKC activity and its interaction with other intrinsic (e.g., T3 and RORα) and extrinsic (see below) signaling pathways remain to be clarified, these results indicate that aPKC mediates the specification of the primary dendrites of PCs by controlling the positioning of the Golgi apparatus, which provides polarized secretory trafficking (Horton et al., 2005).

Extracellular Matrix (ECM) and Granule Cells

Drosophila da neurons and *C. elegans* PVD neurons require extrinsic factors derived from the ECM or neighboring non-neuronal cells to form proper dendritic morphology (Yang & Chien, 2019). In the mammalian central nervous system, the roles of ECM in dendritogenesis are less clear. Chondroitin sulfate proteoglycan (CSPG) is a major component of the ECM. PCs are surrounded by CSPGs, mostly derived from PTPζ, a receptor-type protein tyrosine phosphatase, and phosphacan, the extracellular domain of PTPζ, during dendrite formation. Pleiotrophin (PTN) is considered secreted by Bergman glia in the immature cerebellum (Matsumoto et al., 1994; Wewetzer et al., 1995). When PTN-PTPζ signaling was disrupted in organotypic cultures, PCs showed multiple primary dendrites, some of which were disoriented in the IGL (Tanaka et al., 2003). These findings indicate that astrocytes and the ECM provide cues necessary for proper dendritic orientation and the establishment of single primary dendrites of PCs.

When the number of granule cells is severely reduced, either by spontaneous genetic mutations or by X-ray irradiation to the developing cerebellum, PCs show supernumerary primary dendrites and disrupted planar tree shapes (Berry & Bradley, 1976b; Bradley & Berry, 1978; Caviness & Rakic, 1978). Although these findings indicate the role of granule cells and their axons (PFs) in the formation of single primary dendrites, the contribution of other extrinsic factors, such as changes in the

ECM and the layer structure in the agranular cerebellum, cannot be ruled out. To clarify the role of PFs in the dendritogenesis of PCs, we examined the effect of overexpression of GluD2, which plays an essential role in PF synapse formation by binding to Cbln1 released from PFs (Ibata et al., 2019; Matsuda et al., 2010). We found that in PCs in which GluD2 was overexpressed by IUE, apical dendrites that formed synapses with PFs were not eliminated and remained as multiple apical dendrites (Takeo et al., 2021). These results indicate that stabilization of apical dendrites is regulated by synapse formation with PFs. However, it remains unclear how single primary dendrites are selectively stabilized during the normal development of PCs.

7.3.3 *From Young to Mature PC Stages: Branch Formation and Growth of the Dendritic Tree*

The growth of dendritic trees, accompanied by massive branch formation, mostly occurs after the young PC stage (Fig. 7.1c). This stage is also under the control of RORα, thyroid hormones, and granule cells.

RORα and Thyroid Hormones

Conditional knockdown and overexpression of RORα in PCs at the stellate-cell stage caused a reduction in dendritic branches at the later stages (Takeo et al., 2015). However, the dendritic morphologies under the two conditions differ in several ways. Whereas RORα-knockdown PCs extended apical dendrites straight toward the pia mater (Fig. 7.2a_3), RORα-overexpressing PCs grew curvy dendrites perpendicular to the pia mater (Fig. 7.2a_5). Furthermore, whereas RORα-knockdown PCs had essentially no dendritic spines (Fig. 7.2$b_{1\text{-}3}$), RORα-overexpressing PCs had few filopodia-like protrusions but instead showed an increased number of mushroom-shaped dendritic spines (Fig. 7.2b_4) compared to wild-type PCs (Fig. 7.2b_1). PCs at the late stellate-cell and young PC stages have numerous filopodia-like protrusions on their apical dendrites. These filopodia-like protrusions are thought to act as precursors of dendritic spines and dendritic branches by guiding the direction of branches to extend (Berry & Bradley, 1976a). This hypothesis is consistent with the phenotype of RORα knockdown and overexpressing phenotypes: without RORα, filopodia cannot mature into spines, resulting in multiple primary dendrites with a few branches (Fig. 7.2a_3, b_2), whereas RORα overexpression may accelerate precocious maturation of dendritic spines (Fig. 7.2b_4), reducing filopodia pools required for the formation of apical dendritic branches (Fig. 7.2a_5). Thus, RORα likely promotes the structural maturation of filopodia into mushroom-shaped dendritic spines through its undefined downstream signaling pathways.

TRα1 is expressed ubiquitously in the cerebellum from early embryonic periods through the second postnatal week, whereas TRβ1 was detected predominantly in PCs after P11 in rats (Bradley et al., 1992; Mellström et al., 1991; Strait et al., 1991). TRα1, but not TRβ1, directly mediated T3-dependent dendritic outgrowth in PCs in dissociated cultures (Heuer & Mason, 2003). Overexpression of dominant-negative TRα1 by the pan-neuronal nestin promoter or the GABAergic neuron-specific Ptf1a promoter, but not by the L7/pcp2 promoter, severely reduced the dendritic growth of PCs (Fauquier et al., 2014). Similarly, knock-in mice that express a dominant-negative mutant TRβ1 Δ337 T, which was originally found in patients with hypothyroidism (Usala et al., 1991), exhibited severe impairment of cerebellar development accompanied by reduced PC dendritic growth (Hashimoto et al., 2001). Furthermore, overexpression of a dominant-negative TRβ1 G345R by the L7/pcp2 promoter reduced the dendritic growth of PCs to a much milder degree (Yu et al., 2015). Transgenic mice in which dominant-negative mutants of TRs were expressed by the L7/pcp2 promoter showed the absence of (Fauquier et al., 2014) or milder (Yu et al., 2015) dendritic phenotypes, likely due to the late onset of the L7/pcp2 promoter which is activated during the second postnatal week. Together, these results indicate that the T3-TRs signaling pathway directly regulates the dendritic growth of PCs during the transition from the stellate-cell to young PC stages.

Synapse Formation-Independent and Synapse Formation-Dependent Roles of Granule Cells

The growth factor neurotrophin-3 (NT-3) secreted from granule cells and its receptor TrkC expressed in PCs regulate the dendritic growth of PCs (Joo et al., 2014). Mosaic analysis with MADM showed that *TrkC*$^{-/-}$ PCs (in the *TrkC*$^{+/-}$ background) had shorter dendritic tree heights accompanied by decreased total length and branch points from the third postnatal week. Interestingly, whereas conditional knockout of *TrkC* in all PCs, or *NT-3* in all granule cells, did not affect PC dendritic morphology, conditional knockout of *NT-3* in granule cells rescued the dendritic growth defects caused by sparse deletion of *TrkC* in PCs. These results indicate that neighboring PCs compete for NT3 provided by granule cells to activate TrkC signaling for dendritic growth. Although TrkC has a kinase-independent role in synaptogenesis (Takahashi et al., 2011), its role for dendritic growth relied on its kinase activity (Joo et al., 2014), suggesting that TrkC regulates PC dendritic morphology independently of its synaptogenic activity.

It has been suggested that synapse formation stabilizes and promotes branching of growing dendrites in zebrafish retinotectal neurons (Niell et al., 2004). The effect of synapse formation in granule cells on the dendritogenesis of PCs has been recently investigated by sparsely knocking out *GluD2* (sparse KO, sKO) in PCs via IUE or MADM (Takeo et al., 2021). *GluD2*-sKO PCs showed fewer dendritic branches in the deep region of the ML, with more dendrites in the superficial region of the ML. Consequently, whereas wild-type PCs have a rectangular dendritic tree

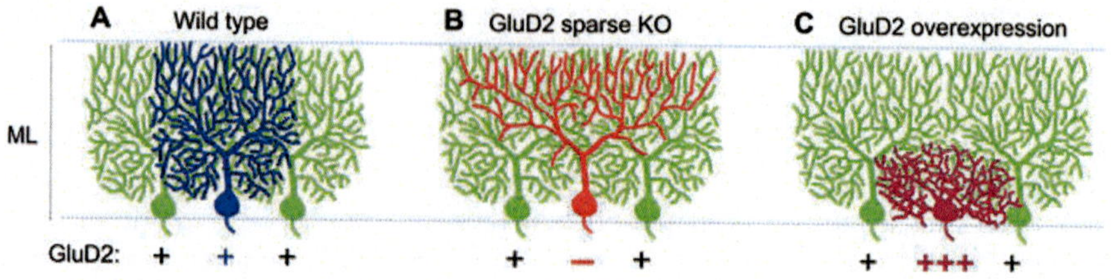

Fig. 7.3 Neighboring PCs compete for GluD2 signaling to grow the dendrites. Schematic showing dendritic morphology of the wild-type PC (**a**), a PC in which GluD2 was sparsely knocked out (**b**) or GluD2 was sparsely overexpressed (**c**)

shape (Fig. 7.3a), *GluD2*-sKO PCs showed an inverted triangular tree shape (Fig. 7.3b). Conversely, PCs that sparsely overexpressed *GluD2* showed phenotypes opposite to those of the *GluD2*-sKO PCs: over-branching in the deep ML and reduced branches in the superficial ML (Fig. 7.3c). The *GluD2*-sKO phenotypes were restored by the expression of the wild-type GluD2, but not the mutant GluD2, which cannot bind Cbln1. Interestingly, this phenotype was not observed in conventional *GluD2* KO mice, in which *GluD2* was knocked out in all PCs (Kaneko et al., 2011) (Fig. 7.4b_1), or in *GluD2*-sKO PCs on the *Cbln1* KO background. These findings indicate that PCs may compete with neighboring PCs for Cbln1 to form synapses with PFs, which anchor dendrites and promote branch formation in the deep ML. *GluD2*-sKO PCs, which lost the competition in the deep ML, likely elongated dendrites in the superficial ML where synaptogenesis may be less competitive. Indeed, a simulation based on the competitive "synaptotrophic" branching hypothesis recapitulated the key features of PC dendritogenesis in wild-type, global-, and sparse- *GluD2* knockout mice (Takeo et al., 2021).

7.3.4 *From Young to Mature PC Stage: Formation of the Planar Dendritic Arbor*

The planar dendritic tree shape is the most characteristic geometrical feature of PC dendrites (Fig. 7.4a_1). Extension of PC dendritic trees in a single plane is thought to play a role in making hundreds of thousands of synapses on PFs with minimal redundancy. Certain types of neurons, such as *Drosophila* da, *C. elegans* PVD, and mammalian amacrine cells, also exhibit planar dendrites (Inberg et al., 2019; Jan & Jan, 2010). Mutagenesis-based in vivo genetic screens have revealed that dendritic morphogenesis in *Drosophila* da neurons is regulated by the surrounding ECM and epidermal cells. *Caenorhabditis elegans* PVD neurons require extrinsic molecules, such as transmembrane proteins in the epidermis or ECM provided by muscle cells to form two-dimensional dendritic arbors (Inberg et al., 2019; Sundararajan et al., 2019; Yang & Chien, 2019). Unlike these neurons that extend their dendrites in the

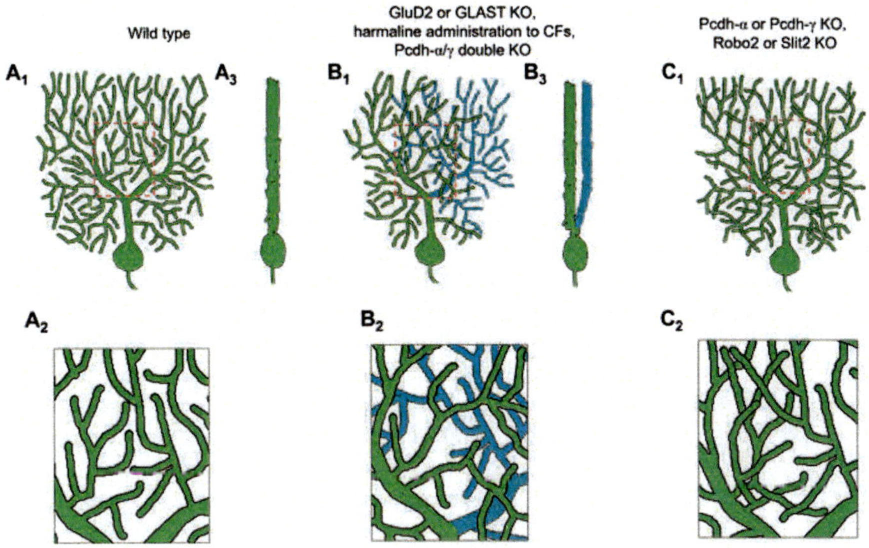

Fig. 7.4 Monoplanar formation and self-avoidance of PC dendrites. (**a**) Schematic showing planar, non-overlapping dendritic morphology (**a**$_1$) of the wild-type PC. The boxed region in **a**$_1$ is enlarged in **a**$_2$. A 90-degree rotation (**a**$_3$) shows the monoplanar dendritic tree shape. (**b**) The multiplanar dendritic phenotype of PCs in conventional *GluD2* KO, *GLAST* KO, wild-type mice in which harmaline was injected to increase CF activities, and *Pcdh-α/γ* double KO. The blue dendritic branch forms a plane different from the green dendrites. The boxed region in **b**$_1$ is enlarged in **b**$_2$. **b**$_3$ shows an image after 90-degree rotation of **b**$_1$. (**c**) Schematic illustration of PC dendritic morphology of *Pcdh-α* KO, *Pcdh-γ* KO, *Robo2* or *Slit2* KO mice. The boxed region in **c**$_1$ is enlarged in **c**$_2$

two-dimensional plane parallel to the pia mater or the body surface, PCs are unique in that they extend dendrites toward the pia mater.

ECM, Bergman Glia, PFs, and CFs

The PTN-PTPζ signaling, provided by Bergman glia and ECM, is also involved in the planar dendritic tree formation of PCs in organotypic slice cultures (Tanaka et al., 2003). PC dendrites have been shown to grow by contacting the radial processes of Bergmann glia in organotypic slice cultures (Lordkipanidze & Dunaevsky, 2005). Thus, Bergmann glia, which extend their radial processes vertically to the pia mater, may provide structural support to restrict the direction of PC dendritic growth in the ML.

If the planar dendritic shape of PCs is evolved to ensure synapse formation with PFs, it would be natural that PFs regulate dendritic planarity. Indeed, PC dendrites were shown to grow in a direction perpendicular to granule cell axons, even in dissociated cultures (Nagata et al., 2006). When GluD2 was overexpressed in sparse

PCs to form more PF synapses, multiple apical dendrites remained and increased the thickness of the dendritic plane. Although general *GluD2* KO PCs showed a normal rectangular shape of dendritic trees (Fig. 7.4b_1), they also showed increased thickness of the dendritic plane (Fig. 7.4$b_{2–3}$) (Kaneko et al., 2011; Takeo et al., 2021), suggesting that the effect of PFs on planar dendritic growth is not mediated by competition between neighboring PCs.

Immature PCs are innervated by multiple CFs, but supernumerary CFs are pruned by the end of the third postnatal week in wild-type mice (Hashimoto & Kano, 2005). When PF synapses are reduced in general *GluD2* KO PCs, multiple CFs remain and innervate the distal dendritic territory that is normally innervated by PFs (Ichikawa et al., 2002). Thus, the impaired dendritic planarity in general *GluD2* KO PCs could be indirectly caused by abnormal CF innervation. In mice wherein a gene encoding the glial glutamate-aspartate transporter (*GLAST*) is disrupted, multiple CFs innervate adult PCs, whereas PFs make normal synapses (Miyazaki et al., 2017). However, *GLAST* KO PCs showed increased thickness of the dendritic plane (Kaneko et al., 2011) (Fig. 7.4b). Furthermore, increased CF activities by repeated administration of harmaline to wild-type mice during the second or third postnatal week caused continued innervation of PCs by multiple CFs and disrupted dendritic planarity (Kaneko et al., 2011) (Fig. 7.4b). These findings indicate that CFs also contribute to the growth of dendritic trees in a single plane.

7.3.5 From Young to Mature PC Stage: Dendritic Self-Avoidance

In young to mature PCs, dendrites extensively divide distally to cover large areas in the ML with minimum crossings. Self-avoidance, a phenomenon in which dendrites of the same neuron avoid each other, is an important property conserved in both vertebrate and invertebrate neurons to maximize the reception of information from afferent connections. This is likely mediated by the recognition of cell-surface molecules expressed on the sibling dendrites. In *Drosophila*, as many as 19,008 isoforms of Dscam1 are generated by alternative splicing of exons and are thought to mediate self-avoidance through isoform-specific homophilic recognition (Zipursky & Grueber, 2013). Mouse DSCAM and DSCAML1 also regulate the self-avoidance of dendrites in a subset of retinal amacrine cells and ganglion cells (Fuerst et al., 2008; Fuerst et al., 2009), although they do not undergo massive alternative splicing as fly Dscam1. In mice, clustered protocadherins (Pcdhs), a different family of cell recognition molecules, have been shown to regulate self-avoidance by a fly Dscam1-like strategy. A total of 58 genes are arranged in gene clusters, *Pcdh-α*, *Pcdh-β*, and *Pcdh-γ*, each consisting of 14, 22, and 22 members, respectively. Pcdh-γ isoforms are reported to form heteromeric tetramers, representing the units of their homophilic trans interactions (Schreiner & Weiner, 2010). Thus, although the number of Pcdh isoforms is much smaller than that of fly Dscam1, combinatorial homophilic interactions likely expand the molecular diversity of Pcdh-based signaling. Single-cell RT-PCR analyses and theoretical considerations have indicated that each PC expresses a total of 15 Pcdh isoforms, 10 randomly (2 Pcdh-α, 4 Pcdh-β, 4 Pcdh-γ) and 5 constitutively (2 Pcdh-α, 3 Pcdh-γ), leading to 12,720 types of cis-tetramers (Yagi, 2012). PC-specific KO of *Pcdh-γ* (Lefebvre et al., 2012) or *Pcdh-α* (Ing-Esteves et al., 2018) resulted in a nearly twofold increase in the number of self-crossings of PC dendrites, suggesting that Pcdhs mediate repulsion of dendrites that belong to the same PC. Furthermore, deletion of both *Pcdh-α* and *Pcdh-γ* caused a sixfold self-crossing of PC dendrites, indicating that Pcdh-α and Pcdh-γ act synergistically (Ing-Esteves et al., 2018). Pcdh-γ has been shown to inhibit focal adhesion kinase (FAK) and protein kinase C (PKC) in the cerebral cortex (Garrett et al., 2012). Although PC-specific deletion of *mTORC2*, an activator of PKC, recapitulated the self-crossing phenotype of *Pcdh-γ* KO PCs (Angliker et al., 2015), deletion of *FAK* in PCs did not cause gross abnormalities in the dendritic arborization of PCs (Watanabe et al., 2008). These findings indicate that although how Pcdhs tetramers are presented on PC dendrites remains to be clarified, combinatorial homophilic interactions of Pcdhs likely contribute to self-avoidance by a mechanism involving local activation of PKC (Fig. 7.4c).

Self-avoidance is often achieved in concert with multiple cell-surface molecules to detect various extracellular cues and to respond to different behaviors of dendrites. Indeed, live imaging of cultured PCs has shown that dendrites undergo two types of self-avoiding behaviors, retraction, and stalling, upon contact with sibling dendrites depending on the developmental stage (Fujishima et al., 2012). Retraction of homoneuronal dendrites is mediated by MTSS, which controls actin nucleator activity in dendritic filopodia, contributing to the loss of proximal branches in young PCs (Kawabata Galbraith et al., 2018). Robo2, a Robo receptor family that provides repulsive cues to regulate axon guidance by binding to Slits, a family of secreted proteins (Andrews et al., 2007) is highly expressed in PC dendrites at P14 in mice. PC-specific deletion of *Robo2* has been shown to cause excessive dendrite self-crossing without affecting arbor size and shape in a manner independent of Pcdh-γ (Gibson et al., 2014). LKB1, a serine-threonine kinase, is active during this period and mediates the self-avoidance of PC dendrites by regulating the dendritic localization of Robo2 (Kuwako & Okano, 2018). Although PC-specific elimination of *Slit2* recapitulated the *Robo2* KO phenotype, mosaic expression of membrane-localized Slit2 via a glycosylphosphatidylinositol anchor, but not diffusible Slit2, restored self-avoidance in *Slit2* KO PCs (Gibson et al., 2014). These results indicate that the Slit2-Robo2 interaction between local dendrites may prevent self-crossing of PC dendrites via repulsive signaling (Fig. 7.4c).

In addition to self-avoidance, dendrites of the same class of neurons typically do not overlap significantly. This phenomenon, referred to as tiling or heteroneuronal avoidance, also relies on the recognition of self and non-self by cell-surface molecules on dendrites (Zipursky & Grueber, 2013). Indeed, PCs wherein both *Pcdh-α* and *Pcdh-γ* genes were deleted not only showed increased self-crossings of dendrites but also extended primary dendrites in multiple planes, suggesting impairment of both homo- and heteroneuronal avoidance (Ing-Esteves et al., 2018)

(Fig. 7.4b). However, reducing Pcdh-γ diversity was shown to increase heteroneuronal avoidance in retinal starburst amacrine cells (Lefebvre et al., 2012). Since PC dendrites located at different sagittal planes are unlikely to express the same Pcdh isoforms, it would be difficult to explain the planar dendritic arrangement by Pcdh-based heteroneuronal avoidance. Rather, severely defective homoneuronal avoidance likely affected the monoplanarity of dendrites in *Pcdh-α/γ* double KO PCs.

7.3.6 Mature PCs: Maintenance of Mature Dendrites

Several molecules involved in PC dendritogenesis remain expressed throughout adult life, but their functional significance in maintaining mature dendrites is not well characterized. Ablation of GluD2 in adult PCs has been shown to cause a gradual reduction in PF synapses accompanied by multiple innervations by CFs (Miyazaki et al., 2010). However, dendritic arborization patterns and their planarity were not affected, indicating the developmental stage-specific role of GluD2. In contrast, deletion of *RORα* by an L7/Pcp2-Cre promoter from the second postnatal week, PC dendrites became atrophic at P28 in vivo (Chen et al., 2013). We showed that tamoxifen-induced knockdown or knockout of RORα at P8 or P21 reduced the number of branches by almost half and a massive reduction of spines at P35 (Fig. 7.2a_6, c, d). Electrophysiological analyses revealed that PF- and CF-evoked synaptic responses were reduced by knockdown of RORα in mature PCs (Takeo et al., 2015). These findings indicate that RORα is essential for the maintenance of mature dendritic complexity and spines in mature PCs.

7.4 Conclusions and Future Perspectives

Recent studies have shed new light on various mechanisms that cooperatively regulate the formation and maintenance of elaborate PC dendrites. Two principles seem to have emerged from these studies. First, the same molecule could exert distinct functions at different stages of dendritogenesis. The most typical utility player is RORα, which is required for the regression of primitive dendrites to form a single PCL by P4, elimination of perisomatic dendrites and maturation of single stem dendrites at the stellate-cell stage by P8, the formation of terminal dendritic branches with massive spines in young PCs after P8, and the maintenance of dendritic complexity and spines in mature PCs (Fig. 7.1c). T3 or PGC-1α also regulates the pruning of supernumerary dendrites at the stellate-cell stage and the subsequent growth of the main dendritic tree at the young PC stage. These findings indicate that RORα and T3 regulate distinct downstream signaling pathways (Anderson, 2008; Serra et al., 2006) depending on the developmental stage. Another non-mutually exclusive possibility is that morphological changes at each developmental stage are causally related. For example, without selecting single primary dendrites, the subsequent outgrowth of dendritic trees would be compromised by the limited supply of membrane resources. Future studies using genetic tools that enable temporally precise in vivo control of gene expression will be warranted to clarify how the same players could perform distinct jobs at each developmental stage of PCs.

Second, competition is another principle of proper dendritogenesis. Competition between homoneuronal dendrites likely regulates the self-avoidance of growing dendrites and their planarity in young and mature PCs. Neighboring PCs also compete with each other for NT-3 provided by granule cells to extend dendritic trees in the ML by activating TrkC signaling independent of synapse formation. Similarly, PCs compete with their neighboring PCs for Cbln1 to form PF synapses, which is necessary to extend the branches in the deep ML. To fully understand the competitive nature of dendritogenesis in vivo, further studies using tools that enable manipulation of gene expression in sparse PCs are warranted.

Another direction of future research includes the role of local neuronal activity in dendritogenesis. Synapses not only serve as anchors to support branching but also secrete neurotransmitters (e.g., glutamate and GABA), trophic factors (e.g., NT-3 and brain-derived neurotrophic factor [BDNF]), and synapse organizers (e.g., Cbln1 and C1ql1 [Kakegawa et al., 2015]) to locally activate receptors expressed on PC dendrites. Importantly, the secretion of these molecules from presynaptic neurons is regulated by neuronal activity. For example, granule cell activities have been reported to promote the growth of PC dendrites by releasing BDNF (Hirai & Launey, 2000; Hisatsune et al., 2006). Similarly, glutamate released from PFs acts on AMPA receptors and metabotropic glutamate receptor 1 on PC dendrites, leading to Ca^{2+} influx from Ca^{2+} channels and Ca^{2+} release from inositol trisphosphate receptor (IP3R), and activation of protein kinase C (PKC) and Ca^{2+}/calmodulin kinase II (CaMKII). Whereas overexpression of CaMKIIβ has been shown to promote dendritic spine formation (Sugawara et al., 2017), PKC reduced dendritic outgrowth (Shimobayashi et al., 2016) and suppressed excess spine formation in PKC-mediated phosphorylation of CaMKIIβ (Sugawara et al., 2017) in PCs in organotypic cerebellar slices. To understand how excitatory synaptic inputs from PFs and CFs and inhibitory inputs from MLIs affect PC dendrite formation and to clarify the contribution mediated by neuronal activities, future studies using optogenetic or chemogenetic tools will be useful to manipulate presynaptic and postsynaptic membrane potentials of each pathway.

RORα deficiency during development is reported to be closely related to the severity of spinocerebellar ataxia type 1 (SCA1) (Serra et al., 2006). Mutations in the gene encoding β-III spectrin, which is crucial for anchoring GluD2 at dendritic spines (Hirai & Matsuda, 1999), cause SCA5 (Ikeda et al., 2006). Similarly, missense mutations in the gene encoding PKCγ cause SCA14 (Chen et al., 2005), and mutations in the gene encoding IP3R cause SCA15/16 (Novak et al., 2010; van de Leemput et al., 2007) and SCA29 (Huang et al., 2012). Although SCAs are neurodegenerative disorders, PCs in SCA patients and corresponding disease model mice show abnormal dendritic architecture, such as multiple primary dendrites, defective planarity, and reduced number of dendritic branches and spines. Indeed, when

RORα was deleted in adult mice, some axons of PCs exhibited torpedoes and rounded swelling of the proximal portion of axons (Takeo et al., 2015) (Fig. 7.2e), which are often observed in pathological conditions (Louis et al., 2014). Future studies are warranted to clarify the extent to which defective dendritic growth contributes to or is caused by the neurodegeneration of PCs in these diseases.

Acknowledgments This work was supported by the CREST from Japan Science and Technology Corporation (M.Y.), Grant-in-Aid for the Ministry of Education, Culture, Sports, Science and Technology of Japan (W.K. and M.Y.), Japan Society for the Promotion of Science (Y-H.T., W.K., M.Y.).

Conflict of Interest The authors declare no conflicts of interest.

References

Anderson, G.W. (2008). Thyroid hormone and cerebellar development. Cerebellum (London, England) *7*, 60–74.

Andrews, W. D., Barber, M., & Parnavelas, J. G. (2007). Slit-Robo interactions during cortical development. *Journal of Anatomy, 211*, 188–198.

Angliker, N., Burri, M., Zaichuk, M., Fritschy, J. M., & Rüegg, M. A. (2015). mTORC1 and mTORC2 have largely distinct functions in Purkinje cells. *The European Journal of Neuroscience, 42*, 2595–2612.

Barnabé-Heider, F., Meletis, K., Eriksson, M., Bergmann, O., Sabelström, H., Harvey, M. A., Mikkers, H., & Frisén, J. (2008). Genetic manipulation of adult mouse neurogenic niches by in vivo electroporation. *Nature Methods, 5*, 189–196.

Berry, M., & Bradley, P. (1976a). The growth of the dendritic trees of Purkinje cells in the cerebellum of the rat. *Brain Research, 112*, 1–35.

Berry, M., & Bradley, P. (1976b). The growth of the dendritic trees of Purkinje cells in irradiated agranular cerebellar cortex. *Brain Research, 116*, 361–387.

Bosman, L. W., & Konnerth, A. (2009). Activity-dependent plasticity of developing climbing fiber-Purkinje cell synapses. *Neuroscience, 162*, 612–623.

Boukhtouche, F., Brugg, B., Wehrlé, R., Bois-Joyeux, B., Danan, J. L., Dusart, I., & Mariani, J. (2010). Induction of early Purkinje cell dendritic differentiation by thyroid hormone requires RORα. *Neural Development, 5*, 18.

Boukhtouche, F., Janmaat, S., Vodjdani, G., Gautheron, V., Mallet, J., Dusart, I., & Mariani, J. (2006). Retinoid-related orphan receptor alpha controls the early steps of Purkinje cell dendritic differentiation. *The Journal of Neuroscience : the Official Journal of the Society for Neuroscience, 26*, 1531–1538.

Bradley, D. J., Towle, H. C., & Young, W. S., 3rd. (1992). Spatial and temporal expression of alpha- and beta-thyroid hormone receptor mRNAs, including the beta 2-subtype, in the developing mammalian nervous system. *The Journal of Neuroscience : the Official Journal of the Society for Neuroscience, 12*, 2288–2302.

Bradley, P., & Berry, M. (1978). The Purkinje cell dendritic tree in mutant mouse cerebellum. A quantitative Golgi study of Weaver and Staggerer mice. *Brain Research, 142*, 135–141.

Callaway, E. M., & Borrell, V. (2011). Developmental sculpting of dendritic morphology of layer 4 neurons in visual cortex: Influence of retinal input. *The Journal of Neuroscience : the Official Journal of the Society for Neuroscience, 31*, 7456–7470.

Caviness, V. S., Jr., & Rakic, P. (1978). Mechanisms of cortical development: A view from mutations in mice. *Annual Review of Neuroscience, 1*, 297–326.

Chen, D. H., Cimino, P. J., Ranum, L. P., Zoghbi, H. Y., Yabe, I., Schut, L., Margolis, R. L., Lipe, H. P., Feleke, A., Matsushita, M., et al. (2005). The clinical and genetic spectrum of spinocerebellar ataxia 14. *Neurology, 64*, 1258–1260.

Chen, X. R., Heck, N., Lohof, A. M., Rochefort, C., Morel, M. P., Wehrlé, R., Doulazmi, M., Marty, S., Cannaya, V., Avci, H. X., et al. (2013). Mature Purkinje cells require the retinoic acid-related orphan receptor-α (RORα) to maintain climbing fiber mono-innervation and other adult characteristics. *The Journal of neuroscience : the official journal of the Society for Neuroscience, 33*, 9546–9562.

Fauquier, T., Chatonnet, F., Picou, F., Richard, S., Fossat, N., Aguilera, N., Lamonerie, T., & Flamant, F. (2014). Purkinje cells and Bergmann glia are primary targets of the TRα1 thyroid hormone receptor during mouse cerebellum postnatal development. *Development (Cambridge, England), 141*, 166–175.

Faustino, L. C., & Ortiga-Carvalho, T. M. (2014). Thyroid hormone role on cerebellar development and maintenance: A perspective based on transgenic mouse models. *Frontiers in Endocrinology, 5*, 75.

Fuerst, P. G., Bruce, F., Tian, M., Wei, W., Elstrott, J., Feller, M. B., Erskine, L., Singer, J. H., & Burgess, R. W. (2009). DSCAM and DSCAML1 function in self-avoidance in multiple cell types in the developing mouse retina. *Neuron, 64*, 484–497.

Fuerst, P. G., Koizumi, A., Masland, R. H., & Burgess, R. W. (2008). Neurite arborization and mosaic spacing in the mouse retina require DSCAM. *Nature, 451*, 470–474.

Fujishima, K., Horie, R., Mochizuki, A., & Kengaku, M. (2012). Principles of branch dynamics governing shape characteristics of cerebellar Purkinje cell dendrites. *Development (Cambridge, England), 139*, 3442–3455.

Fukumitsu, K., Fujishima, K., Yoshimura, A., Wu, Y. K., Heuser, J., & Kengaku, M. (2015). Synergistic action of dendritic mitochondria and creatine kinase maintains ATP homeostasis and actin dynamics in growing neuronal dendrites. *The Journal of Neuroscience : the Official Journal of the Society for Neuroscience, 35*, 5707–5723.

Gao, Y., Perkins, E. M., Clarkson, Y. L., Tobia, S., Lyndon, A. R., Jackson, M., & Rothstein, J. D. (2011). β-III spectrin is critical for development of purkinje cell dendritic tree and spine morphogenesis. *The Journal of Neuroscience : the Official Journal of the Society for Neuroscience, 31*, 16581–16590.

Garrett, A. M., Schreiner, D., Lobas, M. A., & Weiner, J. A. (2012). γ-Protocadherins control cortical dendrite arborization by regulating the activity of a FAK/PKC/MARCKS signaling pathway. *Neuron, 74*, 269–276.

Gibson, D. A., Tymanskyj, S., Yuan, R. C., Leung, H. C., Lefebvre, J. L., Sanes, J. R., Chédotal, A., & Ma, L. (2014). Dendrite self-avoidance requires cell-autonomous slit/robo signaling in cerebellar purkinje cells. *Neuron, 81*, 1040–1056.

Gold, D. A., Baek, S. H., Schork, N. J., Rose, D. W., Larsen, D. D., Sachs, B. D., Rosenfeld, M. G., & Hamilton, B. A. (2003). RORalpha coordinates reciprocal signaling in cerebellar development through sonic hedgehog and calcium-dependent pathways. *Neuron, 40*, 1119–1131.

Hamilton, B. A., Frankel, W. N., Kerrebrock, A. W., Hawkins, T. L., FitzHugh, W., Kusumi, K., Russell, L. B., Mueller, K. L., van Berkel, V., Birren, B. W., et al. (1996). Disruption of the nuclear hormone receptor RORalpha in staggerer mice. *Nature, 379*, 736–739.

Hashimoto, K., Curty, F. H., Borges, P. P., Lee, C. E., Abel, E. D., Elmquist, J. K., Cohen, R. N., & Wondisford, F. E. (2001). An unliganded thyroid hormone receptor causes severe neurological dysfunction. *Proceedings of the National Academy of Sciences of the United States of America, 98*, 3998–4003.

Hashimoto, K., & Kano, M. (2005). Postnatal development and synapse elimination of climbing fiber to Purkinje cell projection in the cerebellum. *Neuroscience Research, 53*, 221–228.

Hatanaka, Y., & Hirata, T. (2020). How do cortical excitatory neurons terminate their migration at the right place? Critical roles of environmental elements. *Frontiers in Cell and Developmental Biology, 8*, 596708.

Hatsukano, T., Kurisu, J., Fukumitsu, K., Fujishima, K., & Kengaku, M. (2017). Thyroid hormone induces PGC-1α during dendritic outgrowth in mouse cerebellar Purkinje cells. *Frontiers in Cellular Neuroscience, 11*, 133.

Heuer, H., & Mason, C. A. (2003). Thyroid hormone induces cerebellar Purkinje cell dendritic development via the thyroid hormone receptor alpha1. *The Journal of Neuroscience : the Official Journal of the Society for Neuroscience, 23*, 10604–10612.

Hirai, H., & Launey, T. (2000). The regulatory connection between the activity of granule cell NMDA receptors and dendritic differentiation of cerebellar Purkinje cells. *The Journal of Neuroscience : the Official Journal of the Society for Neuroscience, 20*, 5217–5224.

Hirai, H., & Matsuda, S. (1999). Interaction of the C-terminal domain of delta glutamate receptor with spectrin in the dendritic spines of cultured Purkinje cells. *Neuroscience Research, 34*, 281–287.

Hisatsune, C., Kuroda, Y., Akagi, T., Torashima, T., Hirai, H., Hashikawa, T., Inoue, T., & Mikoshiba, K. (2006). Inositol 1,4,5-trisphosphate receptor type 1 in granule cells, not in Purkinje cells, regulates the dendritic morphology of Purkinje cells through brain-derived neurotrophic factor production. *The Journal of Neuroscience : the Official Journal of the Society for Neuroscience, 26*, 10916–10924.

Horton, A. C., Rácz, B., Monson, E. E., Lin, A. L., Weinberg, R. J., & Ehlers, M. D. (2005). Polarized secretory trafficking directs cargo for asymmetric dendrite growth and morphogenesis. *Neuron, 48*, 757–771.

Huang, L., Chardon, J. W., Carter, M. T., Friend, K. L., Dudding, T. E., Schwartzentruber, J., Zou, R., Schofield, P. W., Douglas, S., Bulman, D. E., et al. (2012). Missense mutations in ITPR1 cause autosomal dominant congenital nonprogressive spinocerebellar ataxia. *Orphanet Journal of Rare Diseases, 7*, 67.

Ibata, K., Kono, M., Narumi, S., Motohashi, J., Kakegawa, W., Kohda, K., & Yuzaki, M. (2019). Activity-dependent secretion of synaptic organizer Cbln1 from lysosomes in granule cell axons. *Neuron, 102*, 1184–1198.e1110.

Ichikawa, R., Hashimoto, K., Miyazaki, T., Uchigashima, M., Yamasaki, M., Aiba, A., Kano, M., & Watanabe, M. (2016). Territories of heterologous inputs onto Purkinje cell dendrites are segregated by mGluR1-dependent parallel fiber synapse elimination. *Proceedings of the National Academy of Sciences of the United States of America, 113*, 2282–2287.

Ichikawa, R., Miyazaki, T., Kano, M., Hashikawa, T., Tatsumi, H., Sakimura, K., Mishina, M., Inoue, Y., & Watanabe, M. (2002). Distal extension of climbing fiber territory and multiple innervation caused by aberrant wiring to adjacent spiny branchlets in cerebellar Purkinje cells lacking glutamate receptor delta 2. *The Journal of neuroscience : the official journal of the Society for Neuroscience, 22*, 8487–8503.

Ikeda, Y., Dick, K. A., Weatherspoon, M. R., Gincel, D., Armbrust, K. R., Dalton, J. C., Stevanin, G., Dürr, A., Zühlke, C., Bürk, K., et al. (2006). Spectrin mutations cause spinocerebellar ataxia type 5. *Nature Genetics, 38*, 184–190.

Inberg, S., Meledin, A., Kravtsov, V., Iosilevskii, Y., Oren-Suissa, M., & Podbilewicz, B. (2019). Lessons from worm dendritic patterning. *Annual Review of Neuroscience, 42*, 365–383.

Ing-Esteves, S., Kostadinov, D., Marocha, J., Sing, A. D., Joseph, K. S., Laboulaye, M. A., Sanes, J. R., & Lefebvre, J. L. (2018). Combinatorial effects of alpha- and gamma-Protocadherins on neuronal survival and dendritic self-avoidance. *The Journal of Neuroscience : the Official Journal of the Society for Neuroscience, 38*, 2713–2729.

Jan, Y. N., & Jan, L. Y. (2010). Branching out: Mechanisms of dendritic arborization. *Nature Reviews. Neuroscience, 11*, 316–328.

Joo, W., Hippenmeyer, S., & Luo, L. (2014). Neurodevelopment. Dendrite morphogenesis depends on relative levels of NT-3/TrkC signaling. *Science (New York, N.Y.), 346*, 626–629.

Kakegawa, W., Mitakidis, N., Miura, E., Abe, M., Matsuda, K., Takeo, Y. H., Kohda, K., Motohashi, J., Takahashi, A., Nagao, S., et al. (2015). Anterograde C1ql1 signaling is required in order to determine and maintain a single-winner climbing fiber in the mouse cerebellum. *Neuron, 85*, 316–329.

Kalinovsky, A., Boukhtouche, F., Blazeski, R., Bornmann, C., Suzuki, N., Mason, C. A., & Scheiffele, P. (2011). Development of axon-target specificity of ponto-cerebellar afferents. *PLoS Biology, 9*, e1001013.

Kaneko, M., Yamaguchi, K., Eiraku, M., Sato, M., Takata, N., Kiyohara, Y., Mishina, M., Hirase, H., Hashikawa, T., & Kengaku, M. (2011). Remodeling of monoplanar Purkinje cell dendrites during cerebellar circuit formation. *PLoS One, 6*, e20108.

Kapfhammer, J. P. (2004). Cellular and molecular control of dendritic growth and development of cerebellar Purkinje cells. *Progress in Histochemistry and Cytochemistry, 39*, 131–182.

Kawabata Galbraith, K., Fujishima, K., Mizuno, H., Lee, S. J., Uemura, T., Sakimura, K., Mishina, M., Watanabe, N., & Kengaku, M. (2018). MTSS1 regulation of actin-nucleating Formin DAAM1 in dendritic Filopodia determines final dendritic configuration of Purkinje cells. *Cell Rep, 24*, 95–106.e109.

Kim, J., Kwon, N., Chang, S., Kim, K. T., Lee, D., Kim, S., Yun, S. J., Hwang, D., Kim, J. W., Hwu, Y., et al. (2011). Altered branching patterns of Purkinje cells in mouse model for cortical development disorder. *Scientific Reports, 1*, 122.

Kuwako, K. I., & Okano, H. (2018). The LKB1-SIK pathway controls dendrite self-avoidance in Purkinje cells. *Cell Reports, 24*, 2808–2818.e2804.

Lefebvre, J. L., Kostadinov, D., Chen, W. V., Maniatis, T., & Sanes, J. R. (2012). Protocadherins mediate dendritic self-avoidance in the mammalian nervous system. *Nature, 488*, 517–521.

Lefebvre, J. L., Sanes, J. R., & Kay, J. N. (2015). Development of dendritic form and function. *Annual Review of Cell and Developmental Biology, 31*, 741–777.

Lordkipanidze, T., & Dunaevsky, A. (2005). Purkinje cell dendrites grow in alignment with Bergmann glia. *Glia, 51*, 229–234.

Louis, E. D., Kuo, S. H., Vonsattel, J. P., & Faust, P. L. (2014). Torpedo formation and Purkinje cell loss: Modeling their relationship in cerebellar disease. *Cerebellum (London, England), 13*, 433–439.

Matsuda, K., Miura, E., Miyazaki, T., Kakegawa, W., Emi, K., Narumi, S., Fukazawa, Y., Ito-Ishida, A., Kondo, T., Shigemoto, R., et al. (2010). Cbln1 is a ligand for an orphan glutamate receptor delta2, a bidirectional synapse organizer. *Science (New York, N.Y.), 328*, 363–368.

Matsumoto, K., Wanaka, A., Mori, T., Taguchi, A., Ishii, N., Muramatsu, H., Muramatsu, T., & Tohyama, M. (1994). Localization of pleiotrophin and midkine in the postnatal developing cerebellum. *Neuroscience Letters, 178*, 216–220.

Mellström, B., Naranjo, J. R., Santos, A., Gonzalez, A. M., & Bernal, J. (1991). Independent expression of the alpha and beta c-erbA genes in developing rat brain. *Molecular Endocrinology, 5*, 1339–1350.

Miyata, T., Ono, Y., Okamoto, M., Masaoka, M., Sakakibara, A., Kawaguchi, A., Hashimoto, M., & Ogawa, M. (2010). Migration, early axonogenesis, and Reelin-dependent layer-forming behavior of early/posterior-born Purkinje cells in the developing mouse lateral cerebellum. *Neural Development, 5*, 23.

Miyazaki, T., Yamasaki, M., Hashimoto, K., Kohda, K., Yuzaki, M., Shimamoto, K., Tanaka, K., Kano, M., & Watanabe, M. (2017). Glutamate transporter GLAST controls synaptic wrapping by Bergmann glia and ensures proper wiring of Purkinje cells. *Proceedings of the National Academy of Sciences of the United States of America, 114*, 7438–7443.

Miyazaki, T., Yamasaki, M., Takeuchi, T., Sakimura, K., Mishina, M., & Watanabe, M. (2010). Ablation of glutamate receptor GluRδ2 in adult Purkinje cells causes multiple innervation of climbing fibers by inducing aberrant invasion to parallel fiber innervation territory. *The Journal of Neuroscience : the Official Journal of the Society for Neuroscience, 30*, 15196–15209.

Nagata, I., Ono, K., Kawana, A., & Kimura-Kuroda, J. (2006). Aligned neurite bundles of granule cells regulate orientation of Purkinje cell dendrites by perpendicular contact guidance in two-dimensional and three-dimensional mouse cerebellar cultures. *The Journal of Comparative Neurology, 499*, 274–289.

Nakazawa, S., Mizuno, H., & Iwasato, T. (2018). Differential dynamics of cortical neuron dendritic trees revealed by long-term in vivo imaging in neonates. *Nature Communications, 9*, 3106.

Niell, C. M., Meyer, M. P., & Smith, S. J. (2004). In vivo imaging of synapse formation on a growing dendritic arbor. *Nature Neuroscience, 7*, 254–260.

Nishiyama, J., Hayashi, Y., Nomura, T., Miura, E., Kakegawa, W., & Yuzaki, M. (2012). Selective and regulated gene expression in murine Purkinje cells by in utero electroporation. *The European Journal of Neuroscience, 36*, 2867–2876.

Novak, M. J., Sweeney, M. G., Li, A., Treacy, C., Chandrashekar, H. S., Giunti, P., Goold, R. G., Davis, M. B., Houlden, H., & Tabrizi, S. J. (2010). An ITPR1 gene deletion causes spinocerebellar ataxia 15/16: a genetic, clinical and radiological description. *Movement Disorders, 25*, 2176–2182.

Oberdick, J., Smeyne, R. J., Mann, J. R., Zackson, S., & Morgan, J. I. (1990). A promoter that drives transgene expression in cerebellar Purkinje and retinal bipolar neurons. *Science (New York, N.Y.), 248*, 223–226.

Schreiner, D., & Weiner, J. A. (2010). Combinatorial homophilic interaction between gamma-protocadherin multimers greatly expands the molecular diversity of cell adhesion. *Proceedings of the National Academy of Sciences of the United States of America, 107*, 14893–14898.

Serra, H. G., Duvick, L., Zu, T., Carlson, K., Stevens, S., Jorgensen, N., Lysholm, A., Burright, E., Zoghbi, H. Y., Clark, H. B., et al. (2006). RORalpha-mediated Purkinje cell development determines disease severity in adult SCA1 mice. *Cell, 127*, 697–708.

Shima, Y., Kengaku, M., Hirano, T., Takeichi, M., & Uemura, T. (2004). Regulation of dendritic maintenance and growth by a mammalian 7-pass transmembrane cadherin. *Developmental Cell, 7*, 205–216.

Shimobayashi, E., Wagner, W., & Kapfhammer, J. P. (2016). Carbonic anhydrase 8 expression in Purkinje cells is controlled by PKCγ activity and regulates Purkinje cell dendritic growth. *Molecular Neurobiology, 53*, 5149–5160.

Sidman, R. L., Lane, P. W., & Dickie, M. M. (1962). Staggerer, a new mutation in the mouse affecting the cerebellum. *Science (New York, N.Y.), 137*, 610–612.

Smeyne, R. J., Oberdick, J., Schilling, K., Berrebi, A. S., Mugnaini, E., & Morgan, J. I. (1991). Dynamic organization of developing Purkinje cells revealed by transgene expression. *Science (New York, N.Y.), 254*, 719–721.

Soha, J. M., & Herrup, K. (1995). Stunted morphologies of cerebellar Purkinje cells in lurcher and staggerer mice are cell-intrinsic effects of the mutant genes. *The Journal of Comparative Neurology, 357*, 65–75.

Sotelo, C., & Dusart, I. (2009). Intrinsic versus extrinsic determinants during the development of Purkinje cell dendrites. *Neuroscience, 162*, 589–600.

Strait, K. A., Schwartz, H. L., Seybold, V. S., Ling, N. C., & Oppenheimer, J. H. (1991). Immunofluorescence localization of thyroid hormone receptor protein beta 1 and variant alpha 2 in selected tissues: Cerebellar Purkinje cells as a model for beta 1 receptor-mediated developmental effects of thyroid hormone in brain. *Proceedings of the National Academy of Sciences of the United States of America, 88*, 3887–3891.

Sugawara, T., Hisatsune, C., Miyamoto, H., Ogawa, N., & Mikoshiba, K. (2017). Regulation of spinogenesis in mature Purkinje cells via mGluR/PKC-mediated phosphorylation of CaMKIIβ. *Proceedings of the National Academy of Sciences of the United States of America, 114*, E5256–e5265.

Sundararajan, L., Stern, J., & Miller, D. M., 3rd. (2019). Mechanisms that regulate morphogenesis of a highly branched neuron in C. elegans. *Developmental Biology, 451*, 53–67.

Takahashi, H., Arstikaitis, P., Prasad, T., Bartlett, T. E., Wang, Y. T., Murphy, T. H., & Craig, A. M. (2011). Postsynaptic TrkC and presynaptic PTPσ function as a bidirectional excitatory synaptic organizing complex. *Neuron, 69*, 287–303.

Takeo, Y. H. (2016). In utero electroporation of mouse cerebellar Purkinje cells. *Bio-Protocol, 6*, e1835.

Takeo, Y. H., Kakegawa, W., Miura, E., & Yuzaki, M. (2015). RORα regulates multiple aspects of dendrite development in cerebellar Purkinje cells in vivo. *The Journal of Neuroscience : the Official Journal of the Society for Neuroscience, 35*, 12518–12534.

Takeo, Y. H., Shuster, S. A., Jiang, L., Hu, M. C., Luginbuhl, D. J., Rülicke, T., Contreras, X., Hippenmeyer, S., Wagner, M. J., Ganguli, S., et al. (2021). GluD2- and Cbln1-mediated competitive interactions shape the dendritic arbors of cerebellar Purkinje cells. *Neuron, 109*, 629–644.

Tanabe, K., Kani, S., Shimizu, T., Bae, Y. K., Abe, T., & Hibi, M. (2010). Atypical protein kinase C regulates primary dendrite specification of cerebellar Purkinje cells by localizing Golgi apparatus. *The Journal of Neuroscience : the Official Journal of the Society for Neuroscience, 30*, 16983–16992.

Tanaka, M., Maeda, N., Noda, M., & Marunouchi, T. (2003). A chondroitin sulfate proteoglycan PTPzeta /RPTPbeta regulates the morphogenesis of Purkinje cell dendrites in the developing cerebellum. *The Journal of Neuroscience : the Official Journal of the Society for Neuroscience, 23*, 2804–2814.

Tomomura, M., Rice, D. S., Morgan, J. I., & Yuzaki, M. (2001). Purification of Purkinje cells by fluorescence-activated cell sorting from transgenic mice that express green fluorescent protein. *The European Journal of Neuroscience, 14*, 57–63.

Toyoda, S., Kawaguchi, M., Kobayashi, T., Tarusawa, E., Toyama, T., Okano, M., Oda, M., Nakauchi, H., Yoshimura, Y., Sanbo, M., et al. (2014). Developmental epigenetic modification regulates stochastic expression of clustered protocadherin genes, generating single neuron diversity. *Neuron, 82*, 94–108.

Usala, S. J., Menke, J. B., Watson, T. L., Wondisford, F. E., Weintraub, B. D., Bérard, J., Bradley, W. E., Ono, S., Mueller, O. T., & Bercu, B. B. (1991). A homozygous deletion in the c-erbA beta thyroid hormone receptor gene in a patient with generalized thyroid hormone resistance: Isolation and characterization of the mutant receptor. *Molecular Endocrinology, 5*, 327–335.

van de Leemput, J., Chandran, J., Knight, M. A., Holtzclaw, L. A., Scholz, S., Cookson, M. R., Houlden, H., Gwinn-Hardy, K., Fung, H. C., Lin, X., et al. (2007). Deletion at ITPR1 underlies ataxia in mice and spinocerebellar ataxia 15 in humans. *PLoS Genetics, 3*, e108.

Watanabe, F., Miyazaki, T., Takeuchi, T., Fukaya, M., Nomura, T., Noguchi, S., Mori, H., Sakimura, K., Watanabe, M., & Mishina, M. (2008). Effects of FAK ablation on cerebellar foliation, Bergmann glia positioning and climbing fiber territory on Purkinje cells. *The European Journal of Neuroscience, 27*, 836–854.

Weiss, G. M., & Pysh, J. J. (1978). Evidence for loss of Purkinje cell dendrites during late development: A morphometric Golgi analysis in the mouse. *Brain Research, 154*, 219–230.

Wewetzer, K., Rauvala, H., & Unsicker, K. (1995). Immunocytochemical localization of the heparin-binding growth-associated molecule (HB-GAM) in the developing and adult rat cerebellar cortex. *Brain Research, 693*, 31–38.

Williams, D. W., & Truman, J. W. (2005). Remodeling dendrites during insect metamorphosis. *Journal of Neurobiology, 64*, 24–33.

Yagi, T. (2012). Molecular codes for neuronal individuality and cell assembly in the brain. *Frontiers in Molecular Neuroscience, 5*, 45.

Yang, W. K., & Chien, C. T. (2019). Beyond being innervated: The epidermis actively shapes sensory dendritic patterning. *Open Biology, 9*, 180257.

Yu, L., Iwasaki, T., Xu, M., Lesmana, R., Xiong, Y., Shimokawa, N., Chin, W. W., & Koibuchi, N. (2015). Aberrant cerebellar development of transgenic mice expressing dominant-negative thyroid hormone receptor in cerebellar Purkinje cells. *Endocrinology, 156*, 1565–1576.

Zipursky, S. L., & Grueber, W. B. (2013). The molecular basis of self-avoidance. *Annual Review of Neuroscience, 36*, 547–568.

Chapter 8
Physiological Roles of Perineuronal Nets in Cerebellar Functions

Moritoshi Hirono

8.1 Introduction

In the mature central nervous system (CNS), the extracellular matrix enwraps the cell bodies and proximal processes of neurons and forms ladder-like structures. This formation is called a perineuronal net (PNN), which is the fourth most important element for the tetrapartite synapse in addition to presynapses, postsynapses, and glial cells, and serves as a regulator of synaptic functions and plasticity (Chelini et al., 2018; Dityatev & Rusakov, 2011). The major components of PNNs are chondroitin sulfate proteoglycans (CSPGs), tenascin-R, link proteins and hyaluronic acid, and they are synthesized by both neurons and glial cells (Oohashi et al., 2015). CSPGs are proteoglycans, which have a core protein with long chondroitin sulfate chains, which are known to prevent recovery from CNS injuries such as strokes (Quattromani et al., 2018). To date, it has been reported that during brain development, PNNs contribute to the normal maturation of neuronal circuits including fast-spiking parvalbumin-positive neurons (Cabungcal et al., 2013; Reichelt et al., 2019). PNNs morphologically restrict the production of new synapses and the pruning of old synapses and contribute to the regulation of neural plasticity in certain brain areas such as the visual cortex and the amygdala (Carulli et al., 2010; Gogolla et al., 2009; Pizzorusso et al., 2002; Shen, 2018). To remove PNNs, enzymatic or genetic techniques have been widely adopted, and studies using these methods have demonstrated that PNN deletion enhances the formation of memories by facilitating plasticity and encoding new information which occurs by attenuating forgetting or learning information easily (Fawcett et al., 2019; Reichelt et al., 2019; Wang & Fawcett, 2012). Moreover, PNNs have been focused on as the cause of the pathophysiology of brain disorders (Fawcett et al., 2019; Reichelt et al., 2019; Sorg et al., 2016; Wen et al., 2018). Aberrant PNNs have been reported to be associated with

M. Hirono (✉)
Department of Physiology, Wakayama Medical University, Wakayama-shi, Japan
e-mail: mhirono@wakayama-med.ac.jp

H. Mizusawa, S. Kakei (eds.), *Cerebellum as a CNS Hub*,
Contemporary Clinical Neuroscience, https://doi.org/10.1007/978-3-030-75817-2_8

neurodegenerative and neuropsychiatric disorders through abnormal neuroplasticity (Pantazopoulos & Berretta, 2015; Santiago et al., 2018; Suttkus et al., 2016). By contrast, it has not been elucidated whether PNNs can regulate cerebellar neuronal circuits dynamically and functionally, nor how PNNs contribute to the regulation of motor learning.

This chapter focuses on the physiological significance of PNNs, which enwrap the cell bodies of large glutamatergic neurons in the deep cerebellar nuclei (DCN) and synapses between cerebellar Purkinje cells (PCs) and large DCN neurons, in dynamic regulation of GABAergic transmission as well as in cerebellar motor learning. This chapter also introduces recent studies that genetically removed PNNs in order to examine the roles of PNNs in synaptic connections in the DCN and in cerebellar motor learning. These studies suggest that PNNs in the DCN restrict synaptic plasticity that is associated with more learning in adult animals.

8.2 Expression of PNNs in the Cerebellum

PNNs are detected in the brain using labeled *Wisteria floribunda agglutinin* (WFA), which is a lectin that recognizes the *N*-acetylgalactosamine segments of sugar chains in PNNs. In many brain regions, PNNs preferentially enwrap inhibitory parvalbumin-positive neurons, which are highly active and involved in critical periods of brain development, and play a crucial role in the regulation of synaptic plasticity and neuronal function (Cabungcal et al., 2013; Fawcett et al., 2019; Reichelt et al., 2019). By contrast, in the DCN, large and excitatory neurons, which are primarily glutamatergic (Telgkamp & Raman, 2002; Uusisaari et al., 2007), are enwrapped by PNNs in which aggrecan is the predominant CSPG (Bekku et al., 2012; Carulli et al., 2004, 2006, 2007; Foscarin et al., 2011; Zaremba et al., 1990; Zimmermann & Dours-Zimmermann, 2008) (Fig. 8.1a). Large glutamatergic DCN neurons receive direct GABAergic inhibition from PCs (Chan-Palay, 1977; De Zeeuw & Berrebi, 1996; Ito et al., 1964; Obata et al., 1967) and provide cerebellar output to various brain regions including the brainstem, thalamus, and ventral tegmental area (Carta et al., 2019; Kelly & Strick, 2003; Proville et al., 2014). Interestingly, Lugaro cells in the cerebellar cortex, which have spindle-shaped cell bodies underneath the PC layers and receive strong GABAergic inhibition from PCs via their axon collaterals (Dean et al., 2003; Dieudonné & Dumoulin, 2000; Dumoulin et al., 2001; Hirono, 2016; Hirono et al., 2012, 2017; Witter et al., 2016), are also recognized by a monoclonal antibody for agreecan Cat-301 (Crook et al., 2007; Sahin & Hockfield, 1990; Zaremba et al., 1990) (Fig. 8.1b and c). Thus, they could also be surrounded by PNNs.

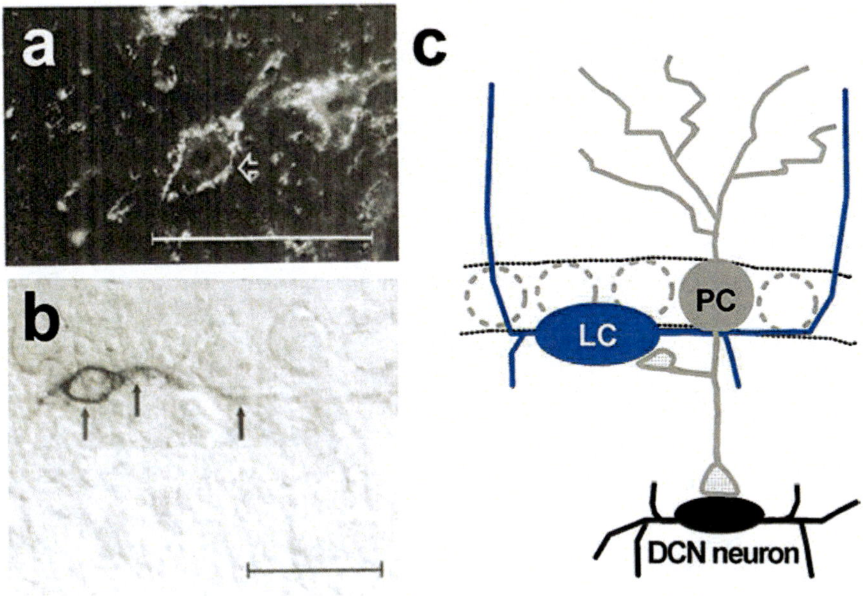

Fig. 8.1 Cerebellar neurons targeted by PC expressing PNN components. A large DCN neuron (**a**) and a Lugaro cell (**b**) are labeled by an antibody for aggrecan, which is a major component of PNNs. Both of these neurons receive strong GABAergic inhibition from PCs (**c**). (**a**) and (**b**) are from Zaremba et al. (1990). (Copyright 1990 Society for Neuroscience)

8.3 Functional Roles of PNNs in GABAergic Transmission in the DCN

To examine the role of PNNs, techniques for the pharmacological or genetic removal of PNNs have been used. Chondroitinase ABC (ChABC), which is a commonly used enzyme that degrades chondroitin sulfate glycosaminoglycans (GAGs), was administrated to mouse cerebellar slices to significantly reduce the intensity of the WFA labeling by 30% around large DCN neurons in our studies. The enzymatic and acute degradation of PNNs caused an increase in the amplitude and a decrease in the paired-pulse ratio of evoked inhibitory postsynaptic currents (IPSCs) recorded from large DCN neurons and also facilitated miniature IPSCs without changing the amplitude, suggesting that acute PNN digestion enhances presynaptic GABA release (Hirono et al., 2018). Under the lower density of PNNs, a decrease in the amplitude of IPSCs evoked by repetitive stimulation at 100 Hz showed steeper, which is similar to that observed in the DCN of juvenile mice (Saitow et al., 2018; Turecek et al., 2016), reflecting a higher propensity for synaptic plasticity and motor learning than adults. Recently, Carulli et al. removed 90% PNNs from large DCN neurons with long-term and stable overexpression of ChABC in the DCN via a lentiviral vector (LV). The LV-ChABC mice indicated that chronic ChABC treatment elicited the formation of new GABAergic terminals and a reduction in density of vesicular glutamate transporter 1 in the DCN, resulting in the reduction of the spontaneous firing of large DCN neurons (Carulli et al., 2020).

Manipulation of specific genes associated with PNNs can degrade PNN components. In mice deficient in each PNN component such as HAPLN1, HAPLN4, and tenascin-C, the number of PC terminals on large DCN neurons was reported to be altered (Bekku et al., 2003, 2012; Foscarin et al., 2011; Stamenkovic et al., 2017). By contrast, the short treatment of mouse cerebellar slices with ChABC did not cause morphological changes of synapses on large DCN neurons; thus, acute enzymatic removal of PNNs had no effect on the size or number of PC axon terminals on large DCN neurons (Hirono et al., 2018). Thus, acute removal of PNNs by ChABC functionally and dynamically increases the release probability of GABA rather than the formation of new presynaptic PC terminals. Therefore, PNNs can regulate presynaptic functions of PC axon terminals in the DCN and could modulate synaptic plasticity and neuronal excitation of large DCN neurons.

8.4 Possible Mechanisms Underlying PNN Regulation of Presynaptic GABA Release

There seem to be four possible mechanisms underlying the facilitation of presynaptic GABA release from PC axon terminals while acute PNN digestion by ChABC as follows: (1) releasing the limitation of availability of extracellular Ca^{2+}, (2) changing the conductance of presynaptic voltage-dependent Ca^{2+} channels, (3) enhancing the GABA release machinery, and (4) increasing the basal intracellular Ca^{2+} concentrations in presynaptic PC terminals. The most possible mechanism is that negatively charged CSPGs buffer Ca^{2+} via electrical interactions (Crank, 1975) and restrict the availability of extracellular Ca^{2+} to PC terminals (Hrabětová et al., 2009; Nicholson & Hrabětová, 2017). The extracellular matrix slows Ca^{2+} diffusion in the tissues, with this effect being reduced by PNN degradation by ChABC. However, the facilitation of presynaptic GABA release by ChABC was unaffected by changes in extracellular Ca^{2+} concentrations or by blocking of voltage-dependent Ca^{2+} channels (Hirono et al., 2018). Therefore, the latter two mechanisms mentioned above could be plausible: PNN degradation could enhance the release machinery (its Ca^{2+} sensitivity and/or the intrinsic activity) or drives intracellular Ca^{2+} release in PC axon terminals. It is possible that CSPG removal alters the activity of type IIa receptor-type protein tyrosine phosphatases (RPTPs), leukocyte common antigen-related (LAR) proteins, PTPσ, and PTPδ, since they are receptors for CSPG-chondroitin sulfates (Fisher et al., 2011; Shen et al., 2009). It has been reported that CSPGs regulate PTPσ and LAR receptors because ChABC-mediated ablation of the carbohydrate chains of CSPGs induces regeneration of axons by the inactivation of PTPσ via clustering (Coles et al., 2015; Lang et al., 2015). Cerebellar PCs express PTPσ and NgR3 as receptors for CSPG-chondroitin sulfates (Brown et al., 2020; Funahashi et al., 2008). Thus, ChABC-mediated chondroitin sulfate

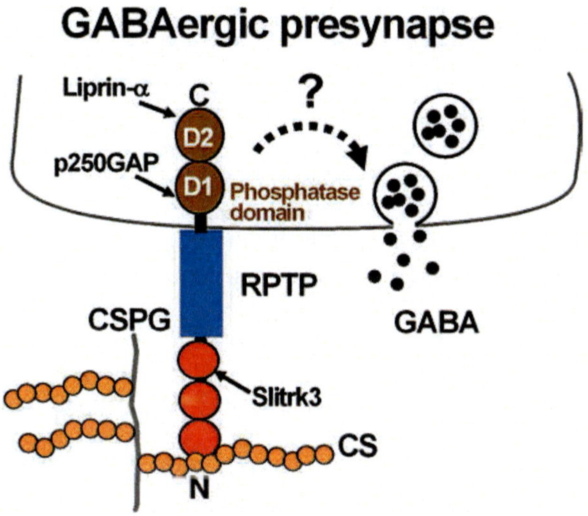

Fig. 8.2 Schematic drawing of a GABAergic presynapse. CSPG binds to RPTP through the chondroitin sulfate binding N-terminal of the RPTP. The intracellular domains of the RPTP may modulate GABA release by directly regulating the release machinery or the intracellular Ca^{2+} concentration via modulation of the activity of phosphatase or by interacting with liprin-α, p250GAP, and Slitrk3 (Südhof, 2012)

digestion, which can inactivate these receptors, may alter the balance between tyrosine phosphorylation and dephosphorylation, leading to increased activation of the release machinery for GABA release in PC axon terminals (Fig. 8.2). Further studies are needed to elucidate the precise mechanism of the new regulation of GABA release by PNNs.

8.5 PNN Depletion Facilitates Rebound Firing in Large DCN Neurons

After the relief of hyperpolarization mediated by PC inhibition, large DCN neurons produce post-inhibitory rebound firing (Aizenman & Linden, 1999; Baumel et al., 2009; Bengtsson et al., 2011; Hoebeek et al., 2010; Llinás & Mühlethaler, 1988; McKay et al., 2005), which contributes to the induction of long-term potentiation (LTP) of excitatory synaptic transmission between mossy fibers and large DCN neurons (McElvain et al., 2010; Person & Raman, 2010; Pugh & Raman, 2006, 2008; Racine et al., 1986). The mossy fiber LTP has been thought to contribute to facilitating action potentials in large DCN neurons (Wu & Raman, 2017; Yarden-Rabinowitz & Yarom, 2017) and to be one of the critical synaptic mechanisms underlying cerebellar motor learning (Attwell et al., 2002; Medina & Mauk, 1999), including ocular reflex adaptation (Kassardjian et al., 2005; Miles & Lisberger,

1981; Okamoto et al., 2011; Shutoh et al., 2006) and eyeblink conditioning (Christian & Thompson, 2003; Kistler & De Zeeuw, 2003; Krupa et al., 1993; Medina & Mauk, 1999; Wetmore et al., 2008). Acute PNN depletion by ChABC administration profoundly facilitated inhibitory GABAergic transmission between PCs and large DCN neurons, which induced an increase in the magnitude of hyperpolarization, thereby augmented the rebound firing in large DCN neurons, suggesting improvements in cerebellar motor learning, as indicated in the next section (Hirono et al., 2018). In contrast to this observation, in vivo electrophysiological recordings from neurons in the interpositus nuclei of LV-ChABC mice indicated that chronic PNN deletion significantly reduced the baseline activity of the neurons (Carulli et al., 2020). This firing reduction of DCN neurons could affect the extent of their rebound firing and is thought to decrease the rebound firing level in DCN neurons during expression of conditioned responses (CRs) (Ten Brinke et al., 2017).

8.6 PNNs in the Interpositus Nuclei Regulate Delay Eyeblink Conditioning

ChABC has been injected directly into the various brain areas, and in vivo experiments with PNN-elimination have demonstrated the importance of PNNs in brain functions such as learning and memory (Pizzorusso et al., 2002; Corvetti & Rossi, 2005; Gogolla et al., 2009; de Vivo et al., 2013; Romberg et al., 2013; Banerjee et al., 2017). Inactivation of the DCN via lesions or pharmacological treatments has been reported to restrict increases in CRs seen during delay eyeblink training (Yeo et al., 1985; Attwell et al., 2002; Ohyama et al., 2006; Boele et al., 2010; Sakamoto and Endo, 2010). Additionally, activation of the interpositus nuclei is required for obtaining CRs of trained mice (Heiney et al., 2014). Thus, to investigate the roles of PNNs in delay eyeblink conditioning, it is reasonable to inject ChABC into the interpositus nuclei and examine its effect on the conditioning. Both the ChABC-injected mice and the LV-ChABC mice exhibited significant enhancements in their CR rates, suggesting that the PNN depletion in the interpositus nuclei facilitates CR acquisition (Carulli et al., 2020; Hirono et al., 2018). Although this facilitation of motor learning is akin to observations in other brain areas, where PNN removal causes structural synaptic reorganization and facilitates memory formation in adult animals (Corvetti & Rossi, 2005; de Vivo et al., 2013; Gogolla et al., 2009; Romberg et al., 2013), a novel role of PNNs in regulating presynaptic GABA release in the DCN is proposed, and a novel mechanistic insight indicating that not only the structural plasticity but also the functional plasticity of synapses regulated by PNNs has been reported to affect behavioral flexibility in adult animals (Hirono et al., 2018). In the DCN, the developmental expression of PNNs attenuates GABAergic transmission at synapses between PCs and large DCN neurons and prevents unnecessary formation of new memories, which may also protect previously encoded memories in the cerebellum during adulthood. Intriguingly, PNNs in the DCN are altered during the acquisition and consolidation of eyeblink conditioning in adult mice (Carulli et al., 2020). Taken together, these findings indicate that modulation of PNNs is crucial for the dynamic regulation of synaptic transmission in the DCN and for fine control of cerebellar motor learning.

8.7 Conclusion

PNNs that wrap CNS neurons contribute to the regulation of synaptic plasticity and neuronal function and are implicated in regulating various brain functions such as learning and memory. PNN depletion enhanced GABAergic transmission in the DCN: chronic PNN removal increased the number of PC terminals contacting large DCN neurons, whereas acute PNN removal functionally facilitated GABA release from the presynaptic PC terminals. Mice that pharmacologically or genetically received ChABC in the interpositus nuclei exhibited a higher CR rate in delay eyeblink conditioning compared to control mice. This evidence suggests that PNN modification makes memories more easily modifiable and stronger, which influences the flexibility of adult cerebellar functions.

Although the cerebellum is classically thought to play a crucial role in motor coordination (Ito, 1984), recent studies suggest that the cerebellum also contributes to a variety of brain functions, including reward prediction (Carta et al., 2019; Heffley et al., 2018; Kostadinov et al., 2019; Wagner et al., 2017) and motor planning (Chabrol et al., 2019; Gao et al., 2018; Wagner et al., 2019). Since PNNs regulate spontaneous firing of large DCN neurons, which project their axons to various brain areas such as the brainstem, thalamus, and ventral tegmental area (Carta et al., 2019; Kelly & Strick, 2003; Proville et al., 2014), alternations of PNN densities in the DCN can regulate not only cerebellar motor learning but also cognitive functions depending on the cerebellum. Thus, future studies on mechanisms underlying the manipulation of specific components in PNNs that are formed in the DCN will be important to unravel the physiology of cerebellar motor learning and cognitive processing.

Acknowledgements This work was supported by JSPS KAKENHI Grant Number JP19K06890.

References

Aizenman, C. D., & Linden, D. J. (1999). Regulation of the rebound depolarization and spontaneous firing patterns of deep nuclear neurons in slices of rat cerebellum. *Journal of Neurophysiology, 82*, 1697–1709.

Attwell, P. J., Ivarsson, M., Millar, L., & Yeo, C. H. (2002). Cerebellar mechanisms in eyeblink conditioning. *Ann NY Acad Sci, 978*, 79–922.

Banerjee, S. B., Gutzeit, V. A., Baman, J., Aoued, H. S., Doshi, N. K., Liu, R., et al. (2017). Perineuronal nets in the adult sensory cortex are necessary for fear learning. *Neuron, 95*, 169–179.

Baumel, Y., Jacobson, G. A., & Cohen, D. (2009). Implications of functional anatomy on information processing in the deep cerebella nuclei. *Frontiers in Cellular Neuroscience, 3*, 14.

Bekku, Y., Saito, M., Moser, M., Fuchigami, M., Maehara, A., Nakayama, M., Kusachi, S., Ninomiya, Y., & Oohashi, T. (2012). Bral2 is indispensable for the proper localization of brevican and the structural integrity of the perineuronal net in the brainstem and cerebellum. *The Journal of Comparative Neurology, 520*, 1721–1736.

Bekku, Y., Su, W. D., Hirakawa, S., Fässler, R., Ohtsuka, A., Kang, J. S., Sanders, J., Murakami, T., Ninomiya, Y., & Oohashi, T. (2003). Molecular cloning of Bral2, a novel brain-specific link protein, and immunohistochemical colocalization with brevican in perineuronal nets. *Molecular and Cellular Neurosciences, 24*, 148–159.

Bengtsson, F., Ekerot, C. F., & Jorntell, H. (2011). *In vivo* analysis of inhibitory synaptic inputs and rebounds in deep cerebellar nuclear neurons. *PLoS One, 6*, e18822.

Boele, H. J., Koekkoek, S. K. E., & De Zeeuw, C. J. (2010). Cerebellar and extracerebellar involvement in mouse eyeblink conditioning: the ACDC model. *Frontiers in Cellular Neuroscience, 3*, 19.

Brown, A. S., Meera, P., Quinones, G., Magri, J., Otis, T. S., Pulst, S. M., & Oro, A. E. (2020). Receptor protein tyrosine phosphatases control Purkinje neuron firing. *Cell Cycle, 19*, 153–159.

Cabungcal, J. H., Steullet, P., Morishita, H., Kraftsik, R., Cuenod, M., Hensch, T. K., & Do, K. Q. (2013). Perineuronal nets protect fast-spiking interneurons against oxidative stress. *Proceedings of the National Academy of Sciences of the United States of America, 110*, 9130–9135.

Carta, I., Chen, C. H., Schott, A. L., Dorizan, S., & Khodakhah, K. (2019). Cerebellar modulation of the reward circuitry and social behavior. *Science, 363*.

Carulli, D., Buffo, A., & Strata, P. (2004). Reparative mechanisms in the cerebellar cortex. *Progress in Neurobiology, 72*, 373–398.

Carulli, D., Pizzorusso, T., Kwok, J. C., Putignano, E., Poli, A., Forostyak, S., Andrews, M. R., Deepa, S. S., Glant, T. T., & Fawcett, J. W. (2010). Animals lacking link protein have attenuated perineuronal nets and persistent plasticity. *Brain, 133*, 2331–2347.

Carulli, D., Rhodes, K. E., Brown, D. J., Bonnert, T. P., Pllack, S. J., Oliver, K., Strata, P., & Fawcett, J. W. (2006). Composition of perineuronal nets in the adult rat cerebellar and the cerebellar origin of their components. *The Journal of Comparative Neurology, 494*, 559–577.

Carulli, D., Rhodes, K. E., & Fawcett, J. W. (2007). Upregulation of aggrecan, link protein 1, and hyaluronan synthases during formation of perineuronal nets in the rat cerebellum. *The Journal of Comparative Neurology, 501*, 83–94.

Carulli, D., Broersen, R., de Winter, F., Muir, E. M., Meškovic, M., de Waal, M., de Vries, S., Boele, H. J., Canto, C. B., De Zeeuw, C. I., & Verhaagen, J. (2020). Cerebellar plasticity and associative memories are controlled by perineuronal nets. *Proc Natl Acad Sci USA, 117*, 6855–6865.

Chabrol, F. P., Blot, A., & Mrsic-Flogel, T. D. (2019). Cerebellar contribution to preparatory activity in motor neocortex. *Neuron, 103*(506–519), e504.

Chan-Palay V (1977) Cerebellar dentate nucleus. Organization, cytology, and transmitters .. : Springer

Chelini, G., Pantazopoulos, H., Durning, P., & Berretta, S. (2018). The tetrapartite synapse: A key concept in the pathophysiology of schizophrenia. *European Psychiatry*.

Christian, K. M., & Thompson, R. F. (2003). Neural substrates of eyeblink conditioning: Acquisition and retention. *Learning & Memory, 10*, 427–455.

Coles, C. H., Jones, E. Y., & Aricescu, A. R. (2015). Extracellular regulation of type Iia receptor protein tyrosine phosphatases: Mechanistic insights from structural analysis. *Seminars in Cell & Developmental Biology, 37*, 98–107.

Corvetti, L., & Rossi, F. (2005). Degradation of chondroitin sulfate proteoglycans induces sprouting of intact Purkinje axons in the cerebellar of the adult rat. *The Journal of Neuroscience, 25*, 7150–7158.

Crank, J. (1975). *The mathematics of diffusion* (2nd ed.). Clarendon Press.

Crook, J. D., Hendrickson, A., Erickson, A., Possin, D., & Robinson, F. R. (2007). Purkinje cell axon collaterals terminate on Cat-301+ neurons in Macaca monkey cerebellum. *Neuroscience, 149*, 834–844.

Dean, I., Robertson, S. J., & Edwards, F. A. (2003). Serotonin drives a novel GABAergic synaptic current recorded in rat cerebellar Purkinje cells: A Lugaro cell to Purkinje cell synapse. *The Journal of Neuroscience, 23*, 4457–4469.

de Vivo, L., Landi, S., Panniello, M., Baroncelli, L., Chiezi, S., Mariotti, L., Spolidoro, M., Pizzorusso, T., Mafffei, L., & Ratto, G. M. (2013). Extracellular matrix inhibits structural and functional plasticity of dendritic spines in the adult visual cortex. *Nature Communications, 4*, 1484.

De Zeeuw, C. I., & Berrebi, A. S. (1996). Individual Purkinje cell axons terminate on both inhibitory and excitatory neurons in the cerebellar and vestibular nuclei. *Ann NY Acad Sci, 781*, 607–610.

Dieudonné, S., & Dumoulin, A. (2000). Serotonin-driven long-range inhibitory connections in the cerebellar cortex. *The Journal of Neuroscience, 20*, 1837–1848.

Dityatev A, Rusakov DA (2011) Molecular signals of plasticity at the tetrapartite synapse. Current Opinion in synapse. Current Opinion in Neurobiology.

Dumoulin, A., Triller, A., & Dieudonné, S. (2001). IPSC kinetics at identified GABAergic and mixed GABAergic and glycinergic synapses onto cerebellar Golgi cells. *The Journal of Neuroscience, 21*, 6045–6057.

Fawcett, J. W., Oohashi, T., & Pizzorusso, T. (2019). The roles of perineuronal nets and the perinodal extracellular matrix in neuronal function. *Nature Reviews. Neuroscience, 20*, 451–465.

Fisher, D., Xing, B., Dill, J., Li, H., Hoang, H. H., Zhao, Z., Yang, X. L., Bachoo, R., Cannon, S., Longo, F. M., Sheng, M., Silver, J., & Li, S. (2011). Leukocyte common antigen-related phosphatase is a functional receptor for chondroitin sulfate proteoglycan axon growth inhibitors. *The Journal of Neuroscience, 31*, 14051–14066.

Foscarin, S., Ponchione, D., Pajaj, E., Leto, K., Gawlak, M., Wilczynski, G. M., Rossi, F., & Carulli, D. (2011). Experience-dependent plasticity and modulation of growth regulatory molecules at central synapses. *PLoS One, 6*, e1666.

Funahashi, S., Hasegawa, T., Nagano, A., & Sato, K. (2008). Differential expression patterns of messenger RNAs encoding nogo receptors and their ligands in the rat central nervous system. *The Journal of Comparative Neurology, 506*, 141–160.

Gao, Z., Davis, C., Thomas, A. M., Economo, N. M., Abrego, A. M., Svoboda, K., De Zeeuw, C. I., & Li, N. (2018). A cortico-cerebellar loop for motor planning. *Nature, 563*, 113–116.

Gogolla, N., Caroni, P., Lüthi, A., & Herry, C. (2009). Perineuronal nets protect fear memories from erasure. *Science, 325*, 1258–1261.

Heffley, W., Song, E. Y., Xu, Z., Taylor, B. N., Hughes, M. A., McKinney, A., Joshua, M., & Hull, C. (2018). Coordinated cerebellar climbing fiber activity signals learned sensorimotor predictions. *Nature Neuroscience, 21*, 1431–1441.

Heiney, S. A., Wohl, M. P., Chettih, S. N., Ruffolo, L. I., & Medina, J. F. (2014). Cerebellar-dependent expression of motor leaning during eyeblink conditioning in head-fixed mice. *The Journal of Neuroscience. 34*, 14845–14853.

Hirono, M., Saitow, F., Kudo, M., Suzuki, H., Yanagawa, Y., Yamada, M., Nagao, S., Konishi, S., & Obata, K. (2012). Cerebellar globular cells receive monoaminergic excitation and monosynaptic inhibition from Purkinje cells. *PLoS One, 7*, e29663.

Hirono, M. (2016). Lugaro cells. In D. Gruol, N. Koibuchi, M. Manto, M. Molinari, J. D. Schmahmann, & Y. Shen (Eds.), *Essentials of cerebellum and cerebellar disorders* (pp. 207–211). Springer.

Hirono, M., Nagao, S., Yanagawa, Y., & Konishi, S. (2017). Monoaminergic modulation of GABAergic transmission onto cerebellar globular cells. *Neuropharmacology, 118*, 79–89.

Hirono, M., Watanabe, S., Karube, F., Fujiyama, F., Kawahara, S., Nagao, S., Yanagawa, Y., & Misonou, H. (2018). Perineuronal nets in the deep cerebellar nuclei regulate GABAergic transmission and delay eyeblink conditioning. *The Journal of Neuroscience, 38*, 6130–6144.

Hoebeek, F. E., Witter, L., Ruigrok, T. J. H., & De Zeeuw, C. I. (2010). Differential olivo-cerebellar cortical control of rebound activity in the cerebellar nuclei. *Proc Natl Acad Sci USA, 107*, 8410–8415.
Hrabětová, S., Masri, D., Tao, L., Xiao, F., & Nicholson, C. (2009). Calcium diffusion enhanced after cleavage of negatively charged components of brain extracellular matrix by chondroitinase ABC. *The Journal of Physiology, 587*, 4029–4049.
Ito, M. (1984). *The cerebellum and neural control.* Raven Press.
Ito, M., Yoshida, M., & Obata, K. (1964). Monosynaptic inhibition of the intracerebellar nuclei induced from the cerebellar cortex. *Experientia, 20*, 575–576.
Kassardjian, C. D., Tan, Y. F., Chung, J.-Y., Heskin, R., Peterson, M. J., & Broussard, D. M. (2005). The site of a memory shifts with consolidation. *The Journal of Neuroscience, 25*, 7979–7985.
Kelly, R. M., & Strick, P. L. (2003). Cerebellar loops with motor cortex and prefrontal cortex of a nonhuman primate. *The Journal of Neuroscience, 23*, 8432–8444.
Kistler, W. M., & De Zeeuw, C. I. (2003). Time windows and reverberating loops: A reverse-engineering approach to cerebellar function. *Cerebellum, 2*, 44–54.
Kostadinov, D., Beau, M., Pozo, M. B., & Hausser, M. (2019). Predictive and reactive reward signals conveyed by climbing fiber inputs to cerebellar Purkinje cells. *Nature Neuroscience, 22*, 950–962.
Krupa, D. J., Thompson, J. K., & Thompson, R. F. (1993). Localization of a memory trace in the mammalian brain. *Science, 260*, 989–991.
Lang, B. T., Cregg, J. M., DePaul, M. A., Tran, A. P., Xu, K., Dyck, S. M., Madalena, K. M., Brown, B. P., Weng, Y. L., Li, S., Karimi-Abdolrezaee, S., Busch, S. A., Shen, Y., & Silver, J. (2015). Modulation of the proteoglycan receptor PTP promotes recovery after spinal cord injury. *Nature, 518*, 404–408.
Llinás, R., & Mühlethaler, M. (1988). Electrophysiology of Guinea-pig cerebellar nuclear cells in the *in vivo* brain stem-cerebellar preparation. *The Journal of Physiology, 404*, 241–258.
McElvain, L. E., Bagnall, M. W., Sakatos, A., & du Lac, S. (2010). Bidirectional plasticity gated by hyperpolarization controls the gain of postsynaptic firing responses at central vestibular nerve synapses. *Neuron, 68*, 763–775.
McKay, B. E., Molineux, M. L., Mehaffey, W. H., & Tuner, R. W. (2005). KV1 K^+ channels control Purkinje cell output to facilitate postsynaptic rebound discharge in deep cerebellar neurons. *The Journal of Neuroscience, 25*, 1481–1492.
Medina, J. F., & Mauk, M. D. (1999). Stimulations of cerebellar motor learning: Computational analysis of plasticity at the mossy fiber to deep nucleus synapse. *The Journal of Neuroscience, 19*, 7140–7151.
Miles, F. A., & Lisberger, S. G. (1981). Plasticity in the vestibule-ocular reflex: A new hypothesis. *Annual Review of Neuroscience, 4*, 273–299.
Nicholson, C., & Hrabětová, S. (2017). Brain extracellular space: The final frontier of neuroscience. *Biophysical Journal, 113*, 2133–2142.
Obata, K., Ito, M., Ochi, R., & Sato, N. (1967). Pharmacological properties of the postsynaptic inhibition by Purkinje cell axons and the action of γ–aminobutyric acid on Deiters neurons. *Experimental Brain Research, 4*, 43–57.
Ohyama, T., Nores, W. L., Medina, J. F., Riusech, F. A., & Mauk, M. D. (2006). Learning-induced plasticity in deep cerebellar nucleus. *The Journal of Neuroscience, 26*, 12656–12663.
Okamoto, T., Endo, S., Shirao, T., & Nagao, S. (2011). Role of cerebellar cortical protein synthesis in transfer of memory trace of cerebellum-dependent motor learning. *The Journal of Neuroscience, 31*, 8958–8966.
Oohashi, T., Edamatsu, M., & Carulli, D. (2015). The hyaluronan and proteoglycan link proteins: Organizers of the brain extracellular matrix and key molecules for neuronal function and plasticity. *Experimental Neurology, 274*, 134–144.
Pantazopoulos, H., & Berretta, S. (2015). In sickness and in health: Perineuronal nets and synaptic plasticity. *Neural Plasticity.*

Person, A. L., & Raman, I. M. (2010). Deactivation of L-type ca current by inhibition controls LTP at excitatory synapses in the cerebellar nuclei. *Neuron, 66*, 550–559.

Pizzorusso, T., Medini, P., Berardi, N., Chierzi, S., Fawcett, J. W., & Maffei, L. (2002). Reactivation of ocular dominance plasticity in the adult visual cortex. *Science, 298*, 1248–1251.

Proville, R. D., Spolidoro, M., Guyon, N., Dugué, G. P., Selimi, F., Isope, P., Popa, D., & Léna, C. (2014). Cerebellum involvement in cortical sensorimotor circuits for the control of voluntary movements. *Nature Neuroscience, 17*, 1233–1239.

Pugh, J. R., & Raman, I. M. (2006). Potentiation of mossy fiber EPSCs in the cerebellar nuclei by NMDA receptor activation followed by postinhibitory rebound current. *Neuron, 51*, 113–123.

Pugh, J. R., & Raman, I. M. (2008). Mechanisms of potentiation of mossy fiber EPSCs in the cerebellar nuclei by coincident synaptic excitation and inhibition. *The Journal of Neuroscience, 28*, 10549–10560.

Quattromani, M. J., Pruvost, M., Guerreiro, C., Backlund, F., Englund, E., Aspberg, A., Jaworski, T., Hakon, J., Ruscher, K., Kaczmarek, L., & Vivien, D. (2018). Extracellular matrix modulation is driven by experience-dependent plasticity during stroke recovery. *Molecular Neurobiology, 55*, 2196–2213.

Racine, R. J., Wilson, D. A., Gingell, R., & Sunderland, D. (1986). Long-term potentiation in the interpositus and vestibular nuclei in the rat. *Experimental Brain Research, 63*, 158–162.

Reichelt, A. C., Hare, D. J., Bussey, T. J., & Saksida, L. M. (2019). Perineuronal nets: Plasticity, protection and therapeutic potential. *Trends in Neurosciences, 42*, 458–470.

Romberg, C., Yang, S., Melani, R., Andrews, M. R., Horner, A. E., Spillantini, M. G., Bussey, T. J., Fawcett, J. W., Pizzorusso, T., & Sakside, L. M. (2013). Depletion of perineuronal nets enhances recognition memory and long-term depression in the perirhinal cortex. *The Journal of Neuroscience, 33*, 7057–7065.

Sahin, M., & Hockfield, S. (1990). Molecular identification of the Lugaro cell in the cat cerebellar cortex. *The Journal of Comparative Neurology, 301*, 575–584.

Saitow, F., Nagano, M., & Suzuki, H. (2018). Developmental changes in serotonergic modulation of GABAergic synaptic transmission and postsynaptic GABAA receptor composition in the cerebellar nuclei. *Cerebellum.*

Sakamoto, T., & Endo, S. (2010). Amygdala, deep cerebellar nuclei and red nucleus contribute to delay eyeblink conditioning in C57BL/6 mice. *The European Journal of Neuroscience. 32*, 1537–1551.

Santiago, A. N., Lim, K. Y., Opendak, M., Sullivan, R. M., & Aoki, C. (2018). Early life trauma increases threat response of peri-weaning rats, reduction of axo-somatic synapses formed by parvalbumin cells and perneuronal net in the basolateral nucleus of amygdala. *The Journal of Comparative Neurology, 526*, 2647–2664.

Shen, H. H. (2018). Perineuronal nets gain prominence for their role in learning, memory, and plasticity. *Proc Natl Acad Sci USA, 115*, 9813–9815.

Shen, Y., Tenney, A. P., Busch, S. A., Horn, K. P., Cuascut, F. X., Liu, K., He, Z., Silver, J., & Flanagan, J. G. (2009). PTPsigma is a receptor for chondroitin sulfate proteoglycan, an inhibitor of neural regeneration. *Science, 326*, 592–596.

Shutoh, F., Ohki, M., Kitazawa, H., Itohara, S., & Nagao, S. (2006). Memory trace of motor learning shifts transsynaptically from cerebellar cortex to nuclei for consolidation. *Neuroscience, 139*, 767–777.

Sorg, B. A., Berretta, S., Blacktop, J. M., Fawcett, J. W., Kitagawa, H., Kwok, J. C. F., & Miquel, M. (2016). Casting a wide net: Role of perineuronal nets in neuronal plasticity. *The Journal of Neuroscience, 36*, 11459–11468.

Stamenkovic, V., Stamenkovic, S., Jaworski, T., Gawlak, M., Jovanovic, M., Jakovcevski, I., Wilczynski, G. M., Kaczmarek, L., Schachner, M., Radenovic, L., & Andjus, P. R. (2017). The extracellular matrix glycoprotein tenascin-C and matrix metalloproteinases modify cerebellar structural plasticity by exposure to an enriched environment. *Brain Structure & Function, 222*, 393–415.

Suttkus, A., Morawski, M., & Arendt, T. (2016). Protective properties of neural extracellular matrix. *Molecular Neurobiology, 53*, 73–82.
Südhof, T. C. (2012). The presynaptic active zone. *Neuron, 75*, 11–25.
Telgkamp, P., & Raman, I. M. (2002). Depression of inhibitory synaptic transmission between Purkinje cells and neurons of the cerebellar nuclei. *The Journal of Neuroscience, 22*, 8447–8457.
Ten Brinke, M. M., Heiney, S. A., Wang, X., Proietti-Onori, M., Boele, H. J., Bakermans, J., Medina, J. F., Gao, Z., & De Zeeuw, C. I. (2017). 2017 dynamic modulation of activity in cerebellar nuclei neurons during pavlovian eyeblink conditioning in mice. *eLife, 6*, e28132.
Turecek, J., Jackman, S. L., & Regehr, W. G. (2016). Synaptic specializations support frequency-independent Purkinje cell output from the cerebellar cortex. *Cell Reports, 17*, 3256–3268.
Uusisaari, M., Obata, K., & Knöpfel, T. (2007). Morphological and electrophysiological properties of GABAergic and non-GABAergic cells in the deep cerebellar nuclei. *Journal of Neurophysiology, 97*, 901–911.
Wagner, M. J., Kim, T. H., Savall, J., Schnitzer, M. J., & Luo, L. (2017). Cerebellar granule cells encode the expectation of reward. *Nature, 544*, 96–100.
Wagner, M. J., Kim, T. H., Kadmon, J., Nguyen, N. D., Ganguli, S., Schnitzer, M. J., & Luo, L. (2019). Shared cortex-cerebellum dynamics in the execution and learning of a motor task. *Cell, 177*(669–682), e624.
Wang, D., & Fawcett, J. (2012). The perineuronal net and the control of CNS plasticity. *Cell and Tissue Research, 349*, 147–160.
Wen, T. H., Binder, D. K., Ethell, I. M., & Razak, K. A. (2018). The perineuronal 'safety' net? Perineuronal net abnormalities in neurological disorders. *Frontiers in Molecular Neuroscience, 11*, 270.
Wetmore, D. Z., Mukamel, E. A., & Schnizer, M. S. (2008). Lock-and-key mechanisms of cerebellar memory recall based on rebound currents. *Journal of Neurophysiology, 100*, 2328–2347.
Witter, L., Rudolph, S., Pressler, R. T., Lahlaf, S. I., & Regehr, W. G. (2016). Purkinje cell collaterals enable output signals from the cerebellar cortex to feed back to Purkinje cells and interneurons. *Neuron, 91*, 312–319.
Wu, Y., & Raman, I. M. (2017). Facilitation of mossy fibre-driven spiking in the cerebellar nuclei by the synchrony of inhibition. *The Journal of Physiology, 595*, 5245–5264.
Yarden-Rabinowitz, Y., & Yarom, Y. (2017). In vivo analysis of synaptic activity in cerebellar nuclei neurons unravels the efficacy of excitatory inputs. *The Journal of Physiology, 595*, 5945–5963.
Yeo, C. H., Hardiman, M. J., & Glickstein, M. (1985). Classical conditioning of the nictitating membrane response of the rabbit. *Experimental Brain Research, 60*, 114–126.
Zaremba, S., Naegele, J. R., Barnstable, C. J., & Hockfield, S. (1990). Neuronal subsets express multiple high-molecular-weight cell-surface glycoconjugates defined by monoclonal antibodies Cat-301 and VC1.1. *The Journal of Neuroscience, 10*, 2985–2995.
Zimmermann, D. R., & Dours-Zimmermann, M. T. (2008). Extracellular matrix of the central nervous system: From neglect to challenge. *Histochemistry and Cell Biology, 130*, 635–653.

Part III
Information Processing in the Cerebellar Neurocircuitry and Its Model

Chapter 9
Roles of Cerebellum-Brainstem Loops in Predictive Optokinetic Eye Velocity Control in Fish, Mice, and Humans

Yutaka Hirata

9.1 Predictive Optokinetic Response (pOKR) in Goldfish

In most vertebrates, a large moving visual scene induces a reflexive eye movement called the optokinetic response (OKR) to stabilize visual images on the retina. In natural environments, such large-field visual motion is often produced by an animal's own movement. When an animal moves, its head moves in reference to the world, causing visual flow in the opposite direction to the head motion. During head motion, another reflexive eye movement, called the vestibulo-ocular reflex (VOR), is primarily induced, and the eyes are counter-rotated in the orbits to stabilize visual flow projected onto the retina. However, the VOR is not perfectly compensatory, resulting in residual visual motion in the opposite direction to the head turn, thereby inducing the OKR. The OKR was shown to be predictive after prolonged exposure to temporally periodic visual motion stimulation first in goldfish (Marsh & Baker, 1997) and later in other animals (Miki et al., 2020). This section summarizes characteristics of predictive OKR (pOKR) identified in various behavioral experiments mainly in goldfish.

Y. Hirata (✉)
Department of Robotic Science and Technology, Chubu University, Kasugai, Japan

Center of Mathematics for Artificial Intelligence and Data Science, Chubu University, Kasugai, Japan

Academy of Emerging Sciences, Chubu University, Kasugai, Japan
e-mail: yutaka@isc.chubu.ac.jp

H. Mizusawa, S. Kakei (eds.), *Cerebellum as a CNS Hub*,
Contemporary Clinical Neuroscience, https://doi.org/10.1007/978-3-030-75817-2_9

9.1.1 General Characteristics

In the laboratory environment, the OKR is induced by moving a large-field visual pattern (typically high-contrast dots or stripes) around an animal whose head is fixed and kept stationary. Time varying eye position data are recorded by video-oculography or a scleral search coil method. Eye velocity is calculated as the first derivative of eye position data. Traditionally, performance of the OKR has been quantified by its gain, defined as eye velocity divided by visual stimulus velocity, and is usually less than 1.0 in naïve animals. Figure 9.1a illustrates horizontal OKR eye velocity of a representative naïve goldfish (Control) responding to bidirectional (CW and CCW) velocity step visual stimulation (gray). Eye velocity gradually builds up to the visual stimulus velocity until the stimulus direction flips, every 8 sec in this example. Figure 9.1b illustrates eye velocity from the same animal 30 minutes after the beginning of repetitive presentation of the same visual stimulation (Trained (Training period)). The trained eye velocity is clearly closer to the stimulus velocity than the control, showing that OKR gain increased, approaching 1.0. This is known as OKR gain adaptation (Ito et al., 1979; Marsh & Baker, 1997; Nagao, 1983). In addition, there is a noticeable change in the eye velocity profile. The trained eye velocity began to decrease prior to the change in stimulus direction, as if the animal anticipated the upcoming change (Fig. 9.1b black arrows). This predictive decrease in eye velocity can be seen more clearly in the averaged traces over different goldfish as shown in Fig. 9.1c (red arrow with a negative slope in contrast to cyan arrow with a positive slope).

When the visual stimulus period was extended after the acquisition of this behavior, the predictive eye velocity decrease persisted even though the visual stimulation continued at a constant velocity (Fig. 9.1d Trained (extended period)). This

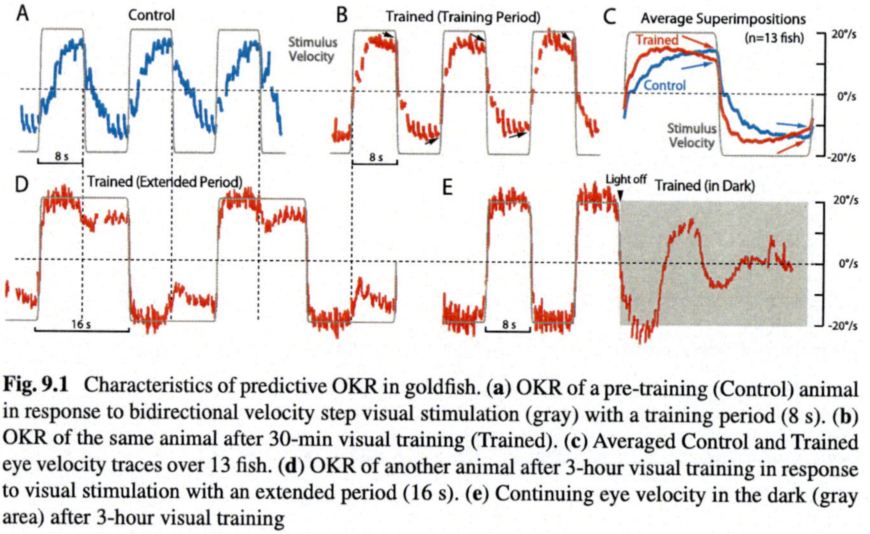

Fig. 9.1 Characteristics of predictive OKR in goldfish. (**a**) OKR of a pre-training (Control) animal in response to bidirectional velocity step visual stimulation (gray) with a training period (8 s). (**b**) OKR of the same animal after 30-min visual training (Trained). (**c**) Averaged Control and Trained eye velocity traces over 13 fish. (**d**) OKR of another animal after 3-hour visual training in response to visual stimulation with an extended period (16 s). (**e**) Continuing eye velocity in the dark (gray area) after 3-hour visual training

predictive eye velocity behavior seen after several hours of training continued, on average, 15 to 30 min before gradually returning toward control behavior. In addition, an alternating eye velocity is observed in the dark when the visual stimulation was suddenly turned off after 3-hour bidirectional training (Fig. 9.1e Trained (in dark)) which is never observed in naïve animals. Duration of this eye velocity, alternation in the dark after the 3-hour training varied from 15 to 60 seconds. When unidirectional (only CW) step velocity visual stimulation was used, eye velocity in the dark after 3-hour training persisted only in the trained direction. In other words, the predictive OKR is direction-selective (Miki et al., 2018). Alternating visual stimulus periods that goldfish could predict have been shown to range from 2 seconds to 128 seconds (Marsh & Baker, 1997).

9.1.2 Prediction of Stimulus Initiation and Termination

In order to achieve timed, predictive, periodic motor output, there may be two different prediction mechanisms involved: one for the stimulus start so as to anticipate the start of movement (initiation) and one for the stimulus end so as to know when to stop (termination). The bidirectional velocity step visual stimulation shown in Fig. 9.1 did not allow the onset and offset timings of the stimulus to be distinguished since offset in one direction was always onset to the other direction. To discriminate visual stimulus onset from offset, unidirectional velocity step visual stimulation in which periodic visual stimulation rotated only in one direction was employed (Miki et al., 2018). As observed in Fig. 9.1a for bidirectional stimulation, Trained eye velocity showed a gradual decrease before the offset of each cycle of the visual stimulation, demonstrating that the visual stimulus offset (termination) is actually predicted. It was also observed that trained eye velocity started to increase before onset (initiation) of the visual stimulation (Miki et al., 2018).

When unidirectional staircase visual stimulation was employed, which stepped up from 10 deg./s to 20 deg./s at 4 seconds after the onset of each stimulation, not only a termination component but also an initiation component presented in the trained eye velocity trace before the stimulus velocity stepped up. Interestingly, trained eye velocity started to increase gradually and actually exceeded the stimulus velocity (10 deg./s) before the visual stimulus stepped up (Miki et al., 2018).

In order to test whether initiation and termination components could be acquired independently, either duration (ON period) or interval (OFF period) was randomized during unidirectional velocity step visual stimulation so that the offset or the onset of the stimulation was not predictable. In fixed duration at variable interval experiments, the stimulus ON duration was constant at 8 seconds, while stimulus OFF duration was randomized between 1 and 15 seconds. After presentation of this stimulus for 3 hours, a termination component was acquired comparable to that after the fixed duration/interval training. However, no initiation component was observed. By contrast, after presentation of variable duration at fixed interval stimulation for 3 hours in which duration of the stimulus ON period was randomized an initiation component was clearly acquired. However, since the stimulus offset could not be predicted, a termination component was not seen. Taken together, these results demonstrated that the initiation and termination predictions can be independently acquired.

9.2 Neuronal Network Subserving OKR

The OKR has been studied in many animal species including rabbits, cats, mice, monkeys, humans, and various fish (Cohen et al., 1973; Collewijn & Grootendorst, 1979; Dieringer et al., 1992; Katoh et al., 1998; Masseck & Hoffmann, 2009; Nagao, 1983; Zee et al., 1976). These studies have demonstrated that basic characteristics such as direct and indirect components constituting the OKR in response to a step of visual stimulus velocity are shared among different species. Goldfish is one of the most thoroughly studied for visual and vestibular behavior, as a structural/functional understanding of its oculomotor neurons and circuitry has been accumulated intensively for the past three decades (Aksay et al., 2000; Aksay et al., 2001; Aksay et al., 2007; Beck et al., 2006; Debowy & Baker, 2011; Graf et al., 1997; Major, Baker, Aksay, Mensh, et al., 2004a; Major, Baker, Aksay, Seung, & Tank, 2004b; Marsh & Baker, 1997; Masseck & Hoffmann, 2009; McElligott et al., 1995; Miki et al., 2018; Miki et al., 2020; Pastor et al., 1991; Pastor et al., 1992; Pastor et al., 1994; Pastor et al., 2019; Soga et al., 2020; Straka et al., 2006). Collectively, these studies have analyzed the visuo-vestibular signal content and anatomical connectivity of accessory optic system neurons (Masseck & Hoffmann, 2009), cerebellar Purkinje cells (Pastor et al., 1997; Straka et al., 2006), pre-cerebellar Area II neurons (Beck et al., 2006), and post-cerebellar vestibular neurons (Pastor et al., 2019). Figure 9.2 summarizes the neuronal network subserving the OKR identified in goldfish, in which multiple loops recursively connect different regions of the brain (dotted loops), with the cerebellum and vestibular nuclei (VN) forming two hubs through which all the loops pass.

9.3 Roles of the Cerebellum in pOKR

The cerebellum has been implicated to be a locus responsible for OKR gain adaptation and predictive motor control in general (Sokolov et al., 2017). Thus, observations of single-unit activity in vestibulo-cerebellar Purkinje cells and oculomotor behavior after cerebellectomy have been performed before, during, and after acquisition of pOKR in goldfish (Miki et al., 2018) to identify its role in acquisition and maintenance of pOKR.

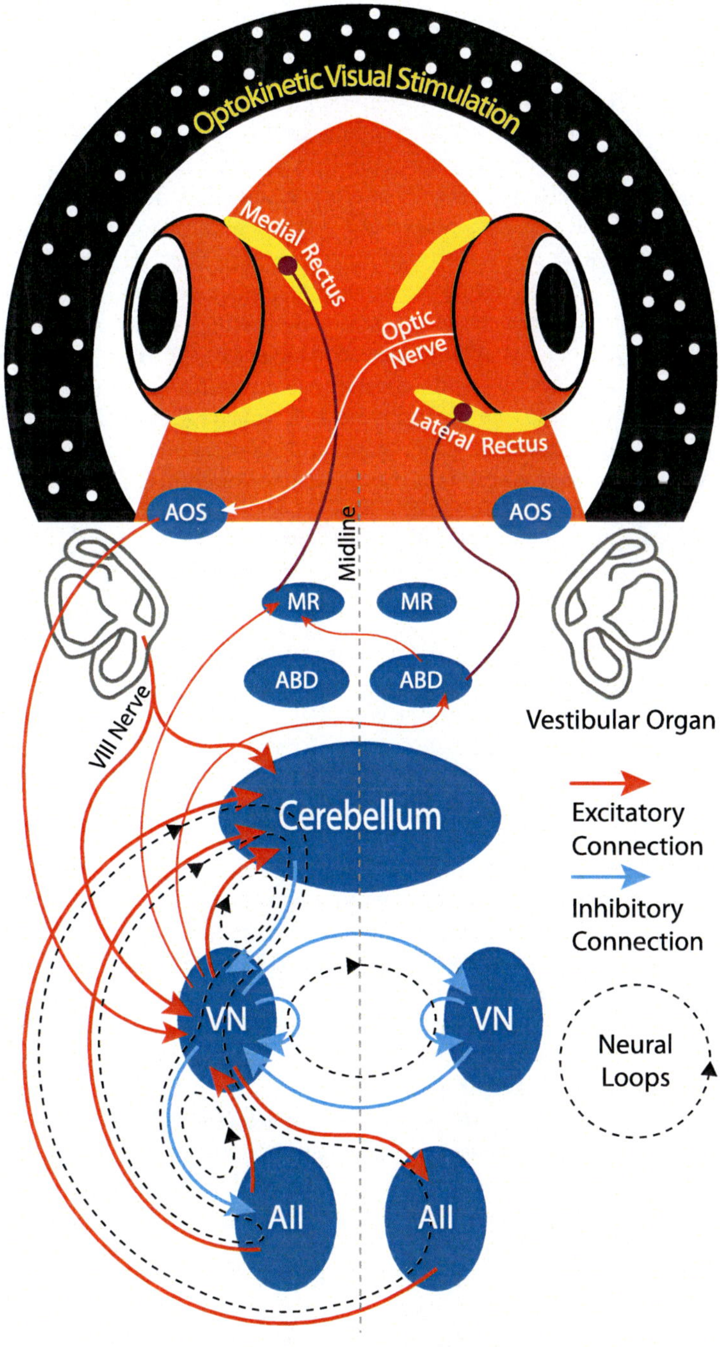

Fig. 9.2 Neuronal circuitry subserving OKR in goldfish. Red and cyan arrows indicate excitatory and inhibitory connections, respectively. Black dotted loops indicate recursive neural loops connecting the brainstem and the cerebellum. (*AOS* accessory optic system, *MR* oculomotor nucleus innervating medial rectus, *ABD* abducens nucleus, *VN* vestibular nucleus, *AII* Area II)

9.3.1 Purkinje Cell Activity during pOKR

The Purkinje cell population recorded in the goldfish vestibulo-cerebellum exhibited firing rates that increased during contraversive eye velocity. According to the terminology proposed by Pastor et al. (1997), these Purkinje cells would be classified as either eye velocity Type II (eII) or head velocity Type I - eye velocity Type II (hIeII). During pOKR visual training, these Purkinje cells showed firing modulation clearly correlated with eye velocity as previously described (Hirata & Highstein, 2001; Pastor et al., 1997). During the extended stimulus period after acquisition of pOKR as in Fig. 9.1d, these neurons increased their firing rate not only with contraversive eye velocity but also with suppressed ipsiversive eye velocity when predictive eye velocity suppression was manifested. Moreover, firing rates in the dark after bidirectional velocity step training in a typical Purkinje cell showed a clear modulation in parallel with eye velocity alternation in the dark as in Fig. 9.1e without any visual stimulation. Collectively these results demonstrate that at least a subset of Purkinje cells in the vestibulo-cerebellum encodes all of the predictive components illustrated in Fig. 9.1 after the acquisition of pOKR (Miki et al., 2018).

9.3.2 Effects of Cerebellectomy before and after Acquisition of pOKR

Acute cerebellectomy after 3 hours of unidirectional velocity step visual training resulted in significant changes in the eye velocity profile. After acute cerebellectomy, overall eye velocity during visual stimulation became slower than that in the post-training condition for intact animals, closely resembling the pre-training (control) eye velocity profile. Initiation and termination components acquired by 3-hour visual training were immediately diminished by cerebellar removal, a procedure that only took only a few minutes (Miki et al., 2018), thereby ruling out extinction as the mechanism. Interestingly, maximum eye velocity (OKR gain) decreased to a lower value than pre-training eye velocity. Eye velocity alternation in the dark acquired during the 3-hour unidirectional velocity step training also was immediately absent after acute cerebellectomy.

In chronically cerebellectomized animals, in which cerebellectomy was performed 1 week prior to training, no clear predictive eye velocity component was observed after 3-hour training, indicating that pOKR was not acquired. Notably, maximum eye velocity reached during the stimulus ON period (OKR gain) appeared significantly smaller in chronically cerebellectomized animals, demonstrating an OKR gain modification (decrease; Miki et al., 2018). The observed OKR gain decrease rather than gain increase demonstrates an adaptation within the brainstem visual circuitry in the absence of cerebellar pathways for sustaining and increasing the OKR gain. None of the cerebellectomy experiments ($N = 8$) showed continuing eye velocity in the dark after 3-hour training (Miki et al., 2018) as observed in cerebellum-intact experiments (Fig. 9.1e).

Taken together, these results demonstrate that the cerebellum is necessary for both acquisition and maintenance of pOKR.

9.4 pOKR in Other Animal Species

A notable characteristic of the cerebellum is a distinct cytoarchitecture and connective anatomy that is evolutionarily highly conserved across vertebrates, from teleosts to primates (Altman & Bayer, 1997; Butler, 1996; Dohaku et al., 2019; Finger, 1983; Hibi & Shimizu, 2012; Nieuwenhuys, 1967). If the cerebellum is sufficient for the acquisition of pOKR, species other than goldfish should be able to acquire the predictive oculomotor behavior. This section summarizes results of behavioral experiments to test this hypothesis in carp, zebrafish, medaka, mice, and humans.

9.4.1 pOKR in Different Fish Species

The eye velocity of the common carp in response to bidirectional velocity step visual stimulation showed a small initial jump followed by a build-up component until the next stimulus direction switch, as observed in goldfish. The initial jump and build-up components have been attributed to the direct and indirect components of OKR, respectively (see below 9.5.1). After the visual training, eye velocity showed a greater initial jump and a faster buildup than those in pre-training control. These features were qualitatively similar to those of goldfish as exemplified in Fig. 9.1a and b. Similarly, trained carps presented a predictive deceleration starting prior to the changes in stimulus direction that was not present in naïve carps (Miki et al., 2020). Carps presented clear predictive deceleration during extended stimulus periods and eye velocity alternation in the dark at the training period, as did goldfish (Fig. 9.1d, e).

Zebrafish and medaka, by contrast, did not show clear changes in their eye velocity profiles after the visual training (Miki et al., 2020). Unlike goldfish and carp, the eye velocity of medaka before training presented a large initial jump and a fast buildup that were unchanged post-training. Eye velocity traces for zebrafish were similar to those of medaka, with slightly greater initial jump and faster buildup in trained than those in control. Furthermore, the maximum trained eye velocities of both zebrafish and medaka are lower than those of goldfish and carp. Previous study (Marsh & Baker, 1997) has shown in goldfish that pOKR can occur at much slower eye velocity (4 deg./s), suggesting that this slower eye velocity in zebrafish and medaka is not related to their inability to acquire predictive decelerations. During extended stimulus periods and in post-training dark conditions, none of zebrafish and medaka tested presented indications of pOKR.

In sum, carps acquired pOKR like goldfish, while zebrafish and medaka did not.

9.4.2 pOKR in Mice and Humans

Similar behavioral experiments in wild-type mice (C57BL/6) indicate so far that they show neither predictive initiation nor termination components during extended period stimulation after a modified version of bidirectional velocity step visual training in which initiation and termination components can be evaluated separately (Yamanaka et al., in prep). Neither did they present eye velocity alternation in the dark after the same visual training, suggesting that mice are not capable of acquiring pOKR.

For human subjects, a head-mount display (HMD: ViVe, HTC) was used to provide large-field visual stimulation to induce a robust OKR. A 3D random-dot stimulation was designed by using a game engine (Unity, Unity Technologies) and displayed in the HMD so that each participant saw its projection on a surrounding transparent cylindrical screen at 1 m (Matsuzawa et al., 2017). Binocular eye positions were recorded in the HMD by using a video eye tracker (VR/AR add-ons, Pupil Labs). In the training session, the visual stimulation was given at 30 deg./s in the rightward direction for 8 seconds and stopped for 4 seconds. This training stimulation was given periodically for 5 minutes and followed immediately by a test session in which left and rightward constant speed rotation at 30 deg./s was given for 16 seconds in each direction as extended period stimulation. Among the 12 participants, 9 presented a clear predictive eye velocity component in the post-training test session, while the rest (3) showed very little predictive component in the test session. These results suggest that majority of humans are capable of acquiring pOKR, while some are not, despite all having an intact cerebellum.

9.5 The Velocity Storage Mechanism as a Possible Determinant of pOKR

We have seen that goldfish, carps, and majority of humans can acquire pOKR while zebrafish, medaka, mice, and some human subjects cannot, although they all share the basic cerebellar neuronal cytoarchitecture. These results suggest that there must be another determinant for acquiring pOKR. As a candidate, there is another visually evoked, reflexive eye movement similar to pOKR in the sense that it persists in the dark following visual stimulation as in Fig. 9.1e, called the optokinetic afternystagmus (OKAN). This section summarizes characteristics of OKAN as a manifestation of a conceptual neuronal mechanism called velocity storage and highlights a paradoxical relationship between OKAN and pOKR.

9.5.1 The Velocity Storage Mechanism (VSM)

With each change of visual stimulus direction, both control and trained eye velocity showed a rapid initial jump followed by gradual buildup (Fig. 9.1). Classically, these two features of OKR are associated with distinct anatomical pathways. The rapid initial jump is thought to be generated by the OKR "direct pathway," while the gradual buildup involves the OKR "indirect pathway" (Fuchs & Mustari, 1993; Waespe & Henn, 1985). The latter pathway is also considered the locus of the VSM, a neural temporal integrator that charges and discharges velocity signals in a manner analogous to a capacitor in an electrical circuit (Cohen et al., 1977).

One of the putative functions of the VSM, among others (Raphen, 2020), is to compensate for imperfect semicircular canal output so as to improve estimates of head velocity in the central nervous system (Karmali, 2019; Laurens & Angelaki, 2011). The output of semicircular canal afferents in response to head motion is a rate-coded signal that falls between head acceleration and head velocity (Fig. 9.3①), and its dynamics can be approximated by a transfer function $SC(s)$ as described in Eq. 9.1:

$$SC(s) = \frac{s}{s+a} \tag{9.1}$$

where s is the Laplace operator. If the VSM is a perfect integrator ($H = 0$) with a gain a and its output is added to semicircular canal output, the entire transfer function $G(s)$ including the semicircular canal and the VSM can be described as Eq. 9.2:

$$G(s) = SC(s) + SC(s)VSM(s) = \frac{s}{s+a} + \frac{s}{s+a}\frac{a}{s} = 1 \tag{9.2}$$

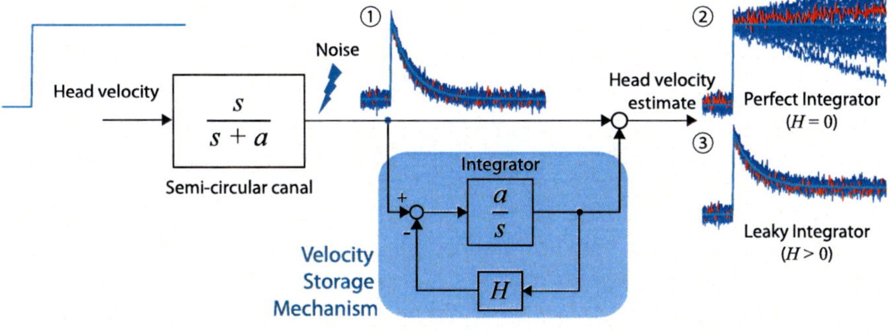

Fig. 9.3 Putative function of the velocity storage mechanism (VSM). ①: Output of a semicircular canal in response to velocity step head rotation. Cyan trace, without noise; blue traces, with 20 different noise sequences with the same mean (0) and standard deviation; red trace, average of blue traces. ②, ③: Head velocity estimates when the VSM is a perfect integrator ($H = 0$), or a leaky integrator ($H > 0$), respectively. Note that the variability of head velocity estimates when $H = 0$ is much greater than that when $H < 0$

where $VSM(s) = a/s$ is the transfzer function of the VSM operating as a perfect integrator. Because $G(s)$ is unity, accurate estimates of head velocity can be obtained as the output of this system no matter what head velocity inputs (Fig. 9.3 ②, cyan trace). However, a perfect integrator enhances low-frequency components of intrinsic noise contained in semicircular canal output (Fig. 9.3 "Noise"), resulting in unstable VSM output, and therefore unstable estimates of head velocity. Blue traces in Fig. 9.3 ② show simulated head velocity estimates with the VSM operating as a perfect integrator (Fig. 9.3, $H = 0$) when different sets of pseudo-white noise with the identical mean and standard deviation were added at the output of the semicircular canal system. The simulation result confirms that head velocity estimates may overshoot or undershoot with various amounts depending on the noise sequences added. A possible solution to avoid such instability in head velocity estimation is to make the integrator leaky ($H > 0$). Blue traces in Fig. 9.3 ③ show simulated head velocity estimates when the VSM was assumed to be a leaky integrator ($H > 0$). The same sets of noise were used to simulate ② and ③. Notably, the head velocity estimates in ③ are much less variable than those in ② although they always undershoot the correct head velocity in ③. Constantly biased (undershoot) estimates are easier to deal with than those that randomly overshoot or undershoot for the central nervous system to generate stable motor commands unaffected by the noise. These simulation results suggest that a leaky VSM integrator is optimal to obtain stable estimates of head velocity in the face of nonstationary noise contaminating semicircular canal output. In animal behavioral experiments, characteristics of the VSM which are manifested as OKAN (see below) have actually been identified to be leaky.

9.5.2 OKAN and the VSM

The VSM not only conditions the vestibular signal as described above (Fig. 9.3) but also appears to be involved in processing the visual signal for the OKR (via the OKR indirect pathway) as well, manifested as optokinetic after-nystagmus (OKAN). In typical OKR experiments, when velocity step visual stimulation is applied to animals, the eye velocity signal gradually builds up, as if being charged in the VSM, and when the visual stimulation is suddenly turned off, the charged eye velocity signal gradually decreases in the dark, following a decaying exponential function until the velocity is completely discharged. The alternating slow- and fast-phase eye velocity profile produced in the dark is called OKAN, and its slow-phase time constant of decay, or more simply its duration, represents the leakiness (Fig. 9.3, "H") of the VSM as a leaky integrator.

Neuronal circuitry subserving the VSM has been postulated to lie within the brainstem vestibular nuclei (Cohen et al., 2008), and a group of secondary vestibular neurons carrying head velocity information in the vestibular nucleus of monkeys have been shown to modulate their firing rates in a manner highly correlated with OKAN slow-phase eye velocity (Blazquez & Highstein, 2007; Waespe & Henn, 1987; Yakushin et al., 2017). Electrical stimulation in this area modulates the VSM

(Yokota et al., 1992), while midline sections of the commissural connection between bilateral vestibular nuclei abolished velocity storage capability (Katz et al., 1991). The degenerated neurons following midline sections were found to be GABAb-ergic (Cohen et al., 2008; Holstein et al., 1999), suggesting that mutual inhibitory connections between bilateral vestibular nuclei were a key element to form the VSM (Yokota et al., 1992).

9.5.3 OKAN in Animals Capable and Incapable of Acquiring pOKR

OKAN was measured before visual training from the same goldfish, carp, medaka, zebrafish, mice, and humans that participated in the pOKR experiments described above. Identical visual stimulation at a constant velocity (20 deg./s) was given to each animal (except for human subjects. See below) for 1 minute, and then the stimulus was turned off to observe OKAN in the dark. Goldfish and carp showed robust long-lasting OKAN. By contrast zebrafish, medaka, and mice hardly exhibited OKAN (Miki et al., 2020; Yamanaka et al., in prep). Thus, those animals that acquired pOKR also exhibited robust OKAN, while those that did not acquire pOKR all showed very short OKAN. For human subjects, identical visual stimulus (random dots) to that used for their pOKR training was given at a constant speed of 30 deg./s for 1 minute, followed by complete darkness for 1 minute. Each of the 12 subjects showed different durations of OKAN, but the 3 subjects who did not acquire clear pOKR behavior presented very short OKAN, while the rest showed significant durations of OKAN (Matsuzawa et al., 2017).

These results suggest that OKAN duration could be an important factor for determining pOKR with those animals exhibiting longer OKAN acquiring pOKR better. All the normal goldfish tested ($n = 13$) showed robust OKAN although with widely varying durations (range = 9–99 s; median = 32 s; Standard deviation = 25 s). Linear regression of the strength of acquired pOKR (predictive deceleration) as a function of OKAN duration among individual goldfish revealed a slight positive correlation between the two values that did not reach statistical significance. Thus, longer OKAN duration did not necessarily predict stronger pOKR acquisition (Miki et al., 2020).

9.5.4 Effects of VIIIth Nerve Neurectomy on OKAN and pOKR

Vestibular neurectomy has been shown to markedly shorten OKAN duration (Cohen et al., 1973; Collewijn, 1976; Zee et al., 1976). In order to directly explore the uncorrelated relationship between pOKR ability and OKAN duration in the same experimental species, VIIIth nerve neurectomy was carried out in goldfish (Miki et al., 2020). The superior branch of VIIIth nerve containing the horizontal and anterior canal afferents as well as those from utricle was cut bilaterally at least 1–3 days before pOKR experiments. In their home water tanks, neurectomized fish maintained normal postures although swimming behaviors were unstable. The horizontal VORs of the neurectomized fish ($N = 7$) measured before visual training were nearly absent, confirming that horizontal semicircular canal afferents were effectively disconnected. The OKAN of a typical neurectomized goldfish measured before visual training lasted significantly shorter than that of controls. Mean OKAN duration for seven neurectomized goldfish was 4.5 s in contrast to 32.2 s for normal animals.

Neurectomized goldfish generated seemingly normal OKR eye velocity with greater initial jumps and faster build-ups than the normal animals. After 30 minutes of visual training, eye velocity started to decrease before changes in the stimulus direction. The averaged trained eye velocity over seven neurectomized fish clearly started to slow down before changes in the stimulus direction. The strength of acquired pOKR (predictive deceleration value) for control and trained in those neurectomized fish was statistically significant (Miki et al., 2020). More precisely, averaged eye velocity of neurectomized animal before visual training (control) arose (initial jump) more rapidly than that of normal animal and reached a plateau level slightly slower than the maximum velocity of the normal animals. After 3 hours of visual training, trained eye velocity of normal animals arose more rapidly (greater initial jump), reached a greater maximum velocity than control eye velocity, and began to slow down before changes in the visual stimulus direction. Similarly, the trained eye velocity of neurectomized animals jumped more rapidly, reached a greater maximum velocity than control, and slowed down prior to changes in the visual stimulus directions. Namely, neurectomized goldfish acquired pOKR as in normal animals although on average predictive deceleration appeared smaller than in normal fish.

To compare the relationship between OKAN duration and acquired predictive deceleration with that of normal goldfish, data from each of seven neurectomized goldfish were superimposed on those from normal fish. As noted above, OKAN durations were significantly shorter and less variable than those of normal animals. Furthermore, neurectomized goldfish with severely shortened OKAN duration acquired pOKR comparable to about 70% of the normal goldfish population.

Together, these results suggest that OKAN duration as an indicator of the varying leakiness of the VSM among animals does not determine their ability to acquire pOKR. It is thus likely that the existence of a neuronal mechanism within the VSM circuitry, irrespective of its connection with VIIIth nerves, is crucial for the acquisition of pOKR.

9.5.5 OKAN Habituation after the Acquisition of pOKR

OKAN has been observed to habituate (Cohen et al., 1992; Collewijn, 1976; Marsh & Baker, 1997), wherein duration gets shorter when OKAN is repeatedly evoked. OKAN of a typical goldfish in the first measurement lasted longer than 20 sec, while that in the fifth measurement lasted less than 10 seconds. OKAN habituation is direction-selective in that OKAN in the direction repeatedly measured selectively habituated, while that in the opposite direction was unchanged (Miki et al., 2020). Acute cerebellectomy after the fifth OKAN test de-habituated OKAN duration and returned the build-up time constant back to values comparable to those in the first test. In addition, OKAN in cerebellectomized goldfish never habituated although generation of OKAN per se is unaffected (Miki et al., 2020). Furthermore, OKAN after acquisition of pOKR in goldfish was found to be habituated (Miki et al., 2020). Before unidirectional visual training for 3 hours, OKAN following both CW and CCW constant velocity 20 deg./s visual stimulation lasted longer than 50 seconds in typical goldfish. OKAN measured immediately after unidirectional (CW) visual training was habituated with a duration less than 20 s in the untrained direction (CCW). Evaluation of OKAN in the trained direction (CW) after unidirectional visual training was impeded by the alternating eye velocity as the training period continued in the dark (Miki et al., 2020).

9.6 Summary and Conclusion

In summary, the cerebellum is required for acquisition and maintenance of both pOKR and OKAN habituation, although the cerebellum is not required to produce OKR and OKAN per se. Furthermore, only animals that presented robust OKAN acquired pOKR, while OKAN was habituated after the acquisition of pOKR in those animals. Nonetheless, animals presenting longer OKAN did not necessarily acquire pOKR better, and neurectomized goldfish whose OKAN was severely shortened did acquire pOKR.

These results from the series of experiments conducted in the context of the well-defined neuronal connectivity of the OKR in goldfish and other species support the argument that the acquired predictive eye velocity behaviors appear to be dependent on VSM signals from the brainstem to the cerebellum (Fig. 9.2), but the adaptive timing mechanism, itself, originates within the cerebellar circuitry. Namely, a central pattern generator or a clock is formed in the cerebellar neuronal network to resonate with future sensory stimulation pattern from the past sequence of sensory input for the generation of properly timed oculomotor commands. The VSM signals may provide modifiable quantities representing the passage of time. Consequently, the eye movement profile is continuously updated within the cerebellar neuronal circuitry by neural loops first employing and then replacing the returning pre-cerebellar VSM signals arising from the bilaterally connected vestibular nuclei

(Fig. 9.2). The computational algorithm and its neural implementation for this functionality are yet to be addressed by mathematical modeling (Inagaki & Hirata, 2017; Pinzon Morales et al., 2019; Shinji & Hirata, 2020) and then tested by further behavioral and neurophysiological experiments.

References

Aksay, E., Baker, R., Seung, H. S., & Tank, D. W. (2000). Anatomy and discharge properties of pre-motor neurons in the goldfish medulla that have eye-position signals during fixations. *Journal of Neurophysiology, 84*(2), 1035–1049.

Aksay, E., Gamkrelidze, G., Seung, H. S., Baker, R., & Tank, D. W. (2001). In vivo intracellular recording and perturbation of persistent activity in a neural integrator. *Nature Neuroscience, 4*(2), 184–193.

Aksay, E., Olasagasti, I., Mensh, B. D., Baker, R., Goldman, M. S., & Tank, D. W. (2007). Functional dissection of circuitry in a neural integrator. *Nature Neuroscience, 10*(4), 494–504.

Altman, J., & Bayer, S. A. (1997). *Development of the cerebellar system in relation to its evolution, structure, and function*. Press.

Beck, J. C., Rothnie, P., Straka, H., Wearne, S. L., & Baker, R. (2006). Precerebellar hindbrain neurons encoding eye velocity during vestibular and optokinetic behavior in the goldfish. *Journal of Neurophysiology, 96*(3), 1370–1382.

Blazquez, P. M., & Highstein, S. M. (2007). Visual-vestibular interaction in vertical vestibular only neurons. *Neuroreport, 18*(13), 1403–1406.

Butler, A. B. (1996). *Hodos, H.* Comparative Vertebrate Neuroanatomy.

Cohen, B., Uemura, T., & Takemori, S. (1973). Effects of labyrinthectomy on optokinetic nystagmus (OKN) and optokinetic after-nystagmus (OKAN). *International Journal of Equilibrium Research, 3*, 88–93.

Cohen, B., Matsuo, V., & Raphan, T. (1977). Quantitative analysis of the velocity characteristics of optokinetic nystagmus and optokinetic after-nystagmus. *The Journal of Physiology, 270*(2), 321–344.

Cohen, H., Cohen, B., Raphan, T., & Waespe, W. (1992). Habituation and adaptation of the vestibuloocular reflex: A model of differential control by the vestibulocerebellum. *Experimental Brain Research, 90*, 526–538.

Cohen, B., Dai, M., Yakushin, S. B., & Raphan, T. (2008). Baclofen, motion sickness susceptibility and the neural basis for velocity storage. *Progress in Brain Research, 171*, 543–553.

Collewijn, H. (1976). Impairment of optokinetic (after-)nystagmus by labyrinthectomy in the rabbit. *Experimental Neurology, 52*, 146–156.

Collewijn, H., & Grootendorst, A. F. (1979). Adaptation of optokinetic and vestibulo-ocular reflexes to modified visual input in the rabbit. *Progress in Brain Research, 50*, 771–781.

Debowy, O., & Baker, R. (2011). Encoding of eye position in the goldfish horizontal oculomotor neural integrator. *Journal of Neurophysiology, 105*(2), 896–909.

Dieringer, N., Reichenberger, I., & Graf, W. (1992). Differences in optokinetic and vestibular ocular reflex performance in teleosts and their relationship to different life styles. *Brain, Behavior and Evolution, 39*, 289–304.

Dohaku, R., et al. (2019). Tracing of afferent connections in the zebrafish cerebellum using recombinant rabies virus. *Front Neural Circuits, 13*, 30.

Finger, T. E. (1983). The gustatory system in teleost fish. In R. G. Northcutt & R. E. Davis (Eds.), *Fish neurobiology* (pp. 285–319). University of Michigan Press.

Fuchs, A. F., & Mustari, M. J. (1993). The optokinetic response in primates and its possible neuronal substrate. *Reviews of Oculomotor Research, 5*, 343–369.

Graf, W., Spencer, R., Baker, H., & Baker, R. (1997). Excitatory and inhibitory vestibular pathways to the extraocular motor nuclei in goldfish. *Journal Neurosphysiol, 77*(5), 2765–2779.

Hibi, M., & Shimizu, T. (2012). Development of the cerebellum and cerebellar neural circuits. *Developmental Neurobiology, 72*, 282–301.

Hirata, Y., & Highstein, S. M. (2001). Acute adaptation of the vestibuloocular reflex: Signal processing by floccular and ventral parafloccular Purkinje cells. *Journal of Neurophysiology, 85*(5), 2267–2288.

Holstein, G. R., Martinelli, G. P., & Cohen, B. (1999). The ultrastructure of GABA-immunoreactive vestibular commissural neurons related to velocity storage in the monkey. *Neuroscience, 93*, 171–181.

Inagaki, K., & Hirata, Y. (2017). Computational theory underlying acute Vestibulo-ocular reflex motor learning with cerebellar long-term depression and long-term potentiation. *Cerebellum, 16*(4), 827–839.

Ito, M., Jastreboff, P. J., & Miyashita, Y. (1979). Adaptive modification of the rabbit's horizontal vestibulo-ocular reflex during sustained vestibular and optokinetic stimulation. *Experimental Brain Research, 37*(1), 17–30.

Karmali, F. (2019). The velocity storage time constant: Balancing between accuracy and precision. *Progress in Brain Research, 248*, 269–276.

Katoh, A., Kitazawa, H., Itohara, S., & Nagao, S. (1998). Dynamic characteristics and adaptability of mouse vestibulo-ocular and optokinetic response eye movements and the role of the flocculo-olivary system revealed by chemical lesions. *Proceedings of the National Academy of Sciences of the United States of America, 95*, 7705–7710.

Katz, E., Vianney de Jong, J. M., Buettner-Ennever, J., & Cohen, B. (1991). Effects of midline medullary lesions on velocity storage and the vestibulo-ocular reflex. *Experimental Brain Research, 87*, 505–520.

Laurens, J., & Angelaki, D. E. (2011). The functional significance of velocity storage and its dependence on gravity. *Experimental Brain Research, 210*(3–4), 407–422.

Major, G., Baker, R., Aksay, E., Mensh, B., Seung, H. S., & Tank, D. W. (2004a). Plasticity and tuning by visual feedback of the stability of a neural integrator. *Proceedings of the National Academy of Sciences of the United States of America, 101*(20), 7739–7744.

Major, G., Baker, R., Aksay, E., Seung, H. S., & Tank, D. W. (2004b). Plasticity and tuning of the time course of analog persistent firing in a neural integrator. *Proceedings of the National Academy of Sciences of the United States of America, 101*(20), 7745–7750.

Marsh, E., & Baker, R. (1997). Normal and adapted visuooculomotor reflexes in goldfish. *Journal of Neurophysiology, 77*(3), 1099–1118.

Masseck, O. A., & Hoffmann, K. P. (2009). Question of reference frames: Visual direction-selective neurons in the accessory optic system of goldfish. *Journal of Neurophysiology, 102*(5), 2781–2789.

Matsuzawa, Y., Baker, R., & Hirata, Y. (2017). Human predictive optokinetic eye movement is correlated with the presence of velocity storage: Demonstration in a virtual reality environment. In *Society for Neuroscience*.

McElligott, J. G., Weiser, M., & Baker, R. (1995). Effect of temperature on the normal and adapted vestibulo-ocular reflex in the goldfish. *Journal of Neurophysiology, 74*(4), 1463–1472.

Miki, S., Baker, R., & Hirata, Y. (2018). Cerebellar role in predictive control of eye velocity initiation and termination. *The Journal of Neuroscience, 38*, 10371–10383.

Miki, S., Urase, K., Baker, R., & Hirata, Y. (2020). Velocity storage mechanism drives a cerebellar clock for predictive eye velocity control. *Scientific Reports, 10*(1), 6944.

Nagao, S. (1983). Effects of vestibulocerebellar lesions upon dynamic characteristics and adaptation of vestibulo-ocular and optokinetic responses in pigmented rabbits. *Experimental Brain Research, 53*(1), 36–46.

Nieuwenhuys, R. (1967). Comparative anatomy of the cerebellum. *Progress in Brain Research, 25*, 1–93.

Pastor, A. M., Torres, B., Delgado-Garcia, J. M., & Baker, R. (1991). Discharge characteristics of medical rectus and abducens motor neurons in the goldfish. *Journal of Neurophysiology, 66*(6), 2125–2140.

Pastor, A. M., de la Cruz, R. R., & Baker, R. (1992). Characterization and adaptive modification of the goldfish vestibuloocular reflex by sinusoidal and velocity step vestibular stimulation. *Journal of Neurophysiology, 68*(6), 2003–2015.

Pastor, A. M., de la Cruz, R. R., & Baker, R. (1994). Cerebellar role in adaptation of the goldfish vestibuloocular reflex. *Journal of Neurophysiology, 72*(3), 1383–1394.

Pastor, A. M., De la Cruz, R. R., & Baker, R. (1997). Characterization of Purkinje cells in the goldfish cerebellum during eye movement and adaptive modification of the vestibulo-ocular reflex. *Progress in Brain Research, 114*, 359–381.

Pastor, A. M., Calvo, P. M., de la Cruz, R. R., Baker, R., & Straka, H. (2019). Discharge properties of morphologically identified vestibular neurons recorded during horizontal eye movements in the goldfish. *Journal of Neurophysiology, 121*, 1865–1878.

Pinzon Morales, R. D., Miki, S., Hirata, Y. (2019). A neural mechanism for predictive optokinetic eye movement. 28th annual computational neuroscience meeting, https://doi.org/10.13140/RG.2.2.12852.60804

Raphen, T. (2020). Vestibular, locomotor, and vestibulo-autonomic research: 50 years of collaboration with Bernard Cohen. *Journal of Neurophysiology, 123*(1), 329–345.

Shinji Y, Hirata Y (2020) Neural network model of the optokinetic eye movement configured by the cerebellum-brainstem connectome. Proc. Annual meeting of the Japanese Neural Network Society.

Soga, J., Matsuyama, M., Miura, H., Highstein, S., Baker, R., & Hirata, Y. (2020). Cerebellar roles in frequency competitive motor learning of the Vestibulo-ocular reflex. *Neuroscience, S0306-4522*(20), 30580–30587.

Sokolov, A. A., Miall, R. C., & Ivry, R. B. (2017). The cerebellum: Adaptive prediction for movement and cognition. *Trends in Cognitive Sciences, 21*(5), 313–332.

Straka, H., Beck, J. C., Pastor, A. M., & Baker, R. (2006). Morphology and physiology of the cerebellar vestibulolateral lobe pathways linked to oculomotor function in the goldfish. *Journal of Neurophysiology, 96*(4), 1963–1980.

Waespe, W., & Henn, V. (1985). Cooperative functions of vestibular nuclei neurons and floccular Purkinje cells in the control of nystagmus slow phase velocity: Single cell recordings and lesion studies in the monkey. *Reviews of Oculomotor Research, 1*, 233–250.

Waespe, W., & Henn, V. (1987). Gaze stabilization in the primate. The interaction of the vestibulo-ocular reflex, optokinetic nystagmus, and smooth pursuit. *Reviews of Physiology, Biochemistry and Pharmacology, 106*, 37–125.

Yakushin, S. B., Raphan, T., & Cohen, B. (2017). Coding of velocity storage in the vestibular nuclei. *Frontiers in Neurology, 8*, 10.3389.

Yokota, J., Reisine, H., & Cohen, B. (1992). Nystagmus induced by electrical stimulation of the vestibular and prepositus hypoglossi nuclei in the monkey: Evidence for site of induction of velocity storage. *Experimental Brain Research, 92*, 123–138.

Zee, D. S., Yee, R. D., & Robinson, D. A. (1976). Optokinetic responses in labyrinthine-defective human beings. *Brain Research, 113*, 423–428.

Chapter 10
Fastigial Nucleus Input/Output Related to Motor Control

Mayu Takahashi and Yoshikazu Shinoda

10.1 Introduction

Purkinje cells in the cerebellar cortex send outputs to both vestibular nuclei (VN) and cerebellar nuclei (CN, comprising the fastigial/medial nucleus, FN, interpositus nucleus IP, and dentate/lateral nucleus DN). The pattern of where the output axons of a certain part of the cerebellar cortex terminate was used to separate the cerebellar cortex into three longitudinal subdivisions (Jansen & Brodal, 1940, 1942). These three subdivisions are the medial zone (vermis), intermediate zone, and lateral zone. These cortical zones project to the FN, IP, and DN, respectively. These three longitudinally organized zones were determined by applying the Marchi method after applying lesions to the corresponding parts of the cerebellar cortex (see a review by Voogd et al., 2013). However, further subdivisions of the cerebellar cortex beyond these three broad divisions have since been identified. The further subdivisions are known as zones A, B, C, and D. These zones were identified using cholinesterase staining in cerebellar white matter (Voogd, 1964, 1969). Later, zones A–D were further refined by observing the topography of the corticonuclear and olivocortical projections (Buisseret-Delmas & Angaut, 1993; Groenewegen & Voogd, 1977; Kawamura & Hashikawa, 1979). Using this latter method of dividing the cerebellar cortex based on the corticonuclear projection patterns, there are a total of six longitudinal zones (A, B, C1, C2, C3, and D). These six zones were confirmed via retrograde axonal transport from the cerebellar nuclei (Voogd & Bigaré, 1980) and anterograde axonal tracing methods in the cat (Trott et al., 1998; Trott & Armstrong, 1987). The zones extend across more than one lobule, even across the almost entire rostrocaudal length of the cerebellar cortex. Among them, A and B zones

M. Takahashi (✉) · Y. Shinoda
Department of Systems Neurophysiology, Tokyo Medical and Dental University, Graduate School of Medicine, Tokyo, Japan
e-mail: takahashi.phy1@tmd.ac.jp

H. Mizusawa, S. Kakei (eds.), *Cerebellum as a CNS Hub*,
Contemporary Clinical Neuroscience, https://doi.org/10.1007/978-3-030-75817-2_10

correspond to the vermis, C1–C3 to the intermediate hemispheric cortex, and D to the lateral hemisphere (Voogd et al., 1987; Voogd & Bigaré, 1980).

The inputs to the cerebellar cortex are also organized into longitudinal zones. This has been shown anatomically in the pattern of olivocerebellar projections (Voogd, 1967) and electrophysiologically in the pattern of spinocerebellar inputs (Oscarsson, 1969). Spinal mossy fibers (Voogd, 1969), reticulocerebellar mossy fibers (Chan-Palay, 1977), and pontocerebellar fibers (Serapide et al., 2001) also terminate in a longitudinally organized fashion in the cerebellar cortex. Overall, it seems the cerebellar cortex is organized in an apposition of structurally and functionally distinct longitudinal stripes. Since the cerebellar cortex can be separated into compartments based on distinct input/output patterns, it is inferred that the function of each cerebellar compartment can be determined by the precise pattern of its unique efferent and afferent connections with extracerebellar neural structures. Oscarsson (1969, 1979) proposed a microzonal structure of the cerebellar cortex as a structural–functional unit with its specific efferent and afferent connections, and Ito (1984) extended this idea to a "corticonuclear microcomplex": a structural–functional unit organization of the cerebellum. To understand the function of each part of the cerebellum, it is essential to know how different mossy fiber (MF) and climbing fiber (CF) afferent systems interact with these cortical longitudinal zones and produce specific patterns of responses in cerebellar cortex Purkinje cells. One must also understand how the final cerebellar output signals (which will travel outside the cerebellum) are generated by cells in the cerebellar nuclei. The responses of these cells in the cerebellar nuclei will necessarily involve not only the Purkinje cell signals from the corresponding cerebellar cortical subdivisions but also interaction with nuclear collateral inputs from the same MF and CF afferents (Shinoda & Sugihara, 2013; Takahashi & Shinoda, 2021).

Sir John Eccles et al. (1967) extensively investigated the neural circuits of mossy and climbing fiber inputs and intrinsic connections in the cerebellum, using electrophysiological methods. After studying in Eccles's lab in Australia, Masao Ito returned to Japan and began his own investigations into cerebellum. Ito applied the same intracellular recording method that he had previously applied to motoneurons in the cat spinal cord to investigate cerebellar connections. While searching for the cerebellar synaptic connections of Deiters' nucleus neurons, Ito and colleagues discovered two important cerebellar mechanisms: (a) Purkinje cell inhibition and disinhibition and (b) long-term depression (LTD) synaptic plasticity at cerebellar synapses.

Purkinje cells formed the nodal point of Ito's landmark model of motor control by the cerebellum. To further understand the function of different parts of the cerebellum, Ito and his colleagues investigated the output targets of the cerebellar and vestibular nuclei, including the fastigial nucleus. Enumeration of these synaptic output connections of the cerebellar and vestibular nuclei is one of the major contributions of Ito's group toward understanding the cerebellar control of movement (see Takahashi & Shinoda, 2021). This article will first discuss the detailed organization of inputs and outputs of the cerebellar cortex and cerebellar nuclei. We will especially focus on how the input/output structure of the cerebellum informs our understanding of the function of the cerebellum in terms of the "corticonuclear microcomplex" model. To clarify how the input/output structure gave rise to its separation into microcomplexes, we will present the axonal trajectories of single mossy and climbing fiber neurons and the relationship of the A–D longitudinal zones with aldolase C longitudinal compartments in the cerebellar cortex and nuclei. The second part of the article will summarize neural circuits of the inputs and outputs of the fastigial nucleus in relation to motor control and will discuss outstanding problems in our understanding of the cerebellar circuit.

10.2 General Structure of the Cerebellar Cortex and the Cerebellar and Vestibular Nuclei

10.2.1 Axonal Trajectories of Single Olivocerebellar Neurons

The cerebellar cortex is for the most part homogenous in its structure (Cajal, 1911), although there are minor regional variations. For example, unipolar brush cells are located in abundance in the granule cell layer, particularly in the uvula-nodulus, but not elsewhere (Mugnaini & Floris, 1994). Since the cellular structure is the same throughout the whole cerebellar cortex, local differences cannot be used to differentiate different functional regions, as is common in cerebral cortex. Instead, cerebellar cortex is separated into functional regions based on the pattern of afferent and efferent connections with extracerebellar structures in the CNS. Although there existed anatomical descriptions of the afferent CF and MF systems, the reports were not able to identify the organization of CF and MF projections at the level of single neurons (primarily due to limitations of the Golgi staining method in identifying axons – a problem that holds for the entire CNS –) (Shinoda, 1999). Using a new intracellular staining method with horseradish peroxide (HRP), Kitai et al. (1976) succeeded in staining single electrophysiologically identified Purkinje cells (PCs). However, in order to properly understand the three-dimensional axonal trajectories of single neurons, a further contribution was made by Shinoda and his colleagues. Shinoda and colleagues developed a method to reconstruct the axonal trajectory with serial sections in the cat (Shinoda et al., 1981, 1986). Using this reconstruction method, the first stained single cerebellar afferent fibers were MFs that were functionally identified as MFs responsive to muscle stretch (Krieger et al., 1985). Following this success, the axonal branching patterns of neurons in more cerebellar afferent and efferent systems were investigated using the same HRP staining and reconstruction method (Shinoda et al., 1992). However, intracellular iontophoretic injection of HRP into a cell or an axon is difficult and requires significant technical skill. As an alternative, Shinoda developed a method of extracellular application of biotinylated dextran amine (BDA) (Veenman et al., 1992) to label a small number of neurons or even a single neuron and reconstruct with serial sections the entire trajectories of single cerebellar afferent and efferent neurons. With this method of extracellular application of BDA in the rat, the entire axonal trajectories of single MF neurons were first revealed in lateral reticular nucleus neurons of the rat (Fig. 10.3a) (Wu et al., 1999). Later, the same method revealed the axonal trajectories of single CF neurons in the inferior olive nucleus (Fig. 10.1) (Sugihara et al., 2001).

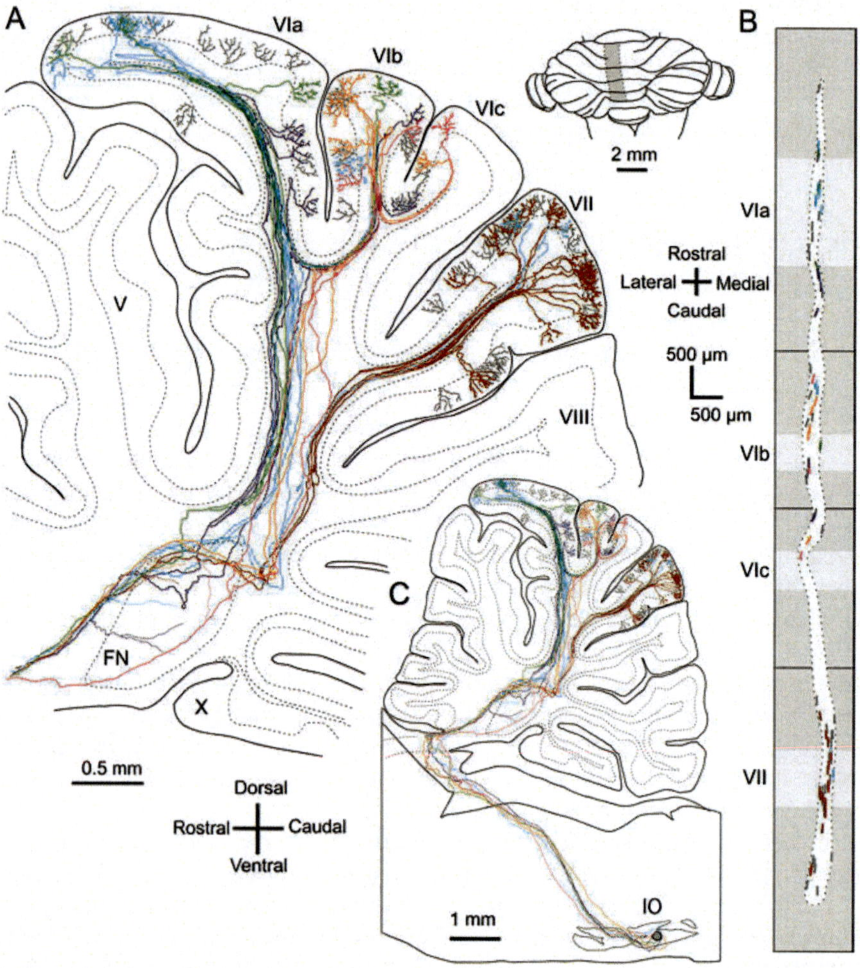

Fig. 10.1 Reconstructions of the entire axonal trajectories of single olivocerebellar neurons that were labeled with biotinylated dextran amine (BDA) injected into the inferior olive in the rat. Climbing fibers originating from small areas in the inferior olive distribute within narrow longitudinal bands in the cerebellar cortex. (**a**) Lateral view of six olivocerebellar axons (colored) reconstructed from 81 serial parasagittal sections. Forty-two climbing fibers arising from these 6 axons, and the other 36 labeled climbing fibers (gray) are included. (**b**) Dorsal view of the distribution of climbing fibers plotted on the unfolded vermal cortex from the midline to the left by 1.3 mm. (**c**) Lateral view under low magnification of the entire axonal trajectories from the injection site in the centromedial portion of the MAO (single injection of 0.01 μl of BDA). Note that the labeled climbing fibers are distributed in a narrow longitudinal band after the small injection of BDA into the localized area of the inferior olive. Colors used for the climbing fibers in b correspond to those used for individual axons in a. Light and dark gray areas in the unfolded scheme represent the cerebellar cortex exposed in the cerebellar surface and hidden in the sulci, respectively. Dotted line indicates the contour of the distribution area. Inset shows the area for the unfolded display (Sugihara et al., 2001)

An example of the application of the BDA method to identifying cerebellar cortex afferents is shown in Fig. 10.1. Axons of single CF neurons spread extensively in a parasagittal plane over a few rostrocaudal lobules (Fig. 10.1a), but mediolaterally their axon terminals are localized in a narrow longitudinal band (Fig. 10.1b). Injection of a small amount of BDA into various localized small areas of the inferior olive (IO) in the rat labeled a small number of olivocerebellar neurons, and one or two narrow longitudinal stripes of their axon terminals were clearly identified in the cerebellar cortex (Sugihara et al., 1999).

10.2.2 Longitudinal Zones A–D and their Relationship with Aldolase C Bands

Although morphologically the cerebellar cortex is homogenous, there does seem to be a difference between the different input–/output-defined zones at the biochemical level of the cells. Richard Hawkes and collaborators used a monoclonal antibody (mab, Q113, zebrin II) to demonstrate the biochemical heterogeneity of Purkinje cells (PCs) in cerebellar cortex (Ahn et al., 1994; Brochu et al., 1990). Although zebrin II was identified as the glycolytic isozyme aldolase C (Aldoc) (Ahn et al., 1994), its functional role in PC heterogeneity remains unknown. However, the cortical distribution pattern of zebrin II positive and negative PCs was composed of thin parasagittal zones containing positive PCs, alternating with thicker zones bearing unstained PCs. This heterogeneity of PCs is very useful as landmarks to determine the cerebellar cortical longitudinal zones morphologically. This is because, as mentioned previously, the cortical longitudinal A–D zones are (morphologically) indistinguishable except for their different patterns of inputs. This relationship between the input-defined A–D longitudinal zones and aldolase C-positive and aldolase C-negative bands was demonstrated by Voogd et al. (2003) and Sugihara and Shinoda (2004). Individual aldolase C-positive and aldolase C-negative stripes in the anterior and posterior halves of the cerebellar cortex are anatomically related such that each stripe in one half of the cortex pairs with one in the other half to form a longitudinal compartment. Approximately 24 such pairs (compartments) have been described and are named using the number of the anterior stripe followed by the number of the posterior stripe (e.g., "compartment 4+//5+"; for details, see Sugihara & Shinoda, 2004). This nomenclature was based on the numeric system developed by Hawkes and Leclerc (1987) combined with an alphabetic system (Voogd et al., 2003; Voogd & Ruigrok, 2004). A plus sign was added for antibody-positive compartments and a negative sign for antibody-negative compartments, with the alphanumeric marker determined by the antibody-positive compartment located immediately medially to that negative compartment. The original studies presumed that rostral and caudal compartments with the same numbering corresponded functionally to each other (Brochu et al., 1990; Hawkes & Leclerc, 1987). However, it turned out that this relationship did not always hold (Sugihara &

Shinoda, 2004). Sugihara and Shinoda (2004) assumed that compartments with PCs of the same aldolase C expression in the anterior and posterior lobes innervated by single olivocerebellar neurons form a common functional unit. Based on this, they then turned to investigate the functional and anatomical relationship between rostral and caudal aspects (Sugihara & Shinoda, 2004). Since compartment a+ (Voogd et al., 2003) was ignored in the original numerical nomenclature of Hawkes and Leclerc (1987), under the older numbering scheme, each rostral compartment ends up being linked with the caudal compartment that is one digit higher (2+//3+, 4+//5+. 5+//6+. 6+//7+) in the revised figure by Sugihara and Shinoda (see the revised links between rostral and caudal aldolase C compartments in Table 2 in Sugihara & Shinoda, 2004). According to this revised nomenclature of aldolase C compartments, conventional zone A occupies the entire vermis and includes multiple aldolase C-positive and aldolase C-negative compartments (see Table 10.1 for details). Conventional zones B, C1, and C3 belong to group IV; zones B and C1 belong to aldolase C-negative compartments (2−//4−, 3−//e1−, respectively), and zone C3 mostly corresponds to 4−//5−. Zone C2 mostly corresponds to 4+//5+. Zone D1 corresponds to 5+//6+ innervated by the ventral principal olive (PO), and lateral zone D2 corresponds to 6+//7+ innervated by the dorsal PO (Voogd et al., 2003).

Taking advantage of the axonal morphology of single olivocerebellar neurons as mentioned above, pairs of rostral and caudal aldolase C compartments can be linked based on the common projection of single OC axons across the rostrocaudal boundary on lobule VIc-crus Ib. By localizing the olivary origins of projection to similar compartments, compartments and olivocerebellar projections can be sorted into five groups (Fig. 10.2 and Table 10.1). Group I comprises positive aldolase C compartments extending from the posterior lobe to the anterior lobe innervated by the principal olive and some neighboring areas. Group II comprises positive compartments localized within the posterior lobe innervated by several medial olivary subnuclei. Group III comprises vermal and central aldolase C-negative compartments innervated by the centrocaudal medial accessory olive. Group IV comprises negative and lightly positive compartments in the hemispheric and the rostral and caudal pars intermedia innervated by the dorsal accessory olive and some neighboring areas. Finally, group V comprises positive aldolase C compartments in the flocculus and nodulus (Sugihara & Shinoda, 2004).

10.2.3 Aldolase C Compartments in the Cerebellar Nucleus

The existence of axon collaterals of olivocerebellar axons to the cerebellar nucleus (CN) was first demonstrated by Gerrits and Voogd (1987), using a reliable modern anatomical staining method. In contrast, the existence of the projections of mossy fibers to the CN had been more controversial (see Shinoda & Sugihara, 2013). However, this controversy was reconciled when Shinoda and colleagues demonstrated via intra-axonal staining with HRP that axon collaterals of single mossy

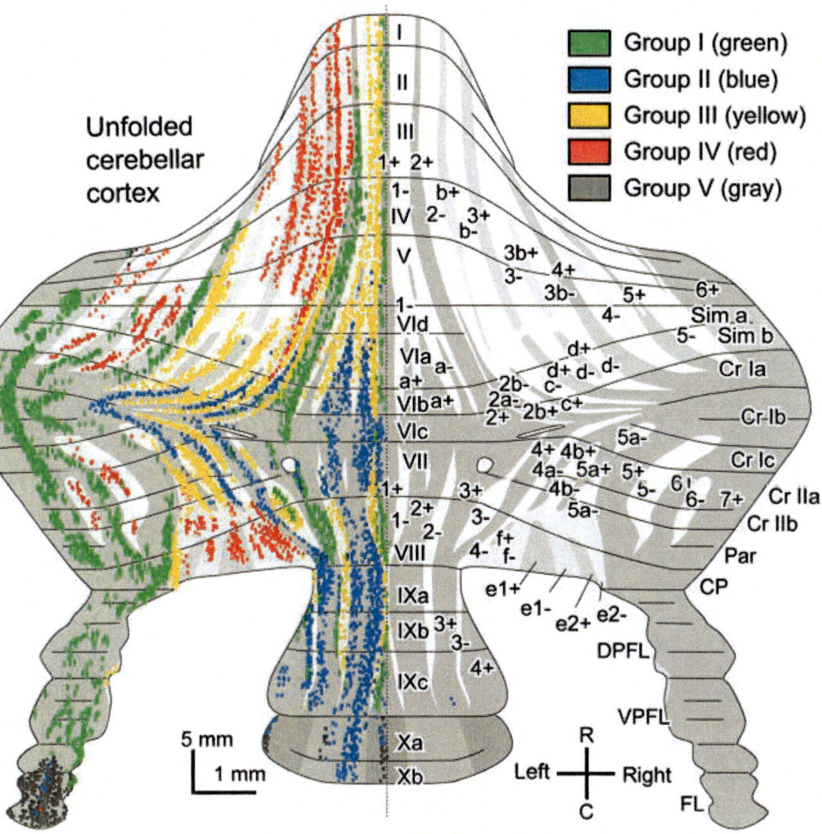

Fig. 10.2 Mapping of labeled climbing fibers in aldolase C compartments on the unfolded scheme of the cerebellar cortex in the rat. Each injection into different parts of the inferior olive was classified into one of five groups based on its projection pattern to aldolase C compartment by referring to the five-group scheme of the olivocerebellar projection (see Table 10.1 for details of groups I–V) (from Sugihara & Shinoda, 2004, 2007)

fiber axons terminate in the cerebellar nucleus (Shinoda et al., 1992). The pontocerebellar mossy fibers identified electrophysiologically were visualized with intracellular injection of HRP, thus demonstrating unequivocally the existence of mossy fiber collaterals to the cerebellar nucleus. Cerebellar nuclear projections from the pontine nucleus and nucleus reticularis tegmenti pontis were soon also reported, using PHA-L tracing in rat (Mihailoff, 1993). Figure 10.3a shows an example of how the morphology of single olivocerebellar axon collaterals in the cerebellar nucleus can be visualized using reconstruction from serial sections (Sugihara et al., 1996). Virtually all olivocerebellar neurons have axon collaterals to the cerebellar nucleus in addition to cortical collaterals terminating in multiple lobules in the parasagittal plane (Sugihara et al., 1999).

Table 10.1 Definition of "groups I–V" and presumed major input to the inferior olive of each group. Refer to Sugihara and Shinoda (2004) for details. The nomenclature of the subnuclei of the inferior olive proposed by Ruigrok and Cella (1995) is used

Group	Inferior olive	Cerebellar cortex	Major input to the inferior olive
I	Subnucleus *a* of the c-MAO, r-MAO, v-PO except for the caudomedial part, d-PO (lateral and rostral IO)	1+//1+, 2+//3+, 4+//5+, 5+//6+, 6+//7+ (positive compartments that extend beyond the primary fissure to the anterior lobe)	Mesodiencephalic
II-a	Caudal subnucleus *c* of the c-MAO, beta, DMCC, caudal DM, caudomedial v-PO (medial IO)	a+//2+ (VIa-b, VIII, IX), 2b+//4+ (positive compartments that do not extend beyond the primary fissure but are located in the posterior lobe)	Vestibular
II-b	Medial and lateral subnucleus *c* of the c-MAO (ventrocaudomedial IO)	a+//2+ (VIc-VII), c+//4b+ 、 d+//5a+ (positive compartments that do not extend beyond the primary fissure but are located in the posterior lobe)	Collicular
III	Subnucleus *b* of the c-MAO (ventrocaudal IO)	1−(med)//1−, 1−(lat)/a−//2−, 2a−//3−,2b−//4a−, c−//4b−, d−//5a−, 3b−//e2− (vermal and central negative compartments)	Somatosensory, vestibular, mesodiencephalic
IV	d-DAO, v-DAO, rostral and central DM (rostrodorsal IO)	2−//4−, b+//f+, b−//f−, 3+//e1+, 3−//e1−, (3b+//e2+), 4−//5−, 5−//6− (negative and lightly positive compartments in the hemisphere and the rostral and caudal pars intermedia)	Somatosensory
V	DC and VLO (dorsocaudal IO)	Flocculus and lobule X (aldolase C-positive)	Visual

Zebrin is able to immunostain somata, axons, and dendrites of Purkinje cells. Labeled axonal bundles of aldolase C-positive Purkinje cells can be traced in serial sections from the cerebellar cortex through the white matter to some parts of the CN where the labeled axons ended in labeled (aldolase C-positive) terminals (Fig. 10.4). In contrast, no somata or dendrites of neurons in the cerebellar (or vestibular) nuclei are clearly labeled by zebrin II. Thus, not only the cerebellar cortex but also the cerebellar nuclei (and vestibular nuclei) can be separated into aldolase C-positive or aldolase C-negative zones based on the presence or absence of labeled Purkinje axon terminals with cell bodies in the corresponding positive or negative zones in the cerebellar cortex. Using this method, clearly demarcated compartments arise in the cerebellar nuclei. We next describe the organization of these cerebellar nuclear compartments.

Based on the presence or absence of labeled Purkinje axon terminals, the cerebellar nuclei also could be divided into aldolase C-positive caudoventral group and aldolase C-negative rostrodorsal group (Fig. 10.4b, c) (Sugihara & Shinoda, 2007).

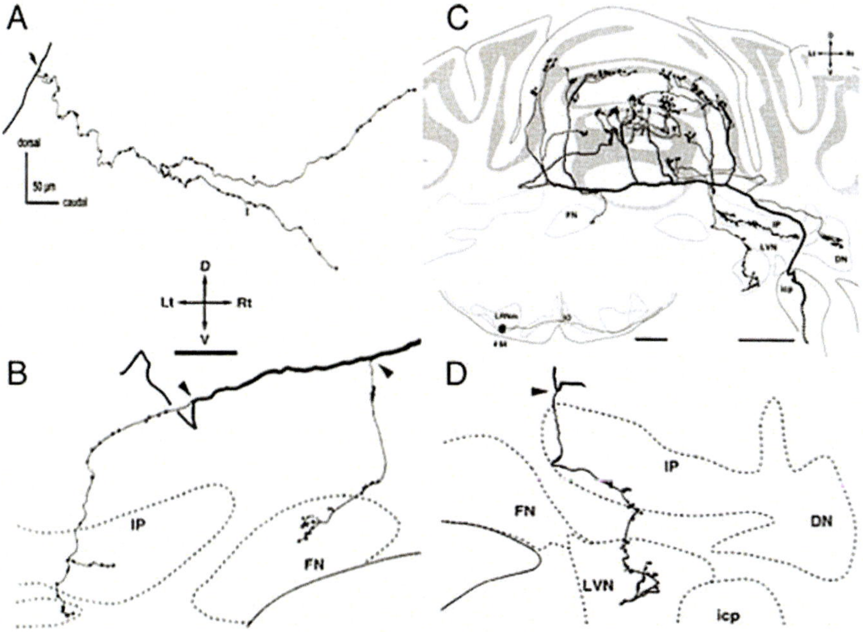

Fig. 10.3 Reconstruction of axonal trajectories of single olivocerebellar neuron (**a**) and single lateral reticular nucleus (LRN) neurons (**b–d**). (**a**) Axon collateral in the fastigial nucleus of a single olivocerebellar axon that projected to vermal lobules VI and VII (Sugihara et al., 1996). (**b**) Collaterals of an LRN axon terminating in the FN and interpositus nucleus (IP). (**c**) Frontal view of a completely reconstructed single axon originating from the middle part of the left LRN. The reconstruction was made by using 96 serial coronal sections. Axon terminals were distributed mainly in lobules II, III, IV, and V of the vermis. Scale bars = 1 mm, respectively. (**d**) Trajectory of a collateral terminating in the lateral vestibular nucleus of the LRN axon in **c**. Arrowheads indicate branching points of collaterals from mossy fiber stem axons. Scale bar = 0.5 mm in (**b**) is also applied to (**d**) (Wu et al., 1999)

Parasagittal sections show that the overall staining patterns in the interpositus and fastigial nuclei were positively stained areas located caudoventrally and negatively stained regions located rostrodorsally with a clear boundary in each nucleus (Fig. 10.4aa, ab). In coronal sections, the ventral parts of the CN were generally aldolase C-positive, consistent with the findings in parasagittal sections (Fig. 10.4b). In each CN, the posterior interpositus nucleus was aldolase C-positive, whereas the anterior interpositus was aldolase C-negative (Fig. 10.4ab, b, c). The FN was aldolase C-positive in the ventrocaudal portion and aldolase C-negative in the rostrodorsal portion (Fig. 10.4aa, b, c). The entire dentate nucleus was aldolase C-positive (Fig. 10.4ac, b, c). The dorsolateral protuberance (DLP), the interstitial cell group (ICG), and the dorsolateral hump (DLH) were aldolase C-negative (Fig. 10.4b, c).

Taking advantage of the selective pattern by which single olivocerebellar axons terminate in the cerebellar cortex and where the same axons' collaterals terminate in the cerebellar nucleus, Sugihara and Shinoda (2007) investigated the functional

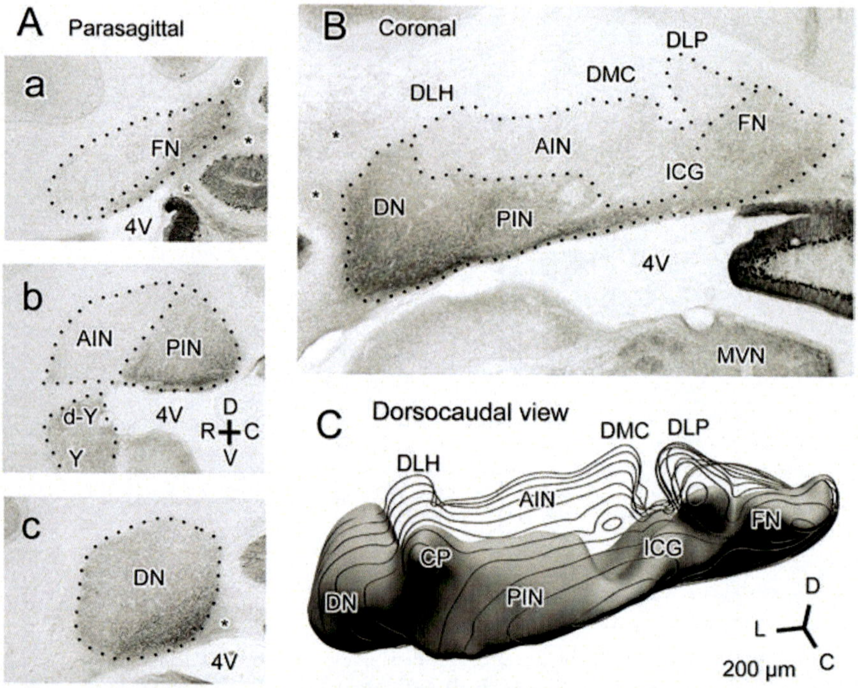

Fig. 10.4 Photomicrographs of aldolase C compartments in the cerebellar nucleus (CN) in the rat. (**a**–**b**) Parasagittal (**a**) and coronal (**b**) sections of the CN labeled with anti-aldolase C antibody. Dotted curves indicate the contour of the CN and borders between aldolase C-positive and aldolase C-negative areas within the nuclei. (**c**) Three-dimensional reconstruction of the aldolase C-positive area (dark areas) in the CN. The wire frames are coronal contours of the CN. Interstitial cell group (ICG); caudal pole (CP); dorsolateral hump (DLH); dentate nucleus (DN); fastigial nucleus (FN); dorsolateral protuberance (DLP); AIN and PIN, anterior and posterior interpositus nucleus (Sugihara & Shinoda, 2007)

relationship between the olivocerebellarly defined compartments of the cerebellar nuclei and the cerebellar cortex. They applied aldolase C expression of Purkinje cell axon terminals to map the compartments of the cerebellar nuclei and compared those compartments with the compartments defined by the olivocerebellar connections. This method revealed the following relationship between the cerebellar cortex and nucleus and the termination of axon collaterals from olivary complex cells. The Purkinje cells located in the aldolase C-positive band in the anterior lobe and the posterior lobe of cerebellar cortex innervated by axons of single olivocerebellar neurons (that is, in anterior/posterior zones corresponding to one another, as described previously) project to the aldolase C-positive nucleus, whereas Purkinje cells in the aldolase C-negative longitudinal band projects to the aldolase C-negative nucleus (Fig. 10.5b).

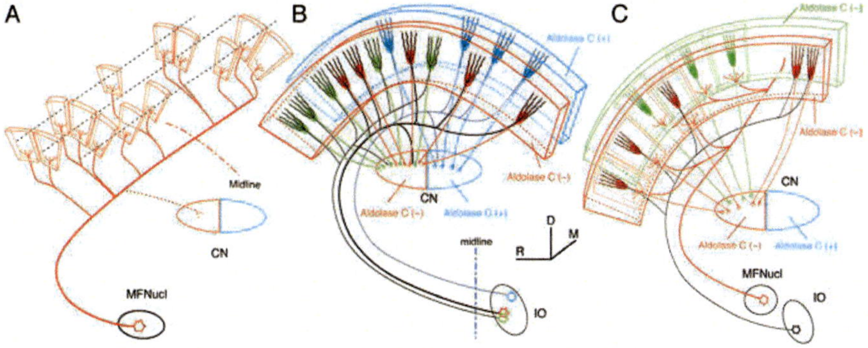

Fig. 10.5 Schematic summary diagrams showing characteristic features of the innervation patterns of a single MF neuron (**a**) and a single climbing fiber neuron (CF) (olivocerebellar neuron) in the cerebellar cortex and CN (**b**) and the relationship between MF and CF longitudinal zones innervated by single MF and CF neurons, respectively (**c**). (**a**) A single MF neuron gives rise to primary collaterals successively from its stem axon, and each primary collateral spreads in a parasagittal plane and terminates in multiple small longitudinal strip-like zones. Such parasagittal zones innervated by individual primary collaterals are arranged successively in a mediolateral direction, often bilaterally. A nuclear collateral may or may not exist in different MF systems. All MFs belonging to a certain mossy fiber nucleus (MF Nucl) project to cortical bands of the same aldolase C expression, namely, either aldolase C-positive or aldolase C-negative cortical bands. (**b**) A single CF neuron (e.g., a blue neuron) gives rise to multiple cortical collaterals that only spread in a parasagittal plane and terminates in a longitudinal zone across one or more lobules. Purkinje cells (PCs) within the longitudinal zone (blue PCs) project to the localized area of a target nucleus where a nuclear axon collateral of the same CF neuron terminates (blue nucleus), and both the PCs and the target nucleus innervated by the same olivocerebellar neuron have the same aldolase C expression pattern (blue, aldolase C-positive). (**c**) The relationship between short band-like zones innervated by MF terminals of a single MF Nucl neuron and longitudinal zones of CF terminals innervated by individual olivary neurons. Single MF neuron (red neuron) with its multiple collaterals spreading mediolaterally innervates PCs of different longitudinal zones of the same aldolase C expression (red and green aldolase C (−) zones) that project to the aldolase C-negative cerebellar nucleus (red) (Shinoda & Sugihara, 2013)

10.2.4 General Structure of the "Microcomplex" of Cerebellar Input–Output Organization

Oscarsson (1969, 1979) proposed a microzonal structure of the cerebellar cortex as a structural–functional unit based on its specific efferent and afferent connections, and Ito (1984) extended this idea to a corticonuclear microcomplex as representing a structural–functional unit organization of the cerebellum. This section introduces how these microcomplexes are demarcated anatomically. It turns out that the microcomplexes can be well defined by both the aldolase C zones we have previously discussed, as well as by the pattern of termination of olivocerebellar axons. There is an additional property that seems to also line up with these other two methods of identification: the axonal trajectory of mossy fiber (MF) neuron. The cerebellar compartmentation we previously presented from the viewpoint of aldolase C expression seems to be well correlated with the zones defined from the viewpoint of olivocerebellar, corticonuclear, and mossy fiber projections (Fig. 10.5c) and form the basic functional units of "microcomplex" proposed by Ito (Shinoda & Sugihara, 2013).

The entire axonal trajectories of single mossy neurons in the cerebellum were first visualized in lateral reticular neurons (Fig. 10.3b–d) (Wu et al., 1999). One of the common features of the branching patterns of single MF neurons is that multiple short longitudinal zones innervated by a single MF axon are arranged mediolaterally in one to a few lobules of the cerebellar cortex. In general, stem axons of MFs run medially toward the midline in the deep white matter rostral and dorsal to the CN and, on their way, give rise to several primary collaterals successively to the cerebellar cortex (Fig. 10.3c). These primary collaterals arise perpendicularly from the stem axons and widely ramify in a rostrocaudal direction mainly in the parasagittal plane of the folium, so that each MF collateral terminates in a relatively narrow longitudinal zone covering one to a few rostrocaudal lobules. It is generally assumed that transverse, lobular subdivisions of the cerebellar cortex are mainly based on the distribution of the MF input, whereas longitudinal divisions represent the output systems of the cerebellar cortex in which both the Purkinje cell zones and their target cerebellar nuclei are innervated by axon collaterals of single climbing fiber neurons in an adjacent area of a subnucleus of the inferior olive (Fig. 10.5b) (Shinoda & Sugihara, 2013). However, the main discontinuities of terminals of single MF axons are mediolateral rather than lobular, and the MF terminals seem to cluster in sagittal short longitudinal strips (Fig. 10.5a). This clustering of terminals of a single MF axon into short sagittal strips may be a common feature of MF projections of different origin (Shinoda & Sugihara, 2013). These branching patterns of single MF axons may account for the "patchy mosaic" or "fractured somatotopy" of granule cell cutaneous receptive fields in which skin surfaces are represented discontinuously in adjacent granule cells (multiple representations of the same receptive fields) (Welker, 1987). Even though the distribution of cutaneous receptive fields looks patchy, the basic organization of the projection of single MF axons in the cerebellar cortex should be regarded as multiple short sagittal strips along longitudinal compartments of aldolase C expression and olivocerebellar projection (Fig. 10.5a) (Sugihara & Shinoda, 2007). The relationship between MF multiple zones and aldolase C-defined longitudinal zones has been examined in several MF systems. The general rule is that MF neurons in each precerebellar nucleus project to longitudinal bands of the identical aldolase C expression. Therefore, single MF neurons project with their multiple primary axon collaterals to multiple longitudinal bands of the same aldolase C expression (Fig. 10.5c). MF projection zones of the lateral reticular nucleus (LRN) are aldolase C-negative in the paramedian lobule (Ruigrok & Cella, 1995). Pontocerebellar MF terminals are located in aldolase C-positive bands (Biswas et al., 2019; Na et al., 2019), whereas most MF terminals originating from the nucleus gracilis and nucleus cuneatus are located in aldolase C-negative bands. MFs of vestibular nucleus neurons terminate in aldolase C-positive zones (Ando et al., 2020). Thus, the cerebellar compartmentation defined from the viewpoint of aldolase C expression seems to be well correlated with the zones defined from the viewpoint of olivocerebellar, corticonuclear, and MF projections (Fig. 10.5c) and form the basic functional units of "microcomplex" proposed by Ito (Shinoda & Sugihara, 2013).

10.2.5 Compartmentalization of the Entire CN and its Correspondence to the Aldolase C Expression Pattern in the CN

This correlation of aldolase C expression with the zones defined via input/output connections is the final piece of the puzzle that allowed pair of rostral and caudal aldolase C compartments to be linked based on the common projection of single OC axons across the rostrocaudal boundary on lobule VIc-crus Ib. Based on the localization of the olivary origins of projection to similar compartments, the compartments and olivocerebellar projections can be sorted into five groups (Fig. 10.2 and Table 10.1); group I, positive aldolase C compartments extending from the posterior lobe to the anterior lobe innervated by the principal olive and some neighboring areas; group II, positive compartments localized within the posterior lobe innervated by several medial olivary subnuclei; group III, vermal and central aldolase C-negative compartments innervated by the centrocaudal medial accessory olive; group IV, negative and lightly positive compartments in the hemispheric and the rostral and caudal pars intermedia innervated by the dorsal accessory olive and some neighboring areas; and group V, positive aldolase C compartments in the flocculus and nodulus (Sugihara & Shinoda, 2004).

To understand how the CN are divided into the termination areas of the five groups, three-dimensional schemes of the termination areas of each group were assembled into a single model of the CN. Figure 10.6 shows this model cut into multiple parasagittal and coronal planes with group organization indicated by colors. Comparison of Fig. 10.4 and 10.6 indicated that the termination areas of groups I (green), II (blue and cyan), and V (gray) are located in an aldolase C-positive caudoventral part of the CN and those of groups III (yellow and orange) and IV (red and pink) are located in an aldolase C-negative rostrodorsal part of the CN (Sugihara & Shinoda, 2007). Indeed, such a relationship was noticed during mapping of nuclear termination areas in individual experiments, because brain sections were double labeled for BDA and aldolase C. The correspondence of termination areas of groups I, II, and V with aldolase C-positive areas was directly shown by comparing dorsocaudal views of the three-dimensional displays of these areas (Fig. 10.4c). The correspondence between the five groups in Fig. 10.2 and aldolase C expression in the CN was consistent with that in the cerebellar cortex, because groups I, II, and V occupied aldolase C-positive compartments in the cerebellar cortex (Fig. 10.2) (Sugihara & Shinoda, 2004). The entire AIN, including the DLH, belonged to group IV (Fig. 10.6ac, af, bb, bd, red) and was generally aldolase C-negative. In the PIN, the main part belonged to aldolase C-positive group I (Fig. 10.6ad, ae, ba, bb,

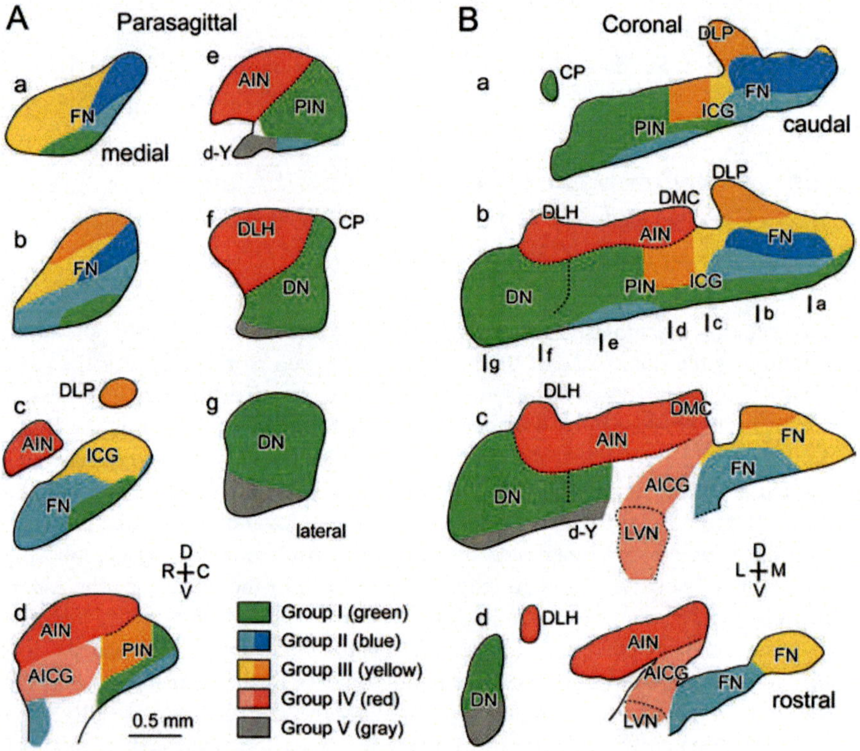

Fig. 10.6 Schematic summary of compartmentalization of the cerebellar nuclei (CN) shown in parasagittal (**a***a*–**a***g*) and coronal (**b***a*–**b***d*) sections of the left CN. Vertical bars in **bb** (*a*–*g*) indicate approximate locations of parasagittal sections in **aa**–**a***g*, respectively. Colors indicate groups I–V defined in Table 10.1 and Fig. 10.2 (green, group I; cyan, group IIa; blue, group IIb; yellow and orange, group III; red and pink, group IV; gray, group V). The drawings are based on the termination areas of olivonuclear collaterals of each group shown in Fig. 10.2. Comparison of the corresponding sections in Fig. 10.4 and this figure indicates that termination areas of groups I, II, and V generally belong to the aldolase C-positive part of the CN and those of groups III and IV belong to the aldolase C-negative part (Sugihara & Shinoda, 2007)

green), but some part was located in the aldolase C-negative dorsomedial portion (group III) (Fig. 10.6ad, ba, bb, orange). The entire DN, which was aldolase C-positive, belonged to group I (Fig. 10.6af, ag, bb–bd, green). The lateral vestibular nucleus (LVN), which was generally aldolase C-negative, belonged to group IV (Fig. 10.6ad, bc, bd, pink). As to the fastigial nuclear subdivisions, the groups I, IIa, and IIb were located in the ventrocaudal FN (Fig. 10.6aa, ab, ba, bb, blue) and were aldolase C-positive, whereas group III areas (Fig. 10.6aa, ab, ba–bd, yellow, orange) occupied the rostrodorsal portion and DLP of the FN, which were aldolase C-negative.

10.3 Input–Output Organization of the Fastigial Nucleus

The cerebellum is known to play a significant role in coordinated control of complex movements. To understand the functional roles of the three longitudinal zones, Chambers and Sprague (1955), and later Thach (as reviewed in Thach et al., 1992), performed lesion experiments and found that among the three zones, the vermis and its underlying fastigial nucleus (FN) were concerned with maintaining posture and walking and its associated body and neck movements. They also found an association with eye movements. The four nuclei can be distinguished in different mammalian species, where they are known as the medial, anterior interposed, posterior interposed, and lateral cerebellar nucleus (Ogawa, 1935). The vermis sends its fibers to the fastigial nucleus. Specifically, the anterior parts of the vermis send axons to the rostral part of the FN and the posterior parts to the caudal parts of the FN. In the original study, it also appears that Purkinje cells of the vermis located in a rostrocaudal direction project to the rostrocaudal parts of the fastigial nucleus, respectively (Jansen & Brodal, 1942). However, the relationship was not so simple, and later anatomical studies revealed finer and complex connections between cortical longitudinal zones and different parts of the FN. The cerebellar efferents from the interpositus and dentate nuclei leave the cerebellum via the superior cerebellar peduncle (SCP) to the contralateral side. Most fibers from the FN, after passing the midline cerebellar white matter in the uncinate fasciculus (the hook bundle), run over the brachium conjunctivum and curve ventromedially into the brainstem on the contralateral side. A minor proportion of the FN runs in the juxtarestiform body, which is just medial to the restiform body, and terminates ipsilaterally in the brainstem (Verhaart, 1964).

To understand the neural circuitry of the fastigial control of complex movements, afferent and efferent projections of the FN have been extensively studied using the silver impregnation method (Carpenter & Nova, 1960; Walberg et al., 1962a, 1962b), the horseradish peroxidase labeling method (cat, Fukushima et al., 1977; Matsushita & Hosoya, 1978), the autoradiographic method (cat, Moolenaar & Rucker, 1976; Sugimoto et al., 1981; monkey, Batton et al., 1977; Asanuma et al., 1983), and the phaseolus vulgaris leucoagglutinin anterograde labeling method (cat, Homma et al., 1995). The cerebellar FN is the main source of cerebellar efferents to the lower brainstem (Thomas et al., 1956). It has been generally accepted in degeneration studies that fibers of the hooked bundle of Russell (Rasmussen, 1933) arise predominantly from the caudal FN and cross within the cerebellum, whereas uncrossed fastigial efferent fibers arise predominantly from the rostral FN and emerge from the cerebellum via the juxtarestiform body (Jansen & Jansen, 1955; Carpenter et al., 1958; Cohen et al., 1958; Walberg et al., 1962a, 1962b; Voogd, 1964; Angaut & Bowsher, 1970). On the other hand, the selective labeling of either rostral or caudal cells in the FN with autoradiographic methods revealed that some cells at all rostrocaudal levels of the FN must give rise to crossed and uncrossed fibers into the brainstem, but fastigioreticular fibers are almost entirely crossed and arise from all parts of the FN (Batton et al., 1977). The densest projection is to the nucleus reticularis gigantocellularis (NRG) from the level of the abducens nucleus to the caudal olivary level of the medulla oblongata. The major terminal regions within the NRG are located rostrally and medially (Batton et al., 1977; Asanuma et al., 1983), although the rostral FN and caudal FN project predominantly to the ventral and dorsal NRG, respectively (Homma et al., 1995). As exemplified above, the details of fastigial efferent projections have historically been rather controversial.

10.3.1 *Efferent System of the Fastigial Nucleus*

Fastigiovestibular Projection

The majority of the crossed and uncrossed fastigovestibular fibers are derived from the caudal and rostral parts of the FN, respectively, in the cat (Cohen et al., 1958). Furthermore, Walberg et al. (1962a, 1962b) showed that fibers from the caudal FN were distributed in the ventral part of the contralateral LVN, whereas fibers from the rostral FN in the dorsal part of the ipsilateral LVN (Brodal et al., 1962). Electrophysiological investigations with intracellular recording techniques revealed that the cerebellar corticovestibular fibers from the anterior lobe had inhibitory synaptic actions in Deiters' neurons (Ito & Yoshida, 1966) and the fastigiovestibular fibers had excitatory synaptic actions in them (Ito et al., 1970). However, the distribution within the Deiters' nucleus of these neurons that receive Purkinje inhibition and/or fastigial excitation has been controversial. This controversy is due in part to the definition of the Deiters' nucleus and is largely due to the complex structures of afferents and efferents around the rostral FN (Brodal, 1981; Voogd, 2016).

The fibers coming from the caudal FN cross the midline through the cerebellar white matter, partly passing through the opposite FN, which forms the hook bundle (uncinate fasciculus of Russell) and bends dorsal to the superior cerebellar peduncle and directs ventrally to the vestibular nuclei and the reticular formation. As pointed out by Alf Brodal (1969), this structure around the rostral FN is unfortunate for experimental studies of neuroanatomy and electrophysiology, because the complicated structure makes it difficult to interpret the results of experimental studies in lesion or stimulation electrophysiological experiments and degeneration or HRP anatomical experiments. Thus, there are several questions regarding FN that remain unsolved, and we summarize these next.

Ito and his colleagues found that ipsilateral fastigial stimulation evoked IPSPs with monosynaptie latencies in the vast majority of Deiters' neurons examined with multipolar stimulating electrodes (Ito & Yoshida, 1966; Ito et al., 1968). However, these IPSPs, mediated by corticovestibular Purkinje cell axons, were absent in those Deiters' neurons located relatively ventrally (Ito et al., 1968, 1969). Figure 10.7 illustrates that EPSPs were induced in a ventral Deiters' neuron from the rostral parts around the bilateral fastigial nuclei (Ito et al., 1970). These EPSPs were evoked at a latency of 0.5–0.9 msec by stimulation at the ipsilateral fastigial region,

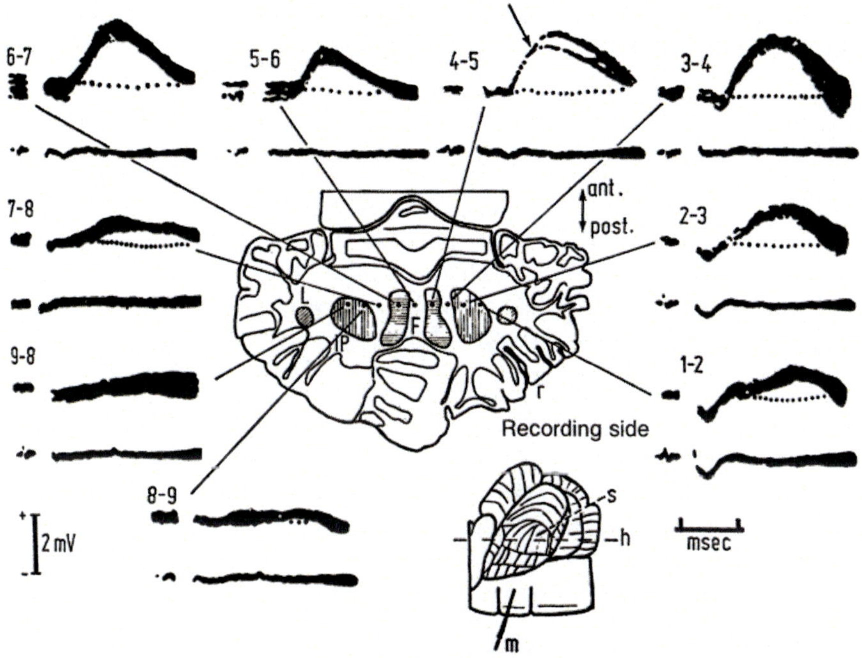

Fig. 10.7 Excitatory postsynaptic potentials (EPSPs) evoked in a Deiters' neuron from the deep nuclear regions of the cat cerebellum. Center diagram indicates a horizontal section of the cerebellum. Positions of nine stimulating spots are indicated, r, right side. Ant., anterior, post., posterior. Records are intracellular potentials evoked in a right ventral Deiters' neuron by applying pulses of 10 volt amplitude to the spots indicated. Upper and lower traces, intracellular and extracellular potentials, respectively. Dotted lines indicate the time course of extracellular controls. Downward arrow in records 4–5 marks the diverging point between the monosynaptic and the superposed delayed EPSPs. Diagram at the bottom shows the left side aspect of the cerebellum and medulla, the directions of inserting stimulating needles (s) and recording microelectrode (m), and the level of the horizontal sectioning (h) of the cerebellum (modified from Ito et al., 1970)

indicating the EPEPs were induced monosynaptically. The effective sites for producing monosynaptie EPSPs in ventral Deiters' neurons extended laterally from the fastigial region not only to the ipsilateral but also to the contralateral side. Small EPSPs from the ipsilateral interpositus stimulation might be due to spread of currents to the fastigial nucleus, but it might also be due to activation of cerebellar afferent fibers passing through or near the interpositus nucleus that terminate on ventral Deiters' neurons. In a majority of descending vestibular nucleus (DVN) neurons, EPSPs were observed in the monosynaptic range during fastigial stimulation. In most ventral Deiters' neurons and DVN neurons, EPSPs were exclusively evoked by stimulation of the bilateral fastigial nuclei, because the EPSPs were not contaminated by Purkinje cell inhibition. Therefore, Ito and colleagues reached the conclusion that the effective of fastigial stimulation on ventral Deiters' and DVN neurons is exclusively excitatory (Ito et al., 1970). However, the question remained as to whether the ipsilateral or contralateral FN has synaptic connections with dorsal Deiters' nucleus neurons. We address each of these in turn.

Projection of the Contralateral FN to the Vestibular Nuclei

To simplify interpretation of the complicated neural circuits of the fastigiovestibular projection, we first address only the connections of the fastigial nucleus to the contralateral vestibular nuclei. Akaike (1983) systematically examined cerebellar corticovestibular and fastigiovestibular projections to the LVN in the cat. In one cat, using a single microelectrode, he mapped synaptic actions evoked form the vermal anterior lobe and the contralateral FN in 85 neurons. These neurons were classified as medial (MVST) and lateral vestibulospinal tract (LVST) neurons by their antidromic responses to stimulation of individual tracts at the spinal cord. Neurons with monosynaptic excitatory input from the contralateral FN were predominantly localized in the ventral half of the LVN and the DVN, although a smaller number of such neurons were also distributed in the dorsal part of the LVN. The ventrally located LVN and DVN neurons with crossed fastigial excitation did not receive ipsilateral Purkinje cell inhibition. These neurons received monosynaptic excitation from the primary vestibular nerve and were either MVST or LVST neurons. These elctrophysiological findings are consistent with the anatomical finding that the fastigial neurons project to the contralateral ventral LVN and DVN (Walberg et al., 1962a, 1962b; Batton et al., 1977). This anatomical connection explains why destruction of the caudal part of the FN results in atonia of the contralateral limbs in decerebrate animals (crossed fastigial atonia) (Moruzzi & Pompeiano, 1956).

However, these findings are not sufficient to determine unequivocally whether the rostral FN neurons project to the contralateral ventral LVN and DVN, because the fibers from the caudal FN pass through the rostral FN to the contralateral side via the uncinate fasciculus. This connection most likely exists, because autoradiographic study (Batton et al., 1977; Asanuma et al., 1983) and PHA-L study (Homma et al., 1995) showed that the rostral FN also projected to the contralateral ventral LVN and DVN. Furthermore, Asanome et al. (1998) injected cholera-toxin b subunit conjugated horseradish peroxidase into the uncinate fasciculus in the midline in the cat, stimulation of which resulted in bilateral and simultaneous augmentation of tonic activities in the neck, lumbar back, fore- and hind limb extensor muscles in a reflexively standing acute decerebrate cat. They found that all retrogradely labeled neurons were scattered symmetrically in the entire rostrocaudal extent of the bilateral fastigial nuclei, indicating that the crossed fastigial efferents originate from not only the caudal but also the rostral FN.

In this experiment, the majority of the dorsal LVN neurons received monosynaptic inhibition from the ipsilateral anterior lobe and were not activated from the ipsilateral primary vestibular nerve. They were LVST neurons. These findings support the interpretation that the LVN should be divided into the dorsal and ventral parts and the dorsal part should be considered as a cerebellar nucleus and the ventral part as a vestibular nucleus (see the discussion in Voogd, 2016). This interpretation is consistent with the anatomical finding that the ventral part of the LVN should be classified as the magnocellular part of the MVN. The Deiters' nucleus is considered to belong to the vestibular nuclear complex (the lateral vestibular nucleus), where large neurons with 30–45μm in longer diameter (Deiters' neurons) are present. Thus, there is no disagreement that the dorsal part of LVN is part of Deiters' nucleus. However, ventral to the dorsal part, neurons with 30μm in longer diameter are present. Brodal's group called this part as ventral LVN in the cat (Brodal & Pompeinano, 1957) and monkey (Brodal, 1984), since they thought this part could be clearly differentiated from the MVN. Voogd's group classified this part as the magnocellular division of the MVN (Epema et al., 1988; Gerrits et al., 1985) to differentiate this part from the more medially located original MVN (parvocellular part of the MVN). The primary vestibular afferents project to these parvocellular and magnocellular parts of the MVN and the DVN, and these neurons do not receive Purkinje cell inhibition from the anterior vermal lobe.

Projection of the Ipsilateral FN to the Vestibular Nuclei

In contrast to the crossed fastigiovestibular projection, the uncrossed fastigiovestibular projection introduces problems about the interpretation of the experimental results. The rostral FN was originally thought to project to ipsilateral dorsal LVN. This interpretation seemed to be based on both anatomical and electrophysiological evidence. However, the experiments do not unambiguously indicate this. It is necessary to determine whether cells of origin responsible for monosynaptic excitation are really located in the rostral FN or whether the monosynaptic excitation could possibly originate from passing fibers from the different cerebellar nuclei. Another possibility is that the axon collateral reflex of mossy fibers and climbing fibers might induce monosynaptic excitation in neurons of the ipsilateral LVN. The neurons in the ventral Deiters' nucleus do not receive Purkinje cell axons from the anterior posterior lobe (Walberg & Jansen, 1961). Dorsal LVN neurons receive strong IPSPs, which makes determination of the presence or absence of monosynaptic EPSPs more difficult, but in ventral LVN neurons, stimulation of the FN usually produces only EPSPs, and these EPSPs are induced usually bilaterally from the fastigial nuclei (Fig. 10.7). On the other hand, in dorsal Deiters' neurons, stimulation of the ipsilateral FN usually evoked monosynaptic EPSPs contaminated by Purkinje cell-induced IPSPs. However, monosynaptic EPSPs can also be evoked from the cerebellar cortex of the anterior and posterior lobes overlying the FN, although the IPSPs arise only from a localized cortical area on the ipsilateral side (B zone) (Ito et al., 1968). The latter EPSPs are presumed to be induced through the cerebellar afferent fibers by a kind of axon reflex. When evoked with brief time intervals, these fastigial and cortical EPSPs interfere with each other in the manner of impulse collision and refractoriness, indicating that a large portion of the fastigial EPSPs (up to two thirds) are indeed induced through certain cerebellar afferent fibers (Ito et al., 1969).

Batton et al. (1977), using a tritiated amino acid autoradiographic technique in the monkey, found the fastigovestibular fibers projected mainly to the ventral part of LVN bilaterally nearly symmetrically. Since the autoradiographic technique provides a method to confirm anterograde labeling of axons of neurons that uptake the material from their cell bodies, this finding suggested that the ipsilateral projection of the rostral FN to the dorsal LVN was not present. Labeled neurons in the rat FN from an injection of a retrograde tracer affecting the entire left-sided output of the nucleus showed that the dorsal part of the left rostral FN was lacking in labeled neurons. However, a small number of the labeled neurons were found in the ventral part of the left rostral LVN (Voogd et al., 2013). In the mouse, the ipsilateral pathway from the FN to the brainstem has been found to originate from glycinergic neurons (Bagnall et al., 2009). The distribution of these glycinergic cells in the ventral and rostral FN in the mouse corresponds closely to the distribution of the neurons with ipsilateral projections in the rat. Therefore, it is most likely that there is no ipsilateral excitatory fastigiovestibular projection to the dorsal LVN. Instead, the anatomical and electrophysiological findings of monosynaptic excitation from the rostral FN to the ipsilateral dorsal LVN reported previously are most likely due to activation (or labeling/staining) of passing fibers of crossed fastigiovestibular efferents or axon reflex of cerebellar cortical mossy and climbing afferents. One such example of the axon reflex is mossy fibers arising from the central cervical nucleus. Matsushita et al. (1995) showed that the central cervical nucleus in the upper cervical cord sent axons to the cerebellar vermis. Along the way, axon collaterals were extended to the LVN and the FN. Therefore, stimulation of these axon collaterals in the FN might evoke monosynaptic excitation in dorsal LVN neurons on the same side.

Fastigioreticular Projections

In previous physiological studies, much effort has been devoted to revealing the functional role of the fastigial nucleus in controlling muscle tonus, using both lesion and stimulation techniques (Brookhart, 1959; Dow & Moruzzi, 1958; Moruzzi & Pompeiano, 1956; Sprague & Chambers, 1954). Although these studies revealed a complex story, they all concluded that the function of the crossed fastigiobulbar projection system is to facilitate the reticular formation and vestibular nuclei (see Moruzzi & Pompeiano, 1957).

On a silver degeneration method, Cohen et al. (1958) found in the cat that crossed fibers of the hook bundle and uncrossed fibers arose from the caudal and rostral parts of the FN, respectively. Furthermore, the rostral third of the FN was a predominant source of the projection to the ipsilateral reticular formation. Walberg et al. (1962a, 1962b) found in the cat that the entire FN participated in the crossed projection via the hook bundle to the reticular formation. The fibers from the rostral part of the FN made up almost half of all fastigiobulbar fibers and left the cerebellum in the ipsilateral restiform body (Flood & Jansen, 1966).

Ito et al. (1970) recorded intracellular potentials in reticulospinal neurons (RSNs) in the medullary reticular formation and showed large monosynaptic EPSPs evoked by stimulation of the rostral FN. Eccles et al. (1975) stimulated the FN and found increased spike activity at the monosynaptic range in pontine and medullary RSNs. Mori et al. (1998) activated fastigiobulbar fibers in the hook bundle and observed extracellular spike activity at the monosynaptic range in RSNs in the medullary reticular formation. However, in all these cases, stimulation of the hook bundle or the rostral FN could have activated fibers arising from rostral and caudal fastigial neurons, and therefore, none of these studies could determine unambiguously whether or not caudal or rostral FN axons directly terminate on reticular neurons.

Ito's pioneering work provided evidence for the excitatory synaptic action of crossed fastigiobulbar fibers by observations in ventral Deiters' and reticular formation neurons in which Purkinje cell inhibition was not observed (Ito et al., 1970). However, the action of the uncrossed fastigioreticular fibers could not be determined in reticular neurons, because stimulation of the FN on either side evoked monosynaptic EPSPs in reticular neurons, and they could not differentiate whether the passing fibers from the crossed FN or fibers arising from the uncrossed FN were activated in their experiments. As discussed for fastigiovestibular projections, the fastigioreticular fibers, arising from all parts of the fastigial nucleus, are predominantly decussating. Terminals of these fibers are distributed widely in the ponto-medullary reticular formation, mostly in the nucleus reticularis gigantocellularis (NRG), nucleus reticularis pontis caudalis (NRPc), and the paramedian reticular nucleus.

In addition to limb and body movements, eye movements were shown to be controlled by the posterior vermis (Keller et al., 1983; Llinás & Wolfe, 1977; Ritchie, 1976; Ron & Robinson, 1973), and microstimulation and recordings of saccade-related Purkinje cell activity (Kase et al., 1980) revealed that vermal lobule VII and its adjacent lobule VIc are concerned with eye movements. The destruction of Purkinje axons by kainic acid and FN injection of bicuculline suppressed saccades evoked by vermal microstimulation, suggesting that the caudal FN relayed impulses from vermal lobules VI and VII to the brainstem oculomotor circuitry. Consistent with this suggestion, neurons in the caudal FN discharge during saccades (Fuchs et al., 1993; Helmchen et al., 1994; Ohtsuka & Noda, 1991). Therefore, the caudal FN is considered to be concerned with eye movements. However, other studies showed that caudal FN is also involved in head movements in cat (Goffart & Pélisson, 1998; Gruart & Delgado-Garcia, 1994) and monkey (Quinet & Goffart, 2007), because inactivation of the caudal FN produced dysmetria of eye and head movements.

As mentioned above, the general pattern of termination of fastigial fibers in the brainstem appeared to be fairly well understood, but the exact projections of the fastigial oculomotor area to specific target neurons in the brainstem saccade generator remained undetermined. To ameliorate this, Noda and colleagues analyzed afferent and efferent projections of the vermal oculomotor region in the brainstem of the monkey (Yamada & Noda, 1987). Subsequent anatomical studies on the caudal FN have shown that axons of neurons in the caudal one-third of the FN terminate in areas where saccade-related premotor burst neurons are located (Takahashi et al., 2014). Specifically, the axons terminated in the excitatory burst neuron (EBN) area in the paramedian pontine reticular formation (PPRF) and the inhibitory burst neuron (IBN) area in the paramedian pontomedullary reticular formation (Hikosaka & Kawakami, 1977). Among these projections, the most extensive projection is to the contralateral NRG where IBNs are located in the monkey (Noda et al., 1990). Noda and colleagues proposed that at least a part of the labeling found in the medullary reticular formation immediately caudal to the abducens nucleus may correspond to the so-called IBN area. Thus, based primarily on these anatomical findings, it is generally assumed that fastigioreticular neurons in the caudal FN terminate on contralateral IBNs via the hook bundle (Kojima et al., 2008; Scudder & McGee, 2003). However, the anatomical methods used cannot unambiguously indicate whether the target neurons of the caudal oculomotor FN are in fact IBNs or whether they are EBNs. This is because the NRPc and NRG also contain RSNs that innervate neck motoneurons (see Shinoda et al., 2006).

While anatomical data could not determine whether axon terminals of caudal fastigial neurons distributed in the NRPc and NRG actually terminate on EBNs and IBNs or RSNs, intracellular recordings of postsynaptic potentials in the target neurons of caudal fastigial axons could potentially settle this question. The type of cell (IBN, EBN) could be identified electrophysiologically by their antidromic responses to stimulation of their destinations (Takahashi & Shinoda, 2018). In spite of many anatomical studies on fastigioreticular projections, there have been far fewer intracellular recording studies on synaptic inputs to their target neurons in the brainstem. Reticular neurons (Ito et al., 1970) and vestibular nucleus neurons (Furuya et al., 1976) have been identified as targets of fastigioreticular neurons. Ito et al. (1970) analyzed the effects of systematic electrical stimulation of the cerebellar nuclei on reticular neurons in the medulla oblongata of the cat. Stimulation of the FN on either side produced monosynaptic excitation of medullary reticular neurons, most of which were RSNs that projected to the spinal cord. However, their stimulation sites in the cerebellar nucleus were located in the rostral FN, where both output neurons in the rostral FN and axons from the caudal FN can be activated. In addition, reticular neurons were recorded rather caudally at the level of the inferior olive where IBNs are not located (Sugiuchi et al., 2005).

Takahashi et al. (2014) investigated neural connections from the oculomotor FN to brainstem neurons in the saccade generator via intracellular recordings. After identifying the oculomotor FN where microstimulation evoked saccades to the contralateral side, Takahashi and colleagues injected dextran biocytin and plotted the axon terminals of the FN in the brainstem (Fig. 10.8). They restricted their injection site to the caudoventral part of the FN and found that the labeled terminals were distributed in the NRPc and NRG only on the contralateral side. Figure 10.9 shows an example of a neuron penetrated in the NRG just caudal to the left abducens nucleus. This neuron received excitation from both the contralateral (Fig. 10.9d, 2–4) and ipsilateral caudal FN (Fig. 10.9d, 5–7). EPSPs were much larger from the contralateral FN than from the ipsilateral FN and were larger from the rostroventral part (sites 2 and 5) than from the more caudodorsal sites in the caudal FNs. The

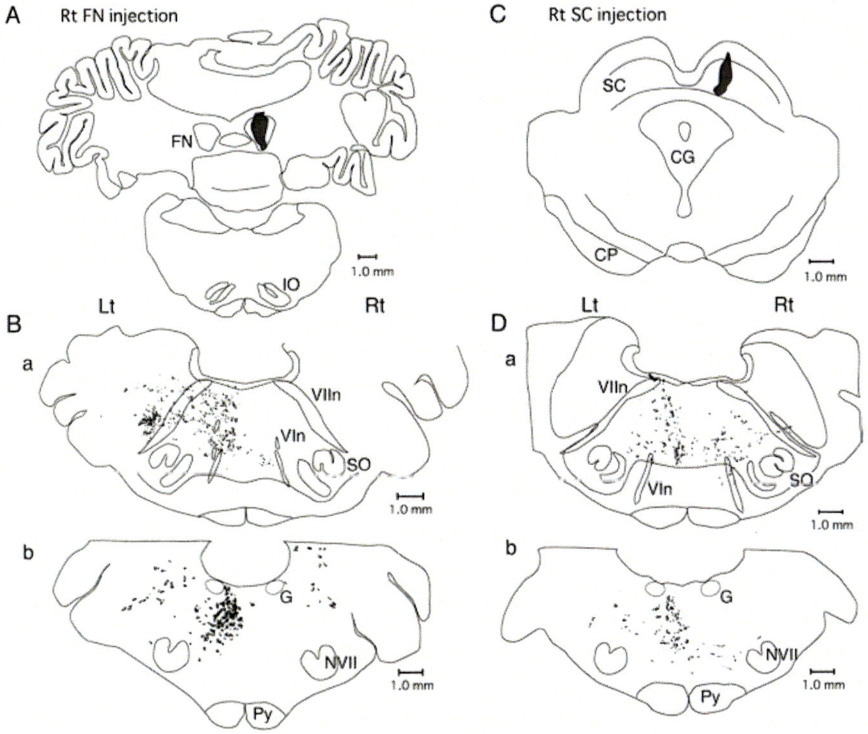

Fig. 10.8 Brainstem distribution of axon terminals of neurons in the caudal fastigial nucleus (FN) (**a** and **b**) and the superior colliculus (SC) (**c** and **d**). (**a**) Frontal section of the cerebellum showing a site at which dextran–biotin (DB) (0.35μl) was injected into the right (Rt) caudal FN (cFN) (black area). (**b**) Distribution of anterogradely labeled axon terminals in the brainstem after the injection of DB into the FN shown in (**a**). All labeled axon terminals in two consecutive 70μm thick sections are plotted in representative frontal sections 1.0 mm rostral to the rostral border of the abducens nucleus (VI) (*a*) and 0.5 mm caudal to the caudal border of the VI (*b*). Each dot indicates one axon terminal in (**b**) and (**d**). (**c**) Frontal section of the midbrain showing a site at which DB (1.0μl) was injected into the SC (black area). (**d**) Distribution of anterogradely labeled axon terminals in the brainstem after the injection of DB into the Rt SC shown in **c**. IO, inferior olive; VIn, abducens nerve; VIIn, facial nerve; SO, superior olive; G, facial genu; NVII, facial nucleus; Py, pyramid; CG, central gray; CP, cerebral peduncle; Lt, left (Takahashi et al., 2014)

ipsilateral EPSPs were almost negligible, while contralateral EPSPs were still large even under weaker stimulus intensities. Based on this, Takahashi et al. (2014) concluded that the ipsilateral EPSPs were due to current spread to the crossed fastigio-reticular fibers (Batton et al. 1977). This pattern of input from the bilateral FNs was typical of NRG neurons. However, these neurons were not activated antidromically from either the ipsilateral abducens nerve (Fig. 10.9c, 1) or the contralateral abducens nucleus (Fig. 10.9c, 14) but were activated antidromically at 1.0 ms from the ipsilateral second cervical cord (Fig. 10.9b). Therefore, these NRG neurons could not be abducens motoneurons nor premotor IBNs. Having excluded these

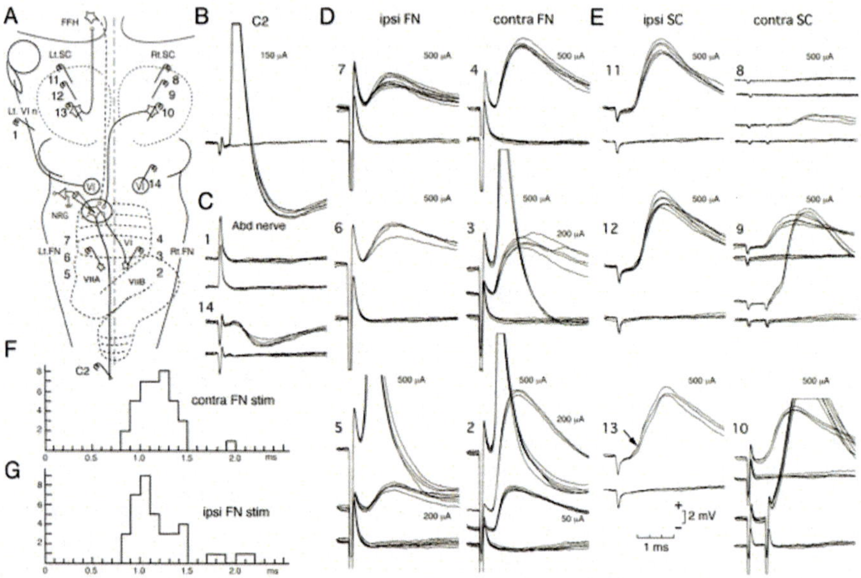

Fig. 10.9 Intracellular records from a neuron in the rostral nucleus reticularis gigantocellularis (NRG). This neuron received excitation from the caudal fastigial nucleus (FN) and the superior colliculus (SC). (**a**) Experimental setup. C2, second cervical cord; FFH, Forel's field H; VI and VIn, abducens nucleus and nerve. (**b**) Antidromic spikes evoked in an all-or-none manner at threshold (150μA) by stimulation of the ipsilateral second cervical spinal cord (C2). (**c**) No antidromic field potentials evoked by stimulation of the Lt VIn (site 1) at 500μA. This neuron was recorded 0.8 mm caudal to the caudal border of the VI. (**d** and **e**) EPSPs evoked from the FNs (**d**) and from the SCs (**e**). (**f** and **g**) Latency histograms of EPSPs in neurons of the rostral NRG evoked by stimulation of the contralateral FN (**f**) and the ipsilateral FN (**g**) (Takahashi et al., 2014)

possibilities, Takahashi and colleagues concluded that the cells must be RSNs that project ipsilaterally to the cervical cord. In fact, these neurons were shown to be antidromically activated from the cervical spinal cord.

To compare synaptic inputs from the caudal oculomotor FN to caudal NRPc neurons and rostral NRG neurons, we have recorded intracellular potentials from NRPc neurons just rostral to the abducens nucleus (Fig. 10.10). Stimulation of the contralateral FN evoked very small EPSPs (Fig. 10.10b, 2), and stimulation of the ipsilateral FN did not evoke any EPSPs (Fig. 10.10b, 5–7). However, we observed large EPSPs after stimulation of the contralateral SC (Fig. 10.10c). We found that caudal NRPc neurons receive only very weak excitation from the contralateral caudal FN and no excitation from the ipsilateral caudal FN (Takahashi et al., 2014). This finding is consistent with the anatomical finding that the caudal fastigial projection to the rostral NRG appears to be considerably stronger than that to the caudal NRPc (Homma et al., 1995; Noda et al., 1990). However, in this experiment, we could not confirm whether fastigioreticular axons arising from the caudal FN directly innervate burst neurons in the brainstem saccade generator on the contralateral side via the hook bundle, as is generally assumed in models of eye movement

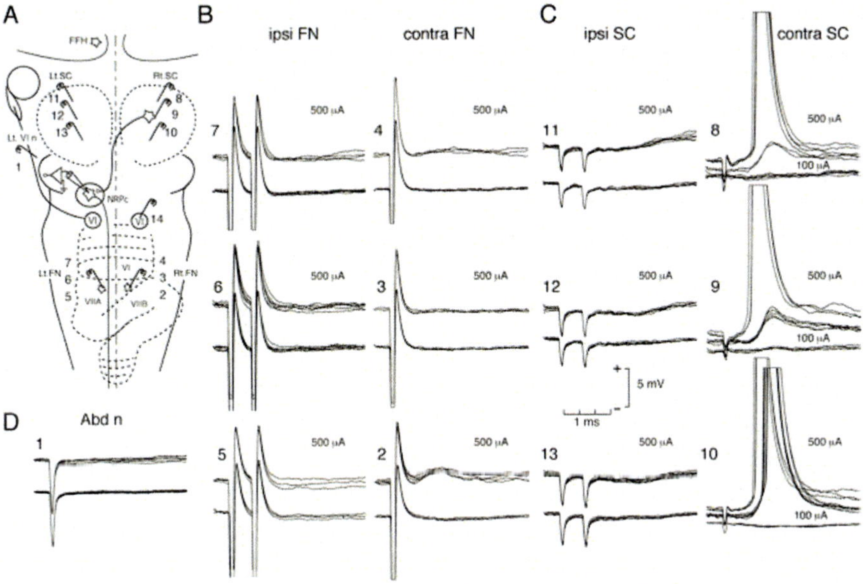

Fig. 10.10 Properties of fastigial and tectal synaptic inputs to a neuron in the nucleus reticularis pontis caudalis (NRPc). (**a**) Experimental setup. (**b**) Properties of postsynaptic potentials evoked by stimulation of the contralateral (sites 2–4) and ipsilateral FN (sites 5–7) at 500μA in an RSN. (**c**) Properties of EPSPs evoked by stimulation of the contralateral (sites 8–10) and ipsilateral SC (sites 11–13) at 500μA in the same neuron as in **b**. Stimulation of the contralateral SC evoked large EPSPs with spikes at 500μA (8–10), and at 100μA, rostral stimulation evoked small EPSPs (8, 9), and caudal stimulation evoked large EPSPs with spikes (10). (**d**) No antidromic spikes or antidromic field potentials were evoked from the ipsilateral VIn in this neuron (Takahashi et al., 2014)

control (Fuchs et al., 1993; Ohtsuka & Noda, 1991). Further experiments are required to reliably conclude that this direct connection really exists by recording intracellular potentials from electrophysiologically identified IBNs and examining the nature of synaptic inputs evoked by stimulation of the caudal oculomotor FN. Confirmation of the presence or absence of this pathway is essential for understanding the neural mechanism of the caudal FN in the cerebellar control of saccadic eye movements.

Fastigiospinal Projections

The direct projection from the FN to the spinal cord was suggested by Thomas et al. (1956), and the clear experimental evidence was first presented in cat by Fukushima et al. (1977) and by Matsushita and Hosoya (1978), who observed labeled neurons in the contralateral FN following HRP injection into the spinal cord. The autoradiographic study by Batton et al. (1977) confirmed these findings. Wilson et al. (1977, 1978) stimulated the rostral part of the FN and recorded intracellular potentials from contralateral motoneurons of biventer cervicis and trapezius muscles in the cat. They evoked short-latency, most likely monosynaptic, EPSPs in these motoneurons. Although they stimulated the rostral FN, this finding does not necessarily prove that only rostral FN neurons are responsible, because fibers from neurons in the caudal FN pass through the rostral FN to the hook bundle. This direct connection of the FN with neck and shoulder motoneurons is one of the ways in which the FN exerts its influence on the spinal cord. In addition to this direct fastigiospinal projection, reticulospinal and vestibulospinal neurons activated from the contralateral FN indirectly influence spinal motoneurons bilaterally (Matsuyama & Jankowska, 2004) via spinal commissural interneurons (Sugiuchi et al., 1992).

Fastigial Projection to the Superior Colliculus

The projection from the FN to the superior colliculus (SC) arises exclusively from the caudal half of the FN and crosses within the cerebellum to form the crossed ascending limb of the uncinate fasciculus. In contrast, the projections from the three other cerebellar nuclei to the SC arise from the almost entire rostrocaudal extent and form a compact fiber bundle within the brachium conjunctivum, which crosses in the brainstem to the contralateral SC. However, the proportion of the SC projections from individual cerebellar nuclei differs among different species. In regard to the fastigial projection, a bilateral projection from the FN is the major component of the cerebellotectal projections in frontal-eyed animals such as the cat (Cohen et al., 1958; Edwards et al., 1979; Roldan & Reinoso-Suarez, 1981) and the monkey (Gonzalo-Ruiz et al., 1988; May et al., 1990). These bilateral projections have been confirmed via autoradiographic study in the cat (Hirai et al., 1982; Kawamura et al., 1982; Sugimoto et al., 1982). In this case, autographic evidence was necessary because in FN degeneration and tracer injection studies, it is not possible to exclude uptake of a tracer by injured passing fibers or interruption of passing fibers from the opposite FN by a lesion. The bilateral projection is likely due to the recrossing of FN axons within the superior colliculi (Takahashi et al., 2010). In contrast, in lateral-eyed animals, the projection to SC is only from the contralateral FN and it is rather sparse. Specifically, it is sparse in the opossum (Walsh & Ebner, 1973), rabbit (Uchida et al., 1983), squirrel (May & Hall, 1986), and rat (Kurimoto et al., 1995). The projection from the FN terminates in the deeper layer IV and layer VI of SC, where tecto-reticulo-spinal neurons give rise to fibers to the contralateral predorsal bundle for control of eye, head, and body movements (Grantyn & Grantyn, 1982; May, 2006; Sparks & Mays, 1981).

Fastigio-Thalamo-Cerebral Projection

The fastigial efferents to the thalamus arise only from the caudal part of the FN and are entirely crossed, leaving from the hook bundle and the ascending brachium conjunctivum (Angaut & Bowsher, 1970; Batton et al., 1977). Terminations of the thalamus are the nucleus ventralis medialis (VM) and the nucleus medial ventralis lateralis in the cat (Angaut & Bowsher, 1970), the ventrolateralis (VL) in the monkey (Kievit & Kuypers, 1972), and the rostral nucleus ventral posterior lateralis in the monkey (Batton et al., 1977). See Jones (1985), Hirai and Jones (1989) and Macchi and Jones (1997) for nomenclature of the thalamic nuclei, because individual neuroanatomists use different delineation and nomenclature of the thalamic nuclei.

The FN of the cerebellum in the cat has been reported to project onto the medial part of the precruciate cortex (area 6) (Nakamura & Matsuda, 1983; Rispal-Padel & Latreille, 1974). In addition to the strong motor cortices projections from the interpositus and dentate nuclei (see Takahashi & Shinoda, 2021), the projection from the cerebellar nuclei to the parietal association (cerebral) cortex was demonstrated by analyzing evoked field potentials in the cerebrum (Sasaki et al., 1972). Laminar field potential analysis suggested that there were two types of thalamic evoked responses, (a) deep and (b) superficial thalamocortical (TC) responses in the cat (Sasaki et al., 1970). The deep TC response, which is composed of a negative potential in the depth of the cortex and a positive potential in the superficial layer of the cortex, corresponds to an augmenting response, whereas the superficial TC response, which is composed of a surface-negative and a deep-positive potential in the cerebral cortex, corresponds to a recruiting response as defined by Jasper (1949). Shinoda and Kakei (1989) found that there are two types of TC neurons; one group of TC axons terminate in the depth (layers III–V) of the cortex, and the other group terminate in the superficial layer (layer I) of the cortex. Sasaki's electrophysiological interpretation was later confirmed anatomically by the anatomical finding (Shinoda et al., 1993; Wannier et al., 1992). Another laminar field potential analysis shows that the FN-evoked field potentials in area 6 are surface-positive and much smaller than the field potentials in area 4 elicited by stimulating the dentate and/or interpositus nuclei (Nakamura & Matsuda, 1983). In contrast, the FN-evoked field potentials in the bilateral anterior ectosylvian sulcus regions (the insular cortex) are surface-negative and most likely via the ventromedial nucleus of the thalamus in the cat (Noda & Oka, 1985). These electrophysiological findings in the cat were recently confirmed morphologically in the mouse using a transsynaptic axonal labeling method investigating the connections from the FN with the cerebral cortex (Fujita et al., 2020). Injection of the tracer into the caudal FN (F4 as identified by Fujita et al., 2020) labeled orthodromically the VM and the medialis dorsalis (MD. Via the VM, terminals were labeled in cortex layer I in the wide parts of the cerebral cortex including insular, sensorimotor, parietal, and cingulate cortices. Meanwhile, via the MD, axon terminals were labeled in cortical layers III–V only in the cingulate and orbitofrontal cortices. In contrast, injection of the tracer into the caudal dorsolateral protuberance labeled the VL, and labeled terminals were found in layers III–V of the cortex (mainly sensorimotor cortex).

It is assumed that one of the important characteristics of the cerebellocerebral loops is that the dentate, interpositus, and fastigial nuclei send outputs mainly back to the same area of the cerebral cortex from which the cerebral inputs derive. However, it is technically difficult to unravel the mutual interactions between different cerebral cortical areas and different cerebellar cortical and nuclear parts. In most anatomical studies, the neural connections between a cerebellar nucleus to the thalamus, from the thalamus to the cerebral cortex, from the cerebral cortex to a precerebellar nucleus, and from a precerebellar nucleus to the cerebellum have been investigated separately, one connection at a time. More recently, however, Sasaki's group has used field potential analysis (Sasaki, 1979), and Strick's group has used a transsynaptic labeling method with an appropriate virus in the monkey (Kelly & Strick, 2013). In addition to its well-known motor function on posture and eye and head movements, the FN has been associated with nonmotor functions such as affective and cognitive functions in human imaging studies (Schmahmann, 1991, 1998; Schmahmann & Sherman, 1998). To understand the functional involvement of the FN in the higher brain functions, it will be necessary to investigate the details of cerebellocerebral circuits involving the FN in nonhuman primates.

10.3.2 *Afferents to the Fastigial Nucleus*

Vestibular Projection to the Fastigial Nucleus

The FN has been reported to receive secondary vestibulocerebellar fibers (Brodal & Torvik, 1957; Carpenter et al., 1959). Carpenter (1960) described fastigial termination of primary vestibulocerebellar afferents, but Brodal and Hϕivik (1964) found a number of such fibers passing through the FN, but could not demonstrate their termination in FN. Using intracellular and extracellular recording from the FN, Precht and Llinás (1968) reported monosynaptic and disynaptic activation of fastigial neurons from the primary vestibular afferents. HRP study showed that second-order vestibular neurons project to the FN (Kotchbkdi & Walberg, 1978; Ruggiero et al., 1977). Fastigial neurons can be influenced by stimulation of the vestibular nerve or labyrinth. The rostral fastigial nucleus contains not only type I but also horizontal type II neurons in the monkey (Gardner & Fuchs, 1975). Many neurons in the rostral FN also respond to tilt in α fashion in the cat (Ghelarducci, 1973).

Neurons in the rostral FN are sensitive to head velocity in the dark (Büttner et al., 1991; Gardner & Fuchs, 1975). These neurons in the rostral FN are called" vestibular-only" neurons since they discharge during vestibular stimulation, but are not modulated during saccadic or smooth pursuit eye movements. In addition to the convergent otholith-canal inputs, rostral FN neurons receive inputs from the neck proprioceptors (Matsushita & Xiong, 2001). This convergence is important for computation of body motion (Merger et al., 1983). Half of the rostral FN neurons encode the motion of the body in space (bimodal neurons), whereas the other half encode the motion of the head in space (unimodal neurons) (Brooks & Cullen, 2009).

There are many horizontal type I neurons in the caudal FN of the cat (Furuya et al., 1975). These neurons in the caudal FN are a part of the transcerebellar commissural inhibition between the bilateral vestibular nuclei (Furuya et al., 1976) in parallel to the commissural inhibition in the brainstem (Shimazu & Precht, 1966). Single axon tracing study of medial vestibular nucleus (MVN) neurons shows that two types of MVN neurons project to the fastigial nucleus: (a) one projects ipsilaterally after giving rise to collaterals to the vestibular nucleus, and (b) the other crosses the midline in the brainstem and projects contralaterally (Ando et al., 2020). These neurons also terminate with their mossy fiber terminals in the uvula and nodulus. Caudal FN neurons receive corticovestibular inhibition from the vermal VIc and VII where Purkinje cells are aldolase C-positive (Sugihara & Shinoda, 2004) and mossy fiber inputs from the pontine nucleus including dorsolateral pontine nucleus and nucleus reticularis tegmenti pontis (Gonzalo-Ruiz & Leichnetz, 1990; Noda et al., 1990).

Spinal Projection to the Fastigial Nucleus

Sensory afferents from the neck are one of the most important components of spinal projections to the vestibular nuclei and the cerebellum. Thick-caliber, peripheral muscle afferents from neck proprioceptors ascend ipsilaterally in the lateral cuneate fascicle to terminate mainly in the external cuneate nucleus; the afferents also supply lamina VI of the spinal cord and the central cervical nucleus (CCN) (Neuhuber & Zenker, 1989). Some of these fibers originate in the CCN at the upper cervical cord and, after crossing the midline in the spinal cord, give rise to collaterals to the MVN, DVN, LVN, and FN on their way to the cerebellar vermis (Matsushita et al., 1995). Lateral reticular nucleus neurons project to the LVN (Ruigrok et al., 1995). They project to the cerebellar vermis of the anterior lobe and also to the LVN and the FN with their axon collaterals (Fig. 10.3b–d) (Wu et al., 1999). Single axon reconstruction studies have shown that non-crossed axons of Clarke's column neurons (NCC) project mainly to multiple bilateral stripes in vermal lobules II–IV and VIII–IX, and the ipsilateral FN, and crossed axons of neurons in the medial ventral horn (CMVH) cross the midline in the thoracic cord and ascend contralaterally to multiple stripes in lobules II–V with a small number of terminals but on their way had abundant collaterals in the MVN, DVN, and FN as well as in the spinal cord and medullary reticular nuclei (Luo et al., 2017). These findings strongly suggest that the axon collateral reflex is responsible for monosynaptic excitation in MVN and DVN neurons on the same side when FN is stimulated electrically.

10.4 Conclusion

Research on higher brain functions without proper understanding of the neural circuits can be likened to building a castle on unstable sand. The fastigial and vestibular nuclei send their outputs sometimes directly or sometimes indirectly to motoneurons. The interpositus nuclei meanwhile send their outputs only indirectly via rubrospinal neurons to spinal and brainstem motoneurons, in addition to the interpositus projections to the cerebral cortex via the thalamus (Brodal, 1981; Ito, 1984; Voogd et al., 2013). The exception is the dentate nucleus, which sends outputs only to the cerebral cortex. This difference in output projection patterns is crucial from the viewpoint of internal representation models (Ito, 1970, 1984). To determine whether the cerebellum uses a forward or an inverse representation model, it is necessary to understand the targets of cerebellar output and how these output signals are processed at their destinations. As new anatomical and physiological methods are brought to bear, we will further develop our understanding of the neural circuits in the CNS. However, as we hope this review makes clear, many textbook descriptions are not necessarily based on reliable experimental evidence, which can lead to wasteful misunderstandings. On December 1–4, 2018, the Fujiwara Seminar on "Cerebellum as a CNS hub – from its evolution to the therapeutic strategy" in honor of Masao Ito was organized by Hirohide Mizusawa and Shinji Kakei in Tokyo Medical and Dental University. More than 60 of the leading scientists and clinicians from across the world were invited and discussed a wide range of topics in relation to the anatomy, physiology, and clinical neurology of the cerebellum. So far as we know, such a large-scale international joint symposium of both cerebellar basic scientists and clinicians has been very rare. The authors often participated in international symposia on the cerebellum with Masao Ito, where neuroscientists engaged in the cerebellar research were numerous. However, Ito complained that systems neurophysiologists engaged in unraveling of neural circuits using intracellular recording and staining techniques and investigating the functions of the cerebellum in higher mammals were on the verge of extinction. Therefore, Masao Ito strongly stated the need for the systems neurophysiological approach in the final chapter of his last monograph "The Cerebellum – Brain for an Implicit Self" (2012). He writes, "There are still deficiencies in our knowledge of the cerebellum's neuronal circuits. Sustained neurobiological efforts are required to clarify such issues at both the cellular/molecular level for analyzing single neurons and the systems level for advancing understanding of how neuronal circuits truly work."

Acknowledgments We thank R. Veale for valuable comments in improving the manuscript. This study was supported by JSPS KAKENHI Grant Number JP19K06937 to MT, Grants-in-Aid for Promotion of Scientific Research from the Naito Foundation to MT and from Brain Science Foundation to MT.

References

Ahn, A. H., Dziennis, S., Hawkes, R., & Herrup, K. (1994). The cloning of zebrin II reveals its identity with aldolase C. *Development, 120*, 2081–2090.

Akaike, T. (1983). Neuronal organization of vestibulo-spinal system in the cat. *Brain Research, 259*, 217–227.

Ando, T., Ueda, M., Luo, Y., & Sugihara, I. (2020). Heterogeneous vestibulocerebellar mossy fiber projections revealed by single axon reconstruction in the mouse. *The Journal of Comparative Neurology, 5*, 1775–1802.

Angaut, P., & Bowsher, D. (1970). Ascending projections of the medial cerebellar (fastigial) nucleus: An experimental study in the cat. *Brain Research, 24*, 49–68.

Asanome, M., Matsuyama, K., & Mori, S. (1998). Augmentation of postural muscle tone induced by the stimulation of the descending fibers in the midline area of the cerebellar white matter in the acute decerebrate cat. *Neuroscience Research, 30*, 257–269.

Asanuma, C., Thach, T., & Jones, E. G. (1983). Brainstem and spinal projections of the deep cerebellar nuclei in the monkey, with observations on the brainstem projections of the dorsal column nuclei. *Brain Research Reviews, 5*, 299–322.

Bagnall, M. W., Zingg, B., Sakatos, A., Moghadam, S. H., Zeilhofer, H. U., & du Lac, S. (2009). Glycinergic projection neurons of the cerebellum. *The Journal of Neuroscience, 29*, 10104–10110.

Batton, R. R., III, Jayaraman, A., Rugiero, D., & Carpenter, M. B. (1977). Fastigial efferent projections in the monkey: An autoradiographic study. *The Journal of Comparative Neurology, 174*, 281–306.

Biswas, M. S., Luo, Y., Sarpong, G. A., & Sugihara, I. (2019). Divergent projections of single pontocerebellar axons to multiple cerebellar lobules in the mouse. *The Journal of Comparative Neurology, 527*, 1966–1985.

Brochu, G., Maler, L., & Hawkes, R. (1990). Zebrin II: A polypeptide antigen expressed selectively by Purkinje cells reveals compartments in rat and fish cerebellum. *The Journal of Comparative Neurology, 291*, 538–552.

Brodal, A. (1969). *Neurological anatomy in relation to clinical medicine* (2nd ed.). Oxford University Press.

Brodal, A. (1981). *Neurological Anatomy in Relation to Clinical Medicine* (3rd ed.). Oxford University Press.

Brodal, A. (1984). The vestibular nuclei in the macaque monkey. *The Journal of Comparative Neurology, 227*, 252–266.

Brodal, A., & Hϕivik, B. (1964). Site and mode of termination of primary vestibulo-cerebellar fibers in the cat. An experimental study with silver impregnation methods. *Archives Italiennes de Biologie, 102*, 1–21.

Brodal, A., Pompeiano, O., & Walberg, F. (1962). *The vestibular nuclei and their connections. Anatomy and Functional Correlations*. The Henderson Trust Lectures.

Brodal, A., & Pompeinano, O. (1957). The vestibular nuclei in the cat. *Journal of Anatomy (London), 91*, 438–454.

Brodal, A., & Torvik, A. (1957). Uber den Ursprung der sekundaren vestibulocerebellaren Fasern bei der Katze. Eine experimenteill-anatomische Studie. *Archiv für Psychiatrie Z und Nervenkrankheiten Gesamte Neurologica, 195*, 550–567.

Brookhart, J. M. (1959). The cerebellum. In *Textbook of physiology, Sect. I. Neurophysiology II* (pp. 1245–1280).

Brooks, J. X., & Cullen, K. E. (2009). Multimodal integration in rostral fastigial nucleus provides an estimate of body movement. *The Journal of Neuroscience, 29*(34), 10499–10511.

Buisseret-Delmas, C., & Angaut, P. (1993). The cerebellar olivo-corticonuclear connections in the rat. *Progress in Neurobiology, 40*, 63–87.

Büttner, U., Fuchs, A. F., Markert-Schwab, G., & Buckmaster, P. (1991). Fastigial nucleus activity in the alert monkey during slow eye and head movements. *Journal of Neurophysiology, 65*, 1360–1371.

Cajal, S. R. (1911). *Histologie du Système Nerveux de l'Homme et des Vertébrés* (Vol. 2). Maloine.

Carpenter, M. B. (1960). Experimental anatomical-physiological studies of the vestibular nerve and cerebellar connections. In G. L. Rasmussen & W. Windfle (Eds.), *Neural Mechanisms of the Auditory and Vestibular Systems* (pp. 279–323). CC Thomas.

Carpenter, M. B., Bard, D. S., & Aling, F. A. (1959). Anatomical connections between the fastigial nuclei the labyrinth and the vestibular nuclei in the cat. *The Journal of Comparative Neurology, 111*, 1–26.

Carpenter, M. B., Brittin, G. M., & Pines, J. (1958). Isolated lesions of the fastigial nuclei in the cat. *The Journal of Comparative Neurology, 109*, 65–89.

Carpenter, M. B., & Nova, H. R. (1960). Descending division of the brachium conjunctivum in the cat: A cerebelloreticular system. *The Journal of Comparative Neurology, 114*, 295–305.

Chambers, W. W., & Sprague, J. M. (1955). Functional localization in the cerebellum. I. Organization in longitudinal cortico-nuclear zones and their contribution to the control of posture, both extrapyramidal and pyramidal. *The Journal of Comparative Neurology, 103*, 105–130.

Chan-Palay, V. (1977). *Cerebellar dentate nucleus. Organization, cytology, and transmitters.* Springer.

Cohen, D., Chambers, W. W., & Sprague, J. M. (1958). Experimental study of the efferent projections from the cerebellar nuclei to the brainstem of the cat. *The Journal of Comparative Neurology, 109*, 233–259.

Dow, R. S., & Moruzzi, G. (1958). *The physiology and pathology of the cerebellum.* The University of Minnesota Press.

Eccles, J. C., Ito, M., & Szentágothai, J. (1967). *The cerebellum as a neuronal machine.* Springer.

Eccles, J. C., Nicoll, R. A., Schwarz, W. F., Táboriková, H., & Willey, T. J. (1975). Reticulospinal neurons with and without monosynaptic inputs from cerebellar nuclei. *Journal of Neurophysiology, 38*, 513–530.

Edwards, S. B., Ginsburgh, C. L., Henkel, C. K., & Stein, B. E. (1979). Sources of subcortical projections to the superior colliculus in the cat. *The Journal of Comparative Neurology, 184*, 309–330.

Epema, A. H., Gerrits, N. M., & Voogd, J. (1988). Commissural and intrinsic connections of the vestibular nuclei in the rabbit: A retrograde labeling study. *Experimental Brain Research, 71*, 129–146.

Flood, S., & Jansen, J. (1966). The efferent fibers of the cerebellar nuclei and their distribution on the cerebellar peduncles in the cat. *Acta Anatomica (Basel), 63*, 137–166.

Fuchs, A. F., Robinson, F. R., & Straube, A. (1993). Role of the caudal fastigial nucleus in saccade generation. I. Neuronal discharge pattern. *Journal of Neurophysiology, 70*, 1723–1740.

Fujita, H., Kodama, T., du Lac, S. (2020). Modular output circuits of the fastigial nucleus for diverse motor and nonmotor functions of the cerebellar vermis. *eLife, 9*, e58613.

Fukushima, K., Peterson, B. W., Uchino, Y., Coulter, J. D., & Wilson, V. J. (1977). Direct fastigiospinal fibers in the cat. *Brain Research, 126*, 538–542.

Furuya, N., Kawano, K., & Shimazu, H. (1975). Functional organization of vestibulofastigial projection in the horizontal semicircular canal system in the cat. *Experimental Brain Research, 24*, 75–87.

Furuya, N., Kawano, K., & Shimazu, H. (1976). Transcerebellar inhibitory interaction between bilateral vestibular nuclei and its modulation by cerebellocortical activity. *Experimental Brain Research, 25*, 447–463.

Gardner, E. P., & Fuchs, A. F. (1975). Single-unit responses to natural vestibular stimuli and eye movements in deep cerebellar nuclei of the alert rhesus monkey. *Journal of Neurophysiology, 38*, 627–649.

Gerrits, N. M., & Voogd, J. (1987). The projection of the nucleus reticularis tegmenti pontis and adjacent regions of the pontine nuclei to the central cerebellar nuclei in the cat. *The Journal of Comparative Neurology, 258*, 52–62.

Gerrits, N. M., Voogd, J., & Magras, I. N. (1985). Vestibular afferents of the inferior olive and the vestibulo-olivo-cerebellar climbing fiber pathway to the flocculus in the cat. *Brain Research, 332*, 325–336.

Ghelarducci, B. (1973). Responses of cerebellar fastigial neurones to tilt. *Pflügers Archiv, 344*, 195–206.

Goffart, L., & Pélisson, D. (1998). Orienting gaze shifts during muscimol inactivation of caudal fastigial nucleus in the cat. I. Gaze dysmetria. *Journal of Neurophysiology, 79*, 1942–1958.

Gonzalo-Ruiz, A., & Leichnetz, G. R. (1990). Afferents of the caudal fastigial nucleus in a New World monkey (Cebus apella). *Experimental Brain Research, 80*, 600–608.

Gonzalo-Ruiz, A., Leichnetz, G. R., & Smith, D. J. (1988). Origin of cerebellar projections to the region of the oculomotor complex, medial pontine reticular formation, and superior colliculus in new world monkeys: A retrograde horseradish peroxidase study. *The Journal of Comparative Neurology, 268*, 508–526.

Grantyn, A., & Grantyn, R. (1982). Axonal patterns and sites of termination of cat superior colliculus neurons projecting in the tecto-bulbo-spinal tract. *Experimental Brain Research, 46*, 243–256.

Groenewegen, H. J., & Voogd, J. (1977). The parasagittal zonation within the olivocerebellar projection. I. Climbing fiber distribution in the vermis of cat cerebellum. *The Journal of Comparative Neurology, 174*, 417–488.

Gruart, A., & Delgado-Garcia, J. M. (1994). Signalling properties of identified deep cerebellar nuclear neurons related to eye and head movements in the alert cat. *The Journal of Physiology, 471*, 37–54.

Hawkes, R., & Leclerc, N. (1987). Antigenic map of the rat cerebellar cortex: The distribution of parasagittal bands as revealed by monoclonal anti-Purkinje cell antibody mabQ113. *The Journal of Comparative Neurology, 256*, 29–41.

Helmchen, C., Straube, A., & Buttner, U. (1994). Saccade-related activity in the fastigial oculomotor region of the macaque monkey during spontaneous eye movements in light and darkness. *Experimental Brain Research, 98*, 474–482.

Hikosaka, O., & Kawakami, T. (1977). Inhibitory reticular neurons related to the quick phase of vestibular nystagmus-their location and projection. *Experimental Brain Research, 27*, 377–396.

Hirai, T., & Jones, E. G. (1989). A new parcellation of the human thalamus on the basis of histochemical staining. *Brain Research, 14*, 1–34.

Hirai, T., Onodera, S., & Kawamura, K. (1982). Cerebellotectal projections studied in cats with horseradish peroxidase or tritiated amino acids axonal transport. *Experimental Brain Research, 48*, 1–12.

Homma, Y., Nonaka, S., Matsuyama, K., & Mori, S. (1995). Fastigiofugal projection to the brainstem nuclei in the cat: An antegrade PHA-L tracing study. *Neuroscience Research, 23*, 89–102.

Ito, M. (1970). Neurophysiological aspects of the cerebellar motor control system. *International Journal of Neurology, 7*, 162–176.

Ito, M. (1984). *The cerebellum and neural control.* Raven.

Ito, M. (2012). *The cerebellum - brain for an implicit self.* PFT Press.

Ito, M., Kawai, N., & Udo, M. (1968). The origin of cerebellar-induced inhibition of Deiters neurones. III. Localization of the inhibitory zone. *Experimental Brain Research, 4*, 310–320.

Ito, M., Kawai, N., Udo, M., & Sato, N. (1969). Axon reflex activation of Deiters' neurones from the cerebellar cortex through collaterals of the cerebellar afferents. *Experimental Brain Research, 8*, 249–268.

Ito, M., Udo, M., Mano, N., & Kawai, N. (1970). Synaptic action of the fastigiobulbar impulses upon neurones in the medullary reticular formation and vestibular nuclei. *Experimental Brain Research, 11*, 29–47.

Ito, M., & Yoshida, M. (1966). The origin of cerebellar-induced inhibition of Deiters' neurones. I. Monosynaptic initiation of the inhibitory postsynaptic potential. *Experimental Brain Research, 2*, 330–349.

Jansen, J., & Brodal, A. (1940). Experimental studies on the intrinsic fibers of the cerebellum. II. The corticonuclear projection. *The Journal of Comparative Neurology, 73*, 267–321.

Jansen, J., & Brodal, A. (1942). Experimental studies on the intrinsic fibers of the cerebellum. III. Cortico-nuclear projection in the rabbit and the monkey. *Norsk Vid Akad Avh 1 Math Nat Kl, 3*, 1–50.

Jansen, J., & Jansen, J. (1955). On the efferent fibers of the cerebellar nuclei in the cat. *The Journal of Comparative Neurology, 102*, 607–632.

Jasper, H. H. (1949). Diffuse projection system: The integrative action of the thalamic recruiting system. *Electroencephalography and Clinical Neurophysiology, 1*, 405–420.

Jones, E. G. (1985). *The thalamus*. Plenum Press.

Kase, M., Miller, D. C., & Noda, H. (1980). Discharges of Purkinje cells and mossy fibres in the cerebellar vermis of the monkey during saccadic eye movements and fixation. *The Journal of Physiology, 300*, 539–555.

Kawamura, K., & Hashikawa, T. (1979). Olivocerebellar projections in the cat studied by means of anterograde axonal transport of labeled amino acids as tracers. *Neuroscience, 4*, 1615–1633.

Kawamura, K., Hattori, S., Higo, S., & Matsuyama, T. (1982). The cerebellar projections to the superior colliculus and pretectum in the cat: An autoradiographic and horseradish peroxidase study. *Neuroscience, 7*, 1673–1689.

Keller, E. L., Slakey, D. P., & Crandall, W. F. (1983). Microstimulation of the primate cerebellar vermis during saccadic eye movements. *Brain Research, 288*, 131–143.

Kelly, R. B., & Strick, P. (2013). Cerebellar loops with motor cortex and prefrontal cortex in a non-human primate. *The Journal of Neuroscience, 23*, 8432–8444.

Kievit, J., & Kuypers, H. G. J. M. (1972). Fastigial cerebellar projections to the ventrolateral nucleus of the thalamus and the organization of the descending pathways. In T. L. Frigyesi, E. Rinvik, & Yahr (Eds.), *Corticothalamic projections and sensorimotor activities* (pp. 91–111). Raven Press.

Kitai, S. T., Kocsis, J. D., Preston, R. J., & Sugimori, M. (1976). Monosyaptic inputs to caudate neurons identified by intracellular injection of horseradish peroxidase. *Brain Research, 109*, 601–606.

Kojima, Y., Iwamoto, Y., Robinson, F. R., Noto, C. T., & Yoshida, K. (2008). Premotor inhibitory neurons carry signals related to saccade adaptation in the monkey. *Journal of Neurophysiology, 99*, 220–230.

Kotchbkdi, N., & Walberg, F. (1978). Cerebellar afferent projections from the vestibular nuclei in the cat: An experimental study with the method of retrograde axonal transport of horseradish peroxidase. *Experimental Brain Research, 31*, 591–604.

Krieger, C., Shinoda, Y., & Smith, A. M. (1985). Labeling of cerebellar mossy fiber afferents with intra-axonal horseradish peroxidase. *Experimental Brain Research, 59*, 414–417.

Kurimoto, Y., Kawaguchi, S., & Murata, M. (1995). Cerebellotectal projection in the rat: Anterograde and retrograde WGA-HRP study of individual cerebellar nuclei. *Neuroscience Research, 22*, 57–71.

Llinás, R., & Wolfe, J. W. (1977). Functional linkage between the electrical activity in the vermal cerebellar cortex and saccadic eye movements. *Experimental Brain Research, 29*, 1–24.

Luo, Y., Patel, R. P., Sarpong, G. A., Sasamura, K., & Sugihara. (2017). Single axonal morphology and termination to cerebellar aldolase C stripes characterize distinct spinocerebellar projection systems originating from the thoracic spinal cord in the mouse. *The Journal of Comparative Neurology, 526*, 681–706.

Macchi, G., & Jones, E. G. (1997). Towards an agreement on terminology of nuclear and subthalamic divisions of the motor thalamus. *Journal of Neurosurgery, 86*, 670–685.

Matsushita, M., Gao, X., & Yanagisawa, H. (1995). Spinovestibular projections in the rat, with particular reference to projections from the central cervical nucleus to the lateral vestibular nucleus. *The Journal of Comparative Neurology, 361*, 334–344.

Matsushita, M., & Hosoya, Y. (1978). The location of spinal projection neurons in the cerebellar nuclei (cerebellospinal tract neurons) of the cat. A study with the horseradish peroxidase technique. *Brain Research, 142*, 237–248.

Matsushita, M., & Xiong, G. (2001). Uncrossed and crossed projections from the upper cervical nuclei in the rat, studied by anterograde axonal tracing. *The Journal of Comparative Neurology, 432*, 101–118.

Matsuyama, K., & Jankowska, E. (2004). Coupling between feline cerebellum (fastigial neurons) and motoneurons innervating hindlimb muscles. *Journal of Neurophysiology, 91*, 1183–1192.
May, P. J. (2006). The mammalian superior: Laminar structure and connections. *Progress in Brain Research, 151*, 321–378.
May, P. J., & Hall, W. C. (1986). The sources of the nigrotectal pathway. *Neuroscience, 19*, 159–180.
May, P. J., Hartwich-Young, R., Nelson, J., Sparks, D. L., & Porter, J. D. (1990). Cerebellotectal pathways in the macaque: Implications for collicular generation of saccades. *Neuroscience, 36*, 305–324.
Merger, T., Nardi, G. L., Becker, W., & Deecke, L. (1983). The role of canal-neck interaction for the perception of horizontal trunk and head rotation. *Experimental Brain Research, 49*, 198–208.
Mihailoff, G. A. (1993). Cerebellar nuclear projections from the pontine nucleus and nucleus reticularis tegmenti pontis demonstrated with PHA-L tracing in the rat. *The Journal of Comparative Neurology, 330*, 130–146.
Moolenaar, G. M., & Rucker, H. K. (1976). Autoradiographic study of brain stem projections from the fastigial pressor areas. *Brain Research, 114*, 492–496.
Mori, S., Matsui, T., Kuze, B., Asanome, M., Nakajima, K., & Matsuyama, K. (1998). Cerebellar-induced locomotion: Reticulospinal control of spinal rhythm generating mechanism in cats. *Annals of the New York Academy of Sciences, 860*, 94–105.
Moruzzi, G., & Pompeiano, O. (1956). Crossed fastigial influence on decerebrate rigidity. *The Journal of Comparative Neurology, 106*, 371–392.
Moruzzi, G., & Pompeiano, O. (1957). Effects of vermal stimulation after fastigial lesion. *Archives Italiennes de Biologie, 95*, 31–55.
Mugnaini, E., & Floris, A. (1994). The unipolar brush cell: A neglected neuron of the mammalian cerebellar cortex. *The Journal of Comparative Neurology, 339*, 174–180.
Na, J., Sugihara, I., & Shinoda, Y. (2019). The entire trajectories of single pontocerebellar axons and their lobular and longitudinal terminal distribution patterns in multiple aldolase C positive compartments of the rat cerebellar cortex. *The Journal of Comparative Neurology, 527*, 2488–2511.
Nakamura, M., & Matsuda, Y. (1983). Re-evaluation of cortical and thalamic responses evoked by stimulation of the cerebellar fastigial nucleus in the cat. *Jap J Physiol, 33*, 215–226.
Neuhuber, W. L., & Zenker, W. (1989). Central distribution of cervical primary afferents in the rat, with emphasis on proprioceptive projections to vestibular, perihypoglossal and upper thoracic spinal nuclei. *The Journal of Comparative Neurology, 280*, 231–253.
Noda, T., & Oka, H. (1985). Fastigial inputs to the insular cortex in the cat: Field potential analysis. *Neuroscience Letters, 53*, 331–336.
Noda, H., Sugita, S., & Ikeda, Y. (1990). Afferent and efferent connections of the oculomotor region of the fastigial nucleus in the macaque monkey. *The Journal of Comparative Neurology, 302*, 330–348.
Ogawa, T. (1935). Beiträge zur vergleichende Anatomie des Zentralnervensystems der Wassersäugetiere. Ueber die Kleinhirnkerne der Pinnipedien und Cetaceen. *Arb Anat Inst Sendai, 17*, 63–136.
Ohtsuka, K., & Noda, H. (1991). Saccadic burst neurons in the oculomotor region of the fastigial nucleus of macaque monkeys. *Journal of Neurophysiology, 65*, 1422–1434.
Oscarsson, O. (1969). Termination and functional organization of the dorsal spino-olivocerebellar path. *Journal of Physiology (London), 200*, 129–149.
Oscarsson, O. (1979). Functional units of the cerebellum-sagittal zones and microzones. *Trends in Neurosciences, 2*, 143–145.
Precht W, Llinás R (1968) Direct vestibular afferents to cat cerebellar nuclei. *Proc XXIV IUPS Vol. VI*, 1063.
Quinet, J., & Goffart, L. (2007). Head-unrestrained gaze shifts after muscimol injection in the caudal fastigial nucleus of the monkey. *Journal of Neurophysiology, 98*, 3269–3283.

Rasmussen, A. T. (1933). Origin and course of the fasciculus uncinatus (Rusell) in the cat, with observations on other fiber tracts arising from the cerebellar nuclei. *The Journal of Comparative Neurology, 57*, 165–197.

Rispal-Padel, L., & Latreille, J. (1974). The organization of projections from the cerebellar nuclei to the contralateral motor cortex in the cat. *Experimental Brain Research, 19*, 36–60.

Ritchie, L. (1976). Effects of cerebellar lesions on saccadic eye movements. *Journal of Neurophysiology, 39*, 1246–1256.

Roldan, M., & Reinoso-Suarez, F. (1981). Cerebellar projections to the superior colliculus in the cat. *The Journal of Neuroscience, 1*, 827–834.

Ron, S., & Robinson, D. A. (1973). Eye movements evoked by cerebellar stimulation in the alert monkey. *Journal of Neurophysiology, 36*, 1004–1022.

Ruggiero, D., Batton, R. R., III, Jayaraman, A., & Carpenter, M. B. (1977). Brainstem afferents to the fastigial nucleus in the cat demonstrated by transport of horseradish peroxidase. *The Journal of Comparative Neurology, 172*, 189–210.

Ruigrok, T. J. H., & Cella, F. (1995). Precerebellar nuclei and red nucleus. In G. Paxinos (Ed.), *The rat nervous system, vol III, brain stem and cerebellum*. Academic Sydney.

Ruigrok, T. J. H., Cella, F., & Voogd, J. (1995). Connections of the lateral reticular nucleus to the lateral vestibular nucleus in the rat. An anterograde tracing study with phaseolus vulgaris leucoagglutinin. *The European Journal of Neuroscience, 7*, 1410–1413.

Sasaki, K. (1979). Cerebro-cerebellar interaction in cats and monkeys. In J. Massion & K. Sasaki (Eds.), *Cerebro-cerebellar interactions* (pp. 105–124). Elsevier.

Sasaki, K., Kawaguchi, S., Matsuda, Y., & Mizuno, N. (1972). Electrophysiological studies on the cerebello-cerebral projections in the cat. *Experimental Brain Research, 16*, 75–88.

Sasaki, K., Staunton, H. P., & Dieckmann, G. (1970). Characteristic features of augmenting and recruiting responses in the cerebral cortex. *Experimental Neurology, 26*, 369–392.

Schmahmann, J. D. (1991). An emerging concept. The cerebellar contribution to higher function. *Archives of Neurology, 48*, 1178–1187.

Schmahmann, J. D. (1998). Dysmetria of thought: Clinical consequences of cerebellar dysfunction on cognition and affect. *Trends Cogn Sci (Regul Ed), 2*, 362–371.

Schmahmann, J. D., & Sherman, J. C. (1998). The cerebellar cognitive affective syndrome. *Brain, 121*(Pt4), 561–579.

Scudder, C. A., & McGee, D. M. (2003). Adaptive modification of saccade size produces correlated changes in the discharges of fastigial nucleus neurons. *Journal of Neurophysiology, 90*, 1011–1026.

Serapide, M. F., Panto, M. R., Parenti, R., Zappala, A., & Cicirata, F. (2001). Multiple zonal projections of the basilar pontine nuclei to the cerebellar cortex of the rat. *The Journal of Comparative Neurology, 430*, 471–484.

Shimazu, H., & Precht, W. (1966). Inhibition of central vestibular neurons from the contralateral labyrinth and its mediating pathway. *Journal of Neurophysiology, 29*, 467–492.

Shinoda, Y. (1999). Visualization of the entire trajectory of long axon axons of single mammalian CNS neurons. *Brain Research Bulletin, 50*, 387–388.

Shinoda, Y., & Kakei, S. (1989). Distribution of terminals of thalamocortical fibers originating from the ventrolateral nucleus of the cat thalamus. *Neuroscience Letters, 96*, 163–167.

Shinoda, Y., Kakei, S., Futami, T., & Wannier, T. (1993). Thalamocortical organization in the cerebello-thalamo-cortical system. *Cerebral Cortex, 3*, 421–429.

Shinoda, Y., Ohgaki, T., & Futami, T. (1986). The morphology of single lateral vestibulospinal tract axons in the lower cervical spinal cord of the cat. *The Journal of Comparative Neurology, 249*, 226–241.

Shinoda, Y., & Sugihara, I. (2013). Axonal trajectories of single climbing and mossy fiber neurons in the cerebellar cortex and nucleus. In M. Manto, D. L. Gruol, & S. J. Koibuchi (Eds.), *Handbook of the cerebellum, and cerebellar disorders, Vol. 1* (pp. 437–467). Springer.

Shinoda, Y., Sugiuchi, Y., Futami, T., & Izawa, R. (1992). Axon collaterals of mossy fibers from the pontine nucleus in the cerebellar dentate nucleus. *Journal of Neurophysiology, 67*, 547–560.

Shinoda, Y., Sugiuchi, Y., Izawa, Y., & Hata, Y. (2006). Long descending motor tract axons and their control of neck and axial muscles. *Progress in Brain Research, 151*, 527–561.

Shinoda, Y., Yokota, J., & Futami, T. (1981). Divergent projection of individual corticospinal axons to motoneurons of multiple muscles in the monkey. *Neuroscience Letters, 23*, 7–12.

Sparks, D. L., & Mays, L. E. (1981). The role of the monkey superior colliculus in the control of saccadic eye movements a current perspective. In A. F. Fuchs & W. Becker (Eds.), *Progress in oculomotor research* (Vol. 12, pp. 137–144). Elsevier North Holland.

Sprague, J. M., & Chambers, W. W. (1954). Control of posture by reticular formation and cerebellum in the intact, anesthetized and unanesthetized and in the decerebrate cat. *Amer J Physiol, 176*, 52–64.

Sugihara, I., & Shinoda, Y. (2004). Molecular, topographic, and functional organization of the cerebellar cortex: A study with combined aldolase C and olivocerebellar labeling. *The Journal of Neuroscience, 24*, 8771–8785.

Sugihara, I., & Shinoda, Y. (2007). Molecular, topographic, and functional organization of the cerebellar nuclei: Analysis by three dimensional mapping of olivocerebellar projection and aldolase C labeling. *The Journal of Neuroscience, 27*, 9696–9710.

Sugihara, I., Wu, H., & Shinoda, Y. (1996). Morphology of axon collaterals of single climbing fibers in the deep cerebellar nuclei of the rat. *Neuroscience Letters, 217*, 33–36.

Sugihara, I., Wu, H.-S., & Shinoda, Y. (1999). Morphology of single olivocerebellar axons labeled with biotinylated dextran amine in the rat. *The Journal of Comparative Neurology, 414*, 131–148.

Sugihara, I., Wu, H., & Shinoda, Y. (2001). The entire trajectories of single olivocerebellar axons in the cerebellar cortex and their contribution to cerebellar compartmentalization. *The Journal of Neuroscience, 21*, 7715–7723.

Sugimoto, T., Mizuno, N., & Itoh, K. (1981). An autoradiographic study on the terminal distribution of cerebellothalamic fibers in the cat. *Brain Research, 215*, 29–47.

Sugimoto, T., Mizuno, N., & Uchida, K. (1982). Distribution of cerebellar fiber terminals in the midbrain visuomotor areas: An autoradiographic study in the cat. *Brain Research, 238*, 353–370.

Sugiuchi, Y., Izawa, Y., Takahashi, M., Na, J., & Shinoda, Y. (2005). Physiological characterization of synaptic inputs to inhibitory burst neurons from the rostral and caudal superior colliculus. *Journal of Neurophysiology, 93*, 697–712.

Sugiuchi, Y., Kakei, S., & Shinoda, Y. (1992). Spinal commissural neurons mediating vestibular input to neck motoneurons in the cat cervical spinal cord. *Neuroscience Letters, 145*, 221–224.

Takahashi, M., & Shinoda, Y. (2018). Brain stem neural circuits of horizontal and vertical saccade systems and their frame of reference. *Neuroscience, 392*, 281–328.

Takahashi, M., & Shinoda, Y. (2021). Neural circuits of inputs and outputs of the cerebellar cortex and nuclei. *Neuroscience, 462*, 70–88.

Takahashi, M., Sugiuchi, Y., & Shinoda, Y. (2010). Topographic organization of excitatory and inhibitory commissural connections in the superior colliculi and their functional roles in saccade generation. *Journal of Neurophysiology, 104*, 3146–3167.

Takahashi, M., Sugiuchi, Y., & Shinoda, Y. (2014). Convergent synaptic inputs from the caudal fastigial nucleus and the superior colliculus onto pontine and pontomedullary reticulospinal neurons. *Journal of Neurophysiology, 111*, 849–867.

Thach, W. T., Goodkin, H. P., & Keating, J. G. (1992). The cerebellum and the adaptive coordination of movement. *Annual Review of Neuroscience, 15*, 403–442.

Thomas, D. M., Kaufman, R. P., Sprague, J. M., & Chambers, W. W. (1956). Experimental studies of the vermal cerebellar projections in the brain stem of the cat (fastigiobulbar tract). *Journal of Anatomy, 90*, 371–385.

Trott, J. R., Apps, R., & Armstrong, D. M. (1998). Zonal organization of cortico-nuclear and nucleo-cortical projections of the paramedian lobule of the cat cerebellum. I. the C1 zone. *Experimental Brain Research, 118*, 298–315.

Trott, J. R., & Armstrong, D. M. (1987). The cerebellar corticonuclear projection from lobule Vb/c of the cat anterior lobe: A combined electrophysiological and autoradiographic study. *Experimental Brain Research, 68*, 339–354.

Uchida, K., Mizuno, N., Sugimoto, T., Itoh, K., & Kudo, M. (1983). Direct projections from the cerebellar nuclei to the superior colliculus in the rabbit: An HRP study. *The Journal of Comparative Neurology, 216*, 319–326.

Veenman, C. L., Reiner, A., & Honig, M. G. (1992). Biotinylated dextran amine as an anterograde tracer for single and double labeling studies. *Journal of Neuroscience Methods, 41*, 239–254.

Verhaart, W. J. C. (1964). *A stereotactic atlas of the brain stem of the cat*. Van Gorcum.

Voogd, J. (1964). *The cerebellum of the cat*. Van Gorcum.

Voogd, J. (1967). Comparative aspects of the structure and fibre connexions of the mammalian cerebellum. *Progress in Brain Research, 25*, 94–135.

Voogd, J. (1969). The importance of fiber connections in the comparative anatomy of the mammalian cerebellum. In R. Llinás (Ed.), *Neurobiology of cerebellar evolution and development* (pp. 493–514). A.M.A.E.R.F. Institute for Biomedical Research.

Voogd, J. (2016). Deiters' nucleus. Its role in cerebellar ideogenesis. *Cerebellum, 15*, 54–66.

Voogd, J., & Bigaré, F. (1980). Topographical distribution of olivary and corticonuclear fibers in the cerebellum: A review. In J. Courville et al. (Eds.), *The inferior Olivary nucleus* (pp. 207–234). New York.

Voogd, J., Hess, D. T., & Marani, E. (1987). The parasagittal zonation of the cerebellar cortex in cat and monkey: Topography, distribution of acetylcholinesterase, and development. In J. S. King (Ed.), *New concepts in cerebellar neurobiology*. Liss.

Voogd, J., Pardoe, J., Ruigrok, T. J., & Apps, R. (2003). The distribution of climbing and mossy fiber collateral branches from the copula pyramidis and the paramedian lobule: Congruence of climbing fiber cortical zones and the pattern of zebrin banding within the rat cerebellum. *The Journal of Neuroscience, 23*, 4645–4656.

Voogd, J., & Ruigrok, T. J. H. (2004). The organization of the corticonuclear and olivocerebellar climbing fiber projections to the rat cerebellar vermis: The congruence of projection zones and the zebrin pattern. *Journal of Neurocytology, 33*, 5–21.

Voogd, J., Shinoda, Y., Ruigorok, T., & Sugihara, I. (2013). Cerebellar nuclei and the inferior olivary nuclei: Organization and connections. In M. Manto, D. L. Gruol, J. Schmahmann, N. Koibuchi, & F. Rossi (Eds.), *Handbook of the cerebellum and cerebellar disorders* (Vol. 1, pp. 377–436). Springer Netherland.

Walberg, F., & Jansen, J. (1961). Cerebellar corticovestibular fibers in the cat. *Experimental Neurology, 3*, 32–52.

Walberg, F., Pompeiano, O., Brodal, A., & Jansen, J. (1962a). The fastigiovestibular projection in the cat. *An experimental study with silver impregnation methods. The Journal of Comparative Neurology, 118*, 49–76.

Walberg, F., Pompeiano, O., Westrum, L. E., & Hauglie-Hanssen, E. (1962b). Fastigioreticular fibers I cat. An experimental study in silver methods. *The Journal of Comparative Neurology, 119*, 187–199.

Walsh, T. M., & Ebner, F. F. (1973). Distribution of cerebellar and somatic lemniscal projections in the ventral nuclear complex of the Virginia opossum. *The Journal of Comparative Neurology, 147*, 427–446.

Wannier, T., Kakei, S., & Shinoda, Y. (1992). Two modes of cerebellar input to the parietal cortex in the cat. *Experimental Brain Research, 90*, 241–252.

Welker, W. (1987). Spatial organization of somatosensory projections to granular cell cerebellar cortex: Functional and connectional implications of fractured somatotopy. In J. S. King (Ed.), *New concepts in cerebellar neurobiology*. Liss.

Wilson, V., Uchino, Y., Maunz, R. A., Susswein, A., & Fukushima, K. (1978). Properties and connections of cat fastigiospinal neuros. *Experimental Brain Research, 32*, 1–17.
Wilson, V., Uchino, Y., Susswein, A., & Fukushima, K. (1977). Properties of direct fastigospinal fibers in the cat. *Brain Research, 126*, 543–546.
Wu, H., Sugihara, I., & Shinoda, Y. (1999). Projection patterns of single mossy fibers originating from the lateral reticular nucleus in the rat cerebellar cortex and nuclei. *The Journal of Comparative Neurology, 411*, 97–118.
Yamada, J., & Noda, H. (1987). Afferent and efferent connections of the oculomotor cerebellar vermis in the macaque monkey. *The Journal of Comparative Neurology, 265*, 224–241.

Chapter 11
Evolution of the Marr-Albus-Ito Model

Tadashi Yamazaki

11.1 Introduction

A theory is a way of thinking about what a given system is and how it behaves. It is grounded on available experimental data, but a good theory will also be consistent with data obtained in the future. Furthermore, a good theory provides experimentally testable predictions to further enhance our understandings of the system.

Around 1970, Marr (1969), Ito (1970), and Albus (1971) conceptualized the computational principle of the cerebellum based on a blueprint of the cerebellar circuit (Eccles et al., 1967). About 10 years later, Ito and his colleagues found the missing piece of the concept experimentally, which was plasticity at parallel fiber-Purkinje cell synapses (Ito et al., 1982; Ito, 1989). Following the discovery of the plasticity that we now know as long-term depression (LTD), the Marr-Albus-Ito model was established. Since then, the model has been analyzed, challenged, and extended to obtain better and deeper understandings of cerebellar computational principles (Ito, 1984, 2012).

The Marr-Albus-Ito model has provided a compass with which to navigate the ocean of cerebellar research for 50 years (Yamazaki & Lennon, 2019). Without the compass, researchers would be easily drawn in front of a huge amount of experimental data. Meanwhile, the model has been gradually evolving to account for new findings and concepts up to date.

The original Marr-Albus-Ito model postulated various important concepts: applications to behavioral studies, granular layer encoding, memory capacity of Purkinje cells, distributed and synergistic synaptic plasticity, internal models, and general computational principles of the cerebellar circuit. For these

T. Yamazaki (✉)
Graduate School of Informatics and Engineering, The University of Electro-Communications, Chofu, Tokyo, Japan
e-mail: contact21@numericalbrain.org

H. Mizusawa, S. Kakei (eds.), *Cerebellum as a CNS Hub*,
Contemporary Clinical Neuroscience, https://doi.org/10.1007/978-3-030-75817-2_11

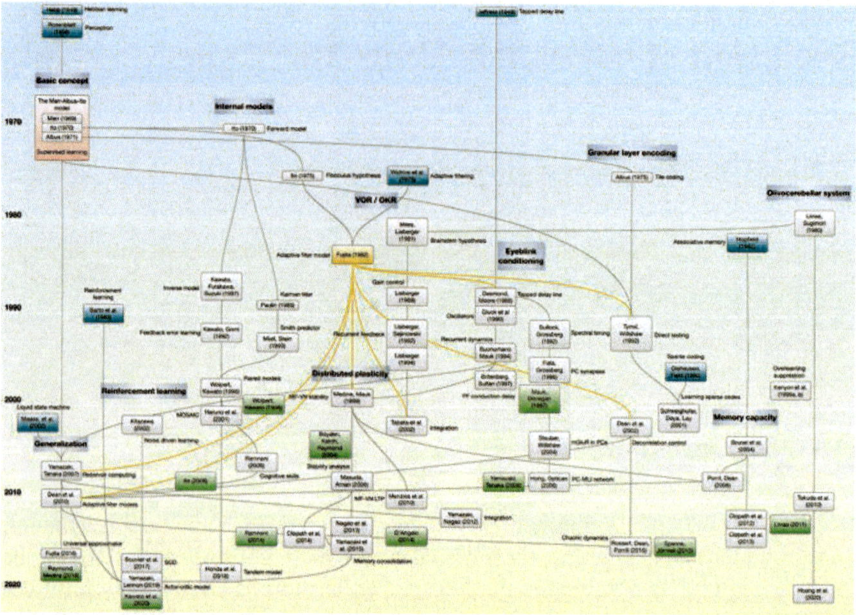

Fig. 11.1 An evolutionary tree of the Marr-Albus-Ito model. Gray boxes represent models, whereas lines represent inheritance relationships. Blue boxes represent conceptual works published outside of the cerebellar research. Green boxes are representative review papers. The pink box indicates the Marr-Albus-Ito model. Please note that the tree is not exhaustive. Only the models necessary to draw the diagram are shown. Abbreviations as in the text

concepts, a number of theoretical models have been proposed. We have summarized the "evolution" of the original model as an evolutionary tree (Fig. 11.1). In this article, we review how the original MAI model has been adapted to account for those concepts.

Meanwhile, computational studies using supercomputers are another important direction for cerebellar research. While theoretical studies aim to extract the principle or "essence" of a given system by eliminating a number of biological details, computational studies aim to replicate the system itself to reproduce the dynamics by incorporating as many details as possible. Both are called models, but their approaches are completely different. For computational studies, another review article will be available (Yamazaki et al., Submitted).

11.2 Evolutionary Tree of the Marr-Albus-Ito Model

To illustrate how the Marr-Albus-Ito model has influenced its successors, we compiled a number of theoretical models that stemmed from the Marr-Albus-Ito model, and created an "evolutionary tree" (Fig. 11.1).

11.2.1 The Marr-Albus-Ito Model

The ancestor is of course the Marr-Albus-Ito model. Although there are fewer than ten types of neurons in the cerebellum (more specifically, in a corticonuclear microcomplex) that constitute a network with recurrent connections, these pioneers focused on a feedforward network composed of Pons (input layer), granule cells (middle layer), and Purkinje cells (output layer) with climbing fiber inputs, while leaving aside the other neurons and connections (Fig. 11.2). The resulting network was assumed to be a cerebellar counterpart of the perceptron (Rosenblatt, 1958), which is the simplest form of supervised learning machine. The Marr-Albus-Ito model, which was also called the perceptron hypothesis, postulated two important observations:

- Granule cells in the middle layer encode afferent inputs from Pons via mossy fibers sparsely in a distributed manner.
- Connection strengths from granule cells to Purkinje cells are adjusted by climbing fibers.

The first idea was called "codon theory" (Marr, 1969) or "expansion recoding" (Albus, 1971), whereas the second one was confirmed about 10 years later by Ito and his colleagues, which we now known as long-term depression (LTD) (Ito et al., 1982). During those 10 years, the parallel fiber-Purkinje cell LTD was a missing piece to establish the Marr-Albus-Ito model. Contrary to the most common synaptic plasticity mechanism called Hebbian learning (Hebb, 1949), in which a synaptic connection between a pair of neurons is updated based on the correlated activity of the pre- and postsynaptic neurons, a synaptic weight is updated based on the

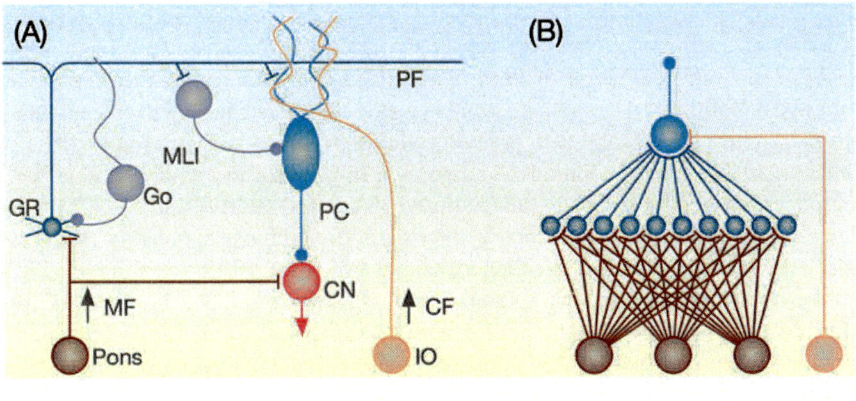

Fig. 11.2 Reduction of the cerebellar circuit to a perceptron. (**a**) Schematic of a corticonuclear microcomplex. Abbreviations: MF mossy fibers, GR granule cells, Go Golgi cells, MLI molecular layer interneurons, PC Purkinje cells, PF parallel fibers, CN cerebellar nuclei, CF climbing fibers, IO inferior olive. (**b**) The Marr-Albus-Ito model known as a perceptron. Only Pons, GR, PC, and IO are considered

correlated activities of the presynaptic parallel fiber and the climbing fiber innervating the same Purkinje cell in LTD.

Moreover, these authors did not forget to discuss the potential roles of other components such as Golgi cells and molecular layer interneurons. The authors first dissected the essential components from other peripheral components, continued to investigate the roles of the other components separately, and finally integrated their roles once again into the main model.

11.2.2 Applications to Behavioral Studies

Eye Movement Control

The Marr-Albus-Ito model was readily applied to behavioral studies. The first attempt was an application to eye movement control such as the vestibulo-ocular reflex (VOR) and optokinetic response (OKR). VOR is an eye movement reflex in which the eyes rotate in the opposite direction compared with the head rotation, whereas OKR is a reflex in which the eyes rotate to the same direction in response to slow movement of the entire visual world. Information on the head and visual world movements are fed by mossy fibers to the cerebellum. When the eye rotation is insufficient against the head or visual world movements, the visual image on the retina slips. This retinal slip provides an "error" signal to Purkinje cells via climbing fibers, which induces learning to adjust the eye movement gain (gain adaptation). In VOR, artificial stimuli could decouple the phases of head rotation and eye rotation while keeping the eye movement gain, which is called phase adaptation.

In the original Marr-Albus-Ito model, granule cells were considered an encoder of spatial input patterns conveyed by mossy fibers. Fujita (1982a) introduced sinusoidal temporal dynamics of mossy fibers representing the head and eye rotations and that of a climbing fiber that represents retinal slip errors. The model also included an implementation of the granular layer network composed of granule cells and a Golgi cell. Due to distributed synaptic weights of mossy fibers and inhibition exerted by the Golgi cell, granule cells exhibited various sinusoidal activity patterns with different amplitudes and phases. In the end, the model successfully reproduced both gain and phase adaptations in VOR and OKR (Fujita, 1982b). The model adopted the notion of adaptive filtering in the field of engineering (Widrow et al., 1975) and therefore was called an adaptive filter model.

The adaptive filter model was a pioneering theoretical model that first put the Marr-Albus-Ito model into action. The model became a stepstone for eyeblink conditioning (section "Eyeblink Conditioning") and general computational principles (Sect. 2.7). In fact, Fujita's original adaptive filter model was later generalized in the context of general computational principles with the same name (Dean et al., 2010). This might produce unnecessary confusion.

After Ito (1975) on the flocculus hypothesis for VOR, another hypothesis on VOR was proposed from outside the Marr-Albus-Ito model (Miles & Lisberger,

1981). Those authors hypothesized that plasticity at mossy fiber synapses on the vestibular nuclei played the prominent role in VOR. In those days, the LTD at parallel fiber synapses on Purkinje cells was a matter of debate. A research group strongly argued that the inferior olive was a timing device that controls motor timing precisely (Llinás & Sugimori, 1980) (Sect. 2.8), but not a subsystem that provided teacher or error signals. Based on the Miles and Lisberger (1981) hypothesis, a series of theoretical models were reported later (Lisberger, 1988; Lisberger & Sejnowski, 1992; Lisberger, 1994), which proposed that Purkinje cells "guide" learning in the downstream neurons of the vestibular nuclei while incorporating recurrent connections from vestibular nuclei to Purkinje cells. Ito immediately responded to the hypothesis (Ito, 1982), and a long-lasting debate over 30 years started (Kandel et al., 2000). Through this debate, the concept of distributed synaptic plasticity within the cerebellum has been gradually developed (Sect. 2.5). In addition, there was an attempt to include the effect of recurrent connections into the Marr-Albus-Ito model (Tabata et al., 2002).

Eyeblink Conditioning

Eyeblink conditioning is a type of classical conditioning, in which an animal receives repeated presentations of a neutral stimulus such as a tone (conditioned stimulus; CS) paired with an aversive stimulus such as an airpuff to the eye (unconditioned stimulus (US)). The animal becomes conditioned to close its eyes in response to the tone (conditioned response (CR)). Moreover, in delay eyeblink conditioning paradigms, the CR is elicited with a delay equal to the interstimulus interval (ISI) between the CS and US onsets (e.g., Mauk and Donegan (1997) for review). The essence of eyeblink conditioning models is how to represent the passage of time during the CS. Most studies attempted to include the timing mechanisms in the granular layer. The first model used tapped delay lines (Desmond & Moore, 1988; Moore et al., 1989), in which neurons are connected in series and activated one by one sequentially. Tapped delay lines are found in superior colliculus for sound localization; this is known as the Jefress model (Jefress, 1948). Fujita's adaptive filter model (section "Eye Movement Control") was also extended and applied (Gluck et al., 1990). In the model, granule cells were assumed to exhibit different sinusoidal temporal activity patterns with various frequencies and phases, which could perform Fourier expansion. Another extension was a spectral timing model (Bullock et al., 1994), which assumed multiple Golgi cells with different membrane time constants and generated transient activities of granule cells with different timings and amplitudes. Buonomano and Mauk (1994) focused on the network dynamics in the granular layer and proposed that granule cells exhibit sparse and chaotic or random activity patterns through recurrent inhibitory connections with Golgi cells. These models led to the refinement of general computational principles of the cerebellum (Sect. 2.7).

Other studies attempted to represent the passage-of-time outside of the granular layer. Braitenberg et al. (1997) proposed that parallel fibers provide delay lines by assuming large conduction delays. The concept of spectral timing models (Bullock et al., 1994) was followed by Fiala et al. (1996), Steuber and Willshaw (2004), and Majoral et al. (2020), which proposed that parallel fiber synapses on Purkinje cells could exhibit various temporal activity patterns lasting for seconds through the activation of metabotropic glutamate receptors. Kotaleski et al. (2002) assumed that biochemical interactions within Purkinje cells produce an increase in protein kinase C (PKC) activation, which could contribute to temporal sensitivity of Purkinje cells lasting for seconds. Hong and Optican (2008) proposed a similar timing mechanism through interactions between Purkinje cells and molecular layer interneurons.

These models provide theoretical support for sparse coding in granule cells (Sect. 2.3), which has long stood in opposition to a similar coding hypothesis (section "Distributed Versus Similar Coding in the Granular Layer").

For timing models, more detailed reviews are provided elsewhere (Yamazaki & Tanaka, 2009).

11.2.3 Granular Layer Encoding

One of the important concepts of the Marr-Albus-Ito model is the granular layer encoding of mossy fiber inputs. After the publication of Albus (1971), Albus published another paper that aimed to apply the cerebellar control mechanisms for engineering applications (Albus, 1975). The model called Cerebellar Model Architecture Control (CMAC) introduced a tile coding scheme within the granular layer. Models for VOR/OKR and eyeblink conditioning assumed various encoding schemes in the granular layer (Sect. 2.2). Tyrrell and Willshaw (1992) examined the possibility and efficiency of the granular layer encoding by a large-scale computer simulation for the first time. Later, the notion of sparse coding (Olshausen & Field, 1996) was introduced for the granular layer encoding, which was realized by anti-Hebbian learning mechanisms (Schweighofer et al., 2001), principal component analysis (PCA) (Dean et al., 2002), and chaotic spatiotemporal dynamics (Rössert et al., 2015). These studies were followed by those for unified gain and timing mechanisms and general computational principles (Sect. 2.7). Recently, the information capacity of sparse coding in the granular layer was examined mathematically (Cayco-Gajic et al., 2017), adding a new approach to an abundance of studies of the information capacity of Purkinje cells (Sect. 2.4). However, not all experimental data support the sparse encoding hypothesis, which is discussed later (section "Distributed Versus Similar Coding in the Granular Layer").

11.2.4 Information Capacity of Purkinje Cells

Another issue of neural encoding is the information encoding by parallel fibers on Purkinje cells. Purkinje cells receive excitatory inputs from about 200,000 parallel fibers, but 80% of them are silent. This massive convergence may play an important functional role in cerebellar computation. The first influential study was reported by Brunel et al. (2004). The authors calculated how many spatial patterns are embedded in parallel fiber synapses on a Purkinje cell while assuming that parallel fibers and Purkinje cells are binary neurons by using the same technique for analyzing associative memory capacity (Hopfield, 1982). Later, the study was extended for temporally correlated input patterns (Clopath et al., 2012) and for analog but not binary neurons (Clopath & Brunel, 2013). Independently from these studies, Porrill and Dean (2008) reported that adaptive filter models using a covariance learning rule could achieve optimal synaptic weights against noisy parallel fiber inputs, suggesting that long-term potentiation (LTP) at parallel fiber-Purkinje cell synapses is also important (Medina & Mauk, 1999).

11.2.5 Distributed Synaptic Plasticity

One of the most active debates on the cerebellum is probably the location of motor memory in the cerebellum. As seen the above (section "Eye Movement Control"), Miles and Lisberger (1981) proposed that mossy fiber-vestibular nuclei synapses store the memory on eye movement gain in VOR, whereas Ito et al. (1982) proposed parallel fiber-Purkinje cell synapses for the memory site (Kandel et al., 2000). Medina and Mauk (1999) built a computer simulation model that has two synaptic plasticity mechanisms for mossy fiber-vestibular nuclei and parallel fiber-Purkinje cell synapses. They found that learned memory on mossy fiber-vestibular nuclei synapses is stable at the resting state if the memory formation is guided by Purkinje cells innervating the nuclei. Dual plasticity models have also been studied in depth mathematically (Masuda & Amari, 2008; Clopath et al., 2014). These models mainly address the formation of motor memory, not the consolidation process of learned memory. Yamazaki et al. (2015) integrated both formation and consolidation processes in a single model and succeeded in reproducing various experimental results including spacing effects.

These studies were accompanied by experimental findings of multiple distributed synaptic plasticity within the cerebellum (e.g., Boyden et al. (2004) and D'Angelo (2014) for review). Among them, plasticity at parallel fiber synapses on molecular layer interneurons is considered a mechanism that could supersede parallel fiber-Purkinje cell LTD. Parallel fiber-molecular layer interneuron synapses undergo LTP with conjunctive activation of a presynaptic parallel fiber and a postsynaptic molecular layer interneuron that could be activated by spillover of glutamate secreted from nearby climbing fibers, whereas molecular layer interneurons inhibit Purkinje cells. This suggests that parallel fiber-molecular layer interneuron LTP could provide the same function for the cerebellar cortex as a supervised learning machine (Jörntell et al., 2010). Yamazaki and Lennon (2019) built a model of the cerebellar cortex that takes both parallel fiber-Purkinje cell LTD and parallel fiber-molecular layer interneuron LTP into account and analyzed the system dynamics. Contrary to the classical hypothesis of the cerebellar cortex as a supervised learning machine, the authors suggested that the cerebellar cortex could act as a reinforcement learning machine. We will discuss this issue in Sect. 2.7.

11.2.6 Internal Models

Ito (1970) already described the role of internal feedbacks from the cerebellum to the cerebral cortex that could act as a forward model. Forward models are well-known in engineering and control theory, and so they were readily adopted in the context of motor control by the cerebellum. A Kalman filter model (Paulin, 1989) and a Smith predictor model (Miall & Stein, 1993) were successful examples. Inverse models, closely related to forward models, were proposed by Kawato et al. (1987). Inverse models are acquired by a feedback error learning scheme (Kawato & Gomi, 1992), which was proposed as a model of cerebro-cerebellar interactions. Furthermore, a general architecture consisting of multiple paired forward and inverse models was proposed (Wolpert & Kawato, 1998) and was finalized as MOSAIC (modular selection and identification for control) architecture (Haruno et al., 2001). The concept has been even adopted for cognitive processes (Ramnani, 2006, 2014). A tandem architecture of forward and inverse models was applied for interpreting adaptation of voluntary movements (Honda et al., 2018). In general, forward and inverse models are called internal models. Internal models may be the most successful of all theoretical studies in the history of cerebellar research. A comprehensive review on internal models has been provided elsewhere (Wolpert et al., 1998).

11.2.7 General Computational Principles

Theoretical models for VOR/OKR (section "Eye Movement Control") and eyeblink conditioning (section "Eyeblink Conditioning") led to generalization of the computational principles. Liquid state machines (Yamazaki & Tanaka, 2007) and generalized adaptive filter models (Dean et al., 2010) were proposed as a general computational principle of the cerebellum as an extension of the Marr-Albus-Ito model. These studies were able to explain gain learning (e.g., VOR adaptation) and timing learning (e.g., eyeblink conditioning) by a single computational principle (Yamazaki & Nagao, 2012). Another theoretical study examined the potential of the cerebellar cortical circuit as a universal functional approximator based on mathematical functional analysis (Fujita, 2016).

Pursuing general computational principles of the cerebellum led to studies that would supersede the classical view of the cerebellum as a supervised learning machine. In supervised learning, learning is driven by teacher or error signals (Raymond & Medina, 2018). However, various different learning schemes would be used as well in the cerebellar cortex (Streng et al., 2018; Hull, 2020). An early attempt was made by Kitazawa (2002), who proposed noise-driven learning at Purkinje cells. The same idea was elaborated recently with the name stochastic gradient descent (SGD), which is a general technique to find optimal solutions used in the field of machine learning (Bouvier et al., 2018). These schemes enable the cerebellar cortex to search optimal solutions autonomously. Further elaboration proposes that the cerebellar cortex is a reinforcement learning machine. In reinforcement learning, an agent (i.e., the cerebellum) acquires an optimal action strategy called a policy for a given environment by maximizing expected future reward through trial and error (Sutton & Barto, 2018). Yamazaki and Lennon (2019) proposed that Purkinje cells and molecular layer interneurons act as an actor and a critic, respectively, in an actor-critic model of reinforcement learning, while assuming that climbing fibers convey reward information.

11.2.8 Olivocerebellar System

In the Marr-Albus-Ito model, inferior olive is considered as the source of teacher or error signals conveyed by climbing fibers that drives learning. Neurons in the inferior olive are connected electrically by gap junctions and exhibit subthreshold oscillation of membrane potentials (Llinás & Sugimori, 1980). A research group has been proposing that inferior olive is not the site for motor learning but a site for controlling motor timing (Welsh et al., 2005) by using temporal dynamics of the subthreshold oscillation (see Llinás (2011) for review). Tokuda et al. (2010) proposed that such subthreshold oscillation combined with chaotic dynamics accelerates learning owing to chaotic resonance mechanisms. A recent theoretical study proposes that gap junctions in the inferior olive constrain motor learning by controlling the degrees of freedom of a system (Hoang et al., 2020). This study could interpret the role of the subthreshold oscillation in the inferior olive, which was the basis of the motor timing hypothesis, within the context of the Marr-Albus-Ito model. Another group studying the olivocerebellar system develops a recurrent network model composed of the inferior olive, Purkinje cells, and cerebellar nuclei via nucleo-olivary connections that could control learning rate and suppress overlearning (Kenyon et al., 1998a, b).

11.3 Perspectives

11.3.1 Summary of the Evolutionary Tree

The evolutionary tree could provide several interesting observations. First, the node that has the largest number of outgoing paths is Fujita (1982a), suggesting that the model is the most influential one from which many followers were born.

On the other hand, the node that has the largest number of incoming paths is Dean et al. (2010), which is also a variant of adaptive filter models. Thus, adaptive filter models could be regarded as a backbone of all theoretical models on the cerebellum. Second, various important concepts have been introduced from the outside of cerebellar research, such as perceptron, adaptive filtering, sparse coding, and reinforcement learning. The "crossover" has improved and expanded the original concept of the Marr-Albus-Ito model continuously for the next generations. In turn, the original concept has not been altered largely against the crossover, indicating the robustness and concreteness of the original concept. Third, a number of excellent review papers have been available in a timely manner. These review papers help researchers to obtain the latest and consistent view of the computational principles of the cerebellum at the age. The evolution will be able to continue further, as long as we regularly incorporate new concepts from the outside and publish many review papers.

11.3.2 Unresolved Issues

Although various issues have been resolved through the evolution of the Marr-Albus-Ito model, there are still several unresolved issues, including:

- Distributed versus a similar coding in the granular layer
- Information representation by mossy fiber and climbing fiber signals
- Local versus global computation revealed by distributed climbing fiber activity

Distributed Versus Similar Coding in the Granular Layer

While the granular layer encoding, in which granule cells receive mossy fibers transmitting different information, has theoretical as well as experimental supports (Billings et al., 2014; Ishikawa et al., 2015; Gilmer & Person, 2017), another group suggests that granule cells receive mossy fibers that represent the same information (Bengtsson & Jörntell, 2009). The scheme, which those authors called "similar coding," may enable granule cells to transmit weak sensory inputs in a graded manner. The similar coding hypothesis suggests that the timing mechanism in eyeblink conditioning (section "Eyeblink Conditioning") exists not in the granular layer but at Purkinje cells, which is supported by experimental observations (e.g., Johansson et al. (2016) for review). To resolve this argument, large-scale, wide field-of-view imaging studies of the granule cells has provided an essential perspective (Knogler et al., 2017; Giovannucci et al., 2017; Wagner et al., 2017). For comprehensive reviews, see Spanne and Jörntell (2015) and Gilmer and Person (2018).

Information Representation by Mossy Fiber and Climbing Fiber Signals

Classical debates on the information conveyed by mossy and climbing fibers were basically on sensory versus motor, because the cerebellum is known as a central locus for motor control. However, recent imaging studies have revealed that various types of information are represented by mossy and climbing fibers. Notably, even reward-related information is represented in both mossy fiber and climbing fiber signals (Badura & Zeeuw, 2017; Gilmer & Person, 2018). Furthermore, anatomical connections with the ventral tegmental area have been found (Carta et al., 2019). These findings suggest the involvement of the cerebellum in reinforcement learning.

Localized Versus Distributed Computation Revealed by Distributed Climbing Fiber Activity

Previous theoretical studies have examined computational capability of a single microcomplex, which is thought of as a functional module of the cerebellum (Ito, 1984). Over the cerebellar surface, a number of microcomplexes are arranged regularly in space to constitute the entire cerebellar circuit. Different microcomplexes could have different functional roles, so for each specific task, a subset of microcomplexes might be employed to achieve the task, while the rest remain silent. In other words, microcomplexes are organized in a task-specific manner. MOSAIC models seem consistent with this observations (Haruno et al., 2001).

However, a striking Ca^{2+} imaging study by Michikawa et al. (2020) has revealed that all microzones are always activated simultaneously, suggesting that all cerebellar modules could function in a holistic manner. This study implies that many rather than a small subset of microcomplexes share functions for a given task. Many questions would arise: how do different microcomplexes share a task? And do they interact with each other to accomplish the task? To address this issue, we must examine how multiple microcomplexes share a task on the fly. This could provide new insights on distributed computations over the entire cerebellum.

11.3.3 Future Directions

A future direction of cerebellar research will be to uncover the role of the cerebellum embedded in the whole brain network for higher-order cognitive functions, for which internal models of mental processes called "mental models" would play essential roles (Ito, 2008, 2012), while sharing the same computational principles with motor functions (Koziol et al., 2012). The interactions will be made through the cerebro-cerebellar communication loop (Allen & Tsukahara, 1974). For such higher- order functions, a whole brain learning mechanism would be necessary including the cerebral cortex, basal ganglia, and cerebellum. In a pioneering study, Doya pointed out different roles of the cerebral cortex, basal ganglia, and cerebellum and argued how these regions interact with each other (Doya, 1999, 2000). In particular, direct interactions between the basal ganglia and cerebellum have been found experimentally (Carta et al., 2019), suggesting that the cerebellum is involved in even social tasks (D'Angelo, 2019). The whole brain learning architecture model has been extended recently (Caligiore et al., 2019).

These studies have assumed that the cerebellum is a supervised learning machine. However, Yamazaki and Lennon (2019) proposed that the cerebellum might act as a reinforcement learning machine by incorporating synaptic plasticity at parallel fiber-molecular layer interneurons as well as conventional parallel fiber-Purkinje cell LTD. Furthermore, the paper proposed that the whole brain could act as a hierarchical deep reinforcement learning machine, where the cerebral cortex stores deep representation of states and actions, the cerebro-basal ganglia loop performs higher reinforcement learning for goal setting and planning by decomposing a global task into a number of smaller subtasks, and the cerebro-cerebellar loop performs lower reinforcement learning that solves the subtasks in parallel (Fig. 11.3). Deep hierarchical reinforcement learning has proven to be a powerful machine learning algorithm (Kulkarni et al., 2016), which would be suitable for higher-order cognitive and social functions realized by the whole brain (Kawato et al., 2021).

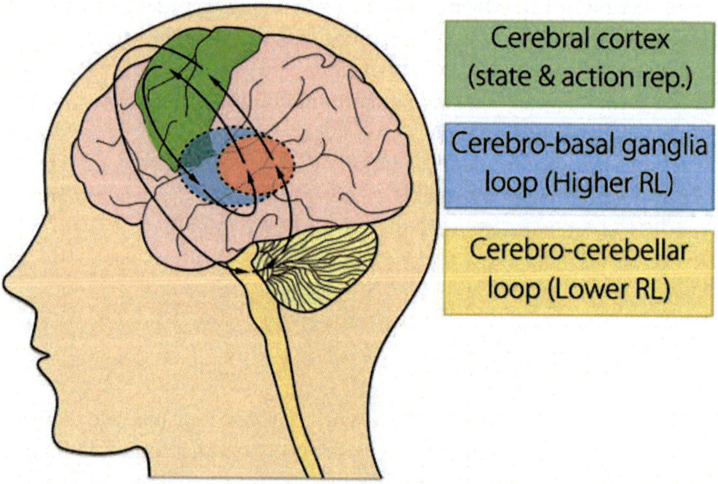

Fig. 11.3 A hypothetical role of cerebro-basal ganglia loop and cerebro-cerebellar loop for hierarchical reinforcement learning (Yamazaki & Lennon, 2019). The cerebral cortex (green) stores deep representation of states and actions. The cerebro-basal ganglia (blue) loop performs higher reinforcement learning that decomposes a global task into a number of smaller subtasks for goal setting and planning. The cerebro-cerebellar (yellow) loop performs lower reinforcement learning to solve subtasks for action execution in parallel. The thalamus (red) would play a certain role in the interactions of the dual loops. Abbreviation: RL reinforcement learning

11.4 Conclusion

The Marr-Albus-Ito model has cultivated a vast research field for theoretical models. The evolution of our knowledge of the cerebellum will continue to expand, including interactions with other brain regions towards understanding the whole brain learning mechanism.

Acknowledgments This article is based on a number of previous and current collaborations with both theoretical and experimental neuroscientists. In particular, we would like to thank Professors Shigeru Tanaka and Soichi Nagao for their mentoring and long-term support on conducting theoretical studies on the cerebellum. We also thank Dr. William Lennon for our international collaboration and stimulating discussions on artificial intelligence and neuroscience. Professor Narender Ramnani and Mr. Ohki Katakura kindly suggested several references. This paper is based on the results obtained by NEDO Next-Generation AI and Robot Core Technologies, and JSPS Kakenhi Grant Number (JP26430009, JP17H06310). Finally, we would like to thank late Dr. Masao Ito for his inspiring leadership and for bringing this mysterious and interesting brain region to the attention of theoretical neuroscientists worldwide.

References

Albus, J. S. (1971). A theory of cerebellar function. *Mathematical Biosciences, 10*, 25–61.

Albus, J. S. (1975). A new approach to manipulator control: The cerebellar model articulation controller (CMAC). *Journal of Dynamics Systems, Measurement, and Control, 97*(3), 220–227.

Allen, G. I., & Tsukahara, N. (1974). Cerebrocerebellar communication systems. *Physiological Reviews, 54*(4), 957–1006.

Badura, A., & Zeeuw, C. I. D. (2017). Cerebellar granule cells: Dense, rich and evolving representa- tions. *Current Biology, 27*(11), R415–R418.

Bengtsson, F., & Jörntell, H. (2009). Sensory transmission in cerebellar granule cells relies on similarly coded mossy fiber inputs. *Proceedings of the National Academy of Sciences, 106*(7), 2389–2394.

Billings, G., Piasini, E., Lörincz, A., Nusser, Z., & Silver, R. A. (2014). Network structure within the cerebellar input layer enables lossless sparse encoding. *Neuron, 83*(4), 960–974.

Bouvier, G., Aljadeff, J., Clopath, C., Bimbard, C., Ranft, J., Blot, A., Nadal, J.-P., Brunel, N., Hakim, V., & Barbour, B. (2018). Cerebellar learning using perturbations. *eLife, 7*, e31599.

Boyden, E. S., Katoh, A., & Raymond, J. L. (2004). Cerebellum-dependent learning: The role of multiple plasticity mechanisms. *Annual Review of Neurosciences, 27*, 581–609.

Braitenberg, V., Heck, D., & Sultan, F. (1997). The detection and generation of sequences as a key to cerebellar function: Experiments and theory. *Behavioral and Brain Sciences, 20*, 229–277.

Brunel, N., Hakim, V., Isope, P., Nadal, J.-P., & Barbour, B. (2004). Optimal information storage and the distribution of synaptic weights: Perceptron versus Purkinje cell. *Neuron, 43*(5), 745–757.

Bullock, D., Fiala, J. C., & Grossberg, S. (1994). A neural model of timed response learning in the cerebellum. *Neural Networks, 7*, 1101–1114.

Buonomano, D. V., & Mauk, M. D. (1994). Neural network model of the cerebellum: Temporal discrimination and the timing of motor responses. *Neural Computation, 6*, 38–55.

Caligiore, D., Arbib, M. A., Miall, R. C., & Baldassarre, G. (2019). The super-learning hypothesis: Integrating learning processes across cortex, cerebellum and basal ganglia. *Neuroscience & Biobehavioral Reviews, 100*, 19–34.

Carta, I., Chen, C. H., Schott, A. L., Dorizan, S., & Khodakhah, K. (2019). Cerebellar modulation of the reward circuitry and social behavior. *Science, 363*(6424), eaav0581t.

Cayco-Gajic, N. A., Clopath, C., & Silver, R. A. (2017). Sparse synaptic connectivity is required for decorrelation and pattern separation in feedforward networks. *Nature Communications, 8*(1), 1116.

Clopath, C., Badura, A., De Zeeuw, C. I., & Brunel, N. (2014). A cerebellar learning model of vestibulo-ocular reflex adaptation in wild-type and mutant mice. *Journal of Neuroscience, 34*(21), 7203–7215.

Clopath, C., & Brunel, N. (2013). Optimal properties of analog perceptrons with excitatory weights. *PLoS Computational Biology, 9*(2), 1–6.

Clopath, C., Nadal, J.-P., & Brunel, N. (2012). Storage of correlated patterns in standard and bistable Purkinje cell models. *PLoS Computational Biology, 8*(4), 1–10.

D'Angelo, E. (2014). The organization of plasticity in the cerebellar cortex: From synapses to control. *Progress in Brain Research, 210*, 31–58.

D'Angelo, E. (2019). The cerebellum gets social. *Science, 363*(6424), 229.

Dean, P., Porrill, J., Ekerot, C.-F., & Jörntell, H. (2010). The cerebellar microcircuit as an adaptive filter: Experimental and computational evidence. *Nature Reviews Neuroscience, 11*, 30–43.

Dean, P., Porrill, J., & Stone, J. (2002). Decorrelation control by the cerebellum achieves oculomotor plant compensation in simulated vestibulo-ocular reflex. *Proceedings of the Royal Society B: Biological Sciences, 269*, 1895–1904.

Desmond, J. E., & Moore, J. W. (1988). Adaptive timing in neural networks: The conditioned response. *Biological Cybernetics, 58*, 405–415.

Doya, K. (1999). What are the computations of the cerebellum, the basal ganglia and the cerebral cortex? *Neural Networks, 12*, 961–974.

Doya, K. (2000). Complementary roles of basal ganglia and cerebellum in learning and motor control. *Current Opinion in Neurobiology, 10*(6), 732–739.

Eccles, J. C., Ito, M., & Szentágothai, J. (1967). *The cerebellum as a neuronal machine*. Springer.

Fiala, J. C., Grossberg, S., & Bullock, D. (1996). Metabotropic glutamate receptor activation in cerebellar Purkinje cells as substrate for adaptive timing of the classically conditioned eyeblink response. *Journal Neuroscience, 16*, 3760–3774.

Fujita, M. (1982a). Adaptive filter model of the cerebellum. *Biological Cybernetics, 45*, 195–206.

Fujita, M. (1982b). Simulation of adaptive modification of the vestibulo-ocular reflex with an adaptive filter model of the cerebellum. *Biological Cybernetics, 45*, 207–214.

Fujita, M. (2016). A theory of cerebellar cortex and adaptive motor control based on two types of universal function approximation capability. *Neural Networks, 75*, 173–196.

Gilmer, J. I., & Person, A. L. (2017). Morphological constraints on cerebellar granule cell combina- torial diversity. *Journal of Neuroscience, 37*(50), 12153–12166.

Gilmer, J. I., & Person, A. L. (2018). Theoretically sparse, empirically dense: New views on cerebellar granule cells. *Trends in Neurosciences, 41*(12), 874–877.

Giovannucci, A., et al. (2017). Cerebellar granule cells acquire a widespread predictive feedback signal during motor learning. *Nature Neuroscience, 20*, 727–734.

Gluck, M. A., Reifsnider, E. S., & Thompson, R. F. (1990). Adaptive signal processing and the cerebellum: Models of classical conditioning and VOR adaptation. In M. A. Gluck & D. E. Rumelhart (Eds.), *Neuroscience and connectionist theory* (pp. 131–186). Erlbaum.

Haruno, M., Wolpert, D. M., & Kawato, M. (2001). Mosaic model for sensorimotor learning and control. *Neural Computation, 13*(10), 2201–2220.

Hebb, D. O. (1949). *The organization of behavior; A neuropsychological theory*. Wiley.

Hoang, H., Lang, E. J., Hirata, Y., Tokuda, I. T., Aihara, K., Toyama, K., Kawato, M., & Schweighofer, N. (2020). Electrical coupling controls dimensionality and chaotic firing of inferior olive neurons. *PLoS Computational Biology, 16*(7), 1–26.

Honda, T., Nagao, S., Hashimoto, Y., Ishikawa, K., Yokota, T., Mizusawa, H., & Ito, M. (2018). Tandem internal models execute motor learning in the cerebellum. *Proceedings of the National Academy of Sciences, 115*(28), 7428–7433.

Hong, S., & Optican, L. M. (2008). Interaction between Purkinje cells and inhibitory interneurons may create adjustable output waveforms to generate timed cerebellar output. *PLoS One, 3*(7), e2770.

Hopfield, J. J. (1982). Neural networks and physical systems with emergent collective computational abilities. *Proceedings of the National Academy of Sciences, 79*, 2554–2558.

Hull, C. (2020). Prediction signals in the cerebellum: Beyond supervised motor learning. *eLife, 9*, e54073.

Ishikawa, T., Shimuta, M., & Häusser, M. (2015). Multimodal sensory integration in single cerebellar granule cells in vivo. *eLife, 4*, e12916.

Ito, M. (1970). Neurophysiological aspects of the cerebellar motor control system. *International Journal of Neurology, 7*(2), 162–176.

Ito, M. (1975). Learning control mechanisms by the cerebellum investigated in the flocculo-vestibulo-ocular system. In D. Tower (Ed.), *The nervous system* (Vol. 1, pp. 245–252). Raven Press.

Ito, M. (1982). Cerebellar control of the vestibulo-ocular reflex–around the flocculus hypothesis. *Annual Review of Neuroscience (Palo Alto, CA), 5*, 275–297.

Ito, M. (1984). *The cerebellum and neural control*. Raven Press.

Ito, M. (1989). Long-term depression. *Annual Review of Neuroscience (Palo Alto, CA), 12*, 85–102.

Ito, M. (2008). Control of mental activities by internal models in the cerebellum. *Nature Reviews Neurology, 9*, 304–313.

Ito, M. (2012). *The cerebellum: Brain for the implicit self*. FT Press.

Ito, M., Sakurai, M., & Tongroach, P. (1982). Climbing fibre induced depression of both mossy fibre responsiveness and glutamate sensitivity of cerebellar purkinje cells. *The Journal of Physiology, 324*, 113–134.

Jefress, L. A. (1948). A place theory of sound localization. *Journal of Comparative and Physiological Psychology, 41*(1), 35–39.

Johansson, F., Hesslow, G., & Medina, J. F. (2016). Mechanisms for motor timing in the cerebellar cortex. *Current Opinion in Behavioral Sciences, 8*, 53–59.

Jörntell, H., Fredrik, B., Schonewille, M., & Zeeuw, C. I. D. (2010). Cerebellar molecular layer interneurons - computational properties and roles in learning. *Trends in Neurosciences, 33*, 524–532.

Kandel, E. R., Schwartz, J. H., & Jessell, T. M. (Eds.). (2000). *Principles of neural science* (4th ed.). McGraw-Hill Medical.

Kawato, M., Furukawa, K., & Suzuki, R. (1987). A hierarchical neural-network model for control and learning of voluntary movement. *Biological Cybernetics, 57*, 169–185.

Kawato, M., & Gomi, H. (1992). A computational model of four regions of the cerebellum based on feedback-error learning. *Biological Cybernetics, 68*, 95–103.

Kawato, M., Ohmae, S., Hoang, H., & Sanger, T. (2021). 50 years since the marr, ito, and albus models of the cerebellum. *Neuroscience, 462*, 151–174.

Kenyon, G. T., Medina, J. F., & Mauk, M. D. (1998a). A mathematical model of the cerebellar-olivary system I: Self-regulating equilibrium of climbing fiber activity. *Journal of Computational Neuroscience, 5*, 17–33.

Kenyon, G. T., Medina, J. F., & Mauk, M. D. (1998b). A mathematical model of the cerebellar-olivary system II: Motor adaptation through systematic disruption of climbing fiber equilibrium. *Journal of Computational Neuroscience, 5*, 71–90.

Kitazawa, S. (2002). Optimization of goal-directed movements in the cerebellum: A random walk hypothesis. *Neuroscience Research, 43*(4), 289–294.

Knogler, L. D., Markov, D. A., Dragomir, E. I., Stih, V., & Portugues, R. (2017). Sensorimotor representations in cerebellar granule cells in larval zebrafish are dense, spatially organized and non-temporally patterned. *Current Biology, 27*(9), 1288–1302.

Kotaleski, J. H., Lester, D., & Blackwell, K. T. (2002). Subcellular interactions between parallel fibre and climbing fibre signals in Purkinje cells predict sensitivity of classical conditioning to interstimulus interval. *Integrative Physiological and Behavioral Science, 37*, 265–292.

Koziol, L. F., et al. (2012). From movement to thought: Executive function, embodied cognition, and the cerebellum. *Cerebellum, 11*(2), 505–525.
Kulkarni, T. D., Narasimhan, K., Saeedi, A., & Tenenbaum, J. (2016). Hierarchical deep reinforcement learning: Integrating temporal abstraction and intrinsic motivation. In D. D. Lee, M. Sugiyama, U. V. Luxburg, I. Guyon, & R. Garnett (Eds.), *Advances in neural information processing systems 29* (pp. 3675–3683). Curran Associates, Inc.
Lisberger, S. G. (1988). The neural basis for learning of simple motor skills. *Science, 242*(4879), 728–735.
Lisberger, S. G. (1994). Neural basis for motor learning in the vestibuloocular reflex of primates. III. Computational and behavioral analysis of the sites of learning. *Journal of Neurophysiology, 72*(2), 974–998.
Lisberger, S. G., & Sejnowski, T. J. (1992). Motor learning in a recurrent network model based on the vestibulo–ocular reflex. *Nature, 360*, 159–161.
Llinás, R., & Sugimori, M. (1980). Electrophysiological properties of in vitro purkinje cell somata in mammalian cerebellar slices. *The Journal of Physiology, 305*, 171–195.
Llinás, R. R. (2011). Cerebellar motor learning versus cerebellar motor timing: The climbing fibre story. *Journal of Physiology, 589*(14), 3423–3432.
Majoral, D., Zemmar, A., & Vicente, R. (2020). A model for time interval learning in the Purkinje cell. *PLoS Computational Biology, 16*(2), e1007601.
Marr, D. (1969). A theory of cerebellar cortex. *Journal of Physiology (London), 202*, 437–470.
Masuda, N., & Amari, S. (2008). A computational study of synaptic mechanisms of partial mem- ory transfer in cerebellar vestibulo-ocular-reflex learning. *Journal of Computational Neuroscience, 24*, 137–156.
Mauk, M. D., & Donegan, N. H. (1997). A model of Pavlovian eyelid conditioning based on the synaptic organization of the cerebellum. *Learning & Memory, 3*, 130–158.
Medina, J. F., & Mauk, M. D. (1999). Simulations of cerebellar motor learning: Computational analysis of plasticity at the mossy fiber to deep nucleus synapse. *Journal of Neuroscience, 19*, 7140–7151.
Miall, R. C., & Stein, J. F. (1993). Is the cerebellum a smith predictor? *Journal of Motor Behavior, 25*(3), 203–216.
Michikawa, T., Yoshida, T., Kuroki, S., Ishikawa, T., Kakei, S., Itohara, S., & Miyawaki, A. (2020). Distributed sensory coding by cerebellar complex spikes in units of cortical segments. *bioRxiv*.
Miles, F., & Lisberger, S. (1981). Plasticity in vestibulo-ocular reflex: A new hypothesis. *Annual Review of Neuroscience (Palo Alto, CA), 4*, 273–299.
Moore, J. W., Desmond, J. E., & Berthier, N. E. (1989). Adaptively timed conditioned responses and the cerebellum: A neural network approach. *Biological Cybernetics, 62*, 17–28.
Olshausen, B. A., & Field, D. J. (1996). Emergence of simple-cell receptive field properties by learning a sparse code for natural images. *Nature, 381*, 607–609.
Paulin, M. (1989). A Kalman filter theory of the cerebellum. In *Dynamic interactions in neural networks: Models and data* (pp. 239–259). Springer.
Porrill, J., & Dean, P. (2008). Silent synapses, LTP, and the indirect parallel-fibre pathway: Computational consequences of optimal cerebellar noise-processing. *PLoS Computational Biology, 4*(5), e1000085.
Ramnani, N. (2006). The primate cortico-cerebellar system. *Nature Reviews Neuroscience, 7*(7), 511–522.
Ramnani, N. (2014). Automatic and controlled processing in the cortico-cerebellar system. In N., R (Ed.), *Cerebellar learning* (Vol. 210, pp. 255–285). Elsevier.
Raymond, J. L., & Medina, J. F. (2018). Computational principles of supervised learning in the cerebellum. *Annual Review of Neuroscience (Palo Alto, CA), 41*, 233–253.
Rosenblatt, M. (1958). The perceptron: A probabilistic model for information storage and organization in the brain. *Psychological Review, 65*, 386–408.

Rössert, C., Dean, P., & Porrill, J. (2015). At the edge of chaos: How cerebellar granular layer network dynamics can provide the basis for temporal filters. *PLoS Computational Biology, 11*, e1004515.

Schweighofer, N., Doya, K., & Lay, F. (2001). Unsupervised learning of granule cell sparse codes enhances cerebellar adaptive control. *Neuroscience, 103*, 35–50.

Spanne, A., & Jörntell, H. (2015). Questioning the role of sparse coding in the brain. *Trends in Neurosciences, 38*(7), 417–427.

Steuber, V., & Willshaw, D. (2004). A biophysical model of synaptic delay learning and temporal pattern recognition in a cerebellar Purkinje cell. *Journal of Computational Neuroscience, 17*, 149–164.

Streng, M. L., Popa, L. S., & Ebner, T. J. (2018). Complex spike wars: A new hope. *The Cerebellum, 17*, 735–746.

Sutton, R. S., & Barto, A. G. (2018). *Reinforcement learning: An introduction* (2nd ed.). MIT Press.

Tabata, H., Yamamoto, K., & Kawato, M. (2002). Computational study on monkey vor adaptation and smooth pursuit based on the parallel control-pathway theory. *Journal of Neurophysiology, 87*, 2176–2189.

Tokuda, I. T., Han, C. E., Aihara, K., Kawato, M., & Schweighofer, N. (2010). The role of chaotic resonance in cerebellar learning. *Neural Networks, 23*(7), 836–842.

Tyrrell, T., & Willshaw, D. (1992). Cerebellar cortex: Its simulation and the relevance of Marr's theory. *Philosophical Transactions of the Royal Society of London. Series B, Biological Sciences, 336*(1277), 239–257.

Wagner, M. J., Kim, T. H., Savall, J., Schnitzer, M. J., & Luo, L. (2017). Cerebellar granule cells encode the expectation of reward. *Nature, 544*, 96–100.

Welsh, J. P., Yamaguchi, H., Zeng, X.-H., Kojo, M., Nakada, Y., Takagi, A., Sugimori, M., & Llinás, R. R. (2005). Normal motor learning during pharmacological prevention of purkinje cell long-term depression. *Proceedings of the National Academy of Sciences, 102*(47), 17166–17171.

Widrow, B., et al. (1975). Adaptive noise cancelling: Principles and applications. *Proceedings of the IEEE, 63*, 1692–1716.

Wolpert, D. M., & Kawato, M. (1998). Multiple paired forward and inverse models for motor control. *Neural Networks, 11*(7–8), 1317–1329.

Wolpert, D. M., Miall, R. C., & Kawato, M. (1998). Internal models in the cerebellum. *Trends in Cogntive Sciences, 2*(9), 338–347.

Yamazaki, T., & Lennon, W. (2019). Revisiting a theory of cerebellar cortex. *Neuroscience Research, 148*(11), 1–8.

Yamazaki, T., & Nagao, S. (2012). A computational mechanism for unified gain and timing control in the cerebellum. *PLoS One, 7*(3), e33319.

Yamazaki, T., Nagao, S., Lennon, W., & Tanaka, S. (2015). Modeling memory consolidation during posttraining periods in cerebellovestibular learning. *Proceedings. National Academy of Sciences. United States of America, 112*, 3541–3546.

Yamazaki, T., & Tanaka, S. (2007). The cerebellum as a liquid state machine. *Neural Networks, 20*, 290–297.

Yamazaki, T., & Tanaka, S. (2009). Computational models of timing mechanisms in the cerebellar granular layer. *Cerebellum, 8*, 423–432.

Part IV
Complex Spikes and Plasticity of the Cerebellar Neurocircuitry

Chapter 12
States Are A-Changing, Complex Spikes Proclaim

Laurentiu S. Popa, Justin D. Aronson, and Timothy J. Ebner

12.1 Internal Models and the Cerebellum

The canonical view of the cerebellum emphasizes its fundamental role in motor control, specifically the production of smooth, continuous movements. More recently, the cerebellum's contributions to non-motor functions have been highlighted, including cognitive and executive processes. The highly stereotypic circuitry of the cerebellum has led to the widely held concept that it provides a uniform computation or transform (Ito, 2008; Schmahmann, 2010; Thach, 2007). A major challenge is to define this common computation and how it is used across all functions and domains.

Multiple hypotheses have been put forth on the essential functions performed by the cerebellum, including motor timing, error detection/correction, and motor learning. Another major set of hypotheses make the case that the cerebellum acquires and stores internal models of the motor system (Imamizu et al., 2000; Kawato, 1999; Pasalar et al., 2006; Popa et al., 2013; Shadmehr & Holcomb, 1997; Shidara et al., 1993; Taylor et al., 2010; Wolpert et al., 1998). Internal models provide neural representations of the input-output relationships of an aspect of the body or the environment to be controlled.

There are two general classes of internal models. Forward models integrate the commands for an action with information about the present state to predict the consequences of that action. By predicting the consequences of a motor act (Miall et al., 1993; Miall & Wolpert, 1996), forward models provide a solution to the problem of producing fast, accurate movements given the long delays in sensory feedback and the low gains of most sensorimotor feedback loops (Kawato, 1999; Wolpert et al., 1998). Predicting the consequences of a motor command has considerable utility, including state estimation, formulating a new motor command without waiting for

L. S. Popa · J. D. Aronson · T. J. Ebner (✉)
Department of Neuroscience, University of Minnesota, Minneapolis, MN, USA
e-mail: popa0003@umn.edu; arons094@umn.edu; ebner001@umn.edu

H. Mizusawa, S. Kakei (eds.), *Cerebellum as a CNS Hub*,
Contemporary Clinical Neuroscience, https://doi.org/10.1007/978-3-030-75817-2_12

the actual sensory feedback, and cancelling sensory reafference due to self-generated movement (Miall & Wolpert, 1996; Wolpert et al., 1995). In addition, forward model predictions can be used for error-based motor learning by generating a sensory prediction error, the difference between the desired and achieved action (Miall & Wolpert, 1996; Shadmehr et al., 2010; Wolpert et al., 1995). Originally termed sensory prediction error, we use the more general term of prediction error (PE), as the prediction is not restricted to the sensory domain. Importantly, PEs drive adaptation of both eye and limb movements (Mazzoni & Krakauer, 2006; Noto & Robinson, 2001; Shadmehr et al., 2010; Wallman & Fuchs, 1998).

In contrast, inverse dynamics models transform a desired outcome or effector state into the necessary motor commands to achieve that state. Considerable evidence suggests the cerebellum does not function as an inverse dynamics model (Bastian, 2006; Izawa et al., 2012; Morton & Bastian, 2006; Pasalar et al., 2006; Tseng et al., 2007). Therefore, this review examines the evidence for the cerebellum as a forward internal model, with a focus on support from electrophysiological studies.

12.2 Cerebellum as a Forward Internal Model

12.2.1 *Patient and Functional Imaging Evidence for the Hypothesis in the Motor Domain*

Support for the forward internal model hypothesis comes from multiple experimental approaches, including patient, functional imaging, noninvasive stimulation, and electrophysiological investigations. In patients with cerebellar disorders, the motor deficits are consistent with those expected for a faulty forward model. These include loss of accuracy due to variability in the motor command (Golla et al., 2008; Xu-Wilson et al., 2009), lack of adaptation to perturbations (Maschke et al., 2004; Taylor et al., 2010; Tseng et al., 2007), and disruptions of predictive adjustments (Bastian, 2006; Morton & Bastian, 2006).

Functional imaging in healthy subjects during motor learning tasks also supports the view that the cerebellum acquires and stores internal models of the motor system (Diedrichsen et al., 2005; Imamizu et al., 2000; Shadmehr & Holcomb, 1997; Tseng et al., 2007). Both the performance errors and sensory prediction errors needed to form and modify internal models result in strong cerebellar activation (Diedrichsen et al., 2005; Flament et al., 1996; Imamizu et al., 2000; Schlerf et al., 2012). Transient cerebellar disruption using transcranial magnetic stimulation induces movement perturbations consistent with the cerebellum making a prediction of the kinematic state of the arm (Miall et al., 2007). Therefore, considerable evidence from human studies, of both normal and disease state, supports the concept of the cerebellum's involvement in prediction, generating PEs, and motor adaptation.

12.2.2 Support from Purkinje Cell Recordings for the Hypothesis in the Motor Domain

Evidence for cerebellar involvement in forward models and in computing PEs is emerging from studies of cerebellar neuronal activity. Within this framework, we proposed that Purkinje cell high-frequency SS firing, modulated by mossy fiber input through the granule cell-parallel fiber network, represents both the predictions of the motor command consequences and the corresponding sensory feedback (Popa et al., 2013, 2016a). For an effector forward model, both the prediction and the sensory feedback are most likely expressed in the effector kinematics domain (Miall & Wolpert, 1996; Wolpert et al., 1995). In a variety of motor behaviors, the SS discharge of Purkinje cells modulates with and is conserved across different behaviors and is correlated with upcoming eye and arm kinematics (for review see (Ebner et al., 2011)). Complementing the feedforward modulation, SSs modulate with limb kinematics during passive movements, providing the needed sensory feedback for computing PEs. Further evidence for a forward model is the observation that SS activity is linked to the kinematic consequences of the motor commands and not to the dynamic output of the motor plant (Pasalar et al., 2006).

Pseudorandom, manual tracking allows decoupling of past and future encoding of kinematics (Hewitt et al., 2011; Popa et al., 2012,; 2017). During pseudorandom tracking in monkeys, Purkinje cell SS firing encodes arm movement kinematics, including position, velocity, and acceleration (Hewitt et al., 2011; Popa et al., 2012, 2017; Streng et al., 2017b). These parametric representations are independent, and individual Purkinje cells simultaneously encode several kinematic parameters. Encoding is primarily linear, with a planar pattern of increases and decreases relative to baseline SS firing. Across the population, these modulation patterns cover the kinematic workspaces. Furthermore, Purkinje cell SS discharge encodes both prediction and feedback of the same kinematic parameters. Approximately, 70% of Purkinje cells exhibit this bimodal, prediction-feedback modulation profile with limb kinematics. This type of representation is consistent with the fundamental properties of a forward model in which predictions based on the motor commands are integrated with known states of the effector.

Another advantage of pseudorandom tracking is that by requiring continuous monitoring and correction of performance errors, it reveals that the SS discharge of Purkinje cells provides robust and complex representations of position error, the difference between hand position and the center of the target (Popa et al., 2012, 2017). Similar to kinematics, position error encoding exhibits both prediction and feedback spanning several hundreds of msec. To better understand the nature of the dual encoding of position error, two perturbations of the visual feedback were incorporated into our pseudorandom tracking (Streng et al., 2018b). The first introduced delays between the hand and cursor movements (Fig. 12.1a), and the second reduced the visual feedback by hiding the cursor when inside the target (Fig. 12.1e). Introducing cursor delays showed that the predictive signals in the SS firing are locked to the hand movement and, therefore, to the motor command, while the

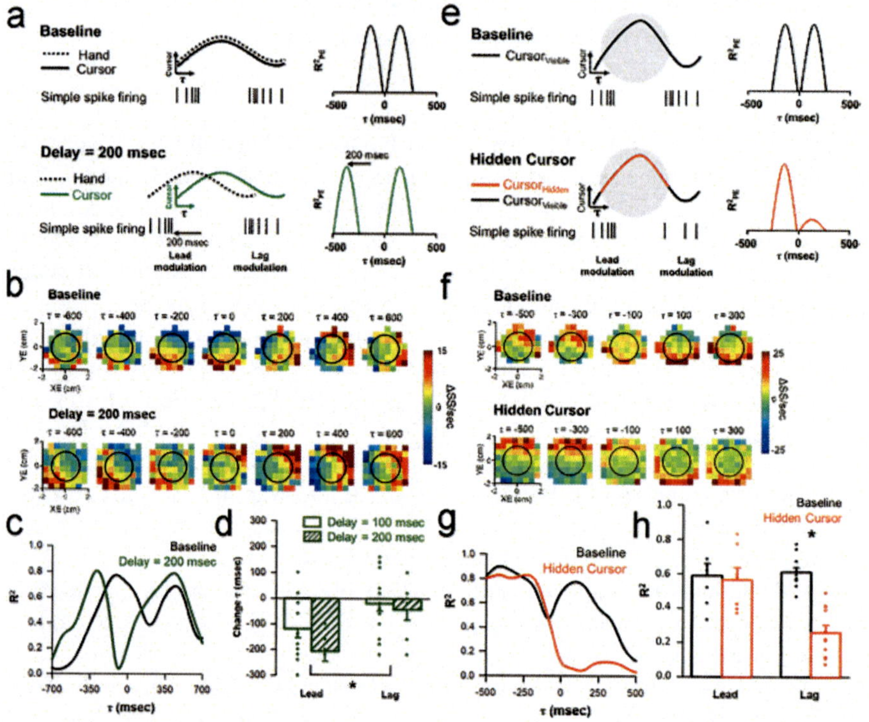

Fig. 12.1 Visual feedback manipulations. (**a**) In the baseline condition, the cursor (continuous trace) and hand (dashed trace) movements are indistinguishable. SS firing both leads (left spike train) and lags (right spike train) the cursor movement as also depicted in the temporal R^2 profile of the SS firing with the position error including a leading ($\tau < 0$) and lagging ($\tau > 0$) peak. In the delay condition, hand movement occurs before the cursor movement by the delay imposed. Based on the forward model hypothesis, the lead SS modulation is time-locked to hand movement and will shift earlier relative to cursor movement. However, the lag modulation should be time-locked to the cursor movement. Therfore, the lead R^2 peak will shift earlier and the lag R^2 peak will remain unchanged. (**b**) Firing maps for an example Purkinje cell with lead and lag SS position error modulation in both baseline (top row) and 200 msec delay (bottom row) conditions at specific lead (negative τ) or lag (positive τ) times. Black circle indicates target edge. (**c**) R^2 temporal profiles for the Purkinje cell shown in b in both baseline (black line) and 200 msec delay (green line) conditions showing the shift in the lead firing with no change in the timing of the feedback SS firing. (**d**) Average change in the timing of lead and lag encoding in the 100 msec (solid white) and 200 msec (hatched green) delay conditions, illustrating a significant shift in the timing of lead SS encoding that matches the cursor delay and no significant change in the timing of the SS lag encoding (* denotes $p = 0.002$). (**e**) In the baseline condition, the cursor (black trace) is visible whether inside or outside of the target (grey circle), while in the hidden condition, the cursor is visible only when outside the target (black trace) and invisible within the target (red segment). For a forward model, the lead SS firing will not be affected, but the lag firing will be reduced as illustrated in both the spike firing R^2 temporal profiles. (**f**) Firing maps for an example Purkinje cell with lead and lag SS encoding of position error in the baseline (top row) and hidden cursor (bottom row) conditions. (**g**) R^2 profiles illustrate the decrease in lagged SS encoding of position error but not the lead SS firing. (**h**) Average Purkinje cell SS encoding strength (R^2) in baseline versus hidden cursor condition shows a significant decrease in the lag but not the lead firing (* denotes $p = 0.0026$). Error bars are SEM. (Modified with permission from (Streng et al., 2018b, https://www.nature.com/articles/s41467-018-03541-0#rightslink))

feedback signals remain locked to the visual input (Fig. 12.1b–d). Hiding the cursor selectively reduces the feedback while conserving the predictive modulation (Fig. 12.1f–h). These results establish that, at the Purkinje cell level, a prediction of the consequence of the motor command is combined with feedback information about the performance. Together with the observation that the predictive feedback pair of signals are biased toward opposing patterns of modulation, we can conclude that single Purkinje cells implement a forward internal model of the task-specific performance errors.

Interestingly, there is no obvious segregation of Purkinje cells into error or kinematic subpopulations. Therefore, the integration of task errors and kinematics occurs at the individual cell level (Popa et al., 2012). This suggests that single Purkinje cells express both a forward internal model of the effector and of the performance errors. A theoretical study has found that having more than one forward model can solve the problem of error credit assignment and preserving optimal models, while updating suboptimal ones (Oh & Schweighofer, 2019). This explains how we can acquire multiple models of motor skills using the same effector internal model.

Moreover, adaptation savings is facilitated by intact task performance feedback when learning a visuomotor rotation during reach movements compared to adaptation in the absence of performance errors (Leow et al., 2020). The results are consistent with the presence of forward models operating in kinematic and task error domains, as we have observed at the Purkinje cell level, that need to simultaneously update to obtain an optimal control solution to a perturbation during tracking.

12.2.3 Cerebellar Circuity Support for the Hypothesis in the Motor Domain

Recent studies have investigated predictions in cerebellar neuronal firing, recording populations of mossy fibers, Purkinje cells, and dentate neurons in a single joint, step-tracking task (Tanaka et al., 2019, 2020; Tomatsu et al., 2016). Purkinje cell SS discharge could be reconstructed as a linear sum of mossy fiber firing and dentate neuronal discharge as the sum of Purkinje cell firing. The reconstruction suggests that dentate neuronal firing contains a prediction of the mossy fiber input from 0 to 200 msec, consistent with a forward internal model at the output stage of the cerebellum (Tanaka et al., 2019; 2020). While not specifically examined, Purkinje cells are also likely to have similar predictive firing in this task, as their SS activity can be reconstructed from the mossy fiber discharge. An important implication is that the predictions are computed by the cerebellar circuitry.

Further support for the cerebellar circuitry performing the required forward model computations involves eye movement signals in the floccular complex (Kim et al., 2019). Smooth pursuit eye movements include torsional components (rotations around gaze direction) that result from the vertical and horizontal rotations and are not due to motor commands. The mossy fiber input to the flocculus is consistent with encoding the motor command and lacks tuning to torsional velocity. However, Purkinje cell SS discharge is tuned to the torsional velocity, suggesting that cerebellar cortical circuitry performs the computation of this particular consequence of a motor command.

For a well-tuned forward model, the predictions should closely match the sensory feedback, effectively cancelling sensory reafferent signals generated by voluntary movements. Conversely, sensory feedback should result in large responses during passive movement in the absence of motor commands and during perturbations when the predictions of the model poorly match the feedback. As expected, rostral fastigial nuclear projection neurons that are the targets of Purkinje cells show greater sensitivity to passive self-motion of the head than to self-generated head movement (Brooks & Cullen, 2013). Further, unexpected loads generate PE signals in the firing of fastigial neurons during self-controlled movements that decrease as the monkeys adapt (Brooks et al., 2015). Together these findings are consistent with the output of an effector forward model that adapts its predictions to minimize sensory prediction errors.

Finally, different motor behaviors will necessitate a spectrum of leads and lags for predictive and feedback signals, respectively. As the cerebellum has roles in controlling diverse behaviors, including interactions among effectors (Bastian et al., 1996; Miall et al., 2001; Serrien & Wiesendanger, 2000; Thach et al., 1992), movement sequences (Molinari et al., 2008; Nixon & Passingham, 2000), and cognitive processes (see below), the associated forward models will use quite different time scales. Purkinje cell discharge has the needed distribution of timing signals for a wide spectrum of behaviors, as the SS firing has highly accurate decoding for leads and lags of up to 500 msec and encoding of feedforward and feedback information up to 2000 msec (Popa et al., 2012, 2017).

12.3 Cerebellum and Error Processing

The studies described above provide compelling evidence that in the cerebellar cortex, Purkinje cell SS firing provides the signals required to compute the PEs for a multitude of behavioral parameters and that PE signals are present in the cerebellar output. However, one of the early and most enduring assumptions regarding cerebellar function is that Purkinje cell complex spikes (CSs), generated by climbing fiber input, are the main conduits of the error signals necessary for motor learning (Albus, 1971; Ito & Kano, 1982; Marr, 1969; Oscarsson, 1980). Although earlier results supported different aspects of this hypothesis (for reviews see (Gao et al., 2012; Ito, 2002)), a number of studies failed to confirm the hypothesis. For example, error signals could not be found, or were limited, in the firing of inferior olivary neurons or in the CS discharge of Purkinje cells (for reviews see (Catz et al., 2005; Llinas, 2013; Popa et al., 2016b; Streng et al., 2018a)). In addition, there are multiple demonstrations of cerebellar learning that is independent of climbing fiber input (Boyden et al., 2004; Hewitt et al., 2015; Ke et al., 2009; Nguyen-Vu et al., 2013; Shin et al., 2014).

The need to re-evaluate the dogma that CSs primarily signal motor errors is highlighted in recent studies showing that CS activity conveys a surprisingly rich and intriguing spectrum of information. In the monkey performing pseudorandom tracking, CSs are only weakly modulated with motor errors (Streng et al., 2017b). Instead, CS discharge is most strongly modulated by kinematic parameters, including position, velocity, and acceleration, with a preference for predictive signaling of these parameters. The predictive property of CSs has been observed with performance-inferred errors in eye movements (Frens et al., 2001; Winkelman & Frens, 2006; Winkelman et al., 2014), with increases prior to anticipatory errors in eye blink conditioning and prior to the conditioned response (Ohmae & Medina, 2015; Ten Brinke et al., 2015).

In addition to prediction, CSs exhibit multifaceted modulation with rewards. During eye movements in the monkey, climbing fiber activity encodes the size of the expected reward during the cue period but not during reward delivery (Wagner & Luo, 2020), arguing for climbing fiber-associative learning beyond error correction. Two-photon (2P) imaging studies in the mouse have demonstrated extensive reward-related CS discharge (Heffley et al., 2018; Heffley & Hull, 2019; Kostadinov et al., 2019), with minimal responses to motor errors (Heffley et al., 2018). Reward-related signaling by climbing fibers includes reward delivery, reward expectations, reward omission, and learned sensory-motor associations. A particularly intriguing finding is that the climbing fiber modulation with reward expectation decreases dramatically with training (Kostadinov et al., 2019), potentially providing insights into the low probability of CS responses observed in highly trained subjects. Therefore, climbing fiber input not only carries a spectrum of information about behavior, but it provides predictions and evaluations of future events.

The realization that CSs do not just encode motor errors requires a reassessment of their role in cerebellar computations. Independent of any error signaling, spontaneous climbing fiber discharge is essential to normal cerebellar function and exerts a powerful control over spontaneous SS firing (Colin et al., 1980; Horn et al., 2013; Llinas et al., 1975; Montarolo et al., 1982). Also, CSs produce a global depolarization of Purkinje cells that alters the processing of subsequent parallel fiber input (for review see (Kitamura & Kano, 2013)). Therefore, we hypothesized that climbing fibers regulate the information in the SS firing. Evaluating this in pseudorandom tracking, we demonstrated that climbing fiber discharge dynamically controls the information present in the SS firing, triggering robust and rapid changes in SS encoding of motor signals in Purkinje cells (Fig. 12.2a–c) (Streng et al., 2017a). The encoding adjustments in SS firing include increases or decreases in the sensitivity to kinematics or position errors. The changes in encoding are tightly coupled to CS occurrence and commonly lead the behavior. At the population level, changes in encoding could serve to optimize behavior as we observed that increases in SS encoding of position error are followed by decreases in error (Fig. 12.2d). Intriguingly, CS-coupled changes in encoding for a given Purkinje cell tend to oppose the encoding "drift" in SS firing occurring independent of CS discharge

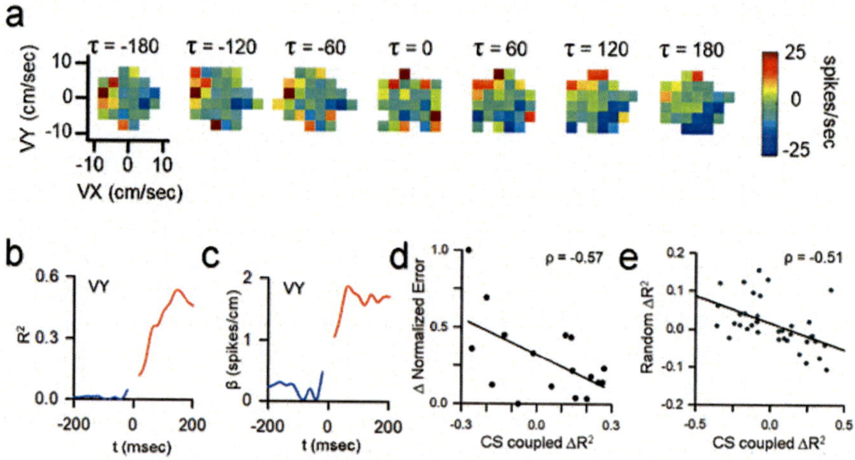

Fig. 12.2 Complex spike-coupled changes in SS encoding. (**a**) Example of Purkinje cell SS firing (mean-subtracted) in relation to hand velocity (VX and VY) demonstrates increased SS modulation after CS occurrence ($t = 0$ msec). (**b**) Increase in SS modulation relative to VY following CSs based on R^2 as a measure of encoding strength, pre- (blue trace) and post-CS (red trace) R^2 plot highlights. (**c**) Corresponding increase in SS sensitivity (β) relative to VY related to CS occurrence (pre-CS blue trace, post-CS red trace). Examples in **a**, **b**, and **c** originate from the same Purkinje cell. (**d**) At the population level, CS-coupled changes in position error are anticorrelated with CS-coupled changes in SS position error encoding (Pearson correlation $\rho = -0.57$, $p = 0.017$). (**e**) At the population level, CS-coupled changes in SS encoding are anticorrelated with changes in SS encoding at random times (i.e., changes independent of CS occurrence, Pearson correlation $\rho = -0.51$, $p = 0.003$). (Modified with permission from (Streng et al., 2017a, https://www.jneurosci.org/content/37/8/1997))

(Fig. 12.2e), thereby stabilizing the information conveyed by the Purkinje cell output. The findings of CS-driven encoding change in the SS firing account for numerous features of climbing fiber input, including why spontaneous climbing fiber input is essential for cerebellar function (for review see (Streng et al., 2018a)).

12.4 Does the Forward Internal Hypothesis Provide a General Model for Cerebellar Function?

Beyond its canonical role in motor control, the cerebellum is engaged in non-motor functions, including cognitive and emotive behaviors. Strong support for the concept comes from bidirectional anatomical and functional connections between the cerebellum and non-motor cerebral cortical regions, cerebellar activation in cognitive, perceptual and emotive tasks, and dysfunction in cerebellar disorders (for reviews see (Baumann et al., 2015; Camus et al., 2015; Ramnani, 2006; Schmahmann,

2004; Strick et al., 2009)). In addition, manipulating cerebellar excitability alters cognitive processes (Ferrucci et al., 2008; Pope & Miall, 2012).

As reviewed above for reward processing, recent animal experiments reveal significant cerebellar impact on non-motor processes, including spatial navigation (Rochefort et al., 2011, 2013), social interaction (Carta et al., 2019), and decision-making (Gao et al., 2018; Wagner & Luo, 2020). Neural recordings demonstrate encoding of cognitive and decision-making processes. In a sensory discrimination, delayed-choice task in which mice were required to plan a future motor movement based on a discriminatory stimulus, optogenetic inactivation in the fastigial nucleus (FN) caused a drop in performance without hampering motor execution (Gao et al., 2018). In zebrafish, cerebellar neurons predict well in advance decisions to change motor behavior (Lin et al., 2020). During optogenetic inhibition in the FN, neuronal firing in the anterior lateral motor (ALM) region decreases in preparatory activity, and optogenetic inhibition of the ALM reduces preparatory activity in the FN, indicating a closed cortico-cerebellar loop. Granule cells exhibit reward-related signals, including anticipation of rewards (Wagner et al., 2017). Simultaneous 2P Ca^{2+} imaging in the premotor cortex and cerebellar granule layer reveals shared, task-encoding features and high correlations in the neural activity between the two regions (Wagner et al., 2019). Chronic imaging reveals that these dynamics co-emerge in the cortex and cerebellum and the correlations between cell activities in both regions increase over time as the mice become more proficient in the task. The presence of non-motor signals suggests the cerebellum may form predictive models of cognitive processes, as it does for sensorimotor processes.

As noted above, there is growing evidence showing that cerebellar involvement in non-motor domains is consistent with the forward model hypothesis. Cerebellar activation patterns related to sentence outcome predictions and to violations of well-established semantic rules are similar to the activation patterns of PEs observed in the motor domain (Lesage et al., 2017; Moberget et al., 2014) and argue that the cerebellar processes in motor control and language are similar and based on forward internal models. Cerebellar transcranial direct current stimulation enhances BOLD signals related to semantic predictions (D'Mello et al., 2017). Disrupting the right cerebellum processing with transcranial magnetic stimulation also provides causal evidence consistent with cerebellar language processing based on forward internal models. For example, cerebellar disruption increases reaction time to predictive language (Lesage et al., 2012) and non-motor impediments to speech production (Runnqvist et al., 2016). Also, in a task requiring subjects to retrieve the information in a given position of a letter sequence maintained in the working memory, cerebellar disruption produces errors consistent with disrupting predictions of a forward internal model operating on the phonological loop information (Sheu et al., 2019). Therefore, cerebellar involvement in language processing has many features in common with a forward model.

12.5 Overall Hypothesis and Summary

One of the remaining controversies of the forward internal model theory regards the functional role of the CS discharge. The traditional view is that CSs provide the error signals postulated by the forward internal model hypothesis used both to control movements and to drive the internal model updates. However, climbing fiber input signals multitude aspects of behavior, at both predictive and feedback timing. Moreover, the low-frequency firing of CSs is rather ill-suited to participate in online motor control, and the apparent stochastic nature of the CS discharge hampers its capability as a teaching signal. Moreover, Purkinje cell SS firing provides for high-fidelity encoding of a spectrum of error signals, as described above. This highlights the need for an alternate hypothesis regarding CS function.

Based on the finding that CSs control SS encoding (Streng et al., 2017a) and that rapid changes in SS encoding are needed as environmental and internal conditions change, we hypothesize that CS discharge directs an internal model selection process (Fig. 12.3). In this view, CSs occur with changes in the behavioral state and engage the appropriate forward models. The combination of excitatory and inhibitory inputs to the inferior olive from the cerebral cortex, mesodiencephalic nuclei,

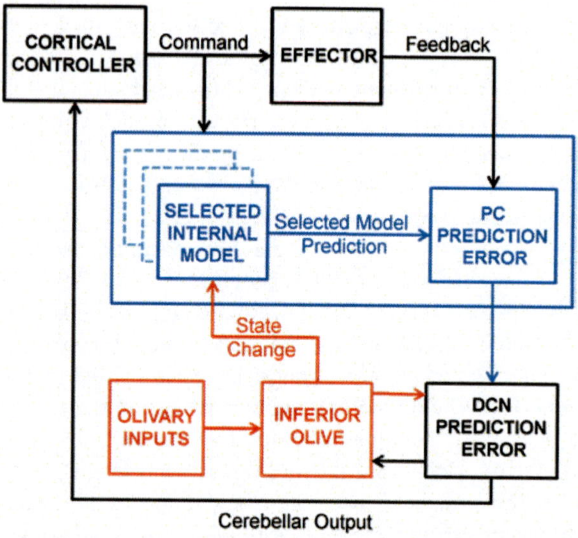

Fig. 12.3 Mechanics of a generalized cerebellar forward internal model. The cerebellar cortex (blue box) encodes a multitude of forward models (blue stack) manifested in the SS modulation and receives the efferent copy from the cortical controller and the subsequent parallel fiber inputs, defining the command feedback. The cerebellar cortex also receives inferior olive inputs associated with behavioral state changes that select the appropriate forward model. The cerebellar cortex computes PEs at the Purkinje cell (PC) level, its sole output projecting on the cerebellar nuclei that in turn provide the cerebellar output as PEs, determined in part by integrating a population of PC inputs

spinal cord, and the cerebellar nuclei could provide the required information about behavioral and internal states at all levels of the CNS (De Zeeuw et al., 1998). The assumption that CSs provide information about state changes can explain the observed heterogeneity of CS encoding, as different experimental paradigms induce state changes in different functional domains. Supporting the relationship between climbing fiber input and state changes is that common drivers of increased CS modulation occur with changes of motor or sensory states, including movement onset, presence of unconditioned stimulus, errors during learning, and internal states such as reward expectations. Also, the low frequency of climbing fiber discharge of around 0.5–2 Hz is similar to the frequency observed for changes in the cortical state (Harris & Thiele, 2011).

In contrast to the dominant view in which the CS role is to provide a reliable teaching signal, this new view suggests that CSs act to select appropriate combinations of internal models among a larger spectrum of models stored by the cerebellar cortex (Fig. 12.3). Under this assumption, CS-driven learning would be determined by the consistency of parallel fiber-Purkinje cell synaptic activity during CS occurrence. As the direction of plasticity at parallel fiber-Purkinje cell synapses is controlled by Purkinje cell intracellular calcium levels that is modulated by both parallel fiber activity and CS occurrence (Coesmans et al., 2004), the consistency in synaptic activity relative to CS occurrence may determine whether long-term depression or potentiation occurs, while randomness will maintain synaptic transmission in the long run. Therefore, we suggest that learning is not contingent on the specific information conveyed either by parallel fibers or climbing fibers; instead it depends on an association between the behavioral context as defined by state changes associated with CSs and control signals conveyed by SS discharge.

As detailed in this review, considerable evidence demonstrates that the cerebellum makes predictions in a wide spectrum of behaviors. However, as opposed to the motor domain in which behavior can be described by well-defined and accurate measures, the non-motor aspects of behavior are more difficult to define and measure, making it challenging to decipher cerebellar involvement. A corollary of the view that the cerebellum provides a uniform computation or transform (Ito, 2008; Schmahmann, 2010; Thach, 2007) would be that the cerebellum provides internal models for non-motor processes similar to those operating in motor control (Imamizu & Kawato, 2009; Ito, 2008; Koziol et al., 2012; Popa et al., 2014). Is it possible to generalize the forward internal model to a framework capable of describing cerebellar processes across the full range of functional domains? Forward models can operate in relation to task-specific performance errors that can be understood as a model of an explicit strategy (Popa et al., 2013, 2014). This provides an indication that the cerebellum can acquire forward internal models of cognitive/executive processes that are linked with cerebral cortical control.

To expand the concept of a forward internal model beyond motor control, we have to generalize the concepts of command and effector to the output of any region of the cerebral cortex (or any other structure) that can send an efferent copy to the cerebellum and the effector as the target of the command, respectively (Fig. 12.3). As for motor behavior, the cerebellum predicts consequences of the command and compares it against other cerebellar inputs, including the output of the effector (i.e., feedback). Rather than the CSs providing error-based teaching signals, in this framework the climbing fiber input acts as a detector of expected or evoked state changes that selects among Purkinje cell internal models.

Prediction is thought to be a dominant mode of CNS function, improving its beliefs about the world by continuously generating predictions from inputs, comparing its predictions with actual results and then acting to reduce PEs (Friston, 2010; Picard & Friston, 2014). In this framework, the CNS is organized into hierarchical generative models in which higher levels provide predictions to lower-level models and the higher levels use PEs from lower levels as inputs to update the predictions (Picard & Friston, 2014; Pickering & Clark, 2014). However, this cognitive architecture raises the question of implementation, with the possibility of relying on a combination of forward and inverse models or relying on only the presence of forward models at each level (Pickering & Clark, 2014). The latter supposition seems the most likely as the cerebellum's extensive reciprocal connections with the cerebral cortex provides the circuitry to implement both top-down predictions and bottom-up PEs based entirely on forward models.

References

Albus, J. S. (1971). A theory of cerebellar function. *Mathematical Biosciences, 10*, 25–61.

Bastian, A. J. (2006). Learning to predict the future: The cerebellum adapts feedforward movement control. *Current Opinion in Neurobiology, 16*, 645–649.

Bastian, A. J., Martin, T. A., Keating, J. G., & Thach, W. T. (1996). Cerebellar ataxia: Abnormal control of interaction torques across multiple joints. *Journal of Neurophysiology, 76*, 492–509.

Baumann, O., Borra, R. J., Bower, J. M., Cullen, K. E., Habas, C., Ivry, R. B., Leggio, M., Mattingley, J. B., Molinari, M., Moulton, E. A., Paulin, M. G., Pavlova, M. A., Schmahmann, J. D., & Sokolov, A. A. (2015). Consensus paper: The role of the cerebellum in perceptual processes. *Cerebellum, 14*, 197–220.

Boyden, E. S., Katoh, A., & Raymond, J. L. (2004). Cerebellum-dependent learning: The role of multiple plasticity mechanisms. *Annual Review of Neuroscience, 27*, 581–609.

Brooks, J. X., & Cullen, K. E. (2013). The primate cerebellum selectively encodes unexpected self-motion. *Current Biology, 23*, 947–955.

Brooks, J. X., Carriot, J., & Cullen, K. E. (2015). Learning to expect the unexpected: Rapid updating in primate cerebellum during voluntary self-motion. *Nature Neuroscience, 18*, 1310–1317.

Camus, S., Ko, W. K., Pioli, E., & Bezard, E. (2015). Why bother using non-human primate models of cognitive disorders in translational research? *Neurobiology of Learning and Memory, 124*, 123–129.

Carta, I., Chen, C. H., Schott, A. L., Dorizan, S., & Khodakhah, K. (2019). Cerebellar modulation of the reward circuitry and social behavior. *Science, 363*, eaav0581.

Catz, N., Dicke, P. W., & Thier, P. (2005). Cerebellar complex spike firing is suitable to induce as well as to stabilize motor learning. *Current Biology, 15*, 2179–2189.

Coesmans, M., Weber, J. T., De Zeeuw, C. I., & Hansel, C. (2004). Bidirectional parallel fiber plasticity in the cerebellum under climbing fiber control. *Neuron, 44*, 691–700.

Colin, F., Manil, J., & Desclin, J. C. (1980). The olivocerebellar system. I. Delayed and slow inhibitory effects: An overlooked salient feature of cerebellar climbing fibers. *Brain Research, 187*, 3–27.

D'Mello, A. M., Turkeltaub, P. E., & Stoodley, C. J. (2017). Cerebellar tDCS modulates neural circuits during semantic prediction: A combined tDCS-fMRI study. *The Journal of Neuroscience, 37*, 1604–1613.

De Zeeuw, C. I., Simpson, J. I., Hoogenraad, C. C., Galjart, N., Koekkoek, S. K., & Ruigrok, T. J. (1998). Microcircuitry and function of the inferior olive. *Trends in Neurosciences, 21*, 391–400.

Diedrichsen, J., Hashambhoy, Y., Rane, T., & Shadmehr, R. (2005). Neural correlates of reach errors. *The Journal of Neuroscience, 25*, 9919–9931.

Ebner, T. J., Hewitt, A. L., & Popa, L. S. (2011). What features of limb movements are encoded in the discharge of cerebellar neurons? *Cerebellum, 10*, 683–693.

Ferrucci, R., Marceglia, S., Vergari, M., Cogiamanian, F., Mrakic-Sposta, S., Mameli, F., Zago, S., Barbieri, S., & Priori, A. (2008). Cerebellar transcranial direct current stimulation impairs the practice-dependent proficiency increase in working memory. *Journal of Cognitive Neuroscience, 20*, 1687–1697.

Flament, D., Ellermann, J. M., Kim, S.-G., Ugurbil, K., & Ebner, T. J. (1996). Functional magnetic resonance imaging of cerebellar activation during the learning of a visuomotor dissociation task. *Human Brain Mapping, 4*, 210–226.

Frens, M. A., Mathoera, A. L., & van der Steen, J. (2001). Floccular complex spike response to transparent retinal slip. *Neuron, 30*, 795–801.

Friston, K. (2010). The free-energy principle: A unified brain theory? *Nature Reviews. Neuroscience, 11*, 127–138.

Gao, Z., van Beugen, B. J., & De Zeeuw, C. I. (2012). Distributed synergistic plasticity and cerebellar learning. *Nature Reviews. Neuroscience, 13*, 619–635.

Gao, Z., Davis, C., Thomas, A. M., Economo, M. N., Abrego, A. M., Svoboda, K., De Zeeuw, C. I., & Li, N. (2018). A cortico-cerebellar loop for motor planning. *Nature, 563*, 113–116.

Golla, H., Tziridis, K., Haarmeier, T., Catz, N., Barash, S., & Thier, P. (2008). Reduced saccadic resilience and impaired saccadic adaptation due to cerebellar disease. *The European Journal of Neuroscience, 27*, 132–144.

Harris, K. D., & Thiele, A. (2011). Cortical state and attention. *Nature Reviews. Neuroscience, 12*, 509–523.

Heffley, W., & Hull, C. (2019). Classical conditioning drives learned reward prediction signals in climbing fibers across the lateral cerebellum. *eLife, 8*, e46764.

Heffley, W., Song, E. Y., Xu, Z., Taylor, B. N., Hughes, M. A., McKinney, A., Joshua, M., & Hull, C. (2018). Coordinated cerebellar climbing fiber activity signals learned sensorimotor predictions. *Nature Neuroscience, 21*, 1431–1441.

Hewitt, A., Popa, L. S., Pasalar, S., Hendrix, C. M., & Ebner, T. J. (2011). Representation of limb kinematics in Purkinje cell simple spike discharge is conserved across multiple tasks. *Journal of Neurophysiology, 106*, 2232–2247.

Hewitt, A. L., Popa, L. S., & Ebner, T. J. (2015). Changes in Purkinje cell simple spike encoding of reach kinematics during adaptation to a mechanical perturbation. *The Journal of Neuroscience, 35*, 1106–1124.

Horn, K. M., Deep, A., & Gibson, A. R. (2013). Progressive limb ataxia following inferior olive lesions. *The Journal of Physiology, 591*, 5475–5489.

Imamizu, H., & Kawato, M. (2009). Brain mechanisms for predictive control by switching internal models: Implications for higher-order cognitive functions. *Psychological Research, 73*, 527–544.

Imamizu, H., Miyauchi, S., Tamada, T., Sasaki, Y., Takino, R., Putz, B., Yoshioka, T., & Kawato, M. (2000). Human cerebellar activity reflecting an acquired internal model of a new tool. *Nature, 403*, 192–195.

Ito, M. (2002). Historical review of the significance of the cerebellum and the role of Purkinje cells in motor learning. *Annals of the New York Academy of Sciences, 978*, 273–288.

Ito, M. (2008). Control of mental activities by internal models in the cerebellum. *Nature Reviews. Neuroscience, 9*, 304–313.

Ito, M., & Kano, M. (1982). Long-lasting depression of parallel fiber-Purkinje cell transmission induced by conjunctive stimulation of parallel fibers and climbing fibers in the cerebellar cortex. *Neuroscience Letters, 33*, 253–258.

Izawa, J., Criscimagna-Hemminger, S. E., & Shadmehr, R. (2012). Cerebellar contributions to reach adaptation and learning sensory consequences of action. *The Journal of Neuroscience, 32*, 4230–4239.

Kawato, M. (1999). Internal models for motor control and trajectory planning. *Current Opinion in Neurobiology, 9*, 718–727.

Ke, M. C., Guo, C. C., & Raymond, J. L. (2009). Elimination of climbing fiber instructive signals during motor learning. *Nature Neuroscience, 12*, 1171–1179.

Kim, G., Laurens, J., Yakusheva, T. A., & Blazquez, P. M. (2019). The macaque cerebellar flocculus outputs a forward model of eye movement. *Frontiers in Integrative Neuroscience, 13*, 12.

Kitamura, K., & Kano, M. (2013). Dendritic calcium signaling in cerebellar Purkinje cell. *Neural Networks, 47*, 11–17.

Kostadinov, D., Beau, M., Blanco-Pozo, M., & Hausser, M. (2019). Predictive and reactive reward signals conveyed by climbing fiber inputs to cerebellar Purkinje cells. *Nature Neuroscience, 22*, 950–962.

Koziol, L. F., Budding, D. E., & Chidekel, D. (2012). From movement to thought: Executive function, embodied cognition, and the cerebellum. *Cerebellum, 11*, 505–525.

Leow, L. A., Marinovic, W., de Rugy, A., & Carroll, T. J. (2020). Task errors drive memories that improve sensorimotor adaptation. *The Journal of Neuroscience, 40*, 3075–3088.

Lesage, E., Morgan, B. E., Olson, A. C., Meyer, A. S., & Miall, R. C. (2012). Cerebellar rTMS disrupts predictive language processing. *Current Biology, 22*, R794–R795.

Lesage, E., Hansen, P. C., & Miall, R. C. (2017). Right lateral cerebellum represents linguistic predictability. *The Journal of Neuroscience, 37*, 6231–6241.

Lin, Q., Manley, J., Helmreich, M., Schlumm, F., Li, J. M., Robson, D. N., Engert, F., Schier, A., Nobauer, T., & Vaziri, A. (2020). Cerebellar neurodynamics predict decision timing and outcome on the single-trial level. *Cell, 180*, 536–551.

Llinas, R. R. (2013). The olivo-cerebellar system: A key to understanding the functional significance of intrinsic oscillatory brain properties. *Frontiers in Neural Circuits, 7*, 96.

Llinas, R., Walton, K., Hillman, D. E., & Sotelo, C. (1975). Inferior olive: Its role in motor learing. *Science, 190*, 1230–1231.

Marr, D. (1969). A theory of cerebellar cortex. *The Journal of Physiology, 202*, 437–470.

Maschke, M., Gomez, C. M., Ebner, T. J., & Konczak, J. (2004). Hereditary cerebellar ataxia progressively impairs force adaptation during goal-directed arm movements. *Journal of Neurophysiology, 91*, 230–238.

Mazzoni, P., & Krakauer, J. W. (2006). An implicit plan overrides an explicit strategy during visuomotor adaptation. *The Journal of Neuroscience, 26*, 3642–3645.

Miall, R. C., & Wolpert, D. M. (1996). Forward models for physiological motor control. *Neural Networks, 9*, 1265–1279.

Miall, R. C., Weir, D. J., Wolpert, D. M., & Stein, J. F. (1993). Is the cerebellum a Smith predictor? *Journal of Motor Behavior, 25*, 203–216.

Miall, R. C., Reckess, G. Z., & Imamizu, H. (2001). The cerebellum coordinates eye and hand tracking movements. *Nature Neuroscience, 4*, 638–644.

Miall, R. C., Christensen, L. O., Cain, O., & Stanley, J. (2007). Disruption of state estimation in the human lateral cerebellum. *PLoS Biology, 5*, e316.

Moberget, T., Gullesen, E. H., Andersson, S., Ivry, R. B., & Endestad, T. (2014). Generalized role for the cerebellum in encoding internal models: Evidence from semantic processing. *The Journal of Neuroscience, 34*, 2871–2878.

Molinari, M., Chiricozzi, F. R., Clausi, S., Tedesco, A. M., De, L. M., & Leggio, M. G. (2008). Cerebellum and detection of sequences, from perception to cognition. *Cerebellum, 7*, 611–615.

Montarolo, P. G., Palestini, M., & Strata, P. (1982). The inhibitory effect of the olivocerebellar input on the cerebellar Purkinje cells in the rat. *The Journal of Physiology, 332*, 187–202.

Morton, S. M., & Bastian, A. J. (2006). Cerebellar contributions to locomotor adaptations during splitbelt treadmill walking. *The Journal of Neuroscience, 26*, 9107–9116.

Nguyen-Vu, T. D., Kimpo, R. R., Rinaldi, J. M., Kohli, A., Zeng, H., Deisseroth, K., & Raymond, J. L. (2013). Cerebellar Purkinje cell activity drives motor learning. *Nature Neuroscience, 16*, 1734–1736.

Nixon, P. D., & Passingham, R. E. (2000). The cerebellum and cognition: Cerebellar lesions impair sequence learning but not conditional visuomotor learning in monkeys. *Neuropsychologia, 38*, 1054–1072.

Noto, C. T., & Robinson, F. R. (2001). Visual error is the stimulus for saccade gain adaptation. *Brain Research. Cognitive Brain Research, 12*, 301–305.

Oh, Y., & Schweighofer, N. (2019). Minimizing precision-weighted sensory prediction errors via memory formation and switching in motor adaptation. *The Journal of Neuroscience, 39*, 9237–9250.

Ohmae, S., & Medina, J. F. (2015). Climbing fibers encode a temporal-difference prediction error during cerebellar learning in mice. *Nature Neuroscience, 18*, 1798–1803.

Oscarsson, O. (1980). Functional organization of olivary projection to the cerebellar anterior lobe. In J. Courville (Ed.), *The inferior olivary nucleus: Anatomy and physiology* (pp. 279–290). Raven.

Pasalar, S., Roitman, A. V., Durfee, W. K., & Ebner, T. J. (2006). Force field effects on cerebellar Purkinje cell discharge with implications for internal models. *Nature Neuroscience, 9*, 1404–1411.

Picard, F., & Friston, K. (2014). Predictions, perception, and a sense of self. *Neurology, 83*, 1112–1118.

Pickering, M. J., & Clark, A. (2014). Getting ahead: Forward models and their place in cognitive architecture. *Trends in Cognitive Sciences, 18*, 451–456.

Popa, L. S., Hewitt, A. L., & Ebner, T. J. (2012). Predictive and feedback performance errors are signaled in the simple spike discharge of individual Purkinje cells. *The Journal of Neuroscience, 32*, 15345–15358.

Popa, L. S., Hewitt, A. L., & Ebner, T. J. (2013). Purkinje cell simple spike discharge encodes error signals consistent with a forward internal model. *Cerebellum, 12*, 331–333.

Popa, L. S., Hewitt, A. L., & Ebner, T. J. (2014). The cerebellum for jocks and nerds alike. *Frontiers in Systems Neuroscience, 8*, 1–13.

Popa, L. S., Streng, M. L., & Ebner, T. J. (2016a). Signaling of predictive and feedback information in Purkinje cell simple spike activity. In D. H. Heck (Ed.), *Neuronal codes of the cerebellum* (pp. 1–25). Elsevier.

Popa, L. S., Streng, M. L., Hewitt, A. L., & Ebner, T. J. (2016b). The errors of our ways: Understanding error representations in cerebellar-dependent motor learning. *Cerebellum, 15*, 93–103.

Popa, L. S., Streng, M. L., & Ebner, T. J. (2017). Long-term predictive and feedback encoding of motor signals in the simple spike discharge of Purkinje cells. *eNeuro, 4*, 0036–17.2017.

Pope, P. A., & Miall, R. C. (2012). Task-specific facilitation of cognition by cathodal transcranial direct current stimulation of the cerebellum. *Brain Stimulation, 5*, 84–94.

Ramnani, N. (2006). The primate cortico-cerebellar system: Anatomy and function. *Nature Reviews. Neuroscience, 7*, 511–522.

Rochefort, C., Arabo, A., Andre, M., Poucet, B., Save, E., & Rondi-Reig, L. (2011). Cerebellum shapes hippocampal spatial code. *Science, 334*, 385–389.

Rochefort, C., Lefort, J. M., & Rondi-Reig, L. (2013). The cerebellum: A new key structure in the navigation system. *Frontiers in Neural Circuits, 7*, 35.

Runnqvist, E., Bonnard, M., Gauvin, H. S., Attarian, S., Trebuchon, A., Hartsuiker, R. J., & Alario, F. X. (2016). Internal modeling of upcoming speech: A causal role of the right posterior cerebellum in non-motor aspects of language production. *Cortex, 81*, 203–214.

Schlerf, J. E., Ivry, R. B., & Diedrichsen, J. (2012). Encoding of sensory prediction errors in the human cerebellum. *The Journal of Neuroscience, 32*, 4913–4922.

Schmahmann, J. D. (2004). Disorders of the cerebellum: Ataxia, dysmetria of thought, and the cerebellar cognitive affective syndrome. *The Journal of Neuropsychiatry and Clinical Neurosciences, 16*, 367–378.

Schmahmann, J. D. (2010). The role of the cerebellum in cognition and emotion: Personal reflections since 1982 on the dysmetria of thought hypothesis, and its historical evolution from theory to therapy. *Neuropsychology Review, 20*, 236–260.

Serrien, D. J., & Wiesendanger, M. (2000). Temporal control of a bimanual task in patients with cerebellar dysfunction. *Neuropsychologia, 38*, 558–565.

Shadmehr, R., & Holcomb, H. H. (1997). Neural correlates of motor memory consolidation. *Science, 277*, 821–825.

Shadmehr, R., Smith, M. A., & Krakauer, J. W. (2010). Error correction, sensory prediction, and adaptation in motor control. *Annual Review of Neuroscience, 33*, 89–108.

Sheu, Y. S., Liang, Y., & Desmond, J. E. (2019). Disruption of cerebellar prediction in verbal working memory. *Frontiers in Human Neuroscience, 13*, 61.

Shidara, M., Kawano, K., Gomi, H., & Kawato, M. (1993). Inverse-dynamics model eye movement control by Purkinje cells in the cerebellum. *Nature, 365*, 50–52.

Shin, S. L., Zhao, G. Q., & Raymond, J. L. (2014). Signals and learning rules guiding oculomotor plasticity. *The Journal of Neuroscience, 34*, 10635–10644.

Streng, M. L., Popa, L. S., & Ebner, T. J. (2017a). Climbing fibers control Purkinje cell representations of behavior. *The Journal of Neuroscience, 37*, 1997–2009.

Streng, M. L., Popa, L. S., & Ebner, T. J. (2017b). Climbing fibers predict movement kinematics and performance errors. *Journal of Neurophysiology, 118*, 1888–1902.

Streng, M. L., Popa, L. S., & Ebner, T. J. (2018a). Complex spike wars: A new hope. *Cerebellum, 17*, 735–746.

Streng, M. L., Popa, L. S., & Ebner, T. J. (2018b). Modulation of sensory prediction error in Purkinje cells during visual feedback manipulations. *Nature Communications, 9*, 1099.

Strick, P. L., Dum, R. P., & Fiez, J. A. (2009). Cerebellum and nonmotor function. *Annual Review of Neuroscience, 32*, 413–434.

Tanaka, H., Ishikawa, T., & Kakei, S. (2019). Neural evidence of the cerebellum as a state predictor. *Cerebellum, 18*, 349–371.

Tanaka, H., Ishikawa, T., Lee, J., & Kakei, S. (2020). The cerebro-cerebellum as a locus of forward model: A review. *Frontiers in Systems Neuroscience, 14*, 19.

Taylor, J. A., Klemfuss, N. M., & Ivry, R. B. (2010). An explicit strategy prevails when the cerebellum fails to compute movement errors. *Cerebellum, 9*, 580–586.

Ten Brinke, M. M., Boele, H. J., Spanke, J. K., Potters, J. W., Kornysheva, K., Wulff, P., IJpelaar, A. C., Koekkoek, S. K., & De Zeeuw, C. I. (2015). Evolving models of Pavlovian conditioning: Cerebellar cortical dynamics in awake behaving mice. *Cell Reports, 13*, 1977–1988.

Thach, W. T. (2007). On the mechanism of cerebellar contributions to cognition. *Cerebellum, 6*, 163–167.

Thach, W. T., Goodkin, H. P., & Keating, J. G. (1992). The cerebellum and the adaptive coordination of movement. *Annual Review of Neuroscience, 15*, 403–442.

Tomatsu, S., Ishikawa, T., Tsunoda, Y., Lee, J., Hoffman, D. S., & Kakei, S. (2016). Information processing in the hemisphere of the cerebellar cortex for control of wrist movement. *Journal of Neurophysiology, 115*, 255–270.

Tseng, Y. W., Diedrichsen, J., Krakauer, J. W., Shadmehr, R., & Bastian, A. J. (2007). Sensory prediction errors drive cerebellum-dependent adaptation of reaching. *Journal of Neurophysiology, 98*, 54–62.

Wagner, M. J., & Luo, L. (2020). Neocortex-cerebellum circuits for cognitive processing. *Trends in Neurosciences, 43*, 42–54.

Wagner, M. J., Kim, T. H., Savall, J., Schnitzer, M. J., & Luo, L. (2017). Cerebellar granule cells encode the expectation of reward. *Nature, 544*, 96–100.

Wagner, M. J., Kim, T. H., Kadmon, J., Nguyen, N. D., Ganguli, S., Schnitzer, M. J., & Luo, L. (2019). Shared cortex-cerebellum dynamics in the execution and learning of a motor task. *Cell, 177*, 669–682.

Wallman, J., & Fuchs, A. F. (1998). Saccadic gain modification: Visual error drives motor adaptation. *Journal of Neurophysiology, 80*, 2405–2416.

Winkelman, B., & Frens, M. (2006). Motor coding in floccular climbing fibers. *Journal of Neurophysiology, 95*, 2342–2351.

Winkelman, B. H., Belton, T., Suh, M., Coesmans, M., Morpurgo, M. M., & Simpson, J. I. (2014). Nonvisual complex spike signals in the rabbit cerebellar flocculus. *The Journal of Neuroscience, 34*, 3218–3230.

Wolpert, D. M., Ghahramani, Z., & Jordan, M. I. (1995). An internal model for sensorimotor integration. *Science, 269*, 1880–1882.

Wolpert, D. M., Miall, R. C., & Kawato, M. (1998). Internal models in the cerebellum. *Trends in Cognitive Sciences, 2*, 338–347.

Xu-Wilson, M., Chen-Harris, H., Zee, D. S., & Shadmehr, R. (2009). Cerebellar contributions to adaptive control of saccades in humans. *The Journal of Neuroscience, 29*, 12930–12939.

Chapter 13
The Quest for a Unifying Framework for the Role of Cerebellar Complex Spikes

Akshay Markanday and Peter Thier

13.1 Introduction

It is well-established that the cerebellum plays a vital role in the fine control and coordination of movements and in motor learning as a means to adjust to new demands. However, the question of how the cerebellum accomplishes this capacity has remained elusive until the publication of David Marr's seminal "theory of cerebellar cortex." In his autobiography (Squire, 1999), Masao Ito describes his frustrating experience at the time when David Marr's theory of cerebellar cortex was not yet available as, "...when we tried to explain a brain function such as learning or motor control on the basis of the brain's neuronal network structures, we found it impossible to do so." As challenging as it may sound, the key to the neural underpinnings of cerebellum-based motor control and learning lies in the understanding of its underlying circuitry and the contribution of its individual components. In fact, the epochal work of David Marr (1969), further elaborated by James Albus (1971) and Masao Ito (1972), collectively known as the Marr-Albus-Ito (*MAI*) hypothesis, is proof of the virtue of rigorous attempts to unravel the functioning of the cerebellar circuitry.

A. Markanday · P. Thier (✉)
Cognitive Neurology Laboratory, Hertie Institute for Clinical Brain Research, Tübingen, Germany
e-mail: thier@uni-tuebingen.de

H. Mizusawa, S. Kakei (eds.), *Cerebellum as a CNS Hub*,
Contemporary Clinical Neuroscience, https://doi.org/10.1007/978-3-030-75817-2_13

13.2 Purkinje Cell Discharge: Simple Spikes and Complex Spikes

The core element of the cerebellar circuitry (Fig. 13.1) is the Purkinje cell (*PC*). *PC* discharge, the only output of the cerebellar cortex, is characterized by the presence of two distinct types of action potentials (Eccles et al., 1966)—simple spikes (*SSs*) and complex spikes (*CSs*). Each of these signals can be traced back to one of the two streams of afferent information, transmitted the via mossy fiber-granule cell-parallel fiber pathway (*PF*) and the climbing fiber (*CF*) pathway, both converging at the level of *SSs*, which represent the collective action of mossy fiber-driven *PFs* modulated by different types of interneurons, are short-duration membrane depolarization events that appear as simple, bi- or triphasic waveforms in extracellular recordings and occur up to several hundred times per second. *CSs*, on the other hand, have a very exotic appearance, not paralleled by action potentials anywhere else in the nervous system. They reflect the direct influence of the *CF* axons originating in the inferior olive (*IO*) neurons. Each *CF*, rising like a creeper along the axis of the *PC*, entwines with the soma and the dendritic tree of the target *PC* to create up to 500 synaptic contacts (Hillman, 1969), thereby forming one of the strongest excitatory synaptic interfaces in the entire brain. *CSs*, fired by the *PC* in response to *CF* activity, are associated with a large influx of Ca^{2+} ions at the *CF-PC* synapses, responsible for a long-lasting depolarization of the targeted *PCs*. Not only their long duration of 4–10 msec is unusual but also their "complex" polyphasic morphologies and their perplexingly low (0.5–2 Hz) firing rates (Davie et al., 2008; Eccles, 1967; Fujita, 1968; Llinás & Sugimori, 1980a, b; Stuart & Häusser, 1994; Thach, 1968).

13.3 The Functional Role of Complex Spikes

Thanks to their more conventional features, the role of *SSs* has been more readily accessible. Many open questions notwithstanding, there is now general consensus that *SSs* encode various movement-related kinematic parameters (Coltz et al., 1999; Dash et al., 2013; Ebner et al., 2011; Fortier et al., 1989, 1993; Frysinger et al., 1984; Fu et al., 1997; Hewitt et al., 2011; Herzfeld et al., 2015, 2018; Ohtsuka & Enoki, 1998; Roitman et al., 2005; Sato & Noda, 1992; Sun et al., 2017; Suzuki & Keller, 1988a; Takagi et al., 2000; Thach, 1968, 1970; Thier et al., 2000; Yamamoto et al., 2007). However, the functional role of *CSs* has turned out to be a tougher nut to crack, a conclusion that might be surprising in view of the fact that already the *MAI* theory had nailed down a highly specific role, well-grounded in a solid body of anatomical and physiological facts already available at that time. According to this theory, the *CS* signals information on motor errors, subsequently corrected by the cerebellum (Albus, 1971; Ito, 1972; Marr, 1969). In other words, the *CS* serves as a

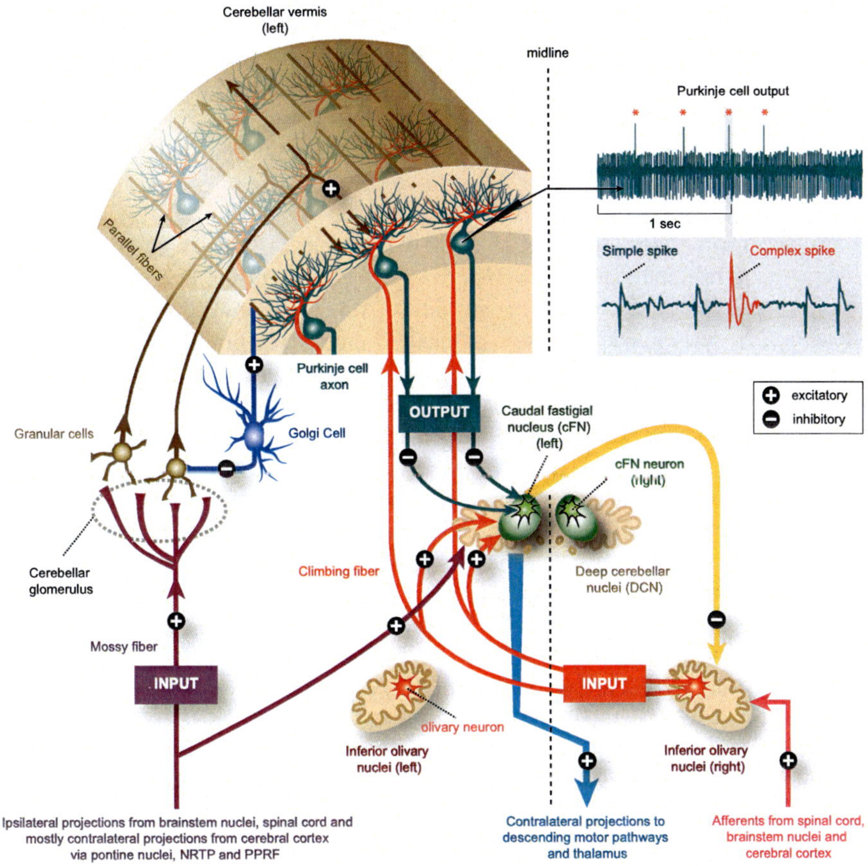

Fig. 13.1 Cerebellar circuitry and its key components. (Modified from Thier and Markanday (2019))

"teacher" struggling to improve motor performance by stimulating motor learning. This is achieved by the strong excitatory influence of *CSs* that causes long-term depression (Ito, 1972, 2001) of *PF-PC* synapses due to the coincidence of the *CS*-based *PC* depolarization and *PF* input. It is this *CS*-mediated adjustment of the coincident *PF* input, ultimately reflected in the *PC's SS* discharge, that is thought to ensure that the movement will reach the desired goal.

This idea has received support from several studies, deploying varying models of cerebellum-based learning such as the adaptation of the vestibulo-ocular reflex (*VOR*) (Ito et al., 1974, 1977; Maekawa & Simpson, 1973), a well-studied example of parametric adjustment. Here the parameters that need to be adjusted are a gain and a phase factor for the translation of a head movement into a compensatory eye movement, able to ensure visual stability by preventing head movement-induced retinal image slip. At that time, perhaps, one might have been tempted to believe that the chapter on the role of *CSs* would be short and conclusive, providing a definite answer to the question of its function. And after all, why should a type of spike, whose sparse firing rates seem to indicate a comparatively limited information capacity, allow room for the development of different interpretations of its role? It is precisely this paradox that explains why many years after the presentation of the *MAI* theory writing the chapter on these intriguing electrophysiological events—*CSs*—continues.

In fact, variant ideas on the role of the *CF* system and observations not easily compatible with the tenets of the *MAI* theory or readily fitting into its conceptual framework emerged already in the seventies. An early example of an alternative view on *CSs* well outside the scope of the *MAI* theory was provided by Llinás and co-workers. They were struck by the fact that neurons in the *IO*, the sources of *CFs*, exhibited a pattern of 8–10 Hz rhythmic and synchronous firing, reminiscent of subliminal tremor of similar frequency that accompanies movements (Lamarre et al., 1971; Lamarre & Mercier, 1971; Villablanca & Riobo, 1970). Considering the clock-like regularity and temporal precision of the firing pattern of these neurons and its resemblance with the tremor, Llinás and co-workers suggested that the *IO-CF* system might be instrumental in coordinating movements by chunking short segments of movement elements (Leznik & Llinas, 2005; Llinas, 1974; Llinas et al., 1974; Llinás & Volkind, 1973). Other studies, unlike the aforementioned studies carried out entirely in reference to the *MAI* theory, led to either partially compatible views, entailing modifications of the *MAI*-concept or results that—at least at the first glance—seemed to contradict the theory.

In this chapter, we will take a look at some of the more recent work on goal-directed eye movements as a convenient model of motor control that allows us to delve more deeply into the controversies on the role of *CSs*. However, rather than listing seemingly incongruent observations and views and endorsing the one or the other, we will try to convince the reader that many of them can be reconciled. We will argue that these views reflect different perspectives on a system transporting information relevant for motor learning that is much richer than hitherto assumed.

Some of the early experiments investigating the role of *PCs* in movement kinematics addressed wrist movements (Thach, 1968), followed by studies of other single joints or the interaction of multiple joints as models of cerebellum-dependent motor control (Ebner et al., 2011; Fortier et al., 1989, 1993; Frysinger et al., 1984; Fu et al., 1997; Hewitt et al., 2011; Roitman et al., 2005; Thach, 1970). However, the large degrees of freedom in the case of multi-joint movements, the particular dynamic challenges of skeletal movements that have to deal with interaction torques and changing loads, the influence of proprioceptive feedback, and the significance of muscular fatigue have impeded the search for an indisputable function of the cerebellum.

13.4 The Advantages of Eye Movement Studies in the Quest for the Function of the Cerebellum

Eye movements are alleviated from the overload of excessive degrees of freedom and the other challenges of skeletal movements. But of course, also eye movements have to strive for precision. This applies particularly to eye movements made by human and nonhuman primates that try to ensure that the advantages of foveal vision for the scrutiny of objects and scenes can be exploited. While saccades are carried out to shift the image of a target that may initially fall onto the peripheral retina to the fovea, smooth-pursuit eye movements (*SPEMs*) are made to stabilize the image in case the target might move at a speed not too fast to evoke a saccade. Given the fact that the fovea has a diameter of only 1°, with its center of highest resolution (the foveola) merely a fraction of that, the amplitude of saccades and the speed of *SPEMs* must be carefully chosen. This is the purpose of saccadic adaptation and smooth-pursuit adaptation (*SPA*), examples of parametric adjustment that ensure that the necessary precision is achieved and maintained by tweaking the relevant movement parameters. For many years scientists have resorted to paradigms that render these parameters inappropriate in a well-controlled manner and unexpected by the subject, necessitating corrections, the subject is not aware of. In one of these paradigms, introduced by McLaughlin (1967), the subject is asked to make a saccade to a target. While the saccade is carried out, the target is moved away to a new location, a shift that is not noticed by the subject because of saccadic suppression (Latour, 1962; Volkmann, 1962). Because the saccade was planned based on the original location of the target, the target is missed, and it is left to a subsequent corrective saccade to finally reach the target. However, if the observer experiences the shift time and again, what one typically sees is that the metric of the primary saccade is gradually adjusted in an attempt to decrease the retinal error and the need to add a corrective saccade. In two popular variants of the task, the target is either moved further out during the primary saccade (Fig. 13.2a) or back in the direction of the fixation point (Fig. 13.2b). In this case, changes in saccade amplitude without concomitant changes in direction are induced. Adaptation results in larger amplitudes ("gain-increase adaptation") or shorter amplitudes ("gain-decrease adaptation"), respectively. In monkeys, it takes a series of a few hundred trials of consistent shifts for adaptation to reach an asymptomatic level, associated with a significant reduction of the retinal error that resulted from the primary saccade. In some individuals, the error may be eliminated completely.

While saccadic adaptation adjusts the amplitude of the targeting saccade, a parametric adaptation of *SPEMs* is needed to avoid that the retinal image of the target slips. *SPEMs* comprise two qualitatively different phases, an early open-loop phase and a subsequent closed-loop phase. The latter can be understood as the product of a typical feedback controller that reduces retinal image slip, induced by the

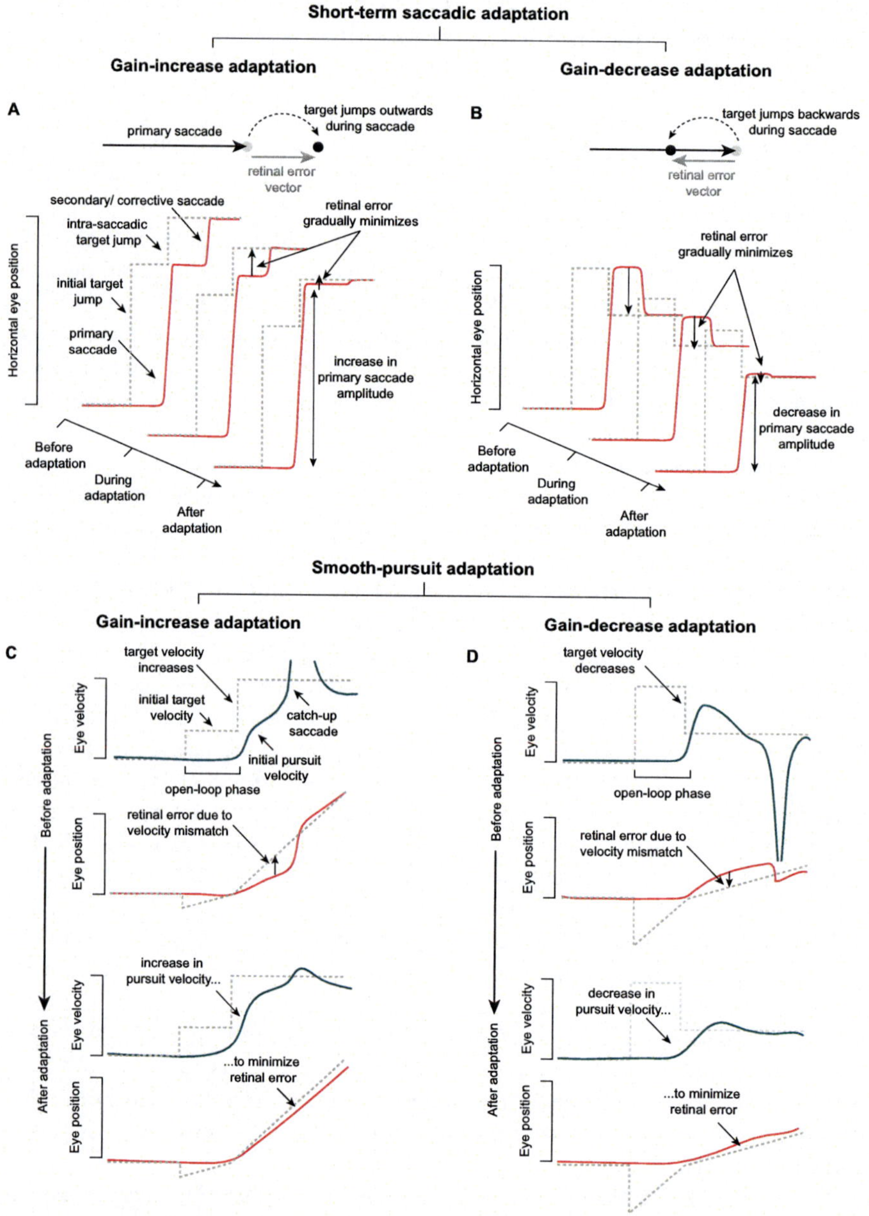

Fig. 13.2 Oculomotor paradigms to test short-term motor learning. (**a, b**) Short-term saccadic adaptation is achieved by consistently introducing intrasaccadic target shifts (dashed gray lines) of the same size, either away from the original target location (gain-increase adaptation, **a**), or backward in the direction of the fixation point (gain-decrease adaptation, **b**), thereby inducing retinal errors which are gradually reduced over many trials of a typical adaptation experiment. A schematic of exemplary trials during the course of gain-increase and gain-decrease adaptation in the horizontal direction is shown in **a** and **b**. In trials before the onset of adaptation, large secondary

difference between the target and eye velocity, by converting the measured slip into a compensatory eye movement. However, the problem is that this conversion has to rely on retinal signals from the past, the information on which becomes available with a delay of some 100 ms, the time incurred by the processing of visual information. Because of the delay, the pursuit eye movement made in the first 100–150 ms after the movement onset will be solely determined by information on the target movement in the 100–150 msec before the onset (labeled as open-loop phase in Fig. 13.2c, d) of the pursuit eye movement, in other words, carried out in an open-loop manner. In order to ensure that the eye velocity will match the target velocity in this early phase, it is essential that an optimal parameter for the feed-forward conversion of target image slip into a *SPEM* is used. *SPA* demonstrates that this is achieved by a parametric adjustment that exploits information on the adequacy of the speed of the initial *SPEM* in a given trial in order to improve future manifestations of initial *SPEM*. In a typical *SPA* paradigm, the target moves at a certain constant speed in one direction until the time the *SPEM* kicks in. At this point in time, the target speed is either increased or decreased (Fig. 13.2c, d, respectively). As the initial *SPEM* is determined by the retinal slip before the change of the target speed, it is most likely inadequate and unable to match the target movement. As a result, the eyes either over- or undershoot the target, depending on the direction of the target velocity step. Yet, provided that the step change in velocity remains consistent and predictable in time over many consecutive trials, the eye velocity during the initial *SPEM* phase is gradually adjusted to precisely track the moving target after the velocity step (Fukushima et al., 1996; Kahlon & Lisberger, 1996). Depending on the direction of the velocity step, the gain of the *SPEM* will be up or downregulated. Although, the outlined paradigms of saccadic adaptation and *SPA* require up to a few hundred trials presented over a couple of tens of minutes in order to reach stable levels of extensive adaptation, clear manifestations of adaptation become apparent much earlier, actually even after single trials (see below). Hence, to refer to these forms of learning as short-term saccadic adaptation (*STSA*) and short-term *SPA*, respectively, is fully justified. The adaption of both, saccades and *SPEMs*, depends critically on a distinct little part of the cerebellum, usually referred to as the oculomotor vermis (Fig. 13.3).

Fig. 13.2 (continued) saccades (eye records shown in magenta) are needed to correct the retinal error (indicated by vertical arrows) entailing insufficient primary saccades. (**c, d**) In smooth-pursuit adaptation (*SPA*), the velocity of the moving target is either increased (**c**) or decreased (**d**) near the end of the open-loop phase of smooth-pursuit eye movements (*SPEMs*). This sudden change in target velocity (dotted lines) induces a retinal slip as shown in the eye position traces (magenta), which is corrected by a "catch-up" saccade, seen as a large change in the eye velocity (turquoise) and position traces. After repeated trials of consistent target velocity change, the eye velocity is adjusted to minimize the retinal error. ((**a, b**) Modified with permission from Catz et al. (2005). (**c, d**) Modified with permission from Dash et al. (2010))

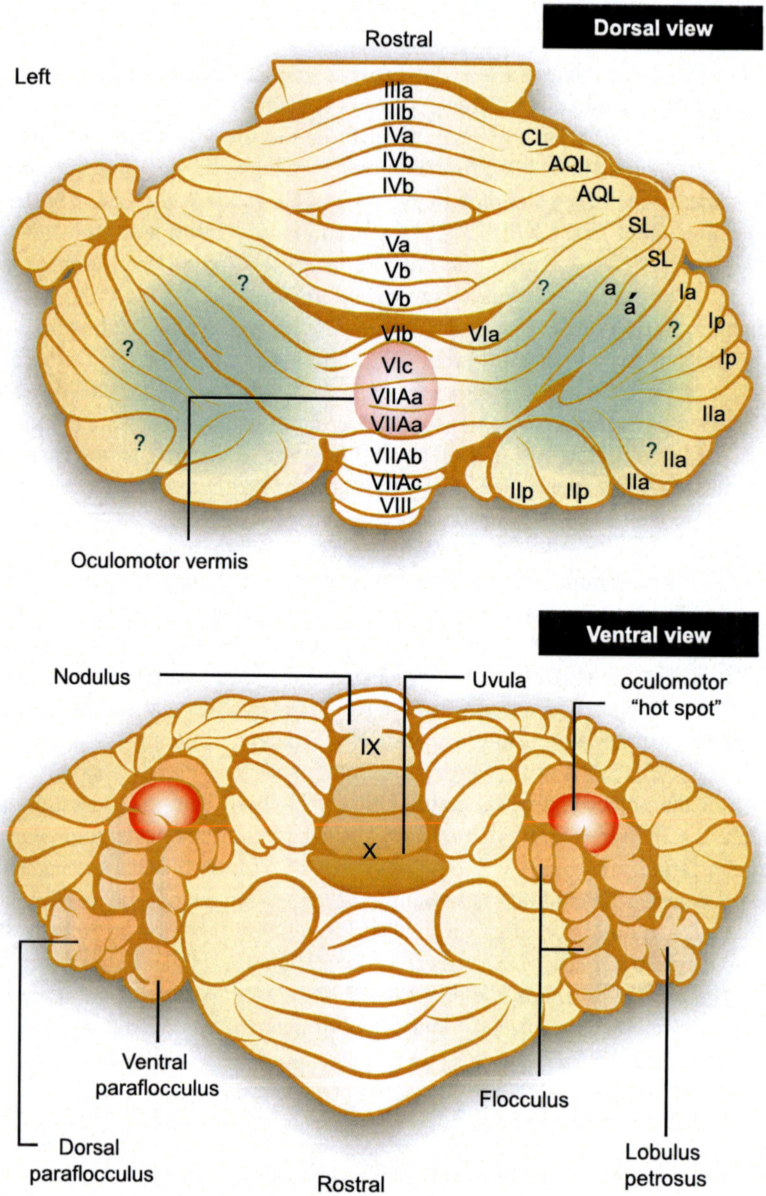

Fig. 13.3 Oculomotor regions of the cerebellar cortex. The anterior surface of the cerebellum (see dorsal view) is formed by the central lobule (*CL*, lobules *III*a, *III*b) and the anterior quadrangular lobe (*AQL*, lobules *IV*a, *IV*b and *V*a, *V*b). In the dorsal midline, they are joined by the oculomotor vermis (*OMV*, pink-shaded region), involved in the parametric control of saccadic and *SPEMs*, that involves the caudal part of the lobule *VI* (*VI*c) and the rostral part of lobule *VII* (*VII*a). On both sides of the *OMV*, a large and relatively ill-defined region (shaded in green) centered on the hemispheric parts of the simple lobule (*SL*)—the hemispheric oculomotor region (*HOR*) has been described as yet another part of cerebellar cortex involved in eye movements and eye movement-related vision (See Thier, 2011 for more details). The uncertainty regarding the exact location of

13.5 The Oculomotor Vermis: A Hub for the Control of Goal-Directed Eye Movements

The cerebellar cortex contains several areas involved in the control of eye movements. A ventral region (Fig. 13.3, ventral view), the "vestibulocerebellum" comprising the flocculus, the neighboring ventral paraflocculus and caudal parts of the midline vermis, the uvula, and the nodulus, is a part of the cerebellum that is essential for the parametric control of vestibular reflexes and their coordination with *SPEMs* (Stone & Lisberger, 1990a, b). The adjoining dorsal paraflocculus and the ill-defined hemispheric oculomotor region (see Fig. 13.3, dorsal view, green-shaded region) are parts of cerebellar cortex that seem to be involved in the control of saccades and *SPEMs*, although the little experimental evidence available does not allow a clearer characterization of their specific functional contributions. Much better understood is the role of the aforementioned small midline region of cerebellar cortex, the "oculomotor vermis" (*OMV*, Fig. 13.3, dorsal view), comprising lobuli *VI*c and *VII*a, first delineated by Noda and co-workers (Noda & Fujikado, 1987a, b) as a low-threshold zone for microstimulation evoked saccades and saccade-related single-unit activity (Fujikado & Noda, 1987). Lesions in cerebellar cortex confined to the *OMV* cause saccadic hypometria (Barash et al., 1999; Ignashchenkova et al., 2009; Ohki et al., 2009; Takagi et al., 1998, 2000; Vahedi et al., 1995), a deficit that is characterized by saccades that undershoot the target. Barash and co-workers (Barash et al., 1999) showed that the experimentally lesioned animals exhibited full recovery from hypometria within a few weeks following the lesion. However, what stayed for the many months of observation, and presumably permanently, was the increased variability in saccade endpoints associated with an inability to adjust saccade amplitudes when tested in a standard *STSA* experiment. Later work involving single-unit recordings and experimental lesions in monkeys could establish that the *OMV* is not only confined to saccades and *STSA* but also involved in the control of *SPEMs* and their learning-based adjustment (Dash et al., 2010, 2011, 2013; Ohtsuka & Enoki, 1998; Sato & Noda, 1992; Sun et al., 2017; Suzuki & Keller, 1988a, b; Takagi et al., 2000). Moreover, not only *PCs* but also individual neurons in the caudal fastigial nucleus, targets of *OMV PCs* that broadcast the information needed by the premotor structures in the brainstem to adjust signals for the control of saccades and *SPEMs*, are known to mediate information on the kinematics of both saccades and *SPEMs* (Sun et al., 2017). This is a surprising fact given the different kinematic parameter spaces of high-velocity saccades and the relatively slow *SPEMs*. The

Fig. 13.3 (continued) its boundaries is denoted by question marks. The ventral view shows the flocculus, the adjoining ventral paraflocculus, the uvula (lobule *X*), and the nodulus (lobule *IX*) which together form the vestibulocerebellum, key players in using vestibular information for the stabilization of gaze in the presence of head movements. The flocculus and ventral paraflocculus are also involved in the control of *SPEMs*. The small region (highlighted in red) in dorsal paraflocculus indicates an oculomotor hot spot that according to Noda and Mikami (1986) seems to prioritize saccade-related information. (Modified from Thier and Markanday (2019))

advantages of this integration remain unclear. How does the *OMV* adjust its *SS* output for the adaptation of brainstem control signals for saccades and *SPEMs* in order to ensure optimal performance, and what relevance does it hold for other learning functions accommodated by the cerebellum? Answers to these questions have been suggested by careful studies of *OMV CSs* as a proxy of information conveyed by the olivocerebellar system.

13.6 The Role of Complex Spikes in OMV-Dependent Oculomotor Learning: Support and Challenges for the Marr-Albus-Ito Theory

Starting out from the *MAI* theory, one would expect that information on retinal errors resulting from inappropriately measured saccades, missing the target, should be conveyed by *CSs*. As the retinal error is annihilated by the corrective saccade, the epoch in which one might expect to see an influence of the error starts with the end of the primary saccade and the onset of the corrective saccade, about 200 ms later. Indeed, Soetedjo and co-workers (Soetedjo & Fuchs, 2006; Soetedjo et al., 2008a, b) reported significant changes in the pattern of *CSs* in this period of time, interpreted as the signal essential for driving learning. They found that the probability of *CS* firing in this period was mostly influenced by the direction of the error, with little or no influence of error magnitude. The notion of *CSs* exhibiting directional tuning for retinal errors was further elaborated by Herzfeld et al. (2015) who reported a distinct topography of direction preferences of *CSs*, with *CS* units preferring leftward errors lying in the right part of the *OMV* and those preferring rightward error on the left side. They, moreover, documented a clear relationship between the preferred direction of *CSs* and their specific consequences on the velocity encoding by *SS* discharge of a given *PC*. While the *SS* discharge increased linearly with saccade speed in both directions, the overall firing of *SSs* was much larger in the *CS's* non-preferred error direction (*CS-OFF*) as compared to the preferred direction of error (*CS-ON*). Based on these observations and the evidence suggesting a monotonic relationship between *SS* population activity and saccade velocity, the same authors later suggested a model for trial-by-trial learning guided by the presence or absence of individual *CSs* that control the eye movements in a bidirectional, "push-pull" manner (Herzfeld et al., 2018). More specifically, they showed that the presence of a single *CS*, elicited by a retinal error in the *CS-ON* direction during the post-saccadic period, reduced the saccade-related *SS* activity of the subsequent trial. Considering the linear relationship between the frequency of *SS* discharge and saccade speed, this *CS*-induced reduction in *SS* discharge reduced the speed of the subsequent primary saccade made in the *CS-OFF* direction. In other words, the force driving the eyes was "pulling" the gaze away from the *CS-OFF*, toward the *CS-ON* direction. Conversely, if the *CS* was absent during the post-saccadic period, the *SS* firing increased in the subsequent trial, thereby increasing the speed of the saccade made in the *CS-OFF* direction, interpreted as a force that "pushed away" the gaze from *CS-ON* (or pulled it toward the *CS-OFF*). In summary, this model of the cerebellar basis of saccadic adaptation rests on two pillars: the first one is the conclusion that *CSs* convey information on retinal errors in a direction specific manner and, furthermore, that changes in *CS* activity determine changes of the cerebellar outflow, based on *SS* activity. The second one is the conclusion that a *SS* population signal encodes saccade velocity. The first conclusion is clearly in line with the central tenet of the *MAI* theory, namely, that the *CF* system initiates learning. The second one is not directly related to it as it addresses the *SS* code used to effect behavioral adjustments. The hypothesis that cerebellar cortex deploys a *SS* population code in order to control saccade kinematics and arguably also other types of movements was suggested much earlier by Thier et al. (2000). Based on their observation of an intriguing link between saccade duration and the duration of a saccade-related *PC* population signal, they argued that the kinematic parameter controlled by the cerebellar cortex might be movement duration. On the other hand, as said before, Herzfeld et al. (2015) suggested that it might be movement velocity whose up- or downregulation leads to larger or smaller saccades. Actually, our recent work on *OMV PC-SS* suggests that the guidance of saccades is based on the control of various metric and kinematic parameters rather than being restricted to just one (Sun et al., 2017; Markanday et al., 2020b). In any case, there is agreement that *SSs* control movement kinematics and, moreover, that the control signal is based on a population signal. The notion of population coding is in good accordance with the anatomical fact that groups of *PC* axons converge on individual target neurons in the deep cerebellar nuclei *(DCN)* (Person & Raman, 2012).

As a matter of fact, attempts to tackle the role of *CSs* in short-term saccadic adaptation started earlier and in contrast to the aforementioned studies, they failed to provide support for the *MAI* theory. We are referring to work published by Catz et al. (2005) who studied two distinct forms of *STSA*—"gain-increase" and "gain-decrease" adaptation—as described earlier. In these experiments the intrasaccadic target jump was constant, either consistently away from or back towards the fixation point. Consequently, Catz et al. observed that saccade amplitude gradually grew with trial number in the case of gain-increase adaptation and shrank if the target shift demanded gain-decrease adaptation. In many cases, at the end of an adaptation block of typically several hundred trials, full adaptation of saccade amplitude and, consequently, elimination of any retinal post-saccadic errors were observed (Fig. 13.4a, b, panels on the right). Assuming that the *CS* firing reflected the size and the direction of error, one should see a clear post-saccadic discharge of *CSs* in the beginning of the adaptation series (i.e., in the absence of adaptation) that would reduce post-saccadic errors. Consequently, one would expect that also the modulation of the *CS* discharge should be strongest in this period. Paralleling the progression of adaptation (and a consecutively decreasing post-saccadic retinal error), this modulation should gradually decrease. However, contrary to this expectation, Catz et al. observed that the CSs fired spontaneously in the beginning of adaptation blocks no matter if gain-increase or gain-decrease adaptation was aimed at (Fig. 13.4a, b). In the case of gain-decrease adaptation (Fig. 13.4b), *CS* firing rates started to grow around the time of the primary saccade all the more, the further the adaptation progressed, reaching a stable maximum at the end of the adaptation series when the error had become minimal or even be completely eliminated. Conversely, in the case of gain-increase adaptation (Fig. 13.4a, also see Fig. 13.4e, f, and g for a single session), they documented a buildup of a suppression of *CS* firing paralleling the progression of adaptation, reaching its maximal level toward the end of the block when adaptation had plateaued. These opposite effects of the two forms of *STSA* on the *CS* firing were most conspicuous at the population level (Fig. 13.4c, d) but also easily discernible with the naked eye at the level of single units (Fig. 13.4a, b and g). The pattern obtained was clearly not compatible with the assumption that *CSs* are sensitive to retinal error. One might argue that pooling *CS* units with opposite preferences for the direction of error in the post-saccadic period might have prevented the emergence of a convincing population signal reflecting the post-saccadic error. While this is a valid objection, it fails to explain the emergence of a clear *CS* signal at the population level, reflecting the direction of adaptation, around the time of the primary saccade. Moreover, also the individual *PC* units whose *CS* activity patterns were analyzed by the authors failed to exhibit signatures of error. An example of *CS* responses of a *PC* unit tested for gain-increase adaptation is given in Fig. 13.4e–j. Figure 13.4h, i shows a scatter plot of saccadic errors before and during adaptation. The pattern of saccadic errors in trials with and without *CS* in the post-saccadic error period, distinguished by color, overlay each other before and during adaptation (Fig. 13.4h, i, red vs. blue dots). The absence of an impact of a post-saccadic error on the *CS* is also indicated by the fact that quantification of mutual information (*MI*) between the *CS* firing and the deviation of eye position from the target did not reach significance, at any time relative to saccade end (Fig. 13.4j, red trace). On the other hand, the *MI* analysis documented that the *CS* firing provided significant information on saccade amplitude (Fig. 13.4j, blue trace). But how is it possible that Soetedjo and Fuchs (2006), based on a similar *STSA* paradigm, found a clear influence of the post-saccadic error on the spiking of *CS* and, on the other hand, failed to note the earlier activity that Catz et al. described? The answer is probably that they confined their analysis to the period starting approximately 70 ms after the end of the primary saccade and did not consider an earlier

Fig. 13.4 (continued) right show the percentage change (±s.e.m) of saccadic gain as a function of adaptation level. **(e, f)** Scatter plots showing the evolution of motor error and saccade amplitude, respectively, in a single gain-increase adaptation session during which an exemplary *CS* unit was recorded. Interval 1 and 2 are separated by the median values of errors **(e)** and amplitudes **(f)**, depicted by the solid red lines. **(g)** Probability of *CS* occurrence within 200 ms from saccade onset as a function of motor error size. Note how the probability drops from $p = 0.1724$ at adaptation onset (maximum error) to $p = 0.0348$ for stable adaptation (minimal error). **(h, i)** Scatter plot comparing motor errors of trials with (red dots) and without (blue dots) a *CS* within a period of 200 ms from saccade onset, before and during adaptation, respectively. **(j)** Information transmission rate of *CSs* of the representative *PC* recorded during the adaptation session (shown in **e** and **f**) for motor error (red trace) and saccade amplitude (blue trace), plotted as a function of time relative to saccade offset (dashed vertical line). Solid and dashed lines depict the significant nonsignificant information carried by *CSs*, respectively. (Modified with permission from Catz et al. (2005))

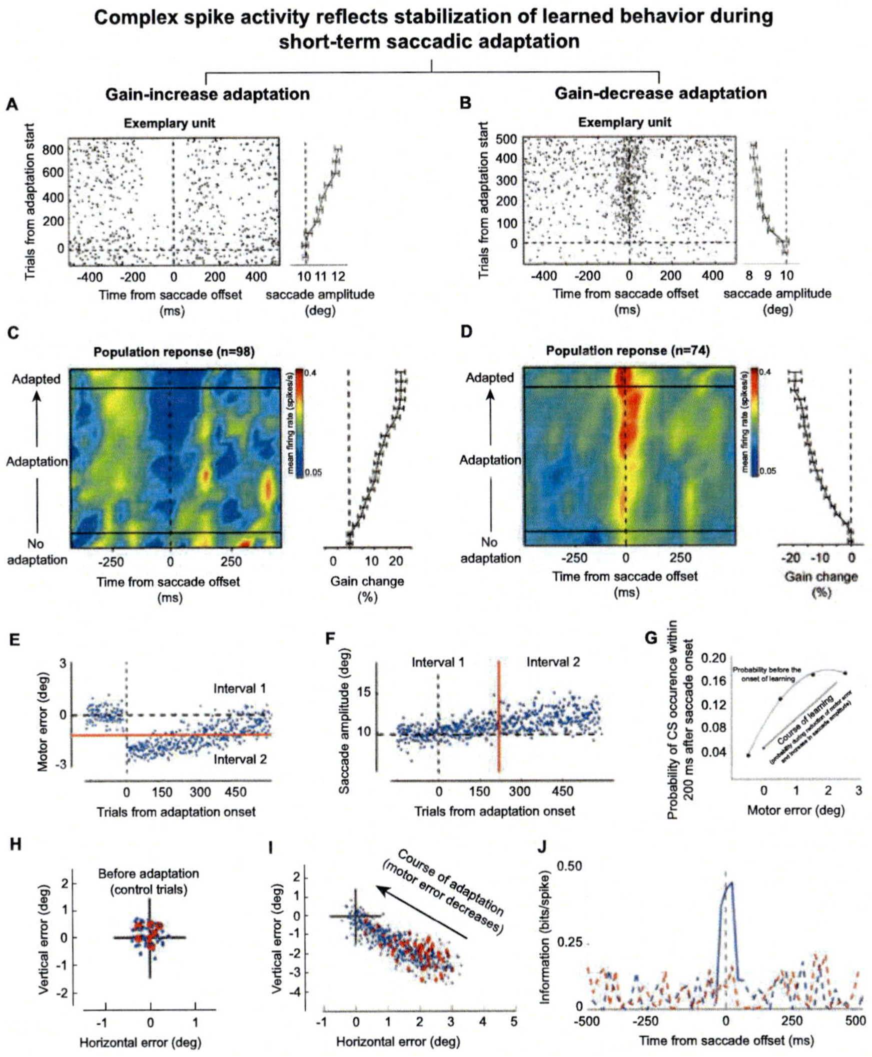

Fig. 13.4 Complex spike activity during short-term saccadic adaptation. (**a**, **b**) Left panels show raster plots of exemplary Purkinje cell (*PC*) units exhibiting a pause (**a**) and a burst (**b**) in complex spike (*CS*) activity around the time of the saccade during gain-increase and gain-decrease adaptation, respectively. Vertical dashed lines represent the time of saccade offset. Horizontal dashed lines represent the onset of adaptation trials. The mean amplitude of saccades (± s.e.m) is shown in the panels on the right. Vertical dashed lines (at 10 deg), representing the size of the initial target jump, serve as a reference for observing changes in saccade amplitudes before and during adaptation. This pause (**c**) and burst (**d**) modulation in *CS* activity is clearly observed in the population responses (shown as heatmaps). Solid horizontal lines represent the onset (no adaptation) and offset (full adaptation) of adaptation trials. Data are aligned to saccade offset (vertical dashed line). Color bars represent the mean *CS* firing rates calculated over 25 ms time bins in *x*-axis and adaptation trial bins in the *y*-axis, corresponding to 1/20th of the trial range from the onset to offset of adaptation. A Gaussian kernel (50 ms *SD*) was used to smoothen the population plot. Panels on the

one. It is less obvious why Catz et al. failed to retrieve information on the post-saccadic error although their analytical window was much wider and the statistical approach used certainly adequate. We will come back to this question after having taken a brief look at learning-based changes of *SPEMs*. As said earlier, *OMV PCs* are not only responsive to saccades but also to *SPEMs*, and lesions of the *OMV* are known to compromise *SPA* (Ohtsuka & Enoki, 1998; Takagi et al., 2000). Do *CSs* fired by *OMV* in conjunction with *SPA* reflect retinal errors, or do they reflect behavioral changes similar to those observed in *STSA*? This was the question addressed by Dash et al. (2010) in a follow-up study. They recorded *PC-CSs* from the *OMV* during "gain-increase" (Fig. va) and "gain-decrease" (Fig. 13.5b) *SPA* by deploying the *SPA* paradigms sketched earlier in Fig. 13.2c, d. They scrutinized *CSs* fired in the early open-loop period, the period in which information on retinal image slip, needed to update the initial *SPEM* velocity, accrues. Although this error was maximal in the first trials of a block, *CSs* were fired randomly without any relation to the task. However, paralleling the progression of adaption, they observed the buildup of a specific pattern of modulation of the *CS* discharge. In the case of "gain-decrease" *SPA*, causing a gradual slowing down of eye velocity in the open-loop phase to align with the pace of the moving target, a likewise gradual increase in *CS* discharge was observed, which reached its maximum at the end of the adaptation series when the velocity errors had become minimal (Fig. 13.5b, e, and f). Conversely, in the case of "gain-increase" *SPA*, a buildup of a drop in *CS* discharge in the open-loop period was documented (Fig. 13.5a, c, and d). In a nutshell, the changes in *CS* activity of *OMV PCs* during *SPA* were found to be completely analogous to the ones observed during *STSA*. In both cases, the *CS* does not reflect the relevant error but the development of suitable behavioral adaptations. Why did they observe this building up of *CS* discharge during motor learning? We argue that this might be a consequence of the predictable nature of the *STSA* and *SPA* paradigms used by Catz et al. and Dash et al., respectively. The subjects in these paradigms could expect a stereotypic movement of the target one trial after another, thus allowing full predictability of the retinal error to be expected in the absence of necessary saccade and *SPA* adjustments, respectively. Although Catz et al. failed to unravel a statistically reliable impact of retinal error on saccade-related response, a look at some of the records obtained by them might suggest weak traces of error in the post-saccadic error period in some of the earlier trials of adaptation sessions (see Fig. 13.4c, left panel). It is tempting to speculate that these traces remained subliminal because of the ability of the experimental subjects to predict the future error and to avoid it, rather than having to wait for its occurrence. This speculation assumes a distribution of complex spikes between two periods, an early period around the primary saccade, allowing complex spikes to cause adjustments of the primary saccade, and a later, post-saccadic period in which unpredicted errors are gauged in order to prompt secondary corrections and to initiate pertinent adjustments of future primary saccades. The higher the predictability of the error, the more complex spikes will be shifted to the early period. This scenario is in accordance with recent work on eyelid conditioning carried out by Ohmae and Medina (2015). Here learning the predictive value of the conditioned stimulus allows a timely eyelid closure that is paralleled by

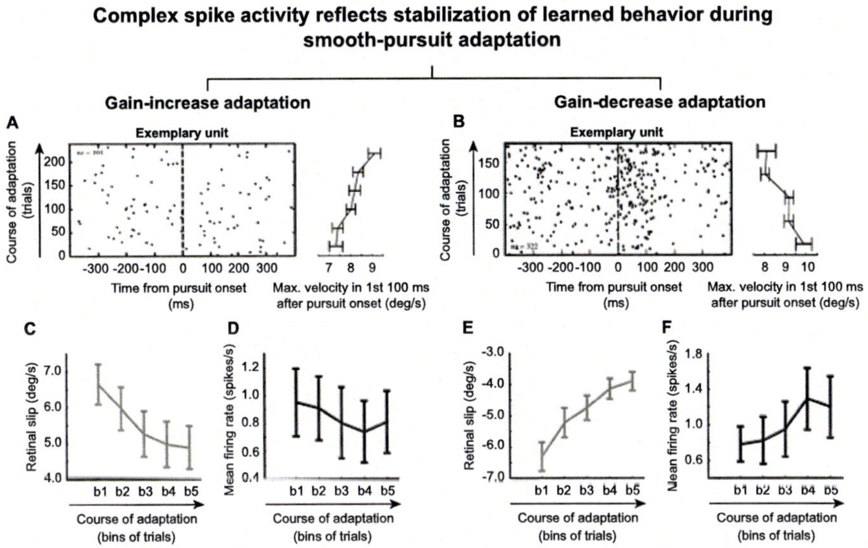

Fig. 13.5 Complex spike responses during smooth-pursuit adaptation. (**a**, **b**) Left panels show raster plots of exemplary *PC* units developing a pause (**a**) and a burst (**b**) in *CS* activity around the time of *SPEM* onset in the course of gain-increase and gain-decrease *SPA*, respectively. Vertical dashed lines represent the time of *SPEM* onset. The gradual change in the mean (± s.e.m) maximum velocity in the open-loop phase (first 100 ms after *SPEM* onset) over the course of adaptation trials is shown in the panels on the right. (**c**, **e**) Mean (± s.e.m) retinal slips as a function of bins of adaptation trials, averaged across all gain-increase ($N = 44$) and gain-decrease ($N = 51$) sessions, respectively. (**d**, **f**) Change in mean (± s.e.m) *CS* firing rate over the course of adaptation. Each adaptation session was divided into five bins containing an equal number of trials. Since the total number of adaptation trials across individual sessions varied the bin size also varied accordingly

a shift of complex spikes to the time of the conditioned stimulus. Prediction involves a subjective element, namely, the varying ability of the subject to extract the information relevant for establishing viable predictions. Individual differences between the monkeys used by Soetedjo et al. and by Catz et al. may explain that the impact of prediction was much stronger in the latter case.

13.7 Complex Spikes Predict Upcoming Errors Based on the Memory of Past Errors

In order to critically assess the relevance of predictability for the occurrence of *CSs*, Junker et al. (2018) embarked on a study of *STSA* in which both the directions of target movement, the presence and absence of error and its type, were randomized (Fig. 13.6a, b). It turned out that in the absence of predictability, *CSs* of *OMV PCs* are indeed strongly influenced by the current error exhibiting a clear directional preference for a specific error vector (Fig. 13.6c; see uppermost raster plot). But

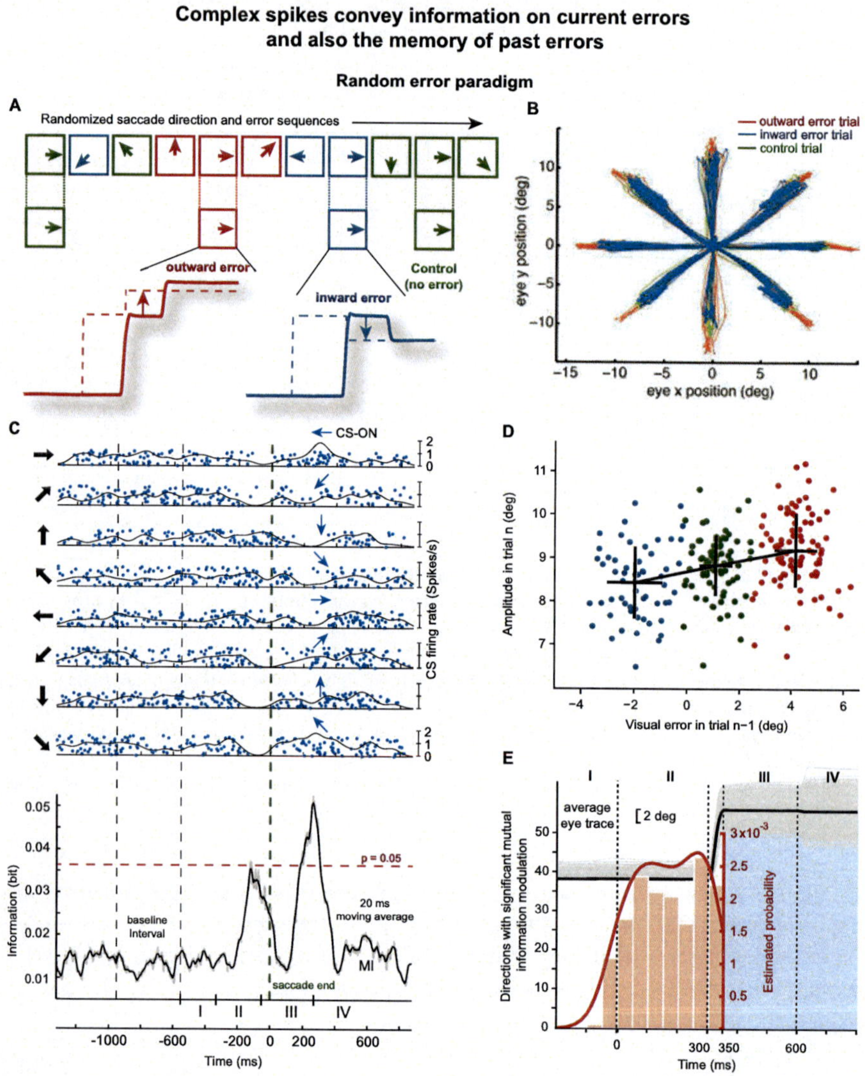

Fig. 13.6 Complex spikes carry a reverberating signal on past errors in a random-error paradigm. (**a**) Random-error paradigm. Primary saccade targets were randomly presented (shown as arrows) in one out of eight equiangular spaced locations. The color of these arrows indicates the direction of intrasaccadic target shifts. It is red, when the target jumped outward; blue, when the target jumped inward; and green, when the target did not jump. (**b**) Saccade trajectories in a 2D space made toward different target locations for an exemplary session. Colors represent the type of trial. (**c**) Upper panels show raster plots of *CSs* recorded from a representative *PC* unit, for saccades made in eight different directions (shown by arrows) followed by inward errors only. Therefore, error vectors are always opposite to the primary saccade vector (blue arrows). Trials were aligned to the primary saccade end (green vertical dashed line). Each raster plot is overlaid by a *CS* density function (solid traces) estimated using a Gaussian kernel of width 40 ms. Note the difference in post-saccadic *CS* activity for inward errors made in different saccade directions, suggesting a clear

furthermore, they also carry a memory trace of past errors. This memory trace may explain the development of more resilient expectations of future errors in case of consistency of past errors, as discussed in the preceding section.

Junker et al. resorted to a paradigm, in which primary saccade targets were randomly presented in one out of the eight possible locations, distributed at equal angular distances (Fig. 13.6a, b). During the target-directed primary saccade, the target was either randomly shifted further away resulting in an "outward error," stepped back toward the starting location causing an "inward error" or remained stationary thereby causing no retinal error as a control. The behavioral analysis revealed that even an individual error in a particular trial n was capable of significantly influencing the amplitude of the subsequent saccade in trial $n + 1$ made in the same direction (Fig. 13.6d), although trials n and $n + 1$ were separated by, on average, some six trials. Outward errors in trial n led to larger saccades in trial $n + 1$ and inward errors to smaller ones. In order to reveal task-related information in the firing of *CSs*, the authors resorted to a *MI* analysis that tried to avoid an overestimation bias by incorporating a Bayesian binning approach (Endres & Foldiak, 2005). The *MI* analysis was applied in a time resolved manner in order to identify significant information in distinct time periods around the primary saccade (Fig. 13.6c, lower panel). One of the periods considered was the post-saccadic or "secondary" retinal error period (labeled as Phase *III* in Fig. 13.6c lower panel and Fig. 13.6e), the time between the end of the primary saccade and the onset of an eventually necessary corrective

Fig. 13.6 (continued) preference for inward (leftward) error following a horizontally rightward primary saccade (topmost panel). Raw mutual information (*MI*) trace (gray line, bottom most panel) for trials in all directions was estimated by calculating *MI* in a 250 ms sliding window and then smoothed with a 20 ms running average filter (black trace). The two vertical dashed lines mark the boundaries of the baseline fixation interval, used for calculating the significant ($p = 0.05$) *MI* threshold (red dashed line). Different phases of a trial that were considered for analysis are indicated by roman letters (*I*–*IV*) at the bottom. *I*, baseline period (fixation); *II*, primary visual error period (from target onset to saccade end); *III* secondary visual error period (primary saccade end to secondary saccade onset); *IV*, post-correction (after secondary saccade end). Note that these intervals are approximate, as individual events may differ from trial to trial. Also note, that the *MI* peaks are most prominent in Phase *II* and Phase *III*. (**d**) Trial-by-trial adaptation in an exemplary session is demonstrated by a scatter plot of saccade amplitudes in trial n as a function of the type (represented by red, blue, and green for outward, inward, and no errors, respectively) and size of visual error in the previous trial $n - 1$ made in a the same (downward) saccade direction. Note that in the case of control trials (green dots), the error observed is due to the natural variability in saccade end points (motor noise). The means (± sd) of each cluster, represented by crosses, were significantly different from each other (*ANOVA*, $p < 0.01$). (**e**) The impact of the type and the size of error in the past trial on *CS* modulation in the given trial in the same direction is demonstrated. The histogram (red bars), depicting the distribution of times of maximally significant trial-by-trial modulation of *MI* of *CS* units, shows a clear accumulation of information on past errors right before the primary saccade. The trial-by-trial analysis was based on the comparison of inward vs. outward error, inward error vs. no error, and outward error vs. no error, in a normalized saccade trial. The blue-shaded region marks those intervals of the current trial in which the *MI* due to current error might have contaminated the information related to the past error. Therefore, the information related to these events cannot be safely segregated during this interval. Individual and average saccade trajectories are shown as thin gray lines and a solid black line, respectively. (Modified with permission from Junker et al. (2018))

saccade. Indeed, more than 40% of the *CS* units studied showed a significant direction dependent *MI* modulation in this period, reflecting either the direction of the past primary saccade or, alternatively, the direction of error (Fig. 13.6c, lower panel). This finding is fully compatible with the notion that the *CF* system conveys information on retinal errors, resulting from insufficient performance. However, sensitivity to retinal information is clearly not confined to the period in which the consequences of the behavior become apparent. This is indicated by the fact that more than 85% of the *CS* units tested exhibited significant *MI* in the preceding, "primary" retinal error period—defined as period between the onset of the target jump and the beginning of a primary saccade (labeled as Phase *II* in Fig. 13.6c and e). In the primary retinal error period, the eyes stay put, while the target image falls onto a peripheral retinal location and the *CS* activity provides information on the direction of the retinal vector connecting the fovea with this peripheral location, determining the metric of the upcoming primary saccade. If we assume that any *CS* signal should have behavioral relevance, we may conclude that it also influences the metric of this saccade. However, what is the need for the *CF* system to influence primary saccades? The answer is that the primary saccades may deviate from the (primary) retinal vector in order to avoid post-saccadic errors due to intrasaccadic target shifts. This is what is shown in Fig. 13.6e, where information on preceding errors modulates the *CS* activity in the period at stake. This figure comprises the result of an analysis in which Junker et al. sorted current trials *n* according to the presence of an inward error, an outward error or the absence of an error in the previous trial $n - 1$ made into the same direction (note that trial $n - 1$ and trial *n* were separated on average by approximately six trials made into other directions). They then looked into the primary retinal error period (Phase *II*) to test whether the *CS* activity in this period reflected information on past errors. To test this, they calculated the *MI* between the inward error and the no-error control group, the *MI* between the outward error and the control group, and, finally, the *MI* information between the inward error and the outward error group for each direction. Significant *MI* for at least one of the three pairs and at least one direction was obtained in 111 out of 128 *PC* units whose *CS* activity exhibited a clear accumulation of significant *MI* peaks right before the primary saccade (Fig. 13.6e, red trace). In other words, *CS* activity before the primary saccade reflects knowledge of a past error that serves as a proxy of an impending future error available to prompt an expedient adjustment of the next saccade, thereby helping to avoid a new error or at least to mitigate it. Given the random nature of the paradigm used, the intriguing fact is that a single error is sufficient to instigate a weak, albeit clear trace of a past error in the *CS* firing before the upcoming saccade, able to predict the need of a pertinent adjustment of the saccade. The more consistent the past experiences of adjustment needs are, the stronger the *CS* modulation seems to get. This is the interpretation of Catz et al.'s finding of a modulation of *CS* firing around the time of the primary saccade, growing in parallel with adaptation and the gradual disappearance of the error. Junker et al.'s study suggests that these changes of *CS* activity associated with the primary saccade influence behavior by way of changes of *SSs* fired before or around the primary saccade onset. The fact that these changes of *SS* activity are later than the memory-based modulation in *CS* firing supports the conclusion that *SSs* serve as a link between the *CS* signal and behavior. However, a word of caution may be needed. *CSs* evoked by post-saccadic errors are tightly linked to changes of *SS* (Herzfeld et al., 2018) which in turn maintain a clear temporal relationship to saccade kinematics (Catz et al., 2008; Herzfeld et al., 2015; Thier et al., 2000). With the qualification that the single-trial adaptation data studied by Junker et al. impeded a reliable characterization of comparable temporal relationships between the *CS* and the resulting behavior, one could gain the impression that the memory-related *CS* activity might be too early and also too dispersed to entail an influence on the eyes. Actually, the same consideration may also hold for the predictive *CS* responses observed in Ohmae and Medina's (2015) study of eyelid conditioning in mice. At the end of the classical eyeblink conditioning paradigm, when the association between with the conditioned stimulus (*LED* signal) and the unconditioned stimulus (air-puffs) had been established, *CSs* no longer fired in response to the delivery of air-puffs, but, rather, the response shifted to an earlier period shortly after the presentation of the *LED* stimulus that predicted the air-puff. However, the eyelid closure (conditioned response) prompted by the *LED* signal followed much later, at the time of the air-puff. How is it possible that the consequences of the predictive *CS* signal are shifted to a behaviorally relevant point in time, rather than giving rise to an immediate response based on prompt changes in *SS* activity? A possible answer may arise from the fact that the optimal interval for the interaction of the *CF* and the *PF* signals causing *LTD* of the *PF* synapse seems to be surprisingly diverse, arguably reflecting the functional needs of particular cerebellar regions (Suvrathan et al., 2016). In the case of the flocculus, it has been argued that *LTD* is induced when the error feedback-driven *CSs* are delayed by 120 ms relative to the *PF* input, a period that matches the delay of visual information. However, in lobuli *V–VI* of the cerebellar vermis of rodents that needs to integrate error feedback provided by different modalities with different delays, the optimal interval between *CS* and *PF* signals is much more diverse. While some *PCs* prefer that the *CF* input may even lead the *PF* signals for optimal *LTD* to emerge, others may require the *CF* signal to be coincident or to be delayed for up to 200 ms relative to the *PF* signals (Chen & Thompson, 1995; Safo & Regehr, 2008). Note, however, also these profoundly delayed *CSs* are able to adjust the *PC-SS* discharge and the behavior by modifying the impact of the *PF* input, earlier in time, that contributed to the occurrence of an error, thereby solving the temporal credit assignment (Sutton & Barto, 1981). Given that *CSs* in the Crus1 region of the rodent cerebellum, mediating eyelid conditioning, have been recently demonstrated to be influenced by a wide range of sensory modalities (Ju et al., 2019), we may speculate that the optimal temporal delays for synaptic plasticity to emerge might vary in this region as well. Hence, the delayed conditioned eyelid closure relative to the *LED*-triggered *CSs*, observed by Ohmae and Medina (2015) in mice, might be mediated by *PCs* preferring a *CF-PF* interval for *LTD* to become optimal, corresponding to the delay of the conditioned reflex needed.

13.8 Speculative Thoughts on the Mechanistic Basis of Predictive Complex Spikes

To address the question of the origin of the predictive *CS* signals found by Junker et al. and earlier by Catz et al., we need to remind us of the modular organization of cerebellar cortex. A zone of five to ten parasagittally aligned *PCs*, innervated by individual *CFs* (Armstrong et al., 1973; Desclin, 1974; Rosina & Provini, 1983) originating from neurons in distinct regions of the *IO*, forms elongated "cerebellar modules" (Voogd, 1969; Voogd & Ruigrok, 1997) that target specific regions in the *DCN* in a similar modular fashion. These target *DCN* neurons send *GABA*ergic projections to the same *IO* neurons that are responsible for driving the response of these *DCN* neurons via the olivo-cortico-nuclear pathway (De Zeeuw et al., 1989). We speculate that changes in the *SS* population discharge, induced by the performance error-related *CS* responses, are transmitted back to the *IO* neurons via the nucleo-olivary projection, thereby inducing a change in the state of *IO* neurons. The arrival of saccade initiating extra-cerebellar inputs may transform this altered state of *IO* neurons into changes in *CF* activity, seen as a signal reverberating a past error acting as predictor of a new error.

13.9 Multiplexing of Information by Complex Spikes

The findings of Junker et al. (2018) clearly demonstrate that *CFs* that encode information on performance errors can shift their responses to another point in time to serve the purpose of predicting future errors based on a memory of past errors. Both error-related *CSs* and predictive *CSs* serve motor learning. Are these two types of signals the only ones relevant for cerebellar learning, or are there further sources of useful information that could be tapped, and if yes, could it be that they are also accommodated by the *CF* system? Moreover, is it conceivable that this system may flexibly select the signals particularly useful in a given situation?

As a matter of fact, a substantial degree of flexibility of the *CF* system is suggested by the aforementioned study of eyeblink conditioning in mice (Ohmae & Medina, 2015). Here, the same *CFs* that responded to the delivery of the unexpected air-puffs (unconditioned stimulus) earlier during the experiment are able to flexibly respond to a conditioned stimulus of completely different modality (either a visual *LED* signal or an auditory tone) well before the occurrence of the unconditioned stimulus, once the association between the conditioned and unconditioned stimulus is established. This is possible because individual *CFs* seem to have access to information from a variety of sensory sources. This is the conclusion of a recent study by Ju et al. (2019) in which two-photon calcium imaging was used to record *CS* activity from the crus1 region in awake mice. The authors demonstrated that *CF* activity could be evoked by a variety of tactile (such as whiskers, lips, and cheek), auditory, and visual stimuli. Access to a wider range of sources of information is also indicated by our recent work on the optimization of saccades (Markanday et al., 2020b), in which we measured the influence of changes in saccade kinematics on the *CS* firing. Changes in saccade kinematics were a consequence of running a relatively simple paradigm, in which monkeys were asked to carry out stereotypic to and fro saccades along the horizontal axis at high frequency while recording *CSs* from the *OMV PCs*. Due to the boring and exhausting nature of the paradigm, the monkeys lose interest over the course of trials associated with a gradual decline in their drive to generate saccades. This "cognitive fatigue" (Prsa et al., 2010), characterized by the progressive decline of saccade velocities, is compensated by the healthy cerebellum by means of upregulating their duration, thereby maintaining the saccade amplitudes within an acceptable range of error (Bahill et al., 1975; BroŽek, 1949; Chen-Harris et al., 2008; Fuchs & Binder, 1983; Straube et al., 1997; Xu-Wilson et al., 2009). Any damage to the cerebellum disrupts this fine velocity duration trade-off mechanism (Markanday et al., 2018) causing an increased scattering of movement end points around the target and, consecutively, a broadening of the distribution of retinal errors. Unlike the intrasaccadic target jumps in standard *STSA* paradigms, that necessitate corrective eye movements, it is the natural variability of saccade end points observed during this fatigue-inducing repetitive saccade paradigm that is responsible for the occurrence of errors, often corrected by secondary saccades. This relatively simple and arguably also more natural paradigm has allowed us to demonstrate the impact of a larger number of behavioral factors influencing the firing of *CSs* than any of the more sophisticated paradigms. This is in support of the aforementioned notion that the *CF* system integrates various streams of information relevant for the optimization of behavior. To start with, the results of this study undeniably confirm the classical view that individual *CSs* encode error-related information with directional specificity. However, the same *CFs* also convey information on the metrics of both primary (macro) and the secondary corrective saccades. More specifically, the saccade-related *CS* firing rate turned out to reflect amplitude in a monotonic format, including the amplitude of the smallest saccades—microsaccades, i.e. saccades of less than 1 deg amplitude, regularly deployed to correct a deficiency of the preceding primary saccade. The same pool of *PCs* also exhibited a transient increase of their instantaneous *CS* firing rate precisely aligned with trial onset. A sharp increase in *CS* firing at this point might be interpreted as a predictor of the beginning of a fresh sequence of well-known events. Finally, it is not only the instantaneous firing rate of *CSs* that was influenced by the information on retinal error, saccade amplitude, and trial onset but also the duration of individual *CSs*. The notion that not only the *CS* firing rate but also the duration, determined by the shape of a *CS*, might be informative seems well justified as they might reflect the state of the *IO* neurons influencing the amount of plasticity at *PC* synapses (Bazzigaluppi et al., 2012; Mathy et al., 2009; Rasmussen et al., 2013; Yang & Lisberger, 2014, 2017). Yet, owing to the challenge of manually labeling start and end points of *CSs*, attempts to test the relevance of *CS* duration for behavior have been few. One such effort was made by Yang and Lisberger (2014, 2017), who were able to demonstrate that changes in *CS* duration were correlated with the degree of motor learning. On the other hand, attempts to establish an analogous relationship between *CS* duration and *STSA* have failed (Herzfeld et al., 2018; Junker et al., 2018). With the availability of a deep neural network tool substantially alleviating the tedious quantification of *CS* timing and morphology without sacrificing reliability (Markanday et al., 2020a), we could now ask if the variables shown to influence the occurrence of *CSs* would also influence *CS* duration. To put it into a nutshell, any informative increase in discharge rates was paralleled by increases in *CS* durations. The bottom line is that *CSs* reflect the impact of a rich spectrum of behaviorally relevant streams of information. These individual streams all affect *CS* firing, albeit at different times. In other words, the *CF* might work like a telecommunication multiplexer that combines various signals for efficient transport. However, the technical system also involves a demultiplexer element that allows the retrieval of a specific informational component from the merged signal stream. But how could the *CF* signal be demultiplexed in order to allow specific chunks of information to have specific effects? An answer might be provided by the preceding discussion of diverse preferred time delays for the interaction between *CF* and *PF* spikes. *PCs* may not differ with respect to the *CF* input, which in every case may reflect the merging of different streams of information. However, they may differ with respect to their preferred interval for spike time-dependent plasticity, the time between *CF* and *PF* activity for optimal *LTD* to emerge. In this way an individual *PC* could be tuned to error, another to saccade metric, and so on. But how could different streams of information be merged in the first place? Given the fact that the adult *PC* receives input from one and not several *CFs* (Eccles, 1967), clearly the merging cannot be a consequence of the convergence of several *CFs*, each mediating a different signal at the level of the *PC*. Hence, the merging of different signals must occur earlier, already at the level of the inferior olive, where an exchange of information between neurons could be accommodated by opening intercellular gap junctions (De Zeeuw et al., 1989; Lefler et al., 2020; Leznik & Llinas, 2005; Llinas et al., 1974; Sotelo et al., 1974). Albeit limited, there is evidence that distinct regions in the *IO* may receive different types of information (Sugihara & Shinoda, 2004). Furthermore, a very recent study (Lefler et al., 2020) using a combination of electrophysiological and computational techniques to determine the functional architecture of the *IO* network suggested that rather than being randomly connected, these neurons appear as distinct clusters. Each of these clusters may form a close network of up to 20 electrically coupled neurons. Hence, we speculate that a putative gap junction-mediated cross-exchange of different streams of information, arriving in distinct neighboring clusters, may allow individual *IO* neurons and their *CF* axons to multiplex information. As it is well-established that *IO* gap conjunctions are subject to control by cerebellar feedback mediated by a *GABA*ergic nucleo-olivary projection (De Zeeuw et al., 1989), one may even speculate that cerebellar cortex is able to determine which signals might be optimally integrated at the level of the *IO* in order to drive motor learning.

13.10 Conclusion: A Unifying Framework for the Role of Complex Spikes

Although originally meant to capture the intricate morphology of an intriguing type of action potential, the *CS*, the attribute "complex" selected for these olivocerebellar signals seems to be even more appropriate when considering the baffling diversity of *CS* patterns emphasized across different studies. In fact, the diversity of results and associated interpretations recently prompted Streng et al. (2018) to metaphorically speak about "complex spike wars" in an attempt to characterize the controversial discussion in the cerebellar community. Yet, this "war" has no champion. The seemingly conflicting findings are in fact pieces of one and the same puzzle. We mentioned a few of these puzzle pieces that have emerged from work on the cerebellar control of goal-directed eye movements and a brief detour into the work on eyeblink conditioning. Other puzzle pieces, which we could not touch upon in the interest of clarity and brevity, have been unraveled by work on the cerebellar control of hand movements (Ebner et al., 1983; Streng et al., 2018). It is much too early to know if all puzzle pieces, the ones known and the others still to be discovered, will in the end fit together. Yet, we are confident that moving forward in our quest for an understanding of the role of the *CF* system will profoundly benefit from relinquishing the standard "either-or" thinking and the willingness to open up for integrative concepts.

References

Albus, J. S. (1971). A theory of cerebellar function. *Mathematical Biosciences, 10*(1–2), 25–61.

Armstrong, D., Harvey, R., & Schild, R. F. (1973). The spatial organisation of climbing fibre branching in the cat cerebellum. *Experimental Brain Research, 18*(1), 40–58.

Bahill, A. T., Clark, M. R., & Stark, L. (1975). The main sequence, a tool for studying human eye movements. *Mathematical Biosciences, 24*(3–4), 191–204.

Barash, S., Melikyan, A., Sivakov, A., Zhang, M., Glickstein, M., & Thier, P. (1999). Saccadic dysmetria and adaptation after lesions of the cerebellar cortex. *Journal of Neuroscience, 19*(24), 10931–10939.

Bazzigaluppi, P., Ruigrok, T., Saisan, P., De Zeeuw, C. I., & De Jeu, M. (2012). Properties of the nucleo-olivary pathway: An in vivo whole-cell patch clamp study. *PLoS One, 7*(9), e46360.

BroŽek, J. (1949). Quantitative criteria of oculomotor performance and fatigue. *Journal of Applied Physiology, 2*(5), 247–260.

Catz, N., Dicke, P. W., & Thier, P. (2005). Cerebellar complex spike firing is suitable to induce as well as to stabilize motor learning. *Current Biology, 15*(24), 2179–2189.

Catz, N., Dicke, P. W., & Thier, P. (2008). Cerebellar-dependent motor learning is based on pruning a Purkinje cell population response. *Proceedings of the National Academy of Sciences, 105*(20), 7309–7314.

Chen, C., & Thompson, R. F. (1995). Temporal specificity of long-term depression in parallel fiber – Purkinje synapses in rat cerebellar slice. *Learning & Memory, 2*(3–4), 185–198.

Chen-Harris, H., Joiner, W. M., Ethier, V., Zee, D. S., & Shadmehr, R. (2008). Adaptive control of saccades via internal feedback. *Journal of Neuroscience, 28*(11), 2804–2813.

Coltz, J., Johnson, M., & Ebner, T. (1999). Cerebellar Purkinje cell simple spike discharge encodes movement velocity in primates during visuomotor arm tracking. *Journal of Neuroscience, 19*(5), 1782–1803.

Dash, S., Catz, N., Dicke, P. W., & Thier, P. (2010). Specific vermal complex spike responses build up during the course of smooth-pursuit adaptation, paralleling the decrease of performance error. *Experimental Brain Research, 205*(1), 41–55.

Dash, S., Catz, N., Dicke, P. W., & Thier, P. (2011). Encoding of smooth-pursuit eye movement initiation by a population of vermal Purkinje cells. *Cerebral Cortex, 22*(4), 877–891. https://doi.org/10.1093/cercor/bhr153

Dash, S., Dicke, P. W., & Thier, P. (2013). A vermal Purkinje cell simple spike population response encodes the changes in eye movement kinematics due to smooth pursuit adaptation. *Frontiers in Systems Neuroscience, 7*, 3.

Davie, J. T., Clark, B. A., & Häusser, M. (2008). The origin of the complex spike in cerebellar Purkinje cells. *Journal of Neuroscience, 28*(30), 7599–7609.

De Zeeuw, C., Holstege, J., Ruigrok, T., & Voogd, J. (1989). Ultrastructural study of the GABAergic, cerebellar, and mesodiencephalic innervation of the cat medial accessory olive: Anterograde tracing combined with immunocytochemistry. *Journal of Comparative Neurology, 284*(1), 12–35.

Desclin, J. C. (1974). Histological evidence supporting the inferior olive as the major source of cerebellar climbing fibers in the rat. *Brain Research, 77*(3), 365–384.

Ebner, T. J., Hewitt, A. L., & Popa, L. S. (2011). What features of limb movements are encoded in the discharge of cerebellar neurons? *The Cerebellum, 10*(4), 683–693.

Ebner, T. J., Yu, Q.-X., & Bloedel, J. R. (1983). Increase in Purkinje cell gain associated with naturally activated climbing fiber input. *Journal of Neurophysiology, 50*(1), 205–219.

Eccles, J. C. (1967). Circuits in the cerebellar control of movement. *Proceedings of the National Academy of Sciences of the United States of America, 58*(1), 336.

Eccles, J. C., Llinás, R., & Sasaki, K. (1966). The action of antidromic impulses on the cerebellar Purkinje cells. *The Journal of Physiology, 182*(2), 316–345. https://doi.org/10.1113/jphysiol.1966.sp007826

Endres, D., & Foldiak, P. (2005). Bayesian bin distribution inference and mutual information. *IEEE Transactions on Information Theory, 51*(11), 3766–3779.

Fortier, P. A., Kalaska, J. F., & Smith, A. M. (1989). Cerebellar neuronal activity related to whole-arm reaching movements in the monkey. *Journal of Neurophysiology, 62*(1), 198–211.

Fortier, P. A., Smith, A. M., & Kalaska, J. F. (1993). Comparison of cerebellar and motor cortex activity during reaching: Directional tuning and response variability. *Journal of Neurophysiology, 69*(4), 1136–1149.

Frysinger, R. C., Bourbonnais, D., Kalaska, J. F., & Smith, A. M. (1984). Cerebellar cortical activity during antagonist cocontraction and reciprocal inhibition of forearm muscles. *Journal of Neurophysiology, 51*(1), 32–49. https://doi.org/10.1152/jn.1984.51.1.32

Fu, Q.-G., Flament, D., Coltz, J., & Ebner, T. (1997). Relationship of cerebellar Purkinje cell simple spike discharge to movement kinematics in the monkey. *Journal of Neurophysiology, 78*(1), 478–491.

Fuchs, A. F., & Binder, M. D. (1983). Fatigue resistance of human extraocular muscles. *Journal of Neurophysiology, 49*(1), 28–34.

Fujikado, T., & Noda, H. (1987). Saccadic eye movements evoked by microstimulation of lobule VII of the cerebellar vermis of macaque monkeys. *The Journal of Physiology, 394*(1), 573–594.

Fujita, Y. (1968). Activity of dendrites of single Purkinje cells and its relationship to so-called inactivation response in rabbit cerebellum. *Journal of Neurophysiology, 31*(2), 131–141.

Fukushima, K., Tanaka, M., Suzuki, Y., Fukushima, J., & Yoshida, T. (1996). Adaptive changes in human smooth pursuit eye movement. *Neuroscience Research, 25*(4), 391–398.

Herzfeld, D. J., Kojima, Y., Soetedjo, R., & Shadmehr, R. (2015). Encoding of action by the Purkinje cells of the cerebellum. *Nature, 526*(7573), 439–442.

Herzfeld, D. J., Kojima, Y., Soetedjo, R., & Shadmehr, R. (2018). Encoding of error and learning to correct that error by the Purkinje cells of the cerebellum. *Nature Neuroscience, 21*(5), 736–743.
Hewitt, A. L., Popa, L. S., Pasalar, S., Hendrix, C. M., & Ebner, T. J. (2011). Representation of limb kinematics in Purkinje cell simple spike discharge is conserved across multiple tasks. *Journal of Neurophysiology, 106*(5), 2232–2247.
Hillman, D. E. (1969). Light and electron microscopical study of the relationships between the cerebellum and the vestibular organ of the frog. *Experimental Brain Research, 9*(1), 1–15.
Ignashchenkova, A., Dash, S., Dicke, P. W., Haarmeier, T., Glickstein, M., & Thier, P. (2009). Normal spatial attention but impaired saccades and visual motion perception after lesions of the monkey cerebellum. *Journal of Neurophysiology, 102*(6), 3156–3168.
Ito, M. (1972). Neural design of the cerebellar motor control system. *Brain Research, 40*(1), 81–84.
Ito, M. (2001). Cerebellar long-term depression: Characterization, signal transduction, and functional roles. *Physiological Reviews, 81*(3), 1143–1195.
Ito, M., Nisimaru, N., & Yamamoto, M. (1977). Specific patterns of neuronal connexions involved in the control of the rabbit's vestibulo-ocular reflexes by the cerebellar flocculus. *The Journal of Physiology, 265*(3), 833–854.
Ito, M., Shiida, T., Yagi, N., & Yamamoto, M. (1974). Visual influence on rabbit horizontal vestibulo-ocular reflex presumably effected via the cerebellar flocculus. *Brain Research, 65*(1), 170–174.
Ju, C., Bosman, L. W., Hoogland, T. M., Velauthapillai, A., Murugesan, P., Warnaar, P., van Genderen, R. M., Negrello, M., & De Zeeuw, C. I. (2019). Neurons of the inferior olive respond to broad classes of sensory input while subject to homeostatic control. *The Journal of Physiology, 597*(9), 2483–2514.
Junker, M., Endres, D., Sun, Z. P., Dicke, P. W., Giese, M., & Thier, P. (2018). Learning from the past: A reverberation of past errors in the cerebellar climbing fiber signal. *PLoS Biology, 16*(8), e2004344.
Kahlon, M., & Lisberger, S. G. (1996). Coordinate system for learning in the smooth pursuit eye movements of monkeys. *Journal of Neuroscience, 16*(22), 7270–7283.
Lamarre, Y., De Montigny, C., Dumont, M., & Weiss, M. (1971). Harmaline-induced rhythmic activity of cerebellar and lower brain stem neurons. *Brain Research, 32*(1), 246–250.
Lamarre, Y., & Mercier, L.-A. (1971). Neurophysiological studies of harmaline-induced tremor in the cat. *Canadian Journal of Physiology and Pharmacology, 49*(12), 1049–1058.
Latour, P. (1962). Visual threshold during eye movements. *Vision Research, 2*(3), 261–262.
Lefler, Y., Amsalem, O., Vrieler, N., Segev, I., & Yarom, Y. (2020). Using subthreshold events to characterize the functional architecture of the electrically coupled inferior olive network. *eLife, 9*, e43560.
Leznik, E., & Llinas, R. (2005). Role of gap junctions in synchronized neuronal oscillations in the inferior olive. *Journal of Neurophysiology, 94*(4), 2447–2456.
Llinas, R. (1974). Motor aspects of cerebellar control. *Physiologist, 17*, 19.
Llinas, R., Baker, R., & Sotelo, C. (1974). Electrotonic coupling between neurons in cat inferior olive. *Journal of Neurophysiology, 37*(3), 560–571.
Llinás, R., & Sugimori, M. (1980a). Electrophysiological properties of in vitro Purkinje cell dendrites in mammalian cerebellar slices. *The Journal of Physiology, 305*(1), 197–213.
Llinás, R., & Sugimori, M. (1980b). Electrophysiological properties of in vitro Purkinje cell somata in mammalian cerebellar slices. *The Journal of Physiology, 305*(1), 171–195.
Llinás, R., & Volkind, R. (1973). The olivo-cerebellar system: Functional properties as revealed by harmaline-induced tremor. *Experimental Brain Research, 18*(1), 69–87.
Maekawa, K., & Simpson, J. (1973). Climbing fiber responses evoked in vestibulocerebellum of rabbit from visual system. *Journal of Neurophysiology, 36*(4), 649–666.
Markanday, A., Bellet, J., Bellet, M. E., Inoue, J., Hafed, Z. M., & Thier, P. (2020a). Using deep neural networks to detect complex spikes of cerebellar Purkinje cells. *Journal of Neurophysiology, 123*(6), 2217–2234.

Markanday, A., Inoue, J., Dicke, P. W., & Thier, P. (2020b). Multiplexing of information by cerebellar complex spikes. Under submission.

Markanday, A., Messner, J., & Thier, P. (2018). A loss of a velocity-duration trade-off impairs movement precision in patients with cerebellar degeneration. *European Journal of Neuroscience, 48*(4), 1976–1989.

Marr, D. (1969). A theory of cerebellar cortex. *The Journal of Physiology, 202*(2), 437–470. https://doi.org/10.1113/jphysiol.1969.sp008820

Mathy, A., Ho, S. S., Davie, J. T., Duguid, I. C., Clark, B. A., & Häusser, M. (2009). Encoding of oscillations by axonal bursts in inferior olive neurons. *Neuron, 62*(3), 388–399.

McLaughlin, S. C. (1967). Parametric adjustment in saccadic eye movements. *Perception & Psychophysics, 2*(8), 359–362.

Noda, H., & Fujikado, T. (1987a). Involvement of Purkinje cells in evoking saccadic eye movements by microstimulation of the posterior cerebellar vermis of monkeys. *Journal of Neurophysiology, 57*(5), 1247–1261.

Noda, H., & Fujikado, T. (1987b). Topography of the oculomotor area of the cerebellar vermis in macaques as determined by microstimulation. *Journal of Neurophysiology, 58*(2), 359–378.

Noda, H., & Mikami, A. (1986). Discharges of neurons in the dorsal paraflocculus of monkeys during eye movements and visual stimulation. *Journal of Neurophysiology, 56*(4), 1129–1146.

Ohki, M., Kitazawa, H., Hiramatsu, T., Kaga, K., Kitamura, T., Yamada, J., & Nagao, S. (2009). Role of primate cerebellar hemisphere in voluntary eye movement control revealed by lesion effects. *Journal of Neurophysiology, 101*(2), 934–947.

Ohmae, S., & Medina, J. F. (2015). Climbing fibers encode a temporal-difference prediction error during cerebellar learning in mice. *Nature Neuroscience, 18*(12), 1798–1803.

Ohtsuka, K., & Enoki, T. (1998). Transcranial magnetic stimulation over the posterior cerebellum during smooth pursuit eye movements in man. *Brain: A Journal of Neurology, 121*(3), 429–435.

Person, A. L., & Raman, I. M. (2012). Purkinje neuron synchrony elicits time-locked spiking in the cerebellar nuclei. *Nature, 481*(7382), 502–505.

Prsa, M., Dicke, P. W., & Thier, P. (2010). The absence of eye muscle fatigue indicates that the nervous system compensates for non-motor disturbances of oculomotor function. *Journal of Neuroscience, 30*(47), 15834–15842.

Rasmussen, A., Jirenhed, D.-A., Zucca, R., Johansson, F., Svensson, P., & Hesslow, G. (2013). Number of spikes in climbing fibers determines the direction of cerebellar learning. *Journal of Neuroscience, 33*(33), 13436–13440.

Roitman, A. V., Pasalar, S., Johnson, M. T., & Ebner, T. J. (2005). Position, direction of movement, and speed tuning of cerebellar Purkinje cells during circular manual tracking in monkey. *Journal of Neuroscience, 25*(40), 9244–9257.

Rosina, A., & Provini, L. (1983). Somatotopy of climbing fiber branching to the cerebellar cortex in cat. *Brain Research, 289*(1–2), 45–63.

Safo, P., & Regehr, W. G. (2008). Timing dependence of the induction of cerebellar LTD. *Neuropharmacology, 54*(1), 213–218.

Sato, H., & Noda, H. (1992). Posterior vermal Purkinje cells in macaques responding during saccades, smooth pursuit, chair rotation and/or optokinetic stimulation. *Neuroscience Research, 12*(5), 583–595.

Soetedjo, R., & Fuchs, A. F. (2006). Complex spike activity of Purkinje cells in the oculomotor vermis during behavioral adaptation of monkey saccades. *Journal of Neuroscience, 26*(29), 7741–7755.

Soetedjo, R., Kojima, Y., & Fuchs, A. (2008a). Complex spike activity signals the direction and size of dysmetric saccade errors. In *Progress in brain research* (Vol. 171, pp. 153–159). Elsevier.

Soetedjo, R., Kojima, Y., & Fuchs, A. F. (2008b). Complex spike activity in the oculomotor vermis of the cerebellum: A vectorial error signal for saccade motor learning? *Journal of Neurophysiology, 100*(4), 1949–1966.

Sotelo, C., Llinás, R., & Baker, R. (1974). Structural study of inferior olivary nucleus of the cat: Morphological correlates of electrotonic coupling. *Journal of Neurophysiology, 37*(3), 541–559.

Squire, L. R. (1999). Masao Ito. In L. R. Squire (Ed.), *The history of neuroscience in autobiography* (Vol. 2, pp. 168–190). Academic Press. https://doi.org/10.1016/S1874-6055(99)80008-5

Stone, L., & Lisberger, S. (1990a). Visual responses of Purkinje cells in the cerebellar flocculus during smooth-pursuit eye movements in monkeys. I. Simple spikes. *Journal of Neurophysiology, 63*(5), 1241–1261.

Stone, L., & Lisberger, S. (1990b). Visual responses of Purkinje cells in the cerebellar flocculus during smooth-pursuit eye movements in monkeys. II. Complex spikes. *Journal of Neurophysiology, 63*(5), 1262–1275.

Straube, A., Fuchs, A. F., Usher, S., & Robinson, F. R. (1997). Characteristics of saccadic gain adaptation in rhesus macaques. *Journal of Neurophysiology, 77*(2), 874–895.

Streng, M. L., Popa, L. S., & Ebner, T. J. (2018). Complex spike wars: A new hope. *The Cerebellum, 17*(6), 735–746.

Stuart, G., & Häusser, M. (1994). Initiation and spread of sodium action potentials in cerebellar Purkinje cells. *Neuron, 13*(3), 703–712.

Sugihara, I., & Shinoda, Y. (2004). Molecular, topographic, and functional organization of the cerebellar cortex: A study with combined aldolase C and olivocerebellar labeling. *Journal of Neuroscience, 24*(40), 8771–8785.

Sun, Z., Smilgin, A., Junker, M., Dicke, P. W., & Thier, P. (2017). The same oculomotor vermal Purkinje cells encode the different kinematics of saccades and of smooth pursuit eye movements. *Scientific Reports, 7*(1), 1–12.

Sutton, R. S., & Barto, A. G. (1981). Toward a modern theory of adaptive networks: Expectation and prediction. *Psychological Review, 88*(2), 135.

Suvrathan, A., Payne, H. L., & Raymond, J. L. (2016). Timing rules for synaptic plasticity matched to behavioral function. *Neuron, 92*(5), 959–967.

Suzuki, D. A., & Keller, E. L. (1988a). The role of the posterior vermis of monkey cerebellum in smooth-pursuit eye movement control. I. Eye and head movement-related activity. *Journal of Neurophysiology, 59*(1), 1–18.

Suzuki, D. A., & Keller, E. L. (1988b). The role of the posterior vermis of monkey cerebellum in smooth-pursuit eye movement control. II. Target velocity-related Purkinje cell activity. *Journal of Neurophysiology, 59*(1), 19–40.

Takagi, M., Zee, D. S., & Tamargo, R. J. (1998). Effects of lesions of the oculomotor vermis on eye movements in primate: Saccades. *Journal of Neurophysiology, 80*(4), 1911–1931.

Takagi, M., Zee, D. S., & Tamargo, R. J. (2000). Effects of lesions of the oculomotor cerebellar vermis on eye movements in primate: Smooth pursuit. *Journal of Neurophysiology, 83*(4), 2047–2062.

Thach, W. (1968). Discharge of Purkinje and cerebellar nuclear neurons during rapidly alternating arm movements in the monkey. *Journal of Neurophysiology, 31*(5), 785–797.

Thach, W. (1970). Discharge of cerebellar neurons related to two maintained postures and two prompt movements. II. Purkinje cell output and input. *Journal of Neurophysiology, 33*(4), 537–547.

Thier, P. (2011). The oculomotor cerebellum. In *The Oxford handbook of eye movements* (pp. 173–193). Oxford University Press.

Thier, P., Dicke, P. W., Haas, R., & Barash, S. (2000). Encoding of movement time by populations of cerebellar Purkinje cells. *Nature, 405*(6782), 72–76.

Thier, P., & Markanday, A. (2019). Role of the vermal cerebellum in visually guided eye movements and visual motion perception. *Annual Review of Vision Science, 5*, 247–268.

Vahedi, K., Rivaud, S., Amarenco, P., & Pierrot-Deseilligny, C. (1995). Horizontal eye movement disorders after posterior vermis infarctions. *Journal of Neurology, Neurosurgery & Psychiatry, 58*(1), 91–94.

Villablanca, J., & Riobo, F. (1970). Electroencephalographic and behavioral effects of harmaline in intact cats and in cats with chronic mesencephalic transection. *Psychopharmacologia, 17*(4), 302–313.

Volkmann, F. C. (1962). Vision during voluntary saccadic eye movements. *JOSA, 52*(5), 571–578.

Voogd, J. (1969). The importance of fiber connections in the comparative anatomy of the mammalian cerebellum. In *Neurology of cerebellar evolution and development* (pp. 493–514).

Voogd, J., & Ruigrok, T. (1997). Transverse and longitudinal patterns in the mammalian cerebellum. In *Progress in brain research* (Vol. 114, pp. 21–37). Elsevier.

Xu-Wilson, M., Zee, D. S., & Shadmehr, R. (2009). The intrinsic value of visual information affects saccade velocities. *Experimental Brain Research, 196*(4), 475–481.

Yamamoto, K., Kawato, M., Kotosaka, S., & Kitazawa, S. (2007). Encoding of movement dynamics by Purkinje cell simple spike activity during fast arm movements under resistive and assistive force fields. *Journal of Neurophysiology, 97*(2), 1588–1599. https://doi.org/10.1152/jn.00206.2006

Yang, Y., & Lisberger, S. G. (2014). Purkinje-cell plasticity and cerebellar motor learning are graded by complex-spike duration. *Nature, 510*(7506), 529–532.

Yang, Y., & Lisberger, S. G. (2017). Modulation of Complex-Spike duration and probability during cerebellar motor learning in visually guided Smooth-Pursuit eye movements of monkeys. *eNeuro, 4*(3) ENEURO.0115-17.2017.

Chapter 14
Role of the Cerebellum in the Acquisition and Consolidation of Memory of Motor Learning

Soichi Nagao

Abbreviations

AMPAR	α-amino-3-hydrooxy-5-methyl-4-isoxazolone propionate receptor
CF	Climbing fiber
CN	Cerebellar nuclei
CS	Conditioned stimuli
EM	Electron microscopy
EMN	Extraocular muscle motor nuclei
FL	Flocculus
HOKR	Horizontal optokinetic response
HVOR	Horizontal vestibulo-ocular reflex
IO	Inferior olive
KI	Knockin
KO	Knockout
LTD	Long-term depression
LTP	Long-term potentiation
MF	Mossy fiber
NO	Nitric monoxide
PC	Purkinje cell
PF	Parallel fiber
US	Unconditioned stimuli
VN	Vestibular nucleus

S. Nagao (✉)
Laboratory for Integrative Brain Function, Nozomi Hospital, Ina, Saitama, Japan

Laboratory for Memory Neuroscience, Tokyo Metropolotan Institute for Gerontology, Itabashi, Tokyo, Japan
e-mail: nagao-so@umin.ac.jp

H. Mizusawa, S. Kakei (eds.), *Cerebellum as a CNS Hub*,
Contemporary Clinical Neuroscience, https://doi.org/10.1007/978-3-030-75817-2_14

14.1 Introduction to Cerebellar Learning Hypothesis

Masao Ito (1928–2018) was a preeminent neuroscientist in the study of the cerebellum (Nagao, 2019; Nagao & Kano, 2019; Nagao et al., 2021). Half a century ago, he and his colleagues discovered that Purkinje cells (PCs), the sole output neurons in the cerebellum, are inhibitory (Ito & Yoshida, 1964) and use γ-aminobutyric acid (GABA) for a neurotransmitter (Obata et al., 1967). In the 1960s, the outline of cerebellar neural network was clarified by electrophysiological and anatomical studies, and how the cerebellum operates in the brain became the subject of research (Eccles et al., 1967).

Around 1970, the two computational neuroscientists inspired by a perceptron neural network model (Rosenblatt, 1958), Marr (1969) in the UK and Albus (1971) in the USA, published their cerebellar learning models by noting the unique anatomical connection of PCs. Two inputs of quite different characteristics converge onto PCs (Fig. 14.1). A large number of parallel fibers (PFs), the axons of granular cells (GRs), convey the signals gathered from wide brain areas via mossy fibers (MFs) and make synapses on distal dendrites of each PC. In contrast, only one or two climbing fibers (CFs), the axons of inferior olive (IO) neurons, make synapses on proximal dendrites of each PC. Both Marr and Albus modeled that PCs learn by modifying the transmission efficacy of PF–PC synapses using the error (teacher) signals fed back via CFs.

14.2 Ocular Reflex Adaptation as an Experimental Paradigm for Motor Learning

To experimentally test their cerebellar learning models, Ito and his colleagues used the paradigm of adaptations of horizontal vestibulo-ocular (HVOR) and optokinetic reflex (HOKR) eye movements (Ito, 1970, 1972, 1974, 1982 and 1984). HVOR is a compensatory eye movement to stabilize the visual image during head movement and tested by the sinusoidal oscillation of the turntable, on which a head-fixed animal is mounted, in the dark. HOKR is a reflex eye movement evoked by the motion of the visual field and tested by the sinusoidal oscillation of a check- or stripe-patterned screen around a head-fixed rabbit or mouse in the light (e.g., Ito & Nagao, 1991). The eye movements evoked by HVOR and HOKR are measured by a TV camera (Batini et al., 1979; Nagao, 1990) or a sclera search-coil system (Fuchs & Robinson, 1966). HVOR and HOKR are quantified from gains determined by comparing the magnitude of the evoked eye movements with that of the turntable or screen movements.

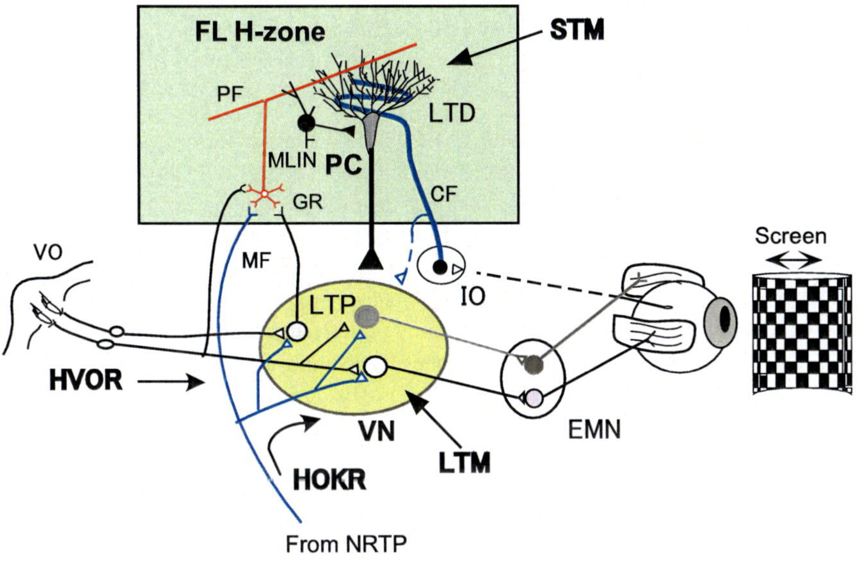

Fig. 14.1 Neural circuitries of the horizontal vestibulo-ocular reflex (HVOR) and optokinetic response (HOKR) eye movements and the sites of memory of adaptations. Note that HVOR and HOKR share the neural circuitry of vestibular nuclei (VN), extraocular motor neurons (EMN), and cerebellar flocculus (FL). H-zone Purkinje cells (PCs) also receive retinal slip signals via climbing fibers (CFs) and directly inhibit the VN neurons that relay HVOR and HOKR. The memory of HOKR gain-up adaptation is initially formed at H-zone parallel fiber (PF)–PC synapses by long-term depression (LTD) and is transferred to vestibular nucleus (VN) neuron synapses by long-term potentiation (LTP) after repetition of short-term adaptation. (Modified from Shutoh et al. (2006)). AOT accessory optic tract, CF climbing fiber, GR granular cell, IO inferior olive, LTM memory of long-term adaptation, MF mossy fiber, MLIN molecular layer inhibitory interneuron, NRTP nucleus reticularis tegmenti pontis, PF parallel fiber, PC Purkinje cell, STM memory of short-term adaptation, VO vestibular organ, VN vestibular nuclei)

Interestingly, training that evokes blurring of vision (retinal slip), an error for HVOR and HOKR, induces adaptation in their gains (Gonshor & Melvill-Jones, 1974; Ito et al., 1974, 1979; Nagao, 1983). For example, when an animal is trained to chronically wear magnifying lenses, which induces a condition equivalent to oscillating the animal and its surrounding dot- or stripe-patterned screen in the out-of-phase direction for hours, HVOR gain increases. When the retinal slip is continuously induced in the head-fixed rabbit (Collewijn & Grootendorst, 1978) and mouse (Katoh et al., 1998) by exposing them to continuous fast screen oscillation for more than 1 h, HOKR gain increases. Adaptations of HVOR and HOKR, as well as the eyeblink conditioning, have been used as an experimental paradigm for motor learning.

14.3 Role of the Cerebellar Flocculus in Ocular Reflex Adaptation

As shown in Fig. 14.1, HVOR and HOKR are driven by not only the vestibular nucleus (VN) and extraocular muscle motor nuclei (EMN) but also the cerebellar flocculus (FL). The H-zone is a microzone in the middle FL (e.g., Ito, 1984; Sugihara & Shinoda, 2004), which is specifically involved in HVOR and HOKR control. H-zone is identified by the effects of microstimulation that induce a slow abduction in the ipsilateral eye (Dufossé et al., 1977; Nagao et al., 1985). H-zone receive vestibular inputs from vestibular ganglia and VN (e.g., Nagao et al., 1997; Osanai et al., 1999) and optokinetic inputs from the nucleus reticularis tegmenti pontis (NRTP) (Kano et al., 1991; Miyashita et al., 1980) via MFs. H-zone PCs directly inhibit the ipsilateral VN neurons that drive EMN (Fukuda et al., 1972; Ito et al., 1977; Kawaguchi, 1985).

Importantly, H-zone PCs receive retinal slip inputs that are necessary for HVOR and HOKR adaptations via CFs (Maekawa & Simpson, 1973). Considering such a unique anatomical connection of FL H-zone to HVOR and HOKR, a hypothesis was proposed that FL H-zone PCs adaptively control HVOR (Ito, 1970, 1972, 1974, 1982 and 1984) and HOKR (Ito, 1984; Ito & Nagao, 1991; Nagao, 1983, 1988) by modifying the vestibular or optokinetic responsiveness at PF–PC synapses using CF retinal slip inputs.

In the 1970s–1990s, Ito and his colleagues demonstrated that both FL and IO are necessary for HVOR and HOKR adaptations by FL (Ito et al. 1982a; Nagao, 1983) and IO (Ito & Miyashita, 1975; Katoh et al., 1998; Shutoh et al., 2006) lesion studies. They also demonstrated that FL H-zone PCs showed simple-spike response that correlated with HVOR or HOKR adaptation in rabbits (Dufossé et al., 1978; Nagao, 1988, 1989) and monkeys (Watanabe, 1985).

14.4 Long-Term Depression (LTD) as a Neural Mechanism of Cerebellar Learning

The models of Marr and Albus and the FL hypothesis of Ito were soon termed collectively as the Marr–Albus–Ito cerebellar learning hypothesis. However, the hypothesis was not accepted by academies, because plasticity of PC synapses underlying cerebellar learning was yet unknown in the 1970s.

In 1982, Ito and his colleagues discovered LTD at PF–PC synapses in the FL (Ito et al. 1982b) and dorsal paraflocculus (Ito & Kano, 1982) using rabbit decerebrated preparations. In both studies, after 5 min of 4 Hz conjunctive electrical stimulation of MFs or PFs with CFs, the simple-spike response of PCs to conjunctively stimulated MFs or PFs decreased for 1 h. The former study showed pharmacologically that the sensitivity of postsynaptic glutamate receptors decreased in LTD. Later, Kano and Kato (1987) confirmed that α-amino-3-hydrooxy-5-methyl-4-isoxazolone propionate receptors (AMPARs) were specifically depressed in LTD. Today, the decrease of recycling AMPARs in the postsynaptic endosome is assumed to underlie LTD (Kim et al., 2017; Kim & Tanaka-Yamamoto, 2019; Parkinson & Hanley, 2018).

Sakurai (1987) developed a method to demonstrate LTD in slice preparations. He showed that after the conjunctive electrical stimulation of PFs and CFs at 4 Hz for 25 s, the PF–PC excitatory postsynaptic potential (EPSP) decreased for 1 h. He also showed that after the electrical stimulation of PFs alone at 4 Hz, not associated with the stimulation of CFs, long-term potentiation (LTP) of PF–PC EPSP occurred for 1 h. Later, Lev-Ram et al. (2002 and 2003) showed that 1–2 Hz PF stimulation-evoked LTP is induced by the postsynaptic mechanism.

14.5 Eyeblink Conditioning and Cerebellar Learning

Delay eyeblink conditioning is cerebellum-dependent timing learning (Longley & Yeo, 2014; Thompson, 1988; Yeo & Hesslow, 1998), in which tone stimuli (conditioned stimuli, CS) are presented repeatedly before and during a brief interval of application of air puff (unconditioned stimuli, US) to the cornea. The animal learns to predict the US timing by referring to CS and blink to only CS. CS and US are conveyed respectively via MFs and CFs to the cerebellar hemisphere, which control the facial muscle motor nucleus via cerebellar nuclei (Yeo et al., 1985a, 1985b, 1985c, 1986). Two computational studies (Yamazaki & Nagao, 2012; Yamazaki & Tanaka, 2007) suggested that PF–PC synapse LTD may induce delay eyeblink conditioning.

14.6 The Cerebellar Learning Hypothesis and Gene-Modified Mouse Studies

The molecular mechanism underlying PF–PC synapse LTD has been examined using slice or culture (Hirano, 1990; Linden & Conner, 1991) preparations. More than 20 molecules, e.g., nitric monoxide (NO), cyclooxygenase-2 (COX-2), protein kinase Cα, etc., were discovered to be involved in LTD induction (Ito, 1989, 2001, 2002, 2006, 2012, Ito et al., 2014, Nagao, 2021), and many strains of mice, which show a deficit in such molecules and consequently show deficient PF–PC synapse LTD, were produced. They have been used to confirm that PF–PC synapse LTD induces ocular reflex adaptations or delay eyeblink conditioning (Yuzaki, 2013).

As shown in Table 14.1, most of the results of PF–PC synapse LTD-deficient mouse studies support the cerebellar learning hypothesis. Schonewille et al. (2011) argued that in three strains [PICK1-knockout (KO), GluR2Delta7-knockin (KI), and GluR2K882A-KI], although the induction of PF–PC synapse LTD tested by an

Table 14.1 PF-PC LTD and short-term motor learning in gene-modified mice

Strains	LTD	Motor Learning	References
mGLuR1-KO	↓	↓	Aiba et al. (1994)
PKCγ-KO	→	→	Kano et al. (1995)
GluD2-KO	↓	↓	Kashiwabuchi et al. (1995)
PKC inhibitor-TG	↓	↓	De Zeeuw et al. (1998)
nNOS-KO	↓	↓	Katoh et al. (2000)
PLCβ4-KO	↓	↓	Miyata et al. (2001)
PKG1-KO	↓	↓	Feil et al. (2003)
CaMKIIα-KO	↓	↓	Hansel et al. (2006)
GS-KO juvenile	↓	↓	Endo et al. (2009)
adult	→	→	
PLA2α-KO	↓	↓	Le et al. (2010)
IP3R1-KO	↓	↓	Sugawara et al. (2013)
patDp/+	↓	↓	Piochon et al. (2014)
PICK1-KO	↓ →	→	Schonewille et al. (2011)
GluRA2K882A-KI	↓ →	→	Yamaguchi et al. (2016)
GluRA2Delta7-KI	↓ →	→	
GluA3-KO	→	↓	Gutierrez-Castellanos et al. (2017)
MLIN-Grin1-KO	↓	↓	Kono et al. (2019)

CaMKIIα Ca^{2+} calmodulin-dependent protein kinase IIα, *GS* G-substrate, *GluA3* type-3 AMPAR, *GluRA2K882A-KI* a knockin harboring a point mutation in the protein kinase C recognition motif of type-2 AMPAR, *GluRA2Delta7-KI* a knockin in which type-2 AMPAR is replaced with a truncated form involving deletion of the last seven amino acids in C-terminus, *GluD2* δ-glutamate rector 2, *nNOS* neuronal nitric oxide synthetase, *IP3RI* inositol 3 phosphate receptor 1, *KO* knockout, *mGLuR1* metabotrophic glutamate receptor type 1, *MLIN-Grin1-KO* molecular layer inhibitory interneuron-specific knockout of a gene encoding an obligatory subunit of N-methyl-D aspartate (NMDA) receptor (GluN1), *pat DP/+*, a mouse model of autism spectrum disorder (ASD) with a paternally inherited human 15q11–13 duplication, *PICK1* protein interacting with PKC 1, *PKCγ* protein kinase C γ subunit, *PKG1* protein kinase G1, *PLA2α* phospholipase A2α, *PLCβ4* phospholipase Cβ4, *TG* transgenic, ↓ impaired, → no change. Note that juvenile (postnatal week 6, PW 6) GS-KO mice showed impaired short-term HOKR adaptation and PF-PC synapse LTD, while adult (PW 12) GS-KO mice showed normal PF-PC synapse LTD and short-term HOKR adaptation, and impaired long-term HOKR adaptation (see Sect. 10 of this Chapter). Also note that although Schonewille et al. (2011) reported that LTD induction was impaired (marked by blue arrows and letters) in three strains of mice (PICK1-KO, GuRA2K882-KI, and GLuRA2Delta7-KI mice), Yamaguchi et al. (2016) showed that LTD induction was normal (marked by black arrows and letters) in the same strains when slightly stronger induction protocols than that of Schonewille et al. (2011) were used

induction protocol (1 Hz PF and 1 Hz CF electrical stimulation for 5 min) was impaired, HVOR and HOKR adaptations were intact. However, Yamaguchi et al. (2016) showed that PF–PC synapse LTD was induced normally in these three strains, when two slightly strong induction protocols (2 Hz PF and 1 Hz CF electrical stimulation for 5 min and 1 Hz PF electrical stimulation and strong PC depolarization for 5 min) were tested. Thus, the causal relationship is preserved between LTD and motor learning in these three strains.

Gutierrez-Castellanos et al. (2017) reported that type-3 AMPAR (GluA3)-KO mice showed normal PF–PC synapse LTD and impaired PF–PC synapse LTP and HVOR gain-up adaptation. At present, this strain is the only one case that does not support the causal relationship between PF–PC synapse LTD and motor learning (Table 14.1). However, deletion of GluA3 may alter the relative roles of the remaining three types of AMPARs in LTD/LTP induction and HVOR adaptation. Further studies are required to correctly interpret their observation.

14.7 The Cerebellar Learning Hypothesis and Pharmacological Studies

Several drugs that block PF–PC synapse LTD have been used to pharmacologically confirm that LTD induces HVOR and HOKR adaptations. N^G-monomethyl-L-arginine (L-NMMA), a blocker of NO synthesis, blocked PF–PC synapse LTD (Crepel & Jaillard, 1990; Ito & Karachot, 1990) in slice preparations and mouse HOKR gain-up adaptation (Katoh et al., 2000; Shutoh et al., 2006) by bilateral FL infusion of 0.5 μl of 20 mM L-NMMA.

Nimesulide, a blocker of COX-2, blocked PF–PC synapse LTD by interfering with the lipid signaling of AMPARs in mouse slice preparations. It blocked mouse HOKR gain-up adaptation by 3 mg/kg (body weight) intraperitoneal (i.p.) application (Le et al., 2010) and HVOR gain-up adaptation in common marmosets by ≥2 mg/kg i.p. application (Anzai & Nagao, 2014).

T-588, which inhibits the intracellular Ca^{2+} concentration increase necessary for LTD induction, blocked PF–PC synapse LTD in rat in vivo preparations (Kimura et al., 2005). Welsh et al. (2005) argued that T-588 did not impair eyeblink conditioning or rotarod motor learning in rats after 3 weeks of oral application. Schonewille et al. (2011) also argued that a single i.p. 10 mg/kg T-588 application did not impair HVOR adaptation in mice. However, Anzai and Nagao (2014) showed that a single i.p. ≥ 3 mg/kg T-588 application blocked HVOR gain-up adaptation in common marmosets. Thus, there may be a significant species difference between marmosets and rodents concerning the ease of T-588 in penetrating the blood–brain barrier or in the transfer from the abdominal cavity to the blood circulation (e.g., Ito et al., 2011).

14.8 LTD Demonstration by New Technology

As the electrophysiological demonstration of PF–PC synapse LTD in in vivo preparations is difficult (Ito et al., 2014), alternative methods using the immuno-electron microscopy (immuno-EM) or optogenetics have been developed today. They have been used to confirm that LTD induces ocular reflex adaptations.

Wang et al. (2014) and Aziz et al. (2014) showed by using anti-AMPAR (type 1–4) antibody and EM that the AMPAR density of freeze-fractured mouse PF–PC synapses decreased by 20% in FL H-zone after 1 h of HOKR gain-up adaptation and returned to baseline levels in 24 h when adaptation recovered. Inoshita and Hirano (2018) showed that in HOKR-adapted mouse FL slices, amplitudes of PF–PC excitatory postsynaptic currents decreased, and PF–PC synapse LTD induction was occluded, compared with slices obtained from non-adapted mice. Eguchi et al. (2020) confirmed their findings using mouse FL slices prepared at physiological temperature by electrophysiology and immuno-EM. These four studies revealed the anatomical or physiological trace of memory of adaptation induced by LTD in FL. Moreover, Kakegawa et al. (2018) showed that the optogenetic blockade of AMPAR endocytosis from postsynaptic membrane blocked PF–PC synapse LTD induction in slice preparations as well as HOKR and HVOR gain-up adaptations in mice.

14.9 Location of Memories of Short- and Long-Term Adaptations

So far, roles of LTD in the short-term HOKR or HVOR adaptations induced by hours of training have been discussed. Here, a question will be addressed on whether the neural mechanism underlying the long-term adaptations induced by days or week of training is the same as that underlying the short-term adaptations. To address on this question, I developed a protocol to quantify HOKR long-term adaptation in mice (Shutoh et al., 2006). Fig. 14.2a shows the adaptation curve induced by 5 days of repetition of HOKR training by continuous fast screen oscillation for 1 h. On the first day, HOKR gain increased by 30%, which recovered to baseline level 24 h after the training. However, the gain at the start of daily training gradually increased by 100% after the 5-day training. This long-term gain increases gradually

Fig. 14.2 (continued) Error bars indicate SEM. (**b**) A box and whisker plot for the effects of micro-infusion of 5% lidocaine into bilateral flocculi (0.5 μl for each) after 4-day HOKR training. $n = 6$. (**c**) Representative eye position traces at the start of daily training on day 1 (initial state, black dotted curve) and day 4 (black solid curve), after the daily training on day 4 (green solid curve), immediately after (red curve) and 24 h after (blue curve) lidocaine infusion are shown. These data were obtained from the same mouse. The inset photograph shows an example of the diffusion of a fluorescent dye mixed with lidocaine. FL flocculus, PF paraflocculus, Calibration bar, 100 μm. (Modified from Shutoh et al. (2006)). *ns* nonsignificant; *, $P < 0.05$; **, $P < 0.01$, Paired T-test)

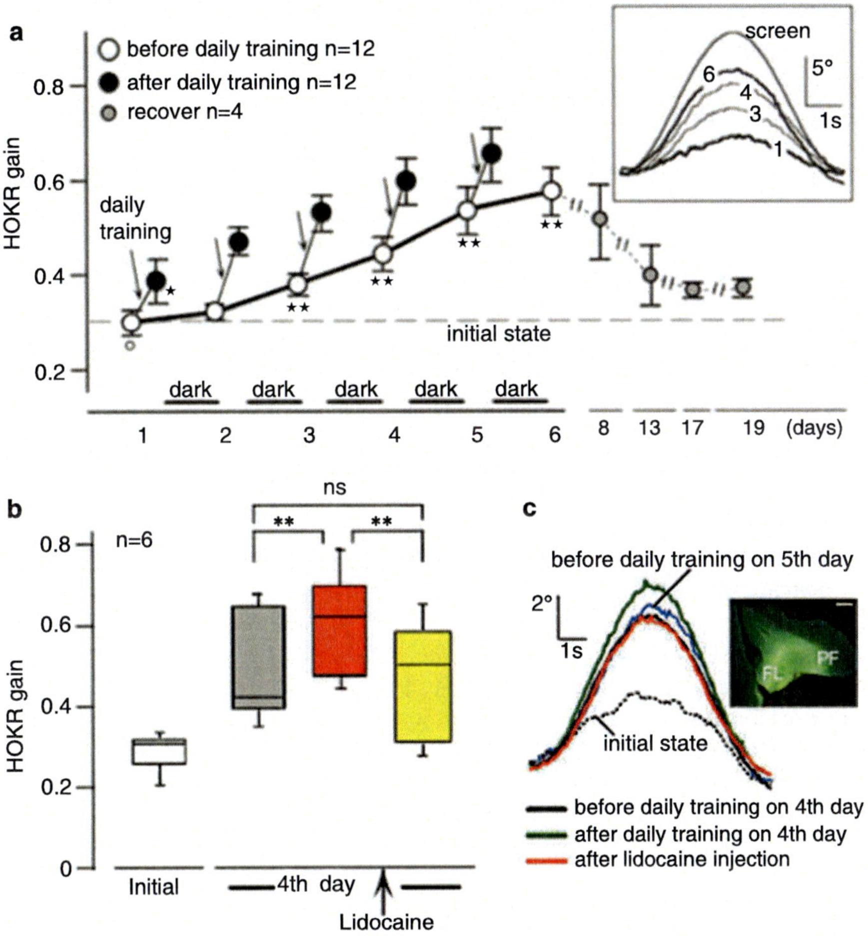

Fig. 14.2 Long-term mouse HOKR adaptation and the effects of bilateral FL shutdown induced by local infusion of lidocaine. (**a**) HOKR gain changes during 5-day HOKR training. Mice with their heads fixed were trained to view the 0.33 Hz-5° (peak-to-peak) checked pattern (check size, 4°) screen oscillating continuously for 1 h daily for 5 days. Eye movements evoked by HOKR during 50 cycles of screen oscillation were recorded using an infrared TV camera before and after the daily training. More than 15 cycles of eye position traces, which were free from saccades and blinks, were averaged and the mean amplitude of eye movements was calculated by a modified Fourier analysis (Jastreboff, 1979). HOKR gain was obtained by comparing the mean amplitude of eye movements with that of screen movements. After daily training, the mice were kept in their home cage under dark condition. After the end of 5-day HOKR training, the mice were reared in their home cages under the normal light condition (12 h in light and 12 h in dark). No screen oscillation was given except for measuring HOKR on day 8, 13, 17, and 19. The inset shows an example of averaged eye position trace over day 1, 3, 4, and 6 in the same mouse. Note that the gain at the start of daily HOKR training gradually increased by 5-day HOKR training, up to two fold at the end of training. The HOKR gain increased by long-term adaptation gradually recovered in 2 weeks. Mean gain before and after 1-h of daily training is shown respectively by black and white circles (n = 12, for each). The recovery of gain after 5-day HOKR training was examined in four mice.

recovered in 10–20 days, when the mice were reared under normal condition without any training. Thus, the gains increased by short- and long-term adaptations are respectively quantified by this HOKR adaptation protocol.

To locate the sites of memory of the short- and long-term adaptations, lidocaine was infused into bilateral FL at the end of the 4-day training. If the memory of adaptation is maintained in FL, lidocaine-induced FL shutdown will erase it, and the adaptation should disappear (e.g., Nagao & Kitazawa, 2003). As shown in Figs. 14.2b and c, lidocaine infusion extinguished the memory of short-term adaptation but hardly affected the memory of long-term adaptation, suggesting that the memory of short-term adaptation is maintained in FL, but not that of long-term adaptation (Shutoh et al., 2006). The same conclusion was obtained from the results of experiments of long-term HVOR adaptations in cats (Kassardjian et al., 2005) and monkeys (Anzai et al., 2010). Difference of memory sites was also suggested in mouse stimulus frequency-dependent HVOR adaptations (Boyden et al., 2006).

Where is the site of the memory of long-term adaptation? Considering the neural circuitries of HVOR and HOKR (Fig. 14.1), the most likely sites of the memory of long-term adaptation are the VN neurons targeted by FL. To confirm this, the field response of VN neurons induced by the electrical stimulation of the horizontal semicircular canal was mapped in the long-term HOKR-adapted mice under trichloroacetaldehyde anesthesia (Shutoh et al., 2006). In these mice, gains increased in not only HOKR but also HVOR. Compared with non-adapted mice, the monosynaptic field response (Precht & Shimazu, 1965) of the long-term adapted mice increased around the medial VN, suggesting the occurrence of LTP in VN neurons. A computational study of Yamazaki et al. (2015) suggested that the plasticity of cerebellar nuclei (CN) or VN neurons (McElvain et al., 2010; Pugh & Raman, 2006; Zhang & Linden, 2006) may underlie the transfer of the memory of long-term HOKR and HVOR adaptations. A recent gene-KO mouse study of Sano et al. (2018) supported this suggestion (see Sect. 14.10).

The transfer of the memory of adaptation is related to the spacing effect, which indicates that the memory acquired through the repetition of learning remains longer than the memory acquired by massed learning (Ebbinghause, 1985). Okamoto, Endo, et al. (2011a) showed that the repetition of 7.5- or 15-min HOKR training at the intervals of 30 min, 1 h, or 1 day (total training time, 1 h) induced the transfer of the memory of adaptation which remained for 2 weeks, suggesting that the repetition of training with an appropriate resting interval induces the transfer of the memory of adaptation.

In summary, the memory of short-term adaptations is initially formed at FL H-zone PF–PC synapses, which is transferred to VN neurons targeted by FL by the repetition of adaptations with an appropriate resting interval. Through memory transfer, the cerebellar cortex may be capable of learning a new motor skill in the same manner at any time, while keeping the old motor memory in VN or CN (Nagao et al., 2013).

14.10 Neural Mechanism Underlying the Transfer of Memory of Adaptation

One may wonder how the FL H-zone PF–PC synapse LTD lead to the LTP of VN neurons targeted by FL in long-term HOKR adaptation. The VN neurons targeted by FL are distributed in only 1–2% of the ipsilateral magno- and parvocellular medial VN, and the adjacent prepositus hypoglossi nucleus neurons (Sato et al., 1987; Sekirinjak et al. 2003; Matsuno et al., 2016). Matsuno et al. (2016) showed by EM that FL H-zone PCs project their axons to the VN neurons not only on their soma and proximal dendrites but also on their distal dendrites, where excitatory synapses of vestibular nerves and inhibitory synapses of FL H-zone PCs are located very closely. In the cerebral cortex pyramidal neurons, inhibitory synapses on the distal dendrites are considered to affect the transmission of neighboring excitatory synapses (Kubota et al., 2007). Such an interaction from inhibitory synapses to neighboring excitatory synapses may underlie LTP of VN neurons.

Several studies consistently suggest that some substances, which are synthesized in FL H-zone PCs during HOKR training and transported to PC axon terminals in VN, may play roles in LTP inductions in VN neurons. First, the pharmacological blockade of de novo protein synthesis in FL during HOKR training blocked the transfer of the memory of HOKR adaptation (Okamoto, Endo, et al., 2011a). Second, the transfer of the memory of HOKR adaptation occurred within 3 h, when the spaced training protocol was used (Okamoto, Endo, et al., 2011a). Three hours is sufficient to transport proteins from mouse FL to VN, the distance of which is 1–2 mm, by fast axonal transport. Third, G-substrate (GS), the substrate of cyclic guanylate monophosphate (cGMP)-dependent protein kinase, is specifically expressed in PCs (Endo et al., 1999). Adult GS-KO mice (Table 14.1) showed normal PF–PC synapse LTD and short-term HOKR adaptation and impaired long-term HOKR adaptation (Endo et al., 2009), suggesting an important role of PCs in the transfer of the memory of long-term adaptation. Fourth, a strain of gene-KO mice showed impaired axonal transport in PCs, MF–CN neuron synapse LTP, and long-term motor leaning (Sano et al., 2018).

G-protein-coupled receptor class 5B (GPRC5B) is an orphan receptor, whose function is unknown at present. GPRC5B develops gradually in the postnatal period in the brain (Fig. 14.3a). Although GPRC5B-KO mice showed no ataxia, their PCs showed age-dependent swellings in distal axons, which contained lots of subcellular organelles including misshapen mitochondria (Figs. 14.3b and c), indicating an impaired axonal transport. In slices obtained from GPRC5B-KO mice, induction of PF–PC synapse LTD was intact (not shown in figures), whereas induction of MF–CN neuron synapse LTP by tetanic MF stimulation followed by rebound depolarization of CN neurons (Pugh & Raman, 2006) was impaired (Fig. 14.3d). In adult GPRC5B-KO mice, the short-term HOKR adaptation was intact (not shown in figures), whereas the long-term HOKR adaptation was impaired (Fig. 14.3e).

Okamoto, Shirao, et al. (2011b) showed that blockade of spontaneous spike discharges of FL PCs after daily HOKR training by bilateral FL muscimol infusions

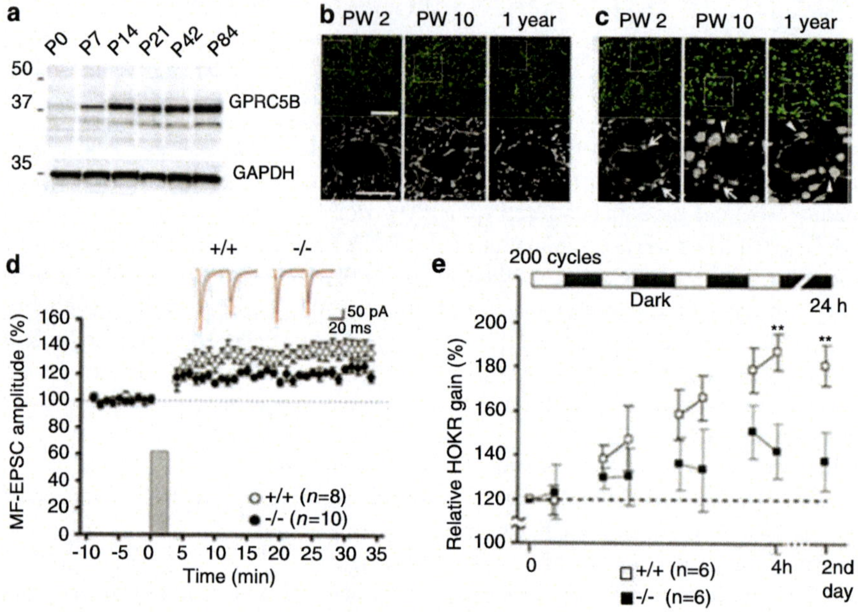

Fig. 14.3 Characteristics of G-protein-coupled receptor class 5B (GPRC5B)-knockout (KO) mice. Loss of GPRC5B impairs morphology of PCs, mossy fiber (MF)–cerebellar nuclear (CN) neuron synapse plasticity, and long-term motor leaning. (**a**) Western blot analysis of postnatal changes in GPRC5B expression in the cerebellum. GPRC5B expression started during the first postnatal day (P 0) and P 7 and increased slowly with age at P 7–P 84. (**b, c**) Photographs of parasagittal sections (thickness, 10–20 μm) of the cerebellum. Sections were labeled with anti-calbindin antibody (green) and counterstained with DAPI (blue) to label nuclei. Calbindin immunopositive swellings of PCs are present at postnatal week (PW) 2, 10, and postnatal year 1 in GPRC5B-KO (**c**) but not in wild-type (**b**) mice. The swellings (arrows or arrow heads) grew larger with increasing age and were immunopositive for Tau1, an axon-specific marker, but not for dendrite-specific microtubule-associated protein 2 (not shown in figures), indicating that the swellings were likely enlargement of PC axons. Square areas marked by white lines in the upper half are shown in the lower half. (**d**) MF–CN neuron synapse LTP tested in slices obtained from mice at PW 2–3. Two electrodes, one for MF stimulation and the other for whole-cell patch-clamp recording from CN neuron, were used. LTP was tested by tetanic stimulation (125 Hz) of MFs, which was immediately followed by rebound depolarization (RD) induced by the injection of hyperpolarizing current into CN neurons via the recording electrode, every 2 s for 2.5 min. Amplitudes of excitatory postsynaptic currents evoked by MF stimulation (MF–EPSC) were measured from CN neurons voltage-clamped at −65 mV. Compared with wild-type mice (+/+, n = 8), LTP was depressed ($P < 0.05$, repeated measure ANOVA) in GPRC5B-KO mice (−/−, n = 10). Insets show representative MF–EPSCs before (black) and after (red) LTP induction. (**e**) Long-term HOKR adaptation tested by a spaced training protocol. At PW 12, wild-type (+/+) and KO (−/−) mice (n = 6, for each) were trained to view 200 cycles of 0.22 Hz-10° screen oscillation with the interval of 1 h for four times. The mice were returned to their home and kept in dark during spaced intervals. HOKR gain change showed significant differences at the end [167% ± 8% (mean ± SEM) for wild-type vs. 114% ± 13% for KO mice; $^{**}P < 0.01$, unpaired T-test] and 24 h after the end (161% ± 8% vs. 116% ± 10%, $^{**}P < 0.01$) of training. Scales, 50 μm (**b**), 20 ms, and 50 pA (**d**). Error bars represent SEM. (Modified from Sano et al. (2018))

impaired the transfer of the memory of HOKR adaptation. Two previous studies showed that PCs release not only GABA but also peptides (e.g., motilin) to VN neurons (Chan-Palay et al., 1982; Todaka et al., 2013). Taken together with these observations, I suggest that proteins or peptides, which are synthesized during HOKR training, transported to PC axon terminals, and released to VN neurons with GABA through spontaneous spike discharges in resting periods, may trigger LTP at neighboring MF–VN neuron synapses underlying the transfer of the memory of long-term gain-up adaptation.

14.11 The Cerebellar Learning Hypothesis: Present and Future

As shown in my review, the Marr–Albus–Ito cerebellar learning hypothesis has been supported for the past 50 years by many studies using computation (Yamazaki et al., 2015; Yamazaki & Nagao, 2012), physiology, pharmacology, gene-modified mice, immuno-EM, and optogenetics. They consistently suggest that PF–PC synapse LTD induces HVOR and HOKR gain-up adaptations. On the other hand, PF–PC synapse LTP is suggested to induce HVOR gain-down adaptation (Boyden & Raymond, 2003; Inagaki & Hirata, 2017; Nagao, 2021). Although several laboratories addressed controversial findings in relation to the hypothesis for past 50 years, the reasons underlying such findings have been clarified except for one study (see Sect. 14.6). Now, the hypothesis is considered as a fundamental mechanism of motor learning (Honda & Ito, 2016; Ito, 2012; Nagao, 2021).

Plasticity of molecular layer inhibitory interneurons in the cerebellar cortex (Jorntell & Ekerot, 2002, 2003; Kano et al., 1992; Rancillac & Crepel, 2004) is suggested to cooperate with PF–PC synapse LTD in HVOR adaptation (Tanaka et al., 2013). Plasticity of VN neurons targeted by FL is suggested to play a role in the transfer of the memory of long-term HOKR (Okamoto, Endo, et al., 2011a; Sano et al., 2018; Shutoh et al., 2006) and HVOR (Anzai et al., 2010; Jang et al., 2020; Kassardjian et al., 2005) adaptations. These two forms of plasticity add new cerebellar learning mechanisms to the hypothesis. Roles of the cerebellar learning in the posttraumatic readjustment of neural circuitry, which are reorganized through axonal regeneration, sprouting, and new synapse formation (Tsukahara et al., 1975), are the target of research in the rehabilitation medicine (Nagao & Ito, 2017).

Ito (1984, 1993, 2001, 2006, 2008, and 2012) expanded the application of cerebellar learning hypothesis from reflexes to voluntary movement and cognitive functions, in which the cerebrocerebellar loop (Kelly & Strick, 2003) operates. In the voluntary movement, the cerebellum is assumed to generate motor command signals through learning of the internal forward (Ito, 1970; Miall et al., 1993) and inverse (Kawato et al., 1987) models of movement (also see Kawato et al., 2021). The internal forward model predicts the sensory consequence of motor command signals, whereas the internal inverse model predicts the motor command signals that generate the appropriate sensory consequence. By using such internal models acquired in the cerebellum, we can automatically touch our finger on the top of our nose without visual guide in the finger-to-nose test (Ito, 1984). Five months before his death, Ito and his colleagues (Honda et al., 2018) proposed a new hypothesis that the cerebellum learns the internal forward and inverse models in tandem during the prism adaptation of human hand-reaching task (Hashimoto et al., 2015). How the cerebellar learning is utilized by the cerebrum should be investigated further in integrative brain functions.

Acknowledgments I thank to late Dr. Masao Ito for his mentorship to my study of the cerebellum and ocular reflexes. I also thank to Dr. Shogo Endo (Tokyo Metropolitan Institute for Gerontology) for his helpful comments on my manuscript and Dr. Fumihiro Shutoh (Maebashi Institute of Technology, Gunma) for his help in the revision of figure.

References

Aiba, A., Kano, M., Chen, C., Stanton, M. E., Fox, G. D., Herrup, K., Zwingman, T. A., & Tonegawa, S. (1994). Deficient cerebellar long-term depression and impaired motor learning in mGluR1 mutant mice. *Cell, 79*, 377–388.

Albus, J. S. (1971). Theory of cerebellar function. *Mathematical Biosciences, 10*, 25–61.

Anzai, M., & Nagao, S. (2014). Motor learning in the common marmosets: Vestibulo-ocular reflex adaptation and its sensitivity to inhibitors of Purkinje cell long-term depression. *Neuroscience Research, 83*, 33–42.

Anzai, M., Kitazawa, H., & Nagao, S. (2010). Effects of reversible pharmacological shutdown of cerebellar flocculus on the memory of long-term horizontal vestibulo-ocular reflex adaptation in monkeys. *Neuroscience Research, 68*, 191–198.

Aziz, W., Wang, W., Kesaf, S., Mohamed, A. A., Fukazawa, Y., & Shigemoto, R. (2014). Distinct kinetics of synaptic structural plasticity, memory formation, and memory decay in massed and spaced learning. *Proceedings of the National Academy of Sciences of the United States of America, 111*, E194–E202.

Batini, C., Ito, M., Kado, R. T., Jastreboff, P. J., & Miyashita, Y. (1979). Interaction between the horizontal vestibulo-ocular reflex and optokinetic response in rabbits. *Experimental Brain Research, 37*, 1–15.

Boyden, E. S., & Raymond, J. L. (2003). Active reversal of motor memories reveals rules governing motor memory. *Neuron, 39*, 1031–1042.

Boyden, E. S., Katoh, A., Pyle, J. L., Chatila, T. A., Tsien, R. W., & Raymond, J. L. (2006). Selective engagement of plasticity mechanisms for motor memory storage. *Neuron, 57*, 823–834.

Chan-Palay, V., Ito, M., Tongroach, P., Sakurai, M., & Palay, S. (1982). Inhibitory effects of motilin, somatostatin, [Leu]enkephalin, [Met]enkephalin, and taurine on neurons of the lateral vestibular nucleus: Interactions with gamma-aminobutyric acid. *Proceedings of the National Academy of Sciences of the United States of America, 79*, 3355–3359.

Collewijn, H., & Grootendorst, A. F. (1978). Adaptation of optokinetic and vestibulo-ocular reflexes to modified visual input in the rabbit. *Progress in Brain Research, 50*, 771–781.

Crepel, F., & Jaillard, D. (1990). Protein kinases, nitric oxide and long-term depression of synapses in the cerebellum. *Neuroreport, 1*, 133–136.

De Zeeuw, C. I., Hansel, C., Bian, F., Koekkoek, S. K., van Alphen, A. M., Linden, D. J., & Oberdick, J. (1998). Expression of a protein kinase C inhibitor in Purkinje cells blocks cerebellar LTD and adaptation of vestibulo-ocular reflex. *Neuron, 20*, 495–508.

Dufossé, M., Miyashita, Y., & Ito, M. (1977). Functional localization in rabbit's cerebellar flocculus determined in relationship with eye movements. *Neuroscience Letters, 5*, 273–277.

Dufossé, M., Ito, M., Jastreboff, P. J., & Miyashita, Y. (1978). A neural correlate in the rabbit's cerebellum to adaptive modification of vestibulo-ocular reflex. *Brain Research, 150*, 611–616.

Ebbinghause, H. E. (1985) Uber das Gedächtnis. Reprinted as memory: A contribution to experimental psychology (Ruger HA, Bussenius C, translators). k: Teachers College-Columbia UP, 1913.

Eccles, J. C., Ito, M., & Szentágothai, J. (1967). *The cerebellum as a neuronal machine*. Springer-Verlag.

Eguchi, K., Velicky, P., Hollergeschwandtner, E., Itakura, M., Fukazawa, Y., Danzi, J. G., & Shigemoto, R. (2020). Advantages of acute slices prepared at physiological temperature in the characterization of synaptic functions. *Front Cell Neuroscience, 14*, 63. https://doi.org/10.3389/fncel.2020.00063

Endo, S., Suzuki, M., Sumi, M., Nairn, A. C., Morita, R., Yamakawa, K., Greengard, P., & Ito, M. (1999). Molecular identification of human G-substrate: A possible downstream component of cyclic GMP-dependent protein kinase. *Proceedings of the National Academy of Sciences of the United States of America, 96*, 2467–2472.

Endo, S., Shutoh, F., Dinh, T. L., Okamoto, T., Ikeda, T., Suzuki, M., Kawahara, S., Yanagihara, D., Yamada, K., Kirino, Y., Hartell, N. A., Yamaguchi, K., Itohara, S., Nairn, A. C., Greengard, P., Nagao, S., & Ito, M. (2009). Dual involvement of G-substrate in motor learning revealed by gene deletion. *Proceedings of the National Academy of Sciences of the United States of America, 106*, 3525–3530.

Feil, R., Hartmann, J., Luo, C., Wolfsgruber, W., Schilling, K., Feil, S., Barski, J. J., Meyer, M., Konnerth, A., De Zeeuw, C. I., & Hofmann, F. (2003). Impairment of LTD and cerebellar learning by Purkinje cell-specific ablation of cGMP-dependent protein kinase I. *The Journal of Cell Biology, 163*, 295–302.

Fuchs, A. F., & Robinson, D. A. (1966). A method for measuring horizontal and vertical eye movement chronically in the monkey. *Journal of Applied Physiology, 21*, 1068–1070.

Fukuda, J., Highstein, S. M., & Ito, M. (1972). Cerebellar inhibitory control of the vestibulo-ocular reflex investigated in rabbit's IIIrd nucleus. *Experimental Brain Research, 14*, 511–526.

Gonshor, A., & Melvill-Jones, G. M. (1974). Extreme vestibulo-ocular adaptation induced by prolonged optical reversal of vision. *Journal of Physiology (London), 256*, 381–414.

Gutierrez-Castellanos, N., Da Silva-Matos, C. M., Zhou, K., Canto, C. B., Renner, M. C., Koene, L. M. C., Ozyildirim, O., Sprengel, R., Kessels, H. W., & De Zeeuw, C. I. (2017). Motor learning requires Purkinje cell synaptic potentiation through activation of AMPA-receptor subunit GluA3. *Neuron, 93*, 409–424.

Hansel, C., de Jeu, M., Belguemunai, A., Houtman, S. H., Buitendijk, G. H., Andreev, D., De Zeeuw, C. I., & Eigersma, Y. (2006). αCaMKII is essential for cerebellar LTD and motor learning. *Neuron, 51*, 835–843.

Hashimoto, Y., Honda, T., Soda, K., Yokota, T., Mizusawa, H., Nagao, S., & Ishikawa, K. (2015). Quantification of human motor learning by prism adaptation of hand-reaching movement. *PLoS One, 10*, e0119376.

Hirano, T. (1990). Effects of postsynaptic depolarization in the induction of synaptic depression between a granule cell and a Purkinje cell in rat cerebellar culture. *Neuroscience Letters, 119*, 145–147.

Honda, T., & Ito, M. (2016). Development from Marr's theory of the cerebellum. In L. M. Vaina & R. E. Passingham (Eds.), *Computational theories and their implication in the brain* (pp. 29–61). Oxford University Press.

Honda, T., Nagao, S., Hashimoto, Y., Ishikawa, K., Yokota, T., Mizusawa, H., & Ito, M. (2018). Tandem internal models execute motor learning in the cerebellum. *Proceedings of the National Academy of Sciences of the United States of America, 115*, 7428–7433.

Inagaki, K., & Hirata, Y. (2017). Computational theory underlying acute vestibulo-ocular reflex motor learning with cerebellar long-term depression and long-term potentiation. *Cerebellum, 16*, 827–839.

Inoshita, T., & Hirano, T. (2018). Occurrence of long-term depression in the cerebellar flocculus during adaptation of optokinetic response. *Elife, 7*, pii: e36209. https://doi.org/10.7554/eLife.36209

Ito, M. (1970). Neurophysiological basis of the cerebellar motor control system. *International Journal of Neurology, 7*, 162–176.
Ito, M. (1972). Neural design of cerebellar motor control system. *Brain Research, 40*, 81–84.
Ito, M. (1974). The control mechanisms of cerebellar motor control system. In D. Schmidt & F. G. Worden (Eds.), *The neuroscience IIIrd study program* (pp. 293–303). MIT Press.
Ito, M. (1982). Cerebellar control of vestibulo-ocular reflex — Around the flocculus hypothesis. *Annual Review of Neuroscience, 5*, 275–296.
Ito, M. (1984). *The cerebellum and neural control.* Raven.
Ito, M. (1989). Long-term depression. *Annual Review of Neuroscience, 12*, 85–102.
Ito, M. (1993). Movement and thought: Identical control mechanisms by the cerebellum. *Trends in Neurosciences, 16*, 448–450.
Ito, M. (2001). Cerebellar long-term depression—characterization, signal transduction and functional roles. *Physiological Reviews, 81*, 1143–1195.
Ito, M. (2002). The molecular organization of cerebellar long-term depression. *Nature Reviews. Neuroscience, 3*, 896–902.
Ito, M. (2006). Cerebellar circuitry as a neuronal machine. *Progress in Neurobiology, 78*, 272–303.
Ito, M. (2008). Control of mental activities by internal models in the cerebellum. *Nature Reviews. Neuroscience, 9*, 304–313.
Ito, M. (2012). *The cerebellum: Brain for an implicit self.* FT Press.
Ito, M., & Kano, M. (1982). Long-lasting depression of parallel fiber–Purkinje cell transmission induced by conjunctive activation of parallel fibers and climbing fibers in the cerebellar cortex. *Neuroscience Letters, 33*, 253–258.
Ito, M., & Karachot, L. (1990). Messengers mediating long-term desensitization in cerebellar Purkinje cells. *Neuroreport, 1*, 129–132.
Ito, M., & Miyashita, Y. (1975). The effect of chronic destruction of inferior olive upon visual modification of vestibulo-ocular reflex. *Proceedings Japan Academy, 51*, 716–720.
Ito, M., & Nagao, S. (1991). Comparative aspects of horizontal ocular reflexes and their cerebellar adaptive control in vertebrates. *Comparative Biochemistry and Physiology, 98C*, 221–228.
Ito, M., & Yoshida, M. (1964). The cerebellar-evoked monosynaptic inhibition of Deiters' neurons. *Experientia, 20*, 515–516.
Ito, M., Shiida, N., Yagi, N., & Yamamoto. (1974). The cerebellar modification of rabbit's horizontal vestibulo-ocular reflex by sustained head rotation combined with visual stimulation. *Proceedings Japan Academy, 50*, 85–90.
Ito, M., Nisimaru, N., & Yamamoto, M. (1977). Specific patterns of neural connections involved in rabbits' vestibulo-ocular reflex. *Journal of Physiology (London), 265*, 833–854.
Ito, M., Jastreboff, P. J., & Miyashita, Y. (1979). Adaptive modification of rabbit's horizontal vestibulo-ocular reflex during sustained vestibular and optokinetic stimulation. *Experimental Brain Research, 37*, 17–30.
Ito, M., Jastreboff, P. J., & Miyashita, Y. (1982a). Specific effects of unilateral lesions in the flocculus upon eye movements in albino rabbits. *Experimental Brain Research, 45*, 233–242.
Ito, M., Sakurai, M., & Tongroach, P. (1982b). Climbing fibre induced both mossy fibre responsiveness and glutamate sensitivity of Purkinje cells. *Journal of Physiology (London), 324*, 113–134.
Ito, K., Uchida, Y., Ohtuski, S., Aizawa, S., Kawakami, H., Katsukura, Y., Kmiiie, J., & Terasaki, T. (2011). Quantitative membrane expression at the blood-brain barrier of adult and younger cynomolgus monkeys. *Journal of Pharmaceutical Sciences, 100*, 3939–3950.
Ito, M., Yamaguchi, K., Nagao, S., & Yamazaki, T. (2014). Long-term depression as a model of cerebellar plasticity. *Progress in Brain Research, 210*, 1–30.
Jang, D. C., Shim, H. G., & Kim, S. J. (2020). Intrinsic plasticity of cerebellar Purkinje cells contributes to motor memory consolidation. *The Journal of Neuroscience, 40*, 4145–4157.
Jastreboff, P. J. (1979). Evaluation and statistical judgement of neural responses to sinusoidal stimulation in cases with superimposed drift and noise. *Biological Cybernetics, 33*, 113–120.
Jorntell, H., & Ekerot, C.-F. (2002). Reciprocal bidirectional plasticity of parallel fiber receptive fields in cerebellar Purkinje cells and their afferent interneurons. *Neuron, 34*, 797–806.

Jorntell, H., & Ekerot, C.-F. (2003). Receptive field plasticity profoundly alters the cutaneous parallel fiber synaptic input to cerebellar interneurons in vivo. *The Journal of Neuroscience, 23*, 9620–9631.

Kakegawa, W., Katoh, A., Narumi, S., Miura, E., Motohashi, J., Takahashi, A., Kohda, K., Fukazawa, Y., Yuzaki, M., & Matsuda, S. (2018). Optogenetic control of synaptic AMPA receptor endocytosis reveals roles of LTD in motor learning. *Neuron, 99*, 985–998.

Kano, M., & Kato, M. (1987). Quisqualate receptors are specifically involved in cerebellar synaptic plasticity. *Nature, 325*, 276–279.

Kano, M., Iino, K., Maekawa, K., & Kano, M.-S. (1991). Optokinetic responses of cells in the nucleus reticularis tegmenti pontis of the pigmented rabbit. *Experimental Brain Research, 87*, 239–244.

Kano, M., Rexhausen, U., Dreessen, J., & Konnerth, A. (1992). Synaptic excitation produces a long-lasting rebound potentiation of inhibitory signals in cerebellar Purkinje cells. *Nature, 356*, 601–604.

Kano, M., Hashimoto, K., Chen, C., Abeliovich, A., Aiba, A., Kurihara, H., Watanabe, M., Inoue, Y., & Tonegawa, S. (1995). Impaired synapse elimination during cerebellar development in PKC gamma mutant mice. *Cell, 83*, 1223–1231.

Kashiwabuchi, N., Ikeda, K., Araki, K., Hirano, T., Shibuki, K., Takayama, C., Inoue, Y., Kutsuwada, T., Yagi, T., Kang, Y., Aizawa, S., & Mishina, M. (1995). Impairment of motor coordination, Purkinje cell synapse formation, and cerebellar long-term depression in GluR delta 2 mutant mice. *Cell, 81*, 245–252.

Kassardjian, C. D., Tan, Y. F., Chung, J. Y., Heskins, R., Petersen, M. J., & Broussard, D. M. (2005). The site of motor memory shifts with consolidation. *The Journal of Neuroscience, 25*, 7979–7985.

Katoh, A., Kitazawa, H., Itohara, S., & Nagao, S. (1998). Dynamic characteristics and adaptability of mouse vestibulo-ocular and optokinetic response eye movements and the role of the flocculo-olivary system revealed by chemical lesions. *Proceedings of the National Academy of Sciences of the United States of America, 95*, 7705–7710.

Katoh, A., Kitazawa, H., Itohara, S., & Nagao, S. (2000). Inhibition of nitric oxide synthesis and gene-knockout of neuronal nitric oxide synthase impaired adaptation of mouse optokinetic eye movements. *Learning and Memory, 7*, 220–226.

Kawaguchi, Y. (1985). Two groups of secondary vestibular neurons mediating horizontal canal signals, probably to the ipsilateral medial rectus muscle, under the influence of cerebellar flocculus in rabbits. *Neuroscience Research, 2*, 434–446.

Kawato, M., Ohmae, S, Hoang, H., & Sanger, T. (2021). 50 years since the Marr, Ito, and Albus models of the cerebellum. *Neuroscience, 462*, 151–174.

Kawato, N., Furukawa, K., & Suzuki, R. (1987). A hierarchical neural-network model for control and learning of voluntary movement. *Biological Cybernetics, 57*, 169–185.

Kelly, R. M., & Strick, P. L. (2003). Cerebellar loops with motor cortex and prefrontal cortex of nonhuman primate. *The Journal of Neuroscience, 23*, 8432–8444.

Kim, T., & Tanaka-Yamamoto, K. (2019). Postsynaptic stability and variability described by a stochastic model of endosomal trafficking. *Frontiers in Cellular Neuroscience, 13*, 72. https://doi.org/10.3389/fncell.2019.00072

Kim, T., Yamamoto, Y., & Tanaka-Yamamoto, K. (2017). Timely regulated sorting from early to late endosome is required to maintain cerebellar long-term depression. *Nature Communications, 8*, 401. https://doi.org/10.1038/s41467-017-00518-3

Kimura, T., Sugimori, M., & Llinás, R. R. (2005). Purkinje cell long-term depression is prevented by T-588, a neuroprotective compound that reduces cytosolic calcium release from intracellular stores. *Proceedings of the National Academy of Sciences of the United States of America, 102*, 17160–17165.

Kono, M., Kakegawa, W., Yoshida, K., & Yuzaki, M. (2019). Interneuronal NMDA receptors regulate long-term depression and motor learning in the cerebellum. *The Journal of Physiology, 597*, 903–920.

Kubota, Y., Hatada, S., Kondo, S., Karube, F., & Kawaguchi, Y. (2007). Neocortical inhibitory terminals innervate dendritic spines targeted by thalamocortical afferents. *The Journal of Neuroscience, 27*, 1139–1150.

Le, T. D., Shirai, Y., Okamoto, T., Tatsukawa, T., Nagao, S., Shimizu, T., & Ito, M. (2010). Lipid signaling in cytosolic phospholipase $A_2\alpha$-cyclooxygenae-2 cascade mediates cerebellar long-term depression and motor learning. *Proceedings of the National Academy of Sciences of the United States of America, 107*, 3198–3203.

Lev-Ram, V., Wong, S. T., Storm, D. R., & Tsien, R. Y. (2002). A new form of cerebellar long-term potentiation is postsynaptic and depends on nitric oxide but not cAMP. *Proceedings of the National Academy of Sciences of the United States of America, 99*, 8389–8394.

Lev-Ram, V., Mehta, S. B., Kleinfeld, D., & Tsien, R. Y. (2003). Reversing cerebellar long-term depression. *Proceedings of the National Academy of Sciences of the United States of America, 100*, 15989–15993.

Linden, D. J., & Conner, J. A. (1991). Long-term depression of glutamate currents in cultured cerebellar Purkinje neurons does not require nitric oxide signaling. *The European Journal of Neuroscience, 4*, 10–15.

Longley, M., & Yeo, C. H. (2014). Distribution of neural plasticity in cerebellum-dependent motor learning. *Progress in Brain Research, 210*, 79–102.

Maekawa, K., & Simpson, J. I. (1973). Climbing fiber responses evoked in vestibulocerebellum of rabbit from visual pathway. *Journal of Neurophysiology, 36*, 649–666.

Marr, D. (1969). A theory of cerebellar cortex. *Journal of Physiology (London), 202*, 437–470.

Matsuno, H., Kudoh, M., Watakabe, A., Yamamori, T., Shigemoto, R., & Nagao, S. (2016). Distribution and structure of synapses on medial vestibular nuclear neurons targeted by cerebellar flocculus Purkinje cells and vestibular nerve in mice: Light and electron microscopy studies. *PLoS One, 11*(10), e0164037.

McElvain, L. E., Bagnall, M. W., Sakatos, A., & du Lac, S. (2010). Bidirectional plasticity gated by hyperpolarization controls the gain of postsynaptic firing responses at central vestibular nerve synapses. *Neuron, 68*, 763–775.

Miall, R. C., Weir, D. J., Wolpert, D. M., & Stein, J. F. (1993). Is the cerebellum a smith predictor? *Journal of Motor Behavior, 25*, 203–216.

Miyashita, Y., Ito, M., Jasareboff, P. J., Maekawa, K., & Nagao, S. (1980). Effect upon eye movements of rabbits induced by severance of mossy fiber visual pathway to the cerebellar flocculus. *Brain Research, 198*, 210–215.

Miyata, M., Kim, H. T., Hashimoto, K., Lee, T. K., Cho, S. Y., Jiang, H., Wu, Y., Jun, K., Wu, D., Kano, M., & Shin, H. S. (2001). Deficient long-term synaptic depression in the rostral cerebellum correlated with impaired motor learning in phospholipase C beta4 mutant mice. *The European Journal of Neuroscience, 13*, 1945–1954.

Nagao, S. (1983). Effects of vestibulocerebellar lesions upon dynamic characteristics and adaptation of vestibulo-ocular and optokinetic responses in pigmented rabbits. *Experimental Brain Research, 53*, 36–46.

Nagao, S. (1988). Behavior of floccular Purkinje cells correlated with adaptation of horizontal optokinetic eye movement response in pigmented rabbits. *Experimental Brain Research, 73*, 489–497.

Nagao, S. (1989). Behavior of floccular Purkinje cells correlated with adaptation of vestibulo-ocular reflex in pigmented rabbits. *Experimental Brain Research, 77*, 531–540.

Nagao, S. (1990). A non-invasive method for eye position recording with an infra-red TV-camera. *Neuroscience Research, 8*, 210–213.

Nagao, S. (2019). Obituary: Masao Ito (1928–2018). *The Cerebellum*. https://doi.org/10.1007/s12311-019-012022-8

Nagao, S. (2021). Ocular reflex adaptation as an experimental model of cerebellar learning — in memory of Masao Ito —. *Neuroscience, 462*, 191–204.

Nagao, S., Hirai, H., Kano, M., & Yuzaki, M. (2021). Masao Ito — A visionary neuroscientist with a passion for the cerebellum. *Neuroscience, 462*, 1–3.

Nagao, S., & Ito, M. (2017). Roles of synaptic plasticity in the functional recovery after brain injury. In L. Petrosini (Ed.), *Contemporary clinical neuroscience, neurobiological and psychological aspects of brain recovery* (pp. 153–181). Springer.

Nagao, S., & Kano, M. (2019). Obituary: Masao Ito (1928–2018). *Neuroscience Research, 141*, 1–3.

Nagao, S., & Kitazawa, H. (2003). Effects of reversible shutdown of the monkey flocculus on the retention of adaptation of horizontal vestibulo-ocular reflex. *Neuroscience, 118*, 563–570.

Nagao, S., Ito, M., & Karachot, L. (1985). Eye fields in the cerebellar flocculus of pigmented rabbits determined with local electrical stimulation. *Neuroscience Research, 3*, 39–51.

Nagao, S., Kitamura, T., Nakamura, N., Hiramatsu, T., & Yamada, J. (1997). Differences of primate flocculus and ventral paraflocculus in mossy and climbing fiber input organization. *The Journal of Comparative Neurology, 382*, 480–498.

Nagao, S., Honda, T., & Yamazaki, T. (2013). Transfer of memory trace of cerebellum-dependent motor learning in human prism adaptation. *Neural Networks, 47*, 72–82.

Obata, K., Ito, M., Ochi, R., & Sato, N. (1967). Pharmacological properties of the postsynaptic inhibition by Purkinje cell axons and the action of gamma-aminobutyric acid on Deiters neuron. *Experimental Brain Research, 4*, 43–57.

Okamoto, T., Endo, S., Shirao, T., & Nagao, S. (2011a). Cerebellar cortical protein synthesis-dependent transfer of memory trace underlies the spacing effect in motor learning. *The Journal of Neuroscience, 31*, 8958–8966.

Okamoto, T., Shirao, T., Shutoh, F., Suzuki, T., & Nagao, S. (2011b). Post-training cerebellar cortical activity plays an important role for consolidation of memory of cerebellum-dependent motor learning. *Neuroscience Letters, 504*, 53–56.

Osanai, R., Nagao, S., Kitamura, T., Kawabata, I., & Yamada, J. (1999). Difference of afferent input organization between the flocculus and paraflocculus in the rat. *Experimental Brain Research, 124*, 248–264.

Parkinson, G. T., & Hanley, G. J. (2018). Mechanisms of AMPA receptor endosomal sorting. *Frontiers in Molecular Neuroscience, 11*, 440. https://doi.org/10.3389/fnmol.2018.00440

Piochon, C., Kloth, A. D., Grasselli, G., Titley, H., Nakayama, H., Hashimoto, K., Wan, V., Simmons DH, Eissa, T., Nakatani, J., Cherskov, A., Miyazaki, T., Watanabe, M., Takumi, T., Kano, M., Wang, S. S.-H., & Hansel, H. (2015). Cerebellar plasticity and motor learning deficits in a copy-number variation mouse model of autism. *Nature Communications*. https://doi.org/10.1038/ncomms6586

Precht, W., & Shimazu, H. (1965). Tonic and kinetic responses of cat's vestibular neurons to horizontal angular acceleration. *Journal of Neurophysiology, 28*, 991–1013.

Pugh, J. R., & Raman, M. (2006). Potentiation of mossy fiber EPSCs in the cerebellar nuclei by NMDA receptor activation followed by postinhibitory rebound current. *Neuron, 51*, 113–123.

Rancillac, A., & Crepel, F. (2004). Synapses between parallel fibers and stellate cells express long-term changes in synaptic efficacy in rat cerebellum. *Journal of Physiology (London), 554*, 707–720.

Rosenblatt, F. (1958). The perceptron: A probabilistic model for information storage and organization in the brain. *Psychological Review, 65*, 386–408.

Sakurai, M. (1987). Synaptic modification of parallel fibre–Purkinje cell transmission in in vitro guinea-pig preparations. *Journal of Physiology (London), 394*, 463–480.

Sano, T., Kohyama-Koganeya, Y., Kinoshita, M. O., Tatsukawa, T., Shimizu, C., Oshima, E., Hama, H., Yamada, K., Le, T. D., Miyawaki, A., Tohyama, K., Nagao, S., & Hirabayashi, Y. (2018). Loss of GPRC5B impairs synapse formation of Purkinje cells with cerebellar nuclear neurons and disrupts cerebellar synaptic plasticity and motor learning. *Neuroscience Research, 136*, 33–47.

Sato, Y., Kanda, K., & Kawasaki, T. (1987). Target neurons of floccular middle zone inhibition in medial vestibular nucleus. *Brain Research, 446*, 225–235.

Schonewille, M., Gao, Z., Boele, H.-J., Veloz, M. F. V., Amerika, W. E., Simek, A. A. M., De Jew, M. T., Steinberg, J. P., Takamiya, K., Hoebeek, F. E., Linden, D. J., Huganir, R. L., &

De Zeeuw, C. I. (2011). Reevaluating the role of LTD in cerebellar motor learning. *Neuron, 70*, 43–50.

Sekirnjak, C., Vissel, B., Bollinger, J., Faulstich, M., & du Lac, S. (2003). Purkinje cell synapses target physiologically unique brainstem neurons. *The Journal of Neuroscience, 23*, 6392–6398.

Shutoh, F., Ohki, M., Kitazawa, H., Itohara, S., & Nagao, S. (2006). Memory trace of motor learning shifts transsynaptically from cerebellar cortex to nuclei for consolidation. *Neuroscience, 139*, 767–777.

Sugawara, T., Hisatsune, C., Le, T. D., Hashikawa, T., Hirono, M., Nagao, S., & Mikoshiba, K. (2013). IP3R1 regulates cerebellar circuits by maintaining the spine and dendritic morphology of Purkinje cells in adult mice. *The Journal of Neuroscience, 33*, 12186–12195.

Sugihara, I., & Shinoda, Y. (2004). Molecular, topographic and functional organization of the cerebellar cortex: A study with combined aldolase C and olivocerebellar labeling. *The Journal of Neuroscience, 24*, 8771–8785.

Tanaka, S., Kawaguchi, S. Y., Shioi, G., & Hirano, T. (2013). Long-term potentiation of inhibitory synaptic transmission onto cerebellar Purkinje neurons contributes to adaptation of vestibulo-ocular reflex. *The Journal of Neuroscience, 33*, 17209–17220.

Thompson, R. F. (1988). The neural basis of basic associative learning of discrete behavioral responses. *Trends in Neurosciences, 11*, 152–155.

Todaka, H., Tatsukawa, T., Hashikawa, T., Yanagawa, Y., Shibuki, K., & Nagao, S. (2013). Heterotrimetric guanosine triphosphate-binding protein-coupled modulatory actions of motilin on K^+ channels and postsynaptic GABA receptors in mouse medial vestibular nuclear neurons. *The European Journal of Neuroscience, 37*, 339–350.

Tsukahara, N., Hultborn, H., Murakami, F., & Fujito, Y. (1975). Electrophysiological study of formation of new synapses and collateral sprouting in red nucleus neurons after partial denervation. *Journal of Neurophysiology, 38*, 1359–1372.

Wang, W., Nakadate, K., Masugi-Tokita, M., Shutoh, F., Aziz, W., Tarusawa, E., Lorincz, A., Molonar, E., Kesef, S., Li, Y.-Q., Fukazawa, Y., Nagao, S., & Shigemoto, R. (2014). Distinct cerebellar engrams in short-term and long-term motor learning. *Proceedings of the National Academy of Sciences of the United States of America, 111*, E188–E193.

Watanabe, E. (1985). Role of the primate flocculus in adaptation of vestibulo-ocular reflex. *Neuroscience Research, 3*, 20–38.

Welsh, J. P., Yamaguchi, H., Zeng, X. H., Kojo, M., Nakada, Y., Takagi, A., Sugimori, M., & Llinás, R. R. (2005). Normal motor learning during pharmacological prevention of Purkinje cell long-term depression. *Proceedings of the National Academy of Sciences of the United States of America, 102*, 17166–17171.

Yamaguchi, K., Itohara, S., & Ito, M. (2016). Reassessment of long-term depression in cerebellar Purkinje cells in mice carrying mutated GluA2 C terminus. *Proceedings of the National Academy of Sciences of the United States of America, 113*, 10192–10197.

Yamazaki, T., & Nagao, S. (2012). A computational mechanism for unified gain and timing control in the cerebellum. *PLoS One, 7*, e33319.

Yamazaki, T., & Tanaka, S. (2007). A spiking network model for passage-of-time representation in the cerebellum. *The European Journal of Neuroscience, 26*, 2279–2292.

Yamazaki, T., Nagao, S., Lennon, W., & Tanaka, S. (2015). Modeling the memory consolidation during post-training periods in cerebellovestibular learning system. *Proceedings of the National Academy of Sciences of the United States of America, 112*, 3541–3546.

Yeo, C. H., & Hesslow, G. (1998). Cerebellum and conditioned reflexes. *Trends in Cognitive Sciences, 2*, 320–330.

Yeo, C. H., Hardman, M. J., & Glickstein, M. (1985a). Classical-conditioning of nictitating response of the rabbit. I. Lesions of the cerebellar nuclei. *Experimental Brain Research, 60*, 87–98.

Yeo, C. H., Hardman, M. J., & Glickstein, M. (1985b). Classical-conditioning of nictitating response of the rabbit. II. Lesions of the cerebellar cortex. *Experimental Brain Research, 60*, 99–113.

Yeo, C. H., Hardman, M. J., & Glickstein, M. (1985c). Classical-conditioning of nictitating response of the rabbit. III. Connections of cerebellar lobule HVI. *Experimental Brain Research, 60*, 114–126.

Yeo, C. H., Hardman, M. J., & Glickstein, M. (1986). Classical-conditioning of nictitating response of the rabbit. IV. Lesions of inferior olive. *Experimental Brain Research, 61*, 81–92.

Yuzaki, M. (2013). Cerebellar LTD vs. motor learning–lessons learned from studying GluD2. *Neural Networks, 47*, 36–41.

Zhang, W., & Linden, D. J. (2006). Long-term depression at the mossy fiber–deep cerebellar nucleus synapse. *The Journal of Neuroscience, 26*, 6935–6944.

Chapter 15
Timing in Purkinje Cells and a Novel Learning Mechanism

Germund Hesslow, Dan-Anders Jirenhed, and Fredrik Johansson

15.1 Introduction: Eyeblink Conditioning

The idea that the cerebellum is important for motor timing has attracted the attention of many investigators for several decades (Braitenberg, 1967; Ivry & Keele, 1989; Jueptner et al., 1996). It has been brought to the forefront after the discovery that classical conditioning of well-timed motor responses is learned by the cerebellum. In eyeblink conditioning, a widely used experimental model of associative learning, an initially neutral conditional stimulus (CS), such as a tone or a light, is repeatedly succeeded by an unconditional stimulus (US), such as an air puff to the cornea, that elicits a blink reflex. After repeated presentations of the CS and US, the subject learns to emit a conditional blink in response to the CS. Eyeblink conditioning is a robust phenomenon that is found in a large number of mammalian species including humans and has very similar properties (Gormezano & Moore, 1969; Kehoe & Macrae, 2002; Mackintosh, 1974).

15.2 Temporal Properties of the CR

The conditional response (CR) is precisely timed. It precedes the US so that it reaches its maximum amplitude close to the time of the (expected) US onset (Gallistel, 1990; Gormezano & Moore, 1969; Kehoe & Macrae, 2002). This holds for CS-US intervals between about a hundred milliseconds and 2 seconds. At shorter intervals, no learning occurs, and at longer intervals learning becomes slow and unreliable and the timing of the CR erratic.

G. Hesslow (✉) · D.-A. Jirenhed · F. Johansson
Department of Experimental Medical Science, Lund University, Lund, Sweden
e-mail: germund.hesslow@med.lu.se

H. Mizusawa, S. Kakei (eds.), *Cerebellum as a CNS Hub*,
Contemporary Clinical Neuroscience, https://doi.org/10.1007/978-3-030-75817-2_15

If the CS-US interval is changed, additional training will result in a CR that is adapted to the new interval. The change does not consist in a gradual move of the CR latency but in the acquisition of a new CR adapted to the new US timing and extinction of the previous CR (Kehoe & Joscelyne, 2005). Subjects that are trained with two different alternating CS-US intervals will often learn double-peaked responses where each peak is timed to one of the two US onset times (Hoehler & Leonard, 1976; Millenson et al., 1977).

The CR latency can be changed by changes in CS parameters. For instance, in a subject that emits reliable and well-timed CR, a sudden increase in CS intensity can cause a considerable shortening of the CR latency. With additional training with the stronger CS, the CR latency will return to the original one (Svensson et al., 1997).

15.3 Pause in Purkinje Cell Firing Drives the CR

After the initial discovery by McCormick et al. (1981) that lesions of the cerebellar output pathway abolished conditional eyeblinks, experimental work by many investigators, using lesions, pharmacological inactivation and electrophysiology, has shown that the main mechanisms underlying eyeblink conditioning reside in the cerebellum (Freeman, 2015; Hesslow & Yeo, 2002). Some investigators have suggested that plasticity in the deep nuclei plays a significant role, but although we cannot exclude a nuclear contribution, it now seems reasonably clear that the most important learning mechanisms are in located the cortex (Longley & Yeo, 2014). A large body of evidence supports a model where the CS is transmitted to the Purkinje cells by mossy fibres and parallel fibres and the US by climbing fibres from the inferior olive.

Consistent with this is the discovery that Purkinje cells, which are known to control the blink (Hesslow, 1994a, 1994b), develop a pause response to the CS during conditioning (Gallistel et al., 2020; Hesslow & Ivarsson, 1994; Jirenhed et al., 2007). This response, the "Purkinje cell CR", has all the known temporal properties of the overt blink CR. It has a latency that corresponds to the CS-US interval, just like the overt CR. The Purkinje cell CR is not learned when the CS-US interval shorter than ~100 ms, and it can be modified just like the overt CR by changes in CS parameters (Jirenhed & Hesslow, 2011; Svensson et al., 2010; Wetmore et al., 2014). Remarkably, when two CS-US intervals are mixed in the way that produces a double-peaked blink (see above), the Purkinje cell develops two pauses in simple spike firing in response to the CS (Jirenhed et al., 2017). For these and other reasons, we concluded that the Purkinje cell CR is the main driver of the overt CR (Jirenhed & Hesslow, 2016).

15.4 Theories of Learning and Timing

Long-term depression (LTD) of parallel fibre to Purkinje cell synapses was the first learning mechanism to be proposed (Marr, 1969) and also demonstrated (Ito et al., 1982) in the cerebellum. Theories about eyeblink conditioning have not surprisingly assumed that the learning is based on Purkinje cell LTD. When the excitatory input from parallel fibres is weakened, inhibitory input from the interneurons would dominate and cause a suppression of the simple spike firing. As pointed out by several authors, however, there is a lot of experimental evidence that would cause us to question the role of LTD in eyeblink conditioning (Hesslow et al., 2013; Schonewille et al., 2011; Welsh et al., 2005).

One serious problem for any theory of learning, which is based on strengthening or weakening of synaptic strength, is how to account for the timing of neural responses and especially the double pause responses mentioned above. To solve this problem, several theories have been proposed where the common idea is that there are inbuilt delays in the afferent pathway that mediate the CS signal, such as the granule cells. There is a family of such "delay-line" models (Desmond & Moore, 1988; Hansel et al., 2001; Mauk & Buonomano, 2004; Medina & Mauk, 2000). For a good review, see Yamazaki and Tanaka (2009). Associative learning in these theories would only take place between elements that are appropriately timed relative to each other. If there are delays in granule cells, for instance, only those parallel fibres to Purkinje cells synapses that are active simultaneously with the climbing fibre input would be depressed. The delays in those parallel fibres would then ensure the correct timing of the Purkinje cell CR. When the appropriately timed parallel fibre signals arrive, before training, glutamatergic excitation of Purkinje cells and GABAergic inhibition via parallel fibre excited interneurons would balance each other out, but after training when LTD had weakened the glutamatergic synapses on the Purkinje cells, the inhibitory effect would dominate and cause a pause in the cells that could release an excitatory signal from the nuclei. Mechanisms of this kind could explain how differently timed CRs might be learned and also how double pause CRs could arise.

15.5 Evidence Against Earlier Theories

We have previously argued that the LTD hypothesis is difficult to reconcile with known facts about the cerebellar circuitry and also that LTD does not have the requisite properties (Hesslow et al., 2013).

For instance, as noted above, conditioning does not occur when the interval between the CS and the US is shorter than about 100 ms. We have tested the effect of using short CS-US intervals on Purkinje cell CRs as well. The results were quite similar to the behavioural findings. With a CS-US interval of 50 ms, no Purkinje cell

CR was acquired. Indeed, the cell learned to *increase* its simple spike firing rate in response to the CS (Wetmore et al., 2014).

LTD, in contrast, is usually reported to be most effective at short delays between pf and cf. input. Most work on LTD has used simultaneous parallel and climbing fibre inputs, so-called conjunctive stimulation, and comparisons of various delays between parallel and climbing fibre stimulation have indicated that LTD is optimal when the delay is close to zero (Ekerot & Kano, 1989; Ito, 2001; Karachot et al., 1995) (cf discussion in (Hesslow et al., 2013).

Another discrepancy between LTD and conditioning concerns the properties of the climbing fibre input. Almost all work on LTD has used single pulses to the climbing fibres to induce LTD. Single climbing fibre impulses in contrast do not cause acquisition of Purkinje cell CRs but rather *extinction* of CRs (Rasmussen et al., 2013). At least three climbing fibre impulses seem to be required for conditioning.

Recently we have obtained even stronger evidence against LTD being the mechanism underlying conditioning.

First, the Purkinje cell pauses do not seem to depend on interneuron inhibition. We used a preparation with decerebrate ferrets in which direct electrical stimulation of parallel fibres and climbing fibres was used as CS and US. Paired CS-US stimulation caused the acquisition of robust Purkinje cell CRs. With so-called "off-beam" stimulation, where parallel fibres are stimulated next to those that excite a particular Purkinje cell, it is possible to obtain interneuron mediated inhibition of the cell without any excitation (Eccles et al., 1967). By injecting very small doses of the GABA receptor antagonist gabazine just next to a Purkinje cell from which we were recording, it was possible to completely abolish a powerful inhibitory response. Yet, this complete removal of interneuron inhibition had no effect at all on the Purkinje cell CR (Johansson et al., 2014).

Second, in a direct test of the LTD hypothesis, parallel fibre excitation of Purkinje cells was found to be same before and after conditioning, that is, LTD did not seem to occur (Johansson et al., 2018).

Third, variable time courses of different parallel fibre inputs, as required by the delay-line models, is not necessary for learned timing. In the setup described in the previous paragraph, we used three different CS-US intervals. The CS was a train of electrical stimuli directly to the parallel fibres. Thus, there could be no delay in the parallel fibre signal. If the timing of the CR had been dependent on such delays, it should not have been possible for the cell to learn differently timed simple spike pauses. Yet, for each of the three intervals, training produced Purkinje cell CRs that reached their maxima close to the corresponding expected US onset, just as is the case when peripheral CSs are used (Johansson et al., 2014).

Fourth, the timing of a Purkinje cell CR changed, when the CS-US interval changed (Jirenhed & Hesslow, 2011; Johansson et al., 2014). It was even possible, when a cell was trained with mixed CS-US intervals, to obtain double Purkinje cell CRs. This means that one cell can learn more than one CR timings. This means that the timing is really learned by the cell and does not result from selection of a subset of Purkinje cells. It also undermines the idea that timing could be achieved by a selection of Purkinje cells with different properties. For a discussion of such timing ideas, for instance, Fiala et al. (1996) or Steuber and Willshaw (2004), see Hesslow et al. (2013).

The unavoidable conclusions are that LTD does not play a major role in conditioning and that the timing mechanism must be intrinsic to the Purkinje cell and does not depend on temporal properties of the input signal.

15.6 Generation of the Purkinje Cell CR-mGluR7 and $K_{ir}3$ Channels

The results described above raises a puzzling question. Parallel fibre release of glutamate has traditionally been assumed to act on ionotropic AMPA/kainate receptors which would be expected to excite the cell. But blocking the GABA receptors has no effect on the Purkinje cell CR, so what is causing the suppression of simple spike firing in the CR?

The obvious candidate would be a metabotropic glutamate receptor of which there are two in Purkinje cells mGluR1 and mGluR7. Local microinjections of antagonists of these receptors indicate that mGluR7, but not mGluR1, is necessary for the generation of a normal Purkinje cell CRs (Johansson et al., 2015).

It is not obvious how activation of mGluR7 could elicit inhibitory responses with learned temporal properties, but a possible mechanism is $K_{ir}3$ channels, also known as GIRK (G protein-coupled inwardly rectifying potassium) channels. These are activated by G protein βγ-subunits from heterotrimeric G protein complexes (Dascal & Kahanovitch, 2015; Niswender & Conn, 2010). $K_{ir}3$ channels are plausibly involved in generating the CR for the following reasons. These channels are composed of subunits that can form multiple combinations. Differences in subunit composition can confer a variety of temporal properties to the channel complex (Wischmeyer et al., 1997). This, together with the fact that the temporal dynamics of the G protein actions are controlled by the regulator of G protein signalling (RGS) family (Doupnik, 2015), would seem to make these channels sufficiently flexible and well suited for generating timed responses in the range of hundreds of milliseconds.

This suggestion has been supported be recent experiments. Local application of the $K_{ir}3$ antagonist tertiapin in the cerebellar cortex suppressed normal performance of Purkinje cell CRs without any detectable effect on spontaneous firing (Johansson & Hesslow, 2020).

15.7 Conclusions

Ever since Cajal (1894) and Hebb (1949) proposed that modification of synaptic strength is the mechanism underlying memory, this idea has completely dominated thinking and research about learning and memory. There have been dissenting voices (Balsam et al., 2010; Gallistel & Matzel, 2013), but they have been few. This is also true of research on motor learning in the cerebellum (Ito et al., 2014; Marr, 1969) where first LTD of parallel fibre to Purkinje cell synapses and later potentiation of other cerebellar synapses (Jörntell et al., 2010; Jörntell & Ekerot, 2002) have been proposed as the basic mechanism. The results summarized above are extremely difficult to reconcile with the Cajal-Hebb hypothesis. Our findings suggest very strongly the existence of an additional intrinsic mechanism, which allows a cell to memorize a temporal interval.

Acknowledgements This work was supported by grants from the Swedish Research Council to F.J. (2016-00127 and 2019-02034) and to G.H. (2018-03191). Additional funding from the Krapperup Foundation to G.H. Additional funding to F.J. from the Swedish Brain Foundation (FO2020-0005), the Royal Swedish Academy of Sciences (ME2019-0048) and the Crafoord Foundation (20200529) as well as the Åke Wiberg, Magnus Bergvall and Segerfalk foundations.

References

Balsam, P. D., Drew, M. R., & Gallistel, C. R. (2010). Time and associative learning. *Comparative Cognition & Behavior Reviews, 5*, 1–22.

Braitenberg, V. (1967). Is the cerebellar cortex a biological clock in the millisecond range? *Progress in Brain Research, 25*, 334–346.

Cajal, S. (1894). La fine structure des centres nerveux. *Proceedings of the Royal Society of London, 55*, 444–468.

Dascal, N., & Kahanovitch, U. (2015). The roles of Gbetagamma and Galpha in gating and regulation of GIRK channels. *International Review of Neurobiology, 123*, 27–85. https://doi.org/10.1016/bs.irn.2015.06.001

Desmond, J. E., & Moore, J. W. (1988). Adaptive timing in neural networks: The conditioned response. *Biology and Cybernetics, 58*(6), 405–415.

Doupnik, C. A. (2015). RGS redundancy and implications in GPCR-GIRK signaling. *International Review of Neurobiology, 123*, 87–116. https://doi.org/10.1016/bs.irn.2015.05.010

Eccles, J. C., Ito, M., & Szentagothai, J. (1967). *The cerebellum as a neuronal machine*. Springer-Verlag.

Ekerot, C. F., & Kano, M. (1989). Stimulation parameters influencing climbing fibre induced long-term depression of parallel fibre synapses. *Neuroscience Research, 6*(3), 264–268.

Fiala, J. C., Grossberg, S., & Bullock, D. (1996). Metabotropic glutamate receptor activation in cerebellar Purkinje cells as substrate for adaptive timing of the classically conditioned eye-blink response. *Journal of Neuroscience, 16*(11), 3760–3774.

Freeman, J. H. (2015). Cerebellar learning mechanisms. *Brain Research, 1621*, 260–269. https://doi.org/10.1016/j.brainres.2014.09.062

Gallistel, C. (1990). *The organization of learning*. Bradford Books/MIT Press.

Gallistel, C. R., & Matzel, L. D. (2013). The neuroscience of learning: Beyond the hebbian synapse. *Annual Review of Psychology, 64*, 169–200. https://doi.org/10.1146/annurev-psych-113011-143807

Gallistel, C, Johansson, F, Jirenhed, D-A, Rasmussen, A, Ricci, M, Hesslow, G. (2020). Quantitative properties of the creation and activation of a cell-intrinsic engram. bioRxiv 20200317995258. https://doi.org/10.1101/2020.03.17.995258.

Gormezano, I., & Moore, J. W. (1969). Classical conditioning. In M. H. Marx (Ed.), *Learning: processes*. Macmillan.

Hansel, C., Linden, D. J., & D'Angelo, E. (2001). Beyond parallel fiber LTD: The diversity of synaptic and non-synaptic plasticity in the cerebellum. *Nature Neuroscience, 4*(5), 467–475. https://doi.org/10.1038/8741987419. [pii].

Hebb, D. O. (1949). *The organization of behavior; a neuropsychological theory*. Wiley.

Hesslow, G. (1994a). Correspondence between climbing fibre input and motor output in eyeblink-related areas in cat cerebellar cortex. *Journal of Physiology (London), 476*(2), 229–244.

Hesslow, G. (1994b). Inhibition of classically conditioned eyeblink responses by stimulation of the cerebellar cortex in the decerebrate cat. *Journal of Physiology (London), 476*(2), 245–256.

Hesslow, G., & Ivarsson, M. (1994). Suppression of cerebellar Purkinje cells during conditioned responses in ferrets. *Neuroreport, 5*(5), 649–652.

Hesslow, G., & Yeo, C. H. (2002). The functional anatomy of skeletal conditioning. In J. W. Moore (Ed.), *A Neuroscientist's guide to classical conditioning* (pp. 86–146). Springer-Verlag.

Hesslow, G., Jirenhed, D.-A., Rasmussen, A., & Johansson, F. (2013). Classical conditioning of motor responses: What is the learning mechanism? *Neural Networks, 47*, 81–87. https://doi.org/10.1016/j.neunet.2013.03.013

Hoehler, F. K., & Leonard, D. W. (1976). Double responding in classical nictitating membrane conditioning with single-CS dual-ISI training. *The Pavlovian Journal of Biological Science, 11*, 180–190.

Ito, M. (2001). Cerebellar long-term depression: Characterization, signal transduction, and functional roles. *Physiological Reviews, 81*(3), 1143–1195.

Ito, M., Sakurai, M., & Tongroach, P. (1982). Climbing fibre induced depression of both mossy fibre responsiveness and glutamate sensitivity of cerebellar Purkinje cells. *Journal of Physiology (London), 324*, 113–134.

Ito, M., Yamaguchi, K., Nagao, S., & Yamazaki, T. (2014). Long-term depression as a model of cerebellar plasticity. *Progress in Brain Research, 210*, 1–30. https://doi.org/10.1016/B978-0-444-63356-9.00001-7

Ivry, R. B., & Keele, S. W. (1989). Timing functions of the cerebellum. *Journal of Cognitive Neuroscience, 1*(2), 136–152. https://doi.org/10.1162/jocn.1989.1.2.136

Jirenhed, D. A., & Hesslow, G. (2011). Learning stimulus intervals--adaptive timing of conditioned purkinje cell responses. *Cerebellum, 10*(3), 523–535. https://doi.org/10.1007/s12311-011-0264-3

Jirenhed, D. A., & Hesslow, G. (2016). Are purkinje cell pauses drivers of classically conditioned blink responses? *Cerebellum, 15*(4), 526–534. https://doi.org/10.1007/s12311-015-0722-4

Jirenhed, D. A., Bengtsson, F., & Hesslow, G. (2007). Acquisition, extinction, and reacquisition of a cerebellar cortical memory trace. *The Journal of Neuroscience, 27*(10), 2493–2502. https://doi.org/10.1523/JNEUROSCI.4202-06.2007

Jirenhed, D.-A., Rasmussen, A., Johansson, F., & Hesslow, G. (2017). Learned response sequences in cerebellar Purkinje cells. *Proceedings of the National Academy of Sciences of the United States of America, 114*(23), 6127–6132. https://doi.org/10.1073/pnas.1621132114

Johansson, F., & Hesslow, G. (2020). Kir3 channel blockade in the cerebellar cortex suppresses performance of classically conditioned purkinje cell responses. *Scientific Reports, 10*(1), 15654. https://doi.org/10.1038/s41598-020-72581-8

Johansson, F., Jirenhed, D. A., Rasmussen, A., Zucca, R., & Hesslow, G. (2014). Memory trace and timing mechanism localized to cerebellar purkinje cells. *Proceedings of the National Academy of Sciences of the United States of America, 111*(41), 14930–14934. https://doi.org/10.1073/pnas.1415371111

Johansson, F., Carlsson, H. A. E., Rasmussen, A., Yeo, C. H., & Hesslow, G. (2015). Activation of a temporal memory in purkinje cells by the mGluR7 receptor. *Cell Reports, 13*(9), 1741–1746. https://doi.org/10.1016/j.celrep.2015.10.047

Johansson, F., Jirenhed, D. A., Rasmussen, A., Zucca, R., & Hesslow, G. (2018). Absence of parallel fibre to purkinje cell LTD during eyeblink conditioning. *Scientific Reports, 8*(1), 14777. https://doi.org/10.1038/s41598-018-32791-7

Jörntell, H., & Ekerot, C.-F. (2002). Reciprocal bidirectional plasticity of parallel fiber receptive fields in cerebellar purkinje cells and their afferent interneurons. *Neuron, 34*, 797–806.

Jörntell, H., Bengtsson, F., Schonewille, M., & De Zeeuw, C. I. (2010). Cerebellar molecular layer interneurons – computational properties and roles in learning. *Trends in Neurosciences, 33*(11), 524–532. https://doi.org/10.1016/j.tins.2010.08.004

Jueptner, M., Flerich, L., Weiller, C., Mueller, S. P., & Diener, H. C. (1996). The human cerebellum and temporal information processing—results from a PET experiment. *Neuroreport, 7*, 2761–2765.

Karachot, L., Kado, R. T., & Ito, M. (1995). Stimulus parameters for induction of long-term depression in in vitro rat purkinje cells. *Neuroscience Research, 21*, 161–168.

Kehoe, E. J., & Joscelyne, A. (2005). Temporally specific extinction of conditioned responses in the rabbit (Oryctolagus cuniculus) nictitating membrane preparation. *Behavioral Neuroscience, 119*, 1011–1022.

Kehoe, E. J., & Macrae, M. (2002). Fundamental behavioral methods and findings in classical conditioning. In J. W. Moore (Ed.), *A neuroscientist's guide to classical conditioning* (pp. 171–231). Springer-Verlag.

Longley, M., & Yeo, C. H. (2014). Distribution of neural plasticity in cerebellum- dependent motor learning. *Progress in Brain Research, 210*, 79–101.

Mackintosh, N. J. (1974). *The psychology of animal learning*. Academic Press.

Marr, D. (1969). A theory of cerebellar cortex. *Journal of Physiology (London), 202*(2), 437–470.

Mauk, M. D., & Buonomano, D. V. (2004). The neural basis of temporal processing. *Annual Review of Neuroscience, 27*, 307–340.

McCormick, D. A., Lavond, D., Clark, G. A., Kettner, R. E., Rising, C. E., & Thompson, R. F. (1981). The engram found? Role of the cerebellum in classical conditioning of nictitating membrane and eyelid responses. *Bulletin of the Psychonomic Society, 18*(3), 103–105.

Medina, J. F., & Mauk, M. D. (2000). Computer simulation of cerebellar information processing. *Nature Neuroscience, 3*, 1205–1211.

Millenson, J. R., Kehoe, E. J., & Gormezano, I. (1977). Classical conditioning of the rabbit's nictitating membrane response under fixed and mixed CS-US intervals. *Learning and Motivation, 8*, 351–366.

Niswender, C. M., & Conn, P. J. (2010). Metabotropic glutamate receptors: Physiology, pharmacology, and disease. *Annual Review of Pharmacology and Toxicology, 50*, 295–322. https://doi.org/10.1146/annurev.pharmtox.011008.145533

Rasmussen, A., Jirenhed, D.-A., Zucca, R., Johansson, F., Svensson, P., & Hesslow, G. (2013). Number of spikes in climbing fibers determines the direction of cerebellar learning. *Journal of Neuroscience, 33*(33), 13436–13440. https://doi.org/10.1523/JNEUROSCI.1527-13.2013

Schonewille, M., Gao, Z., Boele, H. J., Veloz, M. F., Amerika, W. E., Simek, A. A., De Jeu, M. T., Steinberg, J. P., Takamiya, K., Hoebeek, F. E., Linden, D. J., Huganir, R. L., & De Zeeuw, C. I. (2011). Reevaluating the role of LTD in cerebellar motor learning. *Neuron, 70*(1), 43–50. https://doi.org/10.1016/j.neuron.2011.02.044

Steuber, V., & Willshaw, D. (2004). A biophysical model of synaptic delay learning and temporal pattern recognition in a cerebellar purkinje cell. *Journal of Computational Neuroscience, 17*, 149–164.

Svensson, P., Ivarsson, M., & Hesslow, G. (1997). Effect of varying the intensity and train frequency of forelimb and cerebellar mossy fiber conditioned stimuli on the latency of conditioned eye-blink responses in decerebrate ferrets. *Learning & Memory, 4*(1), 105–115.

Svensson, P., Jirenhed, D. A., Bengtsson, F., & Hesslow, G. (2010). Effect of conditioned stimulus parameters on timing of conditioned purkinje cell responses. *Journal of Neurophysiology, 103*(3), 1329–1336. https://doi.org/10.1152/jn.00524.2009

Welsh, J. P., Yamaguchi, H., Zeng, X. H., Kojo, M., Nakada, Y., Takagi, A., Sugimori, M., & Llinas, R. R. (2005). Normal motor learning during pharmacological prevention of Purkinje cell long-term depression. *Proceedings of the National Academy of Sciences of the United States of America, 102*(47), 17166–17171. https://doi.org/10.1073/pnas.0508191102

Wetmore, D. Z., Jirenhed, D.-A., Rasmussen, A., Johansson, F., Schnitzer, M. J., & Hesslow, G. (2014). Bidirectional plasticity of purkinje cells matches temporal features of learning. *Journal of Neuroscience, 34*(5), 1731–1737. https://doi.org/10.1523/JNEUROSCI.2883-13.2014

Wischmeyer, E., Doring, F., Wischmeyer, E., Spauschus, A., Thomzig, A., Veh, R., & Karschin, A. (1997). Subunit interactions in the assembly of neuronal Kir3.0 inwardly rectifying K+ channels. *Molecular and Cellular Neurosciences, 9*(3), 194–206. https://doi.org/10.1006/mcne.1997.0614

Yamazaki, T., & Tanaka, S. (2009). Computational models of timing mechanisms in the cerebellar granular layer. *Cerebellum, 8*, 423–432.

Chapter 16
Contribution of Norepinephrine to Cerebellar Long-Term Depression and Motor Learning

Tomoo Hirano and Takuma Inoshita

16.1 Roles of PF-PN LTD in Motor Learning

Marr (1969) and Albus (1971) proposed that synaptic plasticity at parallel fiber (PF) to Purkinje neuron (PN) synapses in the cerebellar cortex is responsible for motor learning. Ito and his colleagues reported that conjunctive stimulation of PFs and a climbing fiber (CF) depresses PF-PN synaptic transmission for a long term (Ito et al., 1982). This synaptic plasticity is the cerebellar long-term depression (LTD), and its induction mechanisms and roles in motor learning have been intensively studied using various preparations including mutant mice, brain slices, and cultured neurons (Hirano, 1990; Hirano, 2013; Ito, 2001; Linden et al., 1991; Sakurai, 1987).

Ito proposed that adaptation of vestibulo-ocular reflex (VOR), a type of oculomotor reflex, is controlled by the flocculus (Ito, 1982), a small phylogenetically old part of the cerebellar cortex, but also see Miles and Lisberger (1981).Subsequent studies reported that another type of oculomotor reflex, optokinetic response (OKR), also undergoes adaptive changes and is controlled by the flocculus (Ito & Nagao, 1991). Both VOR and OKR are eyeball movements suppressing the image motion on a retina during head movement. In VOR, head turn stimulates the vestibular organs, which cause the eyeball turn in the opposite direction to the head turn. In OKR, motion of a large visual field causes the eyeball turn in the same direction as the image motion. In the case that VOR or OKR cannot suppress the retinal slip effectively, they change so that the retinal slip is minimized, which is called VOR or OKR adaptation. In some animal experiments designed to induce VOR adaptation, rotation of an animal is applied together with sinusoidal rotation of a surrounding screen in the opposite or same direction, which induces gradual increase or decrease of eye movement, respectively, resulting in suppression of the retinal slip. To induce

T. Hirano (✉) · T. Inoshita
Department of Biophysics, Graduate School of Science, Kyoto University, Kyoto, Japan

H. Mizusawa, S. Kakei (eds.), *Cerebellum as a CNS Hub*,
Contemporary Clinical Neuroscience, https://doi.org/10.1007/978-3-030-75817-2_16

OKR adaptation, the surrounding screen is rotated at a relatively high speed without animal rotation, which enhances the eyeball movement.

Many mutant mice deficient in cerebellar LTD showed motor learning failures such as defective VOR and/or OKR adaptation and defective eye-blink conditioning (Aiba et al., 1994; De Zeeuw et al., 1998; Hirano, 2014; Kashiwabuchi et al., 1995; Thompson, 2005). Mutant mice with facilitated LTD induction have also been generated, and they show either enhanced or suppressed motor learning (McConnell et al., 2009; Takeuchi et al., 2008). These studies supported the idea that LTD contributes to motor learning. However, normal motor learning under LTD suppression was reported and challenged the necessity of LTD in motor learning (Schonewille et al., 2011; Welsh et al., 2005). Later, a rebuttal to these challenges was published, demonstrating incomplete suppression of LTD in the previously reported LTD-defective mutant mice (Yamaguchi et al., 2016).

Recently, Inoshita and Hirano (2018) found that after OKR adaptation, unitary postsynaptic responses at PF-PN synapses are decreased, which is in accordance with immunocytochemical results showing that the number of AMPA-type glutamate receptors is decreased after OKR adaptation (Wang et al., 2014). Inoshita and Hirano (2018) also found that LTD is not induced in flocculus slices prepared from a mouse trained for OKR adaptation, implying that LTD has been saturated during the OKR adaptation and additional LTD induction is therefore prevented. These results together suggest that PF-PN LTD occurs during the OKR adaptation process. Thus, at least LTD seems to take place during motor learning.

16.2 Roles of PF-PN LTP in Motor Learning

Repetitive activation of PF alone induces long-term potentiation (LTP) at PF-PN synapses (Hirano, 1990; Sakurai, 1987; Salin et al., 1996). Two types of LTP, one with enhanced presynaptic transmitter release (Hirano, 1991) and the other with enhanced postsynaptic responsiveness to the transmitter glutamate (Lev-Ram et al., 2002), have been reported. Thus, postsynaptic LTP could reverse LTD which is expressed in the postsynaptic membrane (Hirano, 1991; Ito et al., 1982; Linden et al., 1991). Mutant mice deficient in postsynaptic LTP, such as T-type Ca^{2+} channel knockout mice and PN-specific protein phosphatase 2B-knockout mice, show suppressed motor learning such as VOR adaptation, suggesting that PF-PN LTP also contributes to motor learning (Ly et al., 2013; Schonewille et al., 2010).

16.3 Roles of Other Cerebellar Plasticity

Inhibitory GABAergic synapses on a PN also show synaptic plasticity. Repeated CF stimulation potentiates molecular layer inhibitory interneuron (MLIN) to PN synaptic transmission long-term, an effect which was named rebound potentiation (RP)

(Kano et al., 1992). Regulatory mechanisms of RP and its role in motor learning have been intensively studied (Hirano & Kawaguchi, 2014). Mutant mice specifically defective in RP show suppressed VOR adaptation (Tanaka et al., 2013). RP and PF-PN LTD might operate synergistically, since both work to suppress the PN activity in parallel in a manner dependent on the CF activity. Thus, one plasticity mechanism might compensate the other's defects (Hirano, 2013, 2014).

Plasticity has also been reported to occur at cerebellar cortical synapses on neurons other than PNs such as PF to MLIN synapses and mossy fiber to granule neuron (GN) synapses (D'Angelo et al., 2005; Jörntell & Ekerot, 2002). At PF to MLIN synapses, PF stimulation alone induces LTD, and conjunctive stimulation of PF and CF causes LTP (Jörntell & Ekerot, 2002). Thus, similar conditioning stimulation of PF with or without CF stimulation induces changes in the efficacy of PF outputs in opposite directions in two types of postsynaptic neurons PN and MLIN. Considering the sign changes of information occurring at MLIN to PN inhibitory synapses, synaptic plasticity at PF-MLIN synapses also seems to work synergistically with PF-PN synaptic plasticity. Finally, plasticity of intrinsic excitability is induced in a PN and might contribute to motor learning (Titley et al., 2020). Thus, there are multiple types of plasticity in the cerebellar cortex, which might cooperate to improve motor learning performance (Dean et al., 2010; Gao et al., 2012; Hansel et al., 2001; Hirano, 2018).

16.4 NE and Receptors

Norepinephrine (NE) is an aminergic neuromodulator released from neurons in the locus coeruleus (LC) (Sara, 2009; Tully & Bolshakov, 2010). NE-containing neurons in the LC receive inputs from and send outputs to broad regions of the central nervous system, including the cerebellum. Interestingly, it was reported that some PNs send direct outputs to LC neurons (Schwarz et al., 2015).

Receptors for NE are classified into three types, α1-, α2- and β-adrenergic receptors (AR) (Bylund et al., 1994; Philipp & Hein, 2004; Small et al., 2003). β-ARs are further divided into β1, β2, and β3 subtypes. α1- and α2-AR also have subtypes. They are α1A, B, and D and α2A, B, and C. The global knockout mice of each AR subtype have been generated, and their phenotypes have been studied, primarily focusing on cardiovascular regulation (Philipp & Hein, 2004). All ARs are coupled to trimeric G proteins. α1-AR is mainly coupled to Gq protein, which activates phospholipase C producing IP3 and diacylglycerol, which then lead to activation of protein kinase C. β-AR is primarily coupled to Gs protein, which activates adenylyl cyclase-producing cyclic AMP (cAMP), and cAMP activates protein kinase A. In contrast, α2-AR is coupled to Gi protein, which suppresses adenylyl cyclase. Thus, β-AR activity is counteracted by α2-AR activity in general. However, it should be noted that signaling pathways other than those explained above are likely to be activated by each AR subtype (Small et al., 2003).

16.5 Involvement of NE in Learning and Synaptic Plasticity

NE facilitates memory formation, consolidation, maintenance, and retrieval in general (Sara, 2009; Tully & Bolshakov, 2010). Previous studies suggested that when animal encounters a novel object, emotional arousal occurs, and neurons in the locus coeruleus are activated. An increase in NE concentration in an enriched environment was reported in the mouse central nervous system (Naka et al., 2002). NE is released in the hippocampus and amygdala, facilitating LTP and memory formation through β-AR. Maintenance of memory is also supported by NE through protein synthesis, which is presumably enhanced by β-AR, cAMP, and cAMP-dependent transcription factor CREB. Thus, NE and β-AR mediate emotional arousal-triggered enhancement of memory formation and maintenance. Such enhancement of learning and memory also occurs in other regions of the central nervous system such as the cerebral cortex and thalamus (Sara, 2009; Tully & Bolshakov, 2010). The contribution of α1B-AR to spatial learning was also demonstrated using the knockout mice (Spreng et al., 2001).

16.6 Involvement of NE in the Cerebellum-Dependent Motor Learning

NE also contributes to the cerebellum-dependent motor learning paradigms. Lesioning LC neurons impairs acquisition of a locomotor task (Watson & McElligott, 1984), and administration of propranolol, a β-AR antagonist, to the cerebellum of a rabbit or a rat suppresses acquisition of eyeblink conditioning (Gould, 1998; Cartford et al. 2002).

van Neerven et al. (1990) reported that injection of β-AR antagonist into the rabbit flocculus diminishes the gain-up adaptation of VOR, and Tan and Collewijn (1992) found that administration of a β-AR agonist increases the OKR gain. More recently, it was reported that application of NE or β-AR agonist into the mouse flocculus increases the OKR gain and that application of β-AR antagonist suppresses the gain-increase OKR adaptation (Wakita et al., 2017), suggesting that NE supports OKR adaptation through β-AR activation in mice.

16.7 Adrenergic Receptors in the Cerebellum

Morphologically, distribution of each AR subtype has been examined by radio- or fluorescent ligand-binding assays, immunohistochemistry, in situ hybridization, and/or expression of enhanced green fluorescent protein (EGFP) under the control of a specific AR promoter (Minneman et al., 1981; Bondok et al., 1988; Talley et al., 1996; Schambra et al., 2005; Papay et al., 2004, 2006) (Table 16.1). The

Table 16.1 Distribution of AR subtypes in the cerebellar cortex

AR subtypes	Animal	Methods	Neurons	Cellular regions	Authors	Year
β	Chick	FL	PN	S, D	Bondok et al.	1988
β1, β2	Teleost	RL, IH	PN		Zikopolos and Dermon	2005
β1	Rat	IH	PN	S, D	Paschalis et al.	2009
β2	Mouse	IH	PN, GN	S, D	Lippiello et al.	2015
α2A, B	Rat	ISH	PN, GN		Tavares et al.	1996
α1A, B, D, α2A, B, C	Human	ISH	PN, MLIN, GN, GoN		Schmbra et al.	2005
α1A	Mouse	Reporter	PN, GN		Papay et al.	2006

FL fluorescent ligand, *RL* radio-ligand, *IH* immunohistochemistry, *ISH* in situ hybridization, Reporter, EGFP expression under AR -promotor control, *PN* Purkinje neuron, *GN* granule neuron, *MLIN* molecular layer inhibitory interneuron, *GoN* Golgi neuron, *S* soma, *D* dendrite

ligand-binding assays and immunohistochemistry examine distribution of AR proteins, whereas in situ hybridization examines distribution of mRNA in neurons.

Unfortunately, only fragmental information is available about the distribution of each AR subtype in the cerebellar cortex. Ligand-binding studies on β-AR reported that β-AR protein is expressed in PNs in the rat, chicken, and teleost cerebella (Bondok et al., 1988; Minneman et al., 1981; Zikopoulos & Dermon, 2005). Study on the teleost used not only labeled ligands but also immunohistochemistry and reported that both β1- and β2-AR are expressed in PNs (Zikopoulos & Dermon, 2005). Immunohistochemical studies showed that β1-AR is expressed in PN soma and dendrites and also in MLINs in the rat cerebellum (Paschalis et al., 2009) and that β2-AR is expressed in PN soma and dendrites and GNs in the mouse cerebellum (Lippiello et al., 2015). Neither expression of β3-AR in the cerebellar cortex nor expression of β-AR in other neuron types in the cerebellum has been clearly shown.

Schambra et al. (2005) studied expression patterns of mRNAs of α1A-, α1B-, α1D-, α2A, α2B-, or α2C-AR in the human cerebellum by in situ hybridization. PNs strongly express α2A- and α2B-AR, and mRNAs of α1A-, α1B-, and/or α2C-AR are expressed in the cerebellar cortical region-specific manners. MLINs express α1A-, α1B-, and α2A-, α2B-AR mRNAs, but the expression pattern is different between the two types of MLINs, basket and stellate cells. GNs and Golgi neurons (GoNs) express α2A- and α2B-AR mRNAs. Thus, multiple subtypes of α1-AR and α2-AR are expressed in each cerebellar cortical neuron in the human cerebellum. An in situ hybridization study on the rat brain detected α2A- and α2B-AR mRNA in PNs and GNs (Tavares et al., 1996). Expression of EGFP under the control of the α1A- or α1B-AR promotor in the mouse cerebellum was studied (Papay et al., 2004, 2006). α1A- or α1B-AR-related EGFP was expressed in PNs, granular and molecular layers in the cerebellar cortex. The subcellular distribution pattern of AR proteins within each neuron could not been inferred from these studies. In a teleost, expression of α2A- and α2C-ARs was detected only in PNs by ligand-binding and immunohistochemical studies (Zikopoulos & Dermon, 2005). These authors noted that higher β-AR expression and lower α2-AR expression are common in teleost, avian, and mammalian cerebella. However, as far as we know, no information is available about the expression pattern of β-, α1-, or α2-AR in the flocculus.

16.8 NE Regulation of Cerebellar Synaptic Functions

Various effects of NE on excitatory glutamatergic and inhibitory GABAergic synaptic transmissions in the cerebellar cortex have been reported (Table 16.2). Mori-Okamoto and Tatsuno (1988) reported that NE enhances the response to glutamate in cultured chicken PNs through β-AR. In a slice preparation of the vermis of a juvenile mouse, β-AR activation enhances excitatory synaptic transmission and facilitates induction of LTP at PF-PN synapses (Lippiello et al., 2015), although application of β-AR agonist does not significantly affect PF-PN synaptic transmission in slices of the flocculus or vermis of adult mice (Inoshita and Hirano, 2021). Lippiello et al. (2015) also found that activation of α1- or α2-AR suppresses PF-PN synaptic transmission in the juvenile vermis. It is also known that α2-AR activation suppresses the glutamate release from a CF (Carey & Regehr, 2009).

At inhibitory synapses in the cerebellar cortex, activation of β-AR enhances GABA responsiveness of a PN (Cheun & Yeh, 1992; Waterhouse et al., 1982) and also GABA release from MLINs (Kondo & Marty, 1998; Lin et al., 1991; Llano & Gerschenfeld, 1993; Saitow et al., 2000). Activation of α1-AR also enhances inhibitory GABA release to a PN (Herold et al., 2005; Hirono & Obata, 2006). NE is expected to facilitate induction of rebound potentiation, long-lasting potentiation of the MLIN-PN GABAergic synaptic transmission through activation of β-AR and PKA (Kawaguchi & Hirano, 2002; Kitagawa et al., 2009; Sugiyama et al., 2008). Effects of NE on the spontaneous activity of a PN in vivo have also been studied (Guo et al., 2016; Hoffer et al., 1971). NE enhances both excitatory and inhibitory synaptic inputs, and the latter overwhelms the former in mice (Guo et al., 2016). These multiple effects of NE on various cerebellar synapses together with region-specific AR distribution patterns would modulate the neuronal computation in each cerebellar cortical region.

16.9 NE Facilitates LTD Induction Through β-Adrenergic Receptors

Our previous studies showed that LTD at PF-PN synapses occurs in the flocculus during the OKR adaptation process (Inoshita & Hirano, 2018) and that application of NE or a β-AR agonist to the mouse flocculus increases the OKR gain and that application of β-AR antagonist suppresses the OKR adaptation (Wakita et al., 2017). More recently, we found that NE or β-AR agonist facilitates the LTD induction through activation of protein kinase A (Inoshita and Hirano, 2021). It was also

Table 16.2 Synaptic effects of each AR in the cerebellar cortex. Pre or post indicates that pre- or postsynaptic side, respectively, is influenced

Synapses	Animal	Region	Age	AR subtypes	Direct effect	Plasticity	Pre or post	Authors	Year
PF-, CF-PN	Chick	Culture	Juv	β	Up		Post	Mori-Okamoto and Tatsuno	1998
PF-PN	Mouse	Ve	Juv	β	Up	LTP Up		Lippiello et al.	2015
PF-PN	Mouse	Ve	Juv	α1, α2	Down		α1, post; α2, pre and post	Lippiello et al.	2015
PF PN	Mouse	Fl	Ad	β	No effect	LTD up	Post	Inoshita and Hirano	2021
PF-PN	Mouse	Ve	Ad	α1?	Down	LTD occlusion		Inoshita and Hirano	2021
CF-PN	Rat	Ve	Juv	α2	Down		Pre	Carey and Regehr	2009
MLIN-PN	Rat	Ve	Ad	β	Up		Post	Waterhouse et al.	1982
MLIN-PN	Rat	Ve	Ad	β	Up		Pre	Lin et al.	1991
MLIN-PN	Rat	Iso	Neo	β	Up		Post	Cheun and Yeh	1992
MLIN-PN	Rat	Ve	Juv	β	Up		Pre	Llano and Gershenfeld	1993
MLIN-MLIN	Rat	Ve	Juv	β	Up		Pre	Kondo and Marty	1998
MLIN-PN	Rat		Juv	β	Up		Pre	Saitow et al.	2000
MLIN-PN	Rat	Ve	Juv	α1	Up		Pre	Herold et al.	2005
MLIN-PN	Mouse	Ve	Juv	α1	Up		Pre	Hirono and Obata	2006
PN activity	Rat		Ad	β	Down			Hoffer et al.	1971
PN activity	Mouse	Ve	Ad		Down>up			Guo et al.	2016

PF parallel fiber, *CF* climbing fiber, *PN* Purkinje neuron, *MLIN* molecular layer inhibitory interneuron, *Ve* vermis, *Fl* flocculus, *Iso* isolated, *Juv* juvenile, *Ad* adult, *Neo* neonatal

shown that the threshold of LTD induction is lowered without effects on the LTD amplitude. β-AR or NE does not affect the basal PF-PN synaptic transmission in the flocculus. These findings together suggest that NE released in the flocculus contributes to OKR adaptation through facilitation of LTD induction mediated by activation of β-AR and protein kinase A (Fig. 16.1).

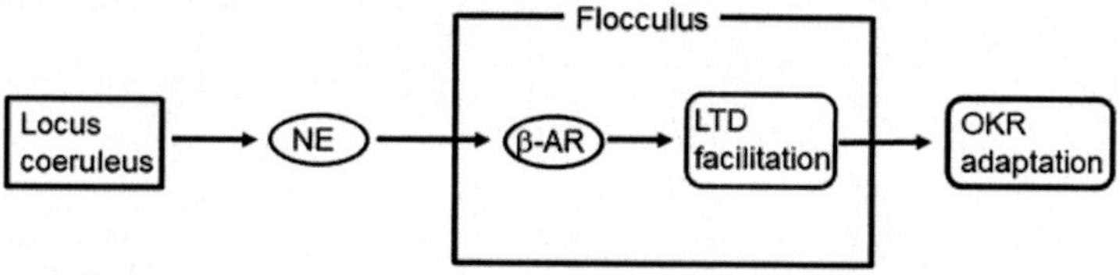

Fig. 16.1 Scheme explaining NE enhancement of OKR adaptation through β-AR-mediated facilitation of LTD

Interestingly, it was also found that in another region of the cerebellar cortex the vermis, NE decreases the basal PF-PN synaptic transmission and occludes the LTD induction independently of β-AR, implying other ARs such as α1-AR might be involved. On the other hand, the effects of a β-AR agonist on PF-PN synaptic responses and LTD induction in the vermis are similar to those in the flocculus. Thus, regulation of PF-PN synaptic transmission by NE differs among cerebellar cortical regions.

16.10 Conclusion

Recent studies suggest that NE may enhance OKR adaptation through facilitation of LTD induction at PF-PN synapses in the flocculus mediated by activation of β-AR, adenylyl cyclase, and protein kinase A. Such NE-dependent facilitation mechanisms of synaptic plasticity in the cerebellar cortex might contribute to motor learning in general. However, different distributions of AR subtypes and regulation of NE release among various cerebellar cortical regions might cause different effects of NE on the respective motor control and learning paradigms.

Acknowledgments The authors thank Drs. S. Kawaguchi and E. Nakajima for helpful comments on this manuscript.

References

Aiba, A., Kano, M., Chen, C., Stanton, M. E., Fox, G. D., Herrup, K., Zwingman, T. A., & Tonegawa, S. (1994). Deficient cerebellar long-term depression and impaired motor learning in mGluR1 mutant mice. *Cell, 7*, 377–388.

Albus, J. (1971). A theory of cerebellar function. *Mathematical Biosciences, 10*, 25–61.

Bondok, A. A., Botros, K. G., & El-Mohandes, E. A. (1988). Fluorescence histochemical study of the localisation and distribution of β-adrenergic receptor sites in the spinal cord and cerebellum of the chicken. *Journal of Anatomy, 160*, 167–174.

Bylund, D. B., Eikenberg, D. C., Hieble, J. P., Langer, S. Z., Lefkowitz, R. J., Minneman, K. P., Molinoff, P. B., Ruffolo, R. R., & Trendelenburg, U. (1994). International union of pharmacology nomenclature of adrenoceptors. *Pharmacological Reviews, 46*, 121–136.

Carey, M., & Regehr, W. (2009). Noradrenergic control of associative synaptic plasticity by selective modulation of instructive signals. *Neuron, 62*, 112–122.

Cartford, M. C., Allgeier, C. A., & Bickford, P. C. (2002). The effects of β-noradrenergic receptor blockade on acquisition of eyeblink conditioning in 3-month-old F344 rats. *Neurobiology of Learning and Memory, 78*, 246–257.

Cheun, J. E., & Yeh, H. H. (1992). Modulation of $GABA_A$ receptor-activated current by norepinephrine in cerebellar Purkinje cells. *Neuroscience, 51*, 951–960.

D'Angelo, E., Rossi, P., Gall, D., Prestori, F., Nieus, T., Maffei, A., & Sola, E. (2005). Long-term potentiation of synaptic transmission at the mossy fiber-granule cell relay of cerebellum. *Progress in Brain Research, 148*, 69–80.

De Zeeuw, C. I., Hansel, C., Bian, F., Koekkoek, S. K., van Alphen, A. M., Linden, D. J., & Oberdick, J. (1998). Expression of a protein kinase C inhibitor in Purkinje cells blocks cerebellar LTD and adaptation of the vestibulo-ocular reflex. *Neuron, 20*, 495–508.

Dean, P., Porrill, J., Ekerot, C. F., & Jörntell, H. (2010). The cerebellar microcircuit as an adaptive filter: Experimental and computational evidence. *Nature Reviews. Neuroscience, 11*, 30–43.

Gao, Z., van Beugen, B. J., & De Zeeuw, C. I. (2012). Distributed synergistic plasticity and cerebellar learning. *Nature Reviews. Neuroscience, 13*, 619–635.

Gould, T. J. (1998). β-Adrenergic involvement in acquisition vs. extinction of a classically conditioned eye blink response in rabbits. *Brain Research, 780*, 174–177.

Guo, A., Feng, J. Y., Li, J., Ding, N., Li, Y. J., Qiu, D. L., Piao, R. L., & Chu, C. P. (2016). Effects of norepinephrine on spontaneous firing activity of cerebellar Purkinje cells in vivo in mice. *Neuroscience Letters, 629*, 262–266.

Hansel, C., Linden, D. J., & D'Angelo, E. (2001). Beyond parallel fiber LTD: The diversity of synaptic and non-synaptic plasticity in the cerebellum. *Nature Neuroscience, 4*, 467–475.

Herold, S., Hecker, C., Deitmer, J., & Brockhaus, J. (2005). α1-adrenergic modulation of synaptic input to Purkinje neurons in rat cerebellar brain slices. *Journal of Neuroscience Research, 82*, 571–579.

Hirano, T. (1990). Depression and potentiation of the synaptic transmission between a granule cell and a Purkinje cell in rat cerebellar culture. *Neuroscience Letters, 119*, 141–144.

Hirano, T. (1991). Differential pre- and postsynaptic mechanisms for synaptic potentiation and depression between a granule cell and a Purkinje cell in rat cerebellar culture. *Synapse, 7*, 321–323.

Hirano, T. (2013). Long-term depression and other synaptic plasticity in the cerebellum. *Proceedings of the Japan Academy, 89*, 183–195.

Hirano, T. (2014). Around LTD hypothesis in motor learning. *Cerebellum, 12*, 645–650.

Hirano, T. (2018). Regulation and interaction of multiple types of synaptic plasticity in a Purkinje neuron and their contribution to motor learning. *Cerebellum, 17*, 756–765.

Hirano, T., & Kawaguchi, S. (2014). Regulation and functional roles of rebound potentiation at cerebellar stellate cell-Purkinje cell synapse. *Frontiers in Cellular Neuroscience, 8*, 42. https://doi.org/10.3389/fncel.2014.00042

Hirono, M., & Obata, K. (2006). α-Adrenoceptive dual modulation of inhibitory GABAergic inputs to Purkinje cells in the mouse cerebellum. *Journal of Neurophysiology, 95*, 700–708.

Hoffer, B. J., Siggins, G. R., & Bloom, F. E. (1971). Studies on norepinephrine-containing afferents to Purkinje cells of rat cerebellum. II. Sensitivity of Purkinje cells to norepinephrine and related substances administered by microiontophoresis. *Brain Research, 25*, 523–534.

Inoshita, T., & Hirano, T. (2018). Occurrence of long-term depression in the cerebellar flocculus during adaptation of optokinetic response. *eLife, 7*, e36209. https://doi.org/10.7554/eLife.36209

Inoshita, T., & Hirano, T. (2021). Norepinephrine facilitates induction of long-term depression through β-adrenergic receptor at parallel fiber-to-Purkinje cell synapses in the flocculus. *Neuroscience, 462*, 141–150.

Ito, M. (1982). Cerebellar control of the vestibulo-ocular reflex--around the flocculus hypothesis. *Annual Review of Neuroscience (Palo Alto, CA), 5*, 275–296.

Ito, M. (2001). Cerebellar long-term depression: Characterization, signal transduction, and functional roles. *Physiological Reviews, 81*, 1143–1195.
Ito, M., & Nagao, S. (1991). Comparative aspects of horizontal ocular reflexes and their cerebellar adaptive control in vertebrates. *Comparative Biochemistry and Physiology. C, 98*, 221–228.
Ito, M., Sakurai, M., & Tongroach, P. (1982). Climbing fibre induced depression of both mossy fibre responsiveness and glutamate sensitivity of cerebellar Purkinje cells. *The Journal of Physiology, 324*, 113–134.
Jörntell, H., & Ekerot, C. F. (2002). Reciprocal bidirectional plasticity of parallel fiber receptive fields in cerebellar Purkinje cells and their afferent interneurons. *Neuron, 34*, 797–806.
Kano, M., Rexhausen, U., Dreessen, J., & Konnerth, A. (1992). Synaptic excitation produces a long-lasting rebound potentiation of inhibitory synaptic signals in cerebellar Purkinje cells. *Nature, 1356*, 601–604.
Kashiwabuchi, N., Ikeda, K., Araki, K., Hirano, T., Shibuki, K., Takayama, C., Inoue, Y., Kutsuwada, T., Yagi, T., Kang, Y., Aizawa, S., & Mishina, M. (1995). Disturbed motor coordination, Purkinje cell synapse formation and cerebellar long-term depression of mice defective in the δ2 subunit of the glutamate receptor channel. *Cell, 81*, 245–252.
Kawaguchi, S., & Hirano, T. (2002). Signaling cascade regulating long-term potentiation of $GABA_A$ receptor responsiveness in cerebellar Purkinje neurons. *The Journal of Neuroscience, 22*, 3969–3976.
Kitagawa, Y., Hirano, T., & Kawaguchi, S. (2009). Prediction and validation of a mechanism to control the threshold for inhibitory synaptic plasticity. *Molecular Systems Biology, 5*, 280. https://doi.org/10.1038/msb.2009.39
Kondo, S., & Marty, A. (1998). Differential effects of noradrenaline on evoked, spontaneous and miniature IPSCs in rat cerebellar stellate cells. *The Journal of Physiology, 509*, 233–243.
Lev-Ram, V., Wong, S., Storm, D., & Tsien, R. (2002). A new form of cerebellar long-term potentiation is postsynaptic and depends on nitric oxide but not cAMP. *Proceedings of the National Academy USA, 99*, 8389–8393.
Lin, A., Freund, R., & Palmer, M. (1991). Ethanol potentiation of GABA-induced electrophysiological responses in cerebellum: Requirement for catecholamine modulation. *Neuroscience Letters, 122*, 154–158.
Linden, D. J., Dickinson, M. H., Smeyne, M., & Connor, J. A. (1991). A long-term depression of AMPA currents in cultured cerebellar Purkinje neurons. *Neuron, 7*, 81–89.
Lippiello, P., Hoxha, E., Volpicelli, F., Duca, G., Tempia, F., & Miniaci, M. (2015). Noradrenergic modulation of the parallel fiber-Purkinje cell synapse in mouse cerebellum. *Neuropharmacology, 89*, 33–42.
Llano, I., & Gerschenfeld, H. (1993). β-adrenergic enhancement of inhibitory synaptic activity in rat cerebellar stellate and Purkinje cells. *The Journal of Physiology, 468*, 201–224.
Ly, R., Bouvier, G., Schonewille, M., Arabo, A., Rondi-Reig, L., Léna, C., Casado, M., De Zeeuw, C. I., & Feltz, A. (2013). T-type channel blockade impairs long-term potentiation at the parallel fiber–Purkinje cell synapse and cerebellar learning. *Proceedings of the National Academy USA, 110*, 20302–20307.
Marr, D. (1969). A theory of cerebellar cortex. *The Journal of Physiology, 202*, 437–470.
McConnell, M. J., Huang, Y. H., Datwani, A., & Shatz, C. J. (2009). H2-K(b) and H2-D(b) regulate cerebellar long-term depression and limit motor learning. *Proceedings of the National Academy USA, 106*, 6784–6789.
Miles, F. A., & Lisberger, S. G. (1981). Plasticity in the vestibulo-ocular reflex: A new hypothesis. *Annual Review of Neuroscience, 4*, 273–299.
Minneman, K. P., Pittman, R. N., & Molinoff, P. B. (1981). β-Adrenergic receptor subtypes: Properties, distribution, and regulation. *Annual Review of Neuroscience (Palo Alto, CA), 4*, 419–461.
Mori-Okamoto, J., & Tatsuno, J. (1988). Effects of noradrenaline on the responsiveness of cultured cerebellar neurons to excitatory amino acids. *Brain Research, 448*, 259–271.

Naka, F., Shiga, T., Yaguchi, M., & Okado, N. (2002). An enriched environment increases noradrenaline concentration in the mouse brain. *Brain Research, 924*, 124–126.

Papay, R., Gaivin, R., McCune, D. F., Rorabaugh, B. R., Macklin, W. B., McGrath, J. C., & Perez, D. M. (2004). Mouse α1B-adrenoreceptor is expressed in neurons and NG2 oligodendrocytes. *The Journal of Comparative Neurology, 478*, 1–10.

Papay, R., Gaivin, R., Jha, A., McCune, D. F., McGrath, J. C., Rodrigo, M. C., Simpson, P. C., Doze, V. A., & Perez, D. M. (2006). Localization of mouse α1A-adrenoreceptor (AR) in the brain: α1AAR is expressed in neurons, GABAergic interneurons, and NG2 oligodendrocyte progenitors. *The Journal of Comparative Neurology, 497*, 209–222.

Paschalis, A., Churchill, L., Marina, N., Kasymov, V., Gourine, A., & Ackland, G. (2009). β1-adrenoceptor distribution in the rat brain: An immunohistochemical study. *Neuroscience Letters, 458*, 84–88.

Philipp, M., & Hein, L. (2004). Adrenergic receptor knockout mice: Distinct functions of 9 receptor subtypes. *Pharmacology & Therapeutics, 101*, 65–74.

Saitow, F., Satake, S., Yamada, J., & Konishi, S. (2000). β-Adrenergic receptor-mediated presynaptic facilitation of inhibitory GABAergic transmission at cerebellar interneuron-Purkinje cell synapses. *Journal of Neurophysiology, 84*, 2016–2025.

Sakurai, M. (1987). Synaptic modification of parallel fibre-Purkinje cell transmission in in vitro Guinea-pig cerebellar slices. *The Journal of Physiology, 394*, 463–480.

Salin, P. A., Malenka, R. C., & Nicoll, R. A. (1996). Cyclic AMP mediates a presynaptic form of LTP at cerebellar parallel fiber synapses. *Neuron, 16*, 797–803.

Sara, S. J. (2009). The locus coeruleus and noradrenergic modulation of cognition. *Nature Reviews. Neuroscience, 10*, 211–223.

Schambra, U. B., Mackensen, G. B., Stafford-Smith, M., Haines, D. E., & Schwinn, D. A. (2005). Neuron specific α-adrenergic receptor expression in human cerebellum: Implications for emerging cerebellar roles in neurologic disease. *Neuroscience, 135*, 507–523.

Schonewille, M., Belmeguenai, A., Koekkoek, S. K., Houtman, S. H., Boele, H. J., van Beugen, B. J., Gao, Z., Badura, A., Ohtsuki, G., Amerika, W. E., Hosy, E., Hoebeek, F. E., Elgersma, Y., Hansel, C., & De Zeeuw, C. I. (2010). Purkinje cell-specific knockout of the protein phosphatase PP2B impairs potentiation and cerebellar motor learning. *Neuron, 67*, 618–628.

Schonewille, M., Gao, Z., Boele, H. J., Veloz, M. F., Amerika, W. E., Simek, A. A., De Jeu, M. T., Steinberg, J. P., Takamiya, K., Hoebeek, F. E., Linden, D. J., Huganir, R. L., & De Zeeuw, C. I. (2011). Reevaluating the role of LTD in cerebellar motor learning. *Neuron, 70*, 43–50.

Schwarz, L. A., Miyamichi, K., Gao, X. J., Beier, K. T., Weissbourd, B., DeLoach, K. E., Ren, J., Ibanes, S., Malenka, R. C., Kremer, E. J., & Luo, L. (2015). Viral-genetic tracing of the input–output organization of a central noradrenaline circuit. *Nature, 524*, 88–92.

Small, K. M., McGraw, D. W., & Liggett, S. B. (2003). Pharmacology and physiology of human adrenergic receptor polymorphisms. *Annual Review of Pharmacology and Toxicology, 43*, 381–411.

Spreng, M., Cotecchia, S., & Schenk, F. (2001). A behavioral study of alpha-1b adrenergic receptor knockout mice: Increased reaction to novelty and selectively reduced learning capacities. *Neurobiology of Learning and Memory, 75*, 214–229.

Sugiyama, Y., Kawaguchi, S., & Hirano, T. (2008). mGluR1-mediated facilitation of long-term potentiation at inhibitory synapses on a cerebellar Purkinje neuron. *The European Journal of Neuroscience, 27*, 884–896.

Takeuchi, T., Ohtsuki, G., Yoshida, T., Fukaya, M., Wainai, T., Yamashita, M., Yamazaki, Y., Mori, H., Sakimura, K., Kawamoto, S., Watanabe, M., Hirano, T., & Mishina, M. (2008). Enhancement of both long-term depression induction and optokinetic response adaptation in mice lacking delphilin. *PLoS One, 3*, e2297. https://doi.org/10.1371/journal.pone.0002297

Talley, E. M., Rosin, D. L., Lee, A., Guyenet, P. G., & Lynch, K. R. (1996). Distribution of alpha2A-adrenergic receptor-like immunoreactivity in the rat central nervous system. *The Journal of Comparative Neurology, 372*, 111–134.

Tan, H. S., & Collewijn, H. (1992). Cholinergic and noradrenergic stimulation in the rabbit flocculus have synergistic facilitatory effects on optokinetic responses. *Brain Research, 586*, 130–134.

Tanaka, S., Kawaguchi, S., Shioi, G., & Hirano, T. (2013). Long-term potentiation of inhibitory synaptic transmission onto cerebellar Purkinje neurons contributes to adaptation of vestibulo-ocular reflex. *The Journal of Neuroscience, 33*, 17209–17220.

Tavares, A., Handy, D. E., Bogdanova, N. N., Rosene, D. L., & Gavras, H. (1996). Localization of α2A- and α2B-adrenergic receptor subtypes in brain. *Hypertension, 27*, 449–455.

Thompson, R. F. (2005). In search of memory traces. *Annual Review of Psychology, 56*, 1–23.

Titley, H. K., Watkins, G. V., Lin, C., Weiss, C., McCarthy, M., Disterhoft, J. F., & Hansel, C. (2020). Intrinsic excitability increase in cerebellar Purkinje cells after delay eye-blink conditioning in mice. *The Journal of Neuroscience, 40*, 2038–2046.

Tully, K., & Bolshakov, V. (2010). Emotional enhancement of memory: How norepinephrine enables synaptic plasticity. *Molecular Brain, 3*, 15.

van Neerven, J., Pomepeiano, O., Collewijn, H., & van der Steen, J. (1990). Injections of β-noradrenergic substances in the flocculus of rabbits affect adaptation of the VOR gain. *Experimental Brain Research, 79*, 249–260.

Wakita, R., Tanabe, S., Tabei, K., Funaki, A., Inoshita, T., & Hirano, T. (2017). Differential regulations of vestibulo-ocular reflex and optokinetic response by α- and β2-adrenergic receptors in the cerebellar flocculus. *Scientific Reports, 7*, 3944. https://doi.org/10.1038/s41598-017-04273-9

Wang, W., Nakadate, K., Masugi-Tokita, M., Shutoh, F., Aziz, W., Tarusawa, E., Lorincz, A., Molnár, E., Kesaf, S., Li, Y.-Q. Q., Fukazawa, Y., Nagao, S., & Shigemoto, R. (2014). Distinct cerebellar engrams in short-term and long-term motor learning. *Proceedings of the National Academy USA, 111*, 188–193.

Waterhouse, B. D., Moises, H. C., Yeh, H. H., & Woodward, D. J. (1982). Norepinephrine enhancement of inhibitory synaptic mechanisms in cerebellum and cerebral cortex: Mediation by beta adrenergic receptors. *The Journal of Pharmacology and Experimental Therapeutics, 221*, 495–506.

Watson, M., & McElligott, J. (1984). Cerebellar norepinephrine depletion and impaired acquisition of specific locomotor tasks in rats. *Brain Research, 296*, 129–138.

Welsh, J. P., Yamaguchi, H., Zeng, X. H., Kojo, M., Nakada, Y., Takagi, A., Sugimori, M., & Llinás, R. R. (2005). Normal motor learning during pharmacological prevention of Purkinje cell long-term depression. *Proceedings of the National Academy USA, 102*, 17166–17171.

Yamaguchi, K., Itohara, S., & Ito, M. (2016). Reassessment of long-term depression in cerebellar Purkinje cells in mice carrying mutated GluA2 C terminus. *Proceedings of the National Academy USA, 113*, 10192–10197.

Zikopoulos, B., & Dermon, C. R. (2005). Comparative anatomy of α2 and β adrenoreceptors in the adult and developing brain of the marine teleost red porgy (*Pagrus Pagrus, Sparidae*): ^{3}H clonidine and ^{3}H dihydroalprenolol quantitative autoradiography and receptor subtypes immunohistochemistry. *The Journal of Comparative Neurology, 489*, 217–240.

Chapter 17
Role of LTD in Cerebellar Motor Learning: The 75th FUJIHARA Seminar "The Cerebellum as a CNS Hub"

Kazuhiko Yamaguchi

17.1 Introduction

Elaborated machine-like neuronal circuit of the cerebellar cortex was revealed 50 years before (Eccles et al., 1967), but its functional interpretation was elusive at that time. Two years later, Marr proposed that this neuronal circuit could work as a learning machine. He proposed a Hebbian-type synaptic plasticity at PF-PC synapse, namely, the synapses from parallel fibers to Purkinje cells were facilitated by the conjunction of presynaptic and climbing fiber (or postsynaptic) activity (Marr, 1969). On the other hand, Albus claimed that in order for the learning process to be stable, pattern storage must be accomplished principally by weakening synaptic weights rather than by strengthening them (Albus, 1971). Ten years later, by means of in vivo recording from rabbit cerebellum, Ito's group experimentally found that the conjunction of PF and CF stimulation induced long-term depression (LTD) of synaptic transmission at PF-PC synapse (Ito et al., 1982). Then, this theory was called as Marr-Albus-Ito's theory of the cerebellar cortex.

The molecular mechanism of LTD was investigated using new in vitro experimental preparations, such as a cerebellar slice (Sakurai, 1987) and cultured cerebellar neurons (Hirano & Ohmori, 1986; Linden & Conner, 1991). Briefly, repetitive firing of PF activates not only AMPA-type glutamate receptor (AMPA-R) but also mGluR1 located at the periphery of postsynaptic region of dendritic spine of PC. Activated mGluR1, coupled with $\alpha_{q/11}$ of GTP-binding protein, activates PLCβ, and then IP_3 and DG were produced. IP_3 bound to IP_3 receptor 1(IP_3R1) of

K. Yamaguchi (✉)
Center for Brain Science, RIKEN, Saitama, Japan

Department of Ultrastructural Research, National Institute of Neuroscience, NCNP, Tokyo, Japan
e-mail: kazuhiko.yamaguchi@ncnp.go.jp

H. Mizusawa, S. Kakei (eds.), *Cerebellum as a CNS Hub*,
Contemporary Clinical Neuroscience, https://doi.org/10.1007/978-3-030-75817-2_17

the endoplasmic reticulum (ER) and released Ca^{2+} from ER. Action potential of CF activates P/Q-type voltage-gated Ca^{2+} channel (VGCC) at PC's dendritic region. Ca^{2+} influx through P/Q Ca^{2+} channel sensitizes IP_3R1 sensitivity to IP_3, then Ca^{2+} release from ER is robustly enhanced, and conjunctive activation of VGCC and mGluR-IP_3 pathways increases $(Ca^{2+})_{in}$ more than an additive level in the dendritic spine (Wang et al., 2000; Kakegawa et al., 2018). This synergistic increase in $(Ca^{2+})_{in}$ activates PKCα, and activated PKCα phosphorylates Ser880 at GluA2 C-terminus. This phosphorylation of Ser880 requires in advance dephosphorylation of Tyr876 by megakaryocyte protein phosphatase (PTPMEG), bound to cytoplasmic terminus of GluR delta 2 (Kohda et al., 2013). Then, AMPA-Rs containing phosphorylated GluA2 are detached from the scaffold protein and internalized with PICK1 in AP2- and clathrin-dependent manners (Wang & Linden, 2000; Kakegawa et al., 2018).

To correlate LTD and motor learning, gene-manipulated animals provided advantageous systems. Gene deletion of mGluR1 caused both deficiency of LTD and conditioned eyeblink (Aiba et al., 1994). In mice lacking Gαq, both LTD and motor coordination were impaired (Hartmann et al., 2004). In PLCβ4-deficient mice, LTD was impaired in rostral cerebellum (lobes 1–6) (Hirono et al., 2001; Miyata et al., 2001), and delay eyeblink conditioning was severely impaired (Kishimoto et al., 2001). Transgenic mouse expressing pseudosubstrate PKC inhibitor only in Purkinje cell showed complete blockade of LTD induction and absence of ability to adapt their vestibulo-ocular reflex (VOR) gain during visuo-vestibular training (De Zeeuw et al., 1998). In GluR delta 2-deficient mice, both LTD and motor coordination were impaired (Kashiwabuchi et al., 1995). Also GluR delta 2 mutant mice failed to change the VOR or optokinetic response (OKR) gain adaptively in response to sustained vestibular and/or visual stimulation (Katoh et al., 2005).

Accordance between impairment of LTD and motor learning in these gene-manipulated animals supported the idea that LTD was the mechanism of the motor learning. Recently, however, discrepancies between LTD and motor learning have been reported in mice with a mutation that targeted the expression of PF-PC LTD by blocking AMPA-R internalization regulated via the phosphorylation of AMPA-Rs. In these mice, motor learning behavior was normal, but no PF-PC LTD was observed (Schonewille et al., 2011). However, the induction of LTD was only attempted using one type of protocol at room temperature. We reexamined slices obtained from these GluA2 K882A and GluA2 Δ7 knock-in mutants at 3–6 months of age. Several types of LTD-inducing protocols at higher temperature were used, and LTD-inducing ability was detected in both types of knock-in mutants (Yamaguchi et al., 2016). These new standard protocols would be necessary especially to detect compensated LTD in mutant animals.

17.2 Materials and Methods

17.2.1 Animals

The knock-in (KI) mice (GluR2K882A and GluR2Δ7) (Fig. 17.1a), which were gifts from R.L. Huganir (Steinberg et al., 2006). Sperms were obtained from both KI mice by ectomy and used for in vitro insemination of oocytes from C57BL/6

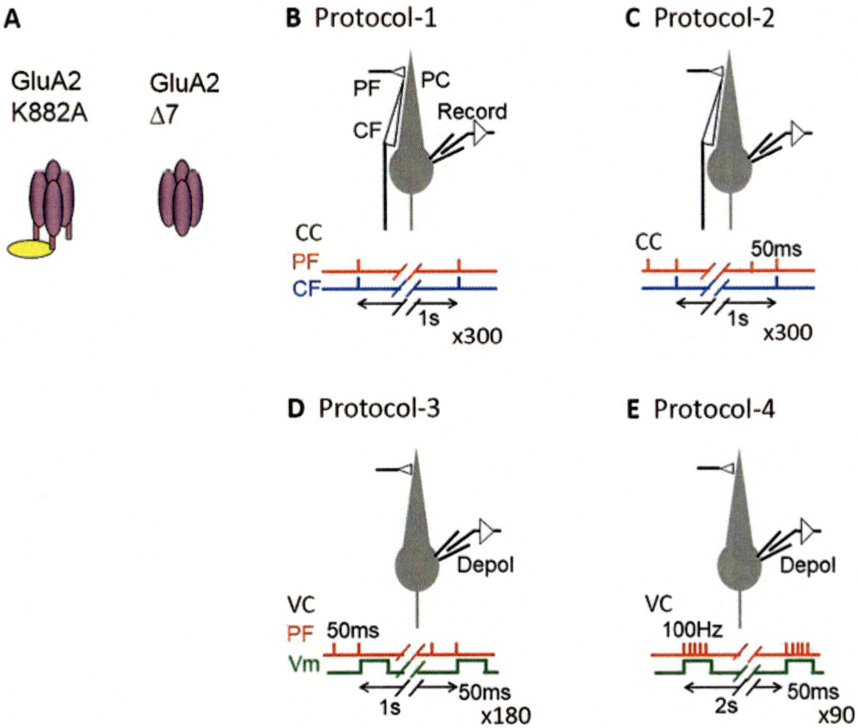

Fig. 17.1 Two types of mutated GluA2 and schematic illustration of LTD-inducing protocols 1–4. (**a**) Mutation of K882 within the intracellular C-terminus of the GluA2 subunit (GluA2 K882A) and the absence of the C-terminal 7 amino acids of GluA2 (GluAΔ7). PKCα cannot phosphorylate S880 of GluA2 K882A, but PIK1 can bind to C-terminus of GluA2 K882A. (**b**) Protocol 1 conjunctive stimulation. 1PF and 1CF stimuli are applied simultaneously 300 times at 1 Hz (5 min) under current-clamp conditions. Electrode for whole-cell recording contains K^+-based internal solution. (**c**) Protocol 2 conjunctive stimulation. 2PF and 1CF stimuli are applied simultaneously 300 times at 1 Hz (5 min) under current-clamp conditions. Electrode contains K^+-based internal solution. (**d**) Protocol 3 conjunctive stimulation. 2PF and somatic depolarization (−70 to 0 mV, 50 ms) are applied 180 times at 1 Hz (3 min) under voltage-clamp conditions, so that the second PF stimulus is applied simultaneously with the beginning of the somatic depolarization. Electrode contains Cs^+-based internal solution. (**e**) Protocol 4 conjunctive stimulation. 5PF at 100 Hz and somatic depolarization are applied 90 times at 0.5 Hz (3 min) under voltage-clamp conditions, simultaneously. Electrode contains Cs^+-based internal solution

mice. The offspring from the fertilized oocytes was proliferated by backcrossing to C57BL/6 mice and then by crossbreeding to obtain homozygous KI mice. Genotypes were determined by PCR amplification of tail DNA samples followed by digestion with restriction enzymes for which unique sites had been engineered (GluR2Δ7, Bgl II; GluR2K882A, Bsl I) (Steinberg et al., 2006). The primers used were Glu L (5'-AACGAATGAAGGTGGCAAAG-3′) and Glu R (5'-AAGAGCTTAAGGACGCGACA-3′). Genotypes were confirmed by second-round genotyping on genomic DNA purified from mouse brain slices used for electrophysiological recording. Mice were kept in the animal facility of RIKEN Brain Science Institute under well-controlled living conditions.

17.2.2 Electrophysiological Studies

All procedures involving animals were approved by the RIKEN committee on the care and use of animals in experiments. Both male and female wild-type mice (C57BL/6, 3–6 months) were used.

All solutions should be made in ultrapure water with free of metals and other impurities. Working artificial cerebrospinal fluid (ACSF) for slice-cutting and recording are made freshly on the day of experiment from a 10 times (x10) stock of ACSF. Bubble the solutions with 5% CO_2/95% O_2 gas mixture before use. The pH of ACSF is adjusted to 7.4 ± 0.1, and osmolarity is adjusted to 315 ± 5 mOsm/kg by adding ultrapure water. ACSF is containing (in mM) 125 NaCl, 3 KCl, 2 $CaCl_2$, 1 $MgSO_4$, 1.25 NaH_2PO_4, 26 $NaHCO_3$, and 20 glucose. For slice-cutting, 50 μl of tetrodotoxin (TTX) (1 μM) was added into the ice-cold ACSF. Using the linear slicer, sagittal slices of vermis were cut with thickness of 300 μm. Slice was used after incubation at 26 °C for, at least, 1 h.

The recording chamber was perfused with oxygenated ACSF containing 100 μM picrotoxin at a rate of 2 ml/min and maintained at 30.0 ± 1.0 °C. The volume of the perfusion bath was 0.5 ml. Whole-cell slice-patch recordings of PCs were performed under an upright microscope with x40 water immersion objective. Patch pipettes were made from borosilicate tubings and filled with a solution containing the following (in mM) (resistance, 2–4 MΩ): for the K^+-based internal solution, 60, KCl; 60, K-gluconate; 0.3, EGTA; 4, $MgCl_2$; 4, ATP; 0.4, GTP; and 30, HEPES (pH 7.2) and, for the Cs^+-based internal solution, 60, CsCl; 40, D-gluconate; 20, TEA-Cl; 0.3, EGTA; 4, $MgCl_2$; 4, ATP; 0.4, GTP; and 30, HEPES (pH 7.2, adjusted using CsOH). Whole-cell recordings were performed using a patch-clamp amplifier (MultiClamp 700A). Recorded signals were filtered at 3 KHz and digitized at 10 KHz. Stimulation and online data acquisition were performed using pCLAMP 9 software (Molecular Devices). Access resistance and input resistance were constantly monitored by applying a small hyperpolarizing voltage step (2 mV, 100 ms). PFs in the molecular layer or CFs in the granule cell layer or molecular cell layer were focally stimulated by applying pulses (duration, 0.1 ms) to a slice through a glass pipette (tip diameter, 5–10 μm) positioned on the surface of a cerebellar slice.

The membrane potential was held at −65 or − 75 mV, and a PF-evoked PF-EPSC was evoked at a frequency of 0.1 Hz as a test response.

17.2.3 *LTD-Inducing Protocols*

Four protocols were used to induce LTD. In protocol 1 and protocol 2, PF and CF were stimulated under current-clamp condition, and in protocol 3 and protocol 4, PF stimulation and somatic depolarization were applied under voltage-clamp condition. Schematic diagrams of protocols 1–4 are shown in Fig. 17.1b–e. In protocol 1 or 2, after switching to the current-clamp mode, LTD was induced by a protocol composed of 300 single (protocol 1) or double (protocol 2) PF stimuli in conjunction with CF stimuli. In the double PF stimuli protocol, the first PF stimulus was followed by a 50 ms interval by a CF stimulus and the second PF stimulus. During the PF-CF conjunctive stimulation, no DC was applied except when there was the need to prevent spontaneous firing. The resting membrane potential was varied between −50 and − 60 mV. Under the voltage-clamp condition using a recording electrode containing the Cs^{+}-based internal solution, two (protocol 3) or five (protocol 4) PF stimuli were applied repeatedly at 1 or 0.5 Hz for 3 min, respectively, coupled with a depolarizing pulse for 50 ms from a holding potential of −75 mV to a potential (0 to +40 mV) at which a few inward current surges were observed, probably representing the generation of dendritic Ca^{2+} spikes.

Statistics To evaluate statistical significances of data between the experimental group and the control group, the *t*-test was used. On the other hand, the statistical significance of intergroup differences among the three genotypes was tested by one-way ANOVA and the Tukey-Kramer post hoc test. Data are represented as mean ± SEM (cell number of tested PCs).

17.3 Results

The two types of mutant, GluA2 K882A and Δ7 KI mice, examined in this study showed seemingly normal daily behavior. Acute slices of cerebellar vermis were obtained from them at 3–6 months of age where motor learning capability was fully developed (Schonewille et al., 2011). Whole-cell recording from PCs under the voltage- or current-clamp condition showed virtually normal features of membrane resistance and PF-EPSCs, simple spikes, and complex spikes in the mutant slices as compared with the wild type (WT). In WT, all types of conjunctive stimulation (protocols 1 to 4) caused LTD, which continued for more than 1 h, but for convenience, the magnitude of LTD was estimated by measuring the amplitude of the depressed PF-EPSC peak at 26–30 min after the onset of conjunctive stimulation, relative to their mean amplitude measured during the 5 min preconjunction period.

Four types of protocols were used in this study to induce cerebellar LTD. In the first two protocols (protocols 1 and 2), conjunction of PF stimulation and CF stimulation was applied under current-clamp condition. In the other two protocols (protocols 3 and 4), CF stimulation was replaced by somatic depolarization under voltage-clamp condition.

Protocol 1 Conjunction of one PF stimulation and one CF stimulation under current-clamp condition was conventionally used in slice preparation (Koekkoek et al., 2005). The shape of complex spike elicited by conjunctive stimulation was similar to that elicited by CF stimulation alone, with the first steep spikelet followed by two to three spikelets (Fig. 17.2a). Firstly, a single-shock stimulus (0.1 ms in duration) was applied to PFs in combination with a single stimulus simultaneously applied to CF and repeated at 1 Hz for 5 min (300 pulses) (Fig. 17.1a). This protocol 1 was used previously as effective in inducing LTD (Karachot et al., 1994; Schonewille et al., 2011). In the present study, protocol 1 was effective in inducing LTD in WT PCs (75.6 ± 4.0%, $n = 6$, $p = 0.002$, t-test) (Fig. 17.2b), but not in Δ7 PCs (103.9 ± 9.6, $n = 6$, $p = 0.898$) (Fig. 17.2d). K882A PCs may appear to exhibit a modest LTD, which, however, is not statistically significant in its magnitude (86.0 ± 7.2, $n = 6$, $p = 0.228$) (Fig. 17.2c). Thus, protocol 1 was effective in inducing LTD only in WT, but not in either K882A or Δ7 PCs (Fig. 17.1e), which is in agreement with the report by Schonewille et al. (2011).

Protocol 2 PFs were stimulated twice at 50 ms intervals, and the second PF stimulus was synchronized with a single CF stimulus (Fig. 17.1c). Similar shape of complex spike was observed during stimulation with protocol 1 (Fig. 17.3a). PF stimulation with 2 pulses was expected to release more Glu and to activate mGluR1 more robustly, because mGluR1 is located at the marginal zone of the synaptic region (Masugi-Tokita et al., 2007). This (2PF + CF) conjunctive stimulation induced significant LTD in K882A PCs (81.0 ± 3.4, $n = 6$, $p = 0.008$) (Fig. 17.3c) comparable to that in WT PCs (82.8 ± 4.95%, $n = 6$, $p = 0.036$) (Fig. 17.3b), there being no statistically significant difference between them ($p = 0.957$, Tukey-Kramer test). However, no LTD was induced in Δ7 PCs (94.8 ± 5.1, $n = 6$, $p = 0.351$) (Fig. 17.3d). Thus, protocol 2 was effective in inducing LTD in both WT and K882A, but not in Δ7 PCs (Fig. 17.3e).

To ensure that protocol 1- and protocol 2-evoked LTD is specifically induced by conjunctive stimulation of PF and CF and not by PF or CF stimulation alone, we tested the effect of applying CF stimuli alone in place of PF-CF conjunctive stimulation. No significant change in EPSC amplitude during 25–29 min was observed in any of the three PC groups, WT PCs (101.1 ± 1.5%, $n = 3$, $p = 0.41$, t-test), K882A PCs (100.5 ± 3.9%, $n = 3$, $p = 0.90$), and Δ7 PCs (102.2 ± 2.8, $n = 3$, $p = 0.48$), as plotted in Figs. 17.2b–d and 17.3b–d as control for conjunctive stimulation. Application of PF alone in place of PF-CF conjunction was shown previously to induce no LTD in WT, K882A, or Δ7 PCs (Schonewille et al., 2011). In this study, we also observed that PF or 2PF induced no LTD, but often long-term potentiation (LTP), as has been known (Schonewille et al., 2011; Lev-Ram et al., 2002).

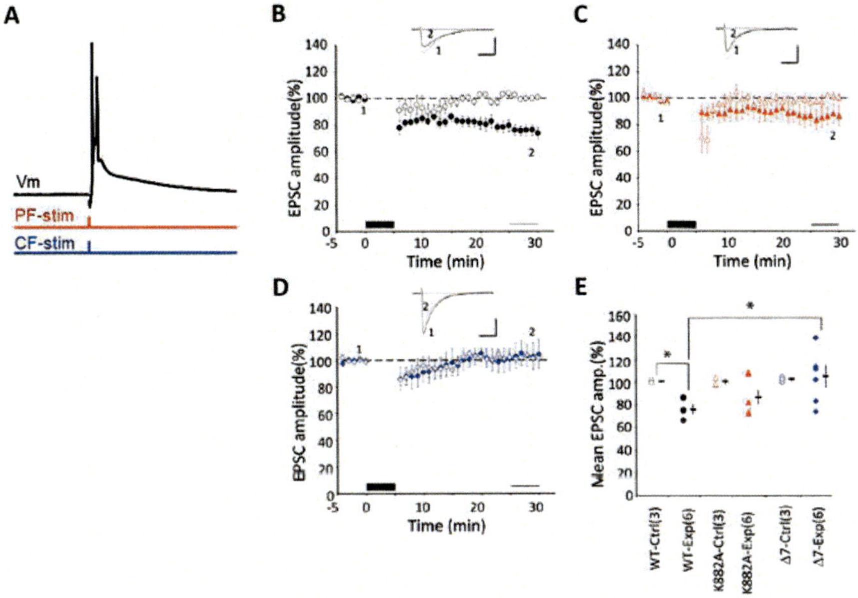

Fig. 17.2 LTD induction by protocol 1 conjunctive stimulation. (**a**) Membrane potential traces of PC elicited by protocol 1 conjunctive stimulation. (**b**) Mean PF-EPSC amplitude recorded from WT PCs before and after protocol 1 conjunctive stimulation (black column at bottom). The horizontal thin line (bottom) indicates the period of 25–29 min after conjunctive stimulation onset, where the amplitudes of PF-EPSCs were compared to estimate LTD. PF-EPSC amplitude was normalized by those recorded before conjunctive stimulation. Filled symbols indicate the mean amplitude of the experimental group. Open symbols indicate the PF-EPSC amplitude of the control group, in which only CF stimulation was applied in place of conjunctive stimulation. Error bars denote SEM. *Inset*: superposed PF-EPSC traces (top) recorded before (marked 1) and 25–29 min after conjunctive stimulation onset (marked 2). Each trace represents the average of six records. *Bars*: 100 ms, 150 pA. (**c**) Similar to **b** but for K882A PCs. (**d**) Similar to **b**, but for Δ7 PCs. (**e**) Summary plot of mean PF-EPSC amplitude recorded during 25–29 min after onset of conjunctive stimulation in WT, K882A, or Δ7 mice. Filled symbols are plots for the test with protocol 1, whereas open symbols are plots for the control measurement where conjunctive stimulation was replaced by 1CF stimulation. The mean and SEM of each group are indicated by a horizontal rod and vertical bar, respectively. Numbers in parentheses represent cell number. $^*p < 0.05$. Intragroup comparison, *t*-test. Intergroup comparison, ANOVA and post hoc Tukey-Kramer test

For the LTD induced by protocol 1 and 2, EPSC amplitude measured during 25–29 min after onset of conjunctive stimulation was widespread. The shape of complex spike was variable from cell to cell. Because the spikelets within the complex spike reflected Ca channel activation (De Schutter & Bower, 1994), whether the shape of complex spike affected LTD amplitude was examined. Relation between sum of voltage peaks 1–4 (Fig. 17.4a) and LTD amplitude in protocol 1 was compared, but there is no correlation ($r = -0.03$) (Fig. 17.4b). Because the first spikelet represents mainly Na component and spikelets 2–4 contain more of the Ca^{2+} component (Swensen & Bean, 2003), the relation between sum of voltage

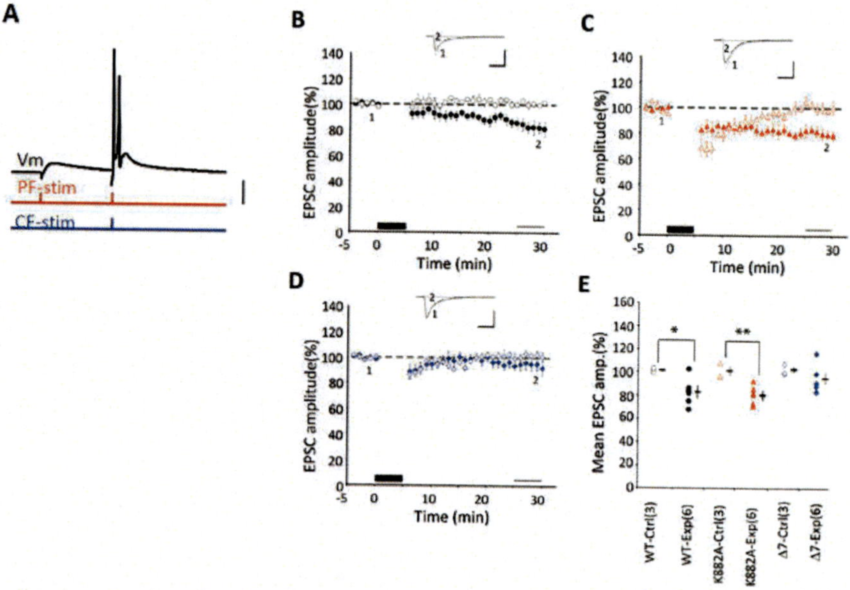

Fig. 17.3 LTD induction by protocol 2 conjunctive stimulation. (**a**) Membrane potential traces of PC elicited by protocol 2 conjunctive stimulation. (**b–e**) Similar to Fig. 17.2b–e, but for conjunction protocol 2, where 2PF and 1CF stimulations were applied conjunctively at 1 Hz for 5 min under current-clamp conditions. In **E**, * $p < 0.05$, ** $p < 0.01$, *t*-test

peaks 2–4 and LTD amplitude was analyzed, and moderate correlation was detected ($r = 0.67$) (Fig. 17.4c). Next, voltage trace of the complex spike was differentiated (Fig. 17.4d), and sum of maximum rate of rises (MRRs) of each spikelet was calculated, and then sum of the MRRs was multiplied by value of each PC's Cm, and this product gave approximate measure of the Ca^{2+} current of each PC during complex spike (Fukuda et al., 1981). Relation between Cm x summed MRRs of spikelets 1–4 and amplitude of LTD was plotted (Fig. 17.4e), but no correlation was detected ($r = 0.17$). The relation between Cm × sum of MRRs of spikelets 2–4 and LTD amplitude showed weak correlation ($r = 0.36$) (Fig. 17.4f).

Protocol 3 Under voltage-clamp condition using Cs^{+}-based internal solution, conjunction of 2PF stimulation and somatic depolarization were applied (Fig. 17.1d). The somatic depolarization was applied 50 ms after the initial PF stimulation, and the second PF stimulation was applied simultaneously with the beginning of the somatic depolarization. Inward current was elicited upon somatic depolarization from −70 to 0 mV. Tail current was also evoked after repolarization. Sometimes, repetitive generation of inward current was observed, which would reflect Ca^{2+} spike activity at the remote dendritic region where the membrane potential was not clamped sufficiently, in spite of using Cs^{+}-based internal solution (Fig. 17.5a) (Steinberg et al., 2006). The patch pipette was filled with a Cs^{+}-based internal solution to improve space clamping along PC dendrites by blocking K^{+}channels. The

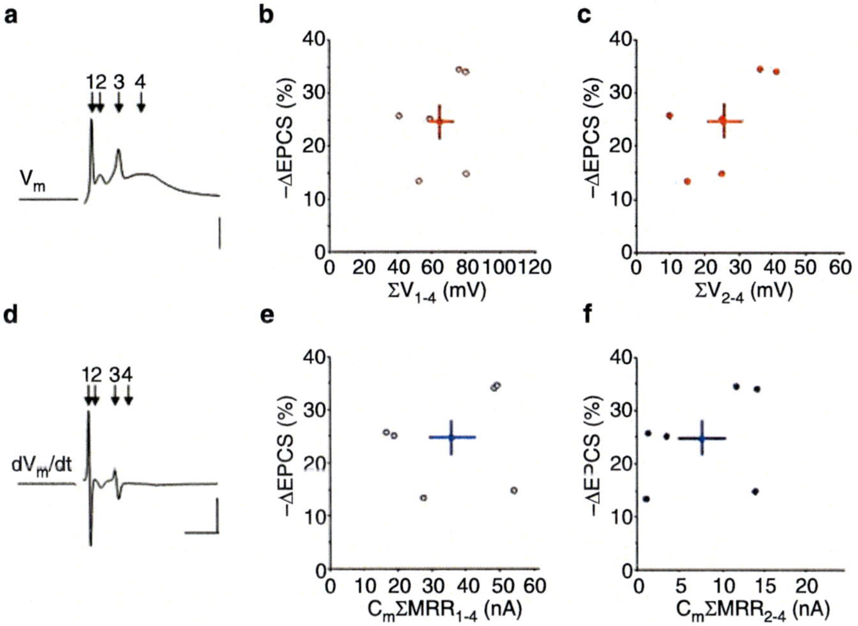

Fig. 17.4 Relationship between spikelet of a complex spike and LTD amplitude. (**a**) Representative trace of a complex spike elicited by protocol 1. Arrows indicate peaks of spikelets (1–4). Bar: 20 mV. (**b**) Relationship between the sum of the amplitude of spikelets (1–4) and LTD amplitude (-ΔEPSC %) ($r = -0.03$). (**c**) Relationship between the sum of the amplitude of spikelets (2–4) and LTD amplitude ($r = 0.67$). (**d**) Representative trace of differentiated complex spikes shown in A. Arrows indicate peaks of dV_m/dt of spikelets. Bars: 5 ms, 50 V/s. (**e**) Relationship between products of the Cm and the sum of the MRR of spikelets (1–4) and amplitude of LTD (-ΔEPSC %) ($r = 0.17$). (**f**) Relationship between the product of the Cm and the sum of the MRR of spikelets (2–4) and amplitude of LTD ($r = 0.36$)

mean **Rm** in cells recorded with the pipette containing the Cs^+-based internal solution was 3–4 times larger than those recorded with a pipette containing a K^+-based solution. This maneuver is expected to ensure efficient activation of voltage-dependent Ca^{2+} channels located at dendritic regions, where PFs form synapses on PCs. With this protocol 3, LTD was induced in WT PCs (64.9 ± 4.3%, $n = 5$, $p = 0.0007$), K882A PCs (72.0 ± 3.8, $n = 7$, $p = 0.003$), and Δ7 PCs (80.0 ± 7.2, $n = 6$, $p = 0.026$) (Fig. 17.5b–d). No significant difference was found between these three groups ($p = 0.151$, one-way ANOVA) (Fig. 17.5e).

Protocol 4 Five PF stimuli at 100 Hz were given simultaneously with the somatic depolarization under voltage-clamp condition (Fig. 17.1e). Repetitive generation of inward current was elicited during depolarization, and tail current was elicited after the repolarization. Timing of repetitive generation of the inward current was not synchronized with PF stimuli (Fig. 17.6a). Sometimes, repetitive generation of the inward current continues after repolarization. This (5PF + Depo) conjunctive stimulation (protocol 4) was effective for inducing LTD in WT PCs (76.0 ± 6.8%,

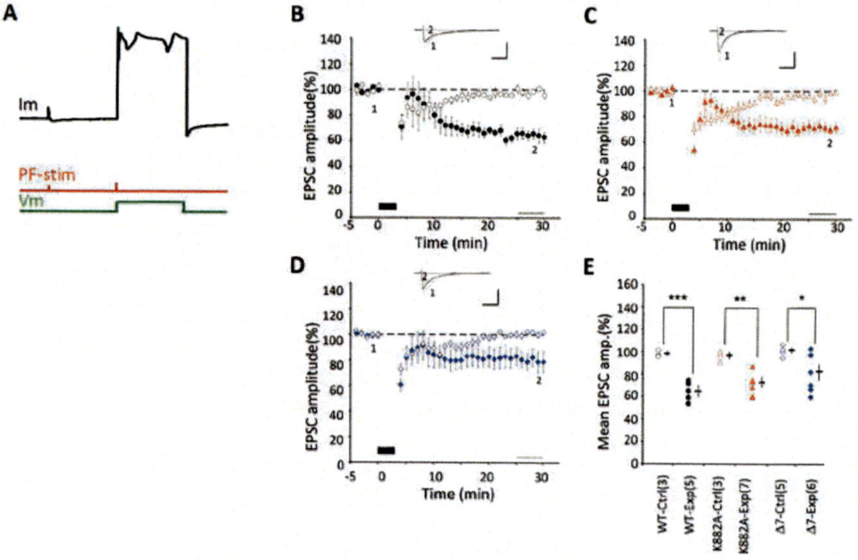

Fig. 17.5 LTD induction by protocol 3 conjunctive stimulation. (**a**) Membrane current traces of PC elicited by protocol 3 conjunctive stimulation. (**b–e**). Similar to Fig. 17.2**b–e**, but for conjunction protocol 3, where 2PF stimulation and one depolarizing pulse were applied conjunctively at 1 Hz for 3 min under voltage-clamp conditions. In **E**, * $p < 0.05$, ** $p < 0.01$, *** $p < 0.001$, *t*-test

$n = 3$, $p = 0.031$), K882A PCs (71.0 ± 2.0%, $n = 4$, $p = 0.00017$), and Δ7 PCs (70.7 ± 7.4%, $n = 3$, $p = 0.013$) (Fig. 17.6b–d). No statistically significant difference was found among the three groups ($p = 0.749$, one-way ANOVA) (Fig. 17.6e). With protocol 3 and protocol 4, CF stimulation was replaced with depolarizing pulses that caused Ca^{2+} entry, which was similar to that evoked by CF responses. As control experiment, we applied 50 ms depolarizing pulses at 1 Hz for 3 min separately from PF stimuli. These depolarizing pulses induced short-term depression lasting for 20 min but did not cause LTD in any of the three PC groups (Figs. 17.5 and 17.6).

Protocol 4 was originally used for young wild-type mouse (P14–21) (Steinberg et al., 2006). They reported that LTD was induced by 30 conjunctive stimulations at 0.5 Hz at RT in wild type, but in neither K882A nor Δ7 PCs (Steinberg et al., 2006). However, when 30 conjunctive stimulations were applied to adult wild-type cerebellar slice (3–6 months) at 30 °C, no LTD was induced (Fig. 17.7a, c). In contrast, when 90 conjunctive stimulations were applied, usual amplitude of LTD was observed in wild type (Fig.17.7b, c). Again, somatic depolarization alone (90 times at 0.5 Hz) did not induce LTD (Fig. 17.5b).

PKCα inhibitor sensitivity It may be questioned whether the LTD induced by replacing CF stimuli with depolarization in protocol 3 or protocol 4 shares a common signal transduction mechanism with the LTD induced by PF-CF conjunctive stimulation in protocol 1 and protocol 2. To test this, we chose Gö6976, a potent PKCα inhibitor, which has been shown to play a crucial role in LTD induction (Xia

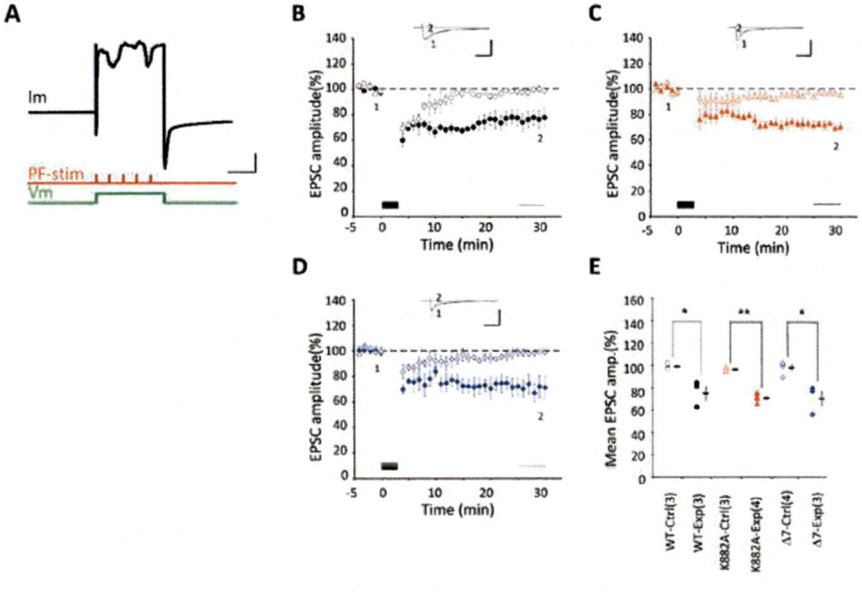

Fig. 17.6 LTD induction by protocol 4 conjunctive stimulation. (**a**) Membrane current trace elicited by protocol 4 conjunctive stimulation. (**b–e**) Similar to Fig. 17.2b–e, but for conjunction protocol 4, where 5PF stimulation and one depolarizing pulse were applied conjunctively at 0.5 Hz for 3 min under voltage-clamp conditions. In (**e**), * $p < 0.05$, ** $p < 0.01$, *t*-test

et al., 2000). When Gö6976 (0.3 μM) was contained in the internal solution, the EPSC amplitude after conjunctive stimulation failed to exhibit LTD in WT PCs either with protocol 2 (105.0 ± 1.7%, $n = 4$, $p = 0.06$, *t*-test) or 3 (97.7 ± 3.6%, $n = 3$, $p > 0.6$) (Fig. 17.8a, b). Hence, as far as the blocking effect of Gö6976 is concerned, the LTD induced by PF-Depo conjunctive stimulation is indistinguishable from that induced by PF-CF conjunctive stimulation. As tested with protocol 3, LTD was also blocked by Gö6976 in K882A PCs (98.5 ± 1.6%, $n = 3$, $p = 0.41$) and Δ7 PCs (99.7 ± 1.7%, $n = 3$, $p = 0.79$) (Fig. 17.7c, d). Hence, PKCα appears to play an essential role in LTD induction also in mutant PCs.

17.4 Discussion

17.4.1 Multiple Protocols for LTD Induction in Mutated Animal

C-terminus of GluA2 was essential for PKC-dependent internalization of AMPA-R containing GluA2, and LTD induction was believed to be impossible in PC expressing mutated GluA2 such as K882A and Δ7 (Steinberg et al., 2006). However, even in animals having these mutated GluA2, some modified protocols could actually

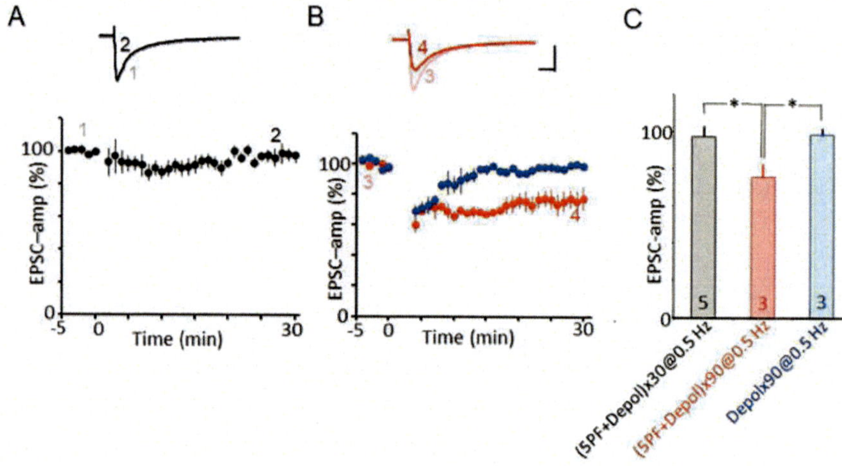

Fig. 17.7 Effect of number of repetitions of conjunctive stimulation on LTD induction by protocol 4. (**a**) Failure of LTD induction by protocol 4 conjunctive stimulation; repetition was 30 times at 0.5 Hz. Mean PF-EPSC amplitude recorded before and after protocol 4 conjunctive stimulation (black column at the bottom). Filled symbol indicates the mean EPSC amplitude. Inset: superimposed PF-EPSC traces (top) were recorded before (marked 1) and 25–29 min after conjunctive stimulation onset (marked 2). Each trace represents the average of six records. Bars: 100 pA, 10 ms. (**b**) Red symbol, LTD induced by protocol 4 conjunctive stimulation; blue symbol, no conjunction stimulation but somatic depolarization was applied 180 times at 0.5 Hz. LTD was not induced. Inset: superimposed PF-EPSC traces (top) were recorded before (marked 3) and 25–29 min after conjunctive stimulation onset (marked 4). Bars: 100 pA, 10 ms. (**c**) Summary plot of mean PF-EPSC amplitude recorded during 25–29 min after onset of conjunctive stimulation. Depol, depolarization. Numerical character in parentheses represents number of cells. ×30 = 30 times, ×90 = 90 times

induce LTD in PC (Yamaguchi et al., 2016). Compared to the simplest protocol, namely, simultaneous stimulation of PF and CF for 5 min at 1 Hz, other protocols were seemed to be more efficient to increase $(Ca^{2+})_{in}$. Thus, some compensated mechanism working in these mutant animals seemed to require higher $(Ca^{2+})_{in}$. Thus, compensatory mechanism of LTD in these mutants might be different from the ordinal molecular mechanism of LTD. However, this compensatory mechanism was indicated to be dependent on PKCα activity as in WT mice, because the selective PKCα inhibitor Gö6976 effectively blocked LTD induction in these mutant mice and WT mice (Fig. 17.5b). It may be that the stronger activation of PKCα by Ca^{2+} drives the downstream molecular machinery that may be modified by functional compensation for the mutations. The mechanism of such compensation is yet to be explored. The simplest possibility would be compensation by GluA3. Note that AMPAR at the PF-PC synapse consists of GluA2 and GluA3 subtypes (Srivastava et al., 1998) and that the GluR3 expression level is normal even in K882A and Δ7 mice (Steinberg et al., 2006). Note also that the GluA3 C-terminus can bind to GRIP/ABP and PICK1 (Bredt & Nicoll, 2003). Suppression of binding between the GluA3 C-terminus and GRIP/ABP through phosphorylation of Ser at

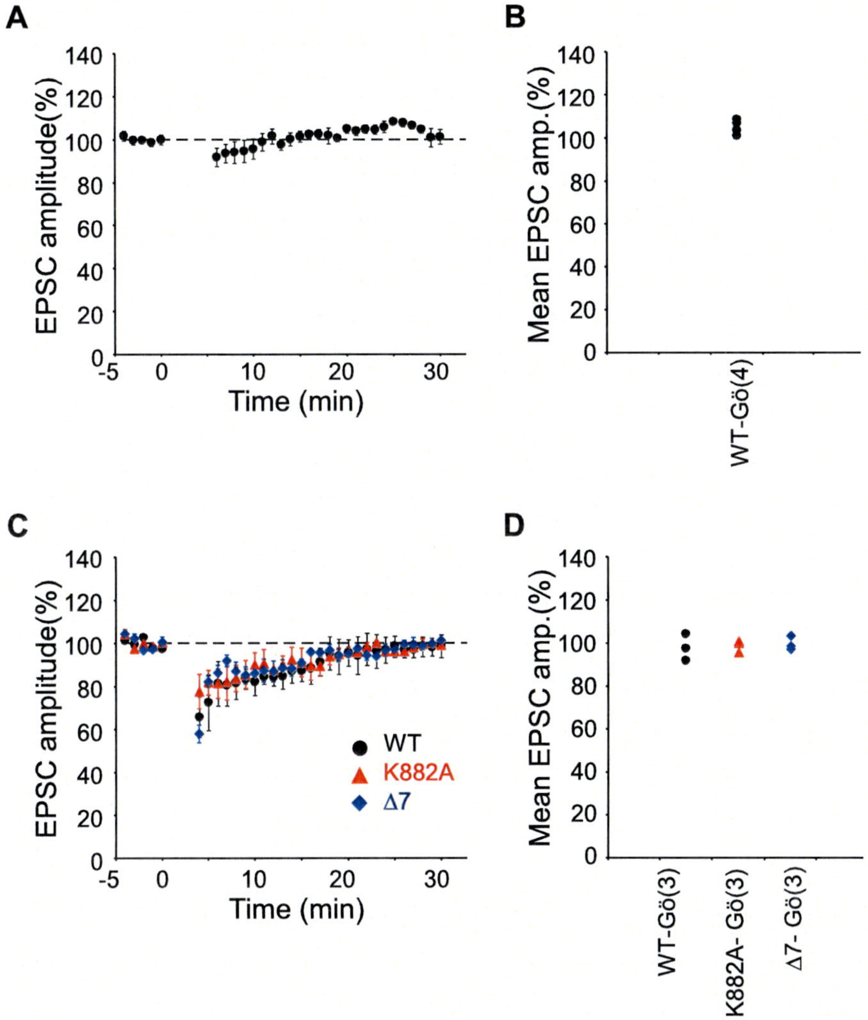

Fig. 17.8 Effects of PKCα blocker on LTD induction. (**a**) Similar to Fig. 17.2b, but obtained under infusion of Gö6976 with protocol 2 in WT mice. (**b**) Summary plot of mean PF-EPSC amplitude recorded during 25–29 min after onset of conjunctive stimulation in WT mice. (**c**) Similar to Fig. 17.5b–d, but obtained with protocol 3 under infusion of Gö6976 in WT, K882A, or Δ7 PCs. (**d**) No significant difference was found in the mean EPSC amplitude between before and after conjunctive stimulation in all three groups ($p > 0.05$, paired *t*-test)

the C-terminus by PKCα seems plausible because the 4-amino-acid sequence at the C-terminus of GluA3 is identical to that of GluA2, including the Ser site. It is interesting to determine whether phosphorylation of Ser at the C-terminus of GluA3 in these mutant mice requires a higher concentration of activated PKCα, just as LTD needs a stronger conjunctive stimulation in these mutant mice. In cultured PCs, even a strong conjunctive stimulation did not induce LTD in cells from K882A or Δ7 mice (Steinberg et al., 2006). The extent of compensation by GluA3 might be low in cultured PCs. Also, compensation by GluA3 might depend on age, because PC in cerebellar slices from young K882A and Δ7 mice (P20–25) showed no LTD even after conjunction of high-frequency PF stimulation and somatic depolarization (0.5 Hz, 1 min) using a Cs^+-containing electrode. However, a similar but longer conjunctive stimulation (3 min) effectively induced LTD in elder mutant mice (3–6 months) (Fig. 17.4).

Compensation in K882A and Δ7 mutant mice might involve scaffold molecules in addition to GRIP1/2 for maintaining GluA2 at the postsynaptic density (PSD). Transmembrane AMPAR regulatory proteins (TARPs) are candidate molecules that deserve attention. In TARP γ-2 (stargazin) gene knockout mice, the content of GluA2 and GluA3 in synaptosome and PSD fractions is markedly reduced, suggesting the involvement of TARPs in the control of AMPAR surface expression at PF-PC synapses (Nomura et al., 2012). TARP γ-2, which is highly phosphorylated in its basal state, was dephosphorylated during chemically induced LTD in cerebellar slices (Yamazaki et al., 2015). The neuronal activity-dependent dephosphorylation of TARP γ-2 by calcineurin was indispensable for cerebellar LTD. Furthermore, PC-specific conditional deletion of TARP γ-2 with a γ-7 KO background impaired motor behaviors. Both GRIP1/2 and TARPs might be involved in stabilization of GluA2/3-containing AMPAR, and dissociation of GluA2/3 from them through direct or indirect action of PKCα might underlie LTD. However, the link between PKCα activation and the dissociation of GluA2/3 from TARP is missing.

An important conclusion from the present results is that LTD remains inducible in the mutants as far as the intensity of conjunctive stimulation is sufficiently increased. Hence, there is no reason to assume that LTD does not occur when these mutants perform motor learning. In this sense, the present results support the Marr-Albus-Ito hypothesis.

17.4.2 Roles of Other Synaptic Plasticities in the Cerebellar Cortex

In the cerebellar cortex, multiple forms of synaptic plasticity at different sites are induced during procedural memory formation (Fig. 17.1) (Gao et al., 2012). In addition to LTD at PF-PC synapse, other several types of synaptic plasticities are reported, such as spike-timing-dependent plasticity (STDP) at mossy fiber (MF)-granule cell (GrC), CF activity-dependent LTP at stellate cell to PC synapse, LTP at PF-PC synapse, and rebound potentiation of basket cell to PC synapse. Roles of these plasticities in motor learning are discussed here.

At the cerebellum input stage, STDP was reported between MF and GrC synapse (Sgritta et al., 2017). This type of synaptic plasticity may take part in cerebellar receptive field reshaping after sensory stimulation. However, there seems no direct feedback signal from the motor performance; thus, this STDP at MF-GrC synapse would be difficult to directly contribute to the motor learning by correcting wrong performance, rather than contributing to fine-tuning of sensory and motor inputs.

As for inhibitory input to PC, GABA-mediated inhibitory synaptic transmission undergoes a long-lasting "rebound potentiation (RP)" after the activation of excitatory CF inputs (Kano et al., 1996). CF activity increases $(Ca^{2+})_{in}$ in PC, and activated Ca^{2+}/calmodulin-dependent protein kinase II (CaMKII), which causes structural alteration of $GABA_AR$-associated protein (GABARAP), subsequently enhances interaction between GABARAP and $GABA_AR$ $\gamma 2$ subunit and shows RP by increasing $GABA_AR$ expression at the inhibitory synaptic site (Kawaguchi & Hirano, 2007). Expression of an inhibitory peptide, which blocks interaction between GABARAP and $GABA_AR$, selectively in PC impairs RP and VOR-adaptation, but not OKR adaptation (Tanaka et al. 2013). The RP obviously had feedback signal from motor performance through CF. Differences between LTD and RP are the following: first, RP is induced in all PC belonging to the same microzone, which is innervated by the same CF (Andersson & Oscarsson, 1978). The number of PCs belonging to one microzone was estimated from number of PCs which synchronously increase $[Ca^{2+}]_{in}$ at the firing of complex spike activated by the same climbing fiber, and estimated number was up to 30 (Ghosh et al., 2013). Thus, the rough number of PC in 1 microzone is considered around 30. Induction of LTD requires combined activities of CF and PF; thus, the number of PC expressing LTD should be far less than microzone size, which should correspond to PC pool size expressing RP simultaneously. Second, RP is induced without regarding the PF activity, which is conveying spatiotemporal pattern of sensory information and motor commands. So, we can in one IO cell, most of PC belonging to this CF's microzone would express RP, and only a small number of PC would express LTD. Surely RP could contribute to the motor learning, because it depends on CF-activity which conveys error signals of motor performance. However, comparing to the LTD, motor learning aquired through RP would be coarse, and RP might contribute at the early stage of learning of a complicated motor performance.

Repetitive stimulation of PF alone at low frequency causes potentiation of PF-PC EPSP (Sakurai, 1987). This postsynaptic type of LTP can reset LTD, and vice versa, LTD can reset LTP; thus, LTP and LTD mutually counterbalance each other (Lev-Ram et al. 2003). LTP of PF-PC synapse is mediated by NO (Lev-Ram et al., 2002), NSF (Kakegawa & Yuzaki 2005), and PP2B (Schonewille et al., 2010). Purkinje cell-selective deletion of PP2B abolishes postsynaptic LTP, whereas LTD was unaffected. The mutants showed impaired "gain-decrease" and "gain-increase" adaptation of VOR (Schonewille et al., 2010). In adaptation of VOR, where relatively small number of PC is involved and the same pool of PCs are engaged in new adaptation, old internal model should be reset by LTP to acquire new internal model.

17.4.3 Significance of LTD in Human Behavior

The significance of cerebellar LTD in motor learning was demonstrated in various mammals; however, significance of LTD in human motor learning is elusive. However, some circumstantial evidences suggested a crucial importance of LTD, compared to other types of synaptic plasticity in the cerebellar cortex. Recently, autoantibodies against neuronal surface proteins such as receptors and ion channels are known to cause neurological symptoms (Lancaster & Dalmau, 2012). However, autoantibodies associated with cerebellar ataxia have different feature from those associated with limbic encephalitis. In limbic encephalitis, anti-NMDA receptor and anti-LGI1 are often detected. LGI1 is important to express and localize voltage-dependent K channel (Kv1) (Hivert et al., 2019), and LGI1 enhanced Kv1 activity (Zhou et al., 2018). Kv1 and LGI1 are expressed in cerebellar neurons; however, association with anti-LGI1 antibody and cerebellar ataxia is rare (Mitoma et al., 2020). Also, NMDA-Rs distribute in cerebellar neurons, and NMDA-R of GrC was indicated to be essential for STDP at the synapse between MF and GrC (Sgritta et al., 2017), but association with immune-mediated cerebellar ataxia is not documented (Mitoma et al., 2020). On the other hand, immune-mediated cerebellar ataxia is associated with anti-mGluR1, anti-GluR delta 2, and anti-voltage-gated Ca channel (VGCC). Association of these autoantibodies with autoimmune limbic encephalitis is not documented (Mitoma et al., 2020). Target surface proteins of these three antibodies are essential to induce LTD in PC. The neuronal circuitry of the mature cerebellum stores acquired parameters of successful coordinated movements, and these internal models are essential for controlling the movement of the body parts without the need for sensory feedback (Ito, 2005, 2008). Acquired internal model would be routinely reset by LTD, in order to adapt to change in internal and external environment. Even if NMDA-R or K channel-related proteins are attacked by autoantibodies, as far as LTD remains normal, distortion of the internal model would be possibly recovered by LTD; however, if essential proteins for LTD induction are attacked by autoantibodies, there would be no way of recovery, and then symptom of cerebellar ataxia appears. As for the cerebellar cortex, PC is a final common output and LTD located at the final regulation step of PC output. Thus, LTD would play a crucial role in regulation of human movement.

References

Aiba, A., Kano, M., Chen, C., Stanton, M. E., Fox, G. D., Herrup, K., Zwingman, T. A., & Tonegawa, S. (1994). Deficient cerebellar long-term depression and impaired motor learning in mGluR1 mutant mice. *Cell, 79*, 377–388.

Albus, J. S. (1971). A theory of cerebellar function. *Mathematical Biosciences, 10*, 25–61.

Andersson, G., & Oscarsson, O. (1978). Climbing fiber microzones in cerebellar vermis and their projection to different groups of cells in the lateral vestibular nucleus. *Experimental Brain Research, 15*, 565–579.

Bredt, D. S., & Nicoll, R. A. (2003). AMPA receptor trafficking at excitatory synapses. *Neuron, 40*, 361–379.

De Schutter, E., & Bower, J. M. (1994). An active membrane model of the cerebellar Purkinje cell II. Simulation of synaptic responses. *Journal of Neurophysiology, 71*, 401–419.

De Zeeuw, C. I., Hansel, C., Bian, F., Koekkoek, S. K., van Alphen, A. M., Linden, D. J., & Oberdick, J. (1998). Expression of a protein kinase C inhibitor in Purkinje cells blocks cerebellar LTD and adaptation of the vestibulo-ocular reflex. *Neuron, 20*, 495–508.

Eccles, J. C., Ito, M., & Szentagothai, J. (1967). *The cerebellum as a neuronal machine*. Springer.

Fukuda, J., Kameyama, M., & Yamaguchi, K. (1981). Breakdown of cytoskeletal filaments selectively reduces Na and Ca spikes in cultured mammal neurones. *Nature, 294*, 82–85.

Gao, Z., van Beugen, B. J., & De Zeeuw, C. I. (2012). Distributed synergistic plasticity and cerebellar learning. *Nature Reviews. Neuroscience, 13*, 619–635.

Ghosh, K. K., Burns, L. D., Cocker, E. D., Nimmerjahn, A., Ziv, Y., El Gamal, A., & Schnitzer, M. J. (2013). Miniaturized integration of a fluorescence microscope. *Nature Methods, 8*, 871–878.

Hartmann, J., Blum, R., Kovalchuk, Y., Adelsberger, H., Kuner, R., Durand, G. M., Miyata, M., Kano, M., Offermanns, S., & Konnerth, A. (2004). Distinct roles of Gαq and Gα11 for Purkinje cell signaling and motor behavior. *The Journal of Neuroscience, 24*, 5119–5130.

Hirano, T., & Ohmori, H. (1986). Voltage-gated and synaptic currents in rat Purkinje cells in dissociated cell cultures. *Proceedings of the National Academy Sciences of the USA, 83*, 1945–1949.

Hirono, M., Sugiyama, T., Kishimoto, Y., Sakai, I., Miyazawa, T., Kishio, M., Inoue, H., Nakao, K., Ikeda, M., Kawahara, S., Kirino, Y., Katsuki, M., Horie, H., Ishikawa, Y., & Yoshioka, T. (2001). Phospholipase Cβ4 and protein kinase Cα and/or protein kinase CβI are involved in the induction of long term depression in cerebellar Purkinje cells. *The Journal of Biological Chemistry, 276*, 45236–45242.

Hivert, B., Marien, L., Agbam, K. N., & Faivre-Sarrailh, C. (2019). ADAM22 and ADAM23 modulate the targeting of the Kv1 channel-associated protein LGI1 to the axon initial segment. *Journal of Cell Science, 132*, jcs219774.

Ito, M., Sakurai, M., & Tongroach, P. (1982). Climbing fibre induced depression of both mossy fibre responsiveness and glutamate sensitivity of cerebellar Purkinje cells. *The Journal of Physiology, 324*, 113–134.

Ito, M. (2005). Bases and implications of learning in the cerebellum — Adaptive control and internal model mechanism. *Progress in Brain Research, 148*, 95–109.

Ito, M. (2008). Control of mental activities by internal models in the cerebellum. *Nature Reviews. Neuroscience, 9*, 304–313.

Kakegawa, W., Katoh, A., Narumi, S., Miura, E., Motohashi, J., Takahashi, A., Kohda, K., Fukazawa, Y., Yuzaki, M., & Matsuda, S. (2018). Optogenetic control of synaptic AMPA receptor endocytosis reveals roles of LTD in motor learning. *Neuron, 99*, 985–998.

Kakegawa, W. & Yuzaki, M. (2005) A mechanism underlying AMPA receptor trafficking during cerebellar long-term potentiation. *Proceedings of the National Academy Science of the USA. 102*. 17846–17851.

Kano, M., Kano, M., Fukunaga, K., & Konnerth, A. (1996). Ca^{2+}-induced rebound potentiation of γ-aminobutyric acid-mediated currents requires activation of Ca^{2+}/calmodulin-dependent kinase II. *Proceedings of the National Academy Sciences of the USA, 93*, 13351–13356.

Karachot, L., Kado, R. T., & Ito, M. (1994). Stimulus parameters for induction of long-term depression in in vitro rat Purkinje cells. *Neuroscience Research, 21*, 161–168.

Kashiwabuchi, N., Araki, K., Hirano, T., Shibuki, K., Takayama, C., Inoue, Y., Kutsuwada, T., Yagi, T., Kang, Y. N., Aizawa, S., & Mishina, M. (1995). Impairment of motor coordination, Purkinje cell synapse formation, and cerebellar long-term depression in GluRδ2 mutant mice. *Cell, 81*, 245–252.

Kawaguchi, S., & Hirano, T. (2007). Sustained structural change of $GABA_A$ receptor-associated protein underlies long-term potentiation at inhibitory synapses on a cerebellar Purkinje neuron. *The Journal of Neuroscience, 27*, 6788–6799.

Kishimoto, Y., Hirono, M., Sugiyama, T., Kawahara, S., Nakao, K., Kishio, M., Katsuki, M., Yoshioka, T., & Kirino, Y. (2001). Impaired delay but normal trace eye blink conditioning in PLCβ4 mutant mice. *Neuroreport, 17*, 2919–2922.

Koekkoek, S. K. E., Yamaguchi, K., Milojkovic, B. A., Dortland, B. R., Ruigrok, T. J. H., Maex, R., De Graaf, W., Smit, A. E., Werf, F. V., Bakker, C. E., Willemsen, R., Ikeda, T., Kakizawa, S., Onodera, K., Nelson, D. L., Mientjes, E., Joosten, M., De Schutter, E., Oostra, B. A., Ito, M., & De Zeeuw, C. L. (2005). Deletion of *FMR1* in Purkinje cells enhances parallel fiber LTD, enlarges spines, and attenuates cerebellar eyelid conditioning in fragile X syndrome. *Neuron, 47*, 339–352.

Katoh, A., Yoshida, T., Himeshima, Y., Mishina, M. & Hirano, T. (2005). Defective control and adaptation of reflex eye movements in mutant mice deficit in either the glutamate receptor delta2 subunit or Purkinje cells. *European Journal of Neuroscience, 21*, 1315–1326.

Kohda, K., Kakegawa, W., Matsuda, S., Yamamoto, T., Hirano, H., & Yuzaki, M. (2013). The δ glutamate receptor gates long-term depression by coordinating interactions between two AMPA receptor phosphorylation sites. *Proceedings of the National Academy Sciences of the USA, 110*, E948–E957.

Lancaster, E., & Dalmau, J. (2012). Neuronal autoantigens-pathogenesis, associated disorders and antibody testing. *Nature Reviews. Neurology, 8*, 380–392.

Lev-Ram, V., Wong, S. T., Storm, D. R., & Tsien, R. Y. (2002). A new form of cerebellar long-term potentiation is postsynaptic and depends on nitric oxide but not cAMP. *Proceedings of the National Academy Sciences of the USA, 99*, 8389–8393.

Lev-Ram, V., Mehta, S. B., Kleinfeld, D., & Tsien, R. Y. (2003). Reversing cerebellar long-term depression. *Proceedings of the National Academy Sciences of the USA, 100*, 15989–15993.

Linden, D. J., & Connor, J. A. (1991). Participation of postsynaptic PKC in cerebellar long-term depression in culture. *Science, 254*, 1656–1659.

Marr, D. (1969). A theory of cerebellar cortex. *The Journal of Physiology, 202*, 437–470.

Masugi-Tokita, M., Tarusawa, E., Watanabe, M., Molnár, E., Fujimoto, K., & Shigemoto, R. (2007). Number and density of AMPA receptors in individual synapses in the rat cerebellum as revealed by SDS-digested freeze-fracture replica labeling. *The Journal of Neuroscience, 27*, 2135–2144.

Mitoma, H., Honnorat, J., Yamaguchi, K., & Manto, M. (2020). Fundamental mechanisms of autoantibody-induced impairments on ion channels and synapses in immune-mediated cerebellar ataxias. *International Journal of Molecular Sciences, 21*, 4936.

Miyata, M., Kim, H. T., Hashimoto, K., Lee, T. K., Cho, S. Y., Jiang, H., Wu, Y. P., Jun, K., Wu, D. Q., Kano, M., & Shin, H. S. (2001). Deficient long-term synaptic depression in the rostral cerebellum correlated with impaired motor learning in phospholipase Cβ4 mutant mice. *The European Journal of Neuroscience, 13*, 1945–1954.

Nomura, T., et al. (2012). Cerebellar long-term depression requires dephosphorylation of TARP in Purkinje cell. *The European Journal of Neuroscience, 35*, 402–410.

Sakurai, M. (1987). Synaptic modification of parallel fibre-Purkinje cell transmission in *in vitro* guinea-pig cerebellar slices. *The Journal of Physiology, 394*, 463–480.

Schonewille, M., Belmeguenai, A., Koekkoek, S. K., Houtman, S. H., Boele, H. J., van Beugen, B. J., Gao, Z., Badura, A., Ohtsuki, G., Amerika, W. E., Hosy, E., Hoebeek, F. E., Elgersma, Y., Hansel, C., & De Zeeuw, C. I. (2010). Purkinje cell-specific knockout of the protein phosphatase PP2B impairs potentiation and cerebellar motor learning. *Neuron, 67*, 618–628.

Schonewille, M., Gao, Z., Boele, H. J., Veloz, M. F. V., Amerika, W. E., Simek, A. A. M., De, J. M. T., Steinberg, J. P., Takamiya, K., Hoebeek, F. E., Linden, D. J., Huganir, R. L., De Zeeuw, C. I., & CI. (2011). Reevaluating the role of LTD in cerebellar motor learning. *Neuron, 70*, 43–50.

Srivastava, S., Osten, P., Vilim, F. S., Khatri, L., Inman, G., States, K. B., Daly, C., DeSouza, S., Abagyan, R., Valtschanoff, J. G., Weinberg, R. J., & Ziff, E. B. (1998). Novel anchorage of GluR2/3 to the postsynaptic density by the AMPA receptor–binding protein ABP. *Neuron, 21*, 581–591.

Steinberg, J. P., Takamiya, K., Shen, Y., Xia, J., Rubio, M. E., Yu, S., Jin, W. Y., Thomas, G. M., Linden, D. J., & Huganiret, R. L. (2006). Targeted in vivo mutations of the AMPA receptor subunit GluR2 and its interacting protein PICK1 eliminate cerebellar long-term depression. *Neuron, 49*, 845–860.

Sgritta, M., Locatelli, F., Soda, T., Prestori, F., & D'Angelo, E. U. (2017). Hebbian spike-timing dependent plasticity at the cerebellar input stage. *The Journal of Neuroscience, 37*, 2809–2823.

Swensen, A. M., & Bean, B. (2003). Ionic mechanisms of burst firing in dissociated Purkinje neurons. *Journal of Neuroscience, 23*(29), 9650–9663.

Tanaka, S., Kawaguchi, S., Shioi, G., & Hirano, T. (2013). Long-term potentiation of inhibitory synaptic transmission onto cerebellar Purkinje neurons contributes to adaptation of vestibulo-ocular reflex. *The Journal of Neuroscience, 33*, 17209–17220.

Wang, S. S., Denk, W., & Hausser, M. (2000). Coincidence detection in single dendritic spines mediated by calcium release. *Nature Neuroscience, 3*, 1266–1273.

Wang, Y. T., & Linden, D. J. (2000). Expression of cerebellar long-term depression requires post-synaptic clathrin-mediated endocytosis. *Neuron, 25*, 635–647.

Xia, J., Chung, H. J., Wihler, C., Huganir, R. L., & Linden, D. J. (2000). Cerebellar long-term depression requires PKC-regulated interactions between GluR2/3 and PDZ domain–containing proteins. *Neuron, 28*, 499–510.

Yamaguchi, K., Itohara, S., & Ito, M. (2016). Reassessment of long-term depression in cerebellar Purkinje cells in mice carrying mutated GluA2 C terminus. *Proceedings of the National Academy Sciences of the USA, 113*, 10192–10197.

Yamazaki, M., Le Pichonc, C. E., Jacksona, A. C., Cerpasa, M., Sakimurab, K., Scearce-Leviec, K., & Nicoll, R. A. (2015). Relative contribution of TARPs γ-2 and γ-7 to cerebellar excitatory synaptic transmission and motor behavior. *Proceedings of the National Academy Sciences of the USA, 112*, E371–E379.

Zhou, L., Zhou, L., Su, L. D., Cao, S. L., Xie, Y. J., Wang, N., Shao, C. Y., Wang, Y. N., Zhou, J. H., Cowell, J. K., & Shen, Y. (2018). Celecoxib ameliorates seizure susceptibility in autosomal dominant lateral temporal epilepsy. *The Journal of Neuroscience, 38*, 3346–3357.

Part V
Cerebro-Cerebellar Loop and Its Contribution of the Cerebellum to Higher Brain Functions

Chapter 18
Neural Predictive Computation in the Cerebellum

Hirokazu Tanaka, Takahiro Ishikawa, and Shinji Kakei

18.1 Introduction

The cerebellum is one of the best-studied structures in the central nervous system from anatomical, neurophysiological, and computational perspectives (Ito, 1984). The cerebellum consists of the vestibulocerebellum, the spinocerebellum, and the cerebrocerebellum, which mainly receive afferent inputs from the vestibular nuclei, the dorsal columns of the spinal cord and the trigeminal nerve, and the cerebral cortex, respectively. The wide range of the afferent signals from cortical, subcortical, and peripheral sources indicates that the cerebellum plays a role in not only motor functions but also cognitive, executive, and emotional functions. Despite the multitude of functions attributed to the cerebellum, the cerebellar cortex possesses a uniform structure composed of several distinct cell types. Neural activities of the cerebellar cells have been well characterized in neurophysiological recordings (De Zeeuw et al., 2011). This uniform structure points to a suggestion that there is a unifying functional principle for the working of the cerebellum (Diedrichsen et al., 2019).

Despite a considerable number of findings on the cerebellum, we still lack the holistic picture of algorithmic understanding of the cerebellum. Although there is a consensus that the cerebellum plays a critical role in motor control, motor learning, and cognitive functions, what and how the cerebellum computes is still open to

H. Tanaka (✉)
Faculty of Information Technology, Tokyo City University, Tokyo, Japan
e-mail: htanaka@tcu.ac.jp

T. Ishikawa
Laboratory for Movement Disorders, Tokyo Metropolitan Institute of Medical Science, Tokyo, Japan

S. Kakei
Department of Anatomy and Physiology, Faculty of Life Sciences, Jissen Women's University, Tokyo, Japan

H. Mizusawa, S. Kakei (eds.), *Cerebellum as a CNS Hub*,
Contemporary Clinical Neuroscience, https://doi.org/10.1007/978-3-030-75817-2_18

debate. From an electrophysiological perspective, the perceptron hypothesis of the cerebellar cortex has been the guiding principle in predicting the synaptic plasticity between parallel fibers and a Purkinje cell (PC) (Ito et al., 1982) and the error coding of the climbing fibers (Kitazawa et al., 1998). From a behavioral perspective, the internal-forward-model hypothesis has explained the results of behavioral experiments both in control subjects and cerebellar patients (Cabaraux et al., 2020). Curiously, the two theories are rarely discussed in the same context. Therefore, the missing link is whether and how the perceptron-like architecture of the cerebellum computes the internal-forward-model prediction. The pioneering work by Marr, Albus, and Ito is often collectively referred to as the Marr-Albus-Ito model. Still, a careful reading of the original papers reveals that Marr and Albus proposed the perceptron model (Marr, 1969; Albus, 1971), whereas Ito proposed the internal forward model (Ito, 1970). This review attempts to provide a coherent understanding of the cerebellar functions based on recent studies, including ours (Tanaka et al., 2019; Tanaka et al., 2020).

This book chapter is organized as follows. First, Sect. 18.2 addresses the problem of temporal delay in sensory feedback signals and introduces the concept of internal models as a possible solution. Notably, an internal forward model anticipates action consequences for motor control and motor learning. Then, Sect. 18.3 presents lines of evidence supporting that the cerebellum performs as an internal forward model for motor prediction. Traditionally, the Marr-Albus theory posits that the cerebellum implements classifier computation, but we contend that the cerebellum may function as a regressor. Our recent investigation on the cerebellar activities supports this view. Finally, Sect. 18.4 expands to tentative computational principles of the cerebellum. We review recent studies reporting linear computation in various stages within the cerebellum and speculate the functional implications of such linear processing. We also assume a possible computational role of the cerebellum in a broader context of the cerebrocerebellar loop. To conclude, Sect. 18.5 poses the remaining questions to promote future studies. Whereas Sects. 18.2 and 18.3 summarize well-accepted views on the cerebellum and overlap with previous reviews, Sects. 18.4 and 18.5 venture into speculations about the computational principles of the cerebellum. This chapter outlines our recent work (Tanaka et al., 2019) and the review articles (Ishikawa et al., 2016; Cabaraux et al., 2020; Tanaka et al., 2020), to which the interested reader is referred to for further details and references.

18.2 Internal Models for Motor Control and Motor Learning

This section addresses the problem of temporal delay in conducting sensory feedback signals from peripheral sensory receptors to the central nervous system. One solution for the delayed feedback problem is an internal model that simulates the musculoskeletal and environmental dynamics. We introduce two types of an internal model: a forward model and an inverse model. Then, we focus on an internal forward model and its predictive computation for guiding motor control and motor learning.

18.2.1 Internal Models for Sensory Delay Compensation

Sensory feedback signals from peripheral sensory receptors contain an inevitable delay in reaching the central nervous system, posing a challenging problem of controlling movements of rapidity and stability. The feedback delay ranges from 10 ms for mice and 100 ms for elephants, mostly from the nerve conduction delay (More et al., 2010; More & Donelan, 2018). Therefore, the brain always observes the past state of the body. Feedback control based on the delayed body state produces oscillatory and even divergent movements (Wolpert & Miall, 1996). In reaching a target, even when the hand reaches the target, delayed sensory feedback signals tell the brain that the hand is still on the way toward the target. The brain then pushes the hand further, thereby overshooting the target. Subsequently, when the hand is pulled back and located on the target, the brain still thinks that the hand is away from the target and pulls the hand toward the body, thereby resulting in oscillatory movements. The fact that animals can conduct rapid movements without oscillations or overshoot indicates that biological motor control implements a mechanism for compensating the delay in sensory feedback signals.

One solution to the sensory-feedback-delay problem is to exploit the physics of the skeletomuscular system and the external world (Wolpert et al., 1998). By simulating the skeletomuscular physics internally in the brain, rapid movements are stably controlled without relying on the delayed sensory feedback signals. The internal simulation of physics is called an internal model. There are two types of internal models: an internal inverse model and an internal forward model (Wolpert & Miall, 1996). An internal inverse model transforms a predesigned, desired trajectory into control signals, so the control signals are generated without relying on the delayed sensory feedback signals (Fig. 18.1a). The inverse-model computation is equivalent to solving the equations of motion in a causally inverse manner (therefore, called an inverse model). This type of control is called feedforward control in control engineering. An internal forward model computes an estimate of the current state of the body by integrating a previous state of the body and an efference copy of motor control signals (Fig. 18.1b). The forward-model computation is equivalent to solving the equations of motion in a causally forward manner (therefore, called a forward model). The estimate of the current state is then used to generate motor commands. This type of control is known as internal feedback control in control engineering. Both internal inverse and forward models exploit the knowledge of musculoskeletal dynamics but in distinctive ways; an internal inverse model derives motor commands from a given desired movement, and an internal forward model estimates a current state from a previous state and a motor command. The internal inverse and forward models can solve the sensory-feedback-delay problem, but whether the biological motor control adopts the inverse-model solution or the forward-model solution is still open to debate. This review focuses on the forward-model solution from now on. More detailed discussions on the computational roles of a forward model within the optimal feedback control model (Todorov & Jordan, 2002; Todorov, 2004) can be found in our previous review (Tanaka et al., 2020).

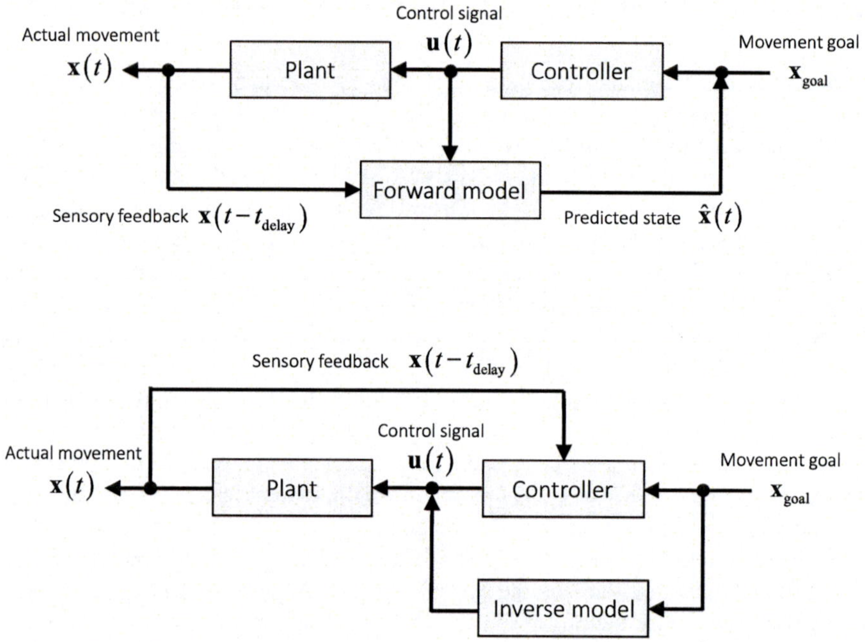

Fig. 18.1 Internal models (forward and inverse). (Top) An internal forward model receives time-delayed sensory feedback $\mathbf{x}(t - t_{\text{delay}})$ and efference copy $\mathbf{u}(t)$ and predicts a current state $\hat{\mathbf{x}}(t)$. The difference between movement goal $\mathbf{x}_{\text{goal}}$ and predicted state $\hat{\mathbf{x}}(t)$ drives a feedback controller to generate a control signal $\mathbf{u}(t)$, which in turn steers the controlled plant. (Bottom) An internal inverse model receives the movement goal $\mathbf{x}_{\text{goal}}$ and then generates the feedforward control signal $\mathbf{u}(t)$ to realize the desired movement. The total control signal consists of the output of the inverse model and the delayed feedback control

18.2.2 Prediction for Motor Control and Motor Learning

An internal forward model solves the sensory-feedback-delay problem by estimating a current state of the body and exerting a feedback controller according to the current estimate. If the current estimate is accurate enough, the internal feedback control should guide rapid movements in a reliable manner. As reviewed above, the delay in sensory feedback signals is of the order of up to 100 ms, so an internal forward model must cover up to such a time scale for online prediction. With the knowledge of body dynamics, the forward model estimates a current state of the body from a previous state of the body communicated via sensory feedback and an efference copy of the motor control signal (Wolpert et al., 1995). Therefore, an internal forward model allows robust motor control under delayed sensory feedbacks (Todorov & Jordan, 2002; Todorov, 2004).

Predictive computation plays an essential role in not only motor control but also motor learning. Learning, in general, occurs when what happens differs from what one expects to happen. The difference between the desired outcome and an actual outcome, known as a prediction error, drives a learning process in improving future performance (Shadmehr et al., 2010). In contrast, another theory, known as feedback-error learning, postulates that a motor correction in a current trial amends feedforward control signals in a next practice (Kawato et al., 1987). Recent studies in human psychophysics support that a prediction error but not a motor correction drives the adaptive process of motor learning (Mazzoni & Krakauer, 2006; Tseng et al., 2007; Izawa & Shadmehr, 2011). First, an ingenious experimental design dissociated prediction error and a target error (an error between a target direction and an actual movement direction) in a visuomotor adaptation by having the subjects aim in a direction different from the target (Mazzoni & Krakauer, 2006). The subjects adapted the moving direction even when the target error was null (i.e., the subjects' movement direction matched with the target direction). The learning curve was explained by not the target error but the prediction error. Another experiment demonstrated that the prediction of an internal forward model adapted to visuomotor adaptation when a visual error was explicitly presented (Izawa & Shadmehr, 2011). One group of subjects received a visual error between the target direction and the movement direction, and the other group of subjects received only a binary variable that indicated a hit or a miss. Whereas the subjects in both groups adapted to the visuomotor rotation successfully, they differed in how the perceived hand position adapted. The subjects who adapted with the visual error reported the hand position along the adapted movement direction, while the subjects who adapted with the binary feedback reported the hand position on the actual location. The altered perception of the hand position reflected the learned dynamics of the internal forward model. These experimental results, taken together, support that learning of the internal forward model by the prediction error is an underlying mechanism of motor learning. Therefore, for motor learning, the brain compares what happens with what is expected to happen and updates motor planning and execution so that the prediction error is reduced.

Recent studies supported the cerebellar predictive computation of reward expectation, conventionally considered as a function of dopaminergic cells in the basal ganglia (Schultz et al., 1997; Schultz, 2000). In reinforcement learning, where a reward or a punishment evaluates an action outcome, not a reward per se but a reward prediction error (i.e., the distinction between an actual reward and an expected reward) drives a learning process. Therefore, a prediction mechanism plays a critical role not only in error-based motor learning but also in reward-based reinforcement learning. The cerebellum seems to participate in a reward-prediction circuit; climbing fiber inputs to a Purkinje cell encode a temporal difference error (Ohmae & Medina, 2015) or an expected reward size (Larry et al., 2019), and cerebellar granule cells (GCs) convey the expectation of reward (Wagner et al., 2017).

18.3 The Cerebellum as an Internal Forward Model

This section reviews extant experimental evidence that the cerebellum functions as an internal forward model. Then, the subtle but important distinction between the perceptron perspective and the internal-forward-model perspective is emphasized. Finally, we explain our recent study demonstrating that the cerebellar output activities at one moment were predictive of the cerebellar input activities in the future.

18.3.1 Cerebellum and Internal Forward Model

Evidence converging from human psychophysics, human neuroimaging, electrophysiology, and clinical observations supports that the cerebellum, especially the cerebrocerebellum, implements the predictive computation of an internal forward model in motor control and motor learning. The anatomical features of the cerebellum, especially the anatomical afferents and efference and the feedforward structure of the cerebellar circuit, make the cerebellum particularly suited for the forward-model computation. We here briefly review evidence supporting the internal-forward-model hypothesis of the cerebellum.

If the cerebellar function is temporally blocked by an artificial stimulation, predictive performance in a task is specifically impeded. The impeded prediction due to cerebellar lesion or stimulation is observed in motor and language tasks, indicating that the cerebellar prediction operates for motor and cognitive functions. Such a lack of prediction is manifested in a psychophysical experiment in which a subject released a ball from one hand and then caught the ball by the other hand (Nowak et al., 2004; Nowak et al., 2007). Healthy controls exhibited anticipating activities in muscles of the receiving hand, whereas cerebellar patients lacked such predictive muscle activities. The predictive computation in the cerebellum is temporally blocked by transcranial magnetic stimulation (Miall et al., 2007). When the cerebellum was stimulated during arm movement, the movement direction deviated as if the hand position had lagged at a slightly past position. The deviation indicated that the lag in the positional estimation was about 138 ms, consistent with the cerebellum's disrupted prediction due to the magnetic stimulation. The cerebellar prediction is not limited in the motor domain but also observed in the cognitive domain (Lesage et al., 2012). When a target of a saccadic movement was predictable from a semantic cue (a verb in this case), the saccadic latency was shortened, indicating the predictive computation from the semantic cue to the target. When repetitive magnetic stimulation disrupted the right cerebellar hemisphere, the predictive shortening in the saccadic latency disappeared, indicating that this part of the cerebellum takes part in the predictive language processing. The cerebellar prediction, therefore, covers not only motor control but also higher cognitive functions.

The PCs are the sole output neurons from the cerebellar cortex to the cerebellar nuclei, and the functional properties have been well studied in a range of sensorimotor tasks. If the cerebellar cortex operates as an internal forward model, PCs' simple spike activity should reflect the current and future kinematics (i.e., position, velocity, and acceleration) rather than the dynamics (i.e., joint torques and muscle tensions). In a circular manual tracking task under varying viscous and elastic loads, the PCs' activity was little affected (Pasalar et al., 2006). Subsequent studies demonstrated that the activity of PCs exhibited a bimodal preference for a future state and a feedback error (Popa et al., 2012, 2013). The state prediction encoded by PCs reflects the computation of an internal forward model, and the feedback error encoded by PCs may be needed for updating an internal forward model. These experimental results show that the cerebellar cortex contains a dual representation of the state prediction and feedback performance about movement kinematics, consistent with the internal-forward-model hypothesis.

An impairment in the cerebellum often leads to a plethora of motor deficits in coordinating movements, collectively known as cerebellar ataxia (Cabaraux et al., 2020). In addition to motor control deficiencies, cerebellar patients often exhibit a diminished ability to adapt to motor perturbations or novel environments. These clinical observations are in line with the two functions of an internal forward model for motor control and motor learning, reviewed above. Regarding the role of the internal forward model in motor control, one symptom in cerebellar ataxia in extremity is dysmetria (Cabaraux et al., 2020). Cerebellar motor ataxias refer to a collection of symptoms, including inaccuracy and clumsiness in control of posture, gait, limbs, and ocular saccades. Among a wide range of manifestations of cerebellar ataxias, dysmetria is a core symptom in cerebellar motor ataxias. Voluntary movements involve multiple degrees of freedom spanning joints and muscles, so a consequence of one body part affects a movement of another body part. An impairment of the prediction of one action consequence leads to uncoordinated and oscillatory movements (see the example of reaching in Sect. 18.2).

In addition, cerebellar patients suffer from an inability to adapt to external perturbations such as visual displacement due to a wedge prism or computer rotation (Martin et al., 1996; Tseng et al., 2007) and viscous force fields (Smith & Shadmehr, 2005). A classic study demonstrated that patients with cerebellar atrophy with lesions of the olivocerebellar system experience an impaired ability in the prism adaptation experiment (Martin et al., 1996). When cerebellar patients were instructed to reach a target and the cursor position was visually rotated, they were not able to compensate for the imposed visual rotation (Tseng et al., 2007). Of note was that those patients made rapid movements without online movement corrections. This finding indicates that the prediction of action consequence, but not the actual movement and correction, contributes to cerebellum-dependent motor adaptation. Deficits in motor adaptation to viscous force fields were observed in cerebellar patients but not in Huntington's disease patients, supporting the concept that the cerebellum but not the basal ganglia is involved explicitly in predicting action consequences.

We emphasize that the internal-forward-model hypothesis was already proposed by Ito (1970) as early as 1970, although the paper is usually not credited appropriately as it deserves. In an early phase of motor learning, the motor cortex must rely on delayed sensory feedback signals, thereby leading to slow and clumsy movements. In a late stage of motor control, the cerebellum simulates the dynamics of the skeletomuscular system and environment to compensate for the delay in sensory feedback. Although Ito did not use the term "an internal forward model," his description precisely defines the cerebellar function as an internal forward model. It is also noteworthy that he correctly depicted a closed loop between the cerebellum and the motor cortex; the anatomical pathway revealed more than 30 years later. Many of Ito's insights in 1970 have turned out to be correct.

18.3.2 Classifier and Regressor Perspectives

In machine learning, supervised learning (i.e., learning based on supervising signal, teacher signal, or error signal) is broadly categorized into classification and regression problems. For a given input, a classifier assigns a discrete output (i.e., a class label), and a regressor computes a continuous output. This distinction between discrete and continuous outputs seems at first technical but has a profound implication on the computational functioning in the cerebellum. Originally 50 years ago, Marr and Albus formulated a theory of the cerebellar cortex by corresponding it to a neural network model called a perceptron (Marr, 1969; Albus, 1971). The Marr-Albus model was inspired by the structural regularity in the cerebellar cortex and hypothesized computational functions specific to individual neuron types. The perceptron model of the cerebellum consists of several assumptions on the functioning of cerebellar cells: mossy fibers (MFs) for network input, GCs for representation expansion, Golgi cells for regularization of input, stellate and basket cells for inverting the sign of input, and PCs for network output. There are three critical hypotheses in the model. First, the GCs that outnumber the MF transform afferent signals into sparse and orthogonal representations in the parallel fibers. Second, the PCs receiving the parallel fiber afferents perform a thresholding computation in the sense of a perceptron. Finally, the climbing fiber input represents a supervising signal to a PC and induces plasticity in the parallel-fiber-to-PC synapses. These hypotheses in the Marr-Albus model have guided both experimental and theoretical studies of the cerebellar cortex. The classifier perspective has dominated the computational study of the cerebellar function (see Cayco-Gajic & Silver (2019) for a recent review). However, it seems peculiar, albeit not inconsistent, that the cerebellum has adopted a perceptron of the classifier with a discrete output for controlling body dynamics of continuous dynamics. We hereby contrast a classifier and a regressor as a computational function of the cerebellum.

In contrast to the perceptron model reviewed above, the internal forward model pioneered by Ito posits that the cerebellum operates as a regressor but not a classifier (Ito, 1970). He clearly stated that the cerebellum implements the dynamics of the external world: "the large loop through the external world may be effectively replaced by an internal one through the cerebellum which would serve as a model of the combination of the spinal motor system, the external world, and the sensory pathways." For this kind of computation, the cerebellar output must be continuous, reflecting the continuous characteristics of the body and the world. Therefore, Ito implicitly indicated that the cerebellum functions as a regressor that performs the continuous computation of dynamics rather than the discrete computation of classification. Our recent study reviewed in the next subsection reported linear transformations and predictions in the cerebellar circuit, thereby supporting the regression computation in the cerebellum (Tanaka et al., 2019).

18.3.3 Neural Evidence for the State Predictor in the Cerebellum

Whereas accumulating evidence points to the cerebellum as a locus of an internal forward model, a direct investigation of the predictive computation had not been demonstrated neurophysiologically. One crucial testable conjecture is that a current output from an internal forward model must contain predictive information about a future input to an internal forward model. Therefore, if the cerebellum is compared to an internal forward model, the current activity of the cerebellar nuclei should predict the future activity of the MFs. Therefore, we proceeded to examine this conjecture using neurophysiological data recoded from a behaving monkey (Tanaka et al., 2019).

We recorded firing rates from 94 MFs, 83 PCs, and 73 dentate nucleus cells (DNCs) while the monkey performed step movements of the wrist toward 1 of the 8 targets (Ishikawa et al., 2014; Tomatsu et al., 2016). We first analyzed the dataset and investigated how the firing rates of one cell population were transformed into the firing rates of a subsequent cell population. The firing rates of PCs were well reconstructed as a weighted linear sum of the firing rates of MFs. Similarly, the firing rates of DNCs were reproduced as a weighted linear sum of the firing rates of PCs and MFs. Besides the linear reconstruction model, we tested nonlinear models (thresholding and quadratic models) and a finite-impulse model. The linear model was the most parsimonious in reconstructing the firing rates, and no nonlinear processing or history dependence was necessary for our dataset. Therefore, we conclude that the cerebellar computation is linear regarding the firing rates.

We then tested the prediction of the internal-forward-model hypothesis by predicting the future inputs to the cerebellum from the current outputs from the cerebellum. If the cerebellum operates as an internal forward model, the cerebellar outputs at one moment should contain the predictive information of the cerebellar inputs in the future. Therefore, we attempted to predict the firing rates of MFs (the cerebellar input) at a future time by reconstructing as a weighted linear sum of the firing rates of DNCs (the cerebellar output) at a current time. The prediction up to a few hundred milliseconds was statistically significant, indicating that the current cerebellar output was predictive of the future cerebellar input. Therefore, the cerebellum performs a predictive transformation of current input into expected input in the future.

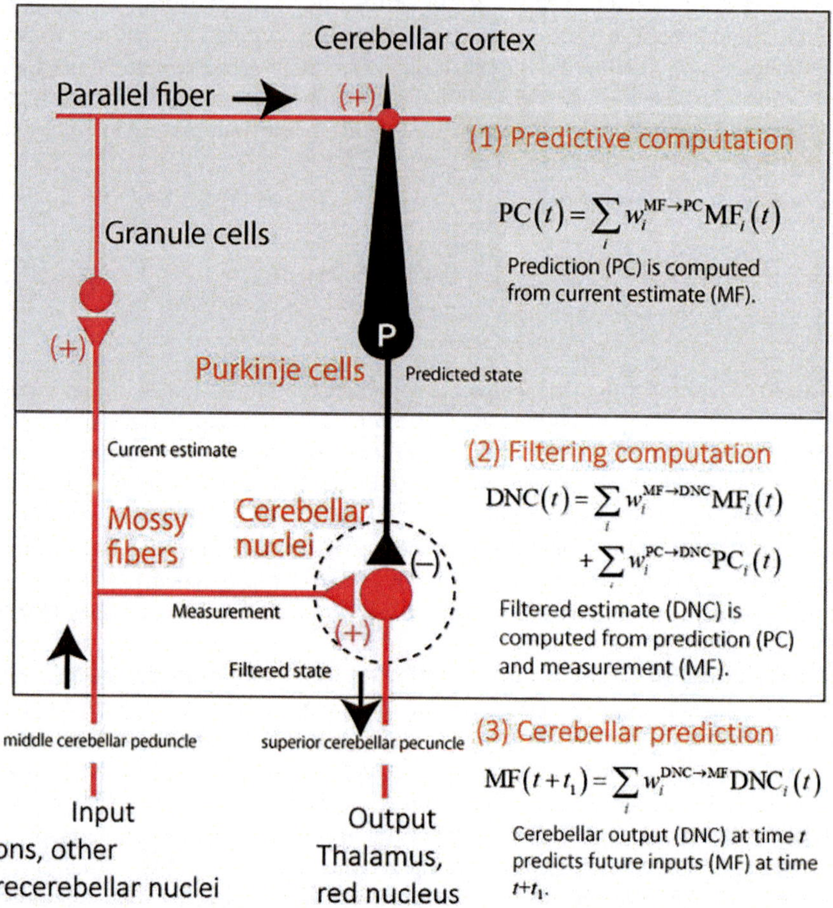

Fig. 18.2 Internal model prediction in the cerebellum (the Kalman filter). Schematics of the cerebellar circuit and corresponding computational steps. (1) Prediction computation. The PCs compute a predicted state from a current estimate conveyed by the MFs. (2) Filtering computation. The predictive state computed by the PCs is integrated with an observation signal conveyed by other MFs for optimal estimation in DNCs. (3) Cerebellar prediction. The current output from the cerebellar circuit (DNCs) can predict future inputs to the cerebellum (MFs). (Figure reproduced from Tanaka et al. (2020))

Our analysis of the neurophysiological dataset revealed the linear reconstruction equations (the PC activities from MF activities and the DNC activities from the MF and PC activities) and the linear prediction equation (the future MF activities from the current DNC activities) (Fig. 18.2). These equations resemble those of the Kalman filter (Kalman, 1960), the iterative Bayesian optimal estimation method for a dynamical system if we assume that the PCs and the DNCs perform a state-prediction step and a filtering step, respectively. The Kalman filter not only predicts a future state (the state-prediction step) but also compares its prediction with a measurement for an optimal estimate (the filtering step). The Kalman filter is a natural extension of an internal forward model. The anatomical inputs to the cerebellar hemisphere support our analogy. The cerebellar hemisphere receives afferents from both the cerebral cortex and the peripheral afferent pathway. We speculate that the cerebellum performs the computation of optimal estimation by combining the predicted state and the incoming sensory feedback signals. Therefore, the cerebellum may function as the Kalman filter beyond the internal forward model. Our analysis of firing rates of the cerebellar cells supports Ito's model of the cerebellum as a regressor rather than Marr and Albus' model of the cerebellum as a classifier.

18.4 Principles of Cerebellar Computation

Based on the evidence for the cerebellum as an internal forward model reviewed in the preceding sections, this section explores theoretical implications about the cerebellar computation. We first argue the computation in the cerebellum, especially emphasizing linear probabilistic computation, and further speculate that the cerebellar predictive computation supports robust and stable functioning in the cerebellar cortex.

18.4.1 Linear Probabilistic Computation in the Cerebellum

Recent developments in recording the neural activities from the cerebellum reveal a novel and an unexpected principle of the cerebellar computation. In particular, calcium imaging allows recording activities from thousands of neurons simultaneously, literally revising the picture of the cerebellar functions. Therefore, what Marr and Albus posited in their models are now possible to examine. In his pioneering paper, Marr stated that "the PC should fire if and only if more than a proportion p of the active parallel fibres have facilitated synapses with it, where p is close to 1," indicating that the PCs perform a nonlinear thresholding operation prerequisite for a classifier (Marr, 1969).

On the contrary to Marr's postulate, one finding is that, in various stages of cerebellar processing, the computation is linear; the activity of output neuron is expressed simply as a weighted sum of activities of input neurons (Tanaka et al., 2019). Several studies reported such linear computation in the cerebellum (Walter & Khodakhah, 2006, 2009; Kennedy et al., 2014; Raymond & Medina, 2018). The firing rate of a PC linearly encoded the strength of GC synaptic input, regardless of the location or temporal pattern of the input (Walter & Khodakhah, 2006). A possible advantage of linear encoding is an increase in the computation capacity of activity patterns that a PC can represent (Walter & Khodakhah, 2009). Another study reported linear computation from MFs to medium ganglion cells in the cerebellum-like structure of electric fish, where linear weighted sums of sparsely and randomly mixed MF inputs reconstructed the membrane potentials of GCs (Kennedy et al., 2014). The reconstructed GC activities exhibited a rich repertoire of temporal bases, which in turn constitute a negative image of sensory inputs.

Besides the linear computation, there is accumulating evidence that cerebellar neurons encode behavioral parameters linearly by MFs (Lisberger & Fuchs, 1978; Laurens et al., 2013), PCs (Shidara et al., 1993; Medina & Lisberger, 2007, 2009; Herzfeld et al., 2015; Chen et al., 2016; Hong et al., 2016; Dugue et al., 2017; Sun et al., 2017), and cerebellar nuclear cells (Ebner et al., 2011; Heiney et al., 2014; Ten Brinke et al., 2017). We note that saccade is encoded by spikes of PCs that were related to the period of occasional pauses (Hong et al., 2016). The findings reviewed above support that both the computation within the cerebellar circuit and the representation of behavioral variables are linear, consistent with the regressor perspective of the cerebellum.

Provided the linear computation in the cerebellum, what is the function of the linear computation? A Bayesian framework of statistical inference offers a unified, systematic method of updating one's belief (posterior probability) by integrating prior knowledge (prior probability) and an incoming observation (likelihood). Specifically, Bayes' theorem states that posterior probability is a product of prior probability and likelihood, divided by marginal probability. It appears, therefore, that the Bayesian probabilistic computation requires multiplicative processing. But the recent studies reviewed above support linear processing in the cerebellum. The multiplicative Bayesian computation seems at odds with the linear computation in the cerebellum. How can the cerebellum perform probabilistic computation based on linear processing?

The key insight is that the brain represents not a probability directly but instead a logarithm of probability and that the logarithm transforms multiplication into summation (Rao, 2004; Yang & Shadlen, 2007; Kira et al., 2015). If neurons encode the logarithm of probabilities of certain events, then the Bayesian computation becomes a weighted sum of firing rates of those neurons. In a probabilistic reasoning task, each figure presented to a monkey was assigned to a probability of obtaining a reward, and neurons in the parietal cortex encoded a logarithm of reward probability by their firing rates (Yang & Shadlen, 2007; Kira et al., 2015). When figures were presented sequentially, the probability of obtaining reward was a product of probabilities of corresponding figures. By encoding logarithmically, the parietal neurons could represent the accumulating probability just by summing their firing rates. Therefore, the logarithmic representation of probability simplifies the neural computation of probabilistic reasoning.

Similarly, the recursive computation of Bayesian inference is reduced to linear computation if logarithms of probabilities are used (Rao, 2004). Our finding of linear computation in the cerebellum, along with other studies reviewed above, bespeaks that the cerebellum also exploits the logarithmic trick to simplify the Bayesian computation. In fact, the Kalman filter is a logarithmic version of a recursive Bayesian filter. Therefore, the cerebellum and the cerebral cortex exploited the logarithmic trick of probabilistic computation long before the invention of the logarithm by John Napier in the seventeenth century!

18.4.2 Complementary Computation in the Cerebrocerebellar Loop

The traditional view of the cerebellar function had been limited to motor control and motor learning, but the revised view has proposed the cerebellar functions in cognitive, executive, and emotional processing (Schmahmann & Caplan, 2006; Schmahmann, 2019). We have so far discussed the cerebellar computation of prediction in the context of motor control, but the predictive computation should have broader utility for cognitive and executive functions. The cerebral cortex and the cerebellum have evolved in conjunction as their volume has increased in a proportional manner (Rilling & Insel, 1998). Also, the cerebellum forms a closed loop not only with the motor cortex but also with the dorsolateral prefrontal cortex (BA46) (Kelly & Strick, 2003). The anatomical evidence suggests a specific function of the cerebrocerebellar loop. But to our knowledge, there is no guiding computational hypothesis. The work of Marr and Albus has guided the study of the cerebellar cortex; the work of Ito again provided a broader perspective of the cerebellum in connection with the cerebral cortex (Ito, 1970).

Computational roles of the cerebellum may best be understood in terms of its connections to other brain areas. The cerebellum is essentially a feedforward circuit with inputs through the MFs and outputs from the cerebellar nuclei. In particular, the cerebrocerebellum receives cortical projections from a cortical area through the pons and projects back to the same cortical area through the thalamus, thereby forming anatomically closed connections called the cerebrocerebellar loop. The cerebellar circuit may well be modeled as a feedforward neural network (Marr, 1969; Albus, 1971). The cerebral cortex, on the other hand, is characterized by recurrent hierarchical connections composed of local and lateral connections (DeFelipe & Jones, 1988), so it is appropriate to model the cerebral cortex as a recurrent neural network. The cerebrocerebellar loop is a structure that has been preserved phylogenetically, but its computational role is still open to debate. One may wonder what computational merits are if a feedforward network and a recurrent network are in tandem.

We contrast the two architectures of neural networks from the perspective of artificial neural networks. First, a feedforward neural network with at least one hidden layer can approximate any functional mapping (Cybenko, 1989; Funahashi, 1989; Hornik et al., 1989) and is relatively straightforward to train with, for example, the backpropagation algorithm (Rumelhart et al., 1986). However, a feedforward neural network can handle only a static mapping but not time-varying dynamics. Next, a recurrent neural network composed of input, recurrent, and output units can approximate any dynamical system (Funahashi & Nakamura, 1993). But its training and control are notoriously difficult because it is difficult to assign a credit of error to recurrent units. Also, a recurrent neural network exhibits a chaotic behavior if recurrent connections are sufficiently forceful. The chaotic behavior provides a rich repertoire of temporal patterns helpful for approximating a dynamical system (Jaeger & Haas, 2004). At the same time, such chaotic behavior is

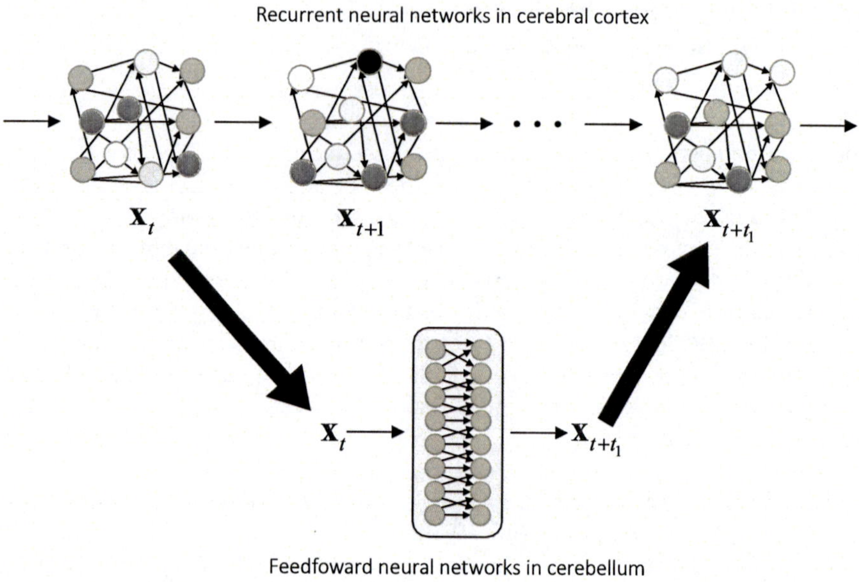

Fig. 18.3 Schematics of the cerebrocerebellar loop and neural network. Proposed mechanism of predictive stabilization of cortical dynamics through the cerebrocerebellar loop. The dynamics of the cerebral cortex and the cerebellum are modeled as a recurrent neural network (top row) and a feedforward neural network (bottom row), respectively. The cerebellum receives a current cortical state at time t and projects back a future cortical state at time $t + t_1$ through the cerebrocerebellar loop. In this proposed mechanism, the cerebellar prediction guides and stabilizes the recurrent dynamics of the cerebral cortex, thereby realizing fast and robust information processing. (Figure reproduced from Tanaka et al. (2020))

difficult to control (Sompolinsky et al., 1988). In sum, a feedforward network can approximate any static function and is easy to train, but cannot handle time-varying signals properly. On the other hand, a recurrent network can approximate any dynamical system of rich temporal behavior, but it is difficult to train and control.

We propose a complementary role of a feedforward neural network of the cerebellum and a recurrent neural network of the cerebral cortex (Fig. 18.3); the cerebellum tames computationally flexible but chaotic dynamics of cortical recurrent neural network by predicting the expected activity of the cerebral cortex. Due to its chaotic characteristic and internal and external noises, a recurrent neural network can behave erratically or even divergently. By iterating its previous activity, a recurrent neural network can amplify a small perturbation exponentially, a typical behavior in a chaotic system. A feedforward network, on the other hand, depends only on the current input and is robust against noisy perturbations. As demonstrated in Sect. 18.3, the cerebellum can predict the expected future input. We propose that the cerebral cortex performs the computation of a dynamical system for processing time-varying sensory and motor functions and that the cerebellum predicts a future state of the cerebral cortex and sends back the prediction for stabilizing and robustifying the cortical activity. The proposed mechanism of prediction-feedback stabilization is in line with the FORCE learning algorithm, where the activity of a recurrent neural network is stabilized by feeding back output units to recurrent units (Sussillo & Abbott, 2009). In our proposed scheme, the cerebellum always monitors the cerebral cortex and collects any deviations in the cortical activity from the cerebellar prediction. The internal-forward-model computation in the cerebellum should apply not only to the motor domain but also to the cognitive domain, as reviewed in Sect. 18.3.

Our proposed function of the cerebrocerebellar loop predicts that through the interactions and mutual learning between the cerebellum and the cerebral cortex, the cerebellar and cortical cells should share the same computational function. Following the prediction, the motor cortical cells and the DNCs share similar directional selectivity and temporal profiles of the firing rates (Tomatsu et al., 2016). A recent imaging study demonstrated that activities of layer 5 pyramidal cells in the neocortex and those of GCs in the cerebellar cortex share task-encoding characteristics acquired during learning, indicating that a critical function of the cerebrocerebellar loop is the propagation of shared neural dynamics (Wagner et al., 2019). The cortical activity acquired in the initial phase of task learning was transferred and copied to the activity of GCs, consistent with our proposed scheme, where the cerebellar function predicts the future cortical state. An additional merit of a feedforward network is that it allows a prediction of multiple time steps without any iteration. To conclude, the cerebellum has evolved to monitor and regulate the activity of the cerebral cortex through the cerebrocerebellar loop.

18.5 Discussion

This review article has attempted to fill the gap in understanding the computational role of the cerebellum and the cerebrocerebellar loop. We emphasize the regressor function for predicting dynamical consequences. Our computational understanding of the cerebellum has progressed since the pioneering studies of Marr, Albus, and Ito. Although we are closing to a unifying understanding of the cerebellar computation, a few outstanding questions need to be answered: (1) computational role of complex spikes of PCs, (2) specific ratios about the cerebellar cells, and (3) linear computation of nonlinear dynamics of the body. We conclude this review article by addressing these unsolved questions.

First, the computational role of climbing fiber inputs to a PC is still an open question. The Marr-Albus theory postulates a supervising signal for a classifier, but its experimental evidence that the climbing fiber encodes a supervising signal is unexpectedly scarce. Complex spikes reflect the climbing fiber input and have a highly stereotyped burst of decrementing spikes. Electrophysiological evidence suggests that complex spikes represent motor errors of voluntary arm movements (Kitazawa et al., 1998) and eye saccadic movements (Herzfeld et al., 2018). However, complex spikes were reported to increase as learning progresses, indicating that complex spikes reflect the consolidation of motor memory rather than a motor error (Catz et al., 2005). Whereas complex spikes encode motor error in part, they also encode other motor-related variables.

Second, the specific ratios about the cerebellar cell numbers have not been addressed in depth. The ratios of MFs, GCs, PCs, and DNCs are 2:1600:1:0.038, so the cerebellum first expands from MF inputs to GCs and compresses in the output of DNCs. Expansion recoding in the Marr-Albus theory refers to the increase in the numbers from MFs to GCs in which neural representation space is expanded through a random projection of MF inputs onto a larger population of granule cells. The expansion plays a key role in sparsifying granule cell activity and increasing pattern separation prior to classification. The sparsification hypothesis of the expansion recoding is challenged by calcium imaging studies of thousands of GCs, which revealed not sparse but rather dense activities of GCs (Gilmer & Person, 2017, 2018). Theoretical considerations suggest the orthogonalization of input representations as a function of the expansion recoding (Cayco-Gajic et al., 2017; Cayco-Gajic & Silver, 2019). The ongoing debate of sparsification and orthogonalization assumes the classifier perspective of the cerebellum. If the cerebellum functions as a regressor rather than a classifier, what is the function of expansion recoding?

More mysterious is the compression ratio of the cerebellar inputs (MFs) to the cerebellar outputs (DNCs). The number of axons in the cerebral peduncle (the cerebellar input from the cerebral cortex) is estimated as 21 million in humans (Tomasch, 1969). The number of axons in the superior cerebellar peduncle (the cerebellar output to the cerebral cortex) is no more than 0.8 million in humans (Heidary & Tomasch, 1969). This ratio indicates that the cerebellum receives massive inputs from the cerebral cortex but projects back only a fraction of the inputs. The Marr-Albus theory addresses the function of the cerebellar cortex but does not provide a whole picture of the cerebellar computation. Why does the cerebellum receive so much and return so little? Expanding our proposed scheme in Sect. 18.4, we speculate that the cerebellum requires the whole cortical activity to predict the future state and that the cerebellar cortex, in return, requires a small fraction of the cerebellar prediction to perform stable and robust computation. This speculation may be tested in a future computational study.

Finally, the mechanism of how the nonlinear dynamics of the body and the world may be expressed in linear neural dynamics is not properly understood. The musculoskeletal system has highly nonlinear dynamics composed of multiple interacting joints and muscles, and controlling such a nonlinear system of large degrees of freedom is a challenge to modern control theory. Biological motor control must face with the large degrees of freedom of the body and corresponding nonlinear dynamics, and we know little about how such dynamics is solved in the cerebellum and the cerebral cortex. We hope that these unresolved problems will promote the future study of the cerebellum in the same way that Marr, Albus, and Ito paved the way for cerebellar research for the following 50 years.

Acknowledgments The authors thank the organizers and the Fujihara seminar participants in December of 2018, which significantly expanded the authors' understanding of the cerebellum. The authors are also grateful to the collaborators, Drs. Yoshikazu Shinoda and Hiroshi Mitoma.

References

Albus, J. S. (1971). A theory of cerebellar function. *Mathematical Biosciences, 10*, 25–61.

Cabaraux, P., Gandini, J., Kakei, S., Manto, M., Mitoma, H., & Tanaka, H. (2020). Dysmetria and errors in predictions: The role of internal forward model. *International Journal of Molecular Sciences, 21*.

Catz, N., Dicke, P. W., & Thier, P. (2005). Cerebellar complex spike firing is suitable to induce as well as to stabilize motor learning. *Current Biology, 15*, 2179–2189.

Cayco-Gajic, N. A., & Silver, R. A. (2019). Re-evaluating circuit mechanisms underlying pattern separation. *Neuron, 101*, 584–602.

Cayco-Gajic, N. A., Clopath, C., & Silver, R. A. (2017). Sparse synaptic connectivity is required for decorrelation and pattern separation in feedforward networks. *Nature Communications, 8*, 1116.

Chen, S., Augustine, G. J., & Chadderton, P. (2016). The cerebellum linearly encodes whisker position during voluntary movement. *eLife, 5*, e10509.

Cybenko, G. (1989). Approximation by superpositions of a sigmoidal function. *Mathematics of Control, Signals and Systems, 2*, 303–314.

De Zeeuw, C. I., Hoebeek, F. E., Bosman, L. W., Schonewille, M., Witter, L., & Koekkoek, S. K. (2011). Spatiotemporal firing patterns in the cerebellum. *Nature Reviews. Neuroscience, 12*, 327–344.

DeFelipe, J., & Jones, E. G. (1988). *Cajal on the cerebral cortex: An annotated translation of the complete wrings*. Oxford University Press.

Diedrichsen, J., King, M., Hernandez-Castillo, C., Sereno, M., & Ivry, R. B. (2019). Universal transform or multiple functionality? Understanding the contribution of the human cerebellum across task domains. *Neuron, 102*, 918–928.

Dugue, G. P., Tihy, M., Gourevitch, B., & Lena, C. (2017). Cerebellar re-encoding of self-generated head movements. *Elife*, 6.

Ebner, T. J., Hewitt, A. L., & Popa, L. S. (2011). What features of limb movements are encoded in the discharge of cerebellar neurons? *Cerebellum, 10*, 683–693.

Funahashi, K.-I. (1989). On the approximate realization of continuous mappings by neural networks. *Neural Networks, 2*, 183–192.

Funahashi, K.-I., & Nakamura, Y. (1993). Approximation of dynamical systems by continuous time recurrent neural networks. *Neural Networks, 6*, 801–806.

Gilmer, J. I., & Person, A. L. (2017). Morphological constraints on cerebellar granule cell combinatorial diversity. *The Journal of Neuroscience, 37*, 12153–12166.

Gilmer, J. I., & Person, A. L. (2018). Theoretically sparse, empirically dense: New views on cerebellar granule cells. *Trends in Neurosciences, 41*, 874–877.

Heidary, H., & Tomasch, J. (1969). Neuron numbers and perikaryon areas in the human cerebellar nuclei. *Acta Anatomica (Basel), 74*, 290–296.

Heiney, S. A., Kim, J., Augustine, G. J., & Medina, J. F. (2014). Precise control of movement kinematics by optogenetic inhibition of Purkinje cell activity. *The Journal of Neuroscience, 34*, 2321–2330.

Herzfeld, D. J., Kojima, Y., Soetedjo, R., & Shadmehr, R. (2015). Encoding of action by the Purkinje cells of the cerebellum. *Nature, 526*, 439–442.

Herzfeld, D. J., Kojima, Y., Soetedjo, R., & Shadmehr, R. (2018). Encoding of error and learning to correct that error by the Purkinje cells of the cerebellum. *Nature Neuroscience, 21*, 736–743.

Hong, S., Negrello, M., Junker, M., Smilgin, A., Thier, P., & De Schutter, E. (2016). Multiplexed coding by cerebellar Purkinje neurons. *Elife*, 5.

Hornik, K., Stinchcombe, M., & White, H. (1989). Multilayer feedforward networks are universal approximators. *Neural Networks, 2*, 359–366.

Ishikawa, T., Tomatsu, S., Tsunoda, Y., Lee, J., Hoffman, D. S., & Kakei, S. (2014). Releasing dentate nucleus cells from Purkinje cell inhibition generates output from the cerebrocerebellum. *PLoS One, 9*, e108774.

Ishikawa, T., Tomatsu, S., Izawa, J., & Kakei, S. (2016). The cerebro-cerebellum: Could it be loci of forward models? *Neuroscience Research, 104*, 72–79.

Ito, M. (1970). Neurophysiological aspects of the cerebellar motor control system. *International Journal of Neurology, 7*, 162–176.

Ito, M. (1984). *The cerebellum and neural control*. Raven Press.

Ito, M., Sakurai, M., & Tongroach, P. (1982). Climbing fibre induced depression of both mossy fibre responsiveness and glutamate sensitivity of cerebellar Purkinje cells. *The Journal of Physiology, 324*, 113–134.

Izawa, J., & Shadmehr, R. (2011). Learning from sensory and reward prediction errors during motor adaptation. *PLoS Computational Biology, 7*, e1002012.

Jaeger, H., & Haas, H. (2004). Harnessing nonlinearity: Predicting chaotic systems and saving energy in wireless communication. *Science, 304*, 78–80.

Kalman, R. E. (1960). A new approach to linear filtering and prediction problems. *Journal of Basic Engineering, 82*(1), 35–45.

Kawato, M., Furukawa, K., & Suzuki, R. (1987). A hierarchical neural-network model for control and learning of voluntary movement. *Biological Cybernetics, 57*, 169–185.

Kelly, R. M., & Strick, P. L. (2003). Cerebellar loops with motor cortex and prefrontal cortex of a nonhuman primate. *The Journal of Neuroscience, 23*, 8432–8444.

Kennedy, A., Wayne, G., Kaifosh, P., Alvina, K., Abbott, L. F., & Sawtell, N. B. (2014). A temporal basis for predicting the sensory consequences of motor commands in an electric fish. *Nature Neuroscience, 17*, 416–422.

Kira, S., Yang, T., & Shadlen, M. N. (2015). A neural implementation of Wald's sequential probability ratio test. *Neuron, 85*, 861–873.

Kitazawa, S., Kimura, T., & Yin, P. B. (1998). Cerebellar complex spikes encode both destinations and errors in arm movements. *Nature, 392*, 494–497.

Larry, N., Yarkoni, M., Lixenberg, A., & Joshua, M. (2019). Cerebellar climbing fibers encode expected reward size. *Elife*, 8.

Laurens, J., Heiney, S. A., Kim, G., & Blazquez, P. M. (2013). Cerebellar cortex granular layer interneurons in the macaque monkey are functionally driven by mossy fiber pathways through net excitation or inhibition. *PLoS One, 8*, e82239.

Lesage, E., Morgan, B. E., Olson, A. C., Meyer, A. S., & Miall, R. C. (2012). Cerebellar rTMS disrupts predictive language processing. *Current Biology, 22*, R794–R795.

Lisberger, S. G., & Fuchs, A. F. (1978). Role of primate flocculus during rapid behavioral modification of vestibuloocular reflex. II. Mossy fiber firing patterns during horizontal head rotation and eye movement. *Journal of Neurophysiology, 41*, 764–777.

Marr, D. (1969). A theory of cerebellar cortex. *The Journal of Physiology, 202*, 437–470.

Martin, T. A., Keating, J. G., Goodkin, H. P., Bastian, A. J., & Thach, W. T. (1996). Throwing while looking through prisms. I. Focal olivocerebellar lesions impair adaptation. *Brain, 119*(Pt 4), 1183–1198.

Mazzoni, P., & Krakauer, J. W. (2006). An implicit plan overrides an explicit strategy during visuomotor adaptation. *The Journal of Neuroscience, 26*, 3642–3645.

Medina, J. F., & Lisberger, S. G. (2007). Variation, signal, and noise in cerebellar sensory-motor processing for smooth-pursuit eye movements. *The Journal of Neuroscience, 27*, 6832–6842.

Medina, J. F., & Lisberger, S. G. (2009). Encoding and decoding of learned smooth-pursuit eye movements in the floccular complex of the monkey cerebellum. *Journal of Neurophysiology, 102*, 2039–2054.

Miall, R. C., Christensen, L. O., Cain, O., & Stanley, J. (2007). Disruption of state estimation in the human lateral cerebellum. *PLoS Biology, 5*, e316.

More, H. L., & Donelan, J. M. (2018). Scaling of sensorimotor delays in terrestrial mammals. *Proceedings of the Biological Sciences, 285*.

More, H. L., Hutchinson, J. R., Collins, D. F., Weber, D. J., Aung, S. K., & Donelan, J. M. (2010). Scaling of sensorimotor control in terrestrial mammals. *Proceedings of the Biological Sciences, 277*, 3563–3568.

Nowak, D. A., Hermsdorfer, J., Rost, K., Timmann, D., & Topka, H. (2004). Predictive and reactive finger force control during catching in cerebellar degeneration. *Cerebellum, 3*, 227–235.
Nowak, D. A., Timmann, D., & Hermsdorfer, J. (2007). Dexterity in cerebellar agenesis. *Neuropsychologia, 45*, 696–703.
Ohmae, S., & Medina, J. F. (2015). Climbing fibers encode a temporal-difference prediction error during cerebellar learning in mice. *Nature Neuroscience, 18*, 1798–1803.
Pasalar, S., Roitman, A. V., Durfee, W. K., & Ebner, T. J. (2006). Force field effects on cerebellar Purkinje cell discharge with implications for internal models. *Nature Neuroscience, 9*, 1404–1411.
Popa, L. S., Hewitt, A. L., & Ebner, T. J. (2012). Predictive and feedback performance errors are signaled in the simple spike discharge of individual Purkinje cells. *The Journal of Neuroscience, 32*, 15345–15358.
Popa, L. S., Hewitt, A. L., & Ebner, T. J. (2013). Purkinje cell simple spike discharge encodes error signals consistent with a forward internal model. *Cerebellum, 12*, 331–333.
Rao, R. P. (2004). Bayesian computation in recurrent neural circuits. *Neural Computation, 16*, 1–38.
Raymond, J. L., & Medina, J. F. (2018). Computational principles of supervised learning in the cerebellum. *Annual Review of Neuroscience, 41*, 233–253.
Rilling, J. K., & Insel, T. R. (1998). Evolution of the cerebellum in primates: Differences in relative volume among monkeys, apes and humans. *Brain, Behavior and Evolution, 52*, 308–314.
Rumelhart, D. E., Hinton, G. E., & Williams, R. J. (1986). Learning representations by back-propagating errors. *Nature, 323*, 533–536.
Schmahmann, J. D. (2019). The cerebellum and cognition. *Neuroscience Letters, 688*, 62–75.
Schmahmann, J. D., & Caplan, D. (2006). Cognition, emotion and the cerebellum. *Brain, 129*, 290–292.
Schultz, W. (2000). Multiple reward signals in the brain. *Nature Reviews. Neuroscience, 1*, 199–207.
Schultz, W., Dayan, P., & Montague, P. R. (1997). A neural substrate of prediction and reward. *Science, 275*, 1593–1599.
Shadmehr, R., Smith, M. A., & Krakauer, J. W. (2010). Error correction, sensory prediction, and adaptation in motor control. *Annual Review of Neuroscience, 33*, 89–108.
Shidara, M., Kawano, K., Gomi, H., & Kawato, M. (1993). Inverse-dynamics model eye movement control by Purkinje cells in the cerebellum. *Nature, 365*, 50–52.
Smith, M. A., & Shadmehr, R. (2005). Intact ability to learn internal models of arm dynamics in Huntington's disease but not cerebellar degeneration. *Journal of Neurophysiology, 93*, 2809–2821.
Sompolinsky, H., Crisanti, A., & Sommers, H.-J. (1988). Chaos in random neural networks. *Physical Review Letters, 61*, 259.
Sun, Z., Smilgin, A., Junker, M., Dicke, P. W., & Thier, P. (2017). The same oculomotor vermal Purkinje cells encode the different kinematics of saccades and of smooth pursuit eye movements. *Scientific Reports, 7*, 40613.
Sussillo, D., & Abbott, L. F. (2009). Generating coherent patterns of activity from chaotic neural networks. *Neuron, 63*, 544–557.
Tanaka, H., Ishikawa, T., & Kakei, S. (2019). Neural evidence of the cerebellum as a state predictor. *Cerebellum, 18*, 349–371.
Tanaka, H., Ishikawa, T., Lee, J., & Kakei, S. (2020). The cerebro-cerebellum as a locus of forward model: A review. *Frontiers in Systems Neuroscience, 14*, 19.
Ten Brinke, M. M., Heiney, S. A., Wang, X., Proietti-Onori, M., Boele, H. J., Bakermans, J., Medina, J. F., Gao, Z., & De Zeeuw, C. I. (2017). Dynamic modulation of activity in cerebellar nuclei neurons during pavlovian eyeblink conditioning in mice. *eLife, 6*.
Todorov, E. (2004). Optimality principles in sensorimotor control. *Nature Neuroscience, 7*, 907–915.

Todorov, E., & Jordan, M. I. (2002). Optimal feedback control as a theory of motor coordination. *Nature Neuroscience, 5*, 1226–1235.

Tomasch, J. (1969). The numerical capacity of the human cortico-pontocerebellar system. *Brain Research, 13*, 476–484.

Tomatsu, S., Ishikawa, T., Tsunoda, Y., Lee, J., Hoffman, D. S., & Kakei, S. (2016). Information processing in the hemisphere of the cerebellar cortex for control of wrist movement. *Journal of Neurophysiology, 115*, 255–270.

Tseng, Y. W., Diedrichsen, J., Krakauer, J. W., Shadmehr, R., & Bastian, A. J. (2007). Sensory prediction errors drive cerebellum-dependent adaptation of reaching. *Journal of Neurophysiology, 98*, 54–62.

Wagner, M. J., Kim, T. H., Savall, J., Schnitzer, M. J., & Luo, L. (2017). Cerebellar granule cells encode the expectation of reward. *Nature, 544*, 96–100.

Wagner, M. J., Kim, T. H., Kadmon, J., Nguyen, N. D., Ganguli, S., Schnitzer, M. J., & Luo, L. (2019). Shared cortex-cerebellum dynamics in the execution and learning of a motor task. *Cell, 177*, 669–682 e624.

Walter, J. T., & Khodakhah, K. (2006). The linear computational algorithm of cerebellar Purkinje cells. *The Journal of Neuroscience, 26*, 12861–12872.

Walter, J. T., & Khodakhah, K. (2009). The advantages of linear information processing for cerebellar computation. *Proceedings of the National Academy of Sciences of the United States of America, 106*, 4471–4476.

Wolpert, D. M., & Miall, R. C. (1996). Forward models for physiological motor control. *Neural Networks, 9*, 1265–1279.

Wolpert, D. M., Ghahramani, Z., & Jordan, M. I. (1995). An internal model for sensorimotor integration. *Science, 269*, 1880–1882.

Wolpert, D. M., Miall, R. C., & Kawato, M. (1998). Internal models in the cerebellum. *Trends in Cognitive Sciences, 2*, 338–347.

Yang, T., & Shadlen, M. N. (2007). Probabilistic reasoning by neurons. *Nature, 447*, 1075–1080.

Chapter 19
The Input-Output Organization of the Cerebrocerebellum as Kalman Filter

Shinji Kakei, Hirokazu Tanaka, Takahiro Ishikawa, Saeka Tomatsu, and Jongho Lee

19.1 Recognition of the Cerebrocerebellum as Loci of Internal Models

The seminal publication of the book on the cerebellum by Eccles et al. (1967) inspired the publication of major theories of the cerebellum and motor control by David Marr (1969), James S. Albus (1971), and Masao Ito (1970). Their legendary papers marked the beginning of the ongoing effort to understand the relationship between the cerebellar neuron circuitry and motor/cognitive control. In particular, Ito (1970) first proposed how the cerebrocerebellum participates in acquiring skilled movements in terms of a control system model (Fig. 19.1a). In voluntary *unskilled* movements (Fig. 19.1a(1)), the initial instruction arising from the association cortex (AC) is transferred to the motor cortex (MX) and transformed into the motor command and relayed to the spinal motor system (SM) through the pyramidal tract (PT). The outcome of the motor command is evaluated by AC using information relayed by the sensory

S. Kakei (✉)
Department of Anatomy and Physiology, Faculty of Life Sciences, Jissen Women's University, Tokyo, Japan
e-mail: kakei-shinji@jissen.ac.jp

H. Tanaka
Faculty of Information Technology, Tokyo City University, Tokyo, Japan

T. Ishikawa
Laboratory for Movement Disorders, Tokyo Metropolitan Institute of Medical Science, Tokyo, Japan

S. Tomatsu
Division of Behavioral Development, National Institute of Physiological Sciences, Okazaki, Aichi, Japan

J. Lee
Faculty of Health Sciences, Komatsu University, Komatsu, Ishikawa, Japan

H. Mizusawa, S. Kakei (eds.), *Cerebellum as a CNS Hub*,
Contemporary Clinical Neuroscience, https://doi.org/10.1007/978-3-030-75817-2_19

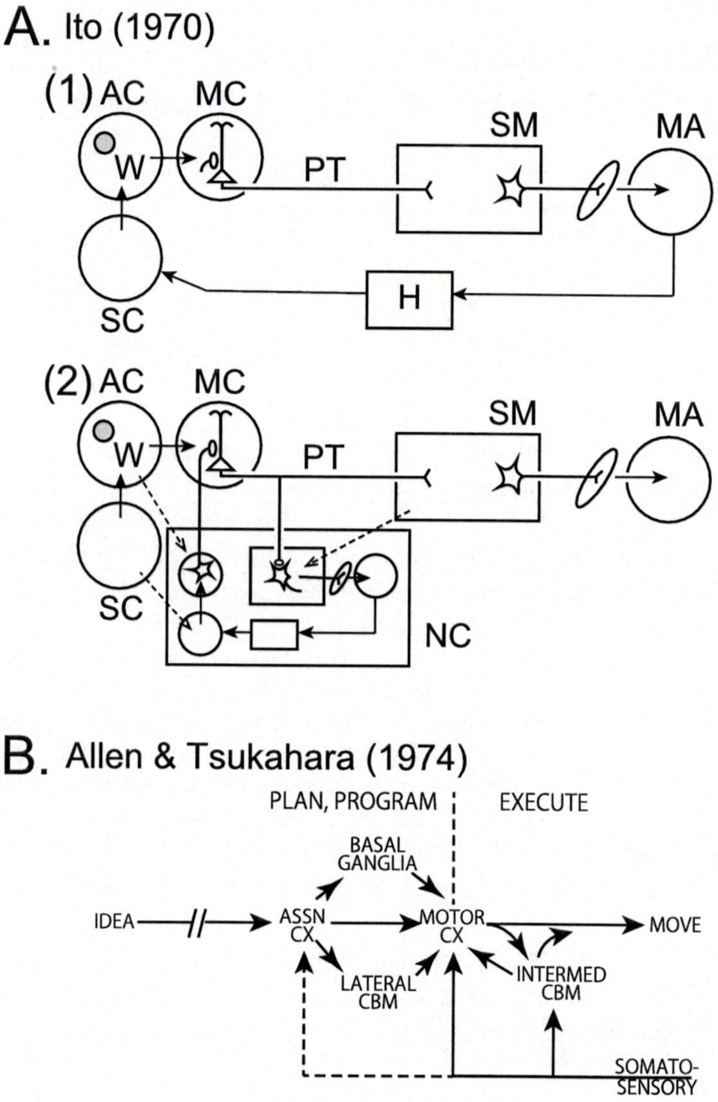

Fig. 19.1 Two types of cerebellar internal models. Model by Ito (1970) (**a**) is consistent with a forward model, while the model by Allen and Tsukahara (1974) (**b**) is consistent with an inverse model. (**a**) Diagram of the possible control system for voluntary movements (Reproduced from Ito (1970), Fig. 19.7). Note the caption is also original. (**1**) Feedback system used in *unskilled* movement. AC cerebral association area, Small gray circle (W) indicates the origin of the *will*. SC cerebral sensory area, MC cerebral motor area, PT pyramidal tract, SM spinal motor system, MA motor activity. H, feedback pathway through the external world. (**2**) Feedforward system formed after learning. NC neocerebellum in which SM, MA, H, SC, and AC in A are indicated in a minimized form (Note Ito (1970) assumed that AC, SC, SM, MA, and H are *all* modeled in *neocerebellum (NC)* (= cerebrocerebellum)). (**b**) Scheme showing proposed roles of several brain structures in movement (Reproduced from Allen and Tsukahara (1974), Fig. 9). Note the caption is also original.

feedback loops (H + SC in Fig. 19.1a). However, with practice, the movement becomes more skilled and predictive while less dependent on the sensory feedback information. In other words, as the learning progresses, the long loop through the *external* world may be effectively replaced by an *internal* loop passing through the cerebrocerebellum (= *neocerebellum* (NC) in Fig. 19.1a), which would serve as a model of the combination of SM, the external world, and the sensory pathways (Fig. 19.1a(2)). According to Ito (1970), it is possible to understand this arrangement as a type of *model inference adaptive control system.* Cerebellar ataxia, such as dysmetria or intention tremor, could be explained as impairment or loss of *the internal model* in the cerebrocerebellum, just as in the stage of unskilled movements before motor learning.

Note that in Ito's model (Fig. 19.1a(2)), the cerebrocerebellum (NC) receives the efference copy from MX and returns its output to the same MX. Therefore, it may play a role that is equivalent to a *forward model.* A forward model provides the controller (i.e., the motor cortex) with a state prediction (Todorov, 2004) to compensate for sensory feedback delays and stabilize movements. To the best of our knowledge, it was the first proposal of a forward model in neuroscience.

Few years after Ito's pioneering paper, Allen and Tsukahara (1974) proposed a different type of cerebrocerebellar organization (Fig. 19.1b) to explain skilled voluntary movements. They envisaged a two-stage planning-execution system between the cerebral cortex and the cerebellum to control voluntary movements (Fig. 19.1b). The schema features the idea that the association cortices (ASSN CX) translate the intention to move into a proper spatiotemporal activation of the motor cortex (MOTOR CX), resulting in the intended movement. ASSN CX that project to the cerebrocerebellum (LATERAL CBM) are among those in the *premotor* circuit. Because LATERAL CBM appears to lack direct sensory inputs, it is more suited for planning the movement than in actual execution and correction of the movement, which was more suitable for the intermediate zone (INTERMED CBM) function. Once the movement has been prepared in ASSN CX, with the help of LATERAL CBM (i.e., the cerebrocerebellum) (Fig. 19.1b), MOTOR CX generates the motor command as the common node. At this point, INTERMED CBM updates the movement based on the difference (i.e., error) between the actual movement and intended movement.

The two schemes (Fig. 19.1a(2), Ito and b, Allen, and Tsukahara) may look similar. But they are exclusive to each other from a functional point of view. In the former scheme (Fig. 19.1a(2)), the cerebrocerebellum provides a long feedback loop model and virtually replaces it, and the cerebrocerebellum resides *outside of the controller.* In contrast, in the latter scheme (Fig. 19.1b), the cerebrocerebellum is *a part of the controller* that translates the intention to move into the motor command. Thus, the cerebrocerebellum in Fig. 19.1b is suitable to play a role that is equivalent to an *inverse model* or a part of it.

Fig. 19.1 (continued) Dashed line represents a pathway of unknown importance. It is proposed that basal ganglia and cerebellar hemisphere are involved with association cortex in programming of volitional movements. At the time that the motor command descends to motoneurons, engaging the movement, the pars intermedia updates the intended movement, based on the motor command and somatosensory description of limb position and velocity on which the movement is to be superimposed. Follow-up correction can be performed by motor cortex when cerebellar hemisphere and pars intermedia do not effectively perform their functions

The two schemes are also exclusive in terms of the neuroanatomical organization between the cerebellum and the cerebral cortex. In the former scheme (Fig. 19.1a(2)), the cerebrocerebellum is connected *reciprocally* with the motor cortex. In contrast, the connectivity in the latter scheme is *non-reciprocal*, collecting its input from the association cortices and returning its output to the motor cortex rather than to the association cortices, the source of the cortical input. Therefore, it is possible to select one from the other by identifying the neuroanatomical organization in theory. Unfortunately, the neuroanatomical techniques available in the 1970s and 1980s, such as the Nauta method or simple neuronal tracers, were not effective enough for this purpose.

19.2 Parallel Organization of the Cerebrocerebellar Communication System Revealed with Transneuronal Tracing Technique

In the 1990s, Peter L. Strick and his colleagues established a revolutionary technique to trace neuron circuitry: a transneuronal tracing technique with neurotropic viruses. It was revolutionary because they enabled the use of transneuronal transport of rabies viruses to reveal the connections of three or more synaptically linked neurons (Kelly & Strick, 2003), which was impossible with conventional neuron tracers. With the new method, analysis of the cerebrocerebellar communication system was within their reach. Their experiments demonstrated that the regions of the cerebellar cortex that receive input from the motor cortex are the same as those that project to the motor cortex. Similarly, the regions of the cerebellar cortex that receive input from area 46 (a part of the prefrontal cortex) are the same as those that project to area 46. Thus, their observations demonstrated that parallel closed-loop circuits represent a fundamental feature of cerebrocerebellar interactions (Fig. 19.2).

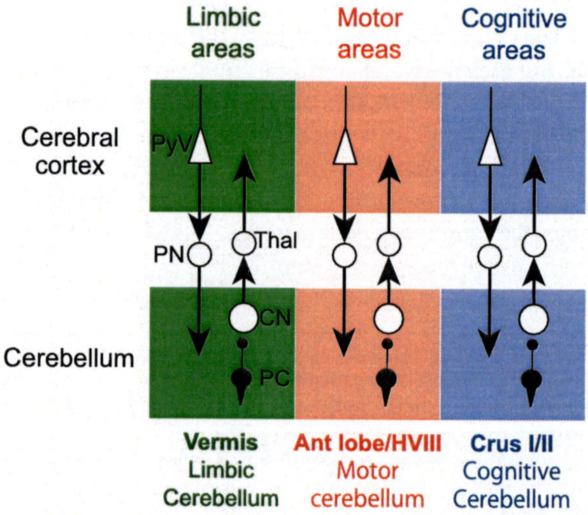

Fig. 19.2 The parallel organization of the cerebrocerebellar loops. The cerebellum contains anatomically separate and functionally distinct motor and non-motor domains. This figure was prepared based on Kelly and Strick (2003). CN cerebellar nuclei cells, PC Purkinje cells, PN pontine nuclei cells, PyV layer V pyramidal cells, Thal thalamus

The closed-loop architecture of the cerebrocerebellar communication system is compatible with Ito's closed-loop scheme (Fig. 19.1a(2)), a forward model. But it is not consistent with Allen and Tsukahara's scheme that assumes an open-loop architecture. The cerebellum integrates its various cortical inputs and returns the output to the heteronymous motor cortex. Looking back, Ito's pioneering forward model hypothesis was almost 30 years ahead of his time, but unfortunately, it somehow remained almost unnoticed until very recently.

19.3 Contests Among Control Laws to Explain Movement Trajectories

In the 1980s and 1990s, there was a vigorous debate about putative control laws that govern movement trajectories in reaching movements. The discussion focused on how the central nervous system selects one specific movement trajectory among an infinite number of possible trajectories that lead to the goal. In other words, the competition was over control laws to reduce excess degrees of freedom. Several candidate theories included minimum jerk theory (Flash & Hogan, 1985), minimum energy theory, minimum mean squared velocity theory, minimum mean squared force theory (Stein et al., 1994), or minimum torque change theory (Uno et al., 1989). Each theory predicts an *ideal* movement trajectory that maximizes or minimizes (optimizes) some *criterion* from the start point to the endpoint. For each ideal trajectory, the causal motor command was determined for the entire path in a feedforward manner. It was also assumed that no noise disturbs the movement's execution because these optimization methods did not make noises into consideration.

19.4 Awareness of Noise in Motor Control

Unfortunately, there certainly *is* noise in the real world. Indeed, Harris and Wolpert (1998) pointed out the critical role of inherent noises to determine the final control signal, i.e., muscle activities. We cannot achieve the *ideal* trajectory however hard we may practice because the control signal is always corrupted with the intrinsic noise. Moreover, there are also various noises or disturbances from the environment. Awareness of these intrinsic and extrinsic noises in motor control dramatically changed our approach to feedforward control. For instance, feedforward control for the entire path makes sense only when everything goes as planned during the movement. In reality, the unpredictable noises force trajectories to deviate from the desired path, increasing uncertainty toward the goal. Therefore, there is no guarantee to optimize the criterion as planned. The best we can hope to do is to maximize or minimize the expected value of the criterion.

19.5 Introduction of Stochastic Optimal Control

Intrinsic and extrinsic noises are a typical condition where "stochastic optimal control" or "optimal control" comes into play. Optimal control was developed initially in engineering to control complex multiple-input, multiple-output systems, which were not amenable to classical control theories (Kirk, 1970). It evaluates the system's random behavior and attempts to optimize responses or stability on the average rather than with assured precision (Stengel, 1994). A stochastic control system performs two functions: first, it controls the system (controller, in Fig. 19.3) and, second, it predicts the current state of the system (estimator, in Fig. 19.3) to provide the best feedback information for the controller (Stengel, 1994). Such an estimator takes efference copy and sensory inputs into account, and it weighs these pieces of information depending on their reliability (i.e., optimally). In modeling practice, one may use a Kalman filter (Kalman & Bucy, 1961), an optimal estimator when the dynamics and sensory measurements are linear and the noise is Gaussian (Todorov, 2004). In Fig. 19.3, the estimator and the controller are in a loop; thus, they can continue to generate time-varying commands recursively without preparing a whole set of motor commands in a feed-forward manner. Then where is the estimator in the central nervous system?

19.6 Difficulty in the Identification of the Cerebellar Forward Model

Previous reviews repeatedly suggested the cerebellum as a potential site of the estimator or forward model mainly based on neuroanatomical data and clinical observations (for instance, Miall et al., 1993; Haggard & Wing, 1995; Wolpert & Miall,

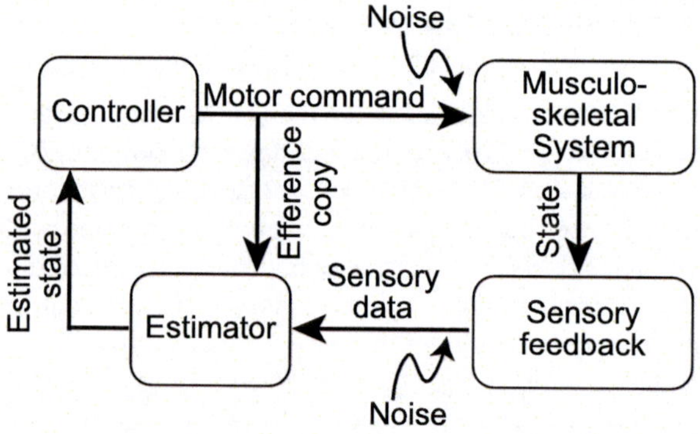

Fig. 19.3 Schematic of closed-loop optimization. (Modified from Todorov (2004)) The estimator needs an efference copy of concurrent motor commands and delayed sensory feedback data in order to compensate for sensory delays. Note that the estimator and controller are in a loop; thus, they can continue to generate time-varying motor commands even when sensory feedback becomes unavailable or unreliable

1996; Bastian, 2006; Ebner & Pasalar, 2008). As mentioned above, a forward model requires two major inputs: (1) a set of sensory feedback signals, which are necessary to update the forward model, and (2) the copy of descending motor commands. These two inputs are integrated in the forward model to generate the state estimate. Indeed, the cerebellum receives both of these inputs. It receives substantial inputs from cortical motor areas via the pontine nuclei (PN) (Brodal and Bijaalie, 2003; Schmahmann et al., 2004), and these inputs represent the efference copy of the descending motor commands (Ishikawa et al., 2014, 2016; Tomatsu et al., 2016). The cerebellum also receives substantial somatosensory inputs directly from the ascending spinocerebellar tracts and indirectly via brain stem nuclei, such as the cuneate nucleus or lateral reticular nucleus. These sensory inputs could provide an update on the state of the motor apparatus. The above argument may appear to support the cerebellar forward model hypothesis. But in reality, it is on insufficient grounds because the two lines of inputs are primarily separated in the cerebellar cortex. The mossy fiber (MF) inputs from the cortical motor areas (via PN) distribute mainly in the hemispheric (i.e., lateral) part (Na et al., 2019), while the sensory MF inputs from the spinal cord or the brain stem nuclei distribute in more rostral and medial part (the anterior lobe and the intermediate zone) (e.g., Wu et al., 1999) of the cerebellar cortex. Therefore, we may expect a convergence of the two MF inputs only in a minor part of the intermediate zone. More importantly, even if the nominal convergence has some role to play, the simple summation of the two MF inputs is *not* consistent with their *asymmetric* roles in the forward model. The efference copy plays an essential role in a state prediction, while the sensory input plays a critical role in an *update* of the prediction, as will be discussed later.

As for the output from a forward model, we expect it to correlate with the future state of the motor apparatus (Wolpert and Miall, 1996). In principle, we should examine the output from the cerebrocerebellum in the dentate nucleus (DN) because it is the sole output node from the cerebrocerebellum. Nevertheless, previous studies tried to address this issue by analyzing the Purkinje cell (PC) activities. Note that PCs' activity represents an *intermediate representation* of the cerebellar circuitry and is not ideal for characterizing the *output* of a forward model. In this regard, few studies are eligible to discuss the output of the cerebellar forward models (Thach, 1975, 1978; Thier & Markanday, 2019).

19.7 Movement Representation in the Cerebrocerebellum

To identify movement representations of a forward model, we need to satisfy two requirements: (1) identification of cerebellar neural elements and (2) identification of movement or sensory coordinate frames for activities of each component. Fortunately, the cerebellum provides an ideal place to achieve the first goal (Ishikawa et al., 2014; Tomatsu et al., 2016). Indeed, it is possible in the cerebellum to isolate single-unit activities of MFs (primary cerebellar inputs), PCs (the sole output from the cerebellar cortex), and DN cells (DNCs) (the sole output from the cerebrocerebellum) (Ishikawa et al., 2014; Tomatsu et al., 2016). Furthermore, it is also possible to achieve the second goal by employing our previous experimental design (Kakei et al., 1999, 2001). With this setup, we recorded activities of MFs, PCs, and DNCs, while monkeys perform wrist movements for eight different directions in two different forearm postures (Ishikawa et al., 2014; Tomatsu et al., 2016). This task design enabled us to dissociate intrinsic coordinate frames from an extrinsic coordinate frame for the wrist movement, depending on the posture-dependent changes in neuron activities. The results revealed distinct steps of movement representation from the input to the output of the cerebrocerebellum.

First, MFs demonstrated temporal and directional properties that were surprisingly similar to those of neurons in the primary motor cortex (M1)/the premotor cortex (PM) (Kakei et al., 1999, 2001). Namely, these MFs relay copies of the M1/PM motor commands to the cerebellum. Besides, their posture-dependent change of directional tuning demonstrated a *bimodal* distribution of shifts in the preferred direction (PD) for the 180° rotation in the forearm posture (Fig. 10a in Tomatsu et al., 2016), much like M1/PM neurons (Tomatsu et al., 2016) – one group with smaller shifts in PD (i.e., *extrinsic-like* neurons) and the other group with larger shifts in PD (*muscle-like* or *joint-like* neurons).

Second, PCs demonstrated much more complex spatiotemporal patterns of activity than MFs. The complexity of PC activities appeared to reflect rapidly changing properties of the peripheral motor apparatus during movement. Also, intricate spatiotemporal patterns of PC activities changed significantly for a change in forearm posture regarding the directional tuning and the gain modulation (Tomatsu et al., 2016). In particular, PCs showed a unimodal distribution of shift in PD that differed from the bimodal distribution of that of MFs (Fig. 10b in Tomatsu et al., 2016). The posture-dependent changes of PC activities indicate that the activities of these PCs encode intrinsic parameters and provide another support that the cerebrocerebellum works as a forward model to predict the state of the motor apparatus (Tomatsu et al., 2016).

Lastly, activities of DNCs, to our great surprise, appeared to *recover* those properties that were typical for MFs (Ishikawa et al., 2014). Namely, DNCs recovered simpler spatiotemporal activity patterns, much like MFs, despite substantial direct inputs from PCs. Also, the posture-dependent shift in PD for DNCs *recovered* a bimodal distribution for the change in the forearm posture (Fig. 19.4a), much like MFs – one group with smaller (i.e., extrinsic-like) shifts in PD and the other group with more extensive (i.e., muscle-like or joint-like) shifts in PD.

Fig. 19.4 (continued) range per 1 ms of the cursor on the monitor controlled by wrist joint movement. See Ishikawa et al. (2014) for the details of the experimental procedures. (**2**)**:** Optimal delay between the movement speed and |ΣSSdec| and |ΣSSinc| for the data shown in A. We calculated the R^2 value for the correlation between them for each 1 ms shift of movement speed from −150 to 50 ms relative to movement onset. Upper panel: R^2 values between the movement speed and |ΣSSdec| for each delay. The value was the h-ighest (= 0.847) when the movement speed profile was shifted by −61 ms (i.e., optimal delay). Lower panel: R^2 values between the movement speed and |ΣSSinc| for each delay. The value was the highest (= 0.732) when the movement speed profile was shifted by −7 ms. (Modified from Ishikawa et al. (2016))

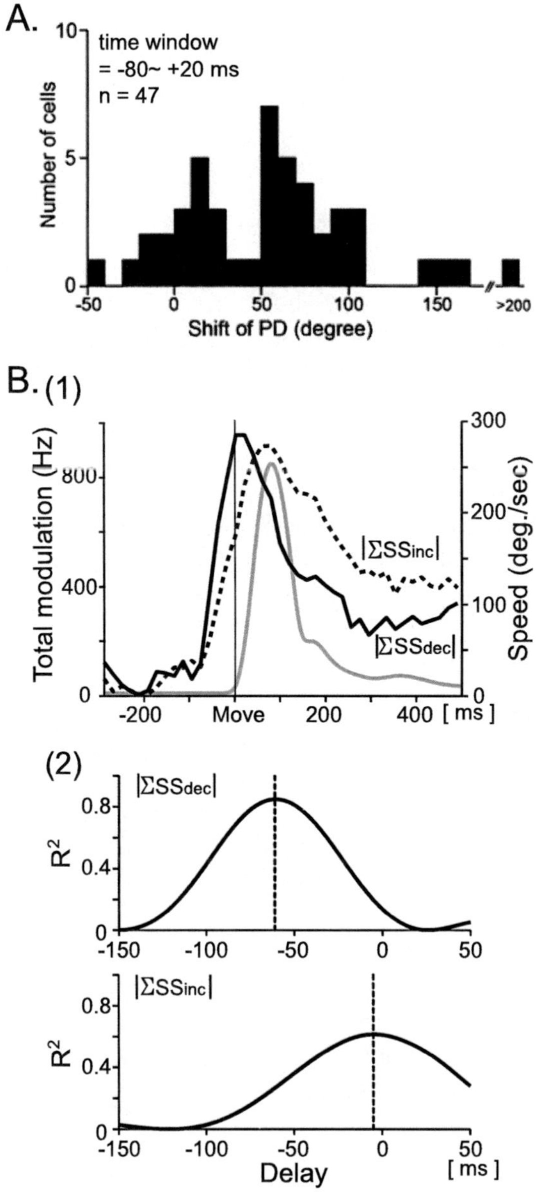

Fig. 19.4 Some spatiotemporal features of dentate nucleus cells (DNCs) and PCs. (**a**) Distribution of shifts in PD from PRO to SUP for DNCs in a time window of −25 to 0 ms relative to movement onset. Bin width = 10°. Note the bimodal distribution (*Ishikawa and Kakei, unpublished data*). (**b**) Correlation between the population modulation of Purkinje cells (PCs) and movement kinematics. **(1)** Temporal patterns of the sum of the decrease (|ΣSSdec|, solid line) and increase (|∑SSinc|, dashed line) of the simple spike (SS) activity of all movement-related PCs and the average speed of the wrist movement (gray line) in a monkey. To obtain |ΣSSdec| and |∑SSinc|, we summed all decreases and increases of SS activity relative to a reference period (200–260 ms before movement onset) separately in each 20 ms bin. The speed profile was calculated from a displacement

In summary, the cerebrocerebellum appears to transform copies of cortical motor commands (i.e., MF inputs) into similar movement representations (i.e., DNC output) through fundamentally distinct representations of PCs in a posture-dependent manner.

19.8 Timing of Movement-Related PC Activities in the Cerebrocerebellum

The timing of the task-related activities of PCs was also compatible with the cerebellar forward model hypothesis. Fig. 19.4b depicts a comparison between the speed profile and PCs' population activity recorded in the cerebrocerebellum of three monkeys during a rapid wrist movement in our recent study (Ishikawa et al., 2014). In this analysis, we summated the increase ($|\Sigma SSinc|$) and decrease ($|\Sigma SSdec|$) of simple spike activity of all movement-related PCs separately. As shown in Fig. 19.4b(2), $|\Sigma SSdec|$ demonstrated the highest correlation with the speed profile of the movement when the speed profile was shifted by −60 ms. Namely, the population activity of PCs *precedes* the actual movement by about 60 ms. Indeed, the lead times of PC activities were comparable to the average onset of muscle activities in the same animals (Tomatsu et al., 2016).

On the other hand, the onset latencies of the PCs lagged behind those of M1 and PMv neurons reported in our previous studies (−97.0 ± 15.3 ms for 44 extrinsic-like M1 neurons, −93.6 ± 20.8 ms for 28 muscle-like M1 neurons, and − 124.3 ± 30.6 ms for 55 extrinsic-like PMv neurons, Kakei et al., 1999, 2001). Therefore, the PCs' population activity *follows* that of the cortical motor command ($p < 0.001$, Mann-Whitney U-test). Thus, PC activities appear to represent the future states of the motor apparatus rather than motor commands or external sensory feedback. Overall, our observations suggest that the cerebrocerebellum could work as a forward model in terms of timing, representation, and transformation of activities.

19.9 System Identification of the Transformation in the Cerebrocerebellum: Its Similarity to Kalman Filter

If the cerebrocerebellum functions as a forward model, it is expected that the current output from DN should contain predictive information about the future MF input. Therefore, in our previous study (Tanaka et al., 2019), we examined the relationship between activities of MFs (cerebellar inputs), PCs (intermediate representation), and DNCs (cerebellar outputs). Briefly, we found that the activities of individual PCs were reconstructed precisely as a weighted sum of those of MFs. Similarly, the activities of individual DNCs were reconstructed strictly as a weighted sum of those of PCs and MFs. We further proved that the activities of DNCs contained predictive information about future MF inputs (Tanaka et al., 2019). Namely, the output from the cerebrocerebellum is capable of predicting 200 ms into the future to compensate for the delay of sensory feedback. We finally note that the linear relationship between MF, PC, and DNC activities resembles an optimal linear estimator known as the Kalman filter (Kalman & Bucy, 1961; Tanaka et al., 2019).

The functional similarity of the cerebellum to the Kalman filter has already been suggested in some previous reviews. Most notably, Paulin (1989, 1997) indicated that the cerebellum could be a neural analog of a Kalman filter. Droulez and Cornílleau-Pérèz (1993) drew attention to the relevance of multisensory integration in the moving organism to the Kalman filter. Nevertheless, the suggested analogy was only at the functional level and lacked correspondence to the cerebellar network. In our study, we demonstrated the three computational steps in the cerebellar circuit that are compatible with the Kalman filter (Tanaka et al., 2019): (1) the PCs compute a predictive state from a current estimate conveyed by the MFs (prediction step); (2) the DNCs combine the predicted state from the PCs and sensory feedback from the MFs (filtering step); and (3) the DNCs represent future activities of MFs (cerebellar prediction).

Note that even a pair of an excitatory granule cell and an inhibitory Golgi cell that receive the same MF input can function as a neural oscillator (Hoppensteadt & Izhikevich, 1997; Wilson & Cowan, 1973). It can show nonlinear input-output organizations and various types of bifurcations of activities depending on system parameters (Izhikevich, 2007). Therefore, these *linear* steps of the cerebellar information processing were unexpected and surprising, considering the complexity of the whole neuron network of the cerebellum.

Overall, the cerebellum appears to perform not only an internal forward model prediction but also an optimal integration of a predicted state and sensory feedback signals, in a way that is equivalent to Kalman filter as summarized below (a) (Tanaka et al., 2019):

$$\hat{X}_{t\backslash t} = \hat{X}_{t\backslash t-1} + K\left(z_t - C\hat{X}_{t\backslash t-1}\right) = \left(I - KC\right)\hat{X}_{t\backslash t-1} + Kz_t \tag{19.a}$$

where a filtered state $\hat{X}_{t\backslash t}$ (DNC output) is generated by combining a predicted state $\hat{X}_{t\backslash t-1}$ (=PC input) and an observed state $\mathbf{z}_t$ (=MF collateral input). We speculate that the weights from PC to DNC and the weights from MF to DNC correspond to the matrices $\mathbf{I} - \mathbf{KC}$ and $\mathbf{K}$ in Eq. (19.a), respectively (Tanaka et al., 2019).

19.10 Morphologic Substrata of the Cerebrocerebellum for Kalman Filter

Nevertheless, we realized that the conventional circuit diagram of the cerebellum (Fig. 19.5a) is *not* compatible with the Kalman filter (a). In this diagram (Fig. 19.5a), a MF projects *both* to PC (via granule cell (GC)) and DNC as collaterals, implying

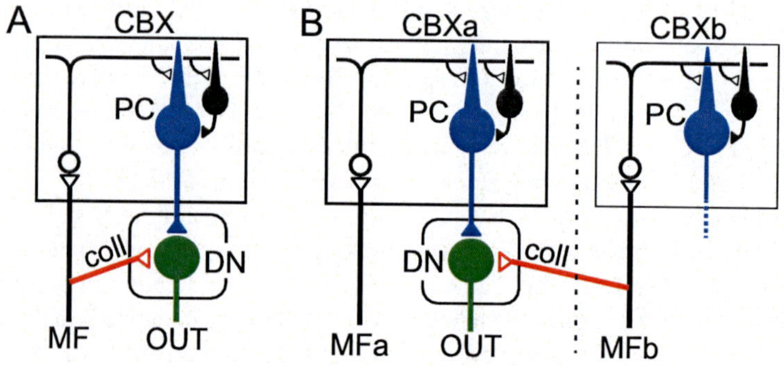

Fig. 19.5 Two schematics of corticonuclear organization. (**a**) Conventional scheme in which the same MF projects to the cerebellar cortex (CBX) and DN, both of which belong to the same corticonuclear complex (Ito, 1984). (**b**) Proposed scheme that one MF (MFa) from pontine nuclei (PN) projects to the cerebellar cortex (i.e., cerebrocerebellum) (CBXa) *without* collateral projection to DN, whereas another separate MF (MFb) projects to DN with a collateral. Note that MFa and MFb have distinct projection areas in the cerebellar cortex, CBXa and CBXb, respectively. Only Scheme B is consistent with the requirements of the Kalman filter model and the latest neuroanatomical data for the cerebrocerebellum. (Adapted from Tanaka et al. (2020) under CC BY license)

that PC and DNC of the same corticonuclear microcomplex (Ito, 1984) share the *same* MF input. In contrast, a Kalman filter (a) requires two distinct MF inputs. One MF input originates from cortical motor areas. It contributes to the prediction step in the cerebellar cortex to generate the *current estimate* ($\hat{X}_{t\backslash t-1}$) (i.e., PC activity). The other MF input conveys sensory feedback input to DNC through the collateral and contributes to the filtering step in DN. Most importantly, the contributions of the two MF inputs in (a) are *asymmetrical* and *uninterchangeable*. Therefore, the neuron circuit (Fig. 19.5a), in which the current estimate and current measurement are indistinguishable (i.e., *interchangeable*), cannot function as a Kalman filter.

Incongruent with the conventional diagram, extant anatomical studies suggest that the cerebrocerebellum receives respective MF inputs to PC and DNC (Fig. 19.5b). The first requirement of the Kalman filter is the cortical MFs project to the cerebrocerebellum *without* collaterals to DNC. Na et al. (2019) recently demonstrated that MFs from PN virtually *lack* collaterals to DNC on their way to the cerebrocerebellum. Namely, the first requirement is satisfied with the input from cortical motor areas to the cerebrocerebellum. The second requirement for the Kalman filter is that MFs conveying sensory input give off collaterals to DN. Indeed, Wu et al. (1999) demonstrated that MFs originated from the lateral reticular nucleus (LRN), which receives strong somatosensory inputs from the spinal cord, have an abundant collateral projection to DN and other cerebellar nuclei on their way to the vermis and the intermediate zone (see Figs. 8, 9, and 10 in Wu et al., 1999). Note that the two MF inputs from PN and LRN have only minor overlap in the cerebellar cortex (Na et al., 2019; Wu et al., 1999). Figure 19.5b summarizes these observations and demonstrates the *asymmetrical* relationship of the two lines of MF inputs, which is consistent with the Kalman filter. We have already pointed out the defect of the symmetrical MF inputs in the previous cerebellar forward model hypothesis (see Section "Difficulty in Identifying the Cerebellar Forward Model" in this article). In this way, the defect has been removed.

Under these anatomical data, we found two functionally distinct populations of MFs in our data (Tanaka et al., 2019). One population of MFs contributed selectively to the reconstruction of PC activities and dominated the prediction step, while the other population of MFs contributed selectively to the reconstruction of DNC activities and dominated the filtering step (Tanaka et al., 2019). The average correlation coefficient between weights of MF–PC and MF–DNC projections was no more than 0.060. A statistical test based on resampling verified that the correlation between the two MF populations was statistically significant ($p < 10^{-5}$). Therefore, we concluded that PCs and DNCs received inputs from distinct populations of MFs, thereby satisfying the Kalman filter model's requirements.

19.11 Inference About the Primordial Operation of the Cerebellum

The "corticonuclear microcomplex" depicted in Fig. 19.5b is most likely specific for the dentate nucleus and corresponding cerebrocerebellum (i.e., the *newer* part of the cerebellum). In contrast, the other older parts of the cerebellar nuclei receive MF inputs in different ways (Ito, 1984). For example, in the vestibular nucleus, neurons are driven primarily by direct (i.e., primary afferent) MF inputs, whereas PCs activated by the same MF inputs exert modulatory action on the nuclear neurons (Fig. 19.5a; see also Fig. 92a in Ito, 1984). In the fastigial nucleus, however, the PC input plays the primary role. At the same time, collaterals of MFs provide a *background excitation* on which PCs can impose efficient bidirectional modulation (Fig. 19.5a; see also Fig. 92b in Ito, 1984). In these phylogenetically *older* cerebellar regions, the corticonuclear microcomplex (Ito, 1984) is *not* consistent with the Kalman filter, where PC and cerebellar nuclear cells share the same MF input. Overall, even if the local neuron circuitry is common for the entire cerebellar cortex, different regions may perform computationally different operations depending on the organization of microcomplex (Ito, 1984: pp. 195–199 and Fig. 92).

Nevertheless, because of the superb "crystal-like" homogeneity of the neuron circuitry, all these regions of the cerebellar cortex most likely hold the *prediction step* in common. Even if the presumed prediction step alone remains suboptimal due to lack of the filtering step, it could still play an invaluable role in improving its owner's survival. Indeed, the cerebellum-like structure of fishes with electroreception systems has been suggested as a neural analog of a dynamical state estimator (Bastian & Zakon, 2005; Paulin, 1989; 1997). According to Paulin, the cerebellum is a *sensory* processing structure with a specific role in the state estimation of dynamical systems. He further suggested that the cerebellum has a common underlying role in sensorimotor, perceptual, and cognitive processes consistent with the state estimator hypothesis (1997). The cerebellar contribution to sensory processing is not surprising if we remember the fact that the cerebellum emerges in the *alar plate* (i.e., sensory domain of the embryonic neural tube) of the rhombencephalon of old jawless fishes (Sugahara et al., 2016). It collects multimodal inputs, including exteroceptive (lateral line, vestibular, acoustic, visual) and somatosensory inputs (Larsell, 1967). The cerebellum has further gained access, in mammals, to cortical information from association areas and motor and sensory areas. Overall, throughout its long history of evolution, the cerebellum has been a unique *hub* to collect afferent, efferent, and finally *internal* (i.e., association) information from the entire brain.

19.12 Extension of Cerebellar Kalman Filter Hypothesis to the Non-motor Cerebrocerebellum

The critical question arises whether the Kalman filter mechanism for the *motor* part of the cerebrocerebellum (Tanaka et al., 2019) generalizes to its *cognitive/affective* part. Our dataset recorded during the motor task may not generalize directly to the cerebellum's contribution to prediction in cognitive/affective domains. Nevertheless, it is possible to search for the *Kalman filter-specific* corticonuclear microcomplex (Fig. 19.5b) in the *non-motor* part of the cerebrocerebellum. There are two requirements: (1) the primary *MFa* input to the cerebrocerebellum is originated from a non-motor cortical area and relayed by PN cells (PNCs) and (2) the filtering *MFb* input is derived from a distinct cortical or subcortical source and relayed by a *non-PN* nucleus with a significant collateral projection to DN (Fig. 19.5b). Requirement (2) is the key because requirement (1) is common for most, if not all, non-motor cortical areas, including prefrontal areas (Schmahmann & Pandya, 1997), parietal association areas (Schmahmann & Pandya, 1989), superior temporal areas (Schmahmann & Pandya, 1991), and occipitotemporal and parahippocampal areas (Schmahmann & Pandya, 1993). There are a few known sources of collateral MF inputs to DN, most notably the lateral reticular nucleus (LRN) (Wu et al., 1999) and the nucleus reticularis tegmenti pontis (NRTP) (Gerrits and Voogd, 1987) in the reticular formation. The LRN receives the main inputs from the spinal cord (Alstermark and Ekerot, 2013) and additional inputs from the sensorimotor areas and the red nucleus (Bruckmoser et al., 1969; Matsuyama and Drew, 1997). The NRTP receives inputs mainly from the sensorimotor areas, the prefrontal areas, and the parietal association areas (Schmahmann et al., 2004). In summary, the Kalman filter model (Fig. 19.5b) is also applicable to the non-motor part of the cerebrocerebellum if the MF collateral to non-motor parts of DN (*MFb*) and MF inputs to the PCs (*MFa*) have distinct sources and *causal* relationship from *MFa* to *MFb*. In that sense, NRTP is a major candidate for the filtering inputs to the non-motor parts of DN (Fig. 19.6, *left*).

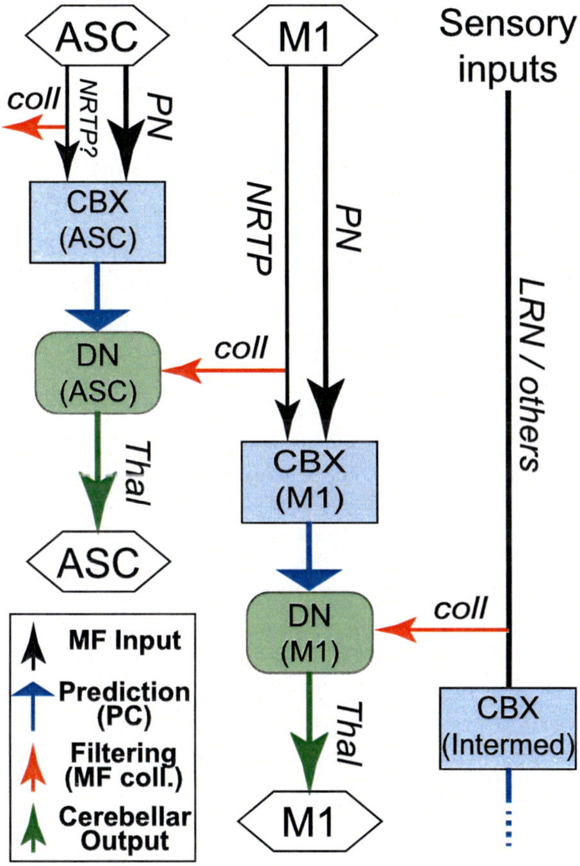

Fig. 19.6 A hypothetical cascade of Kalman filters in the cerebrocerebellar communication loop. The Kalman filter model that predicts M1 activity (*center*) is capable to form a cascade with another Kalman filter that predicts ASC activity (*left*), if M1 sends a filtering input (*coll*) to distinct region of DN (ASC) via NRTP (*center*). Note the collateral input to DN (ASC) from M1 does not project to M1 region of DN (DN (M1)) (see Tanaka et al., 2019). In this way, the filtering input may play a critical role to make two Kalman filters to work together. This model may also explain how parallel forward models in the cerebrocerebellar communication loops function together in a coordinated manner and may provide a partial explanation for unity of mind. ASC association cortex, CBX cerebellar cortex, coll collateral of MFs, LRN lateral reticular nucleus, M1 the primary motor cortex, NRTP nucleus reticularis tegmenti pontis, PN pontine nuclei, Thal thalamus

It should be pointed out that the Kalman filter model that predicts M1 activity (Fig. 19.6, *center*) is capable to form a cascade with another Kalman filter that predicts activity of association cortex (ASC) (Fig. 19.6, *left*), if M1 sends a filtering input (*coll*) to distinct region of DN (ASC) via NRTP (Fig. 19.6, *left*). In this way, the filtering input may play a critical role to make two Kalman filters work together. This model also explains how parallel domains in the cerebrocerebellar communication loops (Kelly & Strick, 2003) are coordinated in a cascadic manner, providing a partial explanation for unity of mind.

19.13 Compressed Prediction of the Cerebellar Internal Model

Few paid attention to the asymmetry of the cerebrocerebellar loop in terms of the number of output neurons. The number of axons in the cerebral peduncle (CP) conveying cortical outputs to PN and other precerebellar nuclei is estimated as 21 million in humans (Tomasch, 1969). In contrast, the number of axons in the return path (i.e., the superior cerebellar peduncle) relaying the cerebellar output to the thalamus is no more than 0.8 million in humans (Heidary & Tomasch, 1969). Namely, the cerebrocerebellum returns its output to the cerebral cortex after significant compression (1:20) (Tanaka et al., 2020). Therefore, the cerebellar output appears to represent a predicted state of cortical activities in a compressed format. The mapping between the cortical output and the cerebellar output may be compatible with a *homomorphism* (https://en.wikipedia.org/wiki/Homomorphism). A homomorphism has a distinguished advantage for an internal model because it enables the model to perform an operation equivalent to the original while using a more simplified representation.

Although there is no consensus on the compressed representation so far (Sanger et al., 2019), the compact (i.e., low-resolution) prediction of the cortical state may help assign more attention to the task currently in focus by minimizing the computational load for the other peripheral tasks. It also reminds us that the cerebellum contributes most to trained and automated repertoires of both motor and cognitive functions with reduced attention.

We cannot spare another important consequence of the relative paucity of DNCs. The MF collateral input to DN (Fig. 19.5b, *coll*) appears far from massive compared (e.g., Wu et al., 1999) to the massive MF input to the cerebrocerebellum (Fig. 19.5b, MFa). Therefore, one may argue that the modest projection of the MF collaterals to DN cannot be effective enough to play such an important function as *filtering* of the Kalman filter. Nevertheless, the limited number of the *target* DNCs appears to help in *amplifying* efficacy of the collateral input.

19.14 Clinical Evidence for the Internal Model Hypothesis of the Cerebellum

Finally, we searched for clinical evidence that supported the cerebellar forward model hypothesis (e.g., Bastian, 2006; Miall et al., 2007). A series of studies from our group confirmed the impaired predictive control in movements of patients with cerebellar ataxia (CA). We first decomposed the muscle activities for the wrist movement into a low-frequency (≤ 0.5 Hz) component (F1) and a high-frequency (>0.5 Hz) component (F2), each of which represented the predictive control and the feedback correction, respectively (Kakei et al., 2019). Then for each component, we identified a recipe of muscle activities by analyzing a relationship between the muscle tension and movement kinematics (the wrist angle $\theta(t)$ and the wrist angular velocity $\dot{\theta}(t)$) weighted by the coefficients of K_r (the elastic term) and B_r (the viscous term) (Kakei et al., 2019; Lee et al., 2012; Mitoma et al., 2016). Importantly, we found that the ratio of B_r/K_r characterized the recipe of muscle activities for each component. In control subjects, the B_r/K_r ratio for the predictive (F1) component demonstrated a higher value (Fig. 19.5 in Kakei et al., 2019) (Fig. 7a), suggesting the velocity control dominance. On the other hand, the B_r/K_r ratio for the corrective (F2) component demonstrated a much smaller value (Fig. 19.5 in Kakei et al., 2019) (Fig. 19.7a), suggesting the role of F2 component in correction of positional errors (Kakei et al., 2019). In contrast, CAs showed a selective decrease of the B_r/K_r ratio for the predictive (F1) component (Fig. 5 in Kakei et al., 2019) (Fig. 19.7a), suggesting poor recruitment of the predictive velocity control and compensatory dependence on the position-dependent pursuit (Kakei et al., 2019). The loss of component-specific differences in the B_r/K_r ratio suggests impairment of predictive control in CA. Indeed, the B_r/K_r ratio decrease correlated with the increase of error in the predictive (F1) movement (Fig. 19.7b) (Kakei et al., 2019). Another critical

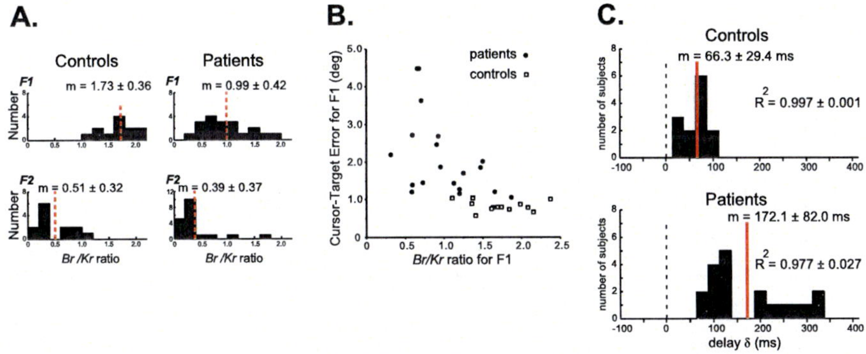

Fig. 19.7 Difference of muscle activity – movement kinematics relationship between controls and cerebellar patients. (**a**) Comparison of the Br/Kr ratios that represent recipe of the motor commands for the F1 and F2 components between the controls and the cerebellar patients. *Controls*: Br/Kr ratios of the control subjects for the F1 component (top) and the F2 component (bottom) ($n = 13$). Note the highly significant difference between the two components. *Patients*: Br/Kr ratios of the patients for the F1 (top) and the F2 (bottom) components ($n = 19$). Note the selective decrease of Br/Kr ratios for the F1 component in the patients. (**b**) Correlation between the Br/Kr ratios for F1 component and cursor-target error for F1 (F1 error, in short). The F1 error is defined as an average error between the target motion and the F1 component of the movement. Note the negative correlation. (**c**) Delay of the predictive (F1) component of the movement relative to the target motion calculated with a cross-correlation analysis for controls ($n = 13$) and patients ($n = 19$). (Adapted from Kakei et al. (2019) under CC BY license)

difference between the control and CA was the increased delay of the predictive (F1) component in CA (Fig. 19.7c). In the control subjects, the predictive (F1) movement lagged the target motion only by 66 ms, which was too small to be a visual feedback delay (i.e., a proof of prediction) (Kakei et al., 2019). In contrast, in patients with CA, the delay increased by more than 100 ms, as much as 172 ms. The increased delay (i.e., 172 ms) is comparable to a visual feedback delay, demonstrating lack of compensation of feedback delay in CA patients. In summary, ataxic movements are consistent with an impairment of a forward model in terms of accuracy and delay of state prediction.

19.14.1 Postscript

The most primitive cerebellum emerged in the alar plate of the rhombencephalon of old jawless fishes as a sensory *hub* to which multimodal sensory inputs converge. The cerebellum later acquired efference copy inputs, which is essential for active sensing. Indeed, we can see its example in the cerebellum-like structure of some fishes that process information from electroreception systems. We speculate that the active sensing evolved to detect causality and finally led to a more sophisticated state prediction in a primitive forward model. Next, in mammals, the cerebellum acquired a strong loop with the cerebral cortex: the cerebrocerebellar communication loop. In this way, the cerebellum developed into the primary hub in the entire CNS. In particular, the acquisition of the DN filtering step evolved the existing dynamic prediction in the cerebellar cortex to a Kalman filter. This revolutionary event gave each region of the cerebrocerebellum a privilege to predict the state of its counterpart in the cerebral cortex, which includes motor areas, parietal association areas, prefrontal association areas, and limbic areas (Ito, 2008). This Kalman filter model also explains how parallel domains in the cerebrocerebellum operate in a cascadic manner and may provide a partial explanation for unity of mind. Finally, we should not forget the morphological asymmetry of the cerebrocerebellar communication loop. Namely, the cerebrocerebellum returns its output from DN back to the cerebral cortex after significant compression (1:20) (Tanaka et al., 2020). The low-resolution prediction of the cortical state may help assign more attention to the task currently in focus by reducing computational load for peripheral tasks. Given the fact that the cerebellum contributes most to trained and automated repertoires with less effort and attention, this asymmetry appears to make perfect sense.

Acknowledgments We thank Profs. Hiroshi Mitoma and Koji Ito for their valuable comments and discussions.

Support This work was supported by Grants-in-Aid from the Ministry of Education, Culture, Sports, Science and Technology in Japan (MEXT) (http://www.mext.go.jp/) (no. 26120003, no. 14580784, no. 15016008, no. 16015212, no. 20033029, and no. 21500319 to SK; no. 25430007, no. 26120005, and no. 16 K12476 to HT; no. 21700229 and no. 24650304 to JL; and no. 24650224 to TI), the Japan Science and Technology Agency (A-STEP) to SK (http://www.jst.go.jp/), the Japan Science and Technology Agency (PRESTO: Intelligent Cooperation and Control) (SK), NBRP "Japanese Monkeys" through the National BioResource Project of MEXT, the JSPS Programs (Program for Advancing Strategic International Networks to Accelerate the Circulation of Talented Researchers, and Embodied-Brain Systems Science) (HT), and the Hitachi-Kurata and the Tateishi Science Foundations (HT). This research was also supported by AMED under grant number 16ek0109048h0003 to SK. The funders had no role in study design, data collection and analysis, decision to publish, or preparation of the manuscript.

References

Albus, J. S. (1971). A theory of cerebellar function. *Mathematical Biosciences, 10*, 25–61.

Allen, G. I., & Tukahara, N. (1974). Cerebrocerebellar communication systems. *Physiological Reviews, 54*, 957–1006.

Alstermark, B., & Ekerot, C. F. (2013). The lateral reticular nucleus: a precerebellar centre providing the cerebellum with overview and integration ofmotor functions at systems level. A new hypothesis. *Journal of Physiology, 591*, 5453–5458.

Bastian, A. J. (2006). Learning to predict the future: The cerebellum adapts feedforward movement control. *Current Opinion in Neurobiology, 16*, 645–649.

Bastian, J., & Zakon, H. H. (2005). Plasticity of sense organs and brain. In T. H. Bullock, C. D. Hopkins, A. N. Popper, & R. R. Fay (Eds.), *Electroception* (pp. 195–228). Springer.

Brodal, P., & Bijaalie, J. G. (2003). Organization of the pontine nuclei. *Neuroscience Research, 13*, 83–118.

Bruckmoser, P., Hepp, M. C., & Wiesendanger, M. (1969). Cortical influence on the lateral reticular nucleus of the cat. *Brain Research, 15*, 556–558.

Droulez, J., & Cornîlleau-Pérèz, V. (1993). Application of the coherence scheme to the multisensory fusion problem. In A. Berthoz (Ed.), *Multisensory control of movement* (pp. 485–501). Oxford University Press.

Ebner, T. J., & Pasalar, S. (2008). Cerebellum predicts the future motor state. *Cerebellum, 7*, 583–588.

Eccles, J. C., Ito, M., & Szentágothai, J. (1967). *The cerebellum as a neuronal machine*. Springer-Verlag.

Flash, T., & Hogan, N. (1985). The coordination of arm movements: an experimentally confirmed mathematical model. *Journal of Neuroscience., 5*, 1688–1703.

Gerrits, N. M., & Voogd, J. (1987). The projection of the nucleus reticularis tegmenti pontis and adjacent regions of the pontine nuclei to the centralcerebellar nuclei in the cat. *Journal of Comparative Neurology, 258*, 52–69.

Haggard, P., & Wing, A. (1995). Coordinate responses following mechanical perturbations of the arm during prehension. *Experimental Brain Research, 102*, 483–494.

Harris, C. M., & Wolpert, D. M. (1998). Signal dependent noise determines motor planning. *Nature, 394*, 780–784.

Heidary, H., & Tomasch, J. (1969). Neuron numbers and perikaryon areas in the human cerebellar nuclei. *Acta Anatomica (Basel), 74*, 290–296.

Hoppensteadt, F. C., & Izhikevich, E. M. (1997). *Weakly connected neural networks*. Springer-Verlag. (Page 23).

Ishikawa, T., Tomatsu, S., Tsunoda, Y., Lee, J., Hoffman, D. S., & Kakei, S. (2014). Releasing dentate nucleus cells from Purkinje cell inhibition generates output from the cerebrocerebellum. *PLoS One, 9*, e108774. https://doi.org/10.1371/journal.pone.0108774

Ishikawa, T., Tomatsu, S., Izawa, J., & Kakei, S. (2016). The cerebro-cerebellum: Could it be loci of forward models? *Neuroscience Research, 104*, 72–79.

Ito, M. (1970). Neurophysiological aspects of the cerebellar motor control system. *International Journal of Neurology, 7*, 162–176.

Ito, M. (1984). *The cerebellum and neural control*. New York: Raven Press.

Ito, M. (2008). Control of mental activities by internal models in the cerebellum. *Nature Reviews. Neuroscience, 9*, 304–313.

Izhikevich, E. M. (2007). *Dynamical Systems in Neuroscience – The geometry of excitability and bursting*. The MIT Press.

Kakei, S., Hoffman, D. S., & Strick, P. L. (1999). Muscle and movement representations in the primary motor cortex. *Science, 285*, 2136–2139.

Kakei, S., Hoffman, D. S., & Strick, P. L. (2001). Direction of action is represented in the ventral premotor cortex. *Nature Neuroscience, 4*, 1020–1025.

Kakei, S., Lee, J., Mitoma, H., Tanaka, H., Manto, M., & Hampe, C. S. (2019). Contribution of the cerebellum to predictive motor control and its evaluation in ataxic patients. *Frontiers in Human Neuroscience, 13*, 216. https://doi.org/10.3389/fnhum.2019.00216

Kalman, R. E., & Bucy, R. S. (1961). New results in linear filtering and prediction. *ASME Journal of Basic Engineering, 83*, 95–108.

Kelly, R. M., & Strick, P. L. (2003). Cerebellar loops with motor cortex and prefrontal cortex of a nonhuman primate. *The Journal of Neuroscience, 23*, 8432–8444.

Kirk, D. E. (1970). *Optimal control theory: An introduction*. Prentice-Hall.

Larsell, O. (1967). *The comparative anatomy and histology of the cerebellum*. University of Minnesota Press.

Lee, J., Kagamihara, Y., Tomatsu, S., & Kakei, S. (2012). The functional role of the cerebellum in visually guided tracking movement. *Cerebellum, 11*, 426–433.

Marr, D. (1969). A theory of cerebellar cortex. *The Journal of Physiology, 202*, 437–470.

Matsuyama, K., & Drew, T. (1997). Organization of the projections from the pericruciate cortex to the pontomedullary brainstem of the cat: a studyusing the anterograde tracer Phaseolus vulgaris-leucoagglutinin. *Journal of Comparative Neurology, 389*, 617–641.

Miall, R. C., Weir, D. J., Wolpert, D. M., & Stein, J. (1993). Is the cerebellum a smith predictor? *Journal of Motor Behavior, 25*, 203–216.

Miall, R. C., Christensen, L. O. D., Cain, O., & Stanley, J. (2007). Disruption of state estimation in the human lateral cerebellum. *PLoS Biology, 5*, e316.

Mitoma, H., Adhikari, K., Aeschlimann, D., Chattopadhyay, P., Hadjivassiliou, M., Hampe, C. S., Honnorat, J., Joubert, B., Kakei, S., Lee, J., Manto, M., Matsunaga, A., Mizusawa, H., Nanri, K., Shanmugarajah, P., Yoneda, M., & Yuki, N. (2016). Consensus paper: Neuroimmune mechanisms of cerebellar ataxias. *Cerebellum, 15*, 213–232.

Na, J., Sugihara, I., & Shinoda, Y. (2019). The entire trajectories of single pontocerebellar axons and their lobular and longitudinal terminal distribution patterns in multiple aldolase C-positive compartments of the rat cerebellar cortex. *The Journal of Comparative Neurology, 527*, 2488–2511.

Paulin, M. (1989). A Kalman filter theory of the cerebellum. In M. A. Arbib & S. Amari (Eds.), *Dynamic interactions in neural networks: Models and data* (pp. 239–259). Springer.

Paulin, M. (1997). Neural representations of moving systems. In J. D. Schmahmann (Ed.), *The cerebellum and cognition* (pp. 515–533). Academic Press.

Sanger, T. D., Yamashita, O., & Kawato, M. (2019). Expansion coding and computation in the cerebellum: 50 years after the Marr-Albus codon theory. *Journal of Physiology, 598*, 913–928.

Schmahmann, J. D., & Pandya, D. N. (1989). Anatomical investigation of projections to the basis pontis from posterior parietal association cortices in rhesus monkey. *The Journal of Comparative Neurology, 289*, 53–73.

Schmahmann, J. D., & Pandya, D. N. (1991). Projections to the basis pontis from the superior temporal sulcus and superior temporal region in the rhesus monkey. *The Journal of Comparative Neurology, 308*, 224–248.

Schmahmann, J. D., & Pandya, D. N. (1993). Prelunate, occipitotemporal, and parahippocampal projections to the basis pontis in rhesus monkey. *The Journal of Comparative Neurology, 337*, 94–112.

Schmahmann, J. D., & Pandya, D. N. (1997). Anatomic organization of the basilar pontine projections from prefrontal cortices in rhesus monkey. *The Journal of Neuroscience, 17*, 438–458.

Schmahmann, J. D., Rosene, D. L., & Pandya, D. N. (2004). Motor projections to the basis pontis in rhesus monkey. *The Journal of Comparative Neurology, 478*, 248–268.

Stein, R. B., Oguztoreli, M. N., & Capaday, C. (1994). What is optimized in muscular movements? In Stengel R. F. *Optimal Control and Estimation*. New York: Dover

Stengel, R. F. (1994). *Optimal control and estimation*. Dover.

Sugahara, F., Pascual-Anaya, J., Oisi, Y., Kuraku, S., Aota, S., Adachi, N., Takagi, W., Hirai, T., Sato, N., Murakami, Y., & Kuratani, S. (2016). Evidence from cyclostomes for complex regionalization of the ancestral vertebrate brain. *Nature, 531*, 97–100.

Tanaka, H., Ishikawa, T., & Kakei, S. (2019). Neural evidence of the cerebellum as a state predictor. *Cerebellum, 18*, 349–371. https://doi.org/10.1007/s12311-018-0996-4

Tanaka, H., Ishikawa, T., Lee, J., & Kakei, S. (2020). The cerebro-cerebellum as a locus of forward model; a review. *Frontiers in Systems Neuroscience, 14*, 19.

Thach, W. T. (1975). Timing of activity in cerebellar dentate nucleus and cerebral motor cortex during prompt volitional movement. *Brain Research, 88*, 233–241.

Thach, W. T. (1978). Correlation of neural discharge with pattern and force of muscular activity, joint position, and direction of intended next movement in motor cortex and cerebellum. *Journal of Neurophysiology, 41*, 654–676.

Thier, P., & Markanday, A. (2019). Role of the vermal cerebellum in visually guided eye movements and visual motion perception. *Annual Review of Vision Science, 5*, 247–268.

Todorov, E. (2004). Optimality principles in sensorimotor control. *Nature Neuroscience, 7*, 907–915.

Tomasch, J. (1969). The numerical capacity of the human cortico-pontocerebellar system. *Brain Research, 13*, 476–484.

Tomatsu, S., Ishikawa, T., Tsunoda, Y., Lee, J., Hoffman, D. S., & Kakei, S. (2016). Information processing in the hemisphere of the cerebellar cortex for control of wrist movement. *Journal of Neurophysiology, 115*, 255–270.

Uno, Y., Kawato, M., & Suzuki, R. (1989). Formation and control of optimal trajectory in human multijoint arm movement: Minimum torque-change model. *Biological Cybernetics, 61*, 89–101.

Wilson, H. R., & Cowan, J. D. (1973). A mathematical theory of the functional dynamics of cortical and thalamic nervous tissue. *Kybernetik, 13*, 55–80.

Wolpert, D. M., & Miall, R. C. (1996). Forward models for physiological motor control. *Neural Networks, 9*, 1265–1279.

Wu, H., Sugihara, I., & Shinoda, Y. (1999). Projection patterns of single mossy fibers originating from the lateral reticular nucleus in the rat cerebellar cortex and nuclei. *The Journal of Comparative Neurology, 411*, 97–118.

Chapter 20
The Cerebellum as a CNS Hub Modulating Autism-Relevant Behaviors

Laura C. Rice and Catherine J. Stoodley

Autism spectrum disorder (autism) is a heterogeneous neurodevelopmental condition that affects approximately 1 in 54 children in the United States (CDC, 2020). Autism is diagnosed based on difficulties in social interaction and communication and the presence of repetitive/stereotyped behaviors and restricted interests (American Psychiatric Association, 2013). Autism can be further characterized by a host of behavioral differences across sensorimotor and cognitive domains, including language, executive function, and behavioral flexibility. Based on the core symptoms, investigations of the neural correlates of autism have focused on networks underpinning theory of mind (or mentalizing, the ability to understand the mental states of others), mirroring/imitation, and processing of social stimuli, including faces. The presence of repetitive behaviors and restricted interests have been associated with sensorimotor networks and differences in the reward system response for social and non-social stimuli. It is not surprising that this complex constellation of behaviors has been associated with atypical structure and function in a number of neural regions. Of these, the cerebellum is one of the most commonly identified neural correlates of autism, and structural and functional differences within the cerebellum have been associated with the full range of autism symptoms.

Autism is highly heritable, with heritability estimates as high as 90%, and polygenic, with several hundred genetic variants of interest (Sandin et al., 2017). It has long been hypothesized that genetically driven aberrations in early development lead to atypical neural network organization in autism (Belmonte et al., 2004; Brambilla, 2003; Donovan & Basson, 2017). Structural differences in autism include unusual brain growth patterns in early life (Courchesne et al., 2001) and complex patterns of regional differences in cellular organization, white matter pathways, and gray matter volume and cortical thickness (for reviews, see Ecker, 2017;

L. C. Rice · C. J. Stoodley (✉)
Department of Neuroscience, American University, Washington, DC, USA
e-mail: laura.blevins@student.american.edu; stoodley@american.edu

H. Mizusawa, S. Kakei (eds.), *Cerebellum as a CNS Hub*,
Contemporary Clinical Neuroscience, https://doi.org/10.1007/978-3-030-75817-2_20

Donovan & Basson, 2017). Researchers have identified differences in network lateralization (Escalante-Mead et al. 2003; Seery et al., 2013), inter- and intra-hemispheric functional connectivity (Lee et al., 2016), and functional connectome hierarchy in autism (Hong et al., 2019). Further evidence supports disruption to specific networks associated with the core symptoms of autism. For example, studies report atypical cross talk between networks involved in mentalizing/theory of mind and mirroring (Fishman et al., 2014); reduced functional connectivity between cortical areas involved in face processing, theory of mind, and the sense of self (Cheng et al., 2015); and differential connectivity within the theory of mind network in autism (Kana et al., 2015). Differences have been reported in other networks supporting autism-associated behaviors, including altered language network lateralization (Seery et al., 2013) and atypical reward system engagement (Choi et al., 2015; Kohls et al., 2018). Disrupted sensorimotor systems (including basal ganglia and cerebellar circuits) have been associated with restricted and repetitive behaviors transdiagnostically (Wilkes & Lewis, 2018).

Postmortem studies have revealed spatial crowding of microcolumns in Brodmann area 9 and wider minicolumns in the primary auditory, auditory association, orbital frontal, and parietal cortex, particularly at younger ages (Donovan & Basson, 2017). These results suggest that developmental differences in cellular organization may underlie atypical system-level neural organization in autism. White matter reductions (Groen et al., 2011), particularly in callosal and subcortical fiber tracts (Shukla et al., 2010), and reduced hemispheric asymmetry of white matter microstructure have been identified in autism (Carper et al., 2016; Postema et al., 2019). Decreased white matter integrity was observed in brain areas involved in social functioning, including the fusiform gyrus, amygdala, and superior temporal gyrus (Jou et al., 2011). There are also reports of decreased lateralization in white matter tracts that connect the hippocampus and amygdala to the fusiform gyrus, reflecting atypical organization of regions involved in the processing of facial emotional expressions, a common behavioral deficit in autism (Travers et al., 2012).

These findings indicate disrupted structural and functional neural networks throughout the brain in autism, but it is not clear how these differences arise during the complex process of pre- and postnatal brain development. It is possible that developmental regional disruptions within the cerebellum could impact the structural and functional development of the cortical circuits that support autism-associated behaviors (Stoodley, 2016; Wang et al., 2014). Here we consider the evidence suggesting that the role of the cerebellum as a CNS hub is an important contributor to the neural bases of autism.

20.1 Cerebellum and Autism

Cerebellar dysfunction in autism is supported by genetic studies, animal models, neuroimaging, and investigations of clinical populations with syndromic forms of autism or cerebellar damage (Courchesne, 1991, 1997; Courchesne et al., 1988; for reviews, Becker & Stoodley, 2013; Wang et al., 2014; D'Mello & Stoodley, 2015; Hulbert & Jiang, 2017; Tsai et al., 2018; Sathyanesan et al., 2019). Reductions in Purkinje cell size and numbers, particularly in the posterior inferior cerebellum, have been reported in postmortem studies of autism (Bauman & Kemper, 2005). Autism risk genes are often expressed in the cerebellum, and the majority of autism mouse models display cerebellar abnormalities (Ellegood et al., 2015). Approximately 50% of patients with tuberous sclerosis complex (TSC) also have an autism diagnosis, and more severe autism symptoms are associated with a higher number of cerebellar tubers in TSC patients (Eluvathingal et al., 2006; Weber et al., 2000). TSC mouse models with targeted cerebellar dysfunction exhibit marked autism features, including deficits in social interaction and increased repetitive behaviors (Tsai et al., 2012). Isolated cerebellar damage can result in an autism diagnosis or autism-like symptoms (for reviews, Becker & Stoodley, 2013; Stoodley & Limperopoulos, 2016; Van Overwalle et al., 2020), and "autism spectrum disorders" are considered part of the neuropsychiatric profile associated with cerebellar dysfunction (Schmahmann et al., 2007). For example, adult patients with cerebellar damage were impaired on an emotion attribution task and reported autism-like behaviors (Hoche et al., 2016), and preterm infants with isolated cerebellar damage are 40 times more likely to be diagnosed with autism, particularly if damage involves the vermis (Limperopoulos et al., 2007). Together with evidence from mouse models showing that primary cerebellar disruption leads to autism-like behaviors (e.g., Badura et al., 2018; Kelly, Escamilla, & Tsai, 2020; Stoodley et al., 2017; Tsai et al., 2012), these findings suggest that developmental disruption of cerebellar circuits has the potential to produce behavioral profiles consistent with an autism diagnosis.

It is important to note that specific cerebellar subregions are engaged during autism-relevant behaviors and show structural and functional differences in individuals with autism (see Fig. 20.1; D'Mello & Stoodley, 2015 for review). The functional subregions of the cerebellum are defined by patterns of anatomical connectivity with different regions of the cerebral cortex and spinal cord (Stoodley & Schmahmann, 2010). The cerebellum forms "closed-loop" circuits with the vast majority of the cerebral cortex and subcortical structures (Buckner et al., 2011; D'Angelo, 2018; Kelly & Strick, 2003; Pisano et al., 2020), by which it can regulate a wide range of behaviors and neural circuits that are relevant to autism (for review, see D'Mello & Stoodley, 2015). Broadly speaking, the cerebellar regions involved in sensorimotor control include the anterior lobe and lobule VIII, whereas the posterolateral hemispheres are anatomically connected to cortical association areas and support a range of cognitive functions, including social cognition (see King et al., 2019; Schmahmann et al., 2019; Stoodley & Schmahmann, 2009; Van Overwalle et al., 2020). The posterior vermis has been implicated in behavioral regulation and flexibility in both humans and animal models (see Kelly, Escamilla, & Tsai, 2020; Schmahmann et al., 2007) and is considered part of limbic circuits underpinning emotional processing. Further evidence for a role of the posterior vermis in features of autism comes from an early investigation of the relationships between visuospatial exploration, stereotyped motor movements, and MRI measures of the cerebellar vermis, whole brain, and frontal lobes in children with autism (Pierce & Courchesne,

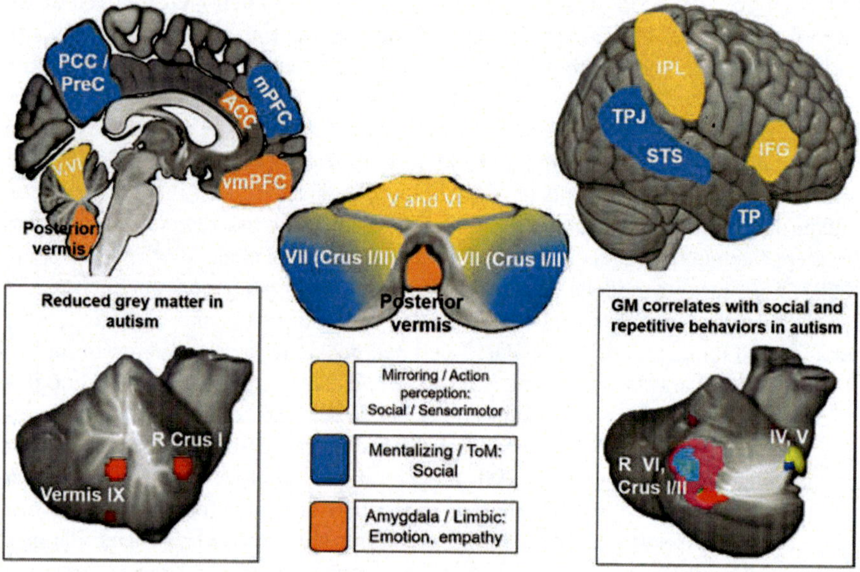

Fig. 20.1 (Top) Supratentorial networks involved in mirroring, mentalizing, and emotion (based on Kennedy & Adolphs, 2012) and corresponding cerebellar regions. Adapted from Stoodley & Tsai, in press. *ACC* anterior cingulate cortex, *IFG* inferior frontal gyrus, *IPL* inferior parietal lobule, *mPFC* medial prefrontal cortex, *PCC/PreC* posterior cingulate cortex/precuneus, *STS* superior temporal sulcus, *TP* temporal pole, *TPJ* temporo-parietal junction, *vmPFC* ventromedial prefrontal cortex. (Bottom left) Reduced gray matter (GM) in autism (red) adapted from Stoodley (2014). (Bottom right) Reduced GM in autism (red) and regions where GM correlated with scores on Autism Diagnostic Observation Schedule (ADOS) Social (violet); ADOS Social Communication (cyan); Autism Diagnostic Interview (ADI) Social Interaction (blue); ADI Restricted, Repetitive, and Stereotyped Behaviors (yellow); ADOS Stereotyped Behaviors and Restricted Interests (green), adapted from D'Mello et al. (2015)

2001). Reduced exploratory behavior and increased stereotyped behaviors were associated with the degree of hypoplasia in vermal lobules VI–VII in the autism group. Both structural neuroimaging meta-analyses (Stoodley, 2014) and prospective studies (e.g., D'Mello et al., 2015) have shown reduced gray matter in cerebellar lobule VII and the posterior vermis in autism (Fig. 20.1). Consistent with connectivity patterns of these regions, gray matter volume in right VII (Crus I/II) correlated with social and communication scores on autism diagnostic scales, and gray matter in the posterior vermis correlated with stereotyped behaviors and restricted interests (D'Mello et al., 2015), suggesting that more severe autism symptoms are associated with reduced gray matter in these cerebellar regions (see Olivito et al., 2018).

Evidence from functional imaging studies support differential cerebellar activation during a range of autism-relevant task paradigms (for reviews, see Becker & Stoodley, 2013; Wang et al., 2014; Sokolov, 2018; Van Overwalle et al., 2020) as well as differences in cerebro-cerebellar functional networks in autism (see D'Mello

& Stoodley, 2015 for review). In neurotypical adults, activity in bilateral hemispheric lobule VII and vermal and paravermal lobule IX was associated with viewing animate versus random movement, and greater activity in left Crus I and lobule VI was associated with the tendency to describe stimuli in social-affective versus motion-related terms (Jack & Pelphrey, 2015). Further, there was preferential effective connectivity between left Crus II and right posterior superior temporal sulcus (pSTS) while viewing animate vs. random stimuli (Jack & Pelphrey, 2015). During a similar task measuring animacy, the Frith-Happe triangle animations, participants with autism had more difficulty with the task and showed decreased activation in left Crus I (Kana et al., 2015), which may underlie behavioral differences in animacy attribution in autism. When performance was equivalent between the autism and neurotypical groups, the autism group showed increased engagement of Crus I bilaterally during a causal attribution task, which may reflect a compensatory mechanism (Kestemont et al., 2016). During a social judgment task, bilateral lobule VII (Crus II) was engaged by the neurotypical group, and the autism group showed significantly less cerebellar activation (bilateral VI and VII) during the social condition relative to gender judgment (Stanfield et al., 2017). These findings suggest that successful engagement of the posterolateral cerebellum is critical for performance on a range of social measures and that structural and functional differences in lobule VII may underlie social deficits in autism.

Disrupted structural (Sivaswamy et al., 2010) and functional (Olivito et al., 2018) connectivity has been reported between the cerebellum and a range of cortical networks in autism. For example, Verly et al. (2014) showed a lack of functional connectivity between the right posterior cerebellum and cerebral cortical language networks in autism, indicating disrupted cerebellar modulation of cortical language circuits. Decreased functional connectivity has also been reported between the left dentate nucleus of the cerebellum and cortical regions involved in the default mode network, which supports social cognition, mentalizing, and higher-order emotional processing (Olivito et al., 2017). Similarly, reduced functional connectivity has been reported between bilateral Crus I and medial regions of the mentalizing network in autism (Van Overwalle et al., 2014), further implicating lobule VII in autism-relevant behaviors, and several studies report reduced connectivity between lobule VII and key social brain hubs, including the temporo-parietal junction, medial prefrontal cortex, and superior temporal sulcus (e.g., D'Mello et al., 2017; Igelström et al., 2017; Kelly, Escamilla, & Tsai, 2020; Olivito et al., 2017, 2018). Consistent with the cerebellar regions showing gray matter reduction in autism, an analysis of whole-brain functional connectivity comparing adults with high-functioning autism and neurotypical controls revealed cerebellar lobules VII and IX as the sites with the greatest difference between groups (Arnold Anteraper et al., 2019). Follow-up analyses in the same cohort revealed that these clusters displayed reduced connectivity with social, emotional, and language brain regions in the autism group. The atypical cross-network connectivity reported in cortical regions in autism is also seen in cerebro-cerebellar circuits, where increased functional connectivity is seen between sensorimotor cortical regions and cerebellar areas that are typically connected with cognitive or default mode cortical networks (e.g., Khan et al., 2015; Verly et al., 2014). These results are consistent with structural imaging findings and provide further evidence that the posterolateral cerebellum – specifically lobule VII – is a consistent neural site of atypical structure and function in autism.

The cerebellum also modulates reward circuitry involving the striatum which is critical to reward learning, an important aspect of developing social behavior (Carta et al., 2019; Kohls et al., 2018; Panasiti et al., 2016). Cerebellar and striatal pathologies are commonly identified in mouse models of autism (Peter et al., 2017), providing further support for reward circuit alterations in autism, in addition to the more common association of these circuits with repetitive behaviors in autism (e.g., Wilkes & Lewis, 2018). Furthermore, the cerebellum is implicated in impulsivity and compulsivity (Miquel et al., 2019) as well as cognitive control (D'Mello et al., 2020), common deficits in autism that may underlie restricted, repetitive, and inflexible behaviors.

In addition to processing of social and reward information, the more traditional role of the cerebellum in sensory perception and sensorimotor processing is also relevant to autism-associated behaviors (for consensus paper, see Baumann et al., 2015). It has been hypothesized that the cerebellar internal forward model is involved in sensory processing particularly for novelty detection (Anderson et al., 2012) and adaptation provided with sensory feedback error (Johnson et al., 2019), and that these processes are atypical in autism. Furthermore, the cerebellum forms part of a network that learns and maintains action-outcome associations through sensory prediction error (Butcher et al., 2017). Cerebellar roles for sensorimotor adaptation have been explored in mouse models using adaptive acceleration of visually evoked smooth eye movements (Kodama & du Lac, 2016), and eye movements, sensorimotor adaptation, and cerebellar-dependent learning may provide potential biomarkers and subphenotypes of autism (Freedman & Foxe, 2018). More specifically, classical eyeblink conditioning, a paradigm which elicits cerebellar-dependent plasticity and learning, may serve as a biomarker for neurodevelopmental disorders including autism (Reeb-Sutherland & Fox, 2015).

20.2 Specific Cerebellar Contribution to Autism

While the cerebellum is a prominent neurobiological correlate of autism, its specific contribution to the etiology and characteristic behaviors of autism remains in question. One hypothesis is that cerebellar dysfunction yields deficits in predictive processing. The ability to predict outcomes of actions is a crucial function of the central nervous system. Such a "prediction network" is posited to comprise cortical-subcortical circuits that support mechanisms of neural prediction (Siman-Tov et al., 2019). The cerebellum has long been associated with predictive processing in the sensorimotor domain (e.g., Miall et al., 1993), and this role has recently been extended to a range of non-motor behaviors (e.g., Sokolov et al., 2017). Accurate prediction is a critical feature of successful social interactions (Tamir & Thornton,

2018; Thornton et al., 2019), and adaptation of internal models to continually optimize behavior could be impaired in autism (Lieder et al., 2019; Kelly, Meng, et al., 2020; Stoodley & Tsai, in press). It has been hypothesized that autism is a disorder of prediction (Pellicano & Burr, 2012; Sinha et al., 2014), and given that the regions of the cerebellum implicated in autism process information relevant to social cognition (Van Overwalle et al., 2014), language (Mariën et al., 2014), emotion (Schmahmann et al., 2007; Strata, 2015), and movement, a failure of cerebellar internal models impacting specific cerebro-cerebellar circuits could explain the broad range of deficits in autism.

Given the cerebellum's uniform cytoarchitecture, it has been proposed that the cerebellum performs a universal computation within functional subregions, also known as the "universal cerebellar transform" (see Schmahmann, 2004; Schmahmann et al., 2019; though this theory has recently been challenged, see Diedrichsen et al., 2019). Various hypotheses have emerged regarding the mechanisms of this universal computation, with a common theme being that the cerebellum builds and hones internal models of movement and "mental models" of thoughts (Ito, 2008), which are trained through feedback and, once optimized, enable outcome prediction (see Ito, 2008; Miall et al., 1993; Raymond & Medina, 2018; Sokolov et al., 2017; Wolpert et al., 1998). This broad predictive modeling mechanism (also known as the internal forward model) can be applied to a range of tasks, from movement kinematics to semantic prediction and social cognition (D'Mello et al., 2020; Lesage et al., 2017; Schmahmann et al., 2019; Sokolov, 2018; Van Overwalle et al., 2014; Van Overwalle et al., 2020). According to this theory, the cerebellum is able to automatize basic and higher-level functions by rapidly detecting disruptions in action and thought sequences and utilizing prediction error to optimize performance (Heleven et al., 2019; Popa et al., 2016).

One way to test the relevance of cerebellar function to predictive processing is through cerebellar neuromodulation. Several studies have shown that the cerebellum is engaged during predictive language processing (D'Mello et al., 2017; Lesage et al., 2017; Moberget et al., 2014) and that modulation of the right posterolateral cerebellum impacts predictive language performance (Pope & Miall, 2012), as well as neural activation patterns and functional connectivity within the language network (D'Mello et al., 2017). In the social domain, several imaging studies have linked cerebellar activation to the sequencing of social interactions (e.g., Heleven et al., 2019; Pu et al., 2020). We have found that cerebellar neuromodulation impacts social learning performance and neural activation patterns in both neurotypical and autistic young adults (Rice et al., 2020). Notably, during a social ball-playing task, neurotypical adults engage cerebellar lobule VII, while adults with autism do not; following neuromodulation targeting right lobule VII, performance significantly improves in adults with autism, and activation increases in right lobule VII (Rice et al., 2020). These findings indicate that individuals with autism do not engage the cerebellum in a typical way during social learning, and when cerebellar engagement increases, social learning performance improves, providing a link between cerebellar predictive processing and social learning in autism.

Cerebellar learning mechanisms are thought to operate at an implicit level (Ito, 2008), and if these are shown to be disrupted, behavioral interventions should aim to explicitly teach information (e.g., social interaction patterns) that typically developing children acquire and adapt implicitly. Indeed, this is already the approach taken by many behavioral interventions for autism, in which explicit instruction is used to overcome atypical learning of the dynamics of social interactions and behaviors. It is possible that developmental disruption of cerebellar implicit learning mechanisms acting on specific cerebro-cerebellar circuits could underpin core social deficits in autism.

An important feature of cerebellar internal models is that they can be adapted as needed based on new information. A recent study suggested that individuals with autism show impaired updating based on new information ("slow updating"; Lieder et al., 2019), which is consistent with disrupted adaptation of internal models of information. It is well-established that the cerebellum is involved in adaptation to sensory feedback (e.g., Johnson et al., 2019; Popa & Ebner, 2018), and it is possible that the "slow updating" seen in autism could be related to inefficient adaptation of cerebellar forward models. This adaptive, context-dependent optimization of behavior is consistent with how Dajani and Uddin (2015) recently defined cognitive flexibility ("the appropriate adjustment of behavior in a changing environment"), suggesting that the inability to adapt internal models could be associated with the inflexible behaviors seen in autism (for reviews, see Kelly, Meng, et al., 2020; Stoodley & Tsai, in press). Indeed, a recent study showed that repetitive and inflexible behaviors could be induced through inhibition of the posterior vermis in mice; critically, activating this region in a mouse model of autism rescued these behaviors (Kelly, Escamilla, & Tsai, 2020).

20.3 Cerebellar Modulation of Network Connectivity

While the internal model hypothesis attempts to establish the specific cerebellar contribution to motor and cognitive control, there is also the question of how the cerebellum exerts its influence on such a wide range of functions. Recent evidence suggests that the cerebellum coordinates the timing of neural oscillations in a site- and frequency-specific manner, whereby cerebellar lobules VI and Crus I differentially represent instantaneous phases and phase differences of local field potentials (oscillations) in the medial prefrontal cortex and dorsal hippocampus CA1 region, respectively (McAfee et al., 2019). Vermal stimulation has been shown to modulate prefrontal cortical activity in a frequency-specific manner, indicating that the cerebellum impacts both local field potential activity in the prefrontal cortex and the synchronization of cortico-cortical networks (Tremblay et al., 2019). The cerebellum may impose coherence between cortical areas (e.g., between S1 and M1 via the ventrolateral thalamus and M1), and different frequency ranges can be dynamically modulated depending on the cerebellar stimulation site and behavioral context (Lindeman et al., 2020). These findings suggest that cerebral cortical oscillations can be modulated by different cerebellar subregions, consistent with the differential functional connectivity patterns evident throughout the cerebellum with core cortical networks (Buckner et al., 2011).

Human neuromodulation-neuroimaging studies offer the opportunity to investigate the impact of primary cerebellar modulation on neural activation and functional connectivity patterns. Altering cerebellar activity with transcranial magnetic stimulation (TMS) or transcranial direct current stimulation (tDCS) has revealed focal and polarity-specific effects on broader cortical regions to which the cerebellum interconnects. For example, Galea and colleagues showed that tDCS targeting a sensorimotor cerebellar region modulated cerebellar brain inhibition (CBI) of contralateral motor cortex, such that cathodal (inhibitory) tDCS decreased CBI and anodal (excitatory) tDCS increased it (Galea et al., 2009). In contrast, modulation targeting cognitive regions of the cerebellum impacts cerebral cortical networks relevant to cognitive functions. For example, functional connectivity increased in the cortical default mode network following intermittent theta-burst TMS targeting right lobule VII, which is functionally connected to this network (Halko et al., 2014). Anodal tDCS targeting right lobule VII modulated activation in right lobule VII during semantic prediction and increased resting-state functional connectivity specifically within reading and language networks (D'Mello et al., 2017). When targeting right lobule VII with tDCS, cerebro-cerebellar functional connectivity changed from the right lobule VII seed, but there were no changes in functional connectivity from adjacent cerebellar lobules (Stoodley et al., 2017; Turkeltaub et al., 2016), suggesting a degree of specificity in the cerebellar modulation of cortical circuits. Consistent with the proposal that the cerebellum modulates cortical coherence, TMS targeting the posterior midline of the cerebellum impacted frontal theta EEG (Schutter & van Honk, 2006). In patients with schizophrenia, TMS targeting the posterior midline of the cerebellum normalized the atypical functional connectivity in a cerebellar-prefrontal network that correlated with negative symptom severity, improving negative symptoms (Brady et al., 2019). These findings suggest that the cerebellum modulates cerebro-cerebellar and cortico-cortical networks in a regionally specific manner.

In animal models, primary regional cerebellar modulation with chemogenetic approaches yields autism-like behaviors, including atypical social approach and novelty as well as restricted and repetitive/inflexible behaviors (Stoodley et al., 2017). In a *Tsc1* autism mouse model, stimulating right Crus I rescued social impairments, but did not alter inflexible behaviors. The medial prefrontal cortex mediates cerebellum-regulated social and inflexible behaviors, and atypical functional connectivity between the medial prefrontal cortex and cerebellum is present in humans with autism as well as multiple autism mouse models (Kelly, Escamilla, & Tsai, 2020; Stoodley et al., 2017). In wild-type mice, inflexible behaviors and social deficits were generated by modulating a circuit involving right Crus I, posterior vermis, deep cerebellar nuclei, ventromedial thalamus, and medial prefrontal cortex. These behaviors were remediated when modulating this circuit in the *Tsc1* autism mouse model (Kelly, Escamilla, & Tsai, 2020).

Finally, it has been proposed that the cerebellum is an important modulator of cortical circuits during development (e.g., Stoodley, 2016; Wang et al., 2014). When the cerebellum is disrupted, particularly during early stages of development, there could be long-term alterations of neural circuits, with significant impacts on behavior (Sathyanesan et al., 2019). Consistent with this idea, perinatal cerebellar lesions have been associated with impaired growth of the contralateral cerebral cortex (e.g., Limperopoulos, 2010). The same phenomena could be true for autism, such that atypical development in specific cerebro-cerebellar circuits leads to the neural and behavioral differences associated with autism (Belmonte et al., 2004). Recent evidence from developmental cerebellar disruption in mice confirms that early disruption of cerebellar right lobule VII leads to later social deficits (Badura et al., 2018; Tsai et al., 2018), providing a potential mechanism by which cerebellar dysfunction contributes to the etiology of autism.

20.4 Cerebellum as a Therapeutic Target for Autism

The extensive connectivity of the cerebellum and its potential to modulate cerebral cortical networks suggest that the cerebellum could be a therapeutic target in autism and other clinical conditions (Grimaldi et al., 2016), particularly schizophrenia (Brady et al., 2019; Escelsior et al., 2019), bipolar disorder (Minichino et al., 2015), and aphasia (Sebastian et al., 2016; Turkeltaub et al., 2016). Stimulation of Purkinje cells in right Crus I rescued social deficits in the *Tsc1* autism mouse model (Stoodley et al., 2017), and stimulation targeting the posterior vermis rescued inflexible behaviors in the same autism model (Kelly et al., 2020). Similarly, in humans, we have shown that tDCS targeting right lobule VII modulates performance and neural activation patterns on a social learning task in neurotypical adults and that cathodal tDCS improves task performance in adults with autism (Rice et al., 2020). Preliminary resting-state results from the same cohort show that reduced functional connectivity between right lobule VII and the medial prefrontal cortex persists into adulthood in autism, and consistent with the improved task performance after cathodal tDCS, resting-state functional connectivity increases from right lobule VII following cathodal tDCS in adults with autism. These findings indicate that structural and functional differences in right lobule VII contribute to social deficits in autism and that cerebellar tDCS targeting this region could increase connectivity in relatively under-connected networks in autism.

Right lobule VII is not the only potential cerebellar target region for altering behaviors in autism. Although right lobule VII stimulation in the *Tsc1* mouse model rescued social behaviors, repetitive grooming and inflexible behaviors did not improve (Stoodley et al., 2017). In humans, repetitive behavior and restricted interest scores correlated with gray matter volume in posterior vermal lobules of the cerebellum (D'Mello et al., 2015) and there is evidence that the posterior vermis is involved in the modulation of attention networks (e.g., Esterman et al., 2017), which could support the flexible allocation of attention during tasks. Stimulation of the posterior vermis rescued inflexible behaviors in *Tsc1* mice on the Y maze task (Kelly et al., 2020), indicating the posterior midline could be targeted to improve behavioral inflexibility. Posterior midline TMS has successfully improved negative symptoms in schizophrenia (Brady et al., 2019), and symptom improvement was associated with increased functional connectivity between the posterior vermis and the prefrontal cortex in this cohort.

It is important to note that cerebellar stimulation was able to rescue social deficits in mice even in adulthood (Kelly, Escamilla, & Tsai, 2020; Stoodley et al., 2017), and currently, there are very few therapeutic options for adults with autism. Pharmacological treatments target psychiatric symptoms, but not the core features of autism. Neuromodulation of the cerebellum could alter the atypical cerebro-cerebellar circuits which underlie these core features. In this way, the role of the cerebellum as a CNS hub could be harnessed to provide novel therapeutic options for individuals with autism.

20.5 Conclusion

Extensive evidence indicates that the cerebellum is a core neuroanatomical correlate of autism. The cerebellar role in building and adapting internal models of information could underpin the deficits in prediction and adaptation in autism, and developmental disruption of the cerebellar modulation of supratentorial networks supporting social and adaptive behaviors could lead to the deficits in social communication and presence of restricted and repetitive behaviors that are the core features of autism.

References

Anderson, S. R., Porrill, J., Pearson, M. J., Pipe, A. G., Prescott, T. J., & Dean, P. (2012). An internal model architecture for novelty detection: Implications for cerebellar and collicular roles in sensory processing. *PLoS One, 7*(9), e44560.

Arnold Anteraper, S., Guell, X., D'Mello, A., Joshi, N., Whitfield-Gabrieli, S., & Joshi, G. (2019 Feb 1). Disrupted cerebrocerebellar intrinsic functional connectivity in young adults with high-functioning autism Spectrum disorder: A data-driven, whole-brain, high-temporal resolution functional magnetic resonance imaging study. *Brain Connectivity., 9*(1), 48–59.

American Psychiatric Association. (2013). *Diagnostic and Statistical Manual of Mental Disorders* (5th ed.).

Badura, A., Verpeut, J. L., Metzger, J. W., Pereira, T. D., Pisano, T. J., Deverett, B., et al. (2018). Normal cognitive and social development require posterior cerebellar activity. *eLife, 7*, e36401.

Bauman, M. L., & Kemper, T. L. (2005). Neuroanatomic observations of the brain in autism: A review and future directions. *International Journal of Developmental Neuroscience, 23*(2–3), 183–187.

Baumann, O., Borra, R. J., Bower, J. M., Cullen, K. E., Habas, C., Ivry, R. B., et al. (2015). Consensus paper: The role of the cerebellum in perceptual processes. *Cerebellum, 14*(2), 197–220.

Becker, E. B. E., & Stoodley, C. J. (2013). Autism Spectrum disorder and the cerebellum. In *International review of neurobiology* (pp. 1–34). Elsevier. Available from: https://linkinghub.elsevier.com/retrieve/pii/B9780124187009000010

Belmonte, M. K., Allen, G., Beckel-Mitchener, A., Boulanger, L. M., Carper, R. A., & Webb, S. J. (2004). Autism and abnormal development of brain connectivity. *The Journal of Neuroscience, 24*(42), 9228–9231.

Brady, R. O., Gonsalvez, I., Lee, I., Öngür, D., Seidman, L. J., Schmahmann, J. D., et al. (2019). Cerebellar-prefrontal network connectivity and negative symptoms in schizophrenia. *Am J Psychiatry., 176*(7), 512–520.

Brambilla, P. (2003). Brain anatomy and development in autism: Review of structural MRI studies. *Brain Research Bulletin, 61*(6), 557–569.

Buckner, R. L., Krienen, F. M., Castellanos, A., Diaz, J. C., & Yeo, B. T. T. (2011). The organization of the human cerebellum estimated by intrinsic functional connectivity. *Journal of Neurophysiology, 106*(5), 2322–2345.

Butcher, P. A., Ivry, R. B., Kuo, S.-H., Rydz, D., Krakauer, J. W., & Taylor, J. A. (2017). The cerebellum does more than sensory prediction error-based learning in sensorimotor adaptation tasks. *J Neurophysiol., 118*(3), 1622–1636.

Carper, R. A., Treiber, J. M., DeJesus, S. Y., & Müller, R.-A. (2016). Reduced hemispheric asymmetry of white matter microstructure in autism Spectrum disorder. *Journal of the American Academy of Child and Adolescent Psychiatry, 55*(12), 1073–1080.

Carta, I., Chen, C. H., Schott, A. L., Dorizan, S., & Khodakhah, K. (2019). Cerebellar modulation of the reward circuitry and social behavior. *Science, 363*(6424).

Centers for Disease Control and Prevention. (2020). Data and statistics on autism spectrum disorder. Available from: https://www.cdc.gov/ncbddd/autism/data.html.

Cheng, W., Rolls, E. T., Gu, H., Zhang, J., & Feng, J. (2015). Autism: Reduced connectivity between cortical areas involved in face expression, theory of mind, and the sense of self. *Brain, 138*(5), 1382–1393.

Choi, U.-S., Kim, S.-Y., Sim, H. J., Lee, S.-Y., Park, S.-Y., Jeong, J.-S., et al. (2015). Abnormal brain activity in social reward learning in children with autism spectrum disorder: An fMRI study. *Yonsei Medical Journal, 56*(3), 705–711.

Courchesne, E. (1991). Neuroanatomic imaging in autism. *Pediatrics, 87*(5 Pt 2), 781–790.

Courchesne, E. (1997). Brainstem, cerebellar and limbic neuroanatomical abnormalities in autism. *Current Opinion in Neurobiology, 7*(2), 269–278.

Courchesne, E., Karns, C. M., Davis, H. R., Ziccardi, R., Carper, R. A., Tigue, Z. D., et al. (2001). Unusual brain growth patterns in early life in patients with autistic disorder: An MRI study. *Neurology, 57*(2), 245–254.

Courchesne, E., Yeung-Courchesne, R., Press, G. A., Hesselink, J. R., & Jernigan, T. L. (1988). Hypoplasia of cerebellar vermal lobules VI and VII in autism. *The New England Journal of Medicine, 318*(21), 1349–1354.

D'Angelo, E. (2018). Physiology of the cerebellum. *Handbook of Clinical Neurology, 154*, 85–108.

D'Mello, A. M., Crocetti, D., Mostofsky, S. H., & Stoodley, C. J. (2015). Cerebellar gray matter and lobular volumes correlate with core autism symptoms. *Neuroimage Clin., 7*, 631–639.

D'Mello, A. M., Gabrieli, J. D. E., & Nee, D. E. (2020). Evidence for hierarchical cognitive control in the human cerebellum. *Curr Biol., 30*(10), 1881–1892.e3.

D'Mello, A. M., & Stoodley, C. J. (2015). Cerebro-cerebellar circuits in autism spectrum disorder. *Frontiers in Neuroscience, 9*, 408.

D'Mello, A. M., Turkeltaub, P. E., & Stoodley, C. J. (2017). Cerebellar tDCS modulates neural circuits during semantic prediction: A combined tDCS-fMRI study. *The Journal of Neuroscience, 37*(6), 1604–1613.

Dajani, D. R., & Uddin, L. Q. (2015). Demystifying cognitive flexibility: Implications for clinical and developmental neuroscience. *Trends in Neurosciences, 38*(9), 571–578.

Diedrichsen, J., King, M., Hernandez-Castillo, C., Sereno, M., & Ivry, R. B. (2019). Universal transform or multiple functionality? Understanding the contribution of the human cerebellum across task domains. *Neuron., 102*(5), 918–928.

Donovan, A. P. A., & Basson, M. A. (2017). The neuroanatomy of autism - a developmental perspective. *Journal of Anatomy, 230*(1), 4–15.

Ecker, C. (2017). The neuroanatomy of autism spectrum disorder: An overview of structural neuroimaging findings and their translatability to the clinical setting. *Autism, 21*(1), 18–28.

Ellegood, J., Anagnostou, E., Babineau, B. A., Crawley, J. N., Lin, L., Genestine, M., et al. (2015). Clustering autism: Using neuroanatomical differences in 26 mouse models to gain insight into the heterogeneity. *Molecular Psychiatry, 20*(1), 118–125.

Eluvathingal, T. J., Behen, M. E., Chugani, H. T., Janisse, J., Bernardi, B., Chakraborty, P., et al. (2006). Cerebellar lesions in tuberous sclerosis complex: Neurobehavioral and neuroimaging correlates. *Journal of Child Neurology, 21*(10), 846–851.

Escalante-Mead P. R., Minshew N. J., & Sweeney J. A. (2003). Abnormal brain lateralization in high-functioning autism. *Journal of Autism and Developmental Disorders, 33*(5), 539–543.

Escelsior, A., Belvederi Murri, M., Calcagno, P., Cervetti, A., Caruso, R., Croce, E., et al. (2019). Effectiveness of cerebellar circuitry modulation in schizophrenia: A systematic review. *The Journal of Nervous and Mental Disease, 207*(11), 977–986.

Esterman, M., Thai, M., Okabe, H., DeGutis, J., Saad, E., Laganiere, S. E., et al. (2017). Network-targeted cerebellar transcranial magnetic stimulation improves attentional control. *Neuroimage., 156*, 190–198.

Fishman, I., Keown, C. L., Lincoln, A. J., Pineda, J. A., & Müller, R.-A. (2014). Atypical cross talk between mentalizing and mirror neuron networks in autism Spectrum disorder. *JAMA Psychiatry., 71*(7), 751.

Freedman, E. G., & Foxe, J. J. (2018). Eye movements, sensorimotor adaptation and cerebellar-dependent learning in autism: Toward potential biomarkers and subphenotypes. *The European Journal of Neuroscience, 47*(6), 549–555.

Galea, J. M., Jayaram, G., Ajagbe, L., & Celnik, P. (2009). Modulation of cerebellar excitability by polarity-specific noninvasive direct current stimulation. *The Journal of Neuroscience, 29*(28), 9115–9122.

Grimaldi, G., Argyropoulos, G. P., Bastian, A., Cortes, M., Davis, N. J., Edwards, D. J., et al. (2016). Cerebellar transcranial direct current stimulation (ctDCS): A novel approach to understanding cerebellar function in health and disease. *The Neuroscientist, 22*(1), 83–97.

Groen, W. B., Buitelaar, J. K., van der Gaag, R. J., & Zwiers, M. P. (2011). Pervasive microstructural abnormalities in autism: A DTI study. *Journal of Psychiatry & Neuroscience, 36*(1), 32–40.

Halko, M. A., Farzan, F., Eldaief, M. C., Schmahmann, J. D., & Pascual-Leone, A. (2014). Intermittent theta-burst stimulation of the lateral cerebellum increases functional connectivity of the default network. *The Journal of Neuroscience, 34*(36), 12049–12056.

Heleven, E., van Dun, K., & Van Overwalle, F. (2019). The posterior cerebellum is involved in constructing social action sequences: An fMRI study. *Scientific Reports, 9*(1), 11110.

Hoche, F., Guell, X., Sherman, J. C., Vangel, M. G., & Schmahmann, J. D. (2016). Cerebellar contribution to social cognition. *Cerebellum, 15*(6), 732–743.

Hong, S.-J., Vos De Wael, R., Rai, B., Lariviere, S., Paquola, C., Valk, S. L., et al. (2019). Atypical functional connectome hierarchy in autism. *Nat Commun, 10*(1), 1022.

Hulbert, S. W., & Jiang, Y.-H. (2017). Cellular and circuitry bases of autism: Lessons learned from the temporospatial manipulation of autism genes in the brain. *Neuroscience Bulletin, 33*(2), 205–218.

Igelström, K. M., Webb, T. W., & Graziano, M. S. A. (2017). Functional connectivity between the temporoparietal cortex and cerebellum in autism spectrum disorder. *Cereb Cortex., 27*(4), 2617–2627.

Ito, M. (2008). Control of mental activities by internal models in the cerebellum. *Nature Reviews. Neuroscience, 9*(4), 304–313.

Jack, A., & Pelphrey, K. A. (2015). Neural correlates of animacy attribution include neocerebellum in healthy adults. *Cerebral Cortex, 25*(11), 4240–4247.

Johnson, J. F., Belyk, M., Schwartze, M., Pinheiro, A. P., & Kotz, S. A. (2019). The role of the cerebellum in adaptation: ALE meta-analyses on sensory feedback error. *Human Brain Mapping, 40*(13), 3966–3981.

Jou, R. J., Jackowski, A. P., Papademetris, X., Rajeevan, N., Staib, L. H., & Volkmar, F. R. (2011). Diffusion tensor imaging in autism spectrum disorders: Preliminary evidence of abnormal neural connectivity. *The Australian and New Zealand Journal of Psychiatry, 45*(2), 153–162.

Kana, R. K., Maximo, J. O., Williams, D. L., Keller, T. A., Schipul, S. E., Cherkassky, V. L., et al. (2015). Aberrant functioning of the theory-of-mind network in children and adolescents with autism. *Molecular Autism, 6*(1), 59.

Kelly, E., Escamilla, C. O., & Tsai, P. T. (2020). Cerebellar dysfunction in autism spectrum disorders: Deriving mechanistic insights from an internal model framework. *Neuroscience, 28*.

Kelly, E., Meng, F., Fujita, H., Morgado, F., Kazemi, Y., Rice, L. C., et al. (2020). Regulation of autism-relevant behaviors by cerebellar–prefrontal cortical circuits. *Nature Neuroscience, 23*(9), 1102–1110.

Kelly, R. M., & Strick, P. L. (2003). Cerebellar loops with motor cortex and prefrontal cortex of a nonhuman primate. *The Journal of Neuroscience, 23*(23), 8432–8444.

Kennedy, D. P., & Adolphs, R. (2012). The social brain in psychiatric and neurological disorders. *Trends in Cognitive Sciences, 16*(11), 559–572.

Kestemont, J., Vandekerckhove, M., Bulnes, L. C., Matthys, F., & Van Overwalle, F. (2016). Causal attribution in individuals with subclinical and clinical autism spectrum disorder: An fMRI study. *Social Neuroscience, 11*(3), 264–276.

Khan, A. J., Nair, A., Keown, C. L., Datko, M. C., Lincoln, A. J., & Müller, R.-A. (2015). Cerebro-cerebellar resting-state functional connectivity in children and adolescents with autism spectrum disorder. *Biological Psychiatry, 78*(9), 625–634.

King, M., Hernandez-Castillo, C. R., Poldrack, R. A., Ivry, R. B., & Diedrichsen, J. (2019). Functional boundaries in the human cerebellum revealed by a multi-domain task battery. *Nature Neuroscience, 22*(8), 1371–1378.

Kodama, T., & du Lac, S. (2016). Adaptive acceleration of visually evoked smooth eye movements in mice. *J Neurosci., 36*(25), 6836–6849.

Kohls, G., Antezana, L., Mosner, M. G., Schultz, R. T., & Yerys, B. E. (2018). Altered reward system reactivity for personalized circumscribed interests in autism. *Molecular Autism, 9*(1), 9.

Lee, J. M., Kyeong, S., Kim, E., & Cheon, K.-A. (2016). Abnormalities of inter- and intra-hemispheric functional connectivity in autism spectrum disorders: A study using the autism brain imaging data exchange database. *Frontiers in Neuroscience, 10*, 191.

Lesage, E., Hansen, P. C., & Miall, R. C. (2017). Right lateral cerebellum represents linguistic predictability. *The Journal of Neuroscience, 37*(26), 6231–6241.

Lieder, I., Adam, V., Frenkel, O., Jaffe-Dax, S., Sahani, M., & Ahissar, M. (2019). Perceptual bias reveals slow-updating in autism and fast-forgetting in dyslexia. *Nature Neuroscience, 22*(2), 256–264.

Limperopoulos, C. (2010). Extreme prematurity, cerebellar injury, and autism. *Seminars in Pediatric Neurology, 17*(1), 25–29.

Limperopoulos, C., Bassan, H., Gauvreau, K., Robertson, R. L., Sullivan, N. R., Benson, C. B., et al. (2007). Does cerebellar injury in premature infants contribute to the high prevalence of long-term cognitive, learning, and behavioral disability in survivors? *Pediatrics, 120*(3), 584–593.

Lindeman, S., Kros, L., Hong, S., Mejias, J. F., Romano, V., Negrello, M., et al. (2020). Cerebellar Purkinje cells can differentially modulate coherence between sensory and motor cortex depending on region and behavior. In *Neuroscience*. Available from: http://biorxiv.org/lookup/doi/10.1101/2020.03.11.986943

Mariën, P., Ackermann, H., Adamaszek, M., Barwood, C. H. S., Beaton, A., Desmond, J., et al. (2014). Consensus paper: Language and the cerebellum: An ongoing enigma. *Cerebellum, 13*(3), 386–410.

McAfee, S. S., Liu, Y., Sillitoe, R. V., & Heck, D. H. (2019). Cerebellar lobulus simplex and crus I differentially represent phase and phase difference of prefrontal cortical and hippocampal oscillations. *Cell Reports, 27*(8), 2328–2334.

Miall, R. C., Weir, D. J., Wolpert, D. M., & Stein, J. F. (1993). Is the cerebellum a smith predictor? *Journal of Motor Behavior, 25*(3), 203–216.

Minichino, A., Bersani, F. S., Bernabei, L., Spagnoli, F., Vergnani, L., Corrado, A., et al. (2015). Prefronto-cerebellar transcranial direct current stimulation improves visuospatial memory, executive functions, and neurological soft signs in patients with euthymic bipolar disorder. *Neuropsychiatric Disease and Treatment, 11*, 2265–2270.

Miquel, M., Nicola, S. M., Gil-Miravet, I., Guarque-Chabrera, J., & Sanchez-Hernandez, A. (2019). A working hypothesis for the role of the cerebellum in impulsivity and compulsivity. *Frontiers in Behavioral Neuroscience, 13*, 99.

Moberget, T., Gullesen, E. H., Andersson, S., Ivry, R. B., & Endestad, T. (2014). Generalized role for the cerebellum in encoding internal models: Evidence from semantic processing. *The Journal of Neuroscience, 34*(8), 2871–2878.

Olivito, G., Clausi, S., Laghi, F., Tedesco, A. M., Baiocco, R., Mastropasqua, C., et al. (2017). Resting-state functional connectivity changes between dentate nucleus and cortical social brain regions in autism Spectrum disorders. *Cerebellum, 16*(2), 283–292.

Olivito, G., Lupo, M., Laghi, F., Clausi, S., Baiocco, R., Cercignani, M., et al. (2018). Lobular patterns of cerebellar resting-state connectivity in adults with autism Spectrum disorder. *The European Journal of Neuroscience, 47*(6), 729–735.

Panasiti, M. S., Puzzo, I., & Chakrabarti, B. (2016). Autistic traits moderate the impact of reward learning on social behaviour. *Autism Research, 9*(4), 471–479.

Pellicano, E., & Burr, D. (2012). When the world becomes "too real": A Bayesian explanation of autistic perception. *Trends in Cognitive Sciences, 16*(10), 504–510.

Peter, S., De Zeeuw, C. I., Boeckers, T. M., & Schmeisser, M. J. (2017). Cerebellar and striatal pathologies in mouse models of autism Spectrum disorder. *Advances in Anatomy, Embryology, and Cell Biology, 224*, 103–119.

Pierce, K., & Courchesne, E. (2001). Evidence for a cerebellar role in reduced exploration and stereotyped behavior in autism. *Biological Psychiatry, 49*(8), 655–664.

Pisano, T. J., Dhanerawala, Z. M., Kislin, M., Bakshinskaya, D., Engel, E. A., Lee, J., et al. (2020). Parallel organization of cerebellar pathways to sensorimotor, associative, and modulatory forebrain. *Neuroscience*. Available from: http://biorxiv.org/lookup/doi/10.1101/2020.03.06.979153

Popa, L. S., & Ebner, T. J. (2018). Cerebellum, predictions and errors. *Frontiers in Cellular Neuroscience, 12*, 524.

Popa, L. S., Streng, M. L., Hewitt, A. L., & Ebner, T. J. (2016). The errors of our ways: Understanding error representations in cerebellar-dependent motor learning. *Cerebellum, 15*(2), 93–103.

Pope, P. A., & Miall, R. C. (2012). Task-specific facilitation of cognition by cathodal transcranial direct current stimulation of the cerebellum. *Brain Stimulation, 5*(2), 84–94.

Postema, M. C., van Rooij, D., Anagnostou, E., Arango, C., Auzias, G., Behrmann, M., et al. (2019). Altered structural brain asymmetry in autism spectrum disorder: Large-scale analysis via the ENIGMA consortium. *Neuroscience*. Available from: http://biorxiv.org/lookup/doi/10.1101/570655

Pu, M., Heleven, E., Delplanque, J., Gibert, N., Ma, Q., Funghi, G., et al. (2020). The posterior cerebellum supports the explicit sequence learning linked to trait attribution. *Cognitive, Affective, & Behavioral Neuroscience, 20*(4), 798–815.

Raymond, J. L., & Medina, J. F. (2018). Computational principles of supervised learning in the cerebellum. *Annual Review of Neuroscience, 41*(1), 233–253.

Reeb-Sutherland, B. C., & Fox, N. A. (2015). Eyeblink conditioning: A non-invasive biomarker for neurodevelopmental disorders. *Journal of Autism and Developmental Disorders, 45*(2), 376–394.

Rice, L. C., D'Mello, A. M., Martin, S. E., & Stoodley, C. J. (2020). The cerebellum modulates the acquisition of social information in autism. *International Society for Autism Research.*

Sandin, S., Lichtenstein, P., Kuja-Halkola, R., Hultman, C., Larsson, H., & Reichenberg, A. (2017). The heritability of autism Spectrum disorder. *Journal of the American Medical Association, 318*(12), 1182.

Sathyanesan, A., Zhou, J., Scafidi, J., Heck, D. H., Sillitoe, R. V., & Gallo, V. (2019). Emerging connections between cerebellar development, behaviour and complex brain disorders. *Nature Reviews. Neuroscience, 20*(5), 298–313.

Schmahmann, J. D. (2004). Disorders of the cerebellum: Ataxia, dysmetria of thought, and the cerebellar cognitive affective syndrome. *The Journal of Neuropsychiatry and Clinical Neurosciences, 16*(3), 367–378.

Schmahmann, J. D., Guell, X., Stoodley, C. J., & Halko, M. A. (2019). The theory and neuroscience of cerebellar cognition. *Annual Review of Neuroscience, 42*, 337–364.

Schmahmann, J. D., Weilburg, J. B., & Sherman, J. C. (2007). The neuropsychiatry of the cerebellum - insights from the clinic. *Cerebellum, 6*(3), 254–267.

Schutter, D. J. L. G., & van Honk, J. (2006). An electrophysiological link between the cerebellum, cognition and emotion: Frontal theta EEG activity to single-pulse cerebellar TMS. *NeuroImage, 33*(4), 1227–1231.

Sebastian, R., Saxena, S., Tsapkini, K., Faria, A. V., Long, C., Wright, A., et al. (2016). Cerebellar tDCS: A novel approach to augment language treatment post-stroke. *Frontiers in Human Neuroscience, 10*, 695.

Seery, A. M., Vogel-Farley, V., Tager-Flusberg, H., & Nelson, C. A. (2013). Atypical lateralization of ERP response to native and non-native speech in infants at risk for autism spectrum disorder. *Developmental Cognitive Neuroscience, 5*, 10–24.

Shukla, D. K., Keehn, B., Lincoln, A. J., & Müller, R.-A. (2010). White matter compromise of callosal and subcortical fiber tracts in children with autism spectrum disorder: A diffusion tensor imaging study. *J Am Acad Child Adolesc Psychiatry., 49*(12), 1269–1278. 1278.e1–2.

Siman-Tov, T., Granot, R. Y., Shany, O., Singer, N., Hendler, T., & Gordon, C. R. (2019). Is there a prediction network? Meta-analytic evidence for a cortical-subcortical network likely subserving prediction. *Neuroscience and Biobehavioral Reviews, 105*, 262–275.

Sinha, P., Kjelgaard, M. M., Gandhi, T. K., Tsourides, K., Cardinaux, A. L., Pantazis, D., et al. (2014). Autism as a disorder of prediction. *Proc Natl Acad Sci USA., 111*(42), 15220–15225.

Sivaswamy, L., Kumar, A., Rajan, D., Behen, M., Muzik, O., Chugani, D., et al. (2010). A diffusion tensor imaging study of the cerebellar pathways in children with autism Spectrum disorder. *Journal of Child Neurology, 25*(10), 1223–1231.

Sokolov, A. A. (2018). The cerebellum in social cognition. *Frontiers in Cellular Neuroscience, 12*, 145.

Sokolov, A. A., Miall, R. C., & Ivry, R. B. (2017). The cerebellum: Adaptive prediction for movement and cognition. *Trends in Cognitive Sciences, 21*(5), 313–332.

Stanfield, A. C., Philip, R. C. M., Whalley, H., Romaniuk, L., Hall, J., Johnstone, E. C., et al. (2017). Dissociation of brain activation in autism and schizotypal personality disorder during social judgments. *Schizophrenia Bulletin, 43*(6), 1220–1228.

Stoodley, C. J. (2014). Distinct regions of the cerebellum show gray matter decreases in autism, ADHD, and developmental dyslexia. *Frontiers in Systems Neuroscience, 8*, 92.

Stoodley, C. J. (2016). The cerebellum and neurodevelopmental disorders. *Cerebellum, 15*(1), 34–37.

Stoodley, C. J., D'Mello, A. M., Ellegood, J., Jakkamsetti, V., Liu, P., Nebel, M. B., et al. (2017). Altered cerebellar connectivity in autism and cerebellar-mediated rescue of autism-related behaviors in mice. *Nature Neuroscience, 20*(12), 1744–1751.

Stoodley, C. J., & Limperopoulos, C. (2016). Structure–function relationships in the developing cerebellum: Evidence from early-life cerebellar injury and neurodevelopmental disorders. *Seminars in Fetal & Neonatal Medicine, 21*(5), 356–364.

Stoodley, C. J., & Schmahmann, J. D. (2009). Functional topography in the human cerebellum: A meta-analysis of neuroimaging studies. *NeuroImage, 44*(2), 489–501.
Stoodley, C. J., & Schmahmann, J. D. (2010). Evidence for topographic organization in the cerebellum of motor control versus cognitive and affective processing. *Cortex, 46*(7), 831–844.
Stoodley, C. J., & Tsai, P. T. (in press). Adapting predictions for social contexts: The cerebellar contribution to social behaviors. *Annu Rev Neurosci. In press.*
Strata, P. (2015). The emotional cerebellum. *Cerebellum, 14*(5), 570–577.
Tamir, D. I., & Thornton, M. A. (2018). Modeling the predictive social mind. *Trends in Cognitive Sciences, 22*(3), 201–212.
Thornton M. A., Weaverdyck M. E., & Tamir D. I. (2019). The social brain automatically predicts others' future mental states. *The Journal of Neuroscience, 39*(1), 140–148.
Travers, B. G., Adluru, N., Ennis, C., Tromp, D. P. M., Destiche, D., Doran, S., et al. (2012). Diffusion tensor imaging in autism spectrum disorder: A review. *Autism Research, 5*(5), 289–313.
Tremblay, S. A., Chapman, C. A., & Courtemanche, R. (2019). State-dependent entrainment of prefrontal cortex local field potential activity following patterned stimulation of the cerebellar vermis. *Frontiers in Systems Neuroscience, 13*, 60.
Tsai, P. T., Hull, C., Chu, Y., Greene-Colozzi, E., Sadowski, A. R., Leech, J. M., et al. (2012). Autistic-like behaviour and cerebellar dysfunction in Purkinje cell Tsc1 mutant mice. *Nature, 488*(7413), 647–651.
Tsai, P. T., Rudolph, S., Guo, C., Ellegood, J., Gibson, J. M., Schaeffer, S. M., et al. (2018). Sensitive periods for cerebellar-mediated autistic-like behaviors. *Cell Rep., 25*(2), 357–367.e4.
Turkeltaub, P. E., Swears, M. K., D'Mello, A. M., & Stoodley, C. J. (2016). Cerebellar tDCS as a novel treatment for aphasia? Evidence from behavioral and resting-state functional connectivity data in healthy adults. *Restorative Neurology and Neuroscience, 34*(4), 491–505.
Van Overwalle, F., Baetens, K., Mariën, P., & Vandekerckhove, M. (2014). Social cognition and the cerebellum: A meta-analysis of over 350 fMRI studies. *NeuroImage, 86*, 554–572.
Van Overwalle, F., Manto, M., Cattaneo, Z., Clausi, S., Ferrari, C., Gabrieli, J. D. E., et al. (2020). Consensus paper: Cerebellum and social cognition. *Cerebellum, 19*(6), 833–868.
Verly, M., Verhoeven, J., Zink, I., Mantini, D., Peeters, R., Deprez, S., et al. (2014). Altered functional connectivity of the language network in ASD: Role of classical language areas and cerebellum. *Neuroimage Clin., 4*, 374–382.
Wang, S. S.-H., Kloth, A. D., & Badura, A. (2014). The cerebellum, sensitive periods, and autism. *Neuron, 83*(3), 518–532.
Weber, A. M., Egelhoff, J. C., McKellop, J. M., & Franz, D. N. (2000). Autism and the cerebellum: Evidence from tuberous sclerosis. *Journal of Autism and Developmental Disorders, 30*(6), 511–517.
Wilkes, B. J., & Lewis, M. H. (2018). The neural circuitry of restricted repetitive behavior: Magnetic resonance imaging in neurodevelopmental disorders and animal models. *Neuroscience and Biobehavioral Reviews, 92*, 152–171.
Wolpert, D. M., Miall, R. C., & Kawato, M. (1998). Internal models in the cerebellum. *Trends in Cognitive Sciences, 2*(9), 338–347.

Part VI
Cerebellar Disorders and their Evaluation

Chapter 21
Cerebellar Reserve: From Theoretical Framework to Therapeutic Strategy

Hiroshi Mitoma and Mario Manto

21.1 Definition of Cerebellar Reserve

Various pathological injuries can inflict damage in the cerebellum, leading to the development of cerebellar ataxias (CAs) impairing motor, oculomotor, and cognitive controls (Manto et al., 2012; Schmahmann & Caplan, 2006). Cerebellar motor ataxias are characterized by impairment of predictive controls of timing and synergy. It is well known that such CAs can show partial and sometimes complete improvement with time (Manto, 2008). This unique feature of reversibility, which is hardly observed in other CNS-related symptoms, was described clearly more than a century ago in the classical paper by the British neurologist Gordon Holmes (Holmes, 1917) where he described the CA recovery process in two patients (pages 514–515, Holmes 1917). One patient had a limited lesion in the lateral lobe, and the CAs disappeared after 58 days. Another patient with a more severe CA, which was induced by a large lesion in the lateral and medial lobes, improved after 71 days. There was no difference in the gait between the two patients during the recovery phase, despite the marked difference in the extent of the two lesions. Here, we define cerebellar reserve as the capacity for compensation and restoration following cerebellar pathological changes (Cendelin et al., 2018; Cendelin et al., 2019; Cendelin & Mitoma, 2018; Mitoma et al., 2018, 2019, 2020; Mitoma & Manto, 2016). This property can result in clinical tolerance to pathology and reversibility after its removal.

H. Mitoma (✉)
Department of Medical Education, Tokyo Medical University, Tokyo, Japan
e-mail: mitoma@tokyo-med.ac.jp

M. Manto
Department of Neurology, CHU-Charleroi, Belgium and Service des Neurosciences, University of Mons, Mons, Belgium
e-mail: mmanto@ulb.ac.be

H. Mizusawa, S. Kakei (eds.), *Cerebellum as a CNS Hub*,
Contemporary Clinical Neuroscience, https://doi.org/10.1007/978-3-030-75817-2_21

The concept of the "reserve" was initially proposed to account for differences in susceptibility of the cerebral cortex to aging and pathological damage (Serra et al., 2018; Serra & Gelfo, 2019; Steffener & Stern, 1822; Stern, 2012, 2017). In this regard, reserve is defined as a moderator between pathology and outcome (Stern, 2012). Stern (2012) proposed three types of reserve: brain (i.e., anatomical) reserve, cognitive (i.e., functional) reserve, and neural (i.e., network-based) reserve (Steffener & Stern, 1822; Stern, 2012, 2017). For the tolerance to aging/pathology, brain reserve focuses on morphological and quantitative natures, such as remaining intact/undamaged neurons and synapses. On the other hand, cognitive reserve is linked to functional activities, e.g., utilization of preexisting cognitive storage. When the integrity of functional brain networks is stressed, the term neural reserve is also used. Thus, compared with this classical meaning, the notion of cerebellar reserve characteristically implies not only resilience to pathologies but also reversibility.

21.2 Two Forms of Cerebellar Reserve Depending on Etiology: Structural Reserve and Functional Reserve

Cerebellar response of compensation and restoration after any insult falls into two forms, depending on the nature of the etiological factor (Mitoma et al., 2020). When the etiology elicits immediate structural damage in a limited area (e.g., in cases of stroke and traumatic injury), the lost cerebellar functions can be compensated for by other areas not affected by the structural loss, and thus it is termed *structural cerebellar reserve*. On the other hand, when the etiology weakens cerebellar neurons (as well as glial cells) in diffuse areas, gradually leading to cell death, the affected lesion itself can replenish vanishing cerebellar functions. Examples of this type of etiologies are immune-mediated cerebellar ataxias (IMCAs), metabolic ataxias, and degenerative ataxias. Cerebellar compensation and restoration occur within the lesion through functional reorganization, and thus it is termed *functional cerebellar reserve*.

Structural Cerebellar Reserve The degree of reversibility after structural damage is assumed to depend on the extension of the lesion. In addition, the reversibility, well- or maladaptation, is influenced by the kind of the lost function. The concept of function-dependent reversibility is based on studies on cerebellar ocular disorders, especially studies using the focal excision models in macaques (Mitoma et al., 2020). Experimental focal lesions in the flocculus/paraflocculus induced impaired reduction in the gain of the ocular pursuit system, deficit in gaze holding and subsequent gaze-evoked nystagmus, as well as abnormal increase in the gain of the vestibulo-ocular reflex (VOR) (Zee et al., 1981). In the immediate postoperative period, the pursuit gain increased over several months' period, and gaze holding function improved several months after the lesion (Mitoma et al., 2020). In contrast, impairment of VOR gain persisted throughout the postoperative course (Mitoma et al., 2020). A similar function-dependent reversibility was observed also in ocular disorders that were induced by restricted lesions in the dorsal vermis. Lesions of the dorsal vermis reduced the gain in saccade and increased its latency and variability (Takagi et al., 1998). A limited recovery of saccade latency and variability was noted at 3–4 months after the lesion phase, but there remained enduring dysmetria (Mitoma et al., 2020). Experimentally induced focal lesions in the dorsal vermis also impaired the smooth pursuit. Excision of a larger area of neural substance from the dorsal vermis led to irreversible loss of the closed-loop pursuit gain, whereas the extent of reversibility in open-loop pursuit gain was determined by the size of the lesion (Mitoma et al., 2020). Based on these observations, Shaikh proposed a hypothesis that reversibility is determined by the proximity of the area to be substituted for the lost function (Mitoma et al., 2020) (Fig. 21.1a). When the backup substitute is anatomically discrete from the primary lesion, the lesion will not affect

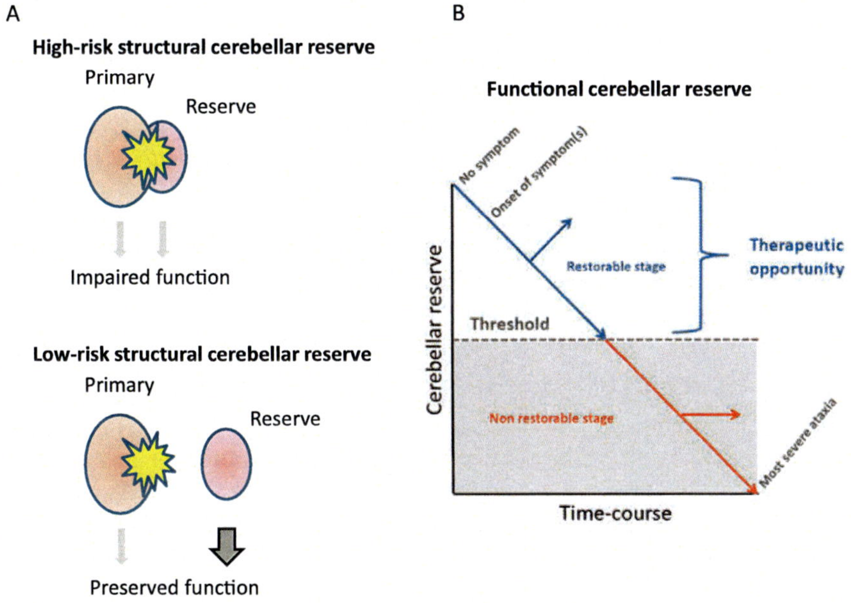

Fig. 21.1 (**a**) When the backup substitute is anatomically discrete from the primary lesion, the lesion will not affect the capacity for reserve ("low-risk structural cerebellar reserve"). On the other hand, when the backup substitute is adjacent to the primary substrate, the backup will be partially involved, leading to impairments in the capacity for reserve ("high-risk structural cerebellar reserve"). (**b**) Schematic diagram of the concept and relationship between *restorable* stage and *functional cerebellar reserve*

the capacity for reserve (low-risk cerebellar reserve). When the backup substitute is adjacent to the primary substrate, the backup will be partially involved, leading to impairment in the capacity for reserve (high-risk cerebellar reserve).

Functional Cerebellar Reserve Partial or full recovery of CAs has been noticed even in patients with diffuse and progressive cerebellar damage, such as IMCAs and metabolic ataxias (Mitoma et al., 2015, 2019; Mitoma & Manto, 2016). Also, motor rehabilitation is effective in enhancing functional restoration in patients with degenerative CAs (Ilg et al., 2014). Thus, the neural structures involved in the restoration process are not only present in the residual damage-free cerebellum with limited and transient lesions but also in the cerebellum with diffuse and progressive lesions. The physiological features of the functional cerebellar reserve are based on clinical observations of therapeutic summaries in IMCAs (Mitoma et al., 2015, 2019; Mitoma & Manto, 2016). Two factors determine the clinical course after immunotherapy.

1. *Response and no response.* Immunotherapy can halt the process of immune progression in certain etiologies but not in others. For example, avoidance of gluten curtails the progression of gluten ataxia (GA), whereas surgical excision of neoplasms and administration of a combination of immunotherapies (e.g., corticosteroids, intravenous immunoglobulins, and immunosuppressants) cannot halt the progression of paraneoplastic cerebellar degenerations (PCDs). These therapeutic benefits reflect the autoimmune pathologies.
2. *Restorable or non-restorable stage.* Induction of immunotherapy can be followed by two different clinical outcomes: patients with early stage of the disease, with normal or mildly atrophied cerebellum, show partial or full recovery. In contrast, patients with advanced stage, most of whom have evident cerebellar atrophy, show no improvement although the CAs remain stable. These features suggest the existence of a threshold that differentiates restorable stage from non-restorable stage. The restorable stage is defined as the period associated with intact cerebellar reversal capacity, i.e., functional cerebellar reserve.

Based on these observations, we designed a scheme for functional cerebellar reserve (Fig. 21.1b). Using this scheme, switching from functional disorders to cell death was clearly identified in anti-GAD ataxia, a subtype of IMCAs (Mitoma et al., 2017a, 2017b), a process likely resembling the pathophysiological background underlying IMCA disease progression from the restorable stage to non-restorable stage. Notably, even in the early stages of degenerative diseases, functional impairment (e.g., synaptic dysfunction or signal flow in cerebellar circuits), that precedes degenerative cell loss, can advance to clinically evident CAs, suggesting generality of the scheme in Fig. 21.1b (Chopra & Shakkottai, 2014; Watson et al., 2015).

21.3 Physiological Mechanisms Underlying Cerebellar Reserve

Experimental studies have demonstrated that cerebellar lesions are followed by substantial recovery even when the lesions are extensive. In rats, both hemicerebellectomy (HCb) and the full cerebellectomy (Cb) are followed by recovery of deficits after a few weeks up to a few months. Limb hyperflexion, wide-based locomotion, and the tendency to side falls are common after cerebellectomy, whereas tremor and body tilt develop after hemicerebellectomy (Federico et al., 2006). Other experimental studies showed that both cortical hemispherectomy and contralateral cerebellar resection were associated with failure of relearning motor tasks (Oulad Ben Taib et al., 2005). Furthermore, injection of kainic acid into the cerebellar nuclei (interpositus/dentate) caused deficits of proximal and distal voluntary movements, but slow recovery of these movements was noted over time. Such compensatory recovery was followed by decompensation when the sensory cortex was also removed (Mackel, 1987). These results suggest that the extra-cerebellar structures or the cerebellar inputs from these areas are prerequisites for relearning of lost cerebellar functions, although it is not clear whether relearning occurs within the cerebellum or in extra-cerebellar structures.

Relearning Within the Cerebellum Three distinguishing features characterize the cerebellar crystal-like neural circuitry (Fig. 21.2). First, about 60% of the entire brain neuron population (85–100 billions) is located in the cerebellum, although the cerebellum constitutes only 10% of brain mass (Colin et al., 2002; Walloe et al., 2014). Second, various forms of synaptic plasticity are involved in updating the

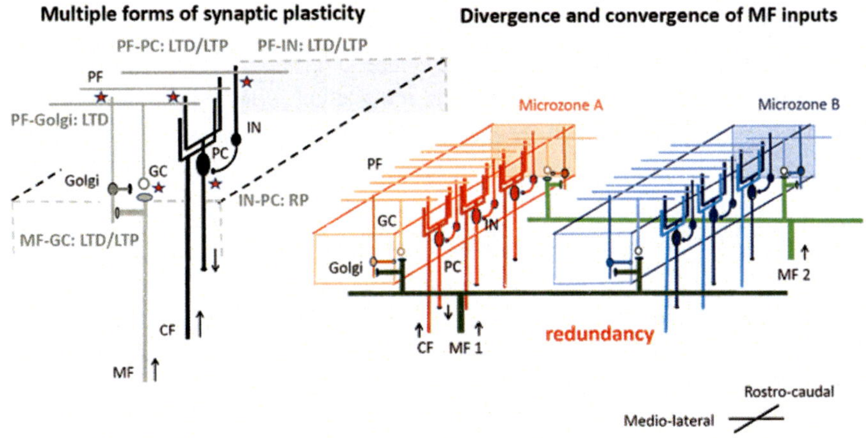

Fig. 21.2 Physiological mechanisms of the concept of cerebellar reserve. (**a**) Many forms of synaptic plasticity are found in the cerebellar cortex. (**b**) The cerebellar functional unit: redundancy of mossy fiber-mediated inputs from the periphery and the cerebral cortex in one microzone. Cited from reference (Mitoma et al., 2018)

internal model after cerebellar damage. Through the updated internal model, the cerebellum can adapt to errors in executed movements and select the desired motor command in a predictive fashion (Kawato et al., 1987). The third feature is the redundancy in cerebellar inputs. Purkinje cells within the narrow rostro-caudal slit act as a functional unit. The mossy fibers (MFs, cerebellar inputs) extend widely in a mediolateral fashion, terminating on the granule cells (Wu et al., 1999). Parallel fibers emerging from granule cells target the dendritic arborization of Purkinje cells and the inhibitory neurons (basket and stellate cells). Based on this architecture, the MFs transmit simultaneously signals from the cerebral cortex or the periphery to multiple functional units. In other words, one cerebellar functional unit receives multiple central and peripheral inputs. This redundant information is used effectively to generate and adapt the internal model thanks to synaptic plasticity. Transcranial direct current stimulation (tDCS) or repetitive transcranial magnetic stimulation (rTMS) is now being used to promote recovery (van Dun et al., 2017), suggesting that activation induced by these apparatuses can be effectively utilized for relearning.

Extra-cerebellar Involvements in Relearning The neurobiological substrates of cerebellar compensation include non-random remodeling of not only cerebellar networks but also the striatum circuits and the cerebello-vestibulo-spinal tracts with structural plasticity. Previous studies showed facilitation of glutamatergic transmission in the contralateral striatum following hemicerebellectomy in rats. Furthermore, pharmacological blockade of N-methyl-d-aspartate (NMDA) receptors with MK-801 impaired the rearrangement of excitatory synapses in the striatum and had a negative impact on compensation from motor disturbances (Centonze et al., 2008). Interestingly, ipsilateral aberrant cerebello-rubral projections developed following neonatal hemicerebellectomy (Naus et al., 1987). This novel pathway mirrored the topographic arrangement of the normal contralateral input at a synaptic level. This aberrant rewiring was explained by the response to deafferentation of the red nucleus, which was not specific to genuine cerebellar lesions. It is interesting to note that the vestibulo-spinal tract also showed remodeling following hemicerebellectomy in newborn rats (Castro & Smith, 1979).

21.4 Assessment of Cerebellar Reserve

From a therapeutic point of view, it is important to estimate how cerebellar reserve is preserved at a particular stage of the disease. The extension of lesions or the presence of atrophy has been generally used for an index in such estimation. However, more direct assessment based on functional analysis is desirable. We have proposed a physiological method using preservation of cerebellar predictive motor controls (Mitoma et al., 2016, 2020). The method is based on the following findings. A feedback signal from the periphery reaches the brain after a certain delay (from 10 to 100 milliseconds). It is well known in engineering that a feedback control based on time-delayed inputs can result in unstable movements, like in CAs (Tanaka et al., 2019). Thus, a forward model that calculates a sensory prediction for an ongoing motor command is embedded in the cerebellum in order to execute stable and dexterous control of compound movements (Tanaka et al., 2019). Kakei and colleagues developed a method by which receipt of motor commands is identified in human by analyzing components of muscle activities in terms of movement kinematics of the wrist joint in smooth task tracking (Kakei et al., 2019; Lee et al., 2012, 2015).

$$\tau(t)=\sum_{i=1}^{4}a_iT_i(t)=M\ddot{\theta}(t)+B\dot{\theta}(t)+K\theta(t) \tag{21.1}$$

where τ(t) represents the wrist joint torque estimated in two ways (middle and right sides of equation), T_i represents the tension of each muscle, and a_i represents the coefficient that converts muscle tension into wrist joint torque. The variables $\theta(t)$, $\dot{\theta}(t)$, and $\theta(t)$ represent the angle, angular velocity, and angular acceleration of the wrist joint, respectively. *M*, *B*, and *K* are the inertia parameter (kgm^2), the viscous coefficient (Nms/rad), and the elastic coefficient (Nm/rad), respectively. The *B/K* ratio can be used as an index of preservation of predictive control, since it represents how much velocity control is weighted relative to position control in muscle activities (Kakei et al., 2019; Lee et al., 2015). For instance, a simple feedback control uses positional information only, resulting in a small value for the *B/K* ratio. In contrast, in order to pursuit a moving target with known velocity and position in a predictive manner, it is necessary to encode both the velocity and position of the target, resulting in a much higher *B/K* ratio. Thus, a decrease in the *B/K* ratio (a decrease in velocity coefficient) suggests a relative lack of predictive control, i.e., a deficient forward model embedded in the cerebellum, in tracking movements (Kakei et al., 2019; Lee et al., 2015).

Using this rationale, we examined the preservation of cerebellar predictive controls in patients with IMCAs at early stage and degenerative CAs, in order to assess the preservation of cerebellar reserve (Mitoma et al., 2016, 2020). Both types of patients showed similar ataxic tracking. Compared with normal subjects, patients with early IMCAs had similar range of *B/K* ratio, whereas patients with degenerative CAs had a much lower *B/K* ratio. Thus, patients with early IMCAs performed the tracing task using predictive control, which was still preserved, but was inaccurate (Fig. 21.3). On the other hand, for patients with degenerative CAs, predictive control was no longer available. The preservation of cerebellar reserve is consistent with the reversibility in early IMCAs.

In conclusion, quantification of cerebellar proper functions, such as predictive motor control, can be a good biomarker for both structural and functional cerebellar reserves (Mitoma et al., 2016, 2020).

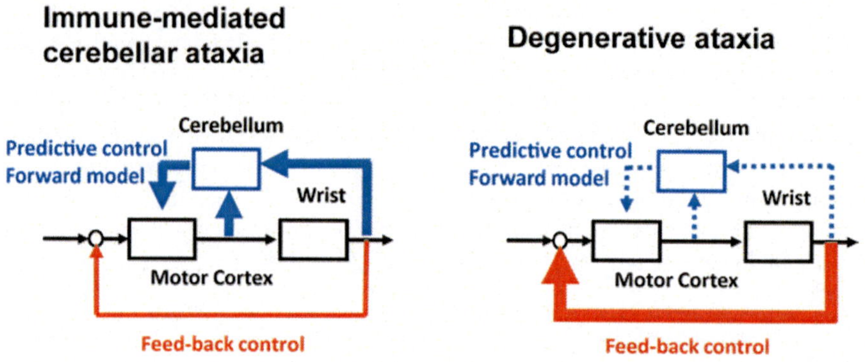

Fig. 21.3 (**a**) Preservation of cerebellar predictive controls in immune-mediated cerebellar ataxias at early stage. (**b**) In contrast, such controls do not exist in degenerative cerebellar ataxias (dotted lines), despite the same degree of ataxic pursuing movements

21.5 Facilitation of Cerebellar Reserve by Environmental Enrichment

Here we show evidence that cerebellar reserve is potentiated by environmental enrichment (EE), suggesting that cerebellar reserve is not rigid or fixed but rather exhibits vast plasticity and can be potentiated by daily routine body activities.

A recent study showed that EE is therapeutically beneficial in autism spectrum disorder (ASD)-like symptoms in BTBR mice (Queen et al., 2020). Juvenile BTBR mice were randomized to standard or EE housing up to 17 weeks, where EE induced metabolic changes (reduced adiposity, increased lean mass, improved glycemic control, and reduction in blood leptin levels) and improved behaviors (decreased anxiety and increased social affiliation) in male mice. On the other hand, the presence of cerebellar abnormalities has been documented in patients with ASD, which include abnormal structures, deregulated neurotransmitters, association of cerebellar motor and cognitive deficits, oxidative stress, and neuroinflammation (Fatemi et al., 2012), suggesting the role of cerebellar circuitry changes in ASD. Taken together, EE-induced improvements in ASD-like symptoms seem to be mediated by compensation and restoration of cerebellar function.

Consistent with this conclusion, recent studies have identified EE-induced changes in the cerebellar circuitry. For example, EE shortened the duration of action potentials in granule cells in rodents and, thus, generated higher firing frequencies, which serve to improve cerebellar controls (Eshra et al., 2019). Furthermore, in the study of Gelfo et al. (2016), rats were exposed to EE followed by hemicerebellectomy, and the effects of EE on dendritic spine density and size of Purkinje cells were examined. All hemicerebellectomized rats showed motor compensation, but standard-reared hemicerebellectomized rats exhibited cognitive impairment, which was almost completely compensated for in the enriched hemicerebellectomized rats. Improvement in the standard-reared hemicerebellectomized animals was associated with changes in the morphology of spines (increased density in the hemisphere coupled with decreased density in the spared hemivermis and enhanced spine area and head diameter in both regions). In contrast, only little improvement with limited compensatory changes in spines was noted in enriched hemicerebellectomized rats (the only exception was reduced density in the hemivermis, a region scarcely sensible to EE influence). These results suggest that EE potentiates cellular cerebellar reserve or remodels the circuit connectivity, which could provide resistant capacities to be employed even in the case of cerebellar lesion and thus make superfluous further post-lesional morphological re-adjustments (Gelfo et al., 2016; Mitoma et al., 2020).

One possible candidate for modulation of the cellular cerebellar reserve is autophagy, a conserved catabolic process that plays an important role in the maintenance of cellular homeostasis and in protection of cells from various insults, including misfolded proteins and damaged organelles (Mitoma et al., 2020; Yamamoto & Yue, 2014). Thus, the process of autophagy is a possible therapeutic target in neurodegenerative diseases. Notably, Buffo et al. (Fucà et al., 2017) found that motor training applied at presymptomatic stages had beneficial effects associated with changes in autophagy in the *tambaleante* mouse model of ataxia.

In conclusion, we propose the hypothesis that environmental enrichment potentiates cerebellar reserve so as to provide resistance to pathologies such as motor and cognitive impairments.

21.6 Conclusion: General Therapeutic Strategies Based on Cerebellar Reserve

Here we propose a general treatment strategy based on cerebellar reserve. In general, the following protocol can be applied irrespective of the etiology (Mitoma et al., 2019, 2020; Mitoma & Manto, 2016):

1. *Cause-cure treatment* should be designed in the initial stages to minimize damage of the cerebellum at the stage when cerebellar impairment is still within the frame of adequate cerebellar reserve (Mitoma et al., 2019; Mitoma & Manto, 2016; Manto et al., 2021). In the case of acute or limited pathology, its extension should be avoided (e.g., early reperfusion therapy in ischemia). On the other hand, in the case of progressive and diffuse pathologies, progression should be minimized or halted (e.g., removal of causes in metabolic CAs and immunotherapies in IMCAs). We expect new molecular therapies that can stop the progression of degenerative CAs to become available in the near future.
2. *Neuromodulation therapy* should be followed to facilitate cerebellar reserve. Neuromodulation therapies include intensive motor rehabilitation or noninvasive cerebellar stimulation (Mitoma et al., 2019; Mitoma & Manto, 2016). In this context, the near future is promising with the introduction of RNA therapies and

neurotransplantation (Mitoma et al., 2019). With regard to the latter, genuine replacement is considered difficult, since cell differentiation and synaptic formation are necessary to establish functional circuitries (Cendelin et al., 2018; Cendelin & Mitoma, 2018; Mitoma et al., 2019). Instead, recent studies have shown that grafted cells rescue surviving cells from neurodegeneration through their trophic effects, providing support to mitochondrial function, modulating neuroinflammation, stimulating endogenous regenerative processes, and facilitating cerebellar compensatory properties thanks to neural plasticity (Mitoma et al., 2019; Mitoma & Manto, 2016). Thus, reinforcement of the cerebellar reserve and prolongation of the restorable stage can be envisioned as future endpoints of neurotransplantation.

Compared with other CNS areas, the cerebellum is endowed with the capacity of restoring functional synapses. Clinicians should not miss this window of treatment opportunity. We have stressed the importance of early diagnosis and treatment by advancing the phrase *Time is Cerebellum* (Mitoma et al., 2018), like *Time is Brain*, a phrase that stresses the importance of early intervention in ischemic brain diseases (Saver, 2006). Further studies on therapeutic strategies based on cerebellar reserve are needed.

References

Castro, A. J., & Smith, D. E. (1979). Plasticity of spinovestibular projections in response to hemicerebellectomy in newborn rats. *Neuroscience Letters, 12*, 69–74.

Cendelin, J., Buffo, A., Hirai, H., Magrassi, L., Mitoma, H., Sherrard, R., et al. (2019). Task force paper on cerebellar transplantation: Are we ready to treat cerebellar disorders with cell therapy? *Cerebellum, 18*, 575–592.

Cendelin, J., & Mitoma, H. (2018). Neurotransplantation therapy. *Handbook of Clinical Neurology, 155*, 379–391.

Cendelin, J., Mitoma, H., & Manto, M. (2018). Neurotransplantation therapy and cerebellar reserve. *CNS & Neurological Disorders - Drug Targets, 17*, 172–183.

Centonze, D., Rossi, S., De Bartolo, P., De Chiara, V., Foti, F., Musella, A., et al. (2008). Adaptations of glutamatergic synapses in the striatum contribute to recovery from cerebellar damage. *The European Journal of Neuroscience, 27*, 2188–2196.

Chopra, R., & Shakkottai, V. G. (2014). The role for alterations in neural activity in the pathogenesis of polyglutamine repeat disorders. *Neurotherapeutics, 11*, 751–763.

Colin, F., Ris, L., & Godaux, E. (2002). Neuroanatomy of the cerebellum. In M. Manto & M. Pandolfo (Eds.), *The cerebellum and its disorders* (pp. 6–29). Cambridge University Press.

van Dun, K., Bodranghien, F., Manto, M., & Mariën, P. (2017). Targeting the cerebellum by noninvasive neurostimulation: A review. *Cerebellum, 16*, 695–741.

Eshra, A., Hirrlinger, P., & Hallermann, S. (2019). Enriched environment shortens the duration of action potential in cerebellar granule cells. *Frontiers in Cellular Neuroscience, 13*, 289.

Fatemi, S. H., Aldinger, K. A., Ashwood, P., Bauman, M. L., Blaha, C. D., Blatt, G. J., et al. (2012). Consensus paper: Pathological role of the cerebellum in autism. *Cerebellum, 11*, 777–807.

Federico, F., Leggio, M. G., Mandolesi, L., & Petrosini, L. (2006). The NMDA receptor antagonist CGS 19755 disrupts recovery following cerebellar lesions. *Restorative Neurology and Neuroscience, 24*, 1–7.

Fucà, E., Guglielmotto, M., Boda, E., Rossi, F., Buffo, L. K., & A. (2017). Preventive motor training but not progenitor grafting ameliorates cerebellar ataxia and deregulated autophagy in tambaleante mice. *Neurobiology of Disease, 102*, 49–59.

Gelfo, F., Florenzano, F., Foti, F., Burello, L., Petrosini, L., & De Bartolo, P. (2016). Lesion-induced and activity-dependent structural plasticity of Purkinje cell dendritic spines in cerebellar vermis and hemisphere. *Brain Structure & Function, 221*, 3405–3426.

Holmes, G. (1917). The symptoms of acute cerebellar injuries due to gunshot injuries. *Brain, 40*, 461–535. (online: https://academic.oup.com/brain/article/40/4/461/273305)

Ilg, W., Bastian, A. J., Boesch, S., Burciu, R. G., Celnik, P., Claaßen, J., et al. (2014). Consensus paper: management of degenerative cerebellar disorders. *Cerebellum, 13*, 248–268.

Kakei, S., Lee, J., Mitoma, H., Tanaka, H., Manto, M., & Hampe, C. S. (2019). Contribution of the cerebellum to predictive motor controls and its evaluation in ataxic patients. *Frontiers in Human Neuroscience.* https://doi.org/10.3389/fnhum.2019.00216

Kawato, M., Furukawa, K., & Suzuki, R. (1987). A hierarchical neural network model for control and learning of voluntary movement. *Biological Cybernetics, 57*, 169–185.

Lee, J., Kagamihara, Y., & Kakei, S. (2015). A new method for functional evaluation of motor commands in patients with cerebellar ataxia. *PLoS One, 10*, e0132983.

Lee, J., Kagamihara, Y., Tomatsu, S., & Kakei, S. (2012). The functional role of the cerebellum in visually guided tracking movement. *Cerebellum, 11*, 426–433.

Mackel, R. (1987). The role of the monkey sensory cortex in the recovery from cerebellar injury. *Experimental Brain Research, 66*, 638–652.

Manto, M. (2008). The cerebellum, cerebellar disorders, and cerebellar research. Two centuries of discoveries. *Cerebellum, 7*, 505–516.

Manto, M., Bower, J. M., Conforto, A. B., Delgado-García, J. M., da Guarda, S. N., Gerwig, M., et al. (2012). Consensus paper: Roles of the cerebellum in motor control—The diversity of ideas on cerebellar involvement in movement. *Cerebellum, 11*, 457–487.

Manto, M., Kakei, S., & Mitoma, H. (2021). The critical need to develop tools assessing cerebellar reserve for the delivery and assessment of non-invasive cerebellar stimulation. *Cerebellum Ataxias, 8*, 2. https://doi.org/10.1186/s40673-020-00126-w

Mitoma, H., Adhikari, K., Aeschlimann, D., Chattopadhyay, P., Hadjivassiliou, M., Hampe, C. S., Honnorat, J., Joubert, B., Kakei, S., Lee, J., Manto, M., Matsunaga, A., Mizusawa, H., Nanri, K., Shanmugarajah, P., Yoneda, M., & Yuki, N. (2016). Consensus paper: Neuroimmune mechanisms of cerebellar ataxias. *Cerebellum, 15*, 213–232.

Mitoma, H., Buffo, A., Gelfo, F., Guell, X., Fucà, E., Kakei, S., et al. (2020). Consensus paper. Cerebellar reserve: From cerebellar physiology to cerebellar disorders. *Cerebellum, 19*, 131–153.

Mitoma, H., Hadjivassiliou, M., & Honnorat, J. (2015). Guidelines for treatment of immune-mediated cerebellar ataxias. *Cerebellum Ataxias., 2*, 14.

Mitoma, H., & Manto, M. (2016). The physiological basis for therapies of cerebellar ataxias. *Therapeutic Advances in Neurological Disorders, 9*, 396–413.

Mitoma, M., Manto, M., & Gandini, J. (2019). Recent advances in the treatment of cerebellar disorders. *Brain Sciences, 10*, E11.

Mitoma, H., Manto, M., & Hampe, C. S. (2017a). Pathogenic roles of glutamic acid decarboxylase 65 autoantibodies in cerebellar ataxias. *Journal of Immunology*, 2913297.

Mitoma, H., Manto, M., & Hampe, C. S. (2017b). Immune-mediated cerebellar ataxias: From bench to bedside. *Cerebellum Ataxias., 4*, 16.

Mitoma, H., Manto, M., & Hampe, C. S. (2018). Time is cerebellum. *Cerebellum, 17*, 387–391.

Naus, C. G., Flumerfelt, B. A., & Hrycyshyn, A. W. (1987). Ultrastructural study of remodeled rubral afferents following neonatal lesions in the rat. *The Journal of Comparative Neurology, 259*, 131–139.

Oulad Ben Taib, N., Manto, M., Pandolfo, M., & Brotchi, J. (2005). Hemicerebellectomy blocks the enhancement of cortical motor output associated with repetitive somatosensory stimulation in the rat. *The Journal of Physiology, 567*(Pt 1), 293–300.

Queen, N. J., Boardman, A. A., Patel, R. S., Siu, J. J., Mo, X., & Cao, L. (2020). Environmental enrichment improves metabolic and behavioral health in the BTBR mouse model of autism. *Psychoneuroendocrinology, 111*, 104476.

Saver, J. L. (2006). Time is brain-quantified. *Stroke, 37*, 263–266.

Schmahmann, J. D., & Caplan, D. (2006). Cognitive, emotion and the cerebellum. *Brain, 129*, 290–292.

Serra, L., & Gelfo, F. (2019). What good is the reserve? A translational perspective for the managing of cognitive decline. *Neural Regeneration Research, 14*, 1219–1220.

Serra, L., Gelfo, F., Petrosini, L., Di Domenico, C., Bozzali, M., & Caltagirone, C. (2018). Rethinking the reserve with a translational approach: Novel ideas on the construct and the interventions. *Journal of Alzheimer's Disease, 65*, 1065–1078.

Steffener, J., & Stern, Y. (1822). Exploring the neural basis of cognitive reserve in aging. *Biochimica et Biophysica Acta, 2012*, 467–473.

Stern, Y. (2012). Cognitive reserve in ageing and Alzheimer's disease. *Lancet Neurology, 11*, 1006–1012.

Stern, Y. (2017). An approach to studying the neural correlates of reserve. *Brain Imaging and Behavior, 11*, 410–416.

Takagi, M., Zee, D. S., & Tamargo, R. J. (1998). Effects of lesions of the oculomotor vermis on eye movements in primate: Saccades. *Journal of Neurophysiology, 80*, 1911–1931.

Tanaka, H., Ishikawa, T., & Kakei, S. (2019). Neural evidence of the cerebellum as a state predictor. *Cerebellum, 18*, 349–371.

Walloe, S., Pakkenberg, B., & Fabricius, K. (2014). Stereological estimation of total cell numbers in the human cerebral and cerebellar cortex. *Frontiers in Human Neuroscience, 8*, 508.

Watson, L. M., Wong, M. M., & Becker, E. B. (2015). Induced pluripotent stem cell technology for modelling and therapy of cerebellar ataxia. *Open Biology, 5*, 150056.

Wu, H. S., Suguhara, I., & Shinoda, Y. (1999). Projection patterns of single mossy fibers originating from the lateral reticular nucleus in the rat cerebellar cortex and nuclei. *The Journal of Comparative Neurology, 411*, 97–118.

Yamamoto, A., & Yue, Z. (2014). Autophagy and its normal and pathogenic states in the brain. *Annual Review of Neuroscience, 37*, 55–78.

Zee, D. S., Yamazaki, A., Butler, P. H., & Gücer, G. (1981). Effects of ablation of flocculus and paraflocculus of eye movements in primate. *Journal of Neurophysiology, 46*, 878–899.

Chapter 22
Prism Adaptation Test (PAT): A Practical and Quantitative Method to Evaluate Cerebellar Function

Hidehiro Mizusawa

22.1 Introduction

Many diseases, such as neoplasm, trauma, congenital malformation, inflammation (including infection), immune-mediated conditions, vascular disorders, intoxication, metabolic disorders, and degeneration (Manto & Pandolfo, 2002), affect the cerebellum. In these situations, we need to evaluate the cerebellar functions to assess the disease's severity and etiology. Neurodegeneration is very difficult to diagnose, evaluate, and cure. The Scale for the Assessment and Rating of Ataxia (SARA) and the International Ataxia Rating Scale (ICARS) are evaluation tools that have been widely used (Schmitz-Hübsch et al., 2006; Trouillas et al., 1997).

However, the SARA score is subjective to the examiner, and its change is very subtle, only approximately one point per year in spinocerebellar ataxia type 6 (SCA 6) (Ashizawa et al., 2013; Jacobi et al., 2011; Moriarty et al., 2016; Yasui et al., 2014). Progression seems very slow in many SCA cases. Changes of one point in SARA could be attributed to the placebo effect (Nishizawa et al., 2020). Therefore, a quantitative method is needed to evaluate cerebellar function. This is very important when we consider clinical trials of neurodegenerative diseases of the cerebellum that belong to the so-called rare diseases (Manto & Pandolfo, 2002).

In addition, many aged people are in an unstable state when walking (Cesari, 2012). This may be due to cerebellar ataxia due to aging, while some cases can be explained by frailty or functional decline of other systems such as the musculoskeletal system (Safe et al., 1992). The cerebellum is known to be sensitive to the aging process; particularly, the vermis shows age-dependent atrophy, comparable to that of the hippocampus in normal aging (Raz et al., 2010). Furthermore, the upper part of the vermis is the most severely affected, during normal aging, as in many

H. Mizusawa (✉)
Neurology, NCNP Hospital, National Center of Neurology and Psychiatry (NCNP), Kodaira, Tokyo, Japan
e-mail: mizusawa@ncnp.go.jp

H. Mizusawa, S. Kakei (eds.), *Cerebellum as a CNS Hub*,
Contemporary Clinical Neuroscience, https://doi.org/10.1007/978-3-030-75817-2_22

degenerative SCAs. This cerebellar function change during aging is very subtle; we need to develop a quantitative method to evaluate these small changes.

The cerebellum is thought to be involved in many functions, including maintaining balance and posture, coordination of voluntary movement, motor learning, and cognitive function (Manto & Pandolfo, 2002). In clinical practice, however, neurological examination is confined to balance and posture and coordination of voluntary movement (Babinski, 1899; Holmes, 1939). The situation could be understood very well based on the name of tests such as SARA and ICARS, suggesting that they involve ataxia almost exclusively. We focused on motor learning because it is considered a fundamental function supporting other motor functions, such as the coordination of voluntary movement (Ito, 1984; Ito, 2011; Lisberger & Thach, 2013).

There are several types of memory, including declarative memory related to the hippocampus and motor memory or memory of motor skills, related to the cerebellum (Ito, 1984; Ito, 2011; Lisberger & Thach, 2013; Manto & Pandolfo, 2002). Motor learning results from practice, a novel experience, or the environment, as a capability for adaptation. Regarding movement coordination, the cerebellum behaves like a computer processor, constantly comparing the actual movement of muscle groups with the motions intended by the motor cortex. The cerebellum plays a major role in adapting and fine-tuning motor programs to make accurate movements through a trial-and-error process. Motor learning memory is acquired through repeated practice and stored in the cerebellum (Shutoh et al., 2006).

Why do we not assess cerebellar motor learning in a clinical setting? There are many motor learning paradigms, including the vestibulo-ocular reflex (VOR), optokinetic response (OKR), eyeblink conditioning, smooth pursuit eye movement, force field perturbation, and visuo-motor perturbation (Gerwig et al., 2007; Shelhamer et al., 1994; Topka et al., 1993; Walker & Zee, 1999; Woodruff-Pak et al., 1996). These paradigms require a relatively large instrumental or special setup to study and have been used only in experimental situations. Perhaps, little attention has been paid to motor learning due to the lack of efficient and feasible methods to routinely assess this important cerebellar function.

The evaluation of motor learning using prism glasses has already been reported in the field of neurophysiology (Baizer & Glickstein, 1974; Rabe et al., 2009; Smith & Shadmehr, 2005; Weiner et al., 1983). Throwing darts is an example (Martin et al., 1996; Norris et al., 2011; Thach et al., 1992). Without prism glasses, the darts landed around the center of the target or the baseline. With prism glasses that shift the visual field to the right, the darts go initially to the right. However, they return to baseline after many trials. Removal of the prism glasses results in movement of the darts to the opposite side (the left side), but again, after many trials, they become closer to the baseline by motor learning or adaptation. This is a normal adaptation pattern. This phenomenon can be understood naturally. However, this normal pattern of adaptation is severely disturbed in patients with cerebellar diseases (Martin et al., 1996).

22.2 Materials and Methods

Dart throwing requires a big examination room and some basic skills, that is, there are good and bad players. Therefore, we developed a system using a hand-reaching task and a touch panel so that the test can be conducted in outpatient clinics (Hashimoto et al., 2015) (Figs. 22.1 and 22.2). The apparatus consists of two personal computers (one for task control and another for data sampling and analysis), a touch screen, a handmade goggle, and an ear sensor. The goggle is outfitted with the electrically controlled shutter, which opens by the pulse-on command voltage (100 V) and closes after pulse-off. A target appears randomly on the screen while the subject touches the sensor.

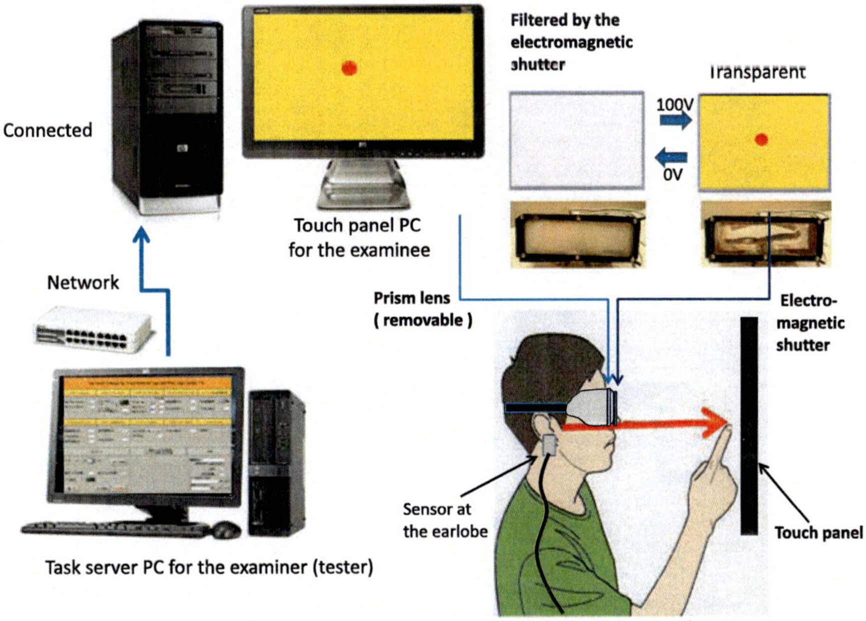

Fig. 22.1 The concept of our touch panel prism adaptation test system. The apparatus consists of two personal computers (one for task control and another for data sampling and analysis), a touch screen, a handmade goggle, and an ear sensor. The goggle is outfitted with the electrically controlled shutter, which opens by the pulse-on command voltage (100 V) and closes after pulse-off. A target appears randomly on the screen while the subject touches the sensor. (*A subject sits in a chair wearing the goggle in a dark room, with his/her head loosely restrained by a chin rest. A touch screen is set 30–50 cm apart, within reach of finger touch, in front of the subject. *At the start, the subject touches its right index finger to the sensor attached at its earlobe. The subject is then asked to touch the target on the screen. *When the subject releases the index finger from the sensor, the shutter changes from transparent (vision on) to translucent (vision off) by the computer. The vision is blocked from the onset to the terminal of movement to avoid eye-hand coordination. *Then, a target on the screen is put off with a beep sound, and the subject is requested to return its index finger from the screen to the sensor. A subject is allowed to see the target for 100 ms after touching the screen)

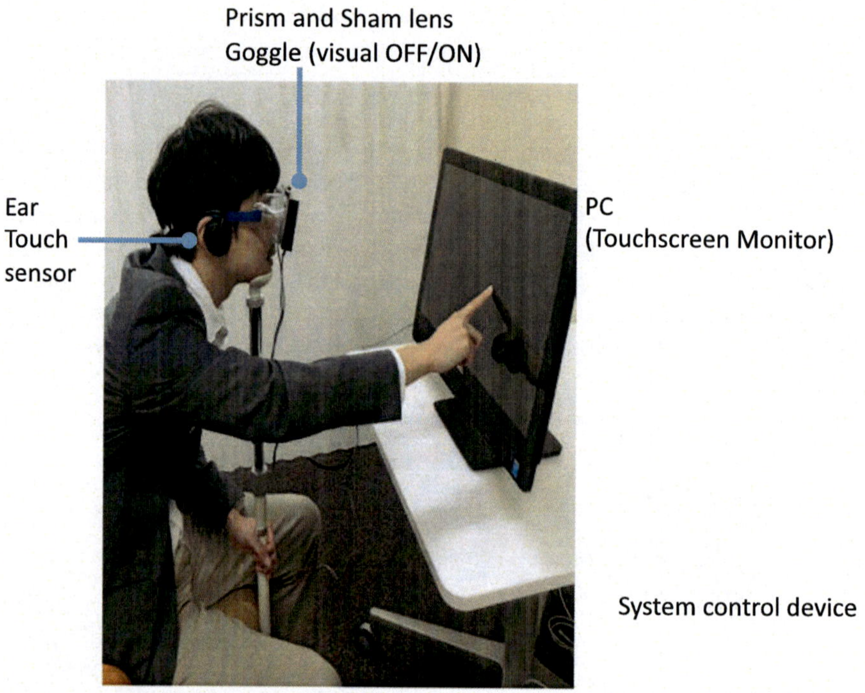

Fig. 22.2 Current prism adaptation test system. Originally, there were two PCs, but now the system has one touchscreen monitor and a control box that is located under the table and cannot be seen. The system is quite concise and can be used in outpatient clinics

The actual procedure is very simple (Fig. 22.3).

After various trials, we chose the following time sequence for a single trial.

The single trial consists of three consecutive tasks:

1. 50 trials with normal vision (*BASELINE*).
2. 100 trials wearing prism glasses shifting the visual field 25° rightward (*PRISM*).
3. 50 trials without the prism glasses (*REMOVAL*).

It takes 15–20 minutes to complete the three sessions.

In normal subjects, the baseline phase shows almost no deviation; wearing the prism lens results in a steep deviation but soon returns to the baseline after repeated trials by motor learning or prism adaptation (Fig. 22.4). Finally, removal of the prism causes a deviation to the opposite side because the adaptation is complete and again quickly returns to the baseline by the second adaptation. In a SCA31 patient, the normal pattern is completely destroyed even in the baseline phase, which is suggestive of dysmetria.

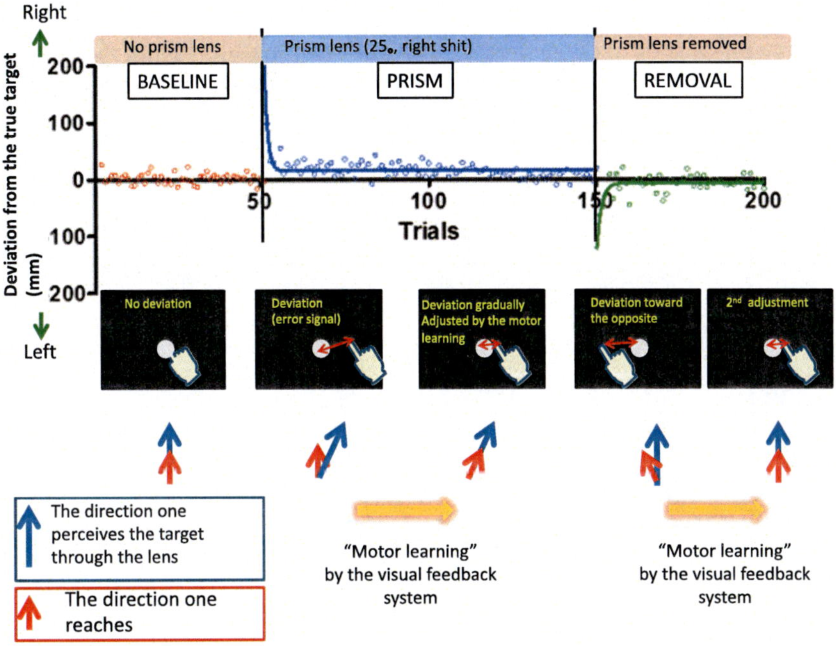

Fig. 22.3 Time sequence of a single test. The whole test consists of three consecutive tasks: (1) 50 trials with normal vision (*BASELINE*), (2) 100 trials wearing prism shifting the visual field 25° rightward (*PRISM*), and (3) 50 trials without the prism (*REMOVAL*). It takes each subject 15–20 min to complete the three sessions

22.3 Data Analysis

First, we defined the correct touch (Fig. 22.5). In the *BASELINE* session, the mean ± 2SD of finger touch in healthy non-elderly subjects was approximately 25 mm from the baseline. Thus, when the finger touch was within 25 mm, the trial was defined as "correct." We think that the last part of prism adaptation and the beginning and last of the removal phase are important, and they may represent acquisition, retention, and extinction of adaptation, namely, motor learning. Acquisition of adaptation was evaluated by the probability of correct touches in the last ten trials of the *PRISM* session, retention of adaptation by the probability of incorrect touches in the initial five trials of the *REMOVAL* session, and extinction by the probability of correct touches of the last ten trials of the *REMOVAL* session. The adaptability index (AI) was calculated using this formula.

There were four groups of subjects (Fig. 22.6): healthy non-elderly subjects (HN, <70 years old, $n = 22$, M/F = 9/13), healthy elderly subjects (HE, ≥ 70 years old, $n = 17$, M/F = 9/8), non-elderly cerebellar ataxia patients (CN, <70 years old, $n = 55$, M/F = 34/21), and elderly patients with cerebellar ataxia (CE, ≥ 70 years old, $n = 13$, M/F = 7/6). The ages of the groups (mean ± SEM, ranges) were as follows: HN, 49.8 ± 2.9, range 28–68; HE, 78.2 ± 1.2, range 72–88; CN, 55.3 ± 1.4,

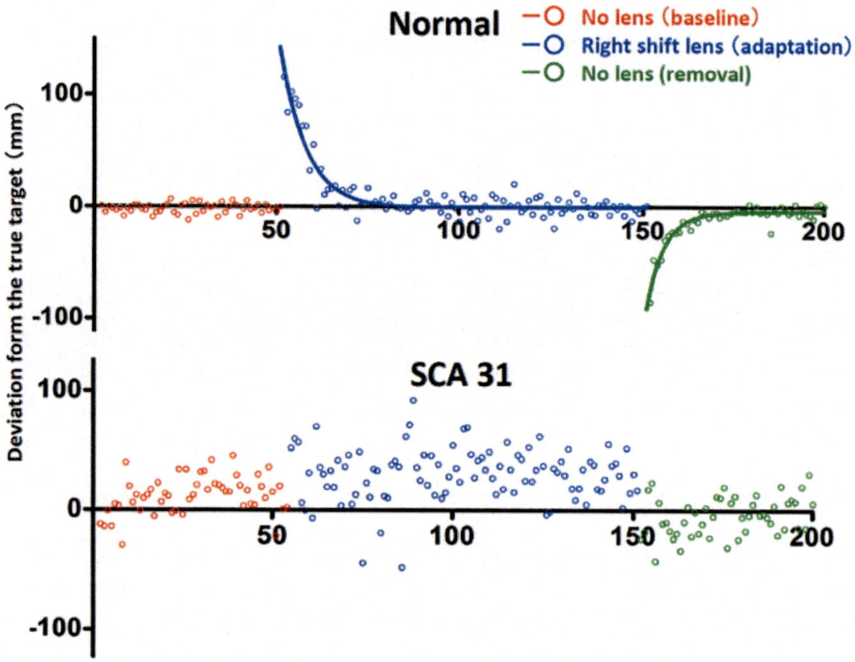

Fig. 22.4 Actual records of touchpoints (deviation from the target). In a normal subject, the baseline phase shows almost no deviation. Wearing the prism glasses results in a steep deviation but gradually returns to the baseline after repeated trials. Finally, removing the prism glasses causes a deviation to the opposite side but again returns to baseline. In a SCA31 patient, the normal pattern is completely destroyed, showing no deviation and no return

range 31–69; and CE, 74.7 ± 1.1, range 70–83. The patients with cerebellar ataxia (n = 68) were diagnosed with SCA2 (n = 1), MJD/SCA3 (n = 13), SCA6 (n = 9), SCA8 (n = 2), SCA31 (n = 12), SCA36 (n = 1), CCA (n = 7), DRPLA (n = 1), and MSA (n = 22 – MSA-C, n = 18; MSA-P, n = 4).

22.4 Results

The actual AI and age of each patient are shown in Fig. 22.6 (Hashimoto et al., 2015). The AI of each group and AI of cerebellar patients were significantly lower than those of control subjects and correlated significantly with SARA and 9HPT scores (Fig. 22.7). In addition, older people had significantly lower AI than younger subjects.

When the AIs of cerebellar ataxia patients and normal subjects were compared, there was very little overlap (Fig. 22.8), suggesting that AI would be able to distinguish cerebellar ataxia patients from normal controls. In fact, AI below 0.68 had a

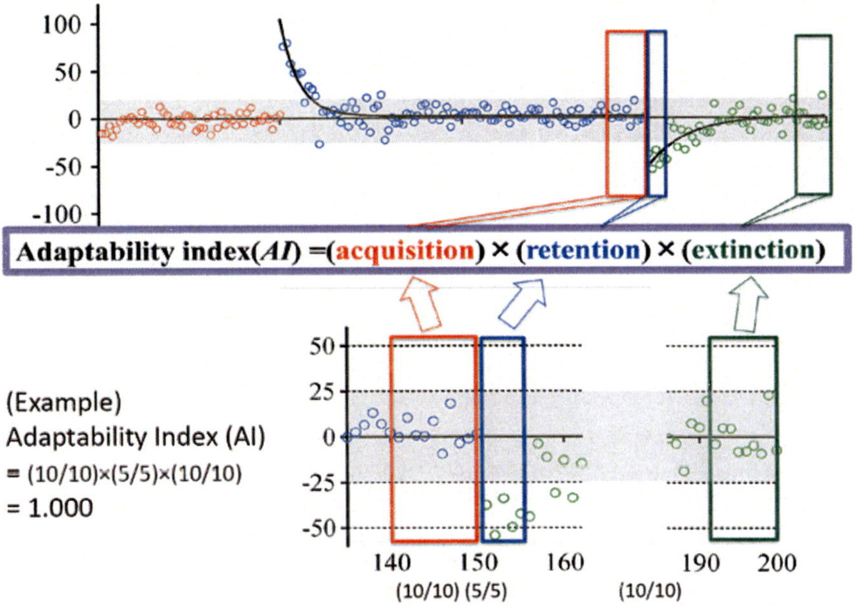

Fig. 22.5 Data analysis, definition of correct touch. In the *BASELINE* session, the mean ± 2SD of finger-touch error in healthy non-elderly subjects is approximately 25 mm. Thus, when the finger-touch error was within 25 mm, the trial was defined as "correct". (*According to the pattern and motor learning theory, we think the last part of prism adaptation and the beginning and end of the removal phase are important. They may represent the acquisition, retention, and extinction of adaptation (motor learning). *The acquisition of adaptation is evaluated by the probability of correct touches in the last ten trials of the *PRISM* session, retention of adaptation by the probability of *incorrect* touches in the initial five trials of the *REMOVAL* session, and extinction by the probability of correct touches of the last ten trials of the *REMOVAL* session. We calculated the adaptability index (AI) using the formula. If a subject has a normal adaptation function, its AI should be close to 1.0, and if adaptation is disturbed, the AI is close to 0)

sensitivity of 98.4% and a specificity of 100% for cerebellar ataxia patients among various receiver operating characteristic (ROC) curves of various indices (Fig. 22.9). As expected, AI progressively decreased with duration (Fig. 22.10).

22.5 Discussion

First, our touch panel prism adaptation test revealed an abnormal pattern very similar to previous studies (Martin et al., 1996), indicating that this system is appropriate for detecting human motor learning function. Results from 39 healthy subjects and 68 patients with degenerative cerebellar ataxia clearly indicate that AI may distinguish such patients with high sensitivity and specificity. Compared with a previous index, the time constant (tau), for example, the ability to discriminate patients from

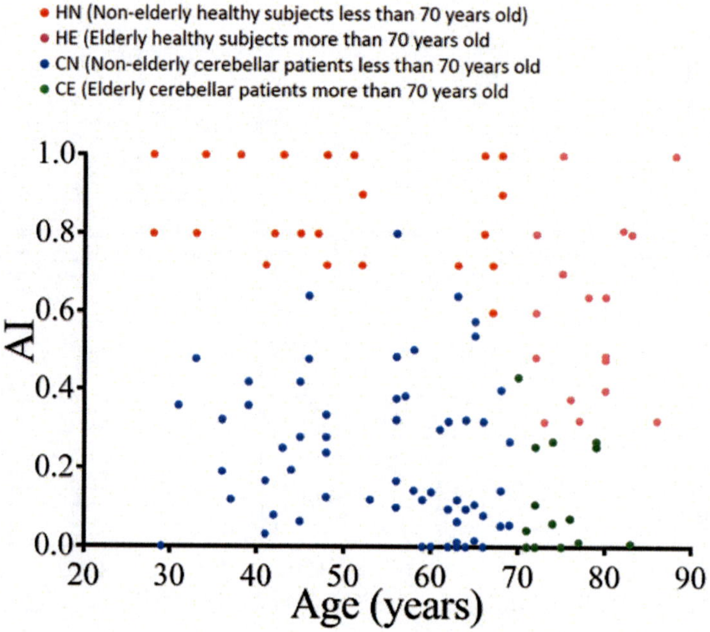

Fig. 22.6 Data of all the subjects plotted by AI and age. Each group occupies a certain area (Hashimoto et al., 2015): red, pink, blue, and green

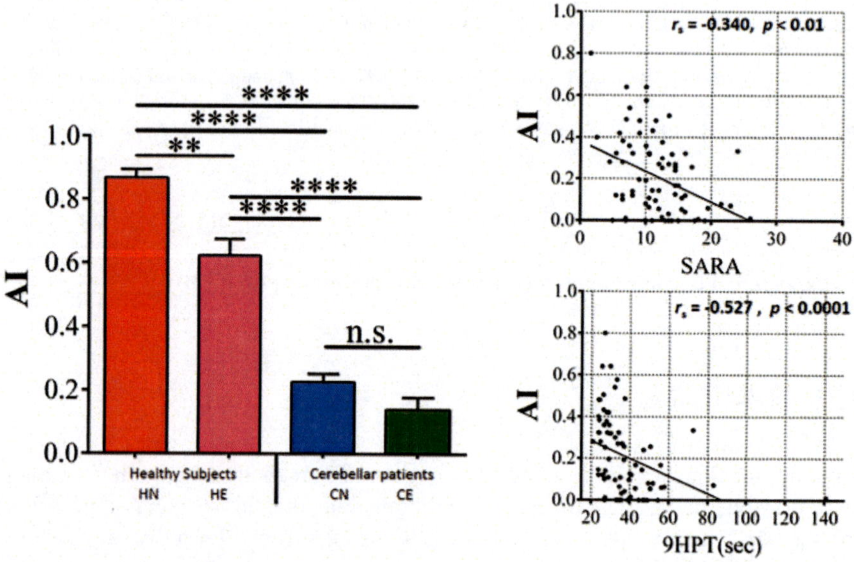

Fig. 22.7 AI of cerebellar ataxia patients is significantly lower than that of normal subjects and correlates significantly with SARA and 9HPT scores. AI also decreased significantly in aged persons among normal subjects (Hashimoto et al., 2015)

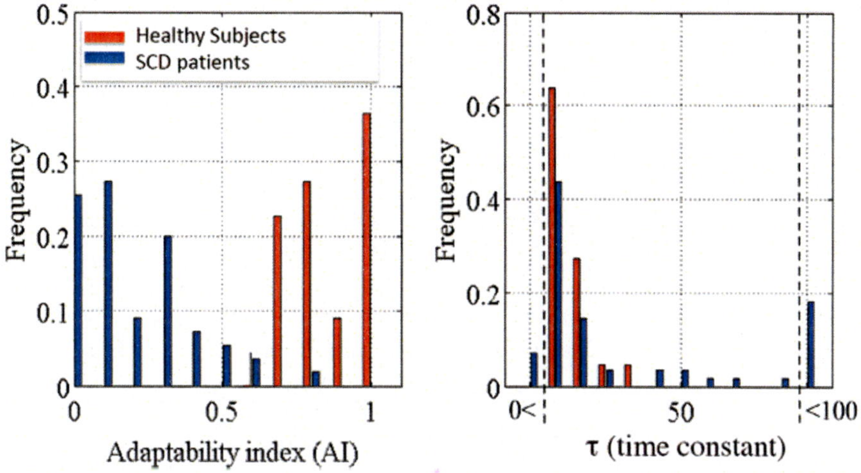

Fig. 22.8 This is another expression of AI of cerebellar ataxia patients and normal subjects. There was very little overlap. A previously used marker tau (time constant) did not distinguish the two groups, suggesting that adequate analysis of data may be very important (Hashimoto et al., 2015)

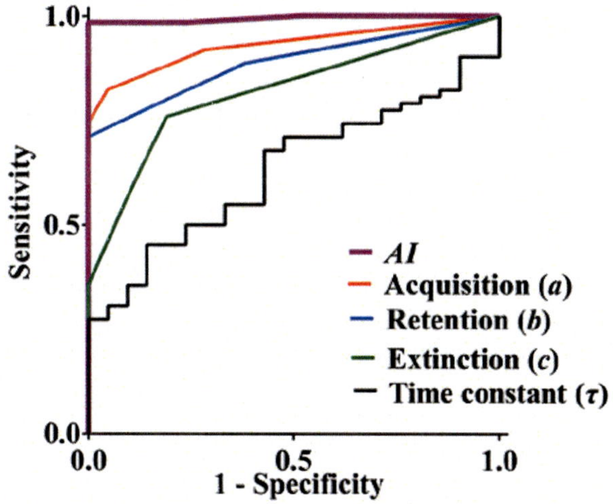

Fig. 22.9 ROC (receiver operating characteristic) curves of various indices. AI below 0.68 shows a sensitivity of 98.4% and a specificity of 100% for cerebellar ataxia patients (Hashimoto et al., 2015)

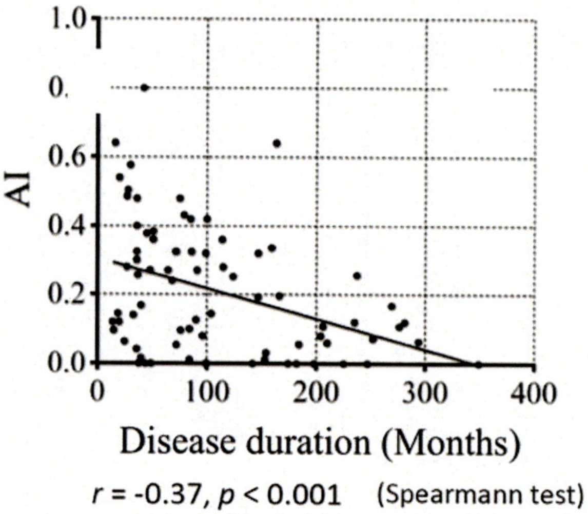

Fig. 22.10 AI correlates with disease duration among cerebellar ataxia patients

controls is quite high in AI, and it is suggested that our estimation of each component of AI is correct.

Motor learning is believed to depend on the long-term depression (LTD) of Purkinje cells, which modifies Purkinje cell discharges and is elicited by error signals coming through climbing fibers from the inferior olivary nucleus during motor performance (Honda et al., 2018; Ito et al., 1982). This short-term memory is located in the cerebellar cortex. In most of our patients, the cerebellar cortex was severely affected, and the results were quite reasonable. This has been proved by our later study comparing PAT and cerebellar atrophy on MRI (Bando et al., 2019). In the near future, we should increase the number of patients and compare the differences among different types of cerebellar degenerative ataxia. Longitudinal alterations are also important for investigation. For these purposes, we have just started a multicenter validation study using this system in Japan. PAT may be useful not only for clinical practice but also for scientific research on the cerebellum.

Acknowledgments This work was performed in collaboration with Doctors Yuji Hashimoto (TMDU, Neurology), Takeru Honda (Riken Brain Science Institute [BSI], Tokyo Metropolitan Medical Institute), Ken Matsumura (TMDU, Neurology), Makoto Nakao (Riken BSI), Kazumasa Soga (TMDU, Neurology), Kazuhiko Katano (Brainway Co., Ltd.), Takanori Yokota (TMDU, Neurology), Soichi Nagao (Riken BSI), and Kinya Ishikawa (TMDU, Neurology) and largely published in 2015 (Hashimoto et al., 2015), with support from the Strategic Research Program for Brain Science, Ministry of Education, Sports, Science and Technology, Japan.

Literature

Ashizawa, T., Figueroa, K. P., Perlman, S. L., Gomez, C. M., Wilmot, G. R., Schmahmann, J. D., Ying, S. H., Zesiewicz, T. A., Paulson, H. L., Shakkottai, V. G., Bushara, K. O., Kuo, S. H., Geschwind, M. D., Xia, G. B., Mazzoni, P., Krischer, J. P., Cuthbertson, D., Holbert, A. R., Ferguson, J. H., Pulst, S. M., & Subramony, S. H. (2013). Clinical characteristics of patients with spinocerebellar ataxias 1, 2, 3 and 6 in the US; a prospective observational study. *Orphanet Journal of Rare Diseases, 8*, 177.

Babinski, J. (1899). De l' asynergie cérébelleuse. *Rev. neurol, 7*, 806–816.

Baizer, J. S., & Glickstein, M. (1974). Role of cerebellum in prism adaptation. *J Physiol Proc, 236*, 34–35.

Bando, K., Honda, T., Ishikawa, K., Takahashi, Y., Mizusawa, H., & Hanakawa, T. (2019). Impaired adaptive motor learning is correlated with cerebellar hemispheric gray matter atrophy in spinocerebellar ataxia patients: A voxel-based morphometry study. *Front Neurol, 10*, 1183. https://doi.org/10.3389/fneur.2019.01183fneur.2019.01183. PMID: 31803128.

Cesari, M. (2012). Frailty and aging. *J Frailty Aging, 1*, 3–6.

Gerwig, M., Kolb, F. P., & Timmann, D. (2007). The involvement of the human cerebellum in eyeblink conditioning. *Cerebellum, 6*, 38–57.

Hashimoto, Y., Honda, T., Matsumura, K., Nakao, M., Soga, K., Katano, K., Yokota, T., Mizusawa, H., Nagao, S., & Ishikawa, K. (2015). Quantitative evaluation of human cerebellum-dependent motor learning through prism adaptation of hand-reaching movement. *PLOS ONE, 10*, e0119376. https://doi.org/10.1371/journal.pone.0119376.eCollection

Holmes, G. (1939). The cerebellum of man. *Brain, 62*, 1–30.

Honda, T., Nagao, S., Hashimoto, Y., Ishikawa, K., Yokota, T., Mizusawa, H., & Ito, M. (2018). Tandem internal models execute motor learning in the cerebellum. *Proc Natl Acad Sci U S A, 115*, 7428–7433.

Ito, M. (1984). *The cerebellum and neural control*. Raven Press.

Ito, M. (2011). *The cerebellum: brain for an implicit self*. FT Press.

Ito, M., Sakurai, M., & Tongroach, P. (1982). Climbing fibre induced depression of both mossy fibre responsiveness and glutamate sensitivity of cerebellar Purkinje cells. *J Physiol, 324*, 113–134.

Jacobi, H., Bauer, P., Giunti, P., Labrum, R., Sweeney, M. G., et al. (2011). The natural history of spinocerebellar ataxia type 1, 2, 3, and 6: a 2-year follow-up study. *Neurology, 77*, 1035–1041.

Lisberger, S. G., & Thach, W. T. (2013). The cerebellum. In E. R. Kandel, J. H. Schwartz, T. M. Jessell, S. A. Siegelbaum, & A. J. Hudspeth (Eds.), *Principles of neural science* (pp. 960–981).

Manto, M.-U., & Pandolfo, M. (2002). *The cerebellum and its disorders*. Cambridge Univ Press.

Martin, T. A., Keating, J. G., Goodkin, H. P., Bastian, A. J., & Thach, W. T. (1996). Throwing while looking through prisms. I. Focal olivocerebellar lesions impair adaptation. *Brain, 119*, 1183–1198.

Moriarty, A., Cook, A., Hunt, H., Adams, M. E., Cipolotti, L., & Giunti, P. (2016). A longitudinal investigation into cognition and disease progression in spinocerebellar ataxia types 1, 2, 3, 6, and 7. *Orphanet J Rare Dis., 11*, 82.

Nishizawa, M., Onodera, O., Hirakawa, A., Shimizu, Y., Yamada, M., & Rovatirelin Study Group. (2020). Effect of rovatirelin in patients with cerebellar ataxia: two randomised double-blind placebo-controlled phase 3 trials. *J Neurol Neurosurg Psychiatry, 91*, 254–262.

Norris, S. A., Hathaway, E. N., Taylor, J. A., & Thach, W. T. (2011). Cerebellar inactivation impairs memory of learned prism gaze-reach calibrations. *J Neurophysiol, 105*, 2248–2259.

Rabe, K., Livne, O., Gizewski, E. R., Aurich, V., Beck, A., Timmann, D., & Donchin, O. (2009). Adaptation to visuomotor rotation and force field perturbation is correlated to different brain areas in patients with cerebellar degeneration. *J Neurophysiol, 101*, 1961–1971.

Raz, N., Ghisletta, P., Rodrigue, K. M., Kennedy, K. M., & Lindenberger, U. (2010). Trajectories of brain aging in middle-aged and older adults: regional and individual differences. *NeuroImage, 51*, 501–511.

Safe, A. F., Cooper, S., & Windsor, A. C. (1992). Cerebellar ataxia in the elderly. *J R Soc Med, 85*, 449–451.

Schmitz-Hübsch, T., du Montcel, S. T., Baliko, L., Berciano, J., Boesch, S., et al. (2006). Scale for the assessment and rating of ataxia: development of a new clinical scale. *Neurology, 66*, 1717–1720.

Shelhamer, M., Tiliket, C., Roberts, D., Kramer, P. D., & Zee, D. S. (1994). Short-term vestibulo-ocular reflex adaptation in humans. II. Error signals. *Exp Brain Res, 100*, 328–336.

Shutoh, F., Ohki, M., Kitazawa, H., Itohara, S., & Nagao, S. (2006). Memory trace of motor learning shifts transsynaptically from cerebellar cortex to nuclei for consolidation. *Neuroscience, 139*, 767–777.

Smith, M. A., & Shadmehr, R. (2005). Intact ability to learn internal models of arm dynamics in Huntington's disease but not cerebellar degeneration. *J Neurophysiol, 93*, 2809–2821.

Thach, W. T., Goodkin, H. P., & Keating, J. G. (1992). The cerebellum and the adaptive coordination of movement. *Annu Rev. Neurosci, 15*, 403–442.

Topka, H., Valls-Solé, J., Massaquoi, S. G., & Hallett, M. (1993). Deficit in classical conditioning in patients with cerebellar degeneration. *Brain, 116*, 961–969.

Trouillas, P., Takayanagi, T., Hallett, M., Currier, R. D., Subramony, S. H., Wessel, K., Bryer, A., Diener, H. C., Massaquoi, S., Gomez, C. M., Coutinho, P., Ben Hamida, M. B., Campanella, G., Filla, A., Schut, L., Timann, D., Honnorat, J., Nighoghossian, N., & Manyam, B. (1997). International Cooperative ataxia Rating Scale for pharmacological assessment of the cerebellar syndrome. The ataxia neuropharmacology committee of the world federation of neurology. *J Neurol Sci, 145*, 205–211.

Walker, M. F., & Zee, D. S. (1999). Directional abnormalities of vestibular and optokinetic responses in cerebellar disease. *Ann NY Acad Sci, 871*, 205–220.

Weiner, M. J., Hallett, M., & Funkenstein, H. H. (1983). Adaptation to lateral displacement of vision in patients with lesions of the central nervous system. *Neurology, 33*, 766–772.

Woodruff-Pak, D. S., Papka, M., & Ivry, R. B. (1996). Cerebellar involvement in eyeblink classical conditioning in humans. *Neuropsychology, 10*, 443–458.

Yasui, K., Yabe, I., Yoshida, K., Kanai, K., Arai, K., Ito, M., Onodera, O., Koyano, S., Isozaki, E., Sawai, S., Adachi, Y., Sasaki, H., Kuwabara, S., Hattori, T., Sobue, G., Mizusawa, H., Tsuji, S., Nishizawa, M., & Nakashima, K. (2014). A 3-year cohort study of the natural history of spinocerebellar ataxia type 6 in Japan. *Orphanet J Rare Dis., 9*, 118.

Part VII
Mechanisms and Models of Spinocerebellar Ataxia and New Treatments

Chapter 23
The Three Cornerstones of Cerebellar Ataxia: Closing the Loop of 200 Years of Cerebellar Research

Pierre Cabaraux, Jordi Gandini, and Mario Manto

23.1 Cerebellum and Cerebellar Research: Cerebellar Scholars from the Eighteenth Century to 2020s

The anatomy of the cerebellum has intrigued investigators for centuries. Although cerebellum was already distinguished from the rest of the brain by Aristotle in the Antiquity, a proper anatomical description of cerebellum was first made by Vincenzo Malacarne in the late eighteenth century in his work entitled "Nuova esposizione della struttura del cervelletto umano, Turin, 1776" (Clarke & O'Malley, 1968; Voogd & Koehler, 2018) (Table 23.1). Malacarne advanced the hypothesis that the number of cerebellar folia was modified by the environment, providing the first approach to the concept of neuroplasticity in the scientific literature (Zanatta et al., 2018). He quantified that the normal cerebellum was composed of about 600 laminae. He also suggested that there was a link between the number of folia and the degree of intelligence. He had observed that the cerebellum was underdeveloped in patients suffering from cretinism. This observation was praised by Charles Bonnet (Zanatta et al., 2018). Malacarne introduced the following terms which are still applied nowadays: vermis, tonsil, pyramid, and uvula. His works were recognized by two famous contemporary anatomists Albrecht von Haller and Felix Vicq d'Azyr, who both dedicated time to understand cerebellar anatomy (Fig. 23.1). From the functional perspective, the description of the motor role of cerebellum was first reported by Luigi Rolando in his ablation studies "Saggio sopra la vera struttura" published from 1807 to 1814. Rolando reported that cerebellar lesions impair

P. Cabaraux · J. Gandini
Department of Neurology, Unité des Ataxies Cérébelleuses, Charleroi, Belgium

M. Manto (✉)
Department of Neurology, Unité des Ataxies Cérébelleuses, Charleroi, Belgium

Department of Neuroscience, University of Mons, Mons, Belgium
e-mail: mmanto@ulb.ac.be

H. Mizusawa, S. Kakei (eds.), *Cerebellum as a CNS Hub*,
Contemporary Clinical Neuroscience, https://doi.org/10.1007/978-3-030-75817-2_23

Table 23.1 Painstaking work of pioneers: two centuries of discoveries

Malacarne (1776)	First description of the human cerebellum
Rolando (1809)	Cerebellar lesions impair posture and voluntary movement
Fodera (1823)	Cerebellectomy causes extensor hypertonia
Flourens (1824)	Cerebellar lesions are linked to muscle incoordination
Magendie (1824)	Cerebellar peduncles contribute to coordination; center of equilibrium
Purkinje (1838)	Histology of the cerebellar cortex
Luciani (1891)	Triad of atonia/asthenia/astasia. Cerebellum is a reinforcer
Lugaro (1894)	Described the elements of the cerebellar cortex
Babinski (1899–1906)	Asynergia, adiadochokinesia, fight against inertia
Sherrington (1900)	Cerebellum regulates a complex proprioceptive system
Holmes (1904–1939)	Cerebellar dysmetria and kinetic tremor
Cajal (1911)	Describes the fine network structure of the cerebellar cortex
Larsell (1937)	Identified 10 lobules (I–X)
Eccles (1963–1964)	Neuronal connectivity in the cerebellar cortex
Voogd (1964–1969)	Longitudinal organization of the cerebellum
Ito (1964)	Inhibitory effect exerted by Purkinje neurons upon cerebellar nuclei
Gilman (1969)	Cerebellum tunes the activity of muscle spindles
Marr and Albus (1969–1971)	Theory of learning
Llinas (1974)	Strong electrotonic coupling between inferior olivary cells
Oscarsson (1976)	Microzone as the functional unit of the cerebellar cortex
Gilbert and Thach (1977)	Increases in complex spike firing rates during motor adaptation
Ito and Kano (1982–1984)	Long-term depression (LTD); concept of microcomplex
Haines (1984)	Reciprocal connections between the hypothalamus and the cerebellum
Mugnaini (1994)	Unipolar brush cells
Schmahmann (1998)	Schmahmann's syndrome (CCAS)

posture and voluntary movement (Fine et al., 2002). Subsequently, in 1823, Michele Fodera in his "Recherches expérimentales sur le système nerveux" observed that cerebellectomy caused extensor hypertonia without paralysis, especially in pigeons and rabbits. Marie Jean-Pierre Flourens (Fine et al., 2002) discussed in 1824 in his "Recherches expérimentales sur les propriétés et les fonctions du système nerveux dans les animaux vertébrés" the consequences of cerebellar lesions on muscle coordination in pigs and dogs. The same year, François Magendie in "Mémoires sur les fonctions de quelques parties du système nerveux" pointed out the specific role of cerebellar peduncles in coordination and proposed the cerebellum as the center of equilibrium (Clarke & O'Malley, 1968; Voogd & Koehler, 2018). After a series of reports on the effects of cerebellar lesions upon motor coordination, a major step was achieved in the morphology of the cerebellum. Indeed, microanatomy of cerebellum was described by Jan Evangelista Purkinje in 1838 in the *Isis Journal of Prague*. These seminal histological studies on the cerebellar led to the identification

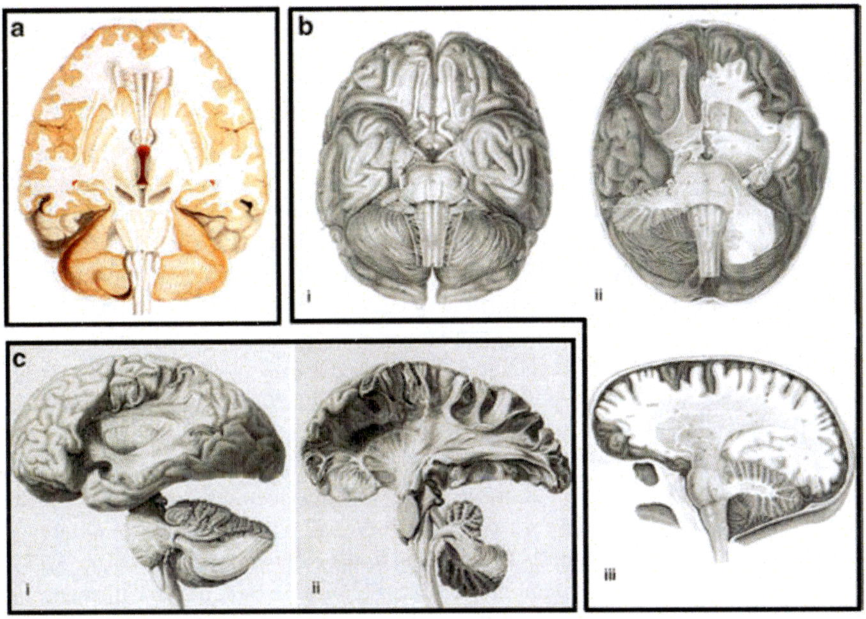

Fig. 23.1 From (Schmahmann, 2016). Depictions of the cerebellum by early anatomists. (**a**) Image from the atlas of Félix Vicq d'Azyr (1786). His Plate IV includes the cerebellum. The image is flipped vertically, as in the atlas the cerebellum is shown at the top. (**b**) Images from the atlas of Franz Joseph Gall and Johann Gaspar Spurzheim (Gall and Spurzheim 1810). (i) Gall and Spurzheim's Plate IV shows the base of the brain with cerebral hemispheres, cerebellum, and brainstem. (ii) Plate XIII shows dissections of the cerebral hemisphere and cerebellum. (iii) Plate X shows cerebral and cerebellar hemispheres partially dissected in the sagittal plane. (**c**) Depictions of white matter dissections of the cerebral hemisphere, cerebellum, and brainstem by Herbert Mayo (1827). (i) Plate III shows dissection of the middle cerebellar peduncle. In (ii), Plate IV, brainstem, and cerebellar dissection with removal of the Middle Cerebellar peduncle (MCP) reveals the inferior and superior cerebellar peduncles

of Purkinje cells. The following decades of the late nineteenth and first half of the twentieth centuries were marked by the description of functional role and anatomic structure of cerebellum as well as clinical observations in animals and human. Luigi Luciani pointed out in 1891 the triad atonia/asthenia/astasia as a consequence of cerebellum insult, describing the cerebellum as a motor reinforcer. Sir Charles Sherrington insisted on a major role of cerebellum as "the head of the proprioceptive system" (Clarke & O'Malley, 1968; Voogd & Koehler, 2018). At the junction between the nineteenth and twentieth century between 1899 and 1906, Joseph Babinski reported asynergia and adiadochokinesia in patients suffering from cerebellum injury (Babinski, 1913). He coined that cerebellar patients exhibit difficulties in adaptation to inertia. In the first half of the twentieth century, Holmes reported on dysmetria and kinetic tremor in patients with cerebellar tumors (Stewart & Holmes, 1904; Voogd & Koehler, 2018). Holmes made meticulous clinical observations in patients during the First World War, pointing the laterality of the deficits.

The helmets of soldiers were poorly designed, rendering the cerebellum particularly vulnerable. In parallel to this clinical progress by pioneers of contemporary neurology, Ernesto Lugaro in 1894 described the elements of cerebellar cortex, and Ramon y Cajal identified the fine network structure of cerebellar cortex, including mossy fibers, glomeruli, parallel fibers, and climbing fibers. Bolk (1906) compared numerous species of mammals and proposed that vermis and hemispheres are folial chains. Olof Larsell set up the actual classification of ten cerebellar lobules around 1937 (Larsell, 1952, 1958; Voogd & Koehler, 2018). Between 1952 and 1954, Larsell provided detailed observations on the cerebellum of rat, cat, pig, monkey, and human, assessing both the developmental phases and adult specimens. Each lobule of the vermis was coined by Roman numerals (I to X), and the hemispheric portion was identified by the prefix H (HII to HX). He underlined that cerebellum is composed of an anterior lobe, a posterior lobe, and a flocculo-nodular lobe (see also Sect. 2). Researchers from the second half of twentieth century focused on the microarchitecture and the fundamental electrophysiological properties of cerebellum circuitry. In the 1960s, Sir John Eccles (1967) discovered the precise neuronal connectivity in the cerebellar cortex, while Jan Voogd pointed out the longitudinal organization of the cerebellum between 1964 and 1969 (Voogd & Koehler, 2018). At the same time, Masao Ito demonstrated the inhibitory effect of Purkinje neurons on cerebellar nuclei by studying the effect of stimulation of anterior vermian part on neurons of Deiters nucleus in cats (Ito et al., 1964). Latency of the inhibitory postsynaptic potentials (IPSPs), the consequence of the monosynaptic interaction between the cerebellar cortex and nucleus, was measured at 0.7–0.9 msec and was found in all cerebellar nuclei. Ugawa confirmed these observations in human using transcranial magnetic stimulation (Ugawa, 2009). The 1960s were marked by the theory of learning proposed by Marr in 1969 followed by Albus in 1971. This theory takes into account the double innervation of Purkinje cells by mossy fibers (via parallel fibers) and climbing fibers from the inferior olive, and was assigned to the cerebellar cortex the ability to learn motor skills. This was a major conceptual step to understand that cerebellum is a major site of plasticity in the CNS. Research in the 1970s led to the discovery of the strong electric coupling between inferior olivary cells and the remote control of Purkinje cells by climbing fibers (Llinas et al., 1974). Subsequently, the cerebellar cortex was divided into microzones, considered as elemental functional units (Oscarsson, 1979). Oscarsson demonstrated that spinocerebellar tracts target the anterior lobe and lobule VIII (spinocerebellum), without projections toward the so-called lateral cerebellum (neocerebellum). The role of complex spike firing rates during motor adaptation was underlined (Gilbert & Thach, 1977). Indeed Gilbert and Thach showed that motor learning takes place in the cerebellum by changes in the strength of transmission of parallel fiber synapses on Purkinje neurons. Later, Masao Ito and Masanobu Kano supported the learning theory proposed by Marr and Albus by theorizing and demonstrating the long-term depression (LTD) model of parallel fiber-Purkinje cell transmission (Ito & Kano, 1982). Consistent with Marr-Albus' theory, motor learning is impaired in animals when pharmacological reagents altering the cascades involved in LTD are applied. Another milestone was reached when the concept of cerebellar microcomplex was proposed (Ito, 1984). Cerebellar cortex is now considered as gathering different classes of Purkinje neurons forming arrays of transverse zones, each divided in parasagittal stripes (Armstrong & Hawkes, 2016). Cerebellum contains more than 200 stripes, and this pattern is conserved across birds and mammals. The architecture of zones and stripes appears early during cerebellar development. The clinical implications of these stripes remain to be discovered. Progress in genetics has led to a better classification of cerebellar ataxias. The advent of next-generation sequencing is now reconfiguring and increasing the spectrum of neurological disorders affecting the cerebellum. Closing the loop of nearly 200 years of research from all over the world, the third cornerstone of cerebellar ataxiology, the cerebellar cognitive syndrome, thanks to Jeremy Schmahmann's work (Schmahmann & Sherman, 1998), was finally suggested and named "Schmahmann's syndrome" few years later, extending to cognitive/affective operations the theory of dysmetria (Manto & Mariën, 2015).

23.2 Brief Functional Anatomy of the Cerebellum: Extracting the Key Elements to Understand the Three Cornerstones of the Clinical Syndrome

Cerebellum is localized in the posterior fossa, connected to the brainstem though three pairs of peduncles, and separated by the fourth ventricle. The cerebellum is composed of a midline part (the cerebellar vermis) and two lateral hemispheres. Cerebellum is divided into three lobes, from top to bottom in an unfolded view: the anterior, the posterior, and the flocculo-nodular lobe. Clinical observations and symptom-lesion studies have led to the subdivision of the cerebellum into (a) a medial portion connected to axial and proximal limb muscles responsible for posture, gait, and slow movements and (b) a lateral portion mainly connected to cerebral cortices (the so-called cerebrocerebellum, lateral cerebellum, or neocerebellum) mainly responsible for fast limb movements (Manto & Habas, 2013; Fig. 23.2).

The subdivision of cerebellum by Larsell into ten lobules, numbering it in Roman nomenclature from I to X (Larsell, 1958; Manto & Habas, 2013), allows to distinguish (a) lobules I to V constituting the anterior lobe, (b) lobules VI to IX gathered in the posterior lobe, and (c) the flocculo-nodular lobe (archicerebellum) corresponding to lobule X (Larsell, 1952, 1958; Manto & Habas, 2013). Lobules are divided into a medial (v, vermian) and a lateral part (H, hemispheric).

Cerebellum shows a highly organized cortex at the periphery, with the three pairs of nuclei embedded in white matter (Manto & Habas, 2013). Cerebellum is organized into microzones (Oscarsson, 1979; Fig. 23.3). Each microzone is composed of a group of approximately 1000 Purkinje cells, all having the same somatotropic receptive field. These neurons are arranged in a long, narrow strip, oriented perpendicular to the cortical folds so that Purkinje cell dendrites are flattened in the same direction as the microzones extend and are crossed by parallel fibers at right angles.

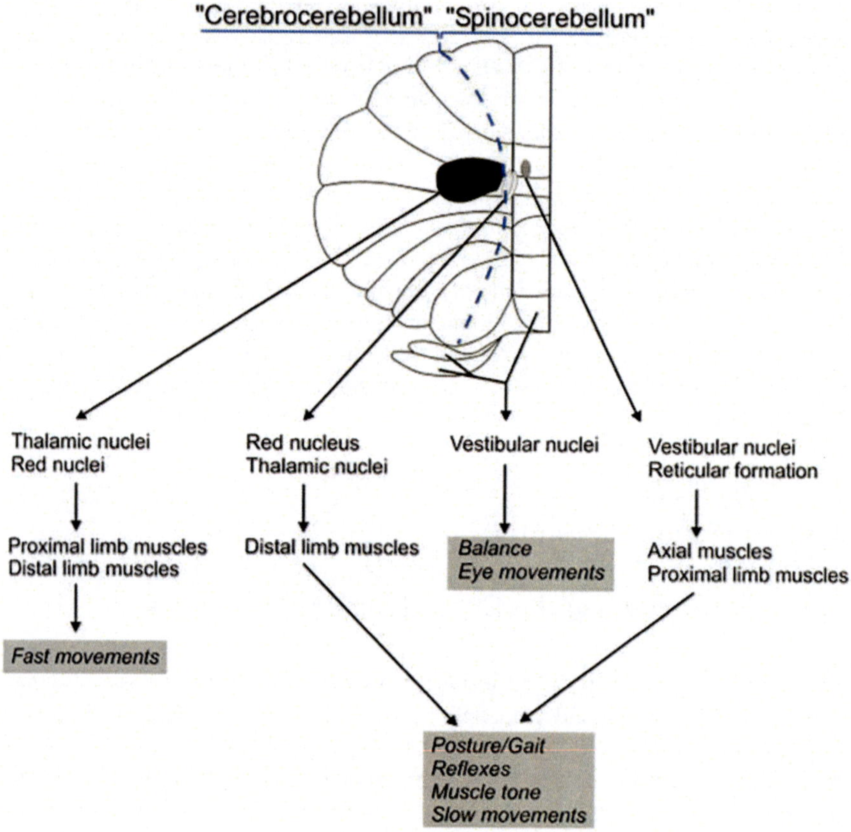

Fig. 23.2 From (Grimaldi et al. 2013). Organisation of cerebellum though its different segment. Cerebellum is subdivided into a medial part, so-called spinocerebellum responsible for posture, gait and slow movements, and a lateral one so-called cerebrocerebellum responsible for fast movement

The climbing fibers' branches (about ten) usually innervate Purkinje cells (there is a 1:1 relationship between one climbing fiber and the dendritic tree of a Purkinje neuron) belonging to the same microzone, and the olivary neurons generating such climbing fibers tend to be coupled by gap junctions. This contributes to synchronizing Purkinje cells within a microzone on a millisecond time scale. The Purkinje cells belonging to a microzone all send their axons to the same small cluster of output cells within the deep cerebellar nuclei (Fig. 23.3 – see modules below). Finally, the axons of basket cells are much longer in the longitudinal direction than in the mediolateral direction (not shown), causing them to be confined largely to a single microzone. Thus, cellular interactions within a microzone are much stronger than those between different microzones, configured perpendicularly to the cortical folds, and receive a homogeneous group of climbing fibers coming from the contralateral inferior olivary neurons. They project through a homogeneous group of

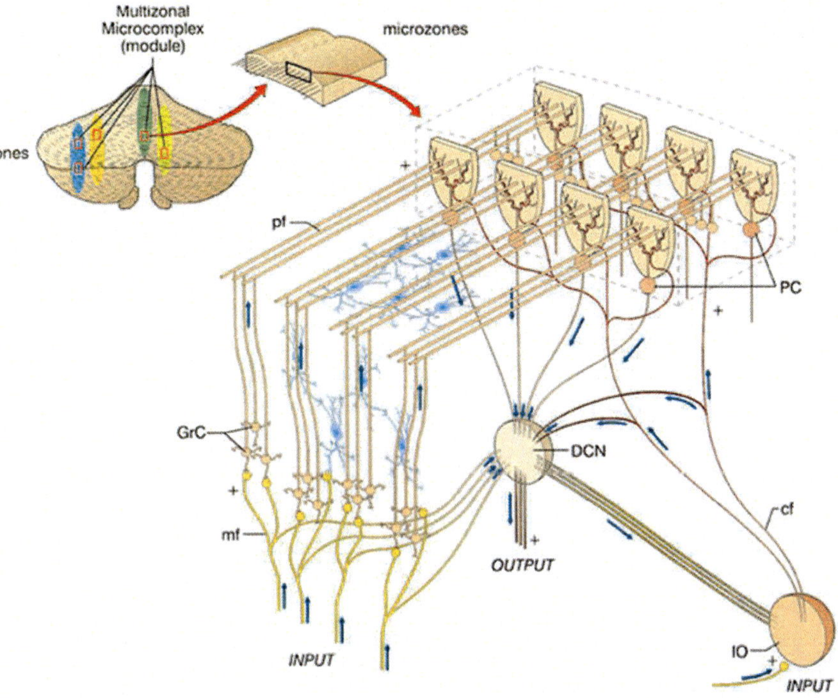

Fig. 23.3 From D'Angelo & Casali, 2013. The modular organization of the cerebellum. The picture shows a flattened view of the cerebellum. Four ideal zones are shown in color, each one containing microzones forming a multizonal microcomplex. The microzones have the basic structure reported in the expansion below, inhibitory interneurons are overlaid in blue

neurons on a deep cerebellar nucleus where they exert a powerful inhibition (D'Angelo & Casali, 2013; Manto & Habas, 2013).

Cerebellum is organized into discrete microcomplexes (Fig. 23.4). Each microcomplex is the recipient of a specific motor error signal from the inferior olive (Ito, 1984; Apps & Garwicz, 2005). Finally, at a macro-scale level, cerebellar modules correspond to a conglomerate of several non-adjacent parasagittal bands of PCs projecting to specific areas of deep cerebellar nuclei and gating segregated projections from the inferior olive (D'Angelo & Casali, 2013). In other words, climbing fibers arising from multiple subunits of olivary nucleus target several microcomplexes from cerebellar cortex, all projecting on the same subunit of cerebellar nuclei cells after having integrated the different inputs. Modules gather clusters of neighboring Purkinje neurons which are coherently active during particular physiological operations, and the state of one of the components affects the state of the other components of the module (Broersen et al., 2016).

Based upon anatomical, neuroimaging, and neuropsychological studies, specific loops running in parallel between the cerebral cortex and cerebellum have been identified (Grimaldi & Manto, 2012). Cerebellum circuitry is indeed segregated into

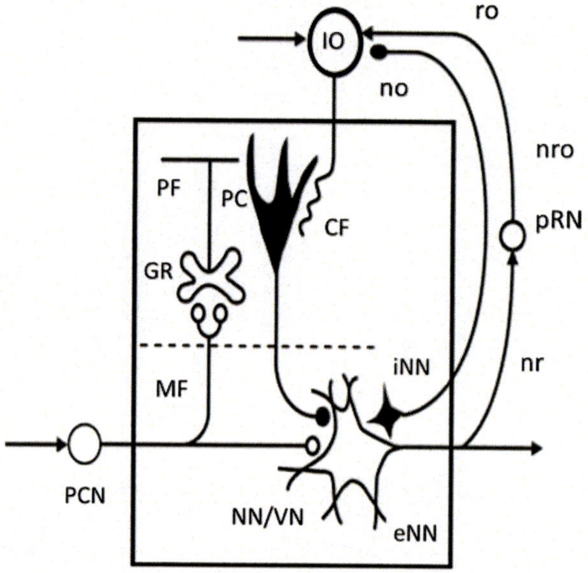

Fig. 23.4 From (Ito, 2013). Basic circuit structure of microcomplex. (Abbreviations: *CF* climbing fiber, *GR* granule cell, *IO* inferior olive, *MF* mossy fiber, *NN/VN* nuclear neuron/vestibular nuclear neuron, *no* nucleoolivary pathway, *nro* nucleorubroolivary pathway, *PC* Purkinje cell, *PCN* precerebellar neuron, *PF* parallel fiber, *pRN* parvocellular red nucleus, *eNN* excitatory nuclear neuron, *iNN* inhibitory nuclear neuron, *ro* rubroolivary pathway, *nr* nucleorubral pathway Inhibitory neurons are in black (others are excitatory))

different functional compartments having a specific connection role with specific cerebral cortex zones (Figs. 23.5 and 23.6).

Output channels from dentate nuclei are segregated: the dentate nucleus contains anatomically distinct motor and nonmotor domains. The neurons targeting motor areas in the cerebral cortex are located in the dorsal parts of the nucleus, whereas the neurons projecting to association cortex are located in the ventral portions of the nucleus (Van Dun et al., 2016). The sensorimotor cerebellum projecting to motor areas (via the dorsal portions of the nucleus) are located in HIV/HV/HVI/HVIII. There is nearly no overlap between the sensorimotor network and the cognitive neocerebellar areas involved in executive control, working memory, attention, flexibility, emotional information, and self-reflection.

Over the last two decades, cerebellum has appeared as a major node of closed loops for:

A. The default mode network comprising BA9/10, BA32, BA7, BA39, BA23, BA29/30, medial temporal lobe and ventral temporal cortex, thalamus, red nucleus, midbrain, lobule IX, and lobule VIIB.
B. The executive network entailing BA45/46, BA9, BA8, BA44/45/46, BA47, BA7, BA39, BA40, caudate nucleus, crus I/II, and lobules VI/VIIB/IX.
C. The salience network which includes BA32, BA24, BA46/47, thalamus, red nucleus, and lobule VI.

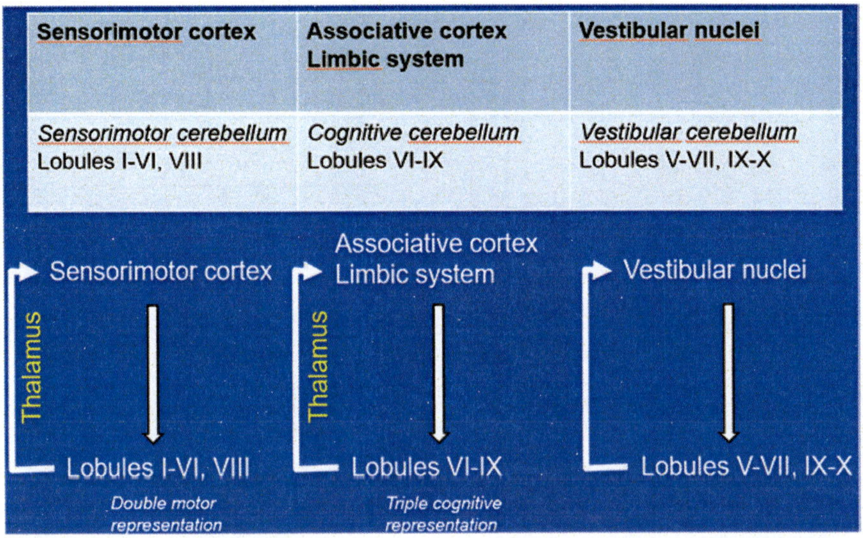

Fig. 23.5 From (Manto et al. 2015). Cerebellar main functions described in ataxiology exhibit distinct connectivity between cerebellar and cerebral cortices

Sensorimotor cerebellum is subdivided into a primary zone corresponding to anterior lobe and lobe VI and a second sensorimotor region located exclusively in lobule VIII.

Brain cortical motor information projects to contralateral cerebellar anterior lobe through the inferior half of the pons. The anterior lobe projects back through the thalamus to cortical motor areas. The so-called cognitive cerebellum is composed of lobules from the posterior lobe (from VI to IX). Fibers from prefrontal cortices project to the medial and rostral pons, while the fibers from posterior areas project to the dorsal and ventral part of the pontine nucleus before making the connection with the posterior cerebellar cortex. This last structure relays then in the thalamus before connecting again to the cerebral cortex. Limbic system in addition, linked to emotion, projects on the vermis.

Finally vestibular nuclei project on flocculo-nodular lobe on three different areas: the flocculus-paraflocculus, the nodulus/ventral uvula (corresponding to lobules IX and X), and the dorsal oculomotor vermis corresponding to lobules V to VII (Grimaldi & Manto, 2012; Manto & Mariën, 2015).

Due to their anatomical connectivity, the vestibular nuclei are at a sensorimotor crossroad (Barmack, 2016). They are influenced by visual information (optokinetic signals reach the vestibular complex from the accessory optic system), proprioceptive signals (proprioceptors in the neck activate vestibular neurons: movements of the head induce reflexive eye movements and postural adjustments), cerebellar

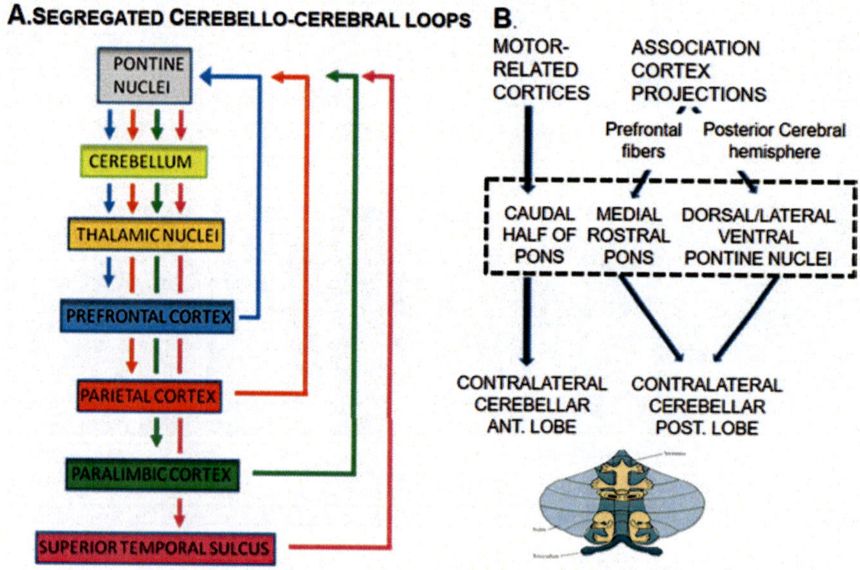

Fig. 23.6 From (Grimaldi et al. 2012). (**a**). Illustration of the segregated loops between the cerebellum and prefrontal cortex, parietal cortex, paralimbic cortex and superior temporal sulcus. (**b**). Topographic distribution of motor-related cortices and association cortex projections the cerebellum. Both motor corticopontine projections and association cortex projections (from prefrontal, posterior parietal, superior temporal, parastriate, parahippocampal and cingulated regions) are somatotopically organized in the pons

signals (mainly from uvula and nodulus and flocculi), and signals emerging from the cerebrum. Vestibular nuclei target the thalamic nuclei, brainstem nuclei, and spinal cord (vestibulospinal tracts). Vestibular projections participate in vestibulocollic reflexes. All these anatomical structures explain the importance of a detailed oculomotor assessment in cerebellar patients. Oculomotor deficits are often prominent in cerebellar patients, such as errors in saccades, pursuit, VOR (vestibular eye movements stabilize gaze during head rotations or translations), fixation (holding an eccentric position of gaze requires a tonic eye position signal), or vergence. The main cerebellar structures involved in oculomotor control can be gathered in three units communicating with the fastigial nuclei: the flocculus/paraflocculus, the nodulus/uvula, and the dorsal vermis (lobules VI–VII).

Functional organization of cerebellum has been studied using mainly two techniques. On one hand, connections between cerebellum and cortical and subcortical areas have been suggested through anatomic resting-state functional connectivity magnetic resonance imaging (fcMRI) work studies. fcMRI relies on temporal correlations between spontaneous low-frequency (0.01–0.1 Hz) fluctuations of the BOLD (blood oxygenation level-dependent) signal between spatially distinct but functionally related cortical and subcortical regions (Biswal et al., 1995; Bernard et al., 2012).

Regions with synchronous spontaneous activity constitute intrinsic connectivity networks (ICNs) and may be linked by mono- or polysynaptic pathways. The ICNs derived from fcMRI studies have been shown to overlap to a large degree with maps of structural connectivity derived from diffusion tensor imaging (DTI) tractography (Stoodley & Schmahmann, 2018; Habas & Manto, 2018). Indeed, these studies have shown that there is an anatomical parcellation of the cerebellar cortex into regions, based on their specific coherence with distinct extracerebellar cortical regions of interest (ROIs), using a probabilistic cerebellar atlas (Diedrichsen et al., 2009; Riedel et al., 2015). On the other hand, neurotropic viral tracers have been successfully used to study the connection between cerebellum and cortical and subcortical areas through works of Peter L. Strick. A disynaptic pathway linking the striatum and deep cerebellar nuclei was first suggested by the injection of rabies virus and the retrograde transneuronal transport from the sensorimotor territory of the striatum to the thalamus and deep cerebellar nuclei, essentially to the dentate nuclei (Hoshi et al., 2005). Later, through the same experimental methods, a disynaptic pathway link between the cerebellar cortex and the subthalamic nuclei (STN) was discovered through the retrograde transneuronal transport of rabies virus from cerebral cortex to pontine nuclei and the STN (Bostan et al., 2010). Strick and his collaborators first arrived to the conclusion that cerebello-cortical loops and cortico-basal ganglia loops interacted only at the cerebral cortex level: outputs of basal ganglia and cerebellum project on distinct thalamic nuclei. However, recent data review suggests interconnection of the basal ganglia and the cerebellum at the subcortical level through disynaptic projection of the STN to the cerebellar cortex and disynaptic connections of the dentate nucleus and the striatum (Bostan & Strick, 2019). Further studies suggested moreover that projections from STN to the cerebellum originate from three functional divisions, subdividing STN in a sensorimotor, an associative, and a limbic part (Pozzi et al., 2016).

23.3 Cerebellum's Role in Motor and Nonmotor Functions

Since centuries, descriptions of clinical features of cerebellar disorder were mainly based upon observation related to cerebellar motor dysfunction. General rules of motor control suggest that optimal strategies are required to perform motion with accuracy, given the highly complex nonlinear biomechanical features of the human body, including the muscles and joints, and the numerous interactions with the environment. The central nervous system (CNS) copes with noise and delays, which are inherent to biology and also motion. The notion of noise in biological signals includes both the input noise and the internal noise. Noise may also fluctuate with time or according to a particular sensorimotor context. Therefore, a high degree of adaptability and modifiability in the operational mechanisms underlying motor control is required, especially for learning procedures. Breaking those rules leads to motor and nonmotor dysfunction of cerebellum, as dysarthria, limb ataxia, and postural/gait deficits, all having in common a lack of anticipation, adaptability, and timing perception impairment (Manto & Habas, 2013). The projection of this specific role of cerebellum to the cognitive-affective field gave rise to the theory of the "dysmetria of thought" proposed by Schmahmann arguing that the mismatch between reality and perceived reality induced a lowering of overall cognitive functions in patients suffering from cerebellar insults (Schmahmann, 1998).

Neuroimaging and neurstimulation studies allowed to suggest cerebellum as a main actor of motor control, thanks to its role in learning, timing, and prediction tasks in motor, cognitive, and emotional/affective function. Moreover, symptom-lesion mapping studies allowed to situate cerebellar cognitive-affective syndrome in cerebellar zones, to differentiate zones specifically attributed to one aspect of Schmahmann's syndrome (Fig. 23.7), and to establish connections with specific cortical areas.

A unique general organization of the cerebellum and cellular architecture, which are highly conserved in vertebrates, is the anatomical substratum of cerebellar electrophysiological property. Simple and repeated fine structure in the cerebellum according to David Marr's works (Marr, 1969) and redundant architecture of cerebellar folia pointed out by Fritz Kahn suggested a highly connected and modulable cerebellar circuitry ideal for motor learning. At a microlevel, research on properties of PCs, their synaptic plasticity, and their connections between each other via the parallel fibers reinforced this concept (D'Angelo, 2018; Bareš et al., 2019).

Patients with cerebellar disorders exhibit impairments on a range of tasks that require precise timing suggesting an involvement of cerebellar cortex in sensorimotor tasks and perceptual tasks requiring precise timing of salient events (Bareš et al.,

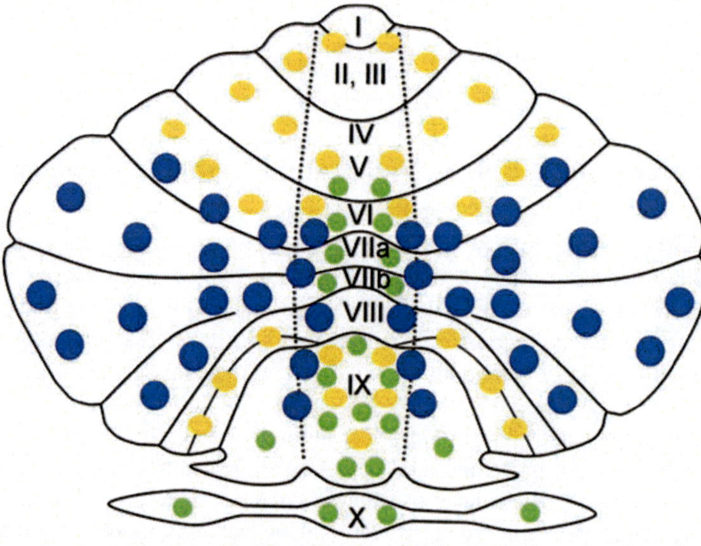

Fig. 23.7 From (Manto et al. 2015). Representation of 10 lobules of the cerebellum (unfolded cerebellum) and areas involved in cerebellar motor syndrome (CMS), vestibulo-cerebellar syndrome (VCS) and cerebellar cognitive affective syndrome (CCAS or Schmahmann syndrome)

2019). Further analyses allowed indeed to point out the role of cerebellar cortex in the representation of temporal relationships in a subsecond range in contrast with the suprasecond range operations of subcortical and cerebral cortices (Bareš et al., 2019). The timing machine concept of cerebellum was born.

There is compelling evidence that cerebellum plays a role in motor and nonmotor predictions. In this context, current dominant theory about cerebellum role uses the internal modern concept. Internal models provide representations of the input-output properties of the motor apparatus to adapt movement control (Manto & Habas, 2013; Lawrenson et al. 2018). Two groups have been hypothesized: the forward model and inverse model.

On one hand, cerebellar circuitry operates as a forward controller and contributes to the predictive mechanisms inherent to sensorimotor control and adaptive behavior (Llinás R et al. 1980; Wolpert et al. 1998; D'Angelo 2018). The hypothetic role of cerebellum of computing both predicted sensory state from previous ones and actual sensory state through efference copies of motor signals in order to generate an adaptation of motor behavior has given birth to the forward internal model (Popa et al. 2014, 2019; D'Angelo 2018). On the other hand, inverse model is the ability of cerebellum to transform a desired outcome or effector state into the necessary commands to achieve that state. This last is considered as the inversion of the forward model (Manto and Habas, 2013; Lawrenson et al. 2018). The cerebellum is also considered as an acquirer and storer of internal models. Moreover, several properties of PCs' simple spike firing during arm movements are consistent with a forward internal model. PCs in lobules IV–VI of the intermediate and lateral cerebellum carry signals related to the position, the direction, the amplitude, and the speed of movement (Lawrenson et al. 2018).

Internal models attribute the special ability to cerebellum to anticipate actions and their consequences to produce and adapt motor sequences. Impairment of this ability was first observed and studied in cerebellar patients by Holmes in 1922, leading to ataxic dysmetria. The concept of limb dysmetria gathers error in trajectory of movement secondary to an abnormal range of force of motion and velocity (Holmes 2007). Hypothesis to explain cerebellar dysmetria would be a bias in internal models of limb dynamics denoting a mismatch between predicted and actual sensory outcomes in cerebellar patients. Underestimation of inertia causes an overshoot, and overestimation of inertia causes an undershoot (Bhanpuri 2014; Lawrenson et al. 2018; Cabaraux et al. 2020). On similar basis, biased internal model of grip force control would produce force overshoot.

Another function of cerebellum in establishing the body state needed for movement has been proposed by researchers over years according data from clinical, experimental and neuroimaging studies (Molinari et al. 2009). Cerebellum compares sequences of movement with previously stored sequences in its microcomplex network as working memory. If sequence stored matches with the actual motor plan, cerebellum will expect of pattern of repetition with those previously memorized to lead to an anticipated response. However, in case of mismatch, an error will be generated, and prediction system will be corrected. In other words, microcomplexes cluster information of efferent copy originating in motor plan generated in area M1 and from incoming sensation. While computing theses different information, they elaborate a comparison between the actual and the predicted sensation. While match between both information will maintain the pattern, mismatch will recalibrate the forward model (Molinari et al. 2009, 2016; Lawrenson et al. 2018). Interestingly, this specific function of cerebellum is compatible with adaptation of the same order regarding cognitive and behavioral function.

One example of errors in motor planning is observed in cerebellar patients showing cerebellar dysarthria. Speech is typically characterized by scanning and explosive features resulting from a lack of coordination between articulation and phonation (Silveri et al. 2016). Dysarthria is observed in particular in patients presenting a progressive loss of neurons and in case of the lesions of the superior portion of the cerebellum. It is presumed that errors in timing are a key mechanism of speech deficits. The hypothesis of the cerebellum holding an "internal clock" involved in both production and perception of sounds would explain dysarthria, keeping in mind that (a) cerebellum contributes to neuronal loops with BA44/45 (Broca's area) and premotor cortex and (b) it is now established that the cerebellum is part of a cerebellum-basal ganglia-cerebellum loop with the subthalamic nucleus projecting to pontine nuclei at the origin of mossy fibers. Cerebellar patients are known to show difficulties to estimate/discriminate the duration of time intervals. Both in children and in adults, "articulatory gestures" would be deficient as a result of the inability to handle feedback signals during articulatory slow movements and to create, store, select, and update internal models during articulatory fast movements (Manto et al. 2018). One has also to consider that speech represents a direct linguistic link between speaker and listener, with a contribution of mirror neurons.

Compelling evidences suggest a role of cerebellum in motion perception, an essential function for many daily life activities from motor learning, driving, and social interaction to body language. According to neurophysiological and lesion research, visual perception of motion associates closed loops between cortical structure as frontal, parietal, and temporal lobe in its superior temporal sulcus mainly in the right hemisphere and the cerebellum. This function, particularly important in social behavior, would set up a bridge between cerebellum and neuropsychiatric disease as schizophrenia and autism described in patient suffering from cerebellar disorder (Baumann et al. 2015). The spectrum of the CCAS/Schmahmann's syndrome (Table 23.2) illustrates that a genuine field of neuropsychiatry of the cerebellum has emerged, with noticeable impacts on daily life in cerebellar patients. Recent works on social cognition (the capacity to understand another person's emotions, intentions, beliefs, and personality traits, based on observed or communicated behaviors) indicate that social interactions are also under cerebellar guidance (Van Overwalle et al. 2019). Again, it is hypothesized that the cerebellar circuitry builds internal action models of human social interactions with the aim of predicting how other people's actions will be performed. Cerebellum would allow to anticipate others' actions. There would be an automatization of our social interactions, with immediate detection of disruptions in action sequences. That would serve as a basis to understand behaviors and to detect violations in social links. Because the fundamental anatomical organization of the cerebellum is highly similar across species

Table 23.2 Main neuropsychiatric features observed in cerebellar patients

Impaired attentional control: Inattentiveness, hyperactivity, rumination, obsessions
Impaired emotional control: Disinhibition, anhedonia, dysphoria
Symptoms suggestive of psychosis: Illogical thought, muted affect, emotional blunting
Social skill set: Irritability, immaturity, difficulties in social interactions
Autism spectrum: Stereotypies, sensory overload

(cerebellar microcomplexes), this hypothesis could have major impacts to understand social interactions in social animals and to establish novel therapeutic strategies in cerebellar patients showing difficulties in social skill set (aggression, opposition, passivity, difficulties to manage social boundaries, difficulties with social cues). The idea of a multipurpose computer designed to smooth voluntary motion is also proposed for mental operations. The context-response linkage that was proposed by Thach is presumed to be valid for the tuning of higher functions. Schmahmann coins the dysmetria of thought and argues that the universal cerebellar transform maintains function around a homeostatic baseline in a given context. The topography of the lesion would dictate the appearance of a deficit.

23.4 Toward an Evaluation of Motor and Nonmotor Cerebellar Dysfunction: From Clinical Scales to Assessment of the Cerebellar Reserve

Subdivision in the three main cerebellar syndromes, motor, vestibulo-cerebellar, and cognitive-affective, is mainly justified by clinical symptoms (Table 23.3) and the highlighting of specific regions impaired by cerebellar disorder.

Scales based on clinical features of cerebellar disorder have been created over years to score cerebellum clinical issues (Bodranghien et al. 2016; Burk et al. 2018) integrating motor, vestibulo-ocular, and cognitive-affective aspects of cerebellum disorders. They constitute a way to assess and follow evolution of the disease over time.

Generally, existing scales evaluate at best one or two aspects of cerebellar syndrome. For example, based upon Schmahmann's studies on cerebellum's contributions on superior function, the cerebellar cognitive-affective syndrome (CCAS) scale evaluates in ataxic patients all aspects of Schmahmann's syndrome, executive, visuospatial, and linguistic functions and affective regulation (Hoche et al. 2018), while the international cooperative ataxia rating scale (ICARS) evaluates general motor function and vestibulo-ocular functions but not the cognitive-affective field of cerebellum (Trouillas et al. 1997).

Separated assessment of the clinical syndrome of cerebellar ataxiology inevitably fails to encompass its different aspects. An assessment of the three cornerstones of cerebellar ataxiology would lead to a better clinical follow-up of patients

Table 23.3 Scales evaluating different aspects of cerebellar syndrome. While cerebellar syndrome may includes symptoms of these three aspects, no scale evaluates all of them together

Syndrome	CMS	VCS	CCAS
Deficits	Ataxic dysarthria Limb dysmetria Postural/Gait ataxia	Dysmetria of saccades Impaired pursuit Nystagmus Impaired VOR Skew deviation	Executive deficits Visuo-spatial deficits Linguistic deficits Affective dysregulation
Scale	ICARS (19 items) SARA (8 items) BARS (5 items) FARS*	ICARS BARS	CCAS rating scale

*Includes a functional staging of ataxia, activities of daily living (ADL), neurological examination and instrumental testing

benefiting therapy and to more realistic results of clinical trials regarding cerebellar disorder treatment.

Given the critical importance of the cerebellar reserve (Mitoma et al. 2020) in the therapies of cerebellar disorders, the discipline of ataxiology now requires the identification of powerful biological, neurophysiological, and neuroimaging biomarkers (structural imaging, functional imaging, isotopic methods). Cerebellar reserve is defined as the capacity of the cerebellum to compensate for tissue lesion or loss of function. When the lesion is due to an acute focal damage such as a stroke, the cerebellar function may be compensated for by other cerebellar areas or by extracerebellar structures. When cerebellar neuronal integrity is gradually affected as observed in neurodegenerative ataxias, the affected area itself might compensate functionally for the slowly progressing cerebellar lesion. Cerebellar reserve is reinforced by environmental enrichment through mechanisms of autophagy and synaptogenesis, highlighting the importance of plasticity potentiated by experience. Biomarkers are essential tools to follow the pathogenic process leading to cerebellar ataxia and to quantify the effects of physical therapy, occupational therapy, speech therapy, drugs, or gene therapies which are now envisioned for real in patients. The revolution brought by MRI to visualize posterior fossa structures now requires an integrative effort to embrace other tools.

23.5 Conclusion

In summary, it is now established that (1) cerebellar circuitry is involved in segregated cerebello-cerebral loops between the cerebellum and cerebral cortex and (2) anatomical connections link the cerebellum to basal ganglia without anatomical relay in the cerebral cortex. The cerebellum is embedded in large-scale brain networks and is critical to perform accurate motor predictions related to body dynamics and environmental events, especially in a context of novelty. The role of cerebellar cortex in inhibiting deep cerebellar nuclei and sculpting of patterns of discharges according to the context is clearly established. The classification of the cerebellar deficits into a cerebellar motor syndrome (**CMS**), a vestibulo-cerebellar syndrome (**VCS**), and **Schmahmann's syndrome** (CCAS – cerebellar cognitive-affective syndrome) is justified not only on a clinical basis but also on neuroanatomical findings. The scientific community has now in hands validated clinical scales. However, these scales analyze only the different aspect of cerebellar ataxiology. There is a need to elaborate a UCARS (universal cerebellar ataxia rating scale) based on CMS, VCS, and CCAS. Efforts are ongoing (SODA scale) to validate a clinical scale dedicated to the VCS. Therapies are being tested, with a focus on one cornerstone only. We can predict that some trials will lead to false-negative results due to a mismatch between the actual needs and the design.

References

Albus, J. A. (1971). A theory of cerebellar function. *Mathematical Biosciences, 10*, 25–60.

Apps, R., & Garwicz, M. (2005). Anatomical and physiological foundations of cerebellar information processing. Nature reviews. *Neuroscience, 6*(4), 297–311. https://doi.org/10.1038/nrn1646

Armstrong, C., & Hawkes, R. (2016). Zones and stripes. In D. Gruol, N. Koibuchi, M. Manto, M. Molinari, J. Schmahmann, & Y. Shen (Eds.), *Essentials of cerebellum and cerebellar disorders*. Springer. https://doi.org/10.1007/978-3-319-24551-5_14

Babinski, J. (1913). *Expose des travaux scientifiques*. Masson.

Bareš, M., Apps, R., Avanzino, L., Breska, A., D'Angelo, E., Filip, P., Gerwig, M., Ivry, R. B., Lawrenson, C. L., Louis, E. D., Lusk, N. A., Manto, M., Meck, W. H., Mitoma, H., & Petter, E. A. (2019). Consensus paper: Decoding the contributions of the cerebellum as a time machine. From neurons to clinical applications. *Cerebellum (London, England), 18*(2), 266–286. https://doi.org/10.1007/s12311-018-0979-5

Barmack, N. H. (2016). Vestibular nuclei and their cerebellar connections. In D. Gruol, N. Koibuchi, M. Manto, M. Molinari, J. Schmahmann, & Y. Shen (Eds.), *Essentials of cerebellum and cerebellar disorders*. Springer. https://doi.org/10.1007/978-3-319-24551-5_8

Baumann, O., Borra, R. J., Bower, J. M., Cullen, K. E., Habas, C., Ivry, R. B., Leggio, M., Mattingley, J. B., Molinari, M., Moulton, E. A., Paulin, M. G., Pavlova, M. A., Schmahmann, J. D., & Sokolov, A. A. (2015). Consensus paper: The role of the cerebellum in perceptual processes. *Cerebellum (London, England), 14*(2), 197–220. https://doi.org/10.1007/s12311-014-0627-7

Bernard, J. A., Seidler, R. D., Hassevoort, K. M., Benson, B. L., Welsh, R. C., Wiggins, J. L., Jaeggi, S. M., Buschkuehl, M., Monk, C. S., Jonides, J., & Peltier, S. J. (2012). Resting state cortico-cerebellar functional connectivity networks: A comparison of anatomical and self-organizing map approaches. *Frontiers in Neuroanatomy, 6*. https://doi.org/10.3389/fnana.2012.00031

Bhanpuri, N. H., Okamura, A. M., & Bastian, A. J. (2014). Predicting and correcting ataxia using a model of cerebellar function. *Brain: A Journal of Neurology, 137*(Pt 7), 1931–1944. https://doi.org/10.1093/brain/awu115

Biswal, B., Yetkin, F. Z., Haughton, V. M., & Hyde, J. S. (1995). Functional connectivity in the motor cortex of resting human brain using echo-planar MRI. *Magnetic Resonance in Medicine, 34*(4), 537–541. https://doi.org/10.1002/mrm.1910340409

Bodranghien, F., Bastian, A., Casali, C., Hallett, M., Louis, E. D., Manto, M., Mariën, P., Nowak, D. A., Schmahmann, J. D., Serrao, M., Steiner, K. M., Strupp, M., Tilikete, C., Timmann, D., & van Dun, K. (2016). Consensus paper: Revisiting the symptoms and signs of cerebellar syndrome. *Cerebellum (London, England), 15*(3), 369–391. https://doi.org/10.1007/s12311-015-0687-3

Bostan, A. C., & Strick, P. L. (2019). The basal ganglia and the cerebellum: Nodes in an integrated network. *Nature reviews. Neuroscience, 19*(6), 338–350. https://doi.org/10.1038/s41583-018-0002-7

Bostan, A. C., Dum, R. P., & Strick, P. L. (2010). The basal ganglia communicate with the cerebellum. *Proceedings of the National Academy of Sciences of the United States of America, 107*(18), 8452–8456. https://doi.org/10.1073/pnas.1000496107

Broersen, R., Winkelman, B. H. J., Ozyildirim, O., & De Zeeuw, C. I. (2016). Physiology of Olivo-cerebellar loops. In D. Gruol, N. Koibuchi, M. Manto, M. Molinari, J. Schmahmann, & Y. Shen (Eds.), *Essentials of cerebellum and cerebellar disorders*. Springer. https://doi.org/10.1007/978-3-319-24551-5_44

Bürk, K., & Sival, D. A. (2018). Scales for the clinical evaluation of cerebellar disorders. *Handbook of Clinical Neurology, 154*, 329–339. https://doi.org/10.1016/B978-0-444-63956-1.00020-5

Cabaraux, P., Gandini, J., Kakei, S., Manto, M., Mitoma, H., & Tanaka, H. (2020). Dysmetria and errors in predictions: The role of internal forward model. *International Journal of Molecular Sciences, 21*(18), E6900. https://doi.org/10.3390/ijms21186900

Clarke, E., & O'Malley, C. D. (1968). *The human brain and spinal cord*. University of California Press.

D'Angelo, E. (2018). *Physiology of the cerebellum. In the cerebellum: From embryology to diagnostic investigations* (pp. 85–108). Elsevier. https://doi.org/10.1016/b978-0-444-63956-1.00006-0

D'Angelo, E., & Casali, S. (2013). Seeking a unified framework for cerebellar function and dysfunction: from circuit operations to cognition. *Frontiers in Neural Circuits*, 6. https://doi.org/10.3389/fncir.2012.00116

Diedrichsen, J., Balsters, J. H., Flavell, J., Cussans, E., & Ramnani, N. (2009). A probabilistic MR atlas of the human cerebellum. *NeuroImage, 46*(1), 39–46. https://doi.org/10.1016/j.neuroimage.2009.01.045

Eccles, J. C., Ito, M., & Szenta Gothai, J. (1967). *The cerebellum as a neuronal machine*. Springer.

Fine, E. J., Ionita, C. C., & Lohr, L. (2002). The history of the development of the cerebellar examination. *Seminars in Neurology, 22*(4), 375–384. https://doi.org/10.1055/s-2002-36759

Gilbert, P. F., & Thach, W. T. (1977). Purkinje cell activity during motor learning. *Brain Research, 128*(2), 309–328. https://doi.org/10.1016/0006-8993(77)90997-0

Grimaldi, G. (2013). Cerebellar Motor Disorders. In M. Manto, J. D. Schmahmann, F. Rossi, D. L. Gruol, & N. Koibuchi (Eds.), *Handbook of the cerebellum and cerebellar disorders*. Springer. https://doi.org/10.1007/978-94-007-1333-8_71

Grimaldi, G., & Manto, M. (2012). Topography of cerebellar deficits in humans. *Cerebellum (London, England), 11*(2), 336–351. https://doi.org/10.1007/s12311-011-0247-4

Habas, C., & Manto, M. (2018). Probing the neuroanatomy of the cerebellum using tractography. In *The cerebellum: From embryology to diagnostic investigations* (pp. 235–249). Elsevier. https://doi.org/10.1016/b978-0-444-63956-1.00014-x

Hoche, F., Guell, X., Vangel, M. G., Sherman, J. C., & Schmahmann, J. D. (2018). The cerebellar cognitive affective/Schmahmann syndrome scale. *Brain: A Journal of Neurology, 141*(1), 248–270. https://doi.org/10.1093/brain/awx317

Holmes, G. (2007). The Croonian lectures on the clinical symptoms of cerebellar disease and their interpretation. Lecture I. 1922. *Cerebellum (London, England), 6*(2), 142–141. https://doi.org/10.1080/14734220701415208

Hoshi, E., Tremblay, L., Féger, J., Carras, P. L., & Strick, P. L. (2005). The cerebellum communicates with the basal ganglia. *Nature Neuroscience, 8*(11), 1491–1493. https://doi.org/10.1038/nn1544

Ito, M., Yoshida, M., & Obata, K. (1964). Monosynaptic inhibition of the cerebellar nuclei induced from the cerebellar cortex. *Experientia, 20*, 575–576.

Ito, M. (1984). The modifiable neuronal network of the cerebellum. *The Japanese Journal of Physiology, 34*(5), 781–792. https://doi.org/10.2170/jjphysiol.34.781

Ito, M. (2013). Error detection and representation in the olivo-cerebellar system. *Frontiers in Neural Circuits, 7*, 1. https://doi.org/10.3389/fncir.2013.00001

Ito, M., & Kano, M. (1982). Long-lasting depression of parallel fiber-Purkinje cell transmission induced by conjunctive stimulation of parallel fibers and climbing fibers in the cerebellar cortex. *Neuroscience Letters, 33*(3), 253–258. https://doi.org/10.1016/0304-3940(82)90380-9

Larsell, O. (1952). The morphogenesis and adult pattern of the lobules and tissues of the cerebellum of the white rat. *The Journal of Comparative Neurology, 97*, 281–356.

Larsell, O. (1958). Lobules of the mammalian and human cerebellum. *The Anatomical Record, 130*, 329–330.

Lawrenson, C., Bares, M., Kamondi, A., Kovács, A., Lumb, B., Apps, R., Filip, P., & Manto, M. (2018). The mystery of the cerebellum: Clues from experimental and clinical observations. *Cerebellum & Ataxias, 5*(1). https://doi.org/10.1186/s40673-018-0087-9

Llinás, R., & Sugimori, M. (1980). Electrophysiological properties of in vitro Purkinje cell somata in mammalian cerebellar slices. *The Journal of Physiology, 305*, 171–195. https://doi.org/10.1113/jphysiol.1980.sp013357

Llinas, R., Baker, R., & Sotelo, C. (1974). Electrotonic coupling between neurons in cat inferior olive. *Journal of Neurophysiology, 37*(3), 560–571. https://doi.org/10.1152/jn.1974.37.3.560

Lugaro, E. (1894). Sulle connessioni tra gli elementi nervosi della corteccia cerebellare con considerazioni generali sul significato fisiologico dei rapporti tragli elementi ner- vosi. *Riv Sper Freniatr Med Leg, 20*, 297–331.

Manto, M. (2018). Cerebellar motor syndrome from children to the elderly. *Handbook of Clinical Neurology, 154*, 151–166. https://doi.org/10.1016/B978-0-444-63956-1.00009-6

Manto, M., & Habas, C. (2013). *Le cervelet, de l'anatomie et la physiologie à la clinique humaine*. Springer.

Manto, M., & Mariën, P. (2015). Schmahmann's syndrome - identification of the third cornerstone of clinical ataxiology. *Cerebellum & Ataxias, 2*, 2. https://doi.org/10.1186/s40673-015-0023-1

Marr, D. (1969). A theory of cerebellar cortex. *The Journal of Physiology, 202*(2), 437–470. https://doi.org/10.1113/jphysiol.1969.sp008820

Mitoma, H., Buffo, A., Gelfo, F., Guell, X., Fucà, E., Kakei, S., Lee, J., Manto, M., Petrosini, L., Shaikh, A. G., & Schmahmann, J. D. (2020). Consensus paper. Cerebellar reserve: From cerebellar physiology to cerebellar disorders. *Cerebellum (London, England), 19*(1), 131–153. https://doi.org/10.1007/s12311-019-01091-9

Molinari, M. (2016). Sequencing. In D. Gruol, N. Koibuchi, M. Manto, M. Molinari, J. Schmahmann, & Y. Shen (Eds.), *Essentials of cerebellum and cerebellar disorders*. Springer. https://doi.org/10.1007/978-3-319-24551-5_54

Molinari, M., Restuccia, D., & Leggio, M. G. (2009). State estimation, response prediction, and cerebellar sensory processing for behavioral control. *Cerebellum, 8*(3), 399–402. https://doi.org/10.1007/s12311-009-0112-x

Oscarsson, O. (1979). Functional units of the cerebellum - sagittal zones and microzones. *Trends in Neurosciences, 2*, 143–145. https://doi.org/10.1016/0166-2236(79)90057-2

Popa, L. S., & Ebner, T. J. (2019). Cerebellum, predictions and errors. *Frontiers in Cellular Neuroscience, 12*, 524. https://doi.org/10.3389/fncel.2018.00524

Popa, L. S., Hewitt, A. L., & Ebner, T. J. (2014). The cerebellum for jocks and nerds alike. *Frontiers in Systems Neuroscience, 8*, 113. https://doi.org/10.3389/fnsys.2014.00113

Pozzi, N. G., Arnulfo, G., Canessa, A., Steigerwald, F., Nickl, R., Homola, G. A., Fato, M. M., Matthies, C., Pacchetti, C., Volkmann, J., & Isaias, I. U. (2016). Distinctive neuronal firing patterns in subterritories of the subthalamic nucleus. *Clinical Neurophysiology: Official Journal of the International Federation of Clinical Neurophysiology, 127*(11), 3387–3393. https://doi.org/10.1016/j.clinph.2016.09.004

Riedel, M. C., Ray, K. L., Dick, A. S., Sutherland, M. T., Hernandez, Z., Fox, P. M., Eickhoff, S. B., Fox, P. T., & Laird, A. R. (2015). Meta-analytic connectivity and behavioral parcellation of the human cerebellum. *NeuroImage, 117*, 327–342. https://doi.org/10.1016/j.neuroimage.2015.05.008

Schmahmann, J. D. (1998). Dysmetria of thought: Clinical consequences of cerebellar dysfunction on cognition and affect. *Trends in Cognitive Sciences, 2*(9), 362–371. https://doi.org/10.1016/s1364-6613(98)01218-2

Schmahmann, J. D. (2016). A brief history of the cerebellum. In D. Gruol, N. Koibuchi, M. Manto, M. Molinari, J. Schmahmann, & Y. Shen (Eds.), *Essentials of cerebellum and cerebellar disorders*. Springer. https://doi.org/10.1007/978-3-319-24551-5_2

Schmahmann, J. D., & Sherman, J. C. (1998). The cerebellar cognitive affective syndrome. *Brain: A Journal of Neurology, 121*(Pt 4), 561–579. https://doi.org/10.1093/brain/121.4.561

Silveri, M. C. (2016). Speech Deficits. In D. Gruol, N. Koibuchi, M. Manto, M. Molinari, J. Schmahmann, & Y. Shen (Eds.), *Essentials of cerebellum and cerebellar disorders*. Springer. https://doi.org/10.1007/978-3-319-24551-5_65

Stewart, T. G., & Holmes, G. M. (1904). Symptomatology of cerebellar tumours. *Brain, 27*, 522–592.

Stoodley, C. J., & Schmahmann, J. D. (2018). Functional topography of the human cerebellum. In *The cerebellum: From embryology to diagnostic investigations* (pp. 59–70). Elsevier. https://doi.org/10.1016/b978-0-444-63956-1.00004-7

Trouillas, P., Takayanagi, T., Hallett, M., Currier, R. D., Subramony, S. H., Wessel, K., Bryer, A., Diener, H. C., Massaquoi, S., Gomez, C. M., Coutinho, P., Ben Hamida, M., Campanella, G., Filla, A., Schut, L., Timann, D., Honnorat, J., Nighoghossian, N., & Manyam, B. (1997). International cooperative Ataxia rating scale for pharmacological assessment of the cerebellar syndrome. The Ataxia neuropharmacology Committee of the World Federation of neurology. *Journal of the Neurological Sciences, 145*(2), 205–211. https://doi.org/10.1016/s0022-510x(96)00231-6

Ugawa, Y. (2009). Can we see the cerebellar activation effect by TMS over the back of the head? *Clinical Neurophysiology: Official Journal of the International Federation of Clinical Neurophysiology, 120*(12), 2006–2007. https://doi.org/10.1016/j.clinph.2009.09.003

Van Dun, K., Manto, M., & Mariën, P. (2016). Cerebello-cerebral feedback projections. In D. Gruol, N. Koibuchi, M. Manto, M. Molinari, J. Schmahmann, & Y. Shen (Eds.), *Essentials of cerebellum and cerebellar disorders*. Springer. https://doi.org/10.1007/978-3-319-24551-5_12

Van Overwalle, F., Manto, M., Leggio, M., & Delgado-García, J. M. (2019). The sequencing process generated by the cerebellum crucially contributes to social interactions. *Medical Hypotheses, 128*, 33–42. https://doi.org/10.1016/j.mehy.2019.05.014

Voogd, J., & Koehler, P. J. (2018). Historic notes on anatomic, physiologic, and clinical research on the cerebellum. *Handbook of Clinical Neurology, 154*, 3–26. https://doi.org/10.1016/B978-0-444-63956-1.00001-1

Wolpert, D. M., Miall, R. C., & Kawato, M. (1998). Internal models in the cerebellum. *Trends in Cognitive Sciences, 2*, 338–347. https://doi.org/10.1016/s1364-6613(98)01221-2

Zanatta, A., Cherici, C., Bargoni, A., Buzzi, S., Cani, V., Mazzarello, P., & Zampieri, F. (2018). Vincenzo Malacarne (1744-1816) and the first description of the human cerebellum. *Cerebellum (London, England), 17*(4), 461–464. https://doi.org/10.1007/s12311-018-0932-7

Chapter 24
Molecular Dissection and Therapeutic Application of SCA1 Pathologies Revealed by Comprehensive Approaches

Hitoshi Okazawa and Hikari Tanaka

24.1 Discovery of Novel SCA1 Pathologies by Comprehensive Analyses

Spinocerebellar ataxia type 1 (SCA1) is a neurodegenerative disease which mainly affects Purkinje cells in the cerebellum and motor neurons in the spinal cord. Since the discovery of causative gene, Ataxin-1 (Atxn1), more than 20 years ago, substantial amount of knowledge about the mechanism has been accumulated. Identification of Atxn1-interacting factors such as capicua (CIC) (Lam et al., 2006) and RNA-binding motif protein 17 (RBM17) as binding proteins to Atxn1 (Lim et al., 2008) indicates the involvement in transcription and splicing in the SCA1 pathology (Zoghbi & Orr, 2009).

Meanwhile, by employing comprehensive analyses, our group has identified key molecules that mediate functional dysregulation caused by mutant Atxn1 protein. First, by yeast two-hybrid screening, we found six clones that interact with polyglutamine (polyQ) tract sequences. One of them was the already known molecule VCP (TERA/p97/VCP), and the others were novel genes such as PQBP1 (polyglutamine-binding protein 1) (Imafuku et al., 1998; Waragai et al., 1999). Functions of PQBP1 have been identified during the last 20 years by our group and other groups (Okazawa, 2018). PQBP1 is involved in transcriptional regulation, RNA splicing, RNA stress, and DNA damage repair (Waragai et al., 1999; Okazawa et al., 2002;

H. Okazawa (✉)
Department of Neuropathology, Medical Research Institute, Tokyo Medical and Dental University, Tokyo, Japan

Center for Brain Integration Research, Tokyo Medical and Dental University, Tokyo, Japan
e-mail: okazawa-tky@umin.ac.jp

H. Tanaka
Department of Neuropathology, Medical Research Institute, Tokyo Medical and Dental University, Tokyo, Japan

H. Mizusawa, S. Kakei (eds.), *Cerebellum as a CNS Hub*,
Contemporary Clinical Neuroscience, https://doi.org/10.1007/978-3-030-75817-2_24

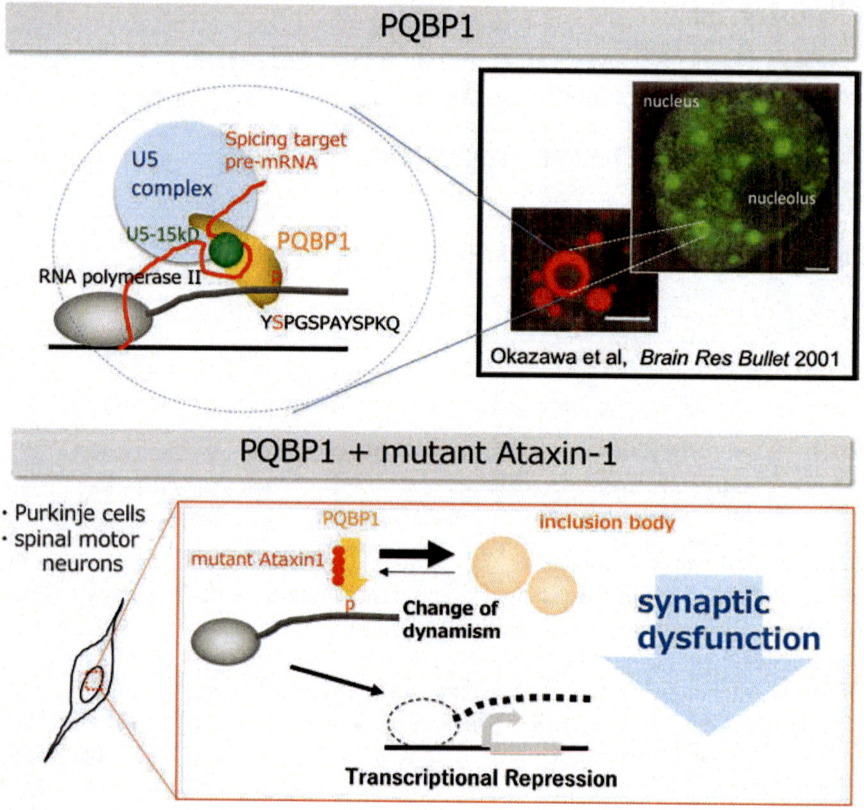

Fig. 24.1 Function of PQBP1. PQBP1 regulates RNA transcription and splicing in nuclear body. And the interaction between mutant Atxn1 and PQBP1 represses transcription and splicing, which eventually leads to synapse dysfunction

Kunde et al., 2011; Mizuguchi et al., 2014; Ito et al., 2015b; Wan et al., 2015; Morchikh et al., 2017) and determines gene expression profiles related to neural stem cell proliferation, neuronal cilia, neurite extension, and synapse function (Ikeuchi et al., 2013; Ito et al., 2015b; Li et al., 2013; Okazawa et al., 2001; Okazawa et al., 2002; Wang et al., 2013; Waragai et al., 1999). Interaction of mutant Atxn1 with PQBP1 basically impairs such multiple functions in neurons including Purkinje cells and spinal motor neurons (Fig. 24.1).

The discovery of HMGB1 (Qi et al., 2007) and VCP (Imafuku et al., BBRC 1998; Waragai et al., Hum Mol Genet 1999) led to identifying that impairment of DNA damage repair is another key pathological event in SCA1 (Ito et al., 2015a; Qi et al., 2007). Comprehensive proteome analysis revealed that HMGB1/2 were decreased in Purkinje cells of SCA1 model mice and that mutant Atxn1 interact with HMGB1/2 to impair the DNA damage repair function (Qi et al., 2007). HMGB1/2 are known to regulate the unwinding or folding DNA structures, and their functional inhibition in nuclei of neurons leads to inhibition of nuclear functions (Alessandra & Bianchi, 2003; Travers, 2003). As mentioned above, VCP was discovered as a binding protein to polyQ tract sequence (Imafuku et al., 1998). Several studies revealed involvement of VCP in DNA damage repair (Acs et al., 2011; Meerang et al., 2011), and interaction of mutant Atxn1 with VCP also leads to impairment of DNA damage repair (Fujita et al., 2013).

24.2 Gene Therapy of SCA1 with HMGB1

We performed a proteome analysis of soluble nuclear proteins from neurons expressing mutant polyQ protein and revealed that HMGB1/2 are decreased in cerebellar neurons expressing mutant Atxn1, which is a mutation of the causative gene of SCA1 and in cerebral neurons expressing huntingtin (Htt), which is a mutation of the causative gene of Huntington's disease (Qi et al., 2007). This result showed that HMGB1/2 proteins were commonly reduced in vulnerable neurons. In addition, similar reductions were observed in vulnerable neurons even before the onset of transgenic mice (R6/2 mice) and knock-in mice (Atxn1-154Q/2Q-KI mice). HMGB1/2 are one of the most abundant proteins in the nucleus, and these are known as essential DNA structural proteins that unwind DNA from histone complex or bent DNA (Alessandra & Bianchi, 2003; Travers, 2003).

Mutant Ataxin-1 or huntingtin binds to HMGB1 and impairs the functions directly or promotes the degradation (Qi et al., 2007). Due to the deficiency of HMGB1, nuclear and possibly cytoplasmic functions including DNA damage repair are impaired (Fig. 24.1) (Qi et al., 2007). Supplementation of HMGB1 suppressed neuronal cell death in *Drosophila* SCA1 model, in which mutant Atxn1 causes degeneration of photoreceptor cells of the complex eye (Qi et al., 2007).

Based on the results from *Drosophila* model, we moved to Atxn1-KI mouse model. We first mated HMGB1-Tg mice with Atxn1-KI mice of the same background C57BL/6 and generated double transgenic mice (Atxn1-KI;HMGB1 mice), and we tested their motor dysfunction and lifespan. In the rotarod test, SCA1 model mice showed motor dysfunction from 5 weeks of age, which continued to decline. On the other hand, in the double transgenic mouse, the shortened rotarod stay time was improved from 7 weeks of age, and the improvement was sustained at least until 21 weeks of age (Ito et al., 2015a). Nuclear DNA damage in Purkinje cells of the mutant Atxn1-KI mice was recovered in the double transgenic mouse. The 50% survival duration was extended from 217 days to 282 days (+30%), and the maximum survival duration was increased from 274 days to 360 days (Ito et al., 2015a).

Moreover, gene therapy using an adeno-associated virus (AAV) vector was effective. HMGB1 was widely and highly expressed in the cerebellar neurons especially in Purkinje cells by a single injection onto the cerebellum surface of 5-week-old SCA1 model mice. Similar to the results in the double transgenic mice, improvement in motor function was observed at 9 and 13 weeks of age in the gene therapy experiment using the AAV vector. Furthermore, the lifespan of mutant Atxn1-KI mice was extended from 217 days to 365.5 days (nearly +70%), and the maximum was extended from 274 days to 448 days (Fig. 24.2) (Ito et al., 2015a). In addition,

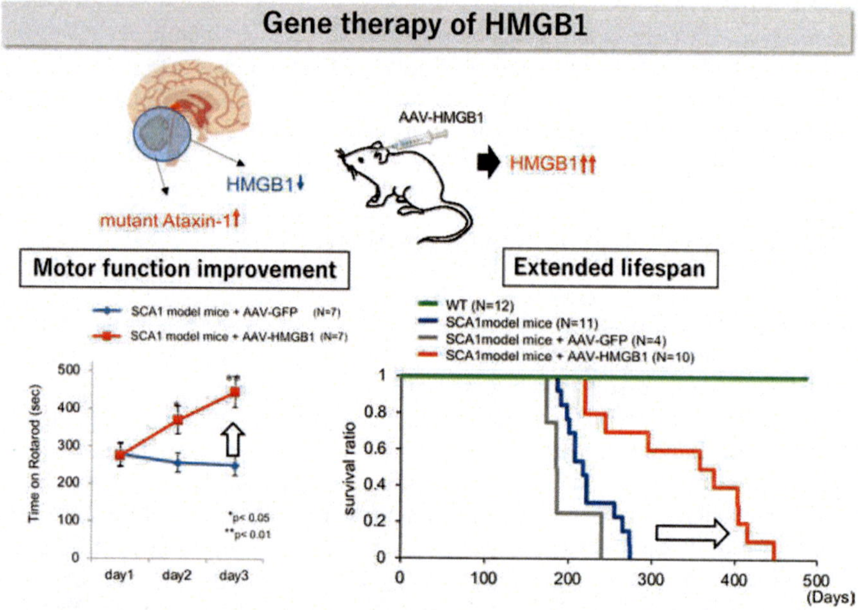

Fig. 24.2 Gene therapy of HMGB1 in SCA1 model mice. HMGB1 is decreased in Purkinje cells and other neurons of mutant Atxn1-KI mice and also human patients. AAV-HMGB1 gene therapy recovers the deficiency of HMGB1 and elongates the lifespan of the model mice. Similar approaches can be used to recover deficiency of other target molecules

it was revealed that HMGB1 also enhances mitochondrial DNA damage repair, suggesting that HMGB1 suppresses neurodegeneration by repairing both nuclear and mitochondrial DNA (Ito et al., 2015a).

24.3 Gene Therapy of SCA1 with RpA1

In addition to the proteome analysis revealing the role of HMGB1/2 in SCA1 pathology, we found from multiple omics analyses that other molecules such as TERA/VCP/p97 and Ku70 involved in DNA damage repair are functionally impaired in polyglutamine disease (Enokido et al., 2010; Fujita et al., 2013; Qi et al., 2007). These proteins are basically involved in non-homologous end joining of DNA double-strand break repair among various types of DNA damage repair. Therefore, we further asked which type of DNA damage repair most significantly contributes to the SCA1 pathology (Barclay et al., 2014). From gene screens with *Drosophila* SCA1 models, we identified that RpA1, a protective protein for naked single-strand DNA in various types of DNA damage, has the largest therapeutic effect on the lifespan-shortening by mutant Atxn1 expression in motor neurons (Barclay et al., 2014). Furthermore, an immunoprecipitation (IP) assay revealed that RpA1 binds to Ataxin-1 and mutant Ataxin-1 binds to RpA1 more strongly (Barclay et al., 2014).

Therefore, we developed gene therapy of mutant Atxn1-KI mice with AAV-RpA1 at the timing of onset, which induced significant improvement of motor function lasting over 50 weeks after injection (Taniguchi et al., 2016). DNA double-strand break, which is the final outcome of various forms of DNA damage, was also ameliorated in cerebellar neurons, and the abnormal patterns of gene expression were also partially corrected (Taniguchi et al., 2016).

24.4 Gene Therapy with PQBP1

Though we have not examined the effect of AAV-PQBP1 on mutant Atxn1-KI model mice, we have unexpectedly experienced the examination in Alzheimer's disease (AD) mouse models (Tanaka et al., 2018). We reached to the conclusion that PQBP1 is also involved in the AD pathology as follows. First, we performed comprehensive phosphoproteome analysis and revealed that SRRM2 phosphorylation occurs at the earliest stage before extracellular Abeta aggregation (Tagawa et al., 2015). The phosphorylation of SRRM2 inhibits its interaction with a chaperone protein TCP1alpha and prevents nuclear translocation of SRRM2 (Tanaka et al., 2018). Given that SRRM2 functions as a scaffold protein in the nucleus, the major target of SRRM2 scaffolding PQBP1 was decreased in the nucleus. The reduction of PQBP1 in the nucleus disturbs proper splicing of synapse-related hnRNA and decreases their mRNA (Ito et al., 2015b), which is just like neurons in patients of Renpenning syndrome (Kalscheuer et al., 2003; Lenski et al., 2004; Lubs et al., 2006; Okazawa, 2018; Stevenson et al., 2005). Therefore, we performed gene therapy against AD model mice with AAV-PQBP1 and found the rescue of phenotypes as expected (Tanaka et al., 2018). The similar approach will be feasible in therapeutics for the SCA1 pathology.

24.5 Developmental Pathology of SCA1

Previous studies have accumulated a great deal of knowledge about pathological conditions caused by gene mutations. However, the relationship between timing and the major pathology is still not unclear, which slows down the progress of disease-modifying therapy. In a previous study, we discovered transcriptional repression-induced atypical cell death (TRIAD) due to RNA polymerase II inhibition and identified a molecule YAPdeltaC that regulates TRIAD (Hoshino et al., 2006). To elucidate the function of YAPdeltaC in SCA1, we used the Tet-ON system for YAPdeltaC expression and examined how the time-specific expression of YAPdeltaC affects the symptoms and survival of Ataxin-1-KI mice. Unexpectedly, expression of YAPdeltaC during development remarkably extended the lifespan, while the expression of YAPdeltaC from adulthood (from 8 weeks old) was less effective (Fujita et al., 2017).

We revealed that YAPdeltaC is a transcriptional co-factor that enhances RORalpha function that regulates gene expression in cerebellar neurons during development collaborating with normal Ataxin-1. On the other hand, mutant Ataxin-1 inhibits the interaction between YAPdeltaC and RORalpha and inhibits gene expression required for maturation of cerebellar neurons (Fujita et al., 2017). These results indicate the role of mutant Atxn1 during the development and the developmental pathology influencing the adult pathology beyond a long time interval.

24.6 Future Prospects

Various comprehensive analyses of our group unraveled novel pathologies of SCA1 such as impairments of DNA damage repair (HMGB1, Ku70, VCP, RpA1, PQBP1) and RNA splicing (PQBP1) and transcription (HMGB1, PQBP1, YAP/YAPdeltaC). A single gene mutation of Atxn1 leads to phenotypes via multiple pathways, which might be "gain of function" or "loss of function." From the viewpoint of mutant Atxn1, it could be described as "gain of function" when mutant Atxn1 deprives some target proteins to inclusion body. However, it could be also described as "loss of function" from the aspect of normal Atxn1. For instance, normal Atxn1 have physiological interaction with YAP/YAPdeltaC or PQBP1 in transcription and splicing regulation. Therefore, prevention of these target proteins from the functional complex by mutant Atxn1 is similar to loss of function of normal Atxn1.

Recently gene-based therapies like gene therapy, antisense oligonucleotide, and genome editing are developing rapidly. Basically, gene-based therapies are classified into two types, upregulation and downregulation of the target gene. Antisense oligonucleotide, therefore, is used for knockdown of mutant Atxn1. However, it has an unexpected pitfall that depletion of Atxn1 might increase the risk of Alzheimer's disease (Crespo-Barreto et al., 2010; Matilla et al., 1998; Suh et al., 2019), and the concern of off-target effect cannot be completely excluded. Actually, currently ongoing clinical trials of antisense therapy targeting selectively mutant Htt mRNA might accelerate brain atrophy of Huntington's disease patients judging from their ventricular size at clinical trial phase 2 (Supplementary Results of ref. (Suh et al., 2019) (Tabrizi et al., 2019). Therefore, upregulating the target whose activity is lost in the pathology of SCA1 might be safer for human patients than downregulating the causative gene.

References

Acs, K., Luijsterburg, M. S., Ackermann, L., Salomons, F. A., Hoppe, T., & Dantuma, N. P. (2011). The AAA-ATPase VCP/p97 promotes 53BP1 recruitment by removing L3MBTL1 from DNA double-strand breaks. *Nature Structural & Molecular Biology, 18*, 1345–1350.

Alessandra, A., & Bianchi, M. E. (2003). HMGB proteins and gene expression. *Current Opinion in Genetics & Development, 2*, 170–178.

Barclay, S. S., Tamura, T., Ito, H., et al. (2014). Systems biology analysis of Drosophila in vivo screen data elucidates core networks for DNA damage repair in SCA1. *Human Molecular Genetics, 23*, 1345–1364.
Crespo-Barreto, J., Fryer, J. D., Shaw, C. A., Orr, H. T., & Zoghbi, H. Y. (2010). Partial loss of ataxin-1 function contributes to transcriptional dysregulation in spinocerebellar ataxia type 1 pathogenesis. *PLoS Genetics, 6*, 1–17.
Enokido, Y., Tamura, T., Ito, H., et al. (2010). Mutant huntingtin impairs Ku70-mediated DNA repair. *The Journal of Cell Biology, 189*, 425–443.
Fujita, K., Mao, Y., Uchida, S., et al. (2017). Developmental YAPdeltaC determines adult pathology in a model of spinocerebellar ataxia type 1. *Nature Communications, 8*, 1864.
Fujita, K., Nakamura, Y., Oka, T., et al. (2013). A functional deficiency of TERA/VCP/p97 contributes to impaired DNA repair in multiple polyglutamine diseases. *Nature Communications, 4*, 1816.
Hoshino, M., Qi, M. L., Yoshimura, N., et al. (2006). Transcriptional repression induces a slowly progressive atypical neuronal death associated with changes of YAP isoforms and p73. *The Journal of Cell Biology, 172*, 589–604.
Ikeuchi, Y., delaTorre-Ubieta, L., Matsuda, T., Steen, H., Okazawa, H., & Bonni, A. (2013). The XLID protein PQBP1 and the GTPase dynamin 2 define a signaling link that orchestrates ciliary morphogenesis in postmitotic neurons. *Cell Reports, 4*, 879–889.
Imafuku, I., Waragai, M., Takeuchi, S., Kanazawa, I., Kawabata, M., Mouradian, M. M., & Okazawa, H. (1998). Polar amino acid-rich sequences bind to polyglutamine tracts. *Biochemical and Biophysical Research Communications, 253*, 16–20.
Ito, H., Fujita, K., Tagawa, K., et al. (2015a). HMGB1 facilitates repair of mitochondrial DNA damage and extends the lifespan of mutant ataxin-1 knock-in mice. *EMBO Molecular Medicine, 7*, 78–101.
Ito, H., Shiwaku, H., Yoshida, C., et al. (2015b). In utero gene therapy rescues microcephaly caused by Pqbp1-hypofunction in neural stem progenitor cells. *Molecular Psychiatry, 20*, 459–471.
Kalscheuer, V. M., Freude, K., Musante, L., et al. (2003). Mutations in the polyglutamine binding protein 1 gene cause X-linked mental retardation. *Nature Genetics, 35*, 313–315.
Kunde, S. A., Musante, L., Grimme, A., Fischer, U., Müller, E., Wanker, E. E., & Kalscheuer, V. M. (2011). The X-chromosome-linked intellectual disability protein PQBP1 is a component of neuronal RNA granules and regulates the appearance of stress granules. *Human Molecular Genetics, 20*, 4916–4931.
Lam, Y. C., Bowman, A. B., Jafar-Nejad, P., et al. (2006). ATAXIN-1 interacts with the repressor Capicua in its native complex to cause SCA1 neuropathology. *Cell, 127*, 1335–1347.
Lenski, C., Abidi, F., Meindl, A., Gibson, A., Platzer, M., Frank Kooy, R., Lubs, H. A., Stevenson, R. E., Ramser, J., & Schwartz, C. E. (2004). Novel truncating mutations in the polyglutamine tract binding protein 1 gene (PQBP1) cause Renpenning syndrome and X-linked mental retardation in another family with microcephaly. *American Journal of Human Genetics, 74*, 777–780.
Li, C., Ito, H., Fujita, K., Shiwaku, H., Qi, Y., Tagawa, K., Tamura, T., & Okazawa, H. (2013). Sox2 transcriptionally regulates Pqbp1, an intellectual disability-microcephaly causative gene, in neural stem progenitor cells. *PLoS One, 8*, 1–15.
Lim, J., Crespo-Barreto, J., Jafar-Nejad, P., Bowman, A. B., Richman, R., Hill, D. E., Orr, H. T., & Zoghbi, H. Y. (2008). Opposing effects of polyglutamine expansion on native protein complexes contribute to SCA1. *Nature, 452*, 713–718.
Lubs, H., Abidi, F. E., Echeverri, R., Holloway, L., Meindl, A., Stevenson, R. E., & Schwartz, C. E. (2006). Golabi-Ito-hall syndrome results from a missense mutation in the WW domain of the PQBP1 gene. *Journal of Medical Genetics, 43*, e30.
Matilla, A., Roberson, E. D., Banfi, S., Morales, J., Armstrong, D. L., Burright, E. N., Orr, H. T., Sweatt, J. D., Zoghbi, H. Y., & Matzuk, M. M. (1998). Mice lacking ataxin-1 display learning deficits and decreased hippocampal paired-pulse facilitation. *The Journal of Neuroscience, 18*, 5508–5516.

Meerang, M., Ritz, D., Paliwal, S., Garajova, Z., Bosshard, M., Mailand, N., Janscak, P., Hubscher, U., Meyer, H., & Ramadan, K. (2011). The ubiquitin-selective segregase VCP/p97 orchestrates the response to DNA double-strand breaks. *Nature Cell Biology, 13*, 1376–1382.

Mizuguchi, M., Obita, T., Serita, T., Kojima, R., Nabeshima, Y., & Okazawa, H. (2014). Mutations in the PQBP1 gene prevent its interaction with the spliceosomal protein U5-15kD. *Nature Communications, 5*, 3822.

Morchikh, M., Cribier, A., Raffel, R., Amraoui, S., Cau, J., Severac, D., Dubois, E., Schwartz, O., Bennasser, Y., & Benkirane, M. (2017). HEXIM1 and NEAT1 long non-coding RNA form a multi-subunit complex that regulates DNA-mediated innate immune response. *Molecular Cell, 67*, 387–399.e5.

Okazawa, H. (2018). PQBP1, an intrinsically disordered/denatured protein at the crossroad of intellectual disability and neurodegenerative diseases. *Neurochemistry International, 119*, 17–25.

Okazawa, H., Rich, T., Chang, A., et al. (2002). Interaction between mutant ataxin-1 and PQBP-1 affects transcription and cell death. *Neuron, 34*, 701–713.

Okazawa, H., Sudol, M., & Rich, T. (2001). PQBP-1 (Np/PQ): A polyglutamine tract-binding and nuclear inclusion-forming protein. *Brain Research Bulletin, 56*, 273–280.

Qi, M. L., Tagawa, K., Enokido, Y., et al. (2007). Proteome analysis of soluble nuclear proteins reveals that HMGB1/2 suppress genotoxic stress in polyglutamine diseases. *Nature Cell Biology, 9*, 402–414.

Stevenson, R. E., Bennett, C. W., Abidi, F., Kleefstra, T., Porteous, M., Simensen, R. J., Lubs, H. A., Hamel, B. C. J., & Schwartz, C. E. (2005). Renpenning syndrome comes into focus. *American Journal of Medical Genetics, 134*(A), 415–421.

Suh, J., Romano, D. M., Nitschke, L., et al. (2019). Loss of Ataxin-1 potentiates Alzheimer's pathogenesis by elevating cerebral BACE1 transcription. *Cell, 178*, 1159–1175.e17.

Tabrizi, S. J., Leavitt, B. R., Landwehrmeyer, G. B., et al. (2019). Targeting huntingtin expression in patients with Huntington's disease. *The New England Journal of Medicine, 380*, 2307–2316.

Tagawa, K., Homma, H., Saito, A., et al. (2015). Comprehensive phosphoproteome analysis unravels the core signaling network that initiates the earliest synapse pathology in preclinical Alzheimer's disease brain. *Human Molecular Genetics, 24*, 540–558.

Tanaka, H., Kondo, K., Chen, X., et al. (2018). The intellectual disability gene PQBP1 rescues Alzheimer's disease pathology. *Molecular Psychiatry, 23*, 2090–2110.

Taniguchi, J. B., Kondo, K., Fujita, K., et al. (2016). RpA1 ameliorates symptoms of mutant ataxin-1 knock-in mice and enhances DNA damage repair. *Human Molecular Genetics, 25*, 4432–4447.

Travers, A. A. (2003). Priming the nucleosome: A role for HMGB proteins? *EMBO Reports, 4*, 131–136.

Wan, D., Zhang, Z. C., Zhang, X., Li, Q., & Han, J. (2015). X chromosome-linked intellectual disability protein PQBP1 associates with and regulates the translation of specific mRNAs. *Human Molecular Genetics, 24*, 4599–4614.

Wang, Q., Moore, M. J., Adelmant, G., Marto, J. A., & Silver, P. A. (2013). PQBP1, a factor linked to intellectual disability, affects alternative splicing associated with neurite outgrowth. *Genes & Development, 27*, 615–626.

Waragai, M., Lammers, C. H., Takeuchi, S., Imafuku, I., Udagawa, Y., Kanazawa, I., Kawabata, M., Mouradian, M. M., & Okazawa, H. (1999). PQBP-1, a novel polyglutamine tract-binding protein, inhibits transcription activation by Brn-2 and affects cell survival. *Human Molecular Genetics, 8*, 977–987.

Zoghbi, H. Y., & Orr, H. T. (2009). Pathogenic mechanisms of a polyglutamine-mediated neurodegenerative disease, spinocerebellar ataxia type 1. *The Journal of Biological Chemistry, 284*, 7425–7429.

Chapter 25
Spinocerebellar Ataxia Type 2

Stefan M. Pulst

25.1 SCA2 Clinical Characteristics

Spinocerebellar ataxia type 2 (SCA2) is a neurodegenerative disease that predominantly affects the cerebellum. While patients with SCA2 show many of the core clinical characteristics that define the SCAs as a group of neurodegenerative disorders, the SCA2 phenotype, when assessed across a large number of individuals, is clinically distinct. Gait ataxia, considered a characteristic of SCA, is the most noticeable symptom and is often the presenting symptom and sign (Luo et al., 2017; Pulst et al., 1993). Onset may also coincide with muscle cramping. Gait ataxia is followed by multiple other symptoms characteristic of cerebellar dysfunction such as appendicular ataxia with instability of stance, dysarthria, and ocular signs. A distinguishing feature of SCA2, not necessarily by its presence, but by its severity, is the slowing of saccadic eye movements (Geschwind et al., 1997; Ashizawa et al., 2013; Gwinn-Hardy et al., 2000). In retrospect, it has become clear that some, but not all, of the families with slow eye movements described by Wadia in India had mutations in the *ATXN2* gene (Wadia & Swami, 1971).

Prominent involvement of basal ganglia as well as upper and lower motor neurons has also been reported in SCA2 patients and led to identification of individuals with pure outlier phenotypes. Attention was first drawn to Taiwanese patients with tremor-predominant L-DOPA-responsive Parkinson disease that were found to have ATXN2 mutations, which were later confirmed in other populations (Gwinn-Hardy et al., 2000; Payami et al., 2003). Molecular studies followed, showing physical interaction of TDP-43 and ATXN2; it was also recognized that a pure ALS phenotypes can occur with *ATXN2* mutations and that long normal repeats in *ATXN2* are risk alleles for ALS (Elden et al., 2010; Tazen et al., 2013; Neuenschwander et al.,

S. M. Pulst (✉)
Department of Neurology, University of Utah, Salt Lake City, UT, USA
e-mail: Stefan.Pulst@hsc.utah.edu

H. Mizusawa, S. Kakei (eds.), *Cerebellum as a CNS Hub*,
Contemporary Clinical Neuroscience, https://doi.org/10.1007/978-3-030-75817-2_25

2014). Genetic variation in *ATXN2* has also been associated with obesity and open-angle glaucoma (Bailey et al., 2016; Figueroa et al., 2009).

25.2 The *ATXN2* Gene and Protein

Nearly two decades after the initial publication of Indian patients, a large population with ataxia and ophthalmoplegia was described in a founder population in the province of Holguin in eastern Cuba (Orozco Diaz et al., 1990). The SCA2 locus was identified in 1993 and mapped to human chromosome 12q24 in several pedigrees, including Cuban individuals (Gispert et al., 1993). In the same year, we mapped a North American family of Italian ancestry to this region (Pulst et al., 1993). This family displayed significant anticipation of age of onset. Using linked genetic markers, effects of potentially late-expressing mutation carriers in the most recent generation were excluded as contributing to any ascertainment bias, thus confirming true earlier onset from generation to generation.

The gene causing SCA2 was identified by three groups almost simultaneously using widely different methodologies (Imbert et al., 1996; Pulst et al., 1996; Sanpei et al., 1996). Most commonly, the *ATXN2* gene has 22 CAG repeats, while ≥33 CAG repeats cause SCA2 (Fernandez et al., 2000). SCA2 is characterized by anticipation with strong inverse correlation between age of onset and CAG repeat length. The CAG repeat expansion is dynamic and unstable during meiosis with a strong propensity for expansion (Figueroa et al., 2017). In Cuban individuals, about half of the variance in age of onset is determined by CAG repeat size, 25% by genetic modifiers and 25% by stochastic factors (Figueroa et al., 2017). Depending on the size of the CAG repeat, ATXN2 repeat expansions can act as recessive, dominant, or risk alleles (reviewed in Pulst, 2018).

25.2.1 Gene Structure, Paralogs, and Orthologs

The *ATXN2* gene consists of 25 exons and spans a total of 147 megabase pairs (Nechiporuk et al., 1997; Sahba et al., 1998). *ATXN2* transcription is regulated in part by the ETS1 transcription factor (Scoles et al., 2012).The *ATXN2* transcript is 4699 bp long including a 162 bp 5'-UTR and a 601 bp 3'-UTR. There are two in-frame start codons, the second one located just 12 bp upstream of the CAG repeat. The predicted molecular weight for ATXN2 is 144 kDa, when translated from the first start codon, and smaller by 17 kDa, when translated from the second one (Scoles et al., 2012).

A smaller approximately 42 kDa ATXN2 fragment was observed in brain extracts from SCA2 patients and SCA2 mice (Huynh et al., 1999; Koyano et al., 1999). A caspase-3 cleavage site at ATXN2 amino acid (aa) 396–399 could explain the origin of this band (Huynh et al., 2000). ATXN2 is a cytoplasmic protein that is phosphorylated with a half-life of ≥21 hours (Turnbull et al., 2004). Its subcellular localization is in part with the trans-Golgi network (Huynh et al., 2003; Turnbull et al., 2004) and the endoplasmic reticulum (Van de Loo et al., 2009). RAN translation appears to play a less important role than in other repeat diseases (Scoles et al., 2015).

Mouse ATXN2 Gene

Screening of a mouse brain cDNA library with multiple human cDNA clones identified several overlapping clones that were sequenced to establish near identity with the human *ATXN2* cDNA (Nechiporuk et al., 1998). Between mouse and human, identity was 91% at the nucleotide and 89% at the amino acid level. Of note, the translation-initiation ATG codon corresponded to the second ATG in the human sequence, and there was only a single polyglutamine instead of the human polyQ repeat. The region flanking the single glutamine was significantly less conserved than the rest of the protein. It is not clear if this lack of conservation has functional implications but could potentially argue for the generation of mouse models using human cDNAs or human BACs as compared with cDNA clones or knock-in models.

C. elegans

C. elegans ataxin-2, designated as Atx-2, has 20.1% amino acid identity and 43.9% similarity to human ATXN2. It encodes a predicted 1026 aa protein. Kiehl and colleagues studied the expression pattern of atx-2 using the endogenous promoter coupled with a GFP expression vector and found widespread expression in the adult worm with strong expression in the muscle and nervous tissue as well as in the embryo (Kiehl et al., 2000). Introduction of interfering dsRNA into larval L4 stage worms of the N2 strain resulted in arrested embryonic development in the offspring of microinjected worms. Although *Atxn2* knockout mice are viable, an essential role of atx-2 for early embryonic development is supported by embryonic lethality of *A2rp/Atxn2l* knockout mice (Key et al., 2020).

Homology searches for ATXN2 identified a gene and gene product with high homology at the nucleotide and predicted amino acid levels (Figueroa & Pulst, 2003). The protein was designated as *A2RP* by us for ataxin-2-related protein and later renamed ATXN2L for ATXN2-like. A2RP/ATXN2L has several isoforms with different C-terminal domains. The longest isoform is composed of 1051 amino acids and has widespread expression in human tissues by Northern and Western blot analyses. A2RP/ATXN2L proteins are highly conserved in evolution with orthologs in mouse, cattle, pig, frog, and plants. Recently, genetic variation in the *ATXN2L* locus was linked to left-handedness in humans (Cuellar-Partida et al., 2020).

The Auburger group generated mice with Crispr/Cas9-mediated deletion of *Atxn2l* exons 5–8, studying homozygotes prenatally and heterozygotes during aging

(Key et al., 2020). Absence of Atxn2L resulted in mid-gestational embryonic lethality, affecting female animals more strongly.

ATXN2 Interactors

Since its discovery in 1996, a large number of proteins and mRNAs have been identified as interactors of ATXN2. ATXN2 interacts with multiple RNA-binding proteins (RBPs) including Staufen-1 (STAU1) (Paul et al., 2018).

Soon after the discovery of the *ATXN2* gene, we embarked on studies aimed at placing ATXN2 into a cellular pathway by identifying interacting proteins. We conducted a yeast two-hybrid (Y2H) screen and identified a novel protein that we designated as Ataxin2-binding protein 1 (A2BP1) (Shibata et al., 2000). This protein was subsequently renamed RBFOX1 to place it in the family of FOX RNA-binding proteins (Gehman et al., 2011). It was first described as fox-1 in *C. elegans* (Skipper et al., 1999). A2BP1/RBFOX1 has RNA-binding motifs and has a highly conserved mouse ortholog (Kiehl et al., 2001). Mouse *A2bp1* is expressed early in development and in multiple tissues including the CNS (Kiehl et al., 2001). It can act as a splicing factor with functions in neural development and CNS disorders.

Another Y2H screen revealed physical interaction between poly(A)-binding protein (PABP) and ATXN2 with verification by coimmunoprecipitation (Ralser et al., 2005). PABP is a component of mammalian stress granules. Colocalization of PABP with ATXN2 and PABP occurred in stress granules in heat-shock-treated COS1 cells providing the first demonstration of ATXN2 in stress granules (Ralser et al., 2005). The same research group also characterized ATXN2 interaction with the DEAD/H-box RNA helicase (DDX6) (Nonhoff et al., 2007).

An interaction between TDP-43 and ATXN2 was first identified by Y2H assays and confirmed by coimmunoprecipitation in HEK293 cells (Elden et al., 2010). This interaction was RNA-dependent. In addition, FUS, another ALS-related protein, was also shown to interact with ATXN2 (Farg et al., 2013).

To investigate a functional connection between Parkinsonism and ATXN2, Huynh and colleagues tested ATXN2 as an interacting protein with Parkin (PARK), an E3 ubiquitin ligase (Huynh et al., 2007). Parkin directly interacted with the wild-type and mutant ATXN2 N-terminal domains in HEK293 cells and ubiquitinated full-length ATXN2 more efficiently than N-terminal fragments. This interaction was independently confirmed in coimmunoprecipitation experiments using cerebellar extracts of ATXN2-CAG42 knock-in mice (Halbach et al., 2015).

The Bezprozvanny group verified that the type 1 receptor for inositol triphosphate (IP3R1) specifically interacted with polyglutamine-expanded ATXN2 protein but only minimally with the normal ATXN2 protein (Liu et al., 2009). They then proceeded to examine the functional consequences of this gain-of-function interaction in cultured PCs and in Pcp2/L7-ATXN2[Q58] mice showing that inhibiting overactive Ca^{++} release from the ER in the presence of mutant ATXN2 improved in vitro and in vivo phenotypes.

In Y2H interaction tests, Ralser et al. showed that ATXN2 interacted with endophilin A1 and endophilin A3 (Ralser et al., 2005). The interaction was mediated by the SH3 domain-binding motif 2 (SBM2) in ATXN2.

A number of studies have identified proteins as ATXN2 interactors by coimmunoprecipitation without further verification of direct interactions. Lastres-Becker et al. (2016) demonstrated that ATXN2 coimmunoprecipitated with TIA1, eIF3B, eIF4G, eIF4A1, and S6 from HEK293 cells treated with or without arsenite (Lastres-Becker et al., 2016). Blokhuis et al. (2016) characterized the ATXN2 interactome in Neuro2A cells using mass spectrometry with validations performed by coimmunoprecipitation (Blokhuis et al., 2016). Key interacting proteins verified in coimmunoprecipitation experiments included Fmrp, Upf1, Caprin1, HuD, Pabpc4, and Dhx9 (Blokhuis et al., 2016). In yeast, the ATXN2 ortholog Pbp1 interacts with PAS kinase 1 (Psk1) in its C-terminal half (DeMille et al., 2015).

Recently, we demonstrated that ATXN2 interacted with Staufen-1 (STAU1) (Paul et al., 2018, 2021). STAU1 is a multifunctional double-stranded RNA-binding protein. We demonstrated that STAU1 protein expression was increased in SCA2 patient-derived fibroblasts, lymphoblasts, and iPSCs and in the cerebella of ATXN2-Q127 transgenic and BAC-ATXN2[Q72] mice. Stau1 overabundance was accompanied by increased presence of stress granules. STAU1 overabundance also leads to an amplification of the unfolded protein response through the PERK-CHOP pathway (Gandelman et al., 2020).

25.2.2 Animal SCA2 Models

We will discuss invertebrate models briefly and then focus on rodent models. ATXN2 is highly conserved in evolution and is similar to poly(A)-binding protein-1 (Pab1)-binding protein (Pbp1) in yeast. We and others quickly embarked on evaluating ATXN2 function in invertebrates. Kiehl and colleagues used RNAi to generate Atx2 deficiency in *C. elegans* and found that it played an essential role in patterning (Kiehl et al., 2000).

Satterfield and Pallanck (2006) showed that fly atx2 assembled with polyribosomes and poly(A)-binding protein (PABP). They pointed to the importance of the N-terminal Lsm/Lsm-associated domain (LsmAD) and the PAM2 motif that mediated physical interaction with PABP. They suggested that Atx2 played a direct role in translational regulation and that mutant ataxin-2 caused neurodegeneration by interfering with the translational regulation of particular mRNAs (Satterfield & Pallanck, 2006). Indeed several years later, Dansithong and colleagues showed in an SCA2 mouse model that ATXN2[Q72] impaired translation of *Rgs8* mRNA (Dansithong et al., 2015).

Fly studies by the Bonini group pointed to an important function of fly atx2 in neurodegeneration. In a screen for modifiers of SCA3 neurodegeneration, they isolated Atx2 and showed that normal activity of Atx2 was critical for SCA3 degeneration, depending in particular on the PAM2 motif that mediates binding of cytoplasmic poly(A)-binding protein (PABP) (Lessing & Bonini, 2008).

We and others have produced multiple SCA2 mouse models, including transgenic and knockout models. Recent reviews describe these mouse lines in detail (Alves-Cruzeiro et al., 2016; Scoles & Pulst, 2018b).

25.2.3 **Pcp2-ATXN2** ***Transgenic Mice***

We generated two transgenic types expressing ATXN2 with mutant repeats of Q58 and Q127 under the control of the Purkinje cell protein 2 (*Pcp2)/L7* promoter (Hansen et al., 2013; Huynh et al., 2000). Rotarod testing demonstrated an ATXN2 dose-dependent motor phenotype for *ATXN2*-Q58 mice first observed at 6 months of age. PCs contained cytoplasmic but not nuclear inclusion bodies. The ATXN2-Q58 mouse was also used in studies demonstrating that dantrolene treatment restored *ATXN2* mouse motor phenotypes (Liu et al., 2009).

The subsequently generated *ATXN2*-Q127 line had an earlier onset at 8 weeks and also showed cytoplasmic inclusions at this age. This line was studied in greater detail using molecular, motor, and electrophysiological techniques (Hansen et al., 2013; Pflieger et al., 2017). Although subtle expression changes were already detected at 4 weeks, more significant changes were seen at the onset of the motor phenotype at 8 weeks. A genome-wide transcriptomic study comparing cerebellar RNAs of wild-type and Q127 mice at 1 day and 3 and 6 weeks of age confirmed these results (Pflieger et al., 2017). The number of differentially expressed genes (DEGs) increased from 138 at day 1 to 458 at 3 weeks and 434 at 6 weeks. Only 3 DEGs were shared across all 3 time points, whereas 87 DEGs were in common at 3 and 6 weeks. Pflieger et al. give specific cutoff expression and significance values (Pflieger et al., 2017). A significant reduction of PC spontaneous firing was also first seen at 8 weeks. These results confirm that this mouse model replicates a salient feature of human SCA2, i.e., an adult-onset neurodegenerative disease with only very minor developmental components (Fig. 25.1).

25.2.4 SCA2 BAC Transgenic Mice

We introduced a BAC containing the entire 176 kb *ATXN2* gene region including 16 kb upstream sequence and 2.5 kb downstream sequence into the mouse germline. Presently, we have two SCA2 BAC lines including alleles expressing ATXN2[Q22] and ATXN2[Q72] (Dansithong et al., 2015). Although the Q22 line has no motor, transcriptomic, or neurophysiological phenotype, it has played a crucial role in the development of gene-targeted therapies. The Q72 line has an onset of a rotarod phenotype at 8 weeks. To cross-validate our different animal models, we examined transcriptomes and found that there was a significant overlap in DEGs seen in the

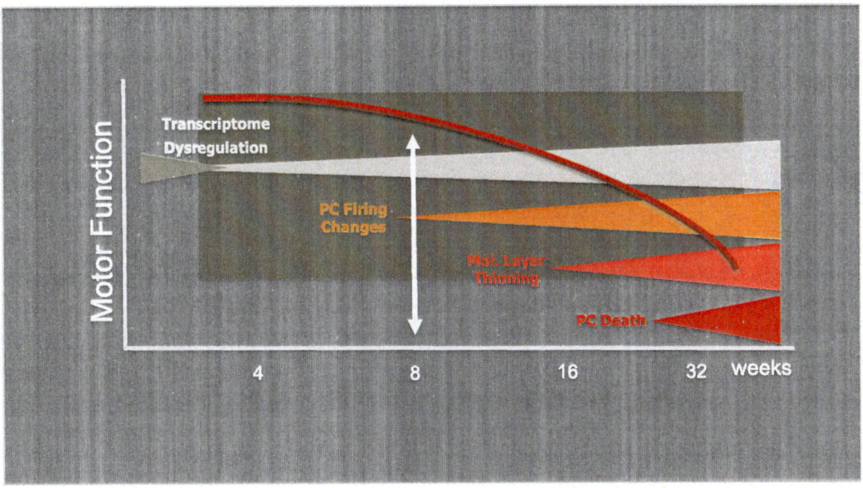

Fig. 25.1 Schematic of time course of molecular, physiologic, and behavioral phenotypes in the Pcp2-ATXN2[Q127] mouse. The curved red line indicates progressive loss of motor coordination beginning at week 8. Note the presence of two sets of transcriptomic changes indicating developmental changes from day 1 to week 3 and neurodegenerative changes afterwards

Q127 line as a result of BAC expression in PCs (Dansithong et al., 2015). Both models also share changes in intrinsic PC excitability (Scoles et al., 2017).

25.2.5 ATXN2-Q75 Transgenic Mice

Aguiar and colleagues generated a transgenic mouse line that expressed ATXN2[Q75] under control of a native human *ATXN2* promoter fragment (Aguiar et al., 2006). Ubiquitous transgene expression was observed, and hemizygous mice were ataxic by 12 weeks of age in rotarod tests as well as abnormal PC morphology. No further reports using this line have been published.

25.2.6 **Atxn2** *Knockout Mice*

After identification of the ATXN2 gene, it was not known whether CAG repeat expansion acted as a dominant gain-of-function allele or also had a loss-of-function component. We therefore generated *Atxn2*-deficient mice for comparison with transgenic mice.

In 2006, we published first results of targeting the *Atxn2* gene by homologous recombination (Kiehl et al., 2006). Mice were viable, but breeding showed significant sex-specific segregation distortion. Knockout mice developed adult-onset hyperphagia and obesity. Of note, heterozygotes showed an intermediate phenotype indicating that *Atxn2* is a highly dosage-sensitive gene. Subsequent studies demonstrated hyperactivity in the open cage and lack of cued and contextual fear conditioning. Long-term potentiation in the amygdala, but not the hippocampus, was impaired, consistent with the observed lack of fear conditioning (Huynh et al., 2009). The phenotype of obesity, hyperactivity, and lack of fear conditioning has been referred to as the 3F phenotype for fat, fearless, and fickle.

Our results were corroborated 2 years later in an independently generated knockout mouse (Lastres-Becker et al., 2008). The Auburger group also showed that *Atxn2* knockout resulted in abnormally low insulin receptor expression.

In contrast to our transgenic lines, recording of PC activity in knockout mice in the acute cerebellar slice did not show significant abnormalities (Pflieger et al., 2017). This was consistent with lack of significant transcriptomic changes in knockout mice. Absence of molecular, physiologic, or morphologic cerebellar neuropathology in *Atxn2*$^{-/-}$ mice supports developing SCA2 therapeutics that target both wild-type and mutant *ATXN2* alleles using antisense oligonucleotides (ASOs).

25.3 Transcriptome Analyses

Our transcriptomic studies of changes in mRNA steady-state levels were a continuous development from semiquantitative immunohistochemistry to quantitative PCR of individual genes and then to transcriptomics. The identification of top dysregulated genes (DEGs) subsequently led to the development of protein assays that allowed us to follow progression of disease and response to therapeutic interventions.

Postmortem biomarkers can be extremely helpful in evaluating onset and progression of disease. Calbindin 28 K is a protein that is highly expressed in PCs and involved in intracellular calcium regulation. Evaluation of immunocytochemical intensity using CALB1 antibodies has been used in many studies of mouse SCA models. In addition to intensity, CALB1 staining can be conveniently used to measure thickness of the molecular layer. The procedure generally involves determining a region of interest on a tissue slide and then measuring intensity with a program such as ImageJ.

Although the primary fissure can help in establishing landmarks, the problems with the above methodology lie in the difficulty to ascertain that identical lobes and locations are examined from mouse to mouse. Furthermore, placement of the region of interest within a given lobule can be highly subjective. In the best of hands, CALB1 staining can be variable even within the same section and even more so between sections.

We therefore began adding a quantitative PCR (qPCR) for *Calb1* to our preclinical evaluations (Hansen et al., 2013). This approach proved to be highly reproducible and allowed us to examine levels at different time points along the disease

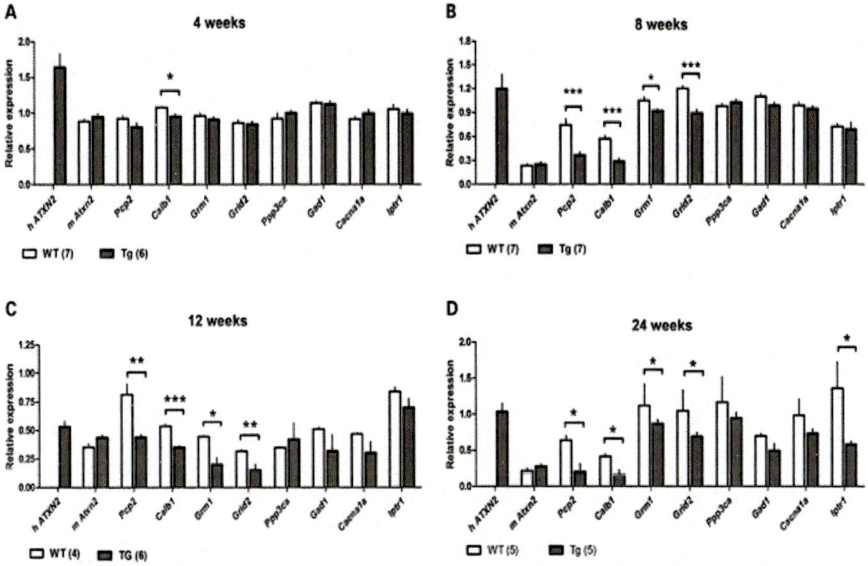

Fig. 25.2 Expression changes of key cerebellar genes including several PC-specific genes at four time points by quantitative PCR data from mouse cerebella: *ATXN2*^{Q127} (gray bars) and WT animals (white bars). Genes tested were human transgene (h*ATXN2*), mouse ataxin-2 (m*Atxn2*), PC protein 2 (*Pcp2*), calbindin 28-k (*Calb1*), metabotropic glutamate receptor 1 (*Grm1*), glutamate receptor ionotropic delta-2 (*Grid2*), calcineurin (*Ppp3ca*), glutamate decarboxylase (*Gad1*), voltage-gated calcium channel Ca$_v$2.1 (*Cacna1a*), and inositol triphosphate receptor 1 (*Itpr1*). *N*, number of mice for each genotype and age group listed in brackets. Gene expression normalized to the *Wasf1* housekeeping gene. **P* < 0.05, ***P* < 0.01, ****P* < 0.001 for Student's two-tailed *t*-test. Error bars represent ±SEM. (Reproduced with permission from Hansen et al., 2013)

process (Fig. 25.2). Levels of *Calb1* were normal at birth, but significantly downregulated as early as 4 weeks of age. Levels declined progressively after that.

Having proven reproducibility with *Calb1*, we went on to select genes from the Allen mouse brain atlas that showed high and specific PC expression, played an important role in PC physiology, or had a known disease association, for evaluation by qPCR. We observed progressive declines in a number of PC-enriched gene products, which in many cases preceded cell loss. We saw a significant reduction in *Pcp2*, *Grm1*, and *Grid2* expression at 8 weeks with continuing decline at later ages (Fig. 25.2). Reduced mRNA expression was not a function of PC death, as PC numbers remained unchanged between WT and *ATXN2*^{Q127} mice until much later (Hansen et al., 2013).

Success with determining mRNA steady-state levels with qPCR prompted us to conduct genome-wide transcriptomic studies of cerebellar RNAs (Dansithong et al., 2015; Pflieger et al., 2017; Scoles et al., 2017). We first used transcriptomics to validate our BAC-ATXN2[Q72] mice in relationship to our Pcp2-ATXN2[Q127] mice (Dansithong et al., 2015). We found that the BAC model very closely followed DEGs in the Pcp2-ATXN2[Q127] model. These genes included DEGs (*Rgs8*, *Pcp2*,

Pcp4, *Cep76*, *Homer3*, and *Fam107b*) that are also significantly reduced at the protein expression level and that respond to ATXN2-ASO therapy (Paul et al., 2018; Scoles et al., 2017).

Gene ontology analysis of top dysregulated pathways in the BAC model identified calcium ion binding, substrate-specific and ion channel activity, and glutamate receptor activity, all of which were shared with the Q127 model (Supplementary Table 4 in Dansithong et al., 2015). On the other hand, a subset of DEGs, designated as class I genes, was not shared with the Q127 model. Several of the class I DEGs (*Grm4*, *Igfbp5*, *Fstl5*, *Snrk*, *D8Ertd82e*, *Dusp5*, *Nab2*, *Btg1*, *Adrbk2*, *Slc25a29*, *Sty12*, *Crhr1*, *Synpr*, *Lrrtm2*, *Rit2*, and *Cabp2*) are predominantly expressed in granule cells. The results utilizing DEGs in addition to examination of microdissected granule cells confirmed that the BAC transgene was expressed in granule cells and that mutant ATXN2 exerted direct effects on granule cell function.

We also used RNA profiling to determine genome-wide changes in early stages of disease in Q127 animals. We found that DEGs were apparent as early as postnatal day 1. Of note, these DEGs at day 1 represented largely developmental cerebellar genes and did not meet cutoff criteria at later stages. These results suggest that the expression of mutant *ATXN2* can lead to subtle developmental abnormalities that are not directly connected to a later developmental transcriptome. Of note, recent studies of aborted human fetuses with Huntington disease show marked developmental abnormalities (Barnat et al., 2020).

Beginning at week 3, we identified a large number of DEGs that were predominantly downregulated, and their downregulation progressed with time. We used transcriptomes at weeks 3 and 6 to employ weighted gene co-expression network analysis (WGCNA). WGCNA can help to identify gene clusters known as co-expression modules, which then facilitate the finding of hub genes that are key drivers of module function and may correspond to biological pathways. We identified several individual modules that represented GTPase-mediated signal transduction, mRNA transport, phosphorylation, and other modules with enrichment of terms related to ER nuclear signaling, regulation of apoptosis, and positive regulation of cell differentiation. Many PC marker genes were also highly connected hub genes (*Calb1*, *Stk17b*, *Pcp4*).

It is interesting to note that transcriptomes of mouse models for different SCAs show significant overlap even if the direct pathogenic events are different. For example, analysis of an *ATXN1* transgenic model shared many key DEGs and pathways with DEGs in our analysis of *ATXN2* models (Ingram et al., 2016). As analyzed in a review by Niewiadomska-Cimicka and colleagues, many PC-enriched genes are also found as DEGs in SCA3, SCA7, and SCA17 (Niewiadomska-Cimicka et al., 2020). Thus, one can identify a small subset of DEGs that appear specific to a given SCA and likely represent the initial insult, for example, interaction with transcription complexes such as RORalpha or capicua or in the case of ATXN2 with cytosolic proteins. It is remarkable, however, that a larger set of DEGS appears to be shared among the different SCAs, potentially similar to the shared changes in intrinsic PC firing. These observations may provide an opportunity for the development of shared therapeutics across the wide spectrum of SCAs and potentially also sporadic ataxias.

In contrast to ATXN2[Q127] transgenic mice, an Atxn2-CAG42 knock-in mouse model displayed ataxic behavior only in the homozygous state at 18 weeks. This was associated with reduced cerebellar expression of poly(A)-binding protein cytoplasmic 1 (*Pabpc1/Pab1*) (Damrath et al., 2012). Cerebellar transcriptomes determined by microarray analysis of symptomatic Atxn2-CAG42 knock-in mice revealed few mRNA changes compared with wild-type (WT) mice, including absence of significant changes for *Calb1*. In a novel model with a knock-in of 100 CAG repeats, the earliest changes were detected at 3 months of age in homozygous knock-in mice showing DEGs in the family of Ca^{++} channels/transporters, IP_3 metabolism (*Plcg1*, *Inpp5a*, *Itpka*), and Ca^{2+}-calmodulin-dependent kinases (*Camk2a*, *Camk4*) (Arsović et al., 2020). The authors hypothesized that the knock-in resulted in non-PC autonomous changes as few changes were observed in the ATXN2[Q58] model examined in a similar vein. These results are at odds with studies of ATXN2[Q127] mice and may relate to different timing and expression levels of transgenes (Pflieger et al., 2017).

We also used transcriptomic analysis to examine whether mutant ATXN2 expression in vivo led to overlap with DEGs in an *Atxn2* loss-of-function mouse model. We compared *ATXN2*$^{+/-}$ and *Atxn2*$^{-/-}$ mice at 10 weeks with Q127 mice at 6 weeks. Overall, either knockdown strain showed only few DEGs at the chosen cutoff values. Heterozygous *Atxn2*$^{+/-}$ mice showed no overlap with transgenic transcriptomes, and knockout mice shared only five DEGs (Pflieger et al., 2017). Combined with a lack of abnormalities of PC firing in knockout animals, these data strongly suggest that mutant ATXN2 does not cause significant loss of function and that knockdown or knockout of *Atxn2* does not lead to major PC dysfunction. Similarly, the Auburger group compared transcriptomes of ATXN2-CAG42 transgenic mice with *Atxn2* knockout mice using a microarray analysis approach. They did demonstrate, however, overlapping abnormalities in calcium homeostasis pathways, although studies are in the end difficult to compare owing to different animal models and techniques used (Halbach et al., 2017).

Our BAC-ATXN2[Q72] mice expressed mtATXN2 in the spinal cord as well as in the brain (Dansithong et al., 2015; Scoles et al., 2020). We therefore used transcriptomes at multiple time points to determine whether this mouse line replicated features seen in other ALS mouse models (Scoles et al., 2020). DEGs defined three interconnected pathways (innate immunity, fatty acid biosynthesis, and cholesterol biosynthesis) in separate modules identified by weighted gene co-expression network analysis. Other key pathways included the complement system and lysosome/phagosome pathways. Of all DEGs in the spinal cord, 13% were also dysregulated in the cerebellum. There was significant overlap with DEGs seen in other ALS mouse models generated by expression of TDP-43 or SOD1 transgenes.

25.4 Changes in Intrinsic Purkinje Cell Activity

Altered PC intrinsic firing has been observed in several mouse models of ataxia including SCA2 (reviewed in Cook et al., 2020). PCs fire at high spontaneous rates, and firing is regular, absent synaptic inputs (reviewed in Meera et al., 2016). Using Pcp2-ATXN2[127Q] mice, we found that PCs exhibited decreased intrinsic firing beginning at 8 weeks of age. This coincided with the first observable deficits in motor behavior on the accelerating rotarod (Hansen et al., 2013). The decrease in PC firing was progressive, but firing regularity was largely unchanged. Similar observations were made in the BAC-ATXN2[Q72] model (Dansithong et al., 2015; Scoles et al., 2020).

Several observations may explain reduced intrinsic firing of PCs related to altered calcium homeostasis, altered ion channel expression, and src kinase signaling. Altered calcium homeostasis may be driven by abnormal physical interaction of mutant ATXN2 and the inositol triphosphate receptor (IP3R) with increased IP3R sensitivity to IP3 resulting in enhanced calcium release from the endoplasmic reticulum (Liu et al., 2009). Reducing calcium release with dantrolene targeting the functionally coupled ryanodine receptor in vitro and in vivo improved PC survival and motor behavior in Pcp2-ATXN2[Q58] mice.

In the acute cerebellar slice, Meera et al. (2017) found that DPHG-mediated stimulation was increased in PCs from Pcp2-ATXN2[Q127] mice as were synaptic mGLur1-mediated calcium transients and excitatory postsynaptic currents (EPSCs). Buffering Ca^{++} at normal resting levels in mutant PCs prevented enhanced mGluR1 function. Consistent with effects of mutant ATXN2 on different limbs of mGluR1 signaling, mGluR1-mediated synaptic currents were enhanced in SCA2 PCs (Fig. 25.3). We postulated a deleterious positive feedback loop that involved elevated intracellular calcium and enhanced mGluR1 function, both contributing to PC dysfunction and subsequent cell loss. The reduced PC firing frequency may represent a homeostatic mechanism to counteract calcium toxicity (Meera et al., 2017).

The Shakkottai laboratory investigated changes of ion channel abundance and physiology in the Pcp2-ATXN2[Q127] mouse (Dell'Orco et al., 2017). At 12 weeks without significant PC atrophy in the Pcp2-ATXN2[Q127] model, they observed a reduction in BK and Kv3.3 transcripts. Reduced function mediated by these channels likely leads to reduced firing frequency due to the loss of membrane repolarization and inability of voltage-gated sodium channels to recover from inactivation. In addition, driven by elevated basal calcium levels, SK-type K^+ conductances are predicted to be tonically active with the resulting membrane hyperpolarization slowing of spontaneous PC firing (Meera et al., 2016). Dell'Orco and colleagues also pointed to different mechanisms at later disease stages when atrophic PCs predominate. At 25 weeks, PCs were able to hyperpolarize the membrane sufficiently through a novel mechanism via K_{ir} currents. This points to the need for potentially different treatment strategies depending on the disease stage.

An unexpected insight into PC firing came from oncology via a knockout mouse of the Missing-in-metastasis 1 (*MTSS1*) gene (Brown et al., 2018). Initially

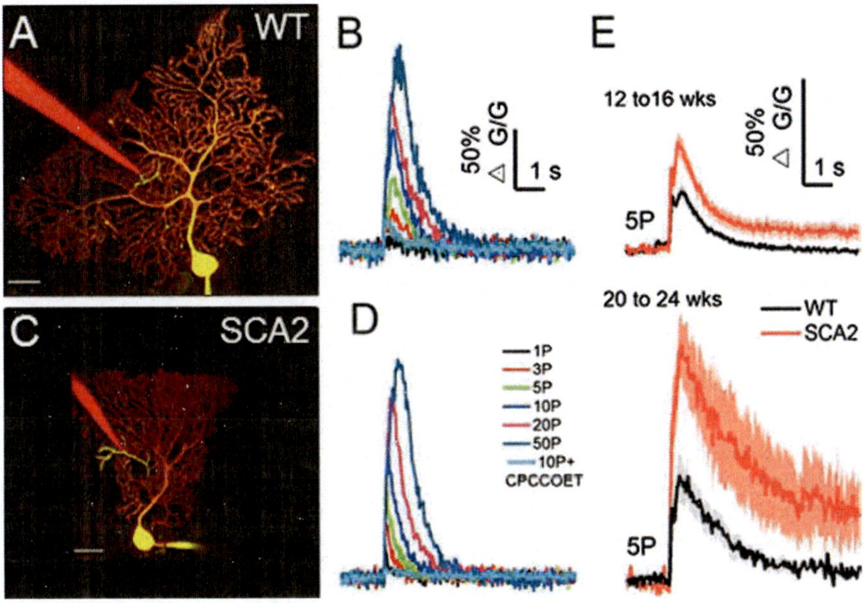

Fig. 25.3 Synaptic mGluR1-mediated calcium transients are significantly increased and prolonged in SCA2 PCs. (**a, c**) Two photon images of 12-week-old wild-type (**a**) and SCA2 (**c**) PCs filled via patch pipette (yellow) with the calcium indicator dye OGB1 (200 μM). A second pipette (red) filled with Alexa Fluor 594 (20 μM) was placed in the dendritic region to minimally stimulate parallel fiber (PF) synaptic inputs. Note the local region of $[Ca^{2+}]$ rises near the tip of the stimulating electrode (green). Scale bars: 25 μm. (**b, d**) Intracellular Ca^{++} signals (ΔG/G) for responses from WT (**b**) and SCA2 (**d**) PCs elicited by stimulation of PFs with 100 Hz trains and indicated number of pulses in the presence of AMPA, NMDA, and $GABA_A$ receptor antagonists. Ca^{++} signals are blocked by the mGluR1 antagonist CPCCOET (cyan traces). (**e**) Mean ± SEM changes in Ca^{++} in response to PF stimulation (five pulses at 100 Hz) across experiments for the indicated age groups and genotypes. WT (black lines) and SCA2 (red lines); lighter lines indicate SEM). (Modified from Meera et al. (2017))

identified in bladder cancer, MTSS1 functions in an evolutionarily conserved signaling cassette to antagonize Src kinase. *MTSS1* knockout mice develop ataxia with early reduction of PC intrinsic firing. Furthermore, *Mtss1* is one of the prominent DEGs in SCA2 mouse models (Dansithong et al., 2015). Src family kinase (SFK) activity was enhanced in a mouse model of SCA2, and reducing SFK signaling using dasatinib rescued PC intrinsic firing (Brown et al., 2018).

In summary, there is converging evidence that dysfunction and PC death are exacerbated by prominent positive feedback mechanisms through effects of elevated basal calcium on mGluR1 coupling to TRPC3 channels and to IP_3R-mediated release of intracellular calcium as well as effects on calcium-dependent potassium (KCa) channels. Effects by mutant ATXN2 on stress granules, the integrated stress response, and src kinases likely add an additional layer of complexity, but also additional targets for therapy.

25.5 RNA-Based SCA2 Therapeutics

Despite identification of many ATXN2 interactors, information on which interactions are targetable and relevant in a given cell type are still lacking. Broadly, two concepts of approaches for developing SCA2 therapeutics were discussed in a recent review on precision medicine for the spinocerebellar ataxias (Bushart et al., 2016). The first approach based on changes in pathways and physiology was discussed above under PC excitability. In this section, we will focus on targeting *ATXN2* directly via antisense oligonucleotides (ASOs). Some ASOs have been approved for human use, and others are in various stages of human clinical trials (reviewed in Scoles & Pulst, 2018a; Scoles & Pulst 2019).

Beginning in 2012 and in collaboration with Ionis Pharmaceuticals, we screened gapmer ASOs in HepG2 cells for their ability to reduce *ATXN2* mRNA abundance. 2'-MOE gapmers are 20 bp in length, are phosphorothioate throughout, and have a 2'-*O*-methoxyethyl (MOE) group on the terminal 5 bps at each end of the oligonucleotide (Rigo et al., 2014; Scoles et al., 2017, 2019). The modifications reduce ASO degradation by nucleases and increase specificity of target mRNA interaction and degradation by RNase-H (Bennett & Swayze, 2010).

We progressed the best ASOs to in vivo studies by ICV injection and measured target engagement in cerebellar tissue. We also examined absence of glial and microglial activation by assessing expression of *Gfap* and *Aif1* (encoding Iba1). The best ASO, designated as ASO7, was taken forward to detailed preclinical evaluation in both our Pcp2-ATXN2-Q127 and in our BAC-ATXN2[Q72] mouse models (Scoles et al., 2017). Symptomatic mice were treated with ASO7 or saline at 8 weeks of age and were tested at different treatment time points on the accelerating rotarod. Both models demonstrated delayed progression of the SCA2 motor phenotypes. At the endpoint (19 weeks of age), we determined the cerebellar expression of ATXN2, Cep76, Fam107b, Homer3, Rgs8, Pcp2, and Pcp4 by qPCR and Western blotting. Human *ATXN2* and mouse *Atxn2* remained significantly reduced even after 11 weeks. The PC marker mRNAs and proteins showed near normalization of mRNA and protein expression. Subsets of mice were tested to determine the effect of the ASO7 on PC physiology in the acute cerebellar slice. Treatment with ASO7 restored the mean PC firing frequency to that observed in age-matched wild-type mice.

In a separate study using BAC-ATXN2[Q72] mice, we analyzed spinal cord transcriptomes and reversal of DEGs after ASO7 treatment (Scoles et al., 2020). This treatment modified DEGs in the innate immunity, the complement system, and the lysosome/phagosome pathways toward levels in wild-type mice. An *ATXN2* ASO, designated as ION541 or BIIB105, is entering a phase 1 clinical trial in sporadic ALS patients and in patients with 30–33 repeats in the *ATXN2* gene.

25.6 Conclusions

Repeat expansion mutations in *ATXN2* cause SCA2, but they are also associated with a wide-ranging set of neurodegenerative disorders. Genetic mechanisms range from dominant to recessive Mendelian alleles to risk alleles for repeats in the range of 30 to 32 glutamines. Numerous ATXN2-interacting proteins have been identified. ATXN2 mouse models strongly support a gain of function for repeat expanded alleles. Progress toward developing drugs targeting calcium signaling and ion channels is accumulating. The opportunity to target *ATXN2* directly using antisense oligonucleotides remains a primary goal of our research group for treating SCA2. Twenty-five years after identification of the *ATXN2* as the gene mutated in SCA2 and 8 years after the first ASO screens, an RNA-based therapy is now in a phase 1 trial.

Acknowledgments I want to thank patients on several continents who have contributed to our understanding of SCA2 and members of my laboratory over the last three decades. Particular thanks go to Dr. Daniel Scoles for a detailed review of this chapter.

Funding This work was supported by grants R01NS33123, R56NS33123, RC4NS073009, UO1 NS103883, and R37NS033123 from the National Institute of Neurological Disorders and Stroke.

References

Aguiar, J., Fernandez, J., Aguilar, A., Mendoza, Y., Vazquez, M., Suarez, J., et al. (2006). Ubiquitous expression of human SCA2 gene under the regulation of the SCA2 self promoter cause specific Purkinje cell degeneration in transgenic mice. *Neuroscience Letters, 392*, 202–206.

Alves-Cruzeiro, J. M., Mendonca, L., Pereira de Almeida, L., & Nobrega, C. (2016). Motor dysfunctions and neuropathology in mouse models of spinocerebellar Ataxia type 2: A comprehensive review. *Frontiers in Neuroscience, 10*, 572.

Ashizawa, T., Figueroa, K. P., Perlman, S. L., Gomez, C. M., Wilmot, G. R., Schmahmann, J. D., Ying, S. H., Zesiewicz, T. A., Paulson, H. L., Shakkottai,, V. G., Bushara, K. O., Kuo, S. H., Geschwind, M. D., Xia, G., Mazzoni, P., Krischer, J. P., Cuthbertson, D., Holbert, A. R., Ferguson, J. H., Pulst, S. M., & Subramony, S. H. (2013). Clinical characteristics of patients with spinocerebellar ataxias 1, 2, 3 and 6 in the US; a prospective observational study. *Orphanet Journal of Rare Diseases, 8*, 177.

Arsović, A., Halbach, M. V., Canet-Pons, J., Esen-Sehir, D., Döring, C., Freudenberg, F., Czechowska, N., Seidel, K., Baader, S. L., Gispert, S., Sen, N. E., & Auburger, G. (2020). Mouse Ataxin-2 Expansion Downregulates CamKII and Other Calcium Signaling Factors, Impairing Granule-Purkinje Neuron Synaptic Strength. *International Journal of Molecular Sciences, 12*(21), 6673.

Bailey, J., Loomis, S., Kang, J., Allingham, R., Gharahkhani, P., Khor, C., et al. (2016). Genome-wide association analysis identifies TXNRD2, ATXN2 and FOXC1 as susceptibility loci for primary open-angle glaucoma. *Nature Genetics, 48*(2), 189–194.

Barnat, M., Capizzi, M., Aparicio, E., Boluda, S., Wennagel, D., Kacher, R., Kassem, R., Lenoir, S., Agasse, F., Braz, B. Y., Liu, J. P., Ighil, J., Tessier, A., Zeitlin, S. O., Duyckaerts, C., Dommergues, M., Durr, A., & Humbert, S. (2020). Huntington's disease alters human neurodevelopment. *Science (New York, N.Y.), 369*(6505), 787–793.

Bennett, C. F., & Swayze, E. E. (2010). RNA targeting therapeutics: Molecular mechanisms of antisense oligonucleotides as a therapeutic platform. *Annual Review of Pharmacology and Toxicology, 50*, 259–293.

Blokhuis, A. M., Koppers, M., Groen, E. J., van den Heuvel, D. M., Dini Modigliani, S., Anink, J. J., et al. (2016). Comparative interactomics analysis of different ALS-associated proteins identifies converging molecular pathways. *Acta Neuropathologica, 132*, 175–196.

Brown, A. S., Meera, P., Altindag, B., Chopra, R., Perkins, E. M., Paul, S., Scoles, D. R., Tarapore, E., Magri, J., Huang, H., Jackson, M., Shakkottai, V. G., Otis, T. S., Pulst, S. M., Atwood, S. X., & Oro, A. E. (2018). MTSS1/Src family kinase dysregulation underlies multiple inherited ataxias. *Proceedings of the National Academy of Sciences of the United States of America, 115*(52), E12407–E12416.

Bushart, D. D., Murphy, G. G., & Shakkottai, V. G. (2016). Precision medicine in spinocerebellar ataxias: Treatment based on common mechanisms of disease. *Annals of Translational Medicine, 4*, 25.

Cook, A. A., Fields, E., & Watt, A. J. (2020). Losing the beat: Contribution of Purkinje cell firing dysfunction to disease, and its reversal. *Neuroscience, S0306–4522*(20), 30377–30378. Advance online publication.

Cuellar-Partida, G., et al. (2020). Genome-wide association study identifies 48 common genetic variants associated with handedness. *Nature Human Behaviour, 5*, 59–70. https://doi.org/10.1038/s41562-020-00956-y

DeMille, D., Badal, B. D., Evans, J. B., Mathis, A. D., Anderson, J. F., & Grose, J. H. (2015). PAS kinase is activated by direct SNF1-dependent phosphorylation and mediates inhibition of TORC1 through the phosphorylation and activation of Pbp1. *Molecular Biology of the Cell, 26*, 569–582.

Damrath, E., Heck, M. V., Gispert, S., Azizov, M., Nowock, J., Seifried, C., et al. (2012). ATXN2-CAG42 sequesters PABPC1 into insolubility and induces FBXW8 in cerebellum of old ataxic knock-in mice. *PLoS Genetics, 8*, e1002920.

Dansithong, W., Paul, S., Figueroa, K. P., Rinehart, M. D., Wiest, S., Pflieger, L. T., et al. (2015). Ataxin-2 regulates RGS8 translation in a new BAC-SCA2 transgenic mouse model. *PLoS Genetics, 11*, e1005182.

Dell'Orco, J. M., Pulst, S. M., & Shakkottai, V. G. (2017). Potassium channel dysfunction underlies Purkinje neuron spiking abnormalities in spinocerebellar ataxia type 2. *Human Molecular Genetics, 26*(20), 3935–3945.

Elden, A. C., Kim, H. J., Hart, M. P., Chen-Plotkin, A. S., Johnson, B. S., Fang, X., Armakola, M., Geser, F., Greene, R., Lu, M. M., Padmanabhan, A., Clay-Falcone, D., McCluskey, L., Elman, L., Juhr, D., Gruber, P. J., Rüb, U., Auburger, G., Trojanowski, J. Q., Lee, V. M., & Gitler, A. D. (2010). Ataxin-2 intermediate-length polyglutamine expansions are associated with increased risk for ALS. *Nature, 466*(7310), 1069–1075.

Farg, M. A., Soo, K. Y., Warraich, S. T., Sundaramoorthy, V., Blair, I. P., & Atkin, J. D. (2013). Ataxin-2 interacts with FUS and intermediate-length polyglutamine expansions enhance FUS-related pathology in amyotrophic lateral sclerosis. *Human Molecular Genetics, 22*, 717–728.

Fernandez, M., McClain, M. E., Martinez, R. A., Snow, K., Lipe, H., Ravits, J., et al. (2000). Late-onset SCA2: 33 CAG repeats are sufficient to cause disease. *Neurology, 55*, 569–572.

Figueroa, K. P., Coon, H., Santos, N., Velazquez, L., Mederos, L. A., & Pulst, S. M. (2017). Genetic analysis of age at onset variation in spinocerebellar ataxia type 2. *Neurology Genetics, 3*(3), e155.

Figueroa, K. P., & Pulst, S. M. (2003). Identification and expression of the gene for human ataxin-2-related protein on chromosome 16. *Experimental Neurology, 184*(2), 669–678.

Figueroa, K. P., Farooqi, S., Harrup, K., Frank, J., O'Rahilly, S., & Pulst, S. M. (2009). Genetic variance in the spinocerebellar ataxia type 2 (ATXN2) gene in children with severe early onset obesity. *PLoS One, 4*(12), e8280.

Gandelman, M., Dansithong, W., Figueroa, K. P., Paul, S., Scoles, D. R., & Pulst, S. M. (2020). Staufen 1 amplifies proapoptotic activation of the unfolded protein response. *Cell Death and Differentiation, 27*(10), 2942–2951.

Gehman, L. T., Stoilov, P., Maguire, J., Damianov, A., Lin, C. H., Shiue, L., Ares, M., Jr., Mody, I., & Black, D. L. (2011). The splicing regulator Rbfox1 (A2BP1) controls neuronal excitation in the mammalian brain. *Nature Genetics, 43*(7), 706–711.

Geschwind, D. H., Perlman, S., Figueroa, C. P., Treiman, L. J., & Pulst, S. M. (1997). The prevalence and wide clinical spectrum of the spinocerebellar ataxia type 2 trinucleotide repeat in patients with autosomal dominant cerebellar ataxia. *American Journal of Human Genetics, 60*, 842–850.

Gispert, S., Twells, R., Orozco, G., Brice, A., Weber, J., Heredero, L., et al. (1993). Chromosomal assignment of the second locus for autosomal dominant cerebellar ataxia (SCA2) to chromosome 12q23-24.1. *Nature Genetics, 4*, 295–299.

Gwinn-Hardy, K., Chen, J. Y., Liu, H. C., Liu, T. Y., Boss, M., Seltzer, W., et al. (2000). Spinocerebellar ataxia type 2 with parkinsonism in ethnic Chinese. *Neurology, 55*, 800–805.

Halbach, M. V., Gispert, S., Stehning, T., Damrath, E., Walter, M., & Auburger, G. (2017). Atxn2 knockout and CAG42-Knock-in cerebellum shows similarly dysregulated expression in calcium homeostasis pathway. *Cerebellum (London, England), 16*(1), 68–81.

Halbach, M. V., Stehning, T., Damrath, E., Jendrach, M., Sen, N. E., Basak, A. N., et al. (2015). Both ubiquitin ligases FBXW8 and PARK2 are sequestrated into insolubility by ATXN2 PolyQ expansions, but only FBXW8 expression is dysregulated. *PLoS One, 10*, e0121089.

Hansen, S. T., Meera, P., Otis, T. S., & Pulst, S. M. (2013). Changes in Purkinje cell firing and gene expression precede behavioral pathology in a mouse model of SCA2. *Human Molecular Genetics, 22*, 271–283.

Huynh, D. P., Del Bigio, M. R., Ho, D. H., & Pulst, S. M. (1999). Expression of ataxin-2 in brains from normal individuals and patients with Alzheimer's disease and spinocerebellar ataxia 2. *Annals of Neurology, 45*, 232–241.

Huynh, D. P., Figueroa, K., Hoang, N., & Pulst, S. M. (2000). Nuclear localization or inclusion body formation of ataxin-2 are not necessary for SCA2 pathogenesis in mouse or human. *Nature Genetics, 26*, 44–50.

Huynh, D. P., Nguyen, D. T., Pulst-Korenberg, J. B., Brice, A., & Pulst, S. M. (2007). Parkin is an E3 ubiquitin-ligase for normal and mutant ataxin-2 and prevents ataxin-2-induced cell death. *Experimental Neurology, 203*, 531–541.

Huynh, D. P., Yang, H. T., Vakharia, H., Nguyen, D., & Pulst, S. M. (2003). Expansion of the polyQ repeat in ataxin-2 alters its Golgi localization, disrupts the Golgi complex and causes cell death. *Human Molecular Genetics, 12*, 1485–1496.

Huynh, D. P., Maalouf, M., Silva, A. J., Schweizer, F. E., & Pulst, S. M. (2009). Dissociated fear and spatial learning in mice with deficiency of ataxin-2. *PLoS One, 4*, e6235.

Imbert, G., Saudou, F., Yvert, G., Devys, D., Trottier, Y., Garnier, J. M., et al. (1996). Cloning of the gene for spinocerebellar ataxia 2 reveals a locus with high sensitivity to expanded CAG/glutamine repeats. *Nature Genetics, 14*, 285–291.

Ingram, M., Wozniak, E., Duvick, L., Yang, R., Bergmann, P., Carson, R., O'Callaghan, B., Zoghbi, H. Y., Henzler, C., & Orr, H. T. (2016). Cerebellar transcriptome profiles of ATXN1 transgenic mice reveal SCA1 disease progression and protection pathways. *Neuron, 89*(6), 1194–1207.

Key, J., Harter, P. N., Sen, N. E., Gradhand, E., Auburger, G., & Gispert, S. (2020). Mid-gestation lethality of Atxn2l-ablated mice. *International Journal of Molecular Sciences, 21*(14), 5124.

Kiehl, T. R., Nechiporuk, A., Figueroa, K. P., Keating, M. T., Huynh, D. P., & Pulst, S. M. (2006). Generation and characterization of Sca2 (ataxin-2) knockout mice. *Biochemical Biophysical Research Communications, 339*, 17–24.

Kiehl, T. R., Shibata, H., Vo, T., Huynh, D. P., & Pulst, S. M. (2001). Identification and expression of a mouse ortholog of A2BP1. *Mammalian Genome: Official Journal of the International Mammalian Genome Society, 12*(8), 595–601.

Kiehl, T. R., Shibata, H., & Pulst, S. M. (2000). The ortholog of human ataxin-2 is essential for early embryonic patterning in C. elegans. *Journal of Molecular Neuroscience, 15*(3), 231–241.

Koyano, S., Uchihara, T., Fujigasaki, H., Nakamura, A., Yagishita, S., & Iwabuchi, K. (1999). Neuronal intranuclear inclusions in spinocerebellar ataxia type 2: Triple-labeling immunofluorescent study. *Neuroscience Letters, 273*, 117–120.

Lastres-Becker, I., Brodesser, S., Lutjohann, D., Azizov, M., Buchmann, J., Hintermann, E., et al. (2008). Insulin receptor and lipid metabolism pathology in ataxin-2 knock-out mice. *Human Molecular Genetics, 17*, 1465–1481.

Lastres-Becker, I., Nonis, D., Eich, F., Klinkenberg, M., Gorospe, M., Kotter, P., et al. (2016). Mammalian ataxin-2 modulates translation control at the pre-initiation complex via PI3K/mTOR and is induced by starvation. *Biochimica et Biophysica Acta, 1862*, 1558–1569.

Lessing, D., & Bonini, N. M. (2008). Polyglutamine genes interact to modulate the severity and progression of neurodegeneration in Drosophila. *PLoS Biology, 6*(2), e29.

Liu, J., Tang, T. S., Tu, H., Nelson, O., Herndon, E., Huynh, D. P., et al. (2009). Deranged calcium signaling and neurodegeneration in spinocerebellar ataxia type 2. *The Journal of Neuroscience, 29*, 9148–9162.

Luo, L., Wang, J., Lo, R. Y., Figueroa, K. P., Pulst, S. M., Kuo, P. H., Perlman, S., Wilmot, G., Gomez, C. M., Schmahmann, J., Paulson, H., Shakkottai, V. G., Ying, S. H., Zesiewicz, T., Bushara, K., Geschwind, M., Xia, G., Subramony, S. H., Ashizawa, T., & Kuo, S. H. (2017). The initial symptom and motor progression in spinocerebellar ataxias. *Cerebellum (London, England), 16*(3), 615–622.

Meera, P., Pulst, S. M., & Otis, T. S. (2016). Cellular and circuit mechanisms underlying spinocerebellar ataxias. *The Journal of Physiology, 594*(16), 4653–4660.

Meera, P., Pulst, S., & Otis, T. (2017). A positive feedback loop linking enhanced mGluR function and basal calcium in spinocerebellar ataxia type 2. *eLife, 6*, e26377.

Nechiporuk, T., Huynh, D. P., Figueroa, K., Sahba, S., Nechiporuk, A., & Pulst, S. M. (1998). The mouse SCA2 gene: cDNA sequence, alternative splicing and protein expression. *Human Molecular Genetics, 7*(8), 1301–1309.

Nechiporuk, T., Nechiporuk, A., Sahba, S., Figueroa, K., Shibata, H., Chen, X. N., Korenberg, J. R., de Jong, P., & Pulst, S. M. (1997). A high-resolution PAC and BAC map of the SCA2 region. *Genomics, 44*, 321–329.

Neuenschwander, A. G., Thai, K. K., Figueroa, K. P., & Pulst, S. M. (2014). Amyotrophic lateral sclerosis risk for spinocerebellar ataxia type 2 ATXN2 CAG repeat alleles: A meta-analysis. *JAMA Neurology, 71*(12), 1529–1534.

Niewiadomska-Cimicka, A., Hache, A., & Trottier, Y. (2020). Gene deregulation and underlying mechanisms in spinocerebellar ataxias with polyglutamine expansion. *Frontiers in Neuroscience, 14*, 571.

Nonhoff, U., Ralser, M., Welzel, F., Piccini, I., Balzereit, D., Yaspo, M. L., et al. (2007). Ataxin-2 interacts with the DEAD/H-box RNA helicase DDX6 and interferes with P-bodies and stress granules. *Molecular Biology of the Cell, 18*, 1385–1396.

Orozco Diaz, O., Nodarse Fleites, A., Cordovés Sagaz, R., & Auburger, G. (1990). Autosomal dominant cerebellar ataxia, clinical analysis of 263 patients from a homogeneous population in Holguín, Cuba. *Neurology, 40*(9), 1369.

Paul, S., Dansithong, W., Figueroa, K. P., Scoles, D. R., & Pulst, S. M. (2018). Staufen1 links RNA stress granules and autophagy in a model of neurodegeneration. *Nature Communications, 9*(1), 3648.

Paul, S., Dansithong W., Figueroa, K. P., Gandelman, M., Scoles D. R., & Pulst S. M. (2021) Staufen1 in Human Neurodegeneration. *Annals of Neurology, 89*, 1114–1128.

Payami, H., Nutt, J., Gancher, S., Bird, T., McNeal, M. G., Seltzer, W. K., Hussey, J., Lockhart, P., Gwinn-Hardy, K., Singleton, A. A., Singleton, A. B., Hardy, J., & Farrer, M. (2003). SCA2 may present as levodopa-responsive parkinsonism. *Movement Disorders: Official Journal of the Movement Disorder Society, 18*(4), 425–429.

Pflieger, L. T., Dansithong, W., Paul, S., Scoles, D. R., Figueroa, K. P., Meera, P., Otis, T. S., Facelli, J. C., & Pulst, S. M. (2017). Gene co-expression network analysis for identifying modules and functionally enriched pathways in SCA2. *Human Molecular Genetics, 26*(16), 3069–3080.

Pulst, S. M. (2018). The complex structure of ATXN2 genetic variation. *Neurology Genetics, 4*(6), e299.

Pulst, S. M., Nechiporuk, A., Nechiporuk, T., Gispert, S., Chen, X. N., Lopes-Cendes, I., et al. (1996). Moderate expansion of a normally biallelic trinucleotide repeat in spinocerebellar ataxia type 2. *Nature Genetics, 14*, 269–276.

Pulst, S. M., Nechiporuk, A., & Starkman, S. (1993). Anticipation in spinocerebellar ataxia type 2. *Nature Genetics, 5*, 8–10.

Ralser, M., Albrecht, M., Nonhoff, U., Lengauer, T., Lehrach, H., & Krobitsch, S. (2005). An integrative approach to gain insights into the cellular function of human ataxin-2. *Journal of Molecular Biology, 346*, 203–214.

Rigo, F., Seth, P. P., & Bennett, C. F. (2014). Antisense oligonucleotide-based therapies for diseases caused by pre-mRNA processing defects. *Advances in Experimental Medicine and Biology, 825*, 303–352.

Sahba, S., Nechiporuk, A., Figueroa, K. P., Nechiporuk, T., & Pulst, S. M. (1998) Genomic structure of the human gene for spinocerebellar ataxia type 2 (SCA2) on chromosome 12q24.1. *Genomics, 47*, 359–364.

Sanpei, K., Takano, H., Igarashi, S., Sato, T., Oyake, M., Sasaki, H., et al. (1996). Identification of the spinocerebellar ataxia type 2 gene using a direct identification of repeat expansion and cloning technique, DIRECT. *Nature Genetics, 14*, 277–284.

Satterfield, T. F., & Pallanck, L. J. (2006). Ataxin-2 and its Drosophila homolog, ATX2, physically assemble with polyribosomes. *Human Molecular Genetics, 15*, 2523–2532.

Scoles, D. R., & Pulst, S. M. (2019) Antisense therapies for movement disorders. *Movement Disorders, 34*, 1112–1119.

Scoles, D. R., Dansithong, W., Pflieger, L. T., Paul, S., Gandelman, M., Figueroa, K. P., Rigo, F., Bennett, C. F., & Pulst, S. M. (2020). ALS-associated genes in SCA2 mouse spinal cord transcriptomes. *Human Molecular Genetics, 29*(10), 1658–1672.

Scoles, D. R., Ho, M. H., Dansithong, W., Pflieger, L. T., Petersen, L. W., Thai, K. K., & Pulst, S. M. (2015). Repeat associated non-AUG translation (RAN translation) dependent on sequence downstream of the ATXN2 CAG repeat. *PLoS One, 10*(6), e0128769.

Scoles, D. R., Meera, P., Schneider, M. D., Paul, S., Dansithong, W., Figueroa, K. P., Hung, G., Rigo, F., Bennett, C. F., Otis, T. S., & Pulst, S. M. (2017). Antisense oligonucleotide therapy for spinocerebellar ataxia type 2. *Nature, 544*(7650), 362–366.

Scoles, D. R., Minikel, E. V., & Pulst, S. M. (2019). Antisense oligonucleotides: A primer. *Neurology Genetics, 5*(2), e323.

Scoles, D. R., Pflieger, L. T., Thai, K. K., Hansen, S. T., Dansithong, W., & Pulst, S. M. (2012). ETS1 regulates the expression of ATXN2. *Human Molecular Genetics, 21*, 5048–5065.

Scoles, D. R., & Pulst, S. M. (2018a). Oligonucleotide therapeutics in neurodegenerative diseases. *RNA Biology, 15*(6), 707–714.

Scoles, D. R., & Pulst, S. M. (2018b). Spinocerebellar Ataxia type 2. *Advances in Experimental Medicine and Biology, 1049*, 175–195.

Skipper, M., Milne, C. A., & Hodgkin, J. (1999). Genetic and molecular analysis of fox-1, a numerator element involved in Caenorhabditis elegans primary sex determination. *Genetics, 151*(2), 617–631.

Shibata, H., Huynh, D. P., & Pulst, S. M. (2000). A novel protein with RNA-binding motifs interacts with ataxin-2. *Human Molecular Genetics, 9*, 1303–1313.

Tazen, S., Figueroa, K., Kwan, J., Goldman, J., Hunt, A., Sampson, J., Gutmann, L., Pulst, S. M., Mitsumoto, H., & Kuo, S. H. (2013). Amyotrophic lateral sclerosis and spinocerebellar ataxia type 2 in a family with full CAG repeat expansions of ATXN2. *JAMA Neurology, 70*(10), 1302–1304.

Turnbull, V. J., Storey, E., Tarlac, V., Walsh, R., Stefani, D., Clark, R., et al. (2004). Different ataxin-2 antibodies display different immunoreactive profiles. *Brain Research, 1027*, 103–116.
Van de Loo, S., Eich, F., Nonis, D., Auburger, G., & Nowock, J. (2009). Ataxin-2 associates with rough endoplasmic reticulum. *Experimental Neurology, 215*, 110–118.
Wadia, N. H., & Swami, R. K. (1971). A new form of heredo-familial spinocerebellar degeneration with slow eye movements (nine families). *Brain, 94*, 359–374.

Chapter 26
Molecular Pathogenesis in Spinocerebellar Ataxia Type 31 (SCA31)

Kinya Ishikawa

26.1 Pentanucleotide TGGAA Repeat Is Tightly Associated with Spinocerebellar Ataxia Type 31 (SCA31)

Spinocerebellar ataxia type 31 (SCA31) is one of a group of disorders that shows progressive cerebellar ataxia as a cardinal symptom with an autosomal-dominant inheritance. This disease was first described when one of us (KI) and his colleagues found their families are mapped to a long arm of chromosome 16 (Nagaoka et al., 2000). We mapped the causative gene to chromosome 16q22.1 (Li et al., 2003), followed by identification of a single-nucleotide exchange (C-to-T) in the 5′ untranslated region of a gene, *PLEKHG4* (also called puratrophin-1), that encodes a protein with a spectrin repeat and Rho-guanine nucleotide exchange factor domain (Ishikawa et al., 2005). However, two affected subjects without this single-nucleotide exchange were subsequently found (Amino et al., 2007; Ohata et al., 2006), indicating that this is a polymorphism rarely found in Japanese population. In fact, many affected individuals across different families shared many rare variants in the critical 2-megabase chromosomal region. Thus, SCA31 was considered to have a strong founder effect (Amino et al., 2007). In support of the notion, SCA31 was not found in any other countries except Japan (Edener et al., 2011; Lee et al., 2007). Fortunately, we were able to narrow down one border of the critical region at this C-to-T in *PLEKHG4* and the other border at rs11640843 with a recombination among affected individuals demonstrating a new 900-kb critical region (Amino et al., 2007).

K. Ishikawa (✉)
Department of Neurology and Neurological Science, Graduate School of Medical and Dental Sciences, Tokyo Medical and Dental University, Tokyo, Japan

The Center for Personalized Medicine for Healthy Aging, Tokyo Medical and Dental University, Tokyo, Japan
e-mail: pico.nuro@tmd.ac.jp

H. Mizusawa, S. Kakei (eds.), *Cerebellum as a CNS Hub*,
Contemporary Clinical Neuroscience, https://doi.org/10.1007/978-3-030-75817-2_26

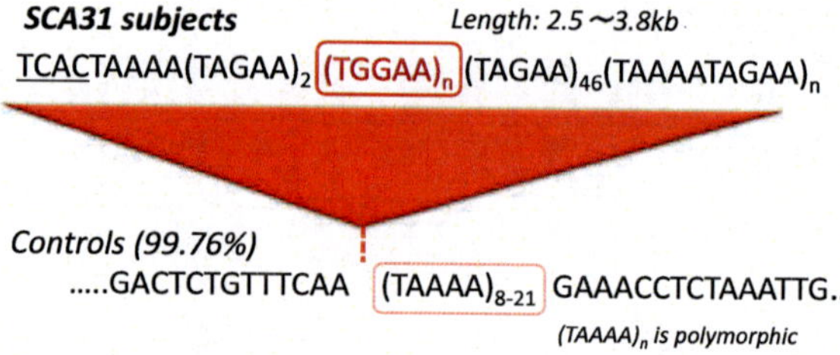

Fig. 26.1 Penta-peptide repeat in the SCA31 locus

By a tiling search of small rearrangement by Northern blot analysis in conjunction with tiling-path shotgun sequencing of the newly set critical SCA31 chromosomal region in 16q22.1, a mutation shared by all family members with founder chromosomes was found to be a 2.5- to 3.8-kb-long insertion (Sato et al., 2009). Cloning and sequencing of this insertion revealed that the internal sequence was a complex pentanucleotide repeat containing $(TGGAA)_n$, $(TAGAA)_n$, $(TAAAA)_n$, and $(TAAAATAGAA)_n$ (Fig. 26.1). The length of this insertion was inversely correlated with the age of onset in SCA31 patients. In contrast, this insertion was not seen in a larger set of control chromosomes with only a few exceptions. The vast majority (99.7%) of Japanese had a short TAAAA repeat of only 8–20 repeats. Sequencing the rare, large insertions in control revealed that the internal sequences were different from those in SCA31 subjects: the sequences were either a long pure stretch of $(TAAAA)_n$ or a complex repeat with $(TAAAA)_n$, $(TAGAA)_n$, and $(TAAAATAGAA)_n$. The $(TGGAA)_n$ was never observed in controls. In addition, presence of TGGAA repeat in SCA31 patients was also replicated in other researchers on different set of patients (Sakai et al., 2010). From these observations, $(TGGAA)_n$ was the only repeat segregating with the phenotype, suggesting its importance in pathogenesis.

The insertion initially appeared to be unrelated to nearby genes, *BEAN1* (brain expressed, associated with Nedd4) and *TK2* (thymidine kinase 2). However, extensive 3'-RACE experiments revealed that these two genes both had multiple downstream exons that had not been deposited in the public databases. This means that the 2.5- to 3.8-kb-long insertion is in an intronic region shared by two different genes, *BEAN1* and *TK2* [8]. While *BEAN1* drives brain-specific expression, *TK2* was expressed in all tissues that we examined. As predicted, the TGGAA repeat was transcribed as UGGAA repeat in SCA31 brains.

Identification of SCA31 repeat clarified clinical picture of SCA31. Typical clinical features can be found in some case reports and cohort studies (Itaya et al., 2018; Nakamura et al., 2017; Sakakibara et al., 2014). Sakakibara and her colleagues (Sakakibara et al., 2014) studied six SCA31 patients. Their average age of onset was 63.8 years. When compared to their own SCA6 patients, they found that SCA31 patients' clinical features were much more confined to cerebellar dysfunction, while their SCA6 patients showed pyramidal tract signs and psychiatric features besides cerebellar ataxia. Itaya and her colleagues described one SCA31 subject who additionally showed blepharospasm (Itaya et al., 2018). Their patient developed dysarthria at his age of 56. Clinical examination at his age of 58 revealed slight ataxia of the trunk and lower limbs as well as dysarthria and blepharospasm. Magnetic resonance imaging of his brain revealed cerebellar atrophy most pronounced in the upper vermis, which is typical for SCA31. Nakamura and his colleagues collected 44 patients with SCA31 and underwent a 4-year prospective study (Nakamura et al., 2017). They evaluated patients yearly using the Scale for the Assessment and Rating of Ataxia (SARA) and the Barthel Index (BI). They showed the annual progression of the SARA score was 0.8 ± 0.1 points/year and that of the BI was −2.3 ± 0.4 points/year (mean ± standard error). Nakamura described that their patients developed ataxic symptoms at 58.5 ± 10.3 years, become wheelchair bound at 79.4 ± 1.7 years, and died at 88.5 ± 0.7 years. This is the first study to show natural course and disease progression of SCA31.

26.2 Founder Effect in SCA31 and Its Implication

As described, the SCA31 shows a strong founder effect. While SCA31 is a common ataxia in Japan, this disease is very rare even in neighboring countries such as Korea (Lee et al., 2007) Taiwan (Lee et al., 2012), and China (Ouyang et al., 2012; Yang et al., 2018). SCA31 was found in Brazilian SCA patients; however, these SCA31 patients were all descendants of Japanese immigrants (Pedroso et al., 2015). In accord with this notion, SCA31 with $(TGGAA)_n$ was never found in the Caucasian SCA families ($n = 320$) in French and German cohorts nor in the 588 healthy control subjects (Ishikawa et al., 2011). Interestingly, nearly 5.5% of the whole SCA and control groups harbored expansions of different pentanucleotide repeats. The most common repeat was $(TACAA)_n$. Other repeats such as $(GAAAA)_n$, $(TGAAA)_n$, and $(TAACA)_n$ were also seen. Unlike the SCA31 insertion consisted of three different repeats $(TGGAA)_n$, $(TAGAA)_n$, and $(TAAAA)_n$, the repeats found in Caucasians were all pure stretches. The most common $(TACAA)_n$ sometimes showed expansion up to 6.5 kb without clinical manifestation.

This observation indicates two things. First, the SCA31 locus in which the vast majority of human beings harbors a short TAAAA repeat (8–20 repeats) can be expanded. The TAAAA can become unstable and expanded as we encountered long $(TAAAA)_n$ in Japanese controls. However, all the expanded sequences in Caucasians were all single-nucleotide mutant versions of TAAAA. For example, $(TACAA)_n$ is a single A-to-C transition form. Such a mutation may make the SCA31 locus unstable. Second, the TGGAA repeat is the only sequence so far that leads to human disease. Considering that the TAGAA gives rise to TGGAA by a single-nucleotide A-to-G transition, it seems likely that the TGGAA repeat, hence SCA31, originates from the TAGAA repeat, which has been seen only in Japanese so far.

26.3 Neuropathology of SCA31

As SCA31 clinically shows progressive ataxia with a purely cerebellar syndrome, and magnetic resonance imaging shows isolated cerebellar atrophy, the neuropathology of SCA31 is expected to be confined to the cerebellar cortex. The first neuropathological study on a 95-year-old female patient who had had SCA31 for 20 years indeed disclosed cerebellar cortical degeneration with Purkinje cell-predominant neuronal loss (Owada et al., 2005) (Fig. 26.2a). What is very unique to SCA31 was that the Purkinje cell often showed shrinkage of its cell body. In addition, an ill-defined amorphous structure was often seen surrounding the Purkinje cell body (Owada et al., 2005). On hematoxylin and eosin staining of the cerebellar tissues, shrunken Purkinje cell body appeared dense pink, while the amorphous structure gave pale pink, resembling the halo of the Lewy body in Parkinson's disease (Fig. 26.2b). Immunohistochemistry against calbindin D28k, a useful Purkinje cell marker, revealed sprouts from the Purkinje cell body (Fig. 26.2c), which are infrequently observed in other conditions. The Purkinje cell's somatic sprout has been described as a pathological hallmark of Menkes' disease, where synaptic

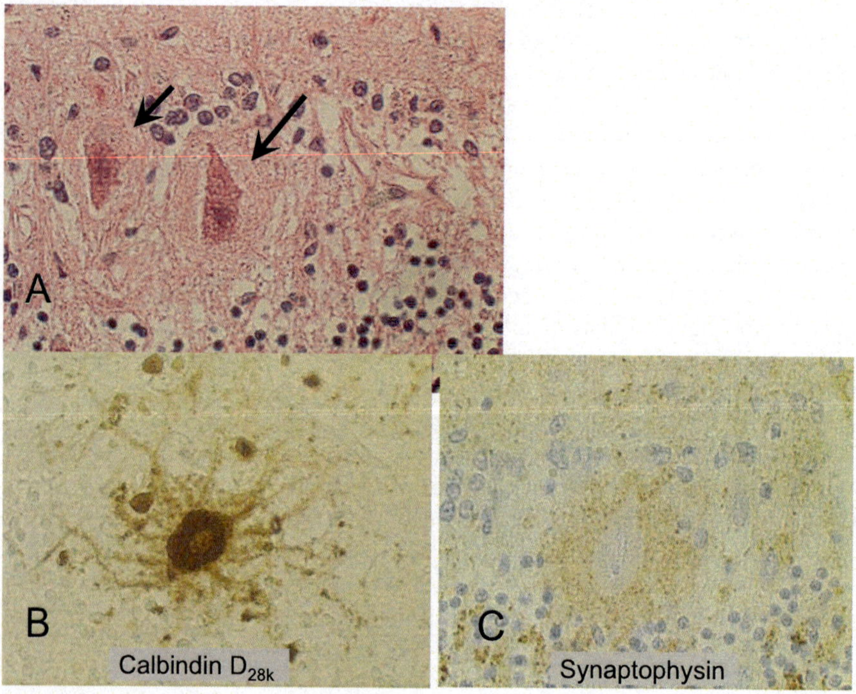

Fig. 26.2 Histopathology of SCA31 patients' Purkinje cells. (**a**) A Purkinje cell body surrounded by the halo-like amorphous materials (arrow). (**b**) Calbindin D_{28k} immunohistochemistry. Numerous calbindin-positive somatic sprouts are seen from the Purkinje cell. (**c**) Increased immunoreactivity against synaptophysin, a presynaptic marker protein, surrounding the Purkinje cell

inputs to the Purkinje cell are known to be dramatically decreased. While SCA31 is similar to Menkes' disease in terms of the Purkinje cell's sprouts, synaptophysin immunohistochemistry disclosed accumulation of presynaptic terminals (Fig. 26.2d). Thus, the Purkinje cell degeneration in SCA31 appears distinct from that in Menkes' disease. The structure of the SCA31 Purkinje cell is also different from other ataxias such as SCA6. Ubiquitin-positive inclusions are sometimes seen in the amorphous structure. As this structure was so remarkable, several brain samples with this figure were suspected to be SCA31 and undergo SCA31 mutation screen, later proven to have SCA31 insertions. One of us (KI) and his mentor, Mizusawa, newly coined a word "halo-like amorphous materials" to this structure as the pathological hallmark of SCA31 (Ishikawa & Mizusawa, 2010) (Fig. 26.2b). As described later in this chapter, the penta-peptide repeat protein specifically translated from the transcript of TGGAA repeat was seen mainly in the amorphous materials.

Yoshida and his colleagues investigated two SCA31 brain samples (Yoshida et al., 2014). They not only found the halo-like amorphous materials in their samples, but they also found that Purkinje cells surrounded by the amorphous materials tend to show bent, elongated, and often folded nuclei with fragmented Golgi apparatus in the cell soma.

From the genetic observation, the $(TGGAA)_n$ transcribed into $(UGGAA)_n$ as BEAN1 transcripts has been implicated an important factor of SCA31 pathogenesis. To gain further insight into the SCA31 pathogenesis, in situ hybridization using RNA probes against $(UGGAA)_n$ or $(UAGAAUAAAA)_n$ was performed. Yusuke Niimi and his colleagues identified RNA foci within SCA31 Purkinje cells' nuclei labeled positive with a locked nucleic acid (LNA)-oligonucleotide $(TTCCA)_5$ probe (Niimi et al., 2013) (Fig. 26.3). Similar RNA foci were also detected by probes

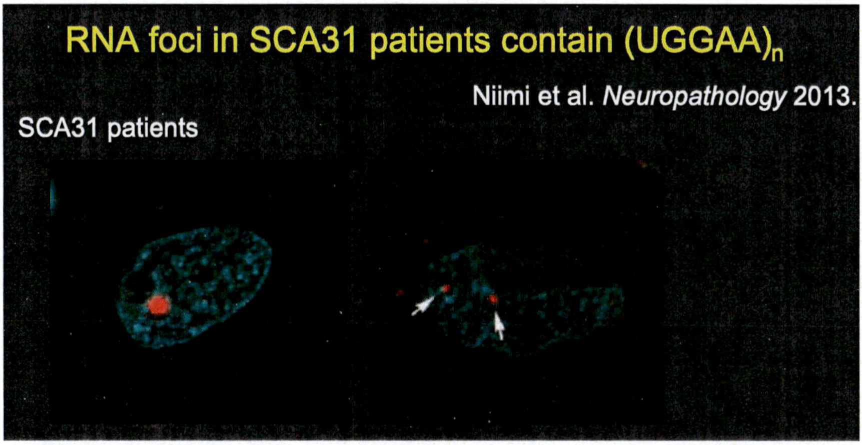

Fig. 26.3 RNA foci in SCA31 patients contain $(UGGAA)_n$
$(UGGAA)_n$ containing RNA foci (red dots indicated by white arrows) in human SCA31 Purkinje cell nuclei. (Niimi et al. (2013))

against $(UAGAAUAAAA)_n$ (Niimi et al., 2013). They also tested whether $(UGGAA)_n$ is toxic than $(UAGAAUAAAA)_n$ in cultured cells. They created transient and stable expression cell systems and found that cell toxicity and formation of RNA foci were both consistently observed upon expression of $(UGGAA)_n$. These observations led us to conclude that $(UGGAA)_n$ could be toxic in cells.

26.4 Dissecting the Molecular Mechanisms of SCA31

In a number of noncoding repeat disorders (Echeverria & Cooper, 2012), RNA-binding proteins are sequestered in RNA foci through which normal functions of those proteins are altered. Such alterations are thought to contribute to their pathogenesis. In line with this, we screened potential $(UGGAA)_n$-binding proteins in vitro by RNA pull-down assay using nuclear fraction of PC12 cells. Mass spectrometry led us to identify various $(UGGAA)_n$-binding candidates, including TDP-43 (TAR DNA-binding protein, 43 kilodalton) (Ishiguro et al., 2017). We further confirmed this binding by Western blotting. We also found FUS and hnRNPs that are all ALS/FTD-linked RNA-binding proteins (RBPs) as well as TDP-43. Our colleagues N. Charlet-Berguerand and C. Sellier independently searched binding proteins using mouse brain lysates and found that these three proteins bind to $(UGGAA)_n$. They also confirmed colocalization of TDP-43 with RNA foci in human SCA31 Purkinje cells (Ishiguro et al., 2017). We therefore decided to focus on TDP-43 to analyze its role in the SCA31 pathology.

To gain an insight into the role of TDP-43 and $(UGGAA)_n$ in the pathogenesis of SCA31, we generated *Drosophila* models of SCA31 to express an expanded $(UGGAA)_n$ repeat RNA (Ishiguro et al., 2017). When the expanded $(UGGAA)_n$ ($n = 80–100$ repeats) were expressed in *Drosophila* eyes, they caused remarkable degeneration depending on their expression level. RNA-fluorescence in situ hybridization (FISH) revealed extensive accumulation of RNA foci consisting of $(UGGAA)_n$ in the eye imaginal discs (Fig. 26.4), consistent with the pathology of SCA31 patients (Niimi et al., 2013). In contrast, control repeat lacking $(UGGAA)_n$ or a short $UGGAA_{22}$ RNA, generated by spontaneous contraction of the TGGAA repeats, did not produce obvious degeneration. We further found that a novel penta-peptide repeat (PPR) protein was produced by a translation from the UGGAA repeat RNA (Fig. 26.4). The PPR protein aggregates were also found in SCA31 human cerebellum, mainly in the amorphous materials surrounding the human Purkinje cell.

To discover the effect of TDP-43 upon binding to $(UGGAA)_n$ in vivo, we crossed flies expressing $(UGGAA)_n$ with those expressing human TDP-43. Co-expression of TDP-43 dramatically suppressed compound eye degeneration in the $(UGGAA)_n$-expressing flies (Fig. 26.5). In contrast, RNA interference (RNAi)-mediated knockdown of endogenous *Drosophila* TDP-43 significantly enhanced

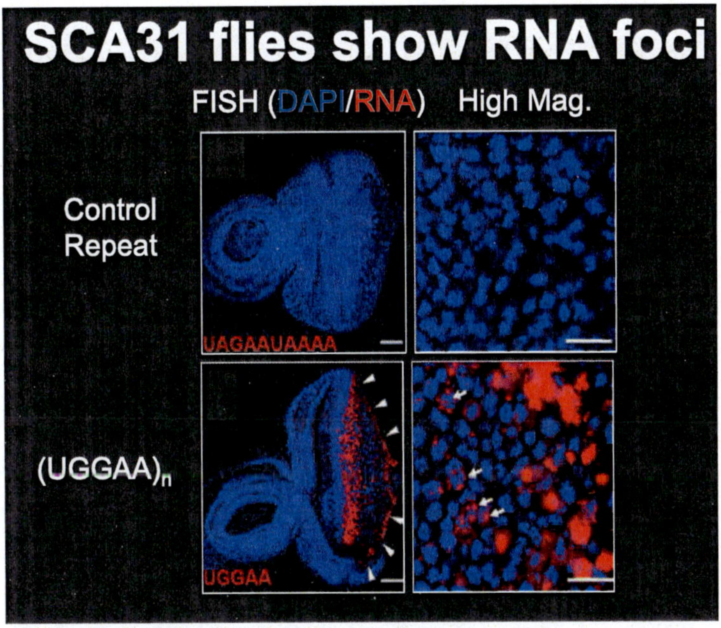

Fig. 26.4 Expression of $(UGGAA)_n$ RNA induces RNA foci formation
Note that control repeat with a different repeat configuration lacking $(UGGAA)_n$ does not induce RNA foci in *Drosophila*. (From reference Ishiguro et al. (2017))

eye degeneration in the $(UGGAA)_n$ flies, indicating a crucial role of TDP-43 in $(UGGAA)_n$-mediated toxicity in vivo (Fig. 26.5). Creating some mutations in the RNA recognition motif (RRM) in TDP-43, which reduced its RNA-binding capacity to $(UGGAA)_n$, failed to suppress $(UGGAA)_n$-mediated eye degeneration. These clearly suggested that the RNA-binding ability of TDP-43 is essential for suppressing $(UGGAA)_n$-mediated toxicity (Fig. 26.5) (Ishiguro et al., 2017). Furthermore, binding of TDP-43 to $(UGGAA)_n$ in *Drosophila* eye caused reduction of both a number of RNA foci and PPR protein aggregations (Fig. 26.6). RT-PCR analyses confirmed that $(UGGAA)_n$ RNA expression levels were not altered by co-expression of TDP-43 (Fig. 26.6), suggesting that TDP-43 neutralize toxicity of $(UGGAA)_n$ via direct binding, not by reducing the RNA expression level. Subsequent circular dichroism (CD) spectroscopy and atomic force microscopy (AFM) analyses done by our colleagues led by C.E. Pearson demonstrated that the binding of TDP-43 to $(UGGAA)_n$ alters structure of $(UGGAA)_n$ and prevents its aggregation, indicating that TDP-43 functions as an RNA chaperone for $(UGGAA)_n$ (Ishiguro et al., 2017; Rajkowitsch et al., 2007). Besides reducing the number of RNA foci, co-expression of TDP-43 significantly reduced the accumulation of PPR proteins in $(UGGAA)_n$-expressing flies (Fig. 26.6).

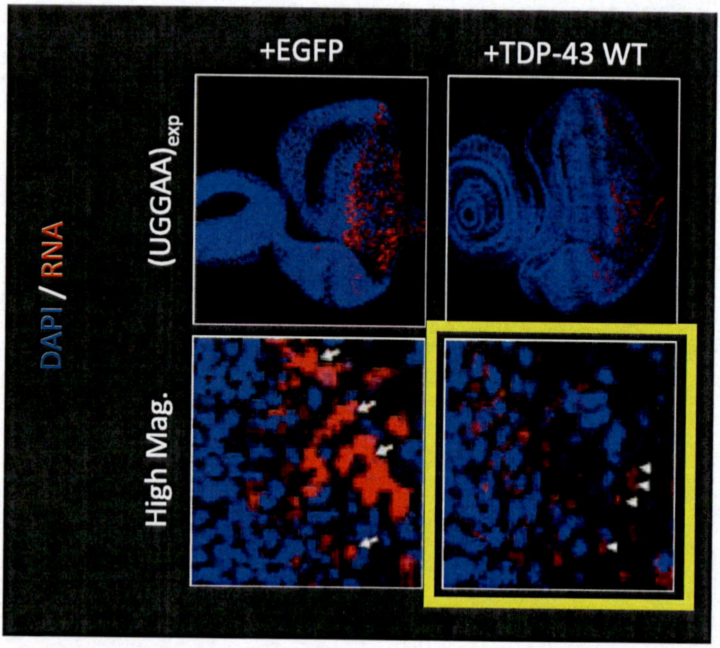

Fig. 26.5 Co-expression of TDP-43 suppresses RNA foci
TDP-43 reduces the number of RNA foci (a yellow square) suppressing toxicity of $(UGGAA)_n$. (From reference Ishiguro et al. (2017))

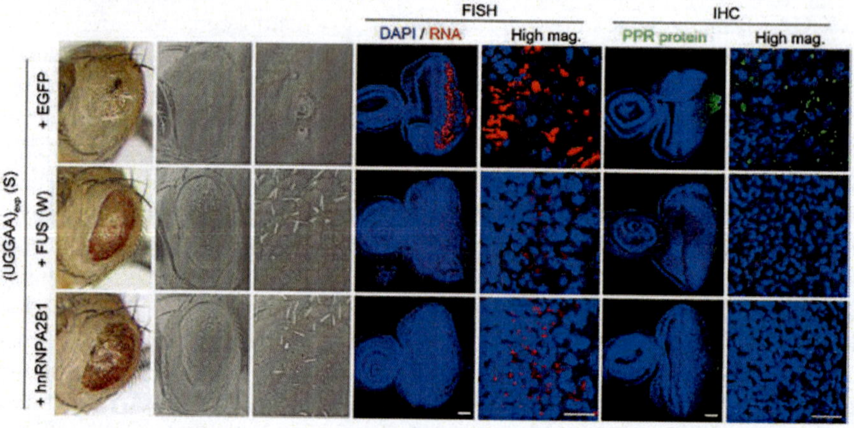

Fig. 26.6 FUS and hnRNPA2/B1 also mitigated $(UGGAA)_n$ toxicity in fly model. Co-expression of FUS or hnRNPA2/B1 also suppresses RNA foci formation and penta-peptide repeat (PPR) protein accumulation as TDP-43 does. (From reference Ishiguro et al. (2017))

26.5 Perspectives

Introducing TDP-43 in SCA31 may provide a fundamental basis for treating SCA31. As TDP-43 is a multifunctional protein with both DNA-binding and RNA-binding properties regulated under a strict protein level, it is critical to clarify the role of TDP-43 in an SCA31 mice model. Further studies would be needed to discover mechanisms that $(TGGAA)_n$ leads to Purkinje cell-predominant neurodegeneration.

Acknowledgments I am deeply in debt to Prof. Hidehiro Mizusawa for leading this study, and I sincerely acknowledge Dr. Nozomu Sato for cloning SCA31 insertion and identifying UGGAA-binding proteins and Dr. Taro Ishiguro and Professor Yoshitaka Nagai for the *Drosophila* study. This work was supported in part by Grants-in-Aid for Scientific Research (A), (B), and (C) and grant on the Research Committee for Ataxic Diseases from the Ministry of Health, Labour and Welfare; by a grant for Practical Research Projects for Rare/Intractable Diseases (15ek0109102h, 16ek0109102h, 17ek0109102h, 18ek0109302h) from the Japan Agency for Medical Research and Development; and by Mitsubishi Foundation, Natural Science.

References

Amino, T., Ishikawa, K., Toru, S., Ishiguro, T., Sato, N., Tsunemi, T., Murata, M., Kobayashi, K., Inazawa, J., Toda, T., & Mizusawa, H. (2007). Redefining the disease locus of 16q22.1-linked autosomal dominant cerebellar ataxia. *Journal of Human Genetics, 52*, 643–649.

Echeverria, G. V., & Cooper, T. A. (2012). RNA-binding proteins in microsatellite expansion disorders: Mediators of RNA toxicity. *Brain Research, 1462*, 100–111.

Edener, U., Bernard, V., Hellenbroich, Y., Gillessen-Kaesbach, G., & Zühlke, C. (2011). Two dominantly inherited ataxias linked to chromosome 16q22.1: SCA4 and SCA31 are not allelic. *Journal of Neurology, 258*(7), 1223–1227.

Ishiguro, T., Sato, N., Ueyama, M., et al. (2017). Regulatory role of RNA chaperone TDP-43 for RNA misfolding and repeat-associated translation in SCA31. *Neuron, 94*(1), 108–124.e7.

Ishikawa, K., & Mizusawa, H. (2010). The chromosome 16q-linked autosomal dominant cerebellar ataxia (16q-ADCA): A newly identified degenerative ataxia in Japan showing peculiar morphological changes of the Purkinje cell. The 50th Anniversary of Japanese Society of Neuropathology Memorial Symposium: Milestones in Neuropathology from Japan. *Neuropathology, 30*, 490–494.

Ishikawa, K., Toru, S., Tsunemi, T., Li, M., Kobayashi, K., Yokota, T., Amino, T., Owada, K., Fujigasaki, H., Sakamoto, M., Tomimitsu, H., Takashima, M., Kumagai, J., Noguchi, Y., Kawashima, Y., Ohkoshi, N., Ishida, G., Gomyoda, M., Yoshida, M., Hashizume, Y., Saito, Y., Murayama, S., Yamonouchi, H., Mizutani, T., Kondo, I., Toda, T., & Mizusawa, H. (2005). An autosomal dominant cerebellar ataxia linked to chromosome 16q22.1 is associated with a single-nucleotide substitution in the 5′ untranslated region of the gene encoding a protein with spectrin repeat and Rho guanine-nucleotide exchange-factor domain. *American Journal of Human Genetics, 77*, 280–296.

Ishikawa, K., Dürr, A., Klopstock, T., et al. (2011). Pentanucleotide repeats at the spinocerebellar ataxia type 31 (SCA31) locus in Caucasians. *Neurology, 77*, 1853–1855.

Itaya, S., Kobayashi, Z., Ozaki, K., Sato, N., Numasawa, Y., Ishikawa, K., Yokota, T., Matsuda, H., & Shintani, S. (2018). Spinocerebellar ataxia type 31 with blepharospasm. *Internal Medicine, 57*(11), 1651–1654.

Lee, P. H., Park, H. Y., Jeoung, S. Y., et al. (2007). 16q-linked autosomal dominant cerebellar ataxia in a Korean family. *European Journal of Neurology, 14*, e16–e17.

Lee, Y. C., Liu, C. S., Lee, T. Y., et al. (2012). SCA31 is rare in the Chinese population on Taiwan. *Neurobiology of Aging, 33*, 426.e23–426.e24.

Li, M., Ishikawa, K., Toru, S., Tomimitsu, H., Takashima, M., Goto, J., Takiyama, Y., Sasaki, H., Imoto, I., Inazawa, J., Toda, T., Kanazawa, I., & Mizusawa, H. (2003). Physical map and haplotype analysis of 16q-linked autosomal dominant cerebellar ataxia (ADCA) type III in Japan. *Journal of Human Genetics, 48*, 111–118.

Nagaoka, U., Takashima, M., Ishikawa, K., et al. (2000). A gene on SCA4 locus causes dominantly-inherited pure cerebellar ataxia. *Neurology, 54*, 1971–1975.

Nakamura, K., Yoshida, K., Matsushima, A., Shimizu, Y., Sato, S., Yahikozawa, H., Ohara, S., Yazawa, M., Ushiyama, M., Sato, M., Morita, H., Inoue, A., & Ikeda, S. I. (2017). Natural history of spinocerebellar ataxia type 31: A 4-year prospective study. *Cerebellum, 16*(2), 518–524.

Niimi, Y., Takahashi, M., Sugawara, E., et al. (2013). Abnormal RNA structures (RNA foci) containing a penta-nucleotide repeat $(UGGAA)_n$ in the Purkinje cell nucleus is associated with spinocerebellar ataxia type 31 pathogenesis. *Neuropathology, 33*, 600–611.

Ohata, T., Yoshida, K., Sakai, H., Hamanoue, H., Mizuguchi, T., Shimizu, Y., Okano, T., Takada, F., Ishikawa, K., Mizusawa, H., Yoshiura, K., Fukushima, Y., Ikeda, S., & Matsumoto, N. (2006). A 16C>T substitution in the 5'UTR of the puratrophin-1 gene is prevalent in autosomal dominant cerebellar ataxia in Nagano. *Journal of Human Genetics, 51*, 461–466.

Ouyang, Y., He, Z., Li, L., et al. (2012). Spinocerebellar ataxia type 31 exists in Northeast China. *Journal of the Neurological Sciences, 316*, 164–167.

Owada, K., Ishikawa, K., Toru, S., et al. (2005). A clinical, genetic, and pathologic study in a family with 16q-linked ADCA type III. *Neurology, 65*, 629–632.

Pedroso, J. L., Abrahao, A., Ishikawa, K., et al. (2015). When should we test patients with familial ataxias for SCA31? A misdiagnosed condition outside Japan? *Journal of the Neurological Sciences, 355*, 206–208.

Rajkowitsch, L., Chen, D., Stampfl, S., et al. (2007). RNA chaperones, RNA annealers and RNA helicases. *RNA Biology, 4*, 118–130.

Sakai, H., Yoshida, K., Shimizu, Y., Morita, H., Ikeda, S., & Matsumoto, N. (2010). Analysis of an insertion mutation in a cohort of 94 patients with spinocerebellar ataxia type 31 from Nagano, Japan. *Neurogenetics, 11*(4), 409–415.

Sakakibara, S., Aiba, I., Saito, Y., Inukai, A., Ishikawa, K., & Mizusawa, H. (2014). Clinical features and MRI findings in spinocerebellar ataxia type 31 (SCA31) comparing with spinocerebellar ataxia type 6 (SCA6). *Rinshō Shinkeigaku, 54*(6), 473–479.

Sato, N., Amino, T., Kobayashi, K., et al. (2009). Spinocerebellar ataxia type 31 (SCA31) is associated with "inserted" pentanucleotide repeat including $(TGGAA)_n$. *American Journal of Human Genetics, 68*, 355–367.

Yang, K., Zeng, S., Liu, Z., et al. (2018). Analysis of spinocerebellar ataxia type 31 related mutations among patients from mainland China. *Zhonghua Yi Xue Yi Chuan Xue Za Zhi, 35*(3), 309–313.

Yoshida, K., Asakawa, M., Suzuki-Kouyama, E., et al. (2014). Distinctive features of degenerating Purkinje cells in spinocerebellar ataxia type 31. *Neuropathology, 34*, 261–267.

Chapter 27
Cerebellar Circuitry of Tremor

Ming-Kai Pan and Sheng-Han Kuo

27.1 Introduction

Tremor is the most common movement disorder symptom and is also the most specific, measurable movement, since tremor is defined by oscillatory and rhythmic movements. This unique feature makes tremor an ideal model to correlate movements with neuronal activity, providing an important window to understand how the movements are generated from the brain (McAuley & Marsden, 2000).

Tremor occurs in a variety of neurological disorders. Essential tremor is the most common tremor disorder, occurring in 4% of the population over the age of 40. The prevalence of essential tremor increases over aging. By the age of 90, as many as 20% of the population will have essential tremor (Louis et al., 1998; Louis & Ferreira, 2010). Therefore, essential tremor is among the most common neurological disorders. Patients with essential tremor have predominant action and postural tremor, at the frequency of 4–12 Hz (Haubenberger & Hallett, 2018; Louis, 2001). Rest tremor sometimes can be seen in severe essential tremor patients (Thenganatt & Louis, 2012). Essential tremor is also a progressive disorder, and tremor becomes lower in frequency but higher in amplitude over time, impairing one's activity of daily living (Fig. 27.1). The treatment choice of essential tremor is rather limited.

M.-K. Pan
Graduate Institute of Pharmacology, College of Medicine, National Taiwan University, Taipei, Taiwan

Neurobiology and Cognitive Science Center, National Taiwan University, Taipei, Taiwan

Molecular Imaging Center, National Taiwan University, Taipei, Taiwan

S.-H. Kuo (✉)
Department of Neurology, Columbia University Medical Center, New York, NY, USA

Initiative for Columbia Ataxia and Tremor, Columbia University Medical Center, New York, NY, USA
e-mail: sk3295@columbia.edu

H. Mizusawa, S. Kakei (eds.), *Cerebellum as a CNS Hub*,
Contemporary Clinical Neuroscience, https://doi.org/10.1007/978-3-030-75817-2_27

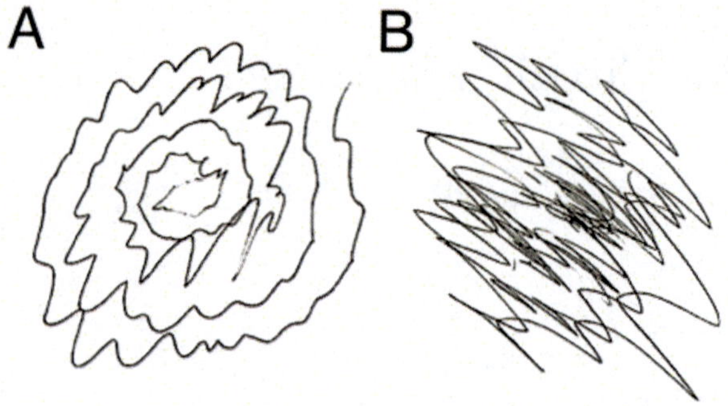

Fig. 27.1 Spiral drawings from patients with essential tremor. (**a**) An example of a spiral drawing from a patient with moderate essential tremor. (**b**) An example of a spiral drawing from a patient with severe essential tremor. Note that spiral drawings from essential tremor patients often have a defined axis

Propranolol, a β-blocker, and primidone, a GABAergic medication, are two first-line pharmacological therapies, but many patients are often non-responders. Even for those with responses to pharmacological therapies, the therapeutic effects are often unsatisfactory (Louis, 2001). Surgical or ablative options targeting at the thalamus can thus be considered and may be very effective (Elias et al., 2016; Zhang et al., 2010). However, the major limitation is the issue of "tolerance," for which tremor might gradually return over time. Therefore, effective and sustainable therapies for essential tremor are urgently needed.

The second most common tremor disorder is tremor associated with Parkinson's disease (i.e., Parkinson tremor). Parkinson's disease is the second most common neurodegenerative disorder, after Alzheimer's disease, occurring in 0.57% over the age of 45 (Marras et al., 2018). As the population ages, the prevalence of Parkinson's disease will likely to rise dramatically. Rest tremor is one of the cardinal features of Parkinson's disease, occurring in 58.2% Parkinson's disease patients. Postural tremor and kinetic tremor are also common (49.7% and 52.3%, respectively) and more commonly seen in patients with severe rest tremor (Gupta et al., 2021). While rest tremor is considered as a symptom related to dopamine deficiency, levodopa treatment is often not satisfactory. Other dopamine agonists may also be tried (Marjama-Lyons & Koller, 2000). In cases refractory to dopamine replacement therapies, surgical options such as deep brain stimulations targeting the thalamus can provide symptomatic benefits.

Dystonic tremor is a tremor associated with dystonia, which is defined by patterned, sustained involuntary movements, and part of dystonic movements may appear to be semi-rhythmic and oscillatory; therefore, these movements are called dystonic tremor (van der Stouwe et al., 2020). Other tremor disorders are quite rare, including cerebellar outflow tremor, Holmes tremor, and orthostatic tremor.

An important aspect for tremor disorders is the overlapping clinical phenotypes. Essential tremor patients can have rest tremor and also sometimes dystonic components (Bhatia et al., 2018), and Parkinson's disease patients can have postural and kinetic tremor (Dirkx et al., 2018; Gupta et al., 2021). Therefore, tremor disorders should be considered to be heterogeneous, and these overlapping clinical features can be the results of the brain circuitry involved. Since tremor is oscillatory and rhythmic, an oscillatory generator or multiple oscillatory generators in the central nervous system are thought to be the main driver(s) for tremor. The olivocerebellum is an intrinsic oscillator; therefore, it has been postulated to play a critical role for tremor disorders.

In this chapter, we will review the current understanding of tremor pathophysiology with a main focus on essential tremor, as the model for kinetic tremor, and Parkinson tremor, as the model for rest tremor. The knowledge will help us broadly understand the cerebellar circuitry of tremor.

27.2 Clinical and Neuroimaging Evidence of the Cerebellar Involvement in Tremor Disorders

The cerebellum has been identified as the main player in tremor disorders. The evidence of the cerebellar involvements comes from multiple converging lines of evidence.

One of the most effective therapies for essential tremor and Parkinson tremor is to implant deep brain stimulators into the ventro-intermediate nucleus of the thalamus that receives the cerebellar output (Fasano & Lozano, 2015). In this part of the thalamus, tremor-related rhythmic neuronal activity can be recorded, suggesting the passing through of such signals from the cerebellum. Delivery of electric currents via deep brain stimulators into the thalamus is thought to suppress the propagation of rhythmic neuronal activity to achieve tremor suppression. Another means to target ventro-intermediate nucleus of the thalamus is to use focused ultrasound-mediated ablation (Elias et al., 2016), which is also highly effective. In addition to ventro-intermediate nucleus of the thalamus, other related areas such as caudal zona incerta and posterior subthalamic area, which also receive cerebellar outflow, are being explored as novel targets for deep brain stimulations.

Another evidence supporting the role of the cerebellum in tremor comes from clinical observations. Patients with essential tremor presumably have defined central oscillators to create rhythmic neuronal activity, which propagates downstream to reach motor cortex. Given the high prevalence of essential tremor, some patients can develop discrete lesions in the central nervous system such as strokes or brain tumors. Interestingly, essential tremor patients can have tremor disappearance and/or reduction with structural lesions at the cerebello-thalamo-cortical loop, suggesting such brain circuitry is critical for tremor (Dupuis et al., 2010).

Structural magnetic resonance imaging has been used to study the brain volume alterations in essential tremor patients. The reduction of the cerebellum volume is often observed (Cerasa & Quattrone, 2016). Specifically, vermal cerebellar atrophy is found, especially in essential tremor patients with both head and arm tremors

(Cerasa & Quattrone, 2016). Furthermore, diffusion tensor imaging indicates white matter loss in the cerebellum and cerebello-thalamic connections in essential tremor patients (Benito-León et al., 2009; Nicoletti et al., 2010; Shin et al., 2008), further supporting involvement of these brain regions. While other parts of the brain have been found to be involved in these structural imaging studies, the abnormalities in the cerebellum are most consistently observed. Another clue for cerebello-thalamo-cortical loop in essential tremor comes from an interesting tractography study to assess the structural connectivity from the effective contacts of deep brain stimulation electrodes in the thalamus. They found that the thalamic regions with active contacts of the electrodes are often in connection with the cerebellum and motor cortex, underscoring the importance of these brain structures in the tremor brain network (Klein et al., 2012). In the functional magnetic resonance imaging, alterations of the connectivity to the cerebellum have been replicated in different studies (Buijink et al., 2015; Fang et al., 2016). Finally, a study using magnetoencephalography to study essential tremor patients also reveals the importance of cerebello-thalamo-cortical loop (Schnitzler et al., 2009).

One of the clinical features for essential tremor is that two-thirds of patients report that their tremor improves with alcohol. Alcohol is a gamma-aminobutyric acid (GABA) agonist; therefore, essential tremor patients might have GABA deficiency. This notion is further corroborated by the first-line therapy for essential tremor, primidone, which is also a GABA agonist. Since the cerebellum has been identified to be a key region for essential tremor, studies have been focusing on measuring GABA concentration in the cerebellum, particularly in the region of dentate nucleus. However, the alterations of GABA concentration are not identified in the cerebellum of essential tremor patients using magnetic resonance spectroscopy by two independent groups (Louis et al., 2018; Tapper et al., 2020). On the other hand, the ratio of GABA to glutamate is different between essential tremor cases and controls, suggesting a broader excitatory and inhibitory imbalance may be important for tremor (Tapper et al., 2020).

The knowledge of Parkinson tremor enjoys recent advances by functional magnetic imaging techniques, coupled with electromyography. The "dimmer-switch" model has been proposed to explain Parkinson tremor. Specifically, the basal ganglia can drive the cerebellum into tremor rhythm; thus, the basal ganglia can serve as the switch for tremor onset, whereas the cerebellum can be the dimmer to modulate tremor amplitude (Helmich et al., 2011). The origin of basal ganglia dysfunction that drives tremor onset comes from dopamine deficiency in the pallidum, and the tremor circuit involves cerebello-thalamo-cortical loop (Dirkx et al., 2016). In addition, dopamine can also directly modulate the ventral intermediate nucleus of the thalamus, providing additional mechanism for the dopamine-responsive Parkinson tremor (Dirkx et al., 2017). Interestingly, certain Parkinson's disease patients' tremor is dopamine responsive, while tremor in others is not, indicating diverse pathophysiology (Zach et al., 2020). While both types of tremor involve the cerebello-thalamo-cortical circuit, dopamine-resistant Parkinson tremor is more related to the cerebellar activity, whereas dopamine-responsive Parkinson tremor is more correlated with activity in the thalamus and secondary somatosensory cortex

(Dirkx et al., 2019). These functional magnetic resonance imaging studies yield tremendous information regarding how the tremor can be generated from the interaction between the basal ganglia and the cerebellum.

Outside of the dopamine system, serotonergic system may also be involved in Parkinson tremor. Specifically, ^{123}I-FP-CIT single-photon emission computed tomography can be used not only to assess the dopamine transporters in the striatum but also to measure serotonin transporter availability in the raphe nuclei. Using this technique, Parkinson patients with tremor have lower serotonin transporter availability than those without tremor (Qamhawi et al., 2015). In addition, raphe serotonin transporter availability is associated with rest tremor constancy and severity (Qamhawi et al., 2015). Interestingly, the ratio of ^{123}I-FP-CIT uptake in raphe nuclei and in putamen is correlated with dopamine responsiveness of tremor (Pasquini et al., 2018), indicating the biochemical balance of these two structures may underlie tremor. Raphe nuclei have wide projections to different parts of the central nervous system. One of the important targets for such serotonergic projects is the cerebellum (Ren et al., 2019) and may modulate cerebellar physiology.

While the above studies demonstrate critical roles of the cerebellum in tremor, few studies investigate the lobule specificity of the cerebellum. The cerebellum can be divided into ten lobules, and there is somatotopic arrangement of these lobules. A study comparing a large number of patients with essential tremor and Parkinson's disease found that lobule VIII volume is negatively correlated with postural tremor severity in essential tremor, whereas lobule IV is positively correlated with rest tremor severity in Parkinson tremor (Lopez et al., 2020). These findings suggest subdomain specificity and lead us to begin understanding the localization of tremor circuitry within the cerebellum.

27.3 Structural Changes in the Cerebellum of Tremor Disorders

Since the cerebellum is heavily implicated in tremor, what are the microstructural changes within the cerebellum that can lead to tremor generation and/or propagation? To answer this question, investigators have been focusing on studying the postmortem human pathology. The cerebellar pathology of essential tremor has recently been extensively studied, and the major pathological alterations have been identified in or centered around Purkinje cells, the principal neurons in the cerebellum.

In 2007, a sentinel paper first identified that there is a modest Purkinje cell loss and Purkinje cell axonal swelling, called torpedoes (Louis et al., 2007). These are common changes in the setting of cerebellar degeneration, raising a question whether essential tremor is a neurodegenerative disorder. These findings have been replicated in multiple follow-up studies using more sophisticated quantitative measures (Louis et al., 2016; Louis et al., 2019). In addition to these two pathological features, essential tremor cerebellum also has increased number of Purkinje cell body displaced into the molecular layer (i.e., heterotopic Purkinje cell body) (Louis et al., 2018), Purkinje cell dendritic swelling (Yu et al., 2012), and reduced Purkinje cell dendritic complexity (Louis et al., 2014). Purkinje cells in essential tremor also undergo tremendous axonal reorganization as their axons often have sprouting, branching, and recurrent collaterals (Babij et al., 2013), which may be secondary to slow, degenerative changes for local micro-circuitry reorganization, contributing to tremor. Basket cells are inhibitory neurons that can modulate Purkinje cell activity by sending dense axonal plexus to wrap around Purkinje cell axonal initial segments. In essential tremor cerebellum, basket cells form very dense axonal plexus onto Purkinje cell axons (Erickson-Davis et al., 2010), and oftentimes, basket cell plexus can be elongated along Purkinje cell axons (Kuo et al., 2013), which are likely to influence Purkinje cell physiology.

One of the most characterized pathologies for essential tremor is climbing fiber-Purkinje cell synaptic pathology. Purkinje cells receive two excitatory inputs, one from climbing fibers and the other from parallel fibers. Climbing fibers form synaptic connections on the thick, proximal Purkinje cell dendrites, whereas parallel fibers form synaptic connections on the thin, spiny branchlets of Purkinje cell dendrites. How climbing fiber and parallel fibers form synaptic connections with Purkinje cells is strictly regulated and is important for cerebellar physiology (Miyazaki et al., 2010; Watanabe, 2008). In essential tremor, there is an increase in abnormal climbing fiber synaptic connections in the parallel fiber synaptic territory on Purkinje cell dendrites and a reduction of climbing fiber synaptic density (Lin et al., 2014). In addition, the climbing fiber-Purkinje cell synaptic pathology occurs in a wide range of essential tremor patients regardless of diverse clinical features (i.e., early onset vs. late onset, with family history vs. without family history) (Lee et al., 2018), indicating that climbing fiber-Purkinje cell synaptic pathology is closely related to the core clinical feature of essential tremor (i.e., tremor). In clinicopathological correlation studies, the amount of the abnormal climbing fiber synapses in the thin, spiny Purkinje cell dendrites *inversely* correlates with tremor severity (Lin et al., 2014; Louis et al., 2015). This finding should be considered in the context of continued dendritic remodeling of Purkinje cells, for which essential tremor patients have loss of Purkinje cell dendritic spines (Louis et al., 2014), further adding to the complexity. Finally, another interesting observation is that chronic thalamic deep brain stimulations may have long-range modulatory effect of cerebellar synaptic pathology, demonstrating potential disease-modifying effects (Kuo et al., 2016).

As mentioned in the previous section, GABA dysfunction has been hypothesized to play a role in essential tremor. Within the cerebellum, Purkinje cells send GABAergic axons to modulate deep cerebellar nucleus neuronal activity. Therefore, GABAergic alterations have been hypothesized at the deep cerebellar nucleus levels. Using receptor-binding autoradiography applied on the postmortem essential tremor cerebellum, a reduction in both $GABA_A$ and $GABA_B$ receptors in the dentate nucleus has been identified (Paris-Robidas et al., 2012). No alterations of either $GABA_A$ or $GABA_B$ receptor levels have been found in the cerebellar cortex (Luo et al., 2012; Paris-Robidas et al., 2012). It is not clear how the above findings fit into the clinical observation of essential tremor patients. A reduction of $GABA_A$ receptor levels is likely to lead to *less responsiveness* to GABAergic agents, such as alcohol or primidone. It is possible that other compensatory mechanisms of extra-synaptic GABA receptors may modulate the cerebellar circuitry for tremor (Handforth et al., 2018), but detailed characterization of GABA axis in the postmortem essential tremor cerebellum is needed.

Many of the postmortem alterations identified in the essential tremor cerebellum can occur in the setting of neurodegeneration. The classical symptom associated with cerebellar degeneration is ataxia. While evidence shows that ataxia and tremor can co-exist (Gan et al., 2017; Lai et al., 2019), these are two distinct types of clinical symptoms. Tremor is rhythmic in nature, whereas ataxia is irregular and lacks rhythm. To sort out the distinct underlying pathology of ataxia and tremor, a study has been conducted to compare the cerebellar pathology of essential tremor and cerebellar ataxia, including spinocerebellar ataxia type 1 and multiple system atrophy. Similar to essential tremor patients, cerebellar ataxia patients have Purkinje cell loss, increased number of Purkinje cell axonal torpedoes, and a reduction of climbing fiber-Purkinje cell synaptic density, but cerebellar ataxia patients have much more severe degree of degeneration (Louis et al., 2019). Could essential tremor thus represent a milder version of cerebellar ataxia? Clinical observations, however, do not support this claim since the majority of essential tremor patients do not evolve into frank ataxia, even in the severe disease stage, whereas cerebellar ataxia patients often do not start out their disease with tremor. Therefore, what is the distinguishing pathological feature between cerebellar ataxia and essential tremor? It turns out that abnormal climbing fiber-Purkinje cell synaptic connections can be a distinguishing feature. Essential tremor has climbing fibers extending to the distal part of the molecular layer, which is the parallel fiber synaptic territory, whereas cerebellar ataxia has climbing fiber loss or regression to the proximal part of the molecular layer (Kuo et al., 2017; Louis et al., 2019). Since climbing fibers are the axons of the inferior olivary neurons, which have intrinsic oscillatory activity, the abnormal climbing fiber wiring of Purkinje cells thus can have disturbed rhythm control of the movements, generating clinical symptoms of ataxia and tremor.

The cerebellum in other tremor disorders, such as Parkinson's disease or dystonia, is much less studied. In a small cohort of Parkinson's disease, no Purkinje cell loss and no reduction in climbing fiber synaptic density have been identified (Louis et al., 2019), suggesting no obvious degenerative changes. Interestingly, abnormal climbing fiber-Purkinje cell synaptic connections in the parallel fiber territory have also been identified in Parkinson's disease cerebellum (Kuo, Lin, et al., 2017), which can possibly serve as a potential pathological substrate to enhance oscillatory neuronal activity. In an unsupervised cluster analysis to categorize different diagnostic categories (essential tremor, Parkinson's disease, spinocerebellar ataxia type 1, and multiple system atrophy), Parkinson's disease with rest tremor tends to have more abnormal climbing fiber-Purkinje cell synaptic connections in the parallel fiber territory than Parkinson's disease without rest tremor (Kuo, Lin, et al., 2017). This observation can possibly support the dimmer-switch model that the primary alterations of Parkinson tremor are originated from the basal ganglia due to dopamine loss, and the cerebellum can have a compensatory alteration to further amplify tremor. More detailed analysis of climbing fiber length and branching complexity in the parallel fiber territory of the molecular layer reveals that Parkinson's disease has shorter and less complex branching when compared to essential tremor (Kuo, Lin, et al., 2017), demonstrating certain specificity of each tremor disorder.

The postmortem human pathology of dystonic tremor has not been sufficiently studied. Even the data on dystonia alone is scarce. The cerebellar pathology of dystonia has been recently studied. Not surprisingly, there is a significant overlap with essential tremor, Parkinson's disease, and healthy controls, demonstrating a heterogeneous group of diseases. Consistently, no severe cerebellar degenerative features have been found (Louis et al., 2019).

Holmes tremor pathology has been identified in a single case of anti-Yo cerebellar ataxia (Rydz et al., 2015). In this case, the patient has fast, progressive cerebellar ataxia, in addition to Holmes tremor. Pathological examination reveals loss of Purkinje cells, dentate nucleus, red nucleus, and inferior olives, with relatively normal striatum, thalamus, and substantia nigra. These findings indicate that degeneration of the Guillain-Mollaret triangle can be the pathological underpinning of Holmes tremor.

27.4 Using Animal Models to Understand Cerebellar Pathophysiology in Tremor

The abovementioned human studies can help us to understand the potential localization and structural changes associated with tremor. However, we still lack a clear understanding of the *causal* relationship of these structural changes and tremor. Therefore, animal models provide an ideal platform to study tremor pathomechanism. In addition, animal models can pave a way for drug screening for tremor disorders. We will discuss commonly used animal models of tremor in this section and how these animal models can provide insight into tremor pathophysiology.

The classical animal model for tremor is harmaline-induced tremor, and a single dose of harmaline can produce tremor in cats, rats, mice, monkeys, and pigs (Cheng et al., 2013; Pan et al., 2018). Harmaline can enhance the gap junction coupling between the inferior olivary neurons, which at baseline have subthreshold membrane potential oscillations at 1–10 Hz. Upon harmaline exposure, the inferior olivary neurons can entrain downstream Purkinje cell to fire synchronously and rhythmically (Llinás, 2013). The synchronous and rhythmic neuronal activity is thought to be the physiological underpinning of tremor, which can propagate through the cerebello-thalamo-cortical loop. Interestingly, harmaline-induced tremor is an action tremor, and when the animals are in complete rest, the tremor disappears or is greatly suppressed (Cheng et al., 2013). This observation suggests that the olivocerebellum is recruited during the action but has less a role at rest.

Harmaline belongs to a group of naturally occurring compounds, β-alkaloids. Elevation of one of these β-alkaloids, called harmane, has been identified in the essential tremor patient blood and brain (Louis et al., 2008; Louis, Benito-León et al., 2013; Louis, Factor-Litvak et al., 2013), indicating a potential environmental toxin exposure can contribute to tremor. β-Alkaloids can be ingested via cooked meats, which can be a dietary source for an increased risk of essential tremor.

As mentioned above, essential tremor patients have abnormal climbing fiber-Purkinje cell synaptic connections in the parallel fiber territory, and this particular pathological feature has been associated with a reduction of the level of cerebellar GluRδ2 (Pan et al., 2020), a master regulator for Purkinje cell synaptic organizer, in the essential tremor cerebellum. A novel mouse model, *Grid2^{dupE3}*, has been established with both molecular (a reduction of the level of cerebellar GluRδ2) and structural (abnormal climbing fiber-Purkinje cell synaptic connections in the parallel fiber territory) similarities (Pan et al., 2020). Interestingly, *Grid2^{dupE3}* mice develop action tremor with no rest tremor, tremor progression with aging, and tremor responsiveness to primidone, propranolol, and alcohol, recapitulating key clinical features of essential tremor patients. Tremor in *Grid2^{dupE3}* mice can be suppressed by silencing either climbing fiber activity or climbing fiber-Purkinje cell synaptic transmission using pharmacology or optogenetic methods (Pan et al., 2020), demonstrating that climbing fiber-Purkinje cell synaptic connections are the origin of tremor. This study provides a strong causal relationship that abnormal climbing fiber wiring identified in essential patients can be the generator of tremor (Fig. 27.2).

Grid2^{dupE3} mice have oscillatory neuronal activity detected by the cerebellar local field potential recording, which comes from the abnormal climbing fiber-Purkinje cell synaptic connections (Pan et al., 2020). In harmaline-induced mouse tremor model, similar oscillatory neuronal activity has also been identified, and such oscillatory activity is coherent with tremor (Fig. 27.3). These data support that oscillatory neuronal activity in the cerebellum can be the pathophysiological underpinning of tremor. Whether the oscillatory neuronal activity in the cerebellum is sufficient to drive tremor still remains an open question. To solve this, a study tested whether

Fig. 27.2 Cerebellar pathology of essential tremor. The postmortem pathological examination of essential tremor cerebellum often shows (**a**) a Purkinje cell axonal torpedo (arrow, Luxol fast blue counter-stained with hematoxylin-eosin); (**b**) abnormal climbing fiber synapses extending to the outer 20% of the molecular layer, which should have been the parallel fiber synaptic territory (arrows, vesicular glutamate transporter type 2 immunohistochemistry); and (**c**) dense basket cell plexus with no Purkinje cell body, "empty baskets" (arrows, calbindin (red) and glutamate decarboxylase (brown) dual immunohistochemistry)

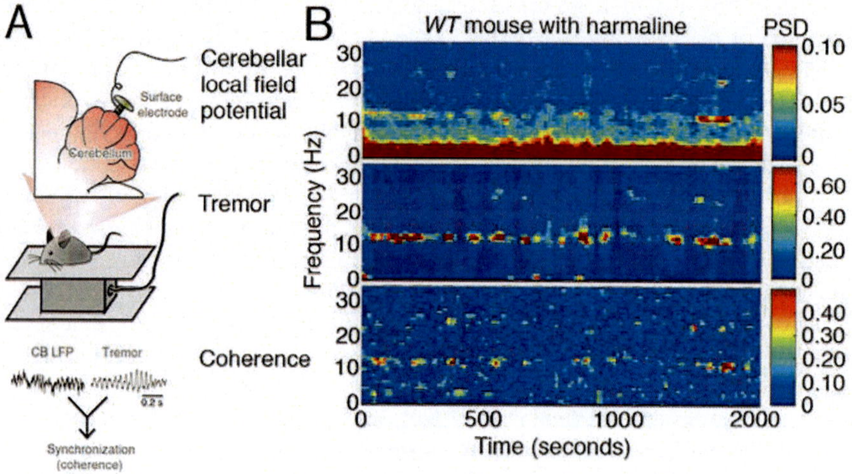

Fig. 27.3 Harmaline-induced mouse tremor. (**a**) A schematic diagram depicting the in vivo recording of cerebellar local field potential and tremor. (**b**) A mouse that receives subcutaneous harmaline injection develops 13 Hz tremor and also 13 Hz cerebellar oscillatory activity. Tremor and cerebellar oscillatory activity have strong coherence

artificially driven oscillatory activity in Purkinje cells can create tremor. When synchronously driving the activity of channelrhodopsin-expressing Purkinje cells into either 10 Hz or 20 Hz in *wild-type* mice, mice will show 10 Hz or 20 Hz tremor, respectively (Pan et al., 2020). This optogenetic mouse model provides evidence that oscillatory neuronal activity in the cerebellum is sufficient for tremor generation and also for the determination of tremor frequency.

Based on the GABA hypothesis of essential tremor, $GABA_A$ receptor α1 subunit knockout mice have been studied. This mouse model develops tremor (Kralic et al., 2005). However, the tremor in this mouse model needs to be measured with tail suspension, rather than in a freely moving setting, raising a question of translatability to essential tremor patients. One of the considerations for $GABA_A$ receptor α1 subunit knockout mice is the genetic background can influence the mouse phenotypes. Some of the strains of $GABA_A$ receptor α1 subunit knockout mice will develop absence and myoclonic seizures (Kuo et al., 2019). To further determine the cerebellar specificity, Purkinje cell-specific $GABA_A$ receptor α1 subunit knockout mice have been recently generated. Tremor in Purkinje cell-specific $GABA_A$ receptor α1 subunit knockout mice is at the similar frequency to the tremor of global $GABA_A$ receptor α1 subunit knockout mice at around 22–25 Hz (Nietz et al., 2020). Purkinje cell-specific $GABA_A$ receptor α1 subunit knockout creates a lack of inhibitory synaptic inputs from interneurons. These studies show that modulation of Purkinje cell activity by interneurons may be important for tremor.

A spontaneous mutant mouse, *waddles*, has been identified as a tremor model. This mouse model is a functional knockout of *Car8* gene. CAR8 is a protein predominantly expressed in Purkinje cells and can influence Purkinje cell activity by modulating the binding of 1,4,5-trisphosphate receptor type 1 to its ligand (Shimobayashi & Kapfhammer, 2018). *Car8*^{wdl} mice not only have tremor but also have dystonia and ataxia, making them a unique model to probe the cerebellar circuitry of three clinical symptoms related to the cerebellum (White et al., 2016). *Car8*^{wdl} mice have tremor age-related tremor progression, and tremor is at 8–12 Hz. In slice physiological recording, *Car8*^{wdl} mice exhibit long pauses of Purkinje cell firing, which increase the coefficient of variation of firing patterns (White et al., 2016). *Car8*^{wdl} mice may be considered as a model for essential tremor with dystonic or ataxia features.

Another recent development of tremor mouse models comes from the conditional deletion of synaptotagmin-2 in the parvalbumin-expressing neurons (Zhou et al., 2020). Synaptotagmin-2 is a fast calcium sensor for neurotransmitter release, and a lack of synaptotagmin-2 can perturb neurotransmitter release. Such genetic manipulation can cause action tremor, which is responsive to alcohol. Since parvalbumin-expressing neurons distribute widely throughout the central nervous system, region-specific, viral-mediated knockout of synaptotagmin-2 in parvalbumin neurons demonstrates the cerebellum, specifically the deep cerebellar nucleus, is important for such tremor. This model offers the knowledge that dysfunction of the presynaptic neurotransmitter release can impact the cerebellar physiology, leading to tremor.

While these animal models provide important insight into tremor generation, several important aspects should be considered when translating animal studies into clinical trial design for tremor patients. First, the method for tremor measurement in animals is critical. Human tremor disorders are rather defined by tremor characteristics during action or at rest. Therefore, tremor measurement in freely moving mice is most ideal to allow analysis to separate tremor during action or at rest. Second, the measurement of tremor responsiveness to pharmacological agents should be standardized. Third, as tremor can be generated in alterations at different levels of cerebellar circuitry, the scientific rationale of tremor animal models should at least be partly based on the human genetic or pathological evidence.

27.5 Physiological Alterations of Tremor Disorders

What is the physiological underpinning of tremor? The olivocerebellum is an intrinsic oscillator and is capable of generating rhythmic movements. As mentioned above, harmaline can induce tremor and the corresponding cerebellar oscillatory activity (Fig. 27.3). Similarly, in *Grid2*^{dupE3} mice with essential tremor-like climbing fiber-Purkinje cell synaptic pathology and essential tremor-like action tremor, cerebellar oscillatory activity can also be recorded (Pan et al., 2020). When using lidocaine micro-infusion into the inferior olivary nucleus to silence climbing fiber activity, both tremor and cerebellar oscillatory activity can be suppressed (Pan et al., 2020), suggesting the role of climbing fibers in tremor and cerebellar oscillatory activity.

The next question is whether cerebellar oscillatory activity can be recorded in essential tremor patients. Using a novel technology, cerebellar electroencephalogram, oscillatory activity at the tremor-range frequency can be observed in essential tremor patients (Pan et al., 2020). Moreover, the intensity of the oscillatory activity correlates with tremor severity, further supporting the relationship between tremor and cerebellar oscillatory activity. Cerebellar oscillatory activity falls within 8–12 Hz in essential tremor patients, which can be confounded by the alpha rhythm coming from the occipital cortex. Therefore, simultaneous recording of occipital cortex and the cerebellum has been performed in essential tremor patients, and it is apparent that occipital alpha rhythm is modulated by the eye opening and eye closing, whereas cerebellar oscillatory activity is not, supporting the distinct brain sources. Another potential confounding factor is muscle artifact, which has a much higher frequency (>50 Hz) than cerebellar oscillatory activity. These studies provide a solid validity of using cerebellar electroencephalogram to detect cerebellar oscillatory activity for essential tremor patients (Pan et al., 2020).

Since the cerebellum is clearly involved in other tremor disorders, whether cerebellar oscillatory activity can be detected in other tremor disorders, such as Parkinson tremor or dystonic tremor, remains to be determined.

27.6 Challenges in the Field

In the past decade, there are great advances in understanding the role of the cerebellum in tremor disorders, from different aspects as outlined above. However, significant challenges exist in the field.

Tremor disorders are likely to be heterogeneous in nature. For example, essential tremor has bimodal age of onset (Deuschl et al., 2015); thus, essential tremor can be further categorized into early-onset and late-onset cases. Early-onset essential tremor cases often have a family history of tremor, whereas late-onset essential tremor cases tend to progress faster. While a neuroimaging study suggests early-onset and late-onset essential tremor cases have different degree of cerebellar involvement (Muthuraman et al., 2015), neuropathological investigation of the cerebellum did not reveal differences of these two subtypes of essential tremor (Kuo et al., 2017; Lee et al., 2018). Note that neuropathology studies can only capture the end stage of the disease, whereas neuroimaging is likely to probe the dynamic changes in the disease process. Therefore, it is possible that these two subtypes of essential tremor have distinct brain circuitries but reach the same endpoint of cerebellar pathology. Similar issues of dynamic regulation during the disease progression should also be considered in Parkinson tremor. Parkinson tremor can occur in the early stage of the disease, and later on, a subset of patients will experience the disappearance of tremor, transforming into akinetic-rigid form of Parkinson's disease. It is possible that the cerebellum can play a role in the early stage of the disease, and as the disease pathology spreads to a wide range of the brain areas, the tremor circuitry is destroyed, leading to tremor disappearance. Further investigations are needed to test these hypotheses.

Another consideration is the co-existing of other movement disorder phenomenology. Essential tremor patients often have subtle features of dystonia, parkinsonism, or ataxia, and sometimes these patients are called to have "essential tremor plus." While this term may be useful to describe these patients, there is no consensus on how essential tremor plus is diagnosed. Among essential tremor plus, dystonic features are most commonly described. Interestingly, both tremor and dystonia can come from the dysfunctional cerebellum. From a harmaline-induced tremor mouse and *Grid2*^{dupE3} mouse models, Purkinje cell rhythmic firing drives real-time rhythmic motor activities (i.e., tremor) (Pan et al., 2020). In mouse models with viral-mediated knockdown of DYT1 or DYT12 in the cerebellum, Purkinje cell burst firing can be induced with corresponding dystonic-like symptoms (Fremont et al., 2014; Fremont et al., 2015; Fremont et al., 2017). These studies suggest that real-time abnormalities of Purkinje cell firing might lead to either tremor or dystonia. Therefore, it is thus not surprising that a disease process can affect Purkinje cell firing patterns at different disease stages or different locations of the cerebellar cortex that can have both tremor and dystonia. Depending on how and when the cerebellum is recruited for motor commands, Purkinje cell can fire at different patterns to generate diverse tremor and dystonia symptoms. Within this framework, dystonic features in a subset of essential tremor patients should have a cerebellar origin. Future studies of different animal models and human physiology recording of tremor and dystonia should help to settle this controversy.

While we begin to understand the importance of cerebello-thalamo-cortical circuit for tremor, other brain regions are likely to be important for tremor and deserve further investigation (Fig. 27.4). For example, dopamine system is the best known modulator for tremor circuit based on the dimmer-switch model. Within the cerebellum, deep cerebellar nucleus send inhibitory projections back to the inferior olivary nucleus (Lang et al., 1996), which regulate the gap junction coupling between inferior olivary neurons. This will have profound influence on the synchrony of Purkinje cell firing and likely to affect tremor presentation. Finally, the complex modulation at the level of motor cortex can play an additional role.

Deep brain stimulations to the thalamus are one of the most effective ways to treat both essential tremor and Parkinson tremor. However, many patients experience "tolerance," meaning the gradual dissipation of therapeutic effects (Fasano & Helmich, 2019). A higher setting of stimulations is often required to achieve therapeutic effects, which usually comes with the price of stimulating neighboring brain structures leading to side effects of dysarthria and paresthesia. The mechanism of tolerance of deep brain stimulations remains unclear. One possibility is that thalamic stimulations only stop the propagation of the oscillatory activity that comes from the cerebellum without addressing the cerebellar oscillatory activity itself. With time, the axonal sprouting of deep cerebellar nucleus to the thalamus can find alternative pathways to allow oscillatory activity to pass through. Further investigations on this topic not only will be important to understand the aberrant neuronal

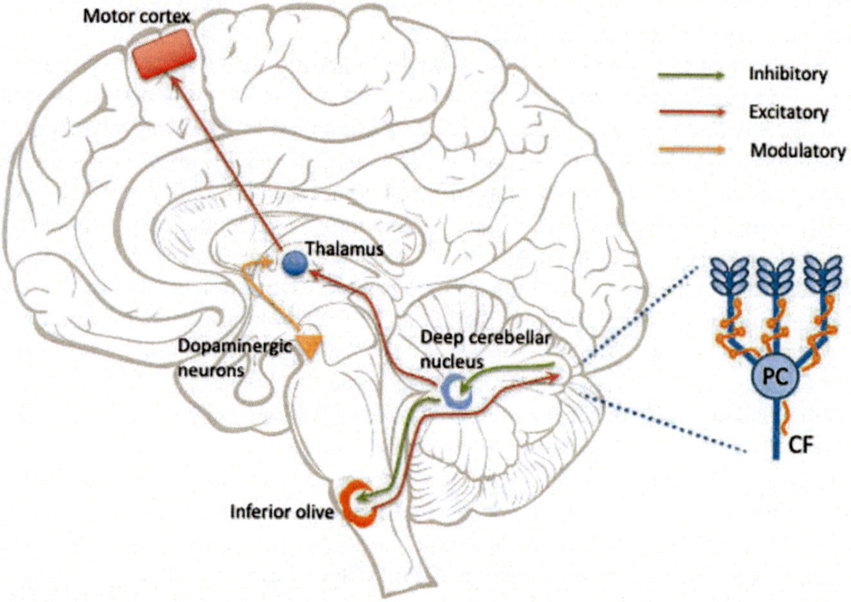

Fig. 27.4 Brain circuitry of tremor. Climbing fiber and Purkinje cell synaptic connection plays an important role for tremor generation, which could be further modulated by the inferior olive and deep cerebellar nucleus. Tremor-related oscillatory activity propagates downstream to the thalamus and motor cortex and can be additionally modulated by the dopamine system

connectivity with chronic deep brain stimulations but also will help to develop novel therapies for this challenging clinical scenario.

27.7 Conclusion

We summarize the importance of the cerebellum in tremor, the most common movement disorder. However, we just begin to understand the specific element within the cerebellum in tremor, which deserves detail investigation to fully understand how the cerebellum generates both normal and abnormal movements, using tremor as a model. Cerebellar synaptic organization and cerebellar oscillatory activity can be promising therapeutic targets for tremor disorders.

References

Babij, R., Lee, M., Cortes, E., Vonsattel, J.-P. G., Faust, P. L., & Louis, E. D. (2013). Purkinje cell axonal anatomy: Quantifying morphometric changes in essential tremor versus control brains. *Brain, 136*, 3051–3061.

Benito-León, J., Alvarez-Linera, J., Hernández-Tamames, J. A., Alonso-Navarro, H., Jiménez-Jiménez, F. J., & Louis, E. D. (2009). Brain structural changes in essential tremor: Voxel-based morphometry at 3-Tesla. *Journal of the Neurological Sciences, 287*, 138–142.

Bhatia, K. P., Bain, P., Bajaj, N., Elble, R. J., Hallett, M., Louis, E. D., Raethjen, J., Stamelou, M., Testa, C. M., & Deuschl, G. (2018). Consensus statement on the classification of tremors. From the task force on tremor of the International Parkinson and Movement Disorder Society. *Movement Disorders, 33*, 75–87.

Buijink, A. W., Broersma, M., van der Stouwe, A. M., van Wingen, G. A., Groot, P. F., Speelman, J. D., Maurits, N. M., & van Rootselaar, A. F. (2015). Rhythmic finger tapping reveals cerebellar dysfunction in essential tremor. *Parkinsonism & Related Disorders, 21*, 383–388.

Cerasa, A., & Quattrone, A. (2016). Linking essential tremor to the cerebellum-neuroimaging evidence. *Cerebellum, 15*, 263–275.

Cheng, M. M., Tang, G., & Kuo, S. H. (2013). Harmaline-induced tremor in mice: Videotape documentation and open questions about the model. *Tremor and Other Hyperkinetic Movements (N Y), 3*, tre-03-205-4668-1.

Deuschl, G., Petersen, I., Lorenz, D., & Christensen, K. (2015). Tremor in the elderly: Essential and aging-related tremor. *Movement Disorders, 30*, 1327–1334.

Dirkx, M. F., den Ouden, H., Aarts, E., Timmer, M., Bloem, B. R., Toni, I., & Helmich, R. C. (2016). The cerebral network of Parkinson's tremor: An effective connectivity fMRI study. *The Journal of Neuroscience, 36*, 5362–5372.

Dirkx, M. F., den Ouden, H. E., Aarts, E., Timmer, M. H., Bloem, B. R., Toni, I., & Helmich, R. C. (2017). Dopamine controls Parkinson's tremor by inhibiting the cerebellar thalamus. *Brain, 140*, 721–734.

Dirkx, M. F., Zach, H., Bloem, B. R., Hallett, M., & Helmich, R. C. (2018). The nature of postural tremor in Parkinson disease. *Neurology, 90*, e1095–e1103.

Dirkx, M. F., Zach, H., van Nuland, A., Bloem, B. R., Toni, I., & Helmich, R. C. (2019). Cerebral differences between dopamine-resistant and dopamine-responsive Parkinson's tremor. *Brain, 142*, 3144–3157.

Dupuis, M. J.-M., Evrard, F. L. A., Jacquerye, P. G., Picard, G. R., & Lermen, O. G. (2010). Disappearance of essential tremor after stroke. *Movement Disorders: Official Journal of the Movement Disorder Society, 25*, 2884–2887.

Elias, W. J., Lipsman, N., Ondo, W. G., Ghanouni, P., Kim, Y. G., Lee, W., Schwartz, M., Hynynen, K., Lozano, A. M., Shah, B. B., Huss, D., Dallapiazza, R. F., Gwinn, R., Witt, J., Ro, S., Eisenberg, H. M., Fishman, P. S., Gandhi, D., Halpern, C. H., Chuang, R., Butts Pauly, K., Tierney, T. S., Hayes, M. T., Cosgrove, G. R., Yamaguchi, T., Abe, K., Taira, T., & Chang, J. W. (2016). A randomized trial of focused ultrasound thalamotomy for essential tremor. *The New England Journal of Medicine, 375*, 730–739.

Erickson-Davis, C. R. C., Faust, P. L. P., Vonsattel, J.-P. G. J., Gupta, S. S., Honig, L. S. L., & Louis, E. D. E. (2010). Hairy baskets associated with degenerative Purkinje cell changes in essential tremor. *Journal of Neuropathology & Experimental Neurology, 69*, 262–271.

Fang, W., Chen, H., Wang, H., Zhang, H., Puneet, M., Liu, M., Lv, F., Luo, T., Cheng, O., Wang, X., & Lu, X. (2016). Essential tremor is associated with disruption of functional connectivity in the ventral intermediate Nucleus--Motor Cortex--Cerebellum circuit. *Human Brain Mapping, 37*, 165–178.

Fasano, A., & Helmich, R. C. (2019). Tremor habituation to deep brain stimulation: Underlying mechanisms and solutions. *Movement Disorders, 34*, 1761–1773.

Fasano, A., & Lozano, A. M. (2015). Deep brain stimulation for movement disorders: 2015 and beyond. *Current Opinion in Neurology, 28*, 423–436.

Fremont, R., Calderon, D. P., Maleki, S., & Khodakhah, K. (2014). Abnormal high-frequency burst firing of cerebellar neurons in rapid-onset dystonia-parkinsonism. *The Journal of Neuroscience, 34*, 11723–11732.

Fremont, R., Tewari, A., Angueyra, C., & Khodakhah, K. (2017). A role for cerebellum in the hereditary dystonia DYT1. *eLife, 6*, e22775.

Fremont, R., Tewari, A., & Khodakhah, K. (2015). Aberrant Purkinje cell activity is the cause of dystonia in a shRNA-based mouse model of Rapid Onset Dystonia-Parkinsonism. *Neurobiology of Disease, 82*, 200–212.

Gan, S. R., Wang, J., Figueroa, K. P., Pulst, S. M., Tomishon, D., Lee, D., Perlman, S., Wilmot, G., Gomez, C. M., Schmahmann, J., Paulson, H., Shakkottai, V. G., Ying, S. H., Zesiewicz, T., Bushara, K., Geschwind, M. D., Xia, G., Subramony, S. H., Ashizawa, T., & Kuo, S. H. (2017). Postural tremor and ataxia progression in spinocerebellar ataxias. *Tremor and Other Hyperkinetic Movements (N Y), 7*, 492.

Gupta, D. K., Marano, M., Zweber, C., Boyd, T., & Kuo, S. H. (2021). Prevalence and relationship of rest and action tremor in Parkinson's Disease. *Tremor and Other Hyperkinetic Movements (N Y), 10*, 58.

Handforth, A., Kadam, P. A., Kosoyan, H. P., & Eslami, P. (2018). Suppression of harmaline tremor by activation of an extrasynaptic GABA(A) receptor: Implications for essential tremor. *Tremor and Other Hyperkinetic Movements (N Y), 8*, 546.

Haubenberger, D., & Hallett, M. (2018). Essential tremor. *The New England Journal of Medicine, 378*, 1802–1810.

Helmich, R. C., Janssen, M. J., Oyen, W. J., Bloem, B. R., & Toni, I. (2011). Pallidal dysfunction drives a cerebellothalamic circuit into Parkinson tremor. *Annals of Neurology, 69*, 269–281.

Klein, J. C., Barbe, M. T., Seifried, C., Baudrexel, S., Runge, M., Maarouf, M., Gasser, T., Hattingen, E., Liebig, T., Deichmann, R., Timmermann, L., Weise, L., & Hilker, R. (2012). The tremor network targeted by successful VIM deep brain stimulation in humans. *Neurology, 78*, 787–795.

Kralic, J. E., Criswell, H. E., Osterman, J. L., O'Buckley, T. K., Wilkie, M. E., Matthews, D. B., Hamre, K., Breese, G. R., Homanics, G. E., & Morrow, A. L. (2005). Genetic essential tremor in gamma-aminobutyric acidA receptor alpha1 subunit knockout mice. *The Journal of Clinical Investigation, 115*, 774–779.

Kuo, S.-H., Tang, G., Louis, E. D., Ma, K., Babji, R., Balatbat, M., Cortes, E., Vonsattel, J.-P. G., Yamamoto, A., Sulzer, D., & Faust, P. L. (2013). Lingo-1 expression is increased in essential tremor cerebellum and is present in the basket cell pinceau. *Acta Neuropathologica, 125*, 879–889.

Kuo, S. H., Lin, C. Y., Wang, J., Liou, J. Y., Pan, M. K., Louis, R. J., Wu, W. P., Gutierrez, J., Louis, E. D., & Faust, P. L. (2016). Deep brain stimulation and climbing fiber synaptic pathology in essential tremor. *Annals of Neurology, 80*, 461–465.

Kuo, S. H., Lin, C. Y., Wang, J., Sims, P. A., Pan, M. K., Liou, J. Y., Lee, D., Tate, W. J., Kelly, G. C., Louis, E. D., & Faust, P. L. (2017). Climbing fiber-Purkinje cell synaptic pathology in tremor and cerebellar degenerative diseases. *Acta Neuropathologica, 133*, 121–138.

Kuo, S. H., Louis, E. D., Faust, P. L., Handforth, A., Chang, S. Y., Avlar, B., Lang, E. J., Pan, M. K., Miterko, L. N., Brown, A. M., Sillitoe, R. V., Anderson, C. J., Pulst, S. M., Gallagher, M. J., Lyman, K. A., Chetkovich, D. M., Clark, L. N., Tio, M., Tan, E. K., & Elble, R. J. (2019). Current opinions and consensus for studying tremor in animal models. *Cerebellum, 18*, 1036–1063.

Kuo, S. H., Wang, J., Tate, W. J., Pan, M. K., Kelly, G. C., Gutierrez, J., Cortes, E. P., Vonsattel, J. G., Louis, E. D., & Faust, P. L. (2017). Cerebellar pathology in early onset and late onset essential tremor. *Cerebellum, 16*, 473–482.

Lai, R. Y., Tomishon, D., Figueroa, K. P., Pulst, S. M., Perlman, S., Wilmot, G., Gomez, C. M., Schmahmann, J. D., Paulson, H., Shakkottai, V. G., Ying, S. H., Zesiewicz, T., Bushara, K., Geschwind, M., Xia, G., Subramony, S. H., Ashizawa, T., & Kuo, S. H. (2019). Tremor in the degenerative cerebellum: Towards the understanding of brain circuitry for tremor. *Cerebellum, 18*, 519–526.

Lang, E. J., Sugihara, I., & Llinas, R. (1996). GABAergic modulation of complex spike activity by the cerebellar nucleoolivary pathway in rat. *Journal of Neurophysiology, 76*, 255–275.

Lee, D., Gan, S. R., Faust, P. L., Louis, E. D., & Kuo, S. H. (2018). Climbing fiber-Purkinje cell synaptic pathology across essential tremor subtypes. *Parkinsonism & Related Disorders, 51*, 24–29.

Lin, C.-Y., Louis, E. D., Faust, P. L., Koeppen, A. H., Vonsattel, J.-P. G., & Kuo, S.-H. (2014). Abnormal climbing fibre-Purkinje cell synaptic connections in the essential tremor cerebellum. *Brain, 137*, 3149–3159.

Llinás, R. R. (2013). The olivo-cerebellar system: A key to understanding the functional significance of intrinsic oscillatory brain properties. *Frontiers in Neural Circuits, 7*, 96.

Lopez, A. M., Trujillo, P., Hernandez, A. B., Lin, Y. C., Kang, H., Landman, B. A., Englot, D. J., Dawant, B. M., Konrad, P. E., & Claassen, D. O. (2020). Structural correlates of the sensorimotor cerebellum in Parkinson's disease and essential tremor. *Movement Disorders, 35*, 1181–1188.

Louis, E. D. (2001). Clinical practice. Essential tremor. *The New England Journal of Medicine, 345*, 887–891.

Louis, E. D., Benito-León, J., Moreno-García, S., Vega, S., Romero, J. P., Bermejo-Pareja, F., Gerbin, M., Viner, A. S., Factor-Litvak, P., Jiang, W., & Zheng, W. (2013). Blood harmane (1-methyl-9H-pyrido[3,4-b]indole) concentration in essential tremor cases in Spain. *Neurotoxicology, 34*, 264–268.

Louis, E. D., Factor-Litvak, P., Liu, X., Vonsattel, J.-P. G., Galecki, M., Jiang, W., & Zheng, W. (2013). Elevated brain harmane (1-methyl-9H-pyrido[3,4-b]indole) in essential tremor cases vs. controls. *Neurotoxicology, 38C*, 131–135.

Louis, E. D., & Ferreira, J. J. (2010). How common is the most common adult movement disorder? Update on the worldwide prevalence of essential tremor. *Movement Disorders: Official Journal of the Movement Disorder Society, 25*, 534–541.

Louis, E. D., Hernandez, N., Dyke, J. P., Ma, R. E., & Dydak, U. (2018). In vivo dentate nucleus gamma-aminobutyric acid concentration in essential tremor vs controls. *Cerebellum, 17*, 165–172.

Louis, E. D., Jiang, W., Pellegrino, K. M., Rios, E., Factor-Litvak, P., Henchcliffe, C., & Zheng, W. (2008). Elevated blood harmane (1-methyl-9H-pyrido[3,4-b]indole) concentrations in essential tremor. *Neurotoxicology, 29*, 294–300.

Louis, E. D., Kerridge, C. A., Chatterjee, D., Martuscello, R. T., Diaz, D. T., Koeppen, A. H., Kuo, S. H., Vonsattel, J. G., Sims, P. A., & Faust, P. L. (2019). Contextualizing the pathology in the essential tremor cerebellar cortex: A patholog-omics approach. *Acta Neuropathologica, 138*, 859–876.

Louis, E. D., Kuo, S. H., Tate, W. J., Kelly, G. C., Gutierrez, J., Cortes, E. P., Vonsattel, J. G., & Faust, P. L. (2018). Heterotopic Purkinje cells: A comparative postmortem study of essential tremor and spinocerebellar ataxias 1, 2, 3, and 6. *Cerebellum, 17*, 104–110.

Louis, E. D., Lee, M., Babij, R., Ma, K., Cortes, E., Vonsattel, J.-P. G., & Faust, P. L. (2014). Reduced Purkinje cell dendritic arborization and loss of dendritic spines in essential tremor. *Brain, 137*, 3142–3148.

Louis, E. D., Ottman, R., & Hauser, W. A. (1998). How common is the most common adult movement disorder? Estimates of the prevalence of essential tremor throughout the world. *Movement Disorders: Official Journal of the Movement Disorder Society, 13*, 5–10.

Louis, E. D., Rabinowitz, D., Choe, M., Tate, W. J., Kelly, G. C., Kuo, S. H., & Faust, P. L. (2016). Mapping Purkinje cell placement along the Purkinje cell layer: An analysis of postmortem tissue from essential tremor patients vs controls. *Cerebellum, 15*, 726–731.

Louis, E. D. E., Faust, P. L. P., Vonsattel, J.-P. G. J., Honig, L. S. L., Rajput, A. A., Robinson, C. A. C., Rajput, A. A., Pahwa, R. R., Lyons, K. E. K., Ross, G. W. G., Borden, S. S., Moskowitz,

C. B. C., Lawton, A. A., & Hernandez, N. N. (2007). Neuropathological changes in essential tremor: 33 cases compared with 21 controls. *Brain, 130*, 3297–3307.

Louis, R. J., Lin, C.-Y., Faust, P. L., Koeppen, A. H., & Kuo, S.-H. (2015). Climbing fiber synaptic changes correlate with clinical features in essential tremor. *Neurology, 84*, 2284–2286. https://doi.org/10.1212/WNL.0000000000001636-0000000000002286

Luo, C., Rajput, A. H., Robinson, C. A., & Rajput, A. (2012). Gamma-aminobutyric acid (GABA)-B receptor 1 in cerebellar cortex of essential tremor. *Journal of Clinical Neuroscience, 19*, 920–921.

Marjama-Lyons, J., & Koller, W. (2000). Tremor-predominant Parkinson's disease. Approaches to treatment. *Drugs & Aging, 16*, 273–278.

Marras, C., Beck, J. C., Bower, J. H., Roberts, E., Ritz, B., Ross, G. W., Abbott, R. D., Savica, R., Van Den Eeden, S. K., Willis, A. W., & Tanner, C. M. (2018). Prevalence of Parkinson's disease across North America. *NPJ Parkinson's Disease, 4*, 21.

McAuley, J. H., & Marsden, C. D. (2000). Physiological and pathological tremors and rhythmic central motor control. *Brain, 123*(Pt 8), 1545–1567.

Miyazaki, T., Yamasaki, M., Takeuchi, T., Sakimura, K., Mishina, M., & Watanabe, M. (2010). Ablation of glutamate receptor GluRdelta2 in adult Purkinje cells causes multiple innervation of climbing fibers by inducing aberrant invasion to parallel fiber innervation territory. *The Journal of Neuroscience, 30*, 15196–15209.

Muthuraman, M., Deuschl, G., Anwar, A. R., Mideksa, K. G., von Helmolt, F., & Schneider, S. A. (2015). Essential and aging-related tremor: Differences of central control. *Movement Disorders, 30*, 1673–1680.

Nicoletti, G., Manners, D., Novellino, F., Condino, F., Malucelli, E., Barbiroli, B., Tonon, C., Arabia, G., Salsone, M., Giofre, L., Testa, C., Lanza, P., Lodi, R., & Quattrone, A. (2010). Diffusion tensor MRI changes in cerebellar structures of patients with familial essential tremor. *Neurology, 74*, 988–994.

Nietz, A., Krook-Magnuson, C., Gutierrez, H., Klein, J., Sauve, C., Hoff, I., Christenson Wick, Z., & Krook-Magnuson, E. (2020). Selective loss of the GABA(Aα1) subunit from Purkinje cells is sufficient to induce a tremor phenotype. *Journal of Neurophysiology, 124*, 1183–1197.

Pan, M. K., Li, Y. S., Wong, S. B., Ni, C. L., Wang, Y. M., Liu, W. C., Lu, L. Y., Lee, J. C., Cortes, E. P., Vonsattel, J. G., Sun, Q., Louis, E. D., Faust, P. L., & Kuo, S. H. (2020). Cerebellar oscillations driven by synaptic pruning deficits of cerebellar climbing fibers contribute to tremor pathophysiology. *Science Translational Medicine, 12*, eaay1769.

Pan, M. K., Ni, C. L., Wu, Y. C., Li, Y. S., & Kuo, S. H. (2018). Animal models of tremor: Relevance to human tremor disorders. *Tremor and Other Hyperkinetic Movements (N Y), 8*, 587.

Paris-Robidas, S., Brochu, E., Sintes, M., Emond, V., Bousquet, M., Vandal, M., Pilote, M., Tremblay, C., Di Paolo, T., Rajput, A. H., Rajput, A., & Calon, F. (2012). Defective dentate nucleus GABA receptors in essential tremor. *Brain, 135*, 105–116.

Pasquini, J., Ceravolo, R., Qamhawi, Z., Lee, J. Y., Deuschl, G., Brooks, D. J., Bonuccelli, U., & Pavese, N. (2018). Progression of tremor in early stages of Parkinson's disease: A clinical and neuroimaging study. *Brain, 141*, 811–821.

Qamhawi, Z., Towey, D., Shah, B., Pagano, G., Seibyl, J., Marek, K., Borghammer, P., Brooks, D. J., & Pavese, N. (2015). Clinical correlates of raphe serotonergic dysfunction in early Parkinson's disease. *Brain, 138*, 2964–2973.

Ren, J., Isakova, A., Friedmann, D., Zeng, J., Grutzner, S. M., Pun, A., Zhao, G. Q., Kolluru, S. S., Wang, R., Lin, R., Li, P., Li, A., Raymond, J. L., Luo, Q., Luo, M., Quake, S. R., & Luo, L. (2019). Single-cell transcriptomes and whole-brain projections of serotonin neurons in the mouse dorsal and median raphe nuclei. *eLife, 8*, e49424.

Rydz, D., Lin, C. Y., Xie, T., Cortes, E., Vonsattel, J. P., & Kuo, S. H. (2015). Pathological findings of anti-Yo cerebellar degeneration with Holmes tremor. *Journal of Neurology, Neurosurgery, and Psychiatry, 86*, 121–122.

Schnitzler, A., Münks, C., Butz, M., Timmermann, L., & Gross, J. (2009). Synchronized brain network associated with essential tremor as revealed by magnetoencephalography. *Movement Disorders: Official Journal of the Movement Disorder Society, 24*, 1629–1635.

Shimobayashi, E., & Kapfhammer, J. P. (2018). Calcium signaling, PKC gamma, IP3R1 and CAR8 link spinocerebellar ataxias and Purkinje cell dendritic development. *Current Neuropharmacology, 16*, 151–159.

Shin, D. H., Han, B. S., Kim, H. S., & Lee, P. H. (2008). Diffusion tensor imaging in patients with essential tremor. *AJNR. American Journal of Neuroradiology, 29*, 151–153.

Tapper, S., Göransson, N., Lundberg, P., Tisell, A., & Zsigmond, P. (2020). A pilot study of essential tremor: Cerebellar GABA+/Glx ratio is correlated with tremor severity. *Cerebellum Ataxias, 7*, 8.

Thenganatt, M. A., & Louis, E. D. (2012). Distinguishing essential tremor from Parkinson's disease: Bedside tests and laboratory evaluations. *Expert Review of Neurotherapeutics, 12*, 687–696.

van der Stouwe, A. M. M., Nieuwhof, F., & Helmich, R. C. (2020). Tremor pathophysiology: Lessons from neuroimaging. *Current Opinion in Neurology, 33*, 474–481.

Watanabe, M. (2008). Molecular mechanisms governing competitive synaptic wiring in cerebellar Purkinje cells. *The Tohoku Journal of Experimental Medicine, 214*, 175–190.

White, J. J., Arancillo, M., King, A., Lin, T., Miterko, L. N., Gebre, S. A., & Sillitoe, R. V. (2016). Pathogenesis of severe ataxia and tremor without the typical signs of neurodegeneration. *Neurobiology of Disease, 86*, 86–98.

Yu, M., Ma, K., Faust, P. L., Honig, L. S., Cortes, E., Vonsattel, J. P., & Louis, E. D. (2012). Increased number of Purkinje cell dendritic swellings in essential tremor. *European Journal of Neurology, 19*, 625–630.

Zach, H., Dirkx, M. F., Roth, D., Pasman, J. W., Bloem, B. R., & Helmich, R. C. (2020). Dopamine-responsive and dopamine-resistant resting tremor in Parkinson disease. *Neurology, 95*, e1461–e1470.

Zhang, K., Bhatia, S., Oh, M. Y., Cohen, D., Angle, C., & Whiting, D. (2010). Long-term results of thalamic deep brain stimulation for essential tremor. *Journal of Neurosurgery, 112*, 1271–1276.

Zhou, M., Melin, M. D., Xu, W., & Südhof, T. C. (2020). Dysfunction of parvalbumin neurons in the cerebellar nuclei produces an action tremor. *The Journal of Clinical Investigation, 130*, 5142–5156.

Chapter 28
Arginine as a Disease-Modifying Therapeutic Candidate for the Polyglutamine Diseases by Stabilizing Polyglutamine Protein Conformation and Inhibiting its Aggregation

Yoshitaka Nagai

28.1 Introduction

The spinocerebellar ataxias (SCAs) are a group of heterogeneous diseases characterized by progressive degeneration of the cerebellum and brainstem, often combined with other brain regions, resulting in various neurological symptoms, such as ataxia, spastic paraplegia, parkinsonism, and peripheral neuropathy (Ashizawa et al., 2018; Klockgether et al., 2019). The SCAs are intractable diseases for which there are only few symptomatic therapies currently available, and thus disease-modifying therapies that delay the disease progression are eagerly awaited. The number of SCA patients in Japan is estimated to be approximately 30,000, and one-third of them are familial forms, whereas two-thirds of them are sporadic forms. Among the familial SCAs in Japan, about 90% show a dominant inheritance, whereas about 10% show a recessive inheritance.

Recent advances in molecular genetics have identified more than 40 causative genes responsible for familial SCAs to date. Among these causative mutations, it should be noted that seven diseases, namely, SCA types 1, 2, 3, 6, 7, and 17 and dentatorubral-pallidoluysian atrophy (DRPLA), share similar types of mutations in each unrelated causative gene, i.e., abnormal expansions of CAG repeat sequences. These CAG repeat sequences are all within the translated region of the causative genes and encode glutamine, thus producing mutant proteins with an abnormally expanded polyglutamine (polyQ) stretch. Therefore, these diseases, together with

Y. Nagai (✉)
Department of Neurology, Kindai University Faculty of Medicine, Osaka-Sayama, Osaka, Japan

Department of Neurotherapeutics, Osaka University Graduate School of Medicine, Suita, Osaka, Japan
e-mail: yoshi.nagai@med.kindai.ac.jp

H. Mizusawa, S. Kakei (eds.), *Cerebellum as a CNS Hub*,
Contemporary Clinical Neuroscience, https://doi.org/10.1007/978-3-030-75817-2_28

Huntington's disease (HD) and spinobulbar muscular atrophy (SBMA) that are also caused by similar gene mutations, are collectively called the polyQ diseases (Table 28.1) (Paulson et al., 2017; Takeuchi & Nagai, 2017).

The polyQ diseases share several clinical and genetic characteristics. The disease-causative genes of the polyQ diseases have neither sequence homology nor any functional similarities, except for the CAG repeat itself. The critical threshold of the CAG repeat length for pathogenic mutations is quite similar among these diseases (about 35–40 repeats), except for SCA6. Most importantly, the CAG repeat length inversely correlates with the age of onset and severity of the disease. Furthermore, most polyQ diseases are inherited in an autosomal-dominant manner except for SBMA (X-linked recessive), and gene dose effects are observed in polyQ disease patients carrying homozygous mutations. Taken together, these facts indicate that the polyQ diseases are caused by gain of toxic function mechanisms that are triggered by the CAG repeat expansion mutation. Indeed, the expression of mutant proteins with an expanded polyQ stretch, non-native proteins fused with an expanded polyQ stretch, or expanded polyQ peptides alone has been shown to induce neurodegeneration in a variety of invertebrate and vertebrate animal models, such as *Caenorhabditis elegans*, *Drosophila*, mice, and monkeys (Burright et al., 1995; Faber et al., 1999; Mangiarini et al., 1996; Tomioka et al., 2017; Warrick et al., 1998; Yang et al., 2008), suggesting that the expanded polyQ stretch itself

Table 28.1 The polyglutamine diseases

Disease	Causative gene	CAG repeat		Affected regions
		Normal	Disease	
Spinocerebellar ataxia type 1 (SCA1)	Ataxin-1	6–39	39–83	Cerebellar cortex, dentate nucleus, brainstem, cerebral cortex
Spinocerebellar ataxia type 2 (SCA2)	Ataxin-2	14–32	32–200	Cerebellar cortex, brainstem, cerebral cortex, peripheral nerves
Spinocerebellar ataxia type 3 (SCA3)/Machado-Joseph disease	Ataxin-3	12–41	55–84	Dentate nucleus, basal ganglia, brainstem, spinal cord
Spinocerebellar ataxia type 6 (SCA6)	α1A calcium channel	4–19	20–33	Cerebellar cortex
Spinocerebellar ataxia type 7 (SCA7)	Ataxin-7	4–35	37–306	Cerebellum, brainstem, retina
Spinocerebellar ataxia type 17 (SCA17)	TATA-binding protein	25–44	46–63	Cerebellum, cerebral cortex, basal ganglia
Dentatorubral-pallidoluysian atrophy (DRPLA)	Atrophin-1	6–36	49–88	Dentate and red nuclei, cerebellum, cerebral cortex brainstem
Huntington's disease (HD)	Huntingtin	6–35	36–180	Caudate nucleus, putamen, cerebral cortex
Spinobulbar muscular atrophy (SBMA)/Kennedy's disease	Androgen receptor	9–36	38–65	Spinal and bulbar motor neurons

plays a pivotal role in the pathogenesis and is sufficient to induce neurodegeneration in the polyQ diseases.

In this review, focusing on the polyQ diseases among the various SCAs, I will introduce our research toward developing disease-modifying therapies for the polyQ diseases targeting polyQ protein misfolding and aggregation.

28.2 Protein Misfolding and Aggregation as a Common Molecular Pathogenesis of the polyQ Diseases and Other Neurodegenerative Diseases

In the molecular pathogenesis of the polyQ diseases, expansions of the polyQ stretch in the disease-causative proteins alter the structural stability of the mutant proteins and induce their misfolding and conformational transition to a β-sheet-dominant structure, resulting in their assembly into soluble oligomers and then subsequently into insoluble amyloid-like fibrillar aggregates (Fig. 28.1) (Nagai et al., 2007; Takeuchi & Nagai, 2017). These polyQ protein aggregates accumulate as inclusion bodies within affected neurons in the brain and spinal cord. However, these inclusion bodies have been shown not always to correlate with neuronal toxicity but rather to decrease the risk of neuronal cell death, implying that inclusion

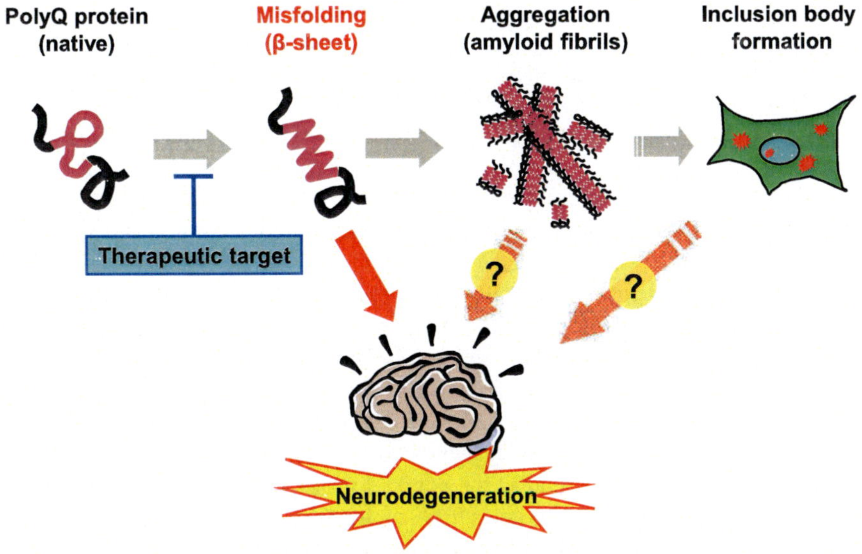

Fig. 28.1 Molecular pathogenesis and therapeutic targets for the polyglutamine diseases. Expansions of the polyQ stretch trigger misfolding of the disease-causative proteins and their conformational transition to a β-sheet-dominant structure, leading to their assembly into insoluble amyloid-like fibrillar aggregates, resulting in their accumulation as inclusion bodies within affected neurons, which eventually cause neurodegeneration

bodies are formed as a protective cellular response against the toxic soluble polyQ proteins (Arrasate et al., 2004; Saudou et al., 1998). Importantly, we and others have demonstrated that soluble β-sheet-rich intermediate species of the expanded polyQ proteins, such as monomers and oligomers, rather than the insoluble amyloid-like fibrillar aggregates, exert cytotoxicity (Legleiter et al., 2010; Miller et al., 2011; Nagai et al., 2007; Nagai & Popiel, 2008; Sathasivam et al., 2010). These protein misfolding and aggregation cascades are commonly implicated in other neurodegenerative diseases, such as Alzheimer's disease (amyloid-β and tau), Parkinson's disease (α-synuclein), and amyotrophic lateral sclerosis (TDP-43), implying a common molecular pathogenesis of a broad range of neurodegenerative diseases (Ross & Poirier, 2005; Soto, 2003).

28.3 Toward Establishment of Potential Therapies for the polyQ Diseases Targeting Protein Misfolding and Aggregation

Toward developing disease-modifying therapies for the polyQ diseases, I decided to target protein misfolding and aggregation, which are thought to be a common molecular pathogenesis of various neurodegenerative diseases, including the polyQ diseases (Fig. 28.1). Based on the hypothesis that molecules selectively binding to the abnormally expanded polyQ stretch would interfere with its misfolding and aggregation, I previously identified polyQ binding peptide 1 (QBP1; SNWKWWPGIFD) by phage display screening (Nagai et al., 2000). Indeed, QBP1 was shown to have high selective affinity for the expanded polyQ stretch, with a dissociation constant (Kd) of 5.7 μM, but not for the normal-length polyQ stretch, as demonstrated by surface plasmon resonance analyses (Okamoto et al., 2009). We then found that QBP1 inhibits the toxic β-sheet conformational transition of the expanded polyQ protein monomer and its subsequent oligomerization and aggregation in vitro (Nagai et al., 2000, 2007; Takahashi et al., 2007). Furthermore, I demonstrated that the expression of QBP1 suppresses inclusion body formation of polyQ proteins and cytotoxicity/neurodegeneration in cellular and *Drosophila* models of the polyQ diseases, indicating the therapeutic effects of QBP1 in vivo (Nagai et al., 2000, 2003). These results demonstrate that inhibition of the toxic β-sheet conformational transition and aggregation of the expanded polyQ protein by QBP1 leads to the suppression of neurodegeneration, establishing a promising target for disease-modifying therapies against the polyQ diseases (Popiel et al., 2013). In an attempt to develop molecular therapies for the polyQ diseases using QBP1, we conjugated QBP1 with protein transduction domains (PTD-QBP1) to increase its cell membrane permeability and successfully demonstrated that the oral administration of PTD-QBP1 suppressed premature death in polyQ disease *Drosophila* models (Popiel et al., 2007). However, PTD-QBP1 showed limited therapeutic effects on

HD model mice upon its peripheral administration (Popiel et al., 2009), probably due to its poor efficiency to penetrate the blood-brain barrier (BBB).

28.4 Identification of Arginine, which Stabilizes Polyglutamine Protein Conformation, as a Disease-Modifying Therapeutic Candidate for the Polyglutamine Diseases

Toward developing disease-modifying therapies for the polyQ diseases using small chemical compounds, which are expected to have high brain permeability, we have been searching for chemical compounds with polyQ aggregation inhibitory activity similar to QBP1. To date, we have performed high-throughput screening of a large-scale chemical compound library (46,000 compounds) using our polyQ protein aggregation turbidity assay (Nagai et al., 2000) and have successfully identified approximately 100 novel polyQ aggregation inhibitor compounds (unpublished).

Among these polyQ aggregation inhibitors, we decided to focus on arginine, because its clinical use has been approved and it has high BBB permeability (Minakawa et al., 2020). Arginine belongs to a group of chemical chaperones, which stabilize the conformation of proteins and suppress their aggregation (Cortez & Sim, 2014). Indeed, we showed that arginine most effectively inhibits polyQ protein aggregation in a dose-dependent manner among the various chemical chaperones that we tested. Importantly, we found that arginine stabilizes the native structure of polyQ proteins and effectively inhibits their conformational transition to a β-sheet-rich structure in vitro. We also confirmed that arginine inhibits the oligomerization of polyQ proteins within cultured cells. We next administered arginine to polyQ disease *Drosophila* models and found that arginine treatment suppresses inclusion body formation of polyQ proteins and neurodegeneration in vivo. We then extended our study to mouse models of the polyQ diseases and found that the oral administration of arginine at dosages equivalent to those approved for humans significantly suppresses the motor impairment in SCA1 model knock-in mice with a Q154 expansion. Neuropathological analyses revealed that arginine administration also suppresses polyQ inclusion body formation and neurodegeneration in SCA1 mice. Most importantly, these therapeutic effects of arginine were also observed even when administered after the disease onset. We also confirmed the effectiveness of arginine in a mouse model of another polyQ disease, SBMA. These results establish the disease-modifying therapeutic effects of arginine against the polyQ diseases, by stabilizing polyQ protein conformation and inhibiting its aggregation (Minakawa et al., 2020; Minakawa & Nagai, 2021). Toward the clinical application of arginine for the polyQ diseases, we are currently conducting a clinical trial of arginine in collaboration with Dr. Onodera at Niigata University, to evaluate the efficacy and safety of long-term arginine administration to SCA6 patients in Japan.

28.5 Future Perspectives

In this review, I introduced our research toward developing therapies for the polyQ diseases among the various SCAs, by targeting protein misfolding and aggregation. Since protein misfolding and aggregation are thought to be a common molecular pathogenesis not only of the polyQ diseases but also of other neurodegenerative diseases, our therapeutic strategy targeting protein misfolding and aggregation can also be applied to a broad range of neurodegenerative diseases, including Alzheimer's disease, Parkinson's disease, and amyotrophic lateral sclerosis (Nagai & Popiel, 2008; Takeuchi & Nagai, 2017). Toward the success of clinical trials of various disease-modifying therapies against neurodegenerative diseases, the development of sensitive disease biomarkers that accurately reflect the disease states and can be used to evaluate therapeutic efficacy is indispensable (Katsuno et al., 2012; Nagai & Minakawa, 2015). Considering that most clinical trials for sporadic neurodegenerative diseases face serious failure because of the difficultly in early diagnosis and early intervention, preventive clinical trials starting before the onset for genetic diseases, such as the polyQ diseases, will possibly open a new avenue for the successful development of disease-modifying therapies against the neurodegenerative diseases.

Acknowledgments I express my sincere thanks to my laboratory members, particularly Dr. Eiko Minakawa (National Center of Neurology and Psychiatry), Dr. Helena Akiko Popiel (Tokyo Medical University), and Mr. Hiroshi Yamane (Asahi Kasei Pharma Corporation), and also to my precious collaborator, Professor Osamu Onodera (Niigata University), for their contribution to our research projects.

Conflicts of Interest Yoshitaka Nagai belongs to an endowed department supported by Nihon Medi-Physics Co., AbbVie GK., Otsuka Pharmaceutical Co., Kyowakai Medical Co., Fujiikai Medical Co., Yukioka Hospital, Osaka Gyoumeikan Hospital, Kyorin Co., and Tokuyukai Medical Co.

Funding This work was funded by Grants-in-Aid for Scientific Research on Priority Areas (Advanced Brain Science Project **14017062**, Research on Pathomechanisms of Brain Disorders **17025026**, and Protein Community **20059023**) and on Innovative Areas (Synapse and Neurocircuit Pathology **23110528** and Brain Protein Aging and Dementia Control **17H05699**) from the Ministry of Education, Culture, Sports, Science, and Technology, Japan; by Grants-in-Aid for Scientific Research (B) (**20390245**) and for Challenging Exploratory Research (**22659172**) from the Japan Society for the Promotion of Science (JSPS), Japan; by Health Labour Sciences Research Grants for Research on Development of New Drugs (H**24**-Soyaku-Sogo-**002**), Research for Intractable Diseases (H**26**-Nanchi-**025**), Research for Persons with Disabilities (H**25**-Shinkei-Kin-**003**), and the Research Committee for Ataxic Diseases (H**23**-Nanchi-**014**) from the Ministry of Health, Labour and Welfare, Japan; by grants for Practical Research Projects for Rare/ Intractable Diseases (JP**16**ek**0109018**, **16**ek**0109048**, JP**19**ek**0109222**, and JP**20**ek**0109459**) and for Project of Translational and Clinical Research Core

Centers (JP**17**lm**0203035** and JP**18**lm**0203071**) from the Japan Agency for Medical Research and Development; by a grant for Core Research for Evolutional Science and Technology (CREST) from the Japan Science and Technology Agency; and by Intramural Research Grants for Neurological and Psychiatric Disorders (**30–3** and **30–9**) from the National Center of Neurology and Psychiatry.

References

Arrasate, M., Mitra, S., Schweitzer, E. S., Segal, M. R., & Finkbeiner, S. (2004). Inclusion body formation reduces levels of mutant huntingtin and the risk of neuronal death. *Nature, 431*(7010), 805–810.

Ashizawa, T., Öz, G., & Paulson, H. L. (2018). Spinocerebellar ataxias: Prospects and challenges for therapy development. *Nature Reviews. Neurology, 14*(10), 590–605.

Burright, E. N., Clark, H. B., Servadio, A., Matilla, T., Feddersen, R. M., Yunis, W. S., et al. (1995). SCA1 transgenic mice: A model for neurodegeneration caused by an expanded CAG trinucleotide repeat. *Cell, 82*(6), 937–948.

Cortez, L., & Sim, V. (2014). The therapeutic potential of chemical chaperones in protein folding diseases. *Prion, 8*(2), 197–202.

Faber, P. W., Alter, J. R., MacDonald, M. E., & Hart, A. C. (1999). Polyglutamine-mediated dysfunction and apoptotic death of a *Caenorhabditis elegans* sensory neuron. *Proceedings of the National Academy of Sciences of the United States of America, 96*(1), 179–184.

Katsuno, M., Tanaka, F., & Sobue, G. (2012). Perspectives on molecular targeted therapies and clinical trials for neurodegenerative diseases. *Journal of Neurology, Neurosurgery, and Psychiatry, 83*(3), 329–335.

Klockgether, T., Mariotti, C., & Paulson, H. L. (2019). Spinocerebellar ataxia. *Nature Reviews. Disease Primers, 5*(1), 24.

Legleiter, J., Mitchell, E., Lotz, G. P., Sapp, E., Ng, C., DiFiglia, M., et al. (2010). Mutant huntingtin fragments form oligomers in a polyglutamine length-dependent manner *in vitro* and *in vivo. Journal of Biological Chemistry, 285*(19), 14777–14790.

Mangiarini, L., Sathasivam, K., Seller, M., Cozens, B., Harper, A., Hetherington, C., et al. (1996). Exon 1 of the HD gene with an expanded CAG repeat is sufficient to cause a progressive neurological phenotype in transgenic mice. *Cell, 87*(3), 493–506.

Miller, J., Arrasate, M., Brooks, E., Libeu, C. P., Legleiter, J., Hatters, D., et al. (2011). Identifying polyglutamine protein species in situ that best predict neurodegeneration. *Nature Chemical Biology, 7*(12), 925–934.

Minakawa, E. N., & Nagai, Y. (2021). Protein aggregation inhibitors as disease-modifying therapies for polyglutamine diseases. *Front Neurosci 15*, 621996.

Minakawa, E. N., Popiel, H. A., Tada, M., Takahashi, T., Yamane, H., Saitoh, Y., et al. (2020). Arginine is a disease modifier for polyQ disease models that stabilizes polyQ protein conformation. *Brain, 143*(6), 1811–1825.

Nagai, Y., Fujikake, N., Ohno, K., Higashiyama, H., Popiel, H. A., Rahadian, J., et al. (2003). Prevention of polyglutamine oligomerization and neurodegeneration by the peptide inhibitor QBP1 in *Drosophila. Human Molecular Genetics, 12*(11), 1253–1259.

Nagai, Y., Inui, T., Popiel, H. A., Fujikake, N., Hasegawa, K., Urade, Y., et al. (2007). A toxic monomeric conformer of the polyglutamine protein. *Nature Structural & Molecular Biology, 14*(4), 332–340.

Nagai, Y., & Minakawa, E. N. (2015). Drug development for neurodegenerative diseases. In K. Wada (Ed.), *Neurodegenerative disorders as systemic diseases* (pp. 183–216). Springer.

Nagai, Y., & Popiel, H. A. (2008). Conformational changes and aggregation of expanded polyglutamine proteins as therapeutic targets of the polyglutamine diseases: Exposed β-sheet hypothesis. *Current Pharmaceutical Design, 14*(30), 3267–3279.

Nagai, Y., Tucker, T., Ren, H., Kenan, D. J., Henderson, B. S., Keene, J. D., et al. (2000). Inhibition of polyglutamine protein aggregation and cell death by novel peptides identified by phage display screening. *Journal of Biological Chemistry, 275*(14), 10437–10442.

Okamoto, Y., Nagai, Y., Fujikake, N., Akiko Popiel, H., Yoshioka, T., Toda, T., et al. (2009). Surface plasmon resonance characterization of specific binding of polyglutamine aggregation inhibitors to the expanded polyglutamine stretch. *Biochemical and Biophysical Research Communications, 378*(3), 634–639.

Paulson, H. L., Shakkottai, V. G., Clark, H. B., & Orr, H. T. (2017). Polyglutamine spinocerebellar ataxias - from genes to potential treatments. *Nature Reviews. Neuroscience, 18*(10), 613–626.

Popiel, H. A., Nagai, Y., Fujikake, N., & Toda, T. (2007). Protein transduction domain-mediated delivery of QBP1 suppresses polyglutamine-induced neurodegeneration *in vivo*. *Molecular Therapy, 15*(2), 303–309.

Popiel, H. A., Nagai, Y., Fujikake, N., & Toda, T. (2009). Delivery of the aggregate inhibitor peptide QBP1 into the mouse brain using PTDs and its therapeutic effect on polyglutamine disease mice. *Neuroscience Letters, 449*(2), 87–92.

Popiel, H. A., Takeuchi, T., Burke, J. R., Strittmatter, W. J., Toda, T., Wada, K., et al. (2013). Inhibition of protein misfolding/aggregation using polyglutamine binding peptide QBP1 as a therapy for the polyglutamine diseases. *Neurotherapeutics, 10*(3), 440–446.

Ross, C. A., & Poirier, M. A. (2005). Opinion: What is the role of protein aggregation in neurodegeneration? *Nature Reviews. Molecular Cell Biology, 6*(11), 891–898.

Sathasivam, K., Lane, A., Legleiter, J., Warley, A., Woodman, B., Finkbeiner, S., et al. (2010). Identical oligomeric and fibrillar structures captured from the brains of R6/2 and knock-in mouse models of Huntington's disease. *Human Molecular Genetics, 19*(1), 65–78.

Saudou, F., Finkbeiner, S., Devys, D., & Greenberg, M. E. (1998). Huntingtin acts in the nucleus to induce apoptosis but death does not correlate with the formation of intranuclear inclusions. *Cell, 95*(1), 55–66.

Soto, C. (2003). Unfolding the role of protein misfolding in neurodegenerative diseases. *Nature Reviews. Neuroscience, 4*(1), 49–60.

Takahashi, Y., Okamoto, Y., Popiel, H. A., Fujikake, N., Toda, T., Kinjo, M., et al. (2007). Detection of polyglutamine protein oligomers in cells by fluorescence correlation spectroscopy. *Journal of Biological Chemistry, 282*(33), 24039–24048.

Takeuchi, T., & Nagai, Y. (2017). Protein Misfolding and aggregation as a therapeutic target for polyglutamine diseases. *Brain Sciences, 7*(10), 128.

Tomioka, I., Ishibashi, H., Minakawa, E. N., Motohashi, H. H., Takayama, O., Saito, Y., et al. (2017). Transgenic monkey model of the polyglutamine diseases recapitulating progressive neurological symptoms. *eNeuro, 4*(2) ENEURO.0250-16.2017.

Warrick, J. M., Paulson, H. L., Gray-Board, G. L., Bui, Q. T., Fischbeck, K. H., Pittman, R. N., et al. (1998). Expanded polyglutamine protein forms nuclear inclusions and causes neural degeneration in *Drosophila*. *Cell, 93*(6), 939–949.

Yang, S. H., Cheng, P. H., Banta, H., Piotrowska-Nitsche, K., Yang, J. J., Cheng, E. C., et al. (2008). Towards a transgenic model of Huntington's disease in a non-human primate. *Nature, 453*(7197), 921–924.

Index

H. Mizusawa, S. Kakei (eds.), *Cerebellum as a CNS Hub*,
Contemporary Clinical Neuroscience, https://doi.org/10.1007/978-3-030-75817-2

R

S

T